# TOWARD ZERO DISCHARGE

# TOWARD ZERO DISCHARGE
## Innovative Methodology and Technologies for Process Pollution Prevention

Edited by

**TAPAS K. DAS**
Washington Department of Ecology
Olympia, WA

**WILEY-INTERSCIENCE**

A JOHN WILEY & SONS, INC., PUBLICATION

*Library of Congress Cataloging-in-Publication Data:*

Toward zero discharge : innovative methodology and technologies for
process pollution prevention / edited by Tapas K. Das.
     p. cm.
    ISBN 0-471-46967-X (cloth : alk. paper)
  1. Factory and trade waste -- Management.  2. Pollution prevention.  3.
Recycling (Waste, etc.)  I. Das, Tapas K.
   TD897.T7 2005
   660′.28′00286--dc22

                                         2004023724

Printed in the United States of America

10  9  8  7  6  5  4  3  2  1

This book is dedicated to my Dad and Mom, Pramatha and Minu, who now live in New York. Their selfless love, common sense, support, inspiration, and dedication have helped me immensely in India and abroad. This book is also dedicated to their grandchildren: Shiva, Nikhil (Niki), Ani, Ria, Apu, Rakhi, Meghasree, and to millions of children across the globe, who will be our champion stewards and caretakers of the Earth.

# Errata

## *Toward Zero Discharge: Innovative Methodology and Technologies for Process Pollution Prevention*

**Tapas K. Das (Editor)**

ISBN 0-471-46967-X    © 2005 John Wiley & Sons, Inc.


## Permissions:

### Reproduced with Permission, Courtesy of Pearson Education:

Section 4.2: (pp. 17-21 Excerpts)) David T. Allen and David R. Shonnard, Green Engineering: Environmentally Conscious Design of Chemical Processes, 1st Edition, © 2002. Reprinted with permission of Pearson Education, Inc., Upper Saddle River, NJ.

### Reproduced with Permission, Courtesy of AIChE:

1. Section 4.5: Mary F. McDaniel, Richard D. Siegel and Roy F. Weston, *Prepare for the Next Phase of Public Involvement*, Chemical Engineering Progress (CEP), 2000, 96(3) 35-40

2. Section 4.6: June C. Bolstridge, *Access Environmental Information on the Internet*, Chemical Engineering Progress (CEP), 2003, 99(1) 50-54

Reprinted with permission of Chemical Engineering Progress (CEP), American Institute of Chemical Engineers, New York, NY

### Reproduced with Permission, Courtesy of Pollution Engineering:

Section 7.1.6: Orest Zacerkowny, *Not a Drop Leaves the Plant*, Pollution Engineering, 19-22, February 2002

Reprinted with permission of Pollution Engineering, Troy, MI

# CONTENTS

**Foreword**                                                                                    xv

**Acknowledgments**                                                                           xvii

**1   INTRODUCTION**                                                                             1

    1.1   Waste as Pollution  /  2
    1.2   Defining Pollution Prevention  /  2
    1.3   The Zero Discharge Paradigm  /  4
    1.4   The Structure of the Book  /  5

## PART I   METHODOLOGY

**2   ZERO DISCHARGE INDUSTRIES**                                                               11

    2.1   Sustainability, Industrial Ecology and Zero Discharge  /  12
    2.2   Why Zero Discharge is Critical to Sustainability  /  15
    2.3   The New Role of Process Engineers and Engineering Firms  /  16
    2.4   Zero Discharge Methodology  /  20
        2.4.1   Analyze Throughput  /  20
        2.4.2   Inventory Inputs and Outputs  /  20
        2.4.3   Build Industrial Clusters  /  21
        2.4.4   Develop Conversion Technologies  /  22
        2.4.5   Designer Wastes  /  22
        2.4.6   Reinvent Regulatory Policies  /  23

2.5   Making the Transition / 23

    2.5.1   Recycling of Materials and Reuse of Products / 24

    2.5.2   Dematerialization / 26

    2.5.3   Investment Recovery / 27

    2.5.4   New Technologies and Materials / 27

    2.5.5   New Mindset / 28

2.6   In the Full Zero Discharge Paradigm / 28

    2.6.1   Opening New Opportunities / 29

    2.6.2   Providing Return on Investment / 30

2.7   Separation—A Key Conversion Technology / 30

2.8   Constraints and challenges / 31

**3   FUNDAMENTALS OF LIFE CYCLE ASSESSMENT**    **35**

3.1   What is Life Cycle Assessment? / 35

    3.1.1   Benefits of Conducting an LCA / 36

    3.1.2   Limitations of LCA / 37

3.2   Conducting LCA / 37

    3.2.1   Goal Definition and Scope / 38

    3.2.2   Life Cycle Inventory / 42

    3.2.3   Life Cycle Impact Assessment / 46

    3.2.4   Life Cycle Interpretation / 53

3.3   LCA and LCI Software Tools / 58

3.4   Evaluating the Life Cycle Environmental Performance of Two Disinfection Technologies / 62

    3.4.1   The Challenge / 62

    3.4.2   The Chlorination (Disinfection) Process / 63

    3.4.3   Dechlorination With Sulfur Dioxide / 64

    3.4.4   UV Disinfection Process / 72

3.5   Case Study: Comparison of LCA of Electricity from Biorenewables and Fossil Fuels / 77

    3.5.1   Results / 77

    3.5.2   Sensitivity Analysis / 82

    3.5.3   Summary and Conclusions / 82

**4   ASSESSMENT AND MANAGEMENT OF HEALTH AND ENVIRONMENTAL RISKS**    **89**

4.1   Health Risk Assessment / 89

    4.1.1   Problem Formulation / 90

    4.1.2   Exposure Assessment / 93

4.1.3   Toxicity Assessment / 101

4.1.4   Risk Characterization / 101

4.2   Assessing the Risks of Some Common Pollutants / 103

    4.2.1   $NO_x$, Hydrocarbons, and VOCs—Ground-Level Ozone / 104

    4.2.2   Carbon Monoxide / 105

    4.2.3   Lead and Mercury / 106

    4.2.4   Particulate Matter / 107

    4.2.5   $SO_2$, $NO_x$, and Acid Deposition / 108

    4.2.6   Air Toxics / 109

4.3   Ecological Risk Assessment / 110

    4.3.1   Technical Aspects of Ecological Problem Formulation / 110

    4.3.2   Ecological Exposure Assessment / 114

    4.3.3   Ecological Effects Assessment / 118

    4.3.4   Additional Components of Ecological Risk Assessments / 120

    4.3.5   Toxicity Testing / 124

    4.3.6   Ecological Risk Characterization / 129

    4.3.7   Uncertainties / 130

4.4   Risk Management / 130

    4.4.1   Valuation of Ecological Resources / 132

    4.4.2   Modeling Risk Management / 134

    4.4.3   Other Considerations for Risk Characterization / 135

    4.4.4   Conceptual Bases for De Minimis Risks / 136

    4.4.5   Ecological Risk Assessment of Chemicals / 137

4.5   Communicating Information on Environmental and Health Risks / 146

    4.5.1   From Concern to Outrage: Determinants of Public
Response / 146

    4.5.2   Sustainable Strategies for Communicating Environmental
and Health Risk / 148

    4.5.3   Case Study: Environmental and Health Risk Communication
Neglected Until After an Accident / 151

    4.5.4   Lessons Learned / 152

4.6   Environmental Information on the Internet / 153

    4.6.1   Internet Sources / 153

    4.6.2   Implications and Limitations of Using the Internet / 156

**5   ECONOMICS OF POLLUTION PREVENTION:
TOWARD AN ENVIRONMENTALLY
SUSTAINABLE ECONOMY                                        161**

5.1   Economics of Pollution Control Technology / 162

    5.1.1   Cost Estimates / 162

5.1.2 Elements of Total Capital Investment / 163

5.1.3 Elements of Total Annual Cost / 165

5.1.4 Economic Criteria for Technology Comparisons / 166

5.1.5 Calculating Cash Flow / 168

5.1.6 Achieving a Responsible Balance / 169

5.2 Best Available Control Technology: Application to Gas Turbine Power Plants / 170

5.2.1 $NO_x$ BACT Review / 171

5.2.2 CO BACT Review: Combustion Turbines and Duct Burners / 176

5.2.3 $PM_{10}$ Control Technologies / 178

5.2.4 VOC Control Technologies / 179

5.3 Case Study: The North Carolina Clean Smokestacks Plan / 179

5.3.1 Consequences of Dirty Air / 180

5.3.2 Technologies and Clean Energy / 181

5.3.3 Costs and Benefits of Reducing Emissions / 182

5.4 Sustainable Economy and The Earth / 183

5.4.1 What is a Sustainable Economy? / 183

5.4.2 Costs of Manufacturing Various Biobased Products / 184

**6 SUSTAINABILITY AND SUSTAINABLE DEVELOPMENT**    **191**

6.1 Introduction / 191

6.1.1 Green Chemistry and Green Engineering / 193

6.2 Sustainable Production and Growth for the Chemical Process Industries / 198

6.2.1 Impact Assessment, Risk Assessment, and Management / 198

6.2.2 Green by Design Process / 199

6.3 Integrating Life Cycle Assessment in Sustainable Product Development / 200

6.4 Biorenewable Energy Resources / 201

6.4.1 Biomass Fuel / 201

6.4.2 Solar-Thermal Power / 203

6.4.3 Wind Power / 203

6.4.4 Hydropower / 205

6.4.5 Wave Energy / 205

6.4.6 Tidal Power / 206

6.5 Applying the Metrics of Sustainability to Transform Business Practices and Public Policy / 206

6.6 Toward a Hydrogen-based Sustainable Economy / 207

6.7 Bio-Based Chemicals and Engineered Materials / 214

6.8 Eco-Efficiency and Eco-industrial Parks / 214

6.9 Process Intensification and Microchannel Reaction / 216

## PART II TECHNOLOGIES AND APPLICATIONS

**7 ZERO DISCHARGE TECHNOLOGY**      **225**

7.1 Zero Water Discharge Systems / 225

    7.1.1 Why Aim for Zero Water Discharge? / 225

    7.1.2 Establishing a Database / 226

    7.1.3 Advantages and Disadvantages / 226

    7.1.4 Design Principles / 227

    7.1.5 Case Study: A Cement Plant in India / 228

    7.1.6 Case Study: DaimlerChrysler's Zero-Discharge Wastewater Treatment Plant in Mexico / 230

    7.1.7 Case Study: Zero Effluent Systems at Formosa Plastic Manufacturing, Texas / 234

7.2 Case Study: Gas Turbine $NO_x$ Emissions at General Electric / 238

    7.2.1 Regulatory Background / 239

    7.2.2 Gas Turbine Emissions / 239

    7.2.3 $NO_x$ Control Technologies / 240

    7.2.4 Discussion / 244

7.3 Questions of Regulatory Policy / 246

7.4 Smokestack Emissions / 247

    7.4.1 Clear Skies Initiative / 248

    7.4.2 The Ebara Process / 249

7.5 Eco-Industrial Parks: Model Zero Discharge Communities / 252

    7.5.1 Industrial Ecology / 252

    7.5.2 What Is an Eco-Industrial Park? / 255

    7.5.3 Eco-Industrial Park Development / 255

    7.5.4 Eco-Industrial Parks—Mini Case Studies in India / 256

    7.5.5 Conclusions / 259

**8 TECHNOLOGIES FOR POLLUTION PREVENTION: AIR**      **263**

8.1 Some On-going Pollution Prevention Technologies / 264

    8.1.1 Destruction of Low Level of VOCs by Electron-Beam-Generated Plasma / 265

    8.1.2 Biofiltration of Organic Vapors / 279

8.1.3   Using Sorbents and Carbon Injection to Remove
Dioxin/Furans and Toxic Metals / 289

8.2   Some Emerging and Innovative Processes / 300

8.2.1   Nonthermal Plasma for Control of Multiple
Pollutants from Coal-Fired Utility Boilers / 300

8.2.2   E-Beam Based Oxidation Treatment for VOC
Contaminated Aqueous Phase Waste Streams / 316

8.2.3   Biodiesel / 330

8.2.4   Fuel Cells for Clean and Efficient Energy / 334

8.2.5   Membrane-Based Removal and Recovery
of VOCs from Air / 352

8.2.6   Carbon Sequestration / 369

**9   TECHNOLOGIES FOR POLLUTION PREVENTION:
WATER**                                                       **383**

9.1   Advances in Wastewater Treatment Technologies / 383

9.1.1   Membrane for Environmental and Pharmaceutical
Applications / 384

9.1.2   Zero Discharge Technologies Based on
Supercritical $CO_2$ and $H_2O$ / 406

9.1.3   Ultraviolet Disinfection / 416

9.2   Some Emerging and Innovative Processes / 432

9.2.1   Brine Concentrators for Recycling
Wastewater / 432

9.2.2   Membrane-Electrode Process / 450

9.2.3   Advances in Pollution Prevention in the Metal
Finishing Industry / 463

9.2.4   Profitable Pollution Prevention in Electroplating:
An In-Process Focused Approach / 476

9.2.5   Advanced Oxidation Technology at Foundries / 491

9.3   Groundwater Quality / 522

9.3.1   How Aquifers Become Contaminated / 522

9.3.2   Investigating New Arsenic Removal
Technologies / 524

**10   TECHNOLOGIES FOR POLLUTION PREVENTION:
SOLID WASTE**                                                 **545**

10.1   Solid Waste Treatment: Some Perspectives
on Recycling / 545

10.1.1   Why Recycle? / 546

10.1.2   What is Recycling? / 546

10.2 Plastic Recycling in a Developing Country:
A Paradoxical Success Story / 550

    10.2.1 Scope of the System in Bangladesh / 552

    10.2.2 Aims and Objectives / 553

    10.2.3 Industrial Ecology and Developing Countries / 554

    10.2.4 Description of the Case Study / 555

    10.2.5 Supply Chain Issues / 558

    10.2.6 Effects of Successive Reprocessing on Recyclate / 565

    10.2.7 Future Scenarios: Mathematical Modeling / 571

    10.2.8 Conclusions / 578

10.3 From Waste to Energy: Catalytic Steam Gasification of
Poultry Litter / 582

    10.3.1 Poultry Litter and Its Uses / 582

    10.3.2 Environmental Effects of Excess Poultry Litter / 583

    10.3.3 Biogas and Direct Combustion / 584

    10.3.4 Coal Gasification Process Concept / 586

    10.3.5 Testing Gasification Processes / 588

    10.3.6 Process Economics / 592

    10.3.7 Toward Zero Discharge in the Poultry Industry / 595

**11 MINIMIZATION OF ENVIRONMENTAL DISCHARGE
THROUGH PROCESS INTEGRATION**     **597**

11.1 Energy Integration and Heat Exchange Networks / 598

11.2 Mass Integration / 600

    11.2.1 End-of-pipe Separation Network Synthesis / 602

    11.2.2 In-Process Separation Network Synthesis / 608

    11.2.3 Multicomponent Nonideal Mass Integration / 614

11.3 Water Optimization and Integration / 622

    11.3.1 Case Study: Minimization of Wastewater Discharge Using
Land Treatment Technology / 624

    11.3.2 Case Study: Optimization of Water Flows in a Kraft Pulp and
Paper Process / 628

11.4 Industrial Applications of Heat Integration and Mass Integration / 638

    11.4.1 Solutia / 638

    11.4.2 General Electric Plastics / 638

11.5 Further Reading / 639

11.6 Final Thoughts / 639

## 12 PROCESS POLLUTION PREVENTION IN THE PULP AND PAPER INDUSTRY 647

12.1 Environmental Management in the Pulp and Paper Industry / 647

12.2 Pollution Prevention in the Pulp and Paper Industry / 648

    12.2.1 Air Pollution / 649

    12.2.2 Effluent Discharges / 653

    12.2.3 Solid Wastes / 655

    12.2.4 Emerging Technologies / 656

12.3 Resource Recovery and Reuse / 658

    12.3.1 Value-Added Chemicals from Pulp Mill Waste Gases / 658

    12.3.2 Recovery and Control of Sulfur Emissions / 660

    12.3.3 Delignification / 660

    12.3.4 Solvent Pulping / 661

    12.3.5 Biopulping: A Review of a Pilot-Plant Project / 661

    12.3.6 New Process for Recovery of Chemicals and Energy from Black Liquor / 665

## 13 PROGRESS TOWARD ZERO DISCHARGE IN PULP AND PAPER PROCESS TECHNOLOGIES 671

13.1 Three Case Studies / 672

    13.1.1 Louisiana-Pacific Corporation: Conversion to Totally Chlorine Free Processing / 672

    13.1.2 The World's First Zero Effluent Pulp Mill at Meadow Lake: The Closed-Loop Concept / 673

    13.1.3 Successful Implementation of a Zero Discharge Program / 676

13.2 Additional Concerns Addressed by Zero Discharge Technology / 678

    13.2.1 Environment Discharges / 678

    13.2.2 Safety / 678

13.3 Other Emission Recovery and Control Processes / 679

    13.3.1 Recovery of Sulfur from Acid Gases / 679

    13.3.2 Greenhouse Gases / 680

    13.3.3 Solid Waste Management and Conserving Energy / 682

13.4 Conclusions / 683

    Epilogue — Final Thoughts / 684

## Index 687

# FOREWORD

Preventing pollution, wherever feasible, has been universally adopted by manufacturing and process industries in preference to waste treatment and control. It has been argued, and in many instances demonstrated, that the costs of complying with regulatory regimes and of future liability is significantly higher over time than eliminating or greatly reducing wastes via process design changes. Zero discharge is a laudable goal of pollution prevention strategies. The practice of pollution prevention or zero discharge, in addition to more efficient and less costly operation, also earns favorable public relation for industry. Avoidance of contributing to local pollution at least satisfies the societal benefit criteria of the sustainability principles.

This book is a compendium of carefully selected chapters that attempt to address the goal of zero discharge of wastes to air, water and land by using advanced tools and techniques, such as life cycle assessment, that analyze and help design environmentally friendly material- and energy-consuming processes. The selected authors are recognized experts in their own right and have instructive stories to tell. The book also presents a series of practical case studies drawn from a range of industrial sectors from around the world that address environmental concerns such as water reuse, renewable energy, and renewable feedstock.

The book consists of thirteen chapters that are divided into Part I, which presents concepts, strategies, evaluation and quantification, and Part II, which explores technologies and applications for pollution prevention and zero discharge, resource conservation and waste minimization. Chapter 13 of Part II highlights a single manufacturing industry, kraft pulp and paper manufacturing, to show how it is slowly but surely achieving zero discharge.

All chapters have been peer-reviewed. The peer review process has led to sharpening of presented concepts, data, and models. The theoretical concepts and laboratory and industrial data presented here will assist the academic and industrial process designers in emulating the presented solutions and developing their own techniques and methods for zero discharge.

Subhas K. Sikdar
Acting Associate Director for Health
National Risk Management Research Laboratory
U.S. Environmental Protection Agency
Cincinnati, Ohio
USA
October 22, 2004

# ACKNOWLEDGMENTS

I want to acknowledge all the contributing authors in the US, England and India, and whose willingness to take time from their very busy schedules and career to share their expertise, guidance, and encouragement have contributed to this book and made this publication possible. My sincere appreciation goes to contributors for your patience and continued support in this project. I also want to acknowledge my teachers, professors and classmates in India, my doctoral and postdoctoral advisors, and mentors in England and United States for their noble efforts, dedications, and integrity to their professions, and exemplary lifestyles. My special thanks to fellow members, officers, and collaborators of the Environmental Division, Sustainable Engineering Forum and other divisions and forums of the American Institute of Chemical Engineers (AIChE). Also, I would like to extend my appreciation to fellow members and officers of the AIChE's Puget Sound Chapter in Seattle and Olympia for their enthusiasms and encouragement.

Without the helping hands of many others, this book publication wouldn't have been successful. My gratitude and special thanks go to Bruce Estus for some figures and graphics for the book, Phyllis Shafer for providing some reference materials from libraries and for checking accuracies of some references, Sonya Kirkendall and Aleta DeBee for their help with word processing, and computer support staff at the Washington Department of Ecology.

I would also like to thank the staff at Wiley who were involved in this project, especially to the editor, Arza Seidel for accepting the idea of the book, and the editorial staff at Wiley, Liam Kuhn, Sarah Harrington, Shirley Thomas, and Kellsee Chu for their continued support and guidance during the preparation of the manuscript and throughout the publishing process. Also, all the reviewers' constructive comments and helpful suggestions were greatly appreciated.

Finally, my very special thanks go to my wife and kids who endured my periodic absence from home, and for burning some incalculable midnight oil while I was working on this book project. Their support, love, and words of encouragement are behind the success of this project.

Tapas K. Das
Olympia, Washington
December 27, 2004

# CHAPTER 1

# INTRODUCTION

TAPAS K. DAS
Washington Department of Ecology, Olympia, Washington 98504, USA

KENNETH L. MULHOLLAND
Kenneth Mulholland & Associate, Inc., 27 Hartech Dr, Wilmington, DE 19807, USA

Our avid interest in environmental issues can be traced directly to awareness that as the world's population continues to expand and to consume natural resources, humanity faces shortages that threaten quality of life in developed areas and elsewhere on the earth, life itself. In attempts to find solutions to these problems, we have created an ever growing inventory of man-made chemicals, ostensibly to improve the quality of life that has in fact contributed to the pollution of our environment. "Pollution prevention," an environmental buzz word since the 1990s, encompasses designing processes that generate no waste to plants that emit only harmless compounds such as pure water.

Zero discharge (ZD) is different from pollution prevention in that it seeks to convert wastes into useful products or valuable resources. Sometimes the conversion of wastes into resources having value to another industry is more efficient than the implementation of pollution prevention techniques. Within the ZD paradigm the goal of resource extraction, refining, or commodity production is approached in much the same way that the mining, pulp and paper, petroleum, and chemical industries go about processing raw materials.

In this book, we will focus on the best available industrial processes, techniques, and technologies that treat waste streams, as well as innovative and emerging processes that have better potential for achieving the highest standards in pollution prevention at the plant level, leading to zero discharge. To move toward ZD via "process pollution prevention" (P3), industries must use processes that deploy materials and energy efficiently enough to neutralize contaminants in the waste stream. The ultimate goal is to remove pollutants from the waste streams and convert them into products or feeds for other processes. Logically then, P3 refers

*Toward Zero Discharge: Innovative Methodology and Technologies for Process Pollution Prevention,* edited by Tapas K. Das.
ISBN 0-471-46967-X   Copyright © 2005 John Wiley & Sons, Inc.

to industrial processes by which materials and energy are efficiently utilized to achieve the end product(s) while reduce or eliminate the creation of pollutants or waste at the source.

## 1.1   WASTE AS POLLUTION

A waste is defined as an unwanted byproduct or damaged, defective, or superfluous material of a manufacturing process. Most often, in its current state, it has or is perceived to have no value. It may or may not be harmful or toxic if released to the environment. Pollution is any release of waste to environment (i.e., any routine or accidental emission, effluent, spill, discharge, or disposal to the air, land or water) that contaminates or degrades the environment.

## 1.2   DEFINING POLLUTION PREVENTION

In this book, we define pollution prevention fairly broadly as any action that prevents the release of harmful materials to the environment. This definition manifests itself in the form of a pollution prevention hierarchy, with safe disposal forms at the base of the pyramid and minimizing the generation of waste at the source at the peak (Figure 1.1).

In contrast, the U.S. Environmental Protection Agency (EPA) definition of pollution prevention recognizes only source reduction, which encompasses only the upper two tiers in the hierarchy—minimize generation and minimize introduction (USEPA, 1992). The EPA describes the seven-level hierarchy of Figure 1.1 as "environment management options." The European Community, on the other hand, includes the entire hierarchy in its definition of pollution prevention. The tiers in the pollution prevention hierarchy are broadly described as follows.

- **Minimize Generation**. Reduce to a minimum the formation of nonsalable byproducts in chemical reaction steps and waste constituents (such as tars, fines, etc.) in all chemical and physical separation steps.
- **Minimize Introduction**. Cut down as much as possible on the amounts of process materials that pass through the system unreacted or are transformed to make waste. This implies minimizing the introduction of materials that are not essential ingredients in making the final product. For examples, plant designers can decide not to use water as a solvent when one of the reactants, intermediates, or products could serve the same function, or they can add air as an oxygen source, heat sink, diluent, or conveying gas instead of large volumes of nitrogen.
- **Segregate and Reuse**. Avoid combining waste streams together with no consideration to the impact on toxicity or the cost of treatment. It may make sense to segregate a low-volume, high toxicity wastewater stream from high-volume, low toxicity wastewater streams. Examine each waste stream at the

**Figure 1.1.** Pollution prevention hierarchy.

source and identify any that might be reused in the process or transformed or reclassified as valuable coproducts.

- **Recycle**. Many manufacturing facilities, especially chemical plants, have internal recycle streams that are considered part of the process. In addition, however, it is necessary to recycle externally such materials as polyester film and bottles, Tyvek envelopes, paper, and spent solvents.
- **Recover Energy Value in Waste**. As a last resort, spent organic liquids, gaseous streams containing volatile organic compounds, and hydrogen gas can be burned for their fuel value. Often the value of energy and resources required to make the original compounds is much greater than that which can be recovered by burning the waste streams for their fuel value.
- **Treat for Discharge**. Before any waste stream is discharged to the environment, measure should be taken to lower its toxicity, turbidity, global warming potential, pathogen-content and so on. Examples include biological wastewater treatment, carbon adsorption, filtration, and chemical oxidation.
- **Safe Disposal**. Render waste streams completely harmless so that they do not adversely impact the environment. In this book, we define this as total conversion of waste constituents to carbon dioxide, water, and nontoxic minerals. An example would be post treatment of a wastewater treatment plant effluent in a private wetland. So-called secure landfills do not fall within this category unless the waste is totally encapsulated in granite.

In this book, we will focus on the lower three tiers of the pollution prevention hierarchy; that is, recovering the energy value in waste, treating for discharge, and arranging for safe disposal. To improve this bottom line, however, businesses should address the upper three tiers first: that is minimize generation, minimize introduction, and segregate and reuse (Mulholland and Dyer, 1999). This is where the real opportunity exists for reducing the volume of wastes to be treated. The volume of the waste stream, in turn, has a strong influence on treatment cost and applicability. Thus useful technologies such as the ultraviolet treatment of groundwater or condensation of volatile organic compounds (VOCs) from air are not economic at large volumetric flow rates. The focus has shifted from "end-of-the-pipe" solutions to more fundamental structural changes in industrial processes.

## 1.3   THE ZERO DISCHARGE PARADIGM

Zero discharge, or something very close to it, is the ultimate goal of P3, while the processes themselves are the tools and pathways to achieve it. Thus industries were to be reorganized into "clusters" in which the wastes or by-products of each industrial process were fully matched with others industries' input requirements, the integrated process would produce no waste of any kind. As described later in the book, this solution is being applied in scattered areas throughout the world, from modern industrial nations such as Denmark to developing countries such as Bangladesh.

Traditionally, pollution control technology processed a "waste" until it was benign enough for discharge into the environment. This was achieved through dilution, destruction, separation or concentration. Within the ZD paradigm, many of these processes will still be applied, but as mentioned earlier, the goal will be resource extraction, refining or commodity production, not simply removal of waste from the premises. Engineering firms will need to develop conversion technologies that create "designer wastes" to meet input specifications of other industries.

Research and development efforts are under way around the world to promote the concept of zero discharge and to work toward it in selected industries. The United Nations University's (UNU) Zero Emissions Research Initiatives (ZERI), headquartered in Tokyo, is a leader in this work (see Chapter 2), with the support of major multinational corporations. One vice president of DuPont has said that whenever the company eliminates a pound of waste, the material most likely ends up in a product. At DuPont, titanium dioxide wastes are converted to high-purity table salt, fertilizer and food-grade carbon dioxide.

Of course almost every manufacturing process generates wastes. For example, brewing beer extracts only 8 to 10% of the nutrients from grains. Pilot studies have been conducted employing spent brewery wastes for aqua-culture and cattle raising. Another example is a flue gas treatment for coal-fired electric plants. Using ammonia and electron beam irradiation, oxides of nitrogen and sulfur are converted into ammonium nitrate and ammonium sulfate for use in fertilizer. Many other ZD processes are being developed, as discussed throughout the book.

## 1.4 THE STRUCTURE OF THE BOOK

What is in this book? The thirteen chapters are divided into Part I, which presents methodology, strategies, evaluation and quantification, and Part II, which explores technologies and applications for pollution prevention and zero discharge.

The subject matter of Chapter 2 is zero discharge itself. Natural resource consumption and waste generation have accelerated tremendously in recent years, placing enormous stress on the delicate ecosystems. Although development is necessary to meet the needs of growing populations and increasing sophisticated societies, it must be sustainable, fostering harmonious interaction between nature and the world's industries, economies and lifestyles. This in turn requires long-term global perspectives, together with more efficient technologies and systems for production, resource conservation and waste minimization. The earth's resources are not boundless, and the challenges just identified must be answered quickly. Fortunately, an answer can be found in ZD, which is no more than applied industrial ecology at the manufacturing level: a practical approach with a concrete method-ology to redesign industrial processes so they have no discharges. Chapter 2 describes this methodology and outlines a path for the transition from wasteful, polluting industrial practices to ZD.

As the name implies, life cycle assessment (LCA) (Chapter 3) evaluates the entire life cycle of a product, process, activity, or service, not just simple economics at the time of delivery. For example, the total environmental impact of a product is a factor, which is sometimes oversimplified. Stakeholders under the LCA concept go beyond the immediate customer and extend to society as a whole, which may be concerned about such issues as natural resource depletion or the impact of degra-dation on the environmental. Companies that subject their operations to LCA, consider the environmental performance of products and processes to be a key issue. Many companies have found it not only responsible but also advantageous to explore ways of moving beyond compliance using pollution prevention strategies and environmental management systems to improve environmental performance. In many cases, LCA leads to better business practices. Chapter 3 illustrates how life-cycle analysis (LCA) can be a very effective tool in quantifying the environmental burdens of a product, process, or activity, looking at the whole cycle from extraction of resources through to recycling or disposal. Case studies are presented, including LCA on biorenewables vs fossil fuels and chlorine vs UV disinfection technologies.

Chapter 4 addresses risk assessment, which is an organized process used to describe and estimate the likelihood of adverse health and environmental impacts from exposures to chemicals released to air, water, and land. Risk assessment is also a systematic, analytical method used to determine the probability of adverse effects. A common application of risk assessment methods is to evaluate human health and ecological impacts of chemical releases to the environment. Information collected from environmental monitoring or modeling is incorporated into models of human or worker activity and exposure, and conclusions on the likelihood of adverse effects are formulated. As such, risk assessment is an important tool for making decisions with environmental and public health consequences, along with economic, societal, technological, and political consequences of a proposed action. This chapter addresses the assessment of risks to human health as well as ecological risks, and, briefly, ecological risk management. In addition, a case study outlines the high costs of failing in the area of risk communication: managers of an oil-and-water separation plant that experienced a brief accidental smoke release learned the heavy penalty of having failed to communicate with the facility's neighbors.

The role of economics in pollution prevention is very important, even as important as the ability to identify technologies changes to the process, new and emerging technologies, zero-discharge technologies, technologies for biobased engineered chemicals, products, renewable energy sources and its associated costs. Chapter 5 shows some methods that can be used to assess the costs of implementing pollution prevention technologies and making, cost comparisons to evaluate the cost-effectiveness of various operations. The concept of best available control technologies is introduced, and we analyze the costs and benefits of manufacturing biobased products. The topics treated illustrate that biobased new development can lead to sustainable economic progress and a healthier planet.

Sustainable development is about creating a business climate in which better goods and services are produced using less energy and materials with no or less waste and pollution. Natural steps and systems are a model for thinking about how to produce, consume and live in sustainable cycles: nature produces little or no waste, relies on free and abundant energy from sun and uses renewable resources. In Chapter 6, we focus on a framework that integrates environmental, social, and economic interests into effective chemical and allied business strategies.

Chapter 7 describes some successful ZD processes and technologies. Case studies presented examples of zero discharge technologies and by-product synergies associated with air pollutants, wastewaters, and solid wastes.

Chapters 8, 9, and 10 address process and technology development tools for achieving pollution prevention at the source for air, water, and solid and hazardous wastes. These chapters summarize some of the best available industrial processes, techniques and technologies that are sustainable and inherently environmentally friendly, and some of the best available control technologies, with process descriptions, theoretical background, and advantages and disadvantages. Each of these chapters introduces some of the innovative processes that are emerging for P3 applications.

Chapter 11 presents a unique framework of design methodologies that are collectively referred to under the general heading of *process integration for the minimization of environmental discharges*. Process integration is a holistic approach to process design and operation that emphasizes the unity of the process and it can be broadly categorized into *mass integration* and *energy integration*. The chemical process industry is facing increased pressure to develop processes and products that are energy efficient, environment-friendly and less expensive. Similarly, the very need to minimize environmental discharges has resulted in the development of "greener" processes. The key to tackling all the challenges of "sustainability" and "industrial ecology" lies in process integration, which involves leveraging all process resources in an optimal fashion to reduce overall cost and increase productivity while simultaneously minimizing energy use and lowering adverse environmental impact. The next decade is likely to witness significant growth in this area, and this chapter is intended to provide the readers with the background necessary to appreciate the value of the process integration approach and to encourage increased and widespread usage of these techniques to improve process and utilities efficiencies, reduce environmental discharges, reduce costs, and optimize industrial processes.

Chapter 12 focuses on P3 efforts in greening the kraft pulp and paper industry, an immense consumer of fossil fuels and process water. The pulp and paper industry has a long history of recycling. The kraft pulping process is unique that it affords the recovery and reuse of most of the chemicals from spent cooking liquors; therefore, this process is a good example of a closed-loop system. In recent years, significant P3 efforts have been focused on this industry, particularly across the North America and Europe. This chapter highlights some of the industry's successes and indicates where it is heading as far as the P3 and ZD issues are concerned.

Chapter 13 highlights a single manufacturing industry, kraft pulp and paper manufacturing, to show how it is slowly but surely achieving ZD status through P3 applications. Our evaluation of the pulp and paper industry's integrated approach toward ZD includes some success stories, in particular, an account of the world's first ZD effluent pulp and paper plant.

A brief epilogue demonstrates that implementation of the methodologies described in the book is being actively pursued all over the world. Governments, foundations, universities, industries, and individuals all have a role to play in this most crucial endeavor.

## REFERENCES

Mulholland, K. L., and Dyer, J. A. (1999). *Pollution Prevention Methodology, Technologies and Practices.* AIChE Press, New York, NY.

U.S. Environmental Protection Agency (1992). *Facility Pollution Prevention Guide.* EPA/600/R-92/088, U.S. EPA, Office of Research and Development, Washington, DC (May).

**PART I**

# METHODOLOGY

# CHAPTER 2

# ZERO DISCHARGE INDUSTRIES

TAPAS K. DAS

Washington Department of Ecology, Olympia, Washington 98504, USA

While there are several practical definitions of zero discharge (ZD) manufacturing, a ZD system is most commonly understood to be one that discharges no waste from a processing and manufacturing site. In such a manufacturing facility (see Figure 2.1) an absolute minimum amount of waste, "ideally zero," is generated and leaves the plant. The only inputs to the facility are the raw materials needed to make salable products and energy. The only outputs are salable products and any byproducts as feedstocks to another plant. The wastes (in air, water, or as solid) or byproducts generated during manufacturing process are recovered using various technologies.

The materials and energy recovered from waste streams either are reused in the plant, or are sold to another plant as feedstock. It is in practice, as well as in theory, possible to isolate some industrial facilities almost completely from the environment by recycling all wastes into materials that can then be manufactured into consumer products. An example of such a facility is a coal-fired power plant. An electron beam-ammonia conversion unit adds ammonia to the effluent gases, which it then irradiates electronically, producing ammonium nitrate and ammonium sulfate that are sold as feedstock to fertilizer manufacturing. The details are given in Chapter 7.

Led by the United Nations University's Zero Emissions Research Initiative (ZERI) in Tokyo, efforts are under way around the world to promote the concept and facilitate the attainment of zero emissions industries. The goal is to modify industrial processes so that services and manufactured goods can be produced without wastes. When wastes cannot be eliminated, technologies are being developed to process the wastes from one industry to serve as the material input for another industry.

The concept of zero emissions was inspired by the business programs of zero defects (total quality management), zero inventory (just-in-time), and zero accidents

*Toward Zero Discharge: Innovative Methodology and Technologies for Process Pollution Prevention*, edited by Tapas K. Das.
ISBN 0-471-46967-X   Copyright © 2005 John Wiley & Sons, Inc.

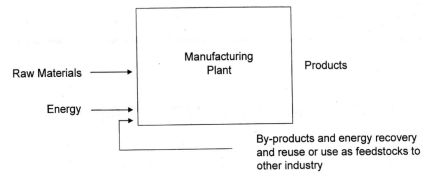

**Figure 2.1.** "Zero" waste manufacturing facility.

(workplace safety), and while its driver is improved business performance, the environmental benefits are also significant.

The basic premise of zero emissions is converting wastes from one industry to the material input of another industry. The application and development of Zero Emissions systems will be the purview of industry, specifically manufacturers and consulting engineering firms. To a degree, the move towards zero emissions is already happening with pollution prevention, waste minimization, and design for the environment. While these systems require further improvement, industries employing them have seen the benefits already.

But it is important to understand that some manufacturing processes inherently produce wastes, even after all reasonable efforts at pollution prevention. Thus in some cases the use of a conversion technology may be more appropriate than a program of pollution prevention: many industrial wastes can be processed to render them viable as material inputs to another industry or to part of an industrial cluster of several connected industries.

## 2.1 SUSTAINABILITY, INDUSTRIAL ECOLOGY AND ZERO DISCHARGE (EMISSIONS)

The concept of *Zero Discharge or Zero Emissions* is the key to sustainable development but is by itself a subset of industrial ecology (Figure 2.2).

Sustainability is defined as "development that meets the needs of the present without compromising the ability of future generations to meet their own needs" (World Commission on Environment and Development, 1987). Sustainability is a worthy vision, but inherently ambiguous, and inescapably expressed in value-laden terms subject to differing ideological interpretations. Accordingly, while the concept provides a useful direction, it is almost impossible to put into operation. Standing alone, it can guide neither technology development or policy formulation.

Industrial ecology is "an approach to the design of industrial products and processes that evaluates these activities through the dual perspectives of product

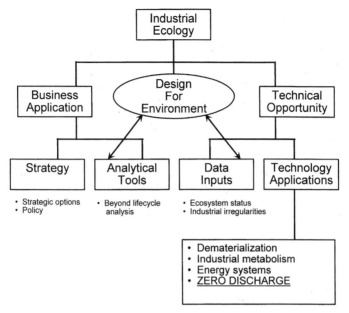

**Figure 2.2.** Zero discharge (emissions) is a subset of industrial ecology. Design for the environment and dematerialization are discussed later in the chapter. Industrial metabolism compromises the energy and value-yielding process essential to economic development.

competitiveness and environmental interactions." The field has been developed, largely academically, over the last 10 years to be "the means by which humanity can deliberately and rationally approach and maintain a desirable carrying capacity" (Graedel and Allenby, 1995). It can be thought of as the science of sustainability for industrial systems — the multidisciplinary study of industrial and economic systems and their linkages with fundamental natural systems.

Even though it is still under development, industrial ecology can provide the theoretical scientific basis upon which understanding, and reasoned improvement, of current practices can be based. It incorporates, among other things, research involving energy supply and use, new materials, new technologies, basic sciences, economics, law, management, and social sciences. It encompasses concurrent engineering, design for the environment (DFE), dematerialization, pollution prevention, waste conversion, waste exchange, waste minimization, and recycling, with *Zero Emissions* as an important subset. Industrial ecology can be a policy tool, but it is neither a policy nor a planning system.

The goal of Zero Emissions is to restructure manufacturing so that there are no wastes. "Zero emissions" is thus applied "industrial ecology" at the manufacturing/ service level, and indeed the terms are often used interchangeably. Note that the emphasis is on the manufacturing level, and not firm or industry level: whereas both "firm" and "industry" imply singular facilities or sectors, wastes are converted most successfully when several facilities are linked in an industrial cluster. As can be

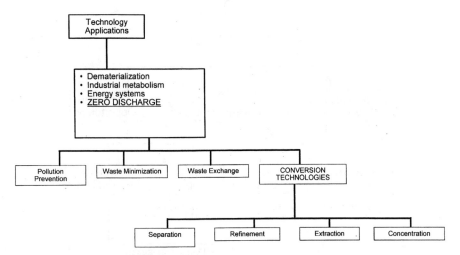

**Figure 2.3.** Zero discharge is supported by an array of tools and methodologies.

seen in Figure 2.3, pollution prevention and waste minimization are part of zero emissions and are technologies to be explored and where possible, optimized. Since it is not always feasible to prevent the generation of wastes early in the industrial cycle, the practice has arisen of trading them after they have been generated. Thus critical components of zero discharge are waste exchange and the conversion of wastes so that they are viable inputs in other sectors. This interaction of systems widens the focus of the waste management effort. Mini-Case Study 2-1 presents a success story often cited in the ZD community.

*Mini-Case Study 2-1: Kalundborg* The evolution of a network of inter-related industries in Kalundborg, Denmark (Figure 2.4), illustrates how a sustainable industrial cluster can operate to achieve an industrial and economic base that functions efficiently (Grann, 1994). A central feature of this system, which has been under development since 1970, is the municipal electric plant (Asnaes), which recaptures surplus steam and distributes it to homes and to the nearby oil refinery, saving 19,000 tons of oil per year. The Statoil refinery removes excess sulfur from its gas, making it suitable for the electric plant to substitute for coal, saving 30,000 tons of coal per year. The removed sulfur is feedstock for the sulfuric acid plant. The electric plant desulfurizes its smoke by using a process that yields calcium sulfate, which in turn is used by Gyproc, a wallboard manufacturer, which obtains 100 percent of its energy requirements from excess gas from Statoil. Statoil sends its wastewater to the electric plant, which purifies it and uses it as boiler feedwater and for cleaning industrial equipment. Excess heat from the power plant warms aquaculture ponds that produce 250 tons of fish per year. The sludge from the ponds, and from the biotechnology plant, becomes fertilizer; fly ash and clinker from the electric plant are sold for use in highway pavement and cement.

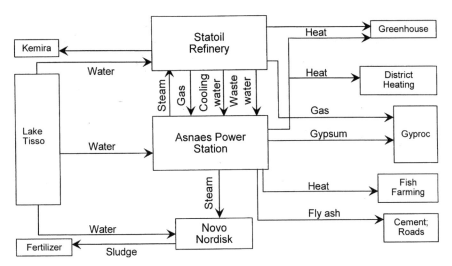

**Figure 2.4.** The industrial cluster at Kalundborg, a coastal community about 75 miles east of Copenhagen, mimics a natural system's use of wastes.

## 2.2 WHY ZERO DISCHARGE IS CRITICAL TO SUSTAINABILITY

Thus in Kalundborg, industry imitates nature — just as one organism's waste is food for another organism, one industry's wastes become another's raw materials. We shall return to Kalundborg in Chapter 7.

To understand why *Zero Discharge* is a critical component of sustainability, it is important to recognize that one principle of sustainability is the efficient and wise use of resources, especially with regard to limiting the amount and type of resource extraction and subsequent pollution loadings. To see how these are related, it may help to think of a cycle with three parts: sources, systems, and sinks.

- The sources include raw materials such as minerals, water, topsoil, and fossil fuels. On this planet, these are limited, but have huge external reserves.
- The systems are our ability to manipulate energy to turn source materials into finished products. Economic and industrial systems are limited only by imagination.
- The sinks are the global waste bins. The ultimate long-term sink is the deep trenches of the oceans; short-term sinks are biosystems such as the atmosphere, rivers, wetlands, and the land. The ability of the sinks to handle wastes is limited; most show adverse effects of pollutant loading in just a few years.

The most restricting rate-limiting component in this three-part model is the sinks. Currently, the depletion of resources is rarely the driving force for resource substitution; instead, change is driven by process innovation to beat the competition, or regulatory intervention imposed from outside. But given the limitations of the sinks, the pressure for modification of an industrial system will increasingly come from the need to reduce loadings to environmental sinks.

To achieve sustainability, the earth's environment must be protected in multiple ways. For example, planners must aim to minimize or eliminate anthropogenic changes to climate, net increases in acidification, loses of topsoil, and withdrawals of fossil water; moreover, biodiversity must be preserved, and buildup of toxic metals and other nonbiodegradable toxics in soils or sediments must be stopped. *Zero discharge* technologies attempt to accomplish these goals by reducing resource extraction and loadings to sinks. The objective is a closed loop in the economic subsystem, so that wastes inevitably created by human activities do not escape to contaminate the environment. *Zero Discharge/Emissions* proponent Gunter Pauli, the founder and former director of ZERI, also notes that in the effort to eliminate waste, *Zero Emissions* "is nothing more than a persistent drive to cut costs." Waste is a form of inefficiency, and an "economic system cannot be considered efficient, or ultimately competitive, if it generates waste" (Pauli, 1996).

*Zero Discharge (or Emissions)* leaves behind the linear "cradle to grave" concept of materials use (Figure 2.5) and embraces a cyclical, "cradle to cradle" vision (Figure 2.6), in which wastes become value-added inputs and the raw materials for other production cycles. This is how natural systems dispose of waste, and according to Pauli, the only way to achieve sustainability.

Meanwhile, the biological sinks are not increasing in capacity. Existing industries will keep operating and generating wastes — some of these wastes, as will be discussed later, containing richer concentrations of recoverable materials than virgin ones. In the interim, there will be a demand for technologies to manage and convert today's wastes into usable feedstocks. Chemical process design engineers and consulting firms will provide focal services to meet this demand through technology development, system integration, and facility operation.

## 2.3   THE NEW ROLE OF PROCESS ENGINEERS AND ENGINEERING FIRMS

Chemical process and product design engineers, environmental engineers, and consulting engineering firms can play a pivotal role as industries move towards the *Zero Emissions* or *Zero Discharge* paradigm, especially firms whose traditional niche has been to treat waste so that it is benign and acceptable for discharge. The role for these engineers in the 21st century is to transform the effluent of one process to serve as the raw material for another process. The new role is not simply facilitating waste exchange; rather, the new jobs include:

- Assessing material flows through the economy and the use of raw materials, water, and energy
- Designing databases with a wider set of information about material flows and manufacturing processes

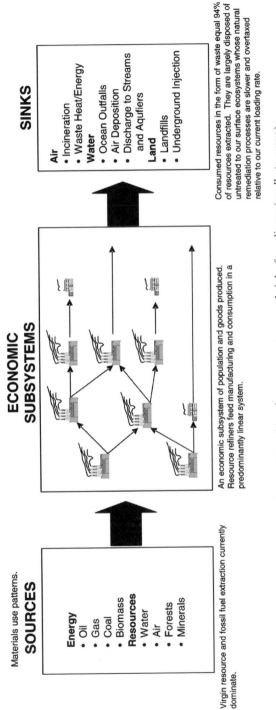

**SOURCES**

Materials use patterns.

**Energy**
• Oil
• Gas
• Coal
• Biomass
**Resources**
• Water
• Air
• Forests
• Minerals

Virgin resource and fossil fuel extraction currently dominate.

**ECONOMIC SUBSYSTEMS**

An economic subsystem of population and goods produced. Resource refiners feed manufacturing and consumption in a predominantly linear system.

**SINKS**

**Air**
• Incineration
• Waste Heat/Energy
**Water**
• Ocean Outfalls
• Air Deposition
• Discharge to Streams and Aquifers
**Land**
• Landfills
• Underground Injection

Consumed resources in the form of waste equal 94% of resources extracted. They are largely disposed of untreated to our surface ecosystems whose natural remediation processes are slower and overtaxed relative to our current loading rate.

**Figure 2.5.** The inter-relationship of sources, systems, and sinks for a linear (cradle to grave).

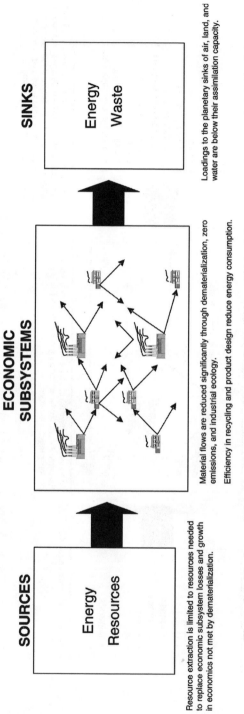

**SOURCES**

Energy

Resources

Resource extraction is limited to resources needed to replace economic subsystem losses and growth in economics not met by dematerialization.

**ECONOMIC SUBSYSTEMS**

Material flows are reduced significantly through dematerialization, zero emissions, and industrial ecology.

Efficiency in recycling and product design reduce energy consumption.

**SINKS**

Energy

Waste

Loadings to the planetary sinks of air, land, and water are below their assimilation capacity.

**Figure 2.6.** The inter-relationship of sources, systems, and sinks for a cyclic (Zero Emissions) materials use pattern.

18

- Working with design firms to understand the production processes of the industries that produce the wastes
- Designing conversion processes
- Identifying purchasers for converted wastes
- Designing material transfer systems to carry wastes to industries that will use them as feedstock
- Identifying industrial clusters and understanding how to fit diverse industries into a successful industrial cluster
- Designing eco-industrial parks and negotiating arrangements that are commercially sound and profitable, yet based on good personal relationships; voluntary, and yet in close collaboration with regulatory agencies

What this means for engineering firms is the need for a broader set of engineering skills and services. As can be seen, consulting engineering firms will find that achieving Zero Emissions entails expertise in areas that have not been part of engineering curriculum, or the professional engineer's exam.

Zero Emissions engineers need to be well trained in design for the environment, concurrent engineering, and industrial engineering, but also able to think and design outside the traditional boundaries of the factory to work in terms of industrial clusters.

Many of the skills and services enumerated above would be applicable to development of an agro-industrial cluster such as the one in Namibia, described in Mini-Case Study 2-2.

*Mini-Case Study 2-2: Beer to Mushrooms: Focusing on the Productivity of Raw Materials*   In the town of Tsumeb in the African desert, Namibian nationals will be implementing Zero Emissions technology at a brewery inaugurated in January of 1997. The three main inputs into beer — grain, water, and energy — are scarce commodities in a developing nation. Brewing uses only 8 to 10 percent of the nutrients in grain, consumes 10 liters of water for every liter of beer produced, and generally requires imported coal, an expensive and polluting energy source.

The lignin-cellulose component of spent grain, which makes up 70 to 80 percent of its bulk, is indigestible to cattle, but is easily broken down by the enzymes of mushrooms. It takes four tons of spent grain to produce one ton of mushrooms, which are a potentially lucrative cash crop for export, because most southern African nations currently import mushrooms. The protein content of the spent grain — up to 26 percent — is used by earthworms, which in turn are fed to chickens and pigs. Processing the waste from the animals in a digester could supply all of the vapor energy required for brewing. Brewery wastewater is high in nutrients but is too alkaline for crops. However, it can be used to grow spirulina, which generate up to 70 percent protein.

The brewery's thermal waste could heat greenhouses or the brewery. These interrelated industries will form an optimal industrial cluster for increasing the productivity of the brewery's raw materials in ways that also produce food for humans that is high in nutritional value.

## 2.4   ZERO DISCHARGE METHODOLOGY

Over the last few years, the members of ZERI have developed a five-step methodology for implementing Zero Emissions. Pauli's *Breakthroughs* (1996) provides a far more comprehensive approach that extends well beyond the manufacturing site. The summary provided here emphasizes the use and impact of a ZD approach at the manufacturing level.

### 2.4.1   Analyze Throughput

The first step towards achieving Zero Discharge and/or Zero Emissions is an in-depth review of the industry to see if total throughput is possible. This means determining whether all material inputs can be found in the final product — if there are no wastes, all inputs must have ended up in the product. One of the few industries where this can occur is cement manufacturing. In the Mini-Case Study 2-2, however, only a small fraction of the nutrients in the grain ends up in beer.

If throughput is not total, the next step is to determine whether the products manufactured can be easily reintegrated into the ecosystem without additional costs for processing, energy, or transportation. However, since this is rarely possible, most industries will not achieve Zero Discharge unilaterally.

Process and product design engineers will probably find their first opportunities by meeting with their traditional clients, analyzing each client's throughput, and looking for opportunities for pollution prevention and waste minimization that the client's in-house experts may not have seen. The analysis would include evaluating products and services presently being produced, processes and materials used, and management of environmental issues including energy efficiency, as well as clarifying the full scope of emissions.

### 2.4.2   Inventory Inputs and Outputs

Once the initial analysis has determined that total throughput is not possible, and that wastes will be generated, the next step is to assess the industry's inputs and outputs, and to inventory all the outputs ("wastes"). A diagram of the inputs and outputs of a system like that of Figure 2.4 is then used to compile basic overview of the company's resources and needs. From this information, design engineers and process specialists can attempt to modify the manufacturing process so that it can become a Zero Emissions system.

Extracting raw materials and processing them imposes significant environmental burdens. An analysis of the industrial metabolism of the product (i.e., its input, materials use, and life cycle expectancy) will help determine the path of least environmental impact. Some materials choices will yield better throughput or by-products that are more suited for use as an input for another industry.

Additional audits and inventories may be needed to determine manufacturing efficiency by percentage of input wasted, to quantify amounts of waste landfill by type of material, to account for amounts of materials collected for recycling, and to identify major emissions of waste heat and the site and amounts of wastewater discharges. Analysis of these outputs may reveal the most effective ways to re-use these outputs and help to determine which industries could use the wastes as raw materials. For example, at the Namibian brewery, spent grain, excess heat, and wastewater all have potential uses in producing food items.

### 2.4.3  Build Industrial Clusters

In sectors that cannot achieve Zero Emissions unilaterally, it may be necessary to build industrial clusters. The input-output analysis leads directly into development of clusters of industries that can use each other's outputs. Developing effective clusters calls for executives to look beyond single industries and make innovative connections among seemingly unrelated potential partners in new industrial clusters. Companies are loathe to implement such changes, however. In addition to concerns about anti-trust regulations, and the need to rely on single vendors for supply, there is fear that relinquishing information about waste stream composition will allow their competitors to deduce proprietary secrets.

Also critical is the geographic location of the client's potential partners, as transportation is a key factor in optimizing waste exchange and use of conversion technologies. The most obvious link in the search for industrial cluster partners will be obtaining industrial input data for other industrial sectors and determining if the client's waste flows could serve (in some converted form) as a material input to another sector. The second place to look for candidate industrial clusters is the historical records of waste exchanges. These material flows will demonstrate which materials being discarded by a sector are of a volume and quality desirable to another sector.

Industries that buy process wastes are taking in nonvirgin material of a grade that may fall short of the purchasers' specification. This is an opportunity for materials blending. For example, plastics can be recycled to make a lumberlike product, but; the grade of such recycled products is not always acceptable as a direct input. If a contaminated waste flow is not of sufficient volume, however, blending in virgin plastics can bring both quality and volume up to manufacturing specifications.

Once the potential partners have been identified, the industrial cluster should be designed and developed. Kalundborg is an excellent example of a cluster that includes heavy industries, while Tsumeb's cluster is based primarily around food production and processing. Elsewhere in the world, industrial clusters have yet to be developed.

China is an industrializing nation that has abundant and cheap supplies of coal, but burning it to generate electricity produces $SO_x$ and $NO_x$. The key to sustainability for China will therefore be development of industrial clusters that link energy, agriculture, and sewage treatment, in the fundamental format for Zero Emission communities. The most effective incentive to develop such clusters is economics, and, unlike conventional $SO_x$ and $NO_x$ treatments, the system for the electron beam/ammonia conversation of these pollutants, has a financial payback of 10 to 15 years. Chapter 7 treats this technology in more detail.

### 2.4.4   Develop Conversion Technologies

The easiest connections for industrial clusters are through a simple, direct waste exchange. The next easiest route is to develop an intermediary process that will take one industry's current waste stream, convert it to a usable form, and transfer it to a purchasing industry. Now we consider briefly the pivotal function of conversion technologies as illustrated by the problems encountered in the paper recycling industry as it works toward attaining Zero Emissions in the United States (see also Chapters 11, 12, and 13).

Paper recycling is quintessentially "green." But current processes used to de-ink paper remove only 70 to 80 percent of the ink particles, leaving recycled papers an unattractive gray. The wastes are a toxic mix of ink, short fibers, coating chemicals, and paper fillers that requires both primary and secondary treatment before disposal. De-inking is both inefficient and expensive, and results in a product that is often higher priced and lower quality.

Under the auspices of ZERI, a conversion technology is being developed that results in 100 percent removal of ink, and three viable outputs. The recaptured ink could be re-used in printing or for making pencils (as is already done with ink from photocopiers). The long fibers could be made into paper again or used in cardboard. The remaining sludgy mixture of short fibers and residues could be dried and used as acoustic insulation inside building walls or as ceiling tiles. The sludge could be used to make shock-absorbent packaging such as egg cartons or replacements for corrugated cardboard.

The industrial cluster built around paper recycling thus includes recapturing ink, making new paper, and making building and packaging materials. Canada, Latvia, and Italy have tested this conversion technology, the steam explosion system. Many cities, states, and national governments, however, require that recycled paper be used in newspapers, and where these regulations are in force, a system that produces a grade of paper better than that needed by newspapers is not likely to be implemented.

### 2.4.5   Designer Wastes

So-called designer wastes can serve as direct feedstock to another sector, or, if properly processed through a conversion technology, as processed feedstock. If the beer industry for example, used a sugar-based cleanser instead of a caustic cleanser for its bottle washing process, the discharge water could serve as a direct feed to fish ponds, without needing any conversion. Yet other solutions will require installation assistance in adjusting or adapting material flows within the client's processes so that its waste output is in acceptable form.

Building industrial clusters may involve working upstream within a facility to modify its production process so that the waste is produced in an acceptable "designer" form for conversion. Or, perhaps downstream, working with the purchaser industry, to help it modify its processes so that it can accept the converted waste.

This means that the involved engineering consultants must become familiar with the products and materials provided by suppliers to the producer and the materials needed by the purchaser of the designer wastes. All parties must have access to a database of specific information about the level of impurities acceptable in each final product. For example, companies that produce basic material inputs (such as plastics, oils, lubricants, and papers) deal in high volumes and some manufacturing processes can tolerate impurities. These companies are already experienced in their own refining processes, and may also be knowledgeable about the manufacturing requirements of their customers, but their expertise may not extend to particular impurities present in the producer's waste stream. The database can provide the information necessary to complete the connections.

### 2.4.6  Reinvent Regulatory Policies

Experienced professionals are probably already aware of how government policies inadvertently inhibit creativity in re-use of wastes. These policies can also inhibit formation of effective industrial clusters. For example, breweries are regulated as an industry and the facilities are generally located in industrial areas. However, to make efficient use of their discharge water in aquaculture, breweries should be located in agricultural zones. Similarly, regulations aimed at providing a market for recycled newsprint need to be revised so that technologies that produce a better grade of paper can flourish — and allow more complete re-use of all the other by-products associated with complete de-inking.

Ironically, Zero Discharge goals are inhibited in the United States by regulations pursuant to the Resource Conservation and Recovery Act of 1976 (RCRA). Transferring "wastes" among the members of industrial clusters is often prohibited by RCRA because its regulatory net entangles all wastes, whether hazardous or not. Its broad scope has the unintended consequences of creating disincentives to invest in recovery technologies and blocking progress toward pollution prevention and recycling. RCRA waste classifications can put kinks in potential closed-loop systems. Indeed, as Robert Herman (1989) put it, "the essence of the environmental crisis is not nearly so much bad actors as the whole, often contradictory structure of incentives in the economy." Regulatory policies need to be reinvented to foster development of breakthrough conversion technologies and encourage cross-sector markets for designer wastes.

## 2.5  MAKING THE TRANSITION

The shift toward a Zero Discharge culture, especially in a world dominated by industrial ecology, will see the development of new products, services, and industries.

Our global economic system depends on extracting massive quantities of materials from the environment — after extraction and processing, the "annual accumulation of active materials embodied in durables, after some allowance for discard and demolition, is probably not more than six percent of the total. The other 94 percent is converted into waste residuals as fast as it is extracted" (Allenby and Richards, 1994; Ayres et al., 1996). In the United States, this means more than 10 tons of "active" mass (excluding fresh water) per person each year.

Of this mass, roughly 75 percent is mineral and nonrenewable, and 25 percent is from biological sources. Of the biological materials, none of the food or fuel becomes part of durable goods and even most timber is burned as fuel or made into pulp and paper products that are disposed of. Of the mineral materials, about 80 percent of the mass of the ores is unwanted impurities, and of the final products, a large portion is processed into consumables and throwaways. Only in the case of nonmetallic minerals is as much as 50 percent of the mass embodied in durable goods such as cement and ceramics (Allenby and Richards, 1994). All of this translates into an estimated more than 12 billion tons of industrial wastes annually in the United States (Allen and Rosselot, 1997).

In addition to materials lost as waste residuals during extraction and processing, finished goods are dissipated/lost because present in concentrations too small to be economically recoverable. Many products are inherently dissipative, and lost with a single normal use. These include packaging, lubricants, solvents, flocculants, anti-freezes, detergents, soaps, bleaches, dyes, paints, paper, cosmetics, pharmaceuticals, fertilizers, pesticides, herbicides, and germicides. From one-half to as much as seven-eighths of the toxic heavy metals used in pigments (red and white lead, cadmium red, chrome green, cobalt blue), in insecticides (arsenic), and in wood preservatives, fungicides, catalysts, and plastic stabilizers are dispersed into the environment beyond economic recoverability. Other materials are lost to uses that are not inherently dissipative, but are so in effect because of the difficulty of recycling. Allenby and Richards (1994) point out that the total elimination of manufacturing wastes probably is an unattainable goal because it would require, in addition to technological advances not yet in place, 100 percent cooperation by consumers.

### 2.5.1 Recycling of Materials and Reuse of Products

A critical element of an interim strategy is enhanced recovery. This can be approached from two directions: re-use of products and recycling of materials. Re-use of products includes return, reconditioning, and remanufacturing. The energy required for re-use and recycling is one of the key factors determining recoverability of a product. The closer the recovered product is to the form it needs to be in for recycling, the less energy is required to make that transformation. From the standpoint of economic development it is worth pointing out that the reconditioning or remanufacturing cycle is relatively; it requires roughly half the energy and twice the labor per physical unit of output.

Recycling materials means closing the loop between the supply of post-consumer waste and the demand for resources for production. Recycling of materials will be the business of the Zero Emissions engineer; re-use of products will also involve the Zero Emissions engineer, but will have lots of front-end work from another professional, the concurrent engineer. Concurrent engineering, which incorporates aspects of industrial engineering, product design, and product manufacturing, is an integrated approach that seeks to optimize materials, assembly, and factory operation. These engineers examine the broader context of a product, including technology for managing the environmental impacts of its transport, intended use, recyclability, and disposal, as well as the environmental consequences of the extraction of the raw material used in its production.

The ultimate fate of all materials is thus dissipation, being discarded, or recycling and recovery. With 94 percent of materials extracted from the environment being converted to wastes, current levels of recovery are clearly not sufficient. Recovery of materials from wastes will reduce the extent of resource extraction (but will not slow the speed of material flows through the economy). Aluminum and lead are two resources currently being heavily recycled, but evidence shows that there is potential for a lot more resources to be recovered from wastes.

Sherwood plots are diagrams that permit the graphic comparison of concentrations of materials in nature against their commodity cost. The sample plot in Figure 2.7 shows that the price for a commodity depends on its concentration in nature before extraction and refining. Figure 2.8, similar plot for concentrations of materials in wastes, again shows a consistent line.

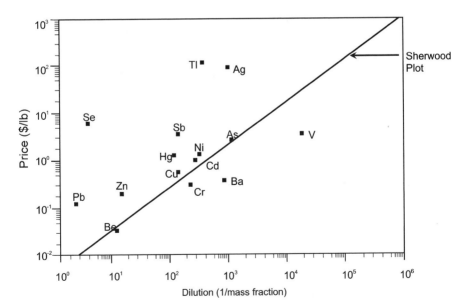

**Figure 2.7.** A Sherwood plot for waste streams: minimum concentration of metal wastes undergoing recycling versus metal prices.

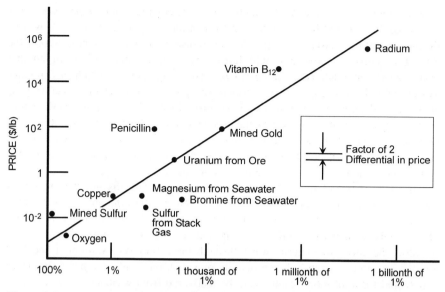

**Figure 2.8.** A Sherwood diagram showing the correlation between the selling price of materials and their degree of dilution in the matrix from which they are separated.

Together, the Sherwood plots demonstrate the recovery potential of materials. The elements plotted above the line in Figure 2.7 should be vigorously recycled because they are present in individual by-products in relatively high concentrations. Lead, zinc, copper, nickel, mercury, arsenic, silver, selenium, antimony, and thallium are more economical to recover from waste than from nature. Extensive waste trading could significantly reduce the quantity of material requiring disposal because resource extraction uses from wastes, not virgin feedstocks (Allen and Behmanesh, 1994).

### 2.5.2 Dematerialization

One critical component of the industrial ecology paradigm is dematerialization. Dematerialization means using less material to make products that perform the same function as predecessors. Sometimes this means smaller or lighter products, but other aspects can include increasing the lifetime of a product or its efficiency. The net effect is a reduction in overall resource extraction. Dematerialization is thus a way to increase the percentage of active materials embodied in durables, and to reduce the percentage that is left as waste residuals.

However, dematerialization has limits in achieving Zero Emissions. We may also need to think of re-materialization — products that may or may not have a lighter weight in their final form, but whose production, use, and subsequent conversion or recyclability fits within the Zero Emissions paradigm. This is demonstrated by the ease and benefit of recycling older model cars versus the newer ones. Nonetheless, economists and engineers point out that optimizing for environmental protection alone means some loss in safety, efficiency, durability, convenience, attractiveness, and price.

### 2.5.3 Investment Recovery

During the transition to Zero Emissions, another early need will be for companies to fill the "decomposer niche," a term for a specialized form of recycling developed by Raymond Cote at Dalhousie University in Ottawa (Cote, 1995). Just as decomposer organisms turn dead animals and vegetable matter into forms that can become food for other animals and plants, decomposer niche companies will "consume" otherwise unusable wastes by processing them into usable feedstock or disassembling equipment and marketing reusable components and materials.

The work of decomposers also can be visualized as investment recovery. Taking a systematic approach to ending waste, investment recovery is a traditional service, according to Cote, "an integrated business process that identified, for redeployment, recycling, or remarketing, non-productive assets generated in the normal course of business." These assets include idle, obsolete, unused, or inoperable equipment, machinery, or facilities; excess raw materials, operating inventories, and supplies; construction debris; equipment and fixtures in facilities scheduled for demolition; off-grade, out of specification, or discontinued products; and process waste (Cote, 1995; 2003).

The goal of investment recovery is to develop strategies and procedures to recapture the highest value from all surplus assets in a company or community. It seeks to reduce operating and disposal costs, prevent disposal of assets as waste, and find markets for redistributing the by-products for increased economic value. One of the operating paradigms for an investment recovery firm would be integration of its functions into a comprehensive strategy for an eco-industrial park. Firms that specialize in this work base their fees on a retainer plus a percent of savings and/ or revenues if there is an incentive.

### 2.5.4 New Technologies and Materials

During transitional stages, existing industries can be identified as potential members of a cluster if minimal design engineering can make them compatible and most of the transfers of materials are occurring in a more basic commodity form, rather than as "designer wastes." Once Zero Emissions has been incorporated at the drawing board level, facilities can be planned to work together in clusters so that the by-products of each enterprise meet the feedstock specifications of other industries in the cluster. This will require innovations in materials and methods alike.

***2.5.4.1 New, Less Toxic Chemicals and Materials*** Examples of new materials include biopolymers, fiber-reinforced composites, high-performance ceramics, and extra-strength concrete. To meet basic environmental requirements, new materials will be biodegradable, nonpolluting, recyclable or convertible; made from renewable resources; have low energy requirements in production and use; and provide a final product with greater strength and durability and lower weight and volume.

**2.5.4.2 Improved Processes**    Consultants can help managers cut costs and create new values by instituting real time monitoring and eliminating inefficiencies in the use of resources all along a product's lifecycle. These inefficiencies include incomplete utilization of material and energy resources, poor process controls, product defects, waste storage costs, discarded packaging, costs passed on to consumers for pollution or low energy efficiency, and the ultimate loss of resources through disposal and dissipative use. Poor resource productivity also can entail costs for waste disposal and regulatory penalties. Methods that are less energy-intensive and more labor-intensive are more sustainable environmentally and socially.

### 2.5.5    New Mindset

As a result of changes in materials and processes, engineering professionals will have to expand the purview of their design parameters. A larger, more integrated design for facilities and manufacturing processes is called for. This is where design for environment comes in: DFE examines the life cycle of the product and considers not only its primary use but the environmental consequences of its production, assembly, testing, servicing, and recycling. In designing an eco-industrial park, for example, manufacturing processes would be linked to material flows and to energy flows. As a result, the design period and overall costs would be higher. Return on investment, however, would be much shorter (see Section 2.6.2).

## 2.6    IN THE FULL ZERO DISCHARGE PARADIGM

Designing Zero Discharge systems requires an expansion of the focus and outputs of the traditional design engineer. Concurrent engineers need to incorporate design for the environment. Industrial engineers need to think in terms of industrial clusters. Environmental engineers need to understand upstream processes better so that they can develop designer wastes. Environmental engineers also need to revamp their processes to begin mimicking resource refining.

Zero discharge engineering firms are expected to be working with design for environment engineers, concurrent engineers, and industrial engineers. They should all be seeking to design wastes, conversion processes, and industrial clusters. Setting the stage for overall product design, the industrial ecology approach assists companies in looking beyond the product to its functionality over its lifecycle. Services and products should be designed and delivered differently as the following six strategic elements of industrial ecology are applied:

- Selection of materials with desired, properties at the outset
- Use of "just in time" materials

- Substitution of processes to eliminate toxic feedstock
- Modification of processes to contain, remove, and treat toxics in waste streams
- Engineering of a robust and reliable process
- Consideration of durability and end of life recyclability

Zero discharge solutions that use conversion technologies should be developed, designed, built, and marketed by the appropriate professionals who understand not only the industrial clusters and the processes involved, but also the upstream and downstream requirements.

## 2.6.1 Opening New Opportunities

As the zero discharge mission gains currency, new opportunities are revealed — to provide cost-saving new design applications, to design new product lines, and to win new customers. These opportunities include:

- Finding cost savings and new revenues in existing operations. Initial cost savings at existing operations should come from pollution prevention and waste minimization, which may already have been optimized. However, new revenues will come from identifying a viable market for the waste stream after it has been converted.
- Entering new markets for existing goods and services. New markets will be entered when cluster partners are identified; for example, brewery specialists will expand into agricultural sectors. The market for handling "designer wastes" is expected to grow significantly, especially for firms specializing in reprocessing. Producers of a wide range of materials processing equipment such as grinding, sifting, sorting, purifying, separating, and packaging will find new markets. However, in some cases the conversion process will be handled by an intermediary company that will alter the wastes mechanically, chemically, or biologically to meet customer specifications.
- Developing new technologies, processes, and materials. Many of the pollution control firms will begin partnering with the upstream commodity producers (e.g., petroleum, chemical, and mining companies) and learning their refining techniques.
- Supporting the organizational changes, and technical and information needs of a Zero Emissions-based economy. This will be a business opportunity primarily for those offering skills in informational and organizational systems.
- Integrating technologies and methods into innovative new systems. As Zero Emissions expands, professionals from different sectors will connect to benefit from each other's skills and experience.
- Developing the infrastructure for eco-industrial parks. Requires equipment to channel the flow of materials, water, or heat between plants and communities. Civil engineering firms that specialize in urban design and infrastructure systems should find great opportunities in providing the integrated system designs.

## 2.6.2 Providing Return on Investment

A simple economic metric, return on investment (ROI) is a quick measure of when an item will pay for itself. The time between the initiation of on investment and the achievement of ROI is called the payback period. Pollution control technology, which is not traditionally viewed as an investment able to generate a return on investment, is usually measured in terms of lowest available cost to meet regulatory guidelines. If wastes are viewed as materials, however, as in ZD, the whole picture changes. Some have joked about giving all materials produced in a facility a product name and an advertising budget and/discontinuing any "product" that does not sell.

Engineers have historically been compensated based on overall project cost. Incentives need to be shifted to designs that reduce material and energy flows. Compensation based on energy efficiency is being implemented in some projects and it successful because energy efficiency can be measured in a single unit (joules of energy saved). While material flows are not as easy to quantify in a single unit, waste recovery systems can provide an excellent return on investment, as illustrated in Mini-Case Study 2-3.

*Mini-Case Study 2-3: Recovery of Wastes from Palm Oil Extraction Yields High Return on Investment* Recovery of wastes from agro-industries is an extremely promising aspect of Zero Emissions. One of ZERI's first projects focuses on recovering all of the solid, liquid, gaseous, and thermal wastes from the Golden Hope Plantation in Malaysia, the largest oil palm plantation in the world. With the commitment of Meta Epsi, a large engineering group with substantial interests in palm oil plantations, operation of the pilot project for the total use of palm oil biomass commenced in the summer of 1996.

The pilot project uses steam explosion to provide for conversion of biomass into recoverable fibers, with a goal of re-using the spent seeds, bunches, leaves, and trunks that Golden Hope used to pay to have disposed of. It costs $50 Malaysian (M$50) to produce one ton of commercially usable fiber, which can be sold for approximately M$350. Products made from the fiber include MDF board, stuffing for car seats, and bedding for medical use. Just one of the mills built to process waste fiber generates pre-tax profits of approximately M$7.8 million (about $3 million U.S.).

## 2.7 SEPARATION — A KEY CONVERSION TECHNOLOGY

Conversion technologies are the reverse of two of the traditional control technologies — destruction and dilution. Destruction technology usually operates as incineration and is profitable for exothermic waste streams, but gasification systems not sufficiently developed for use in extracting base elements or simple gases. "Dilution as the solution to pollution" is inefficient, very wasteful of materials does nothing to hinder buildup of toxics in the environment. However, one already familiar control technology — separation — offers potential for developing conversion technologies.

Separation relates to the removal or isolation of components from process streams. Separation technology includes extraction, refinement, and concentration — also the technologies of materials extraction and refinement industries. Separation technology is a critical element of any Zero Emissions conversion technology because of its applicability to a broad set of materials and its ability to segregate materials and achieve desirable purity levels. The range of technologies included in separation is diverse (see Table 2.1).

In a liquid matrix or process stream, the technologies of primary interest are tied to the removal of suspended solids, dissolved solids, miscible and immiscible liquids, and dissolved gases. For gaseous process streams, the technologies relate to suspended solids removal, miscible gas stream, and suspended liquids.

Most notably, the advances in membrane technologies show incredible promise in separation and refining of low concentration material streams. Membrane technologies include reverse osmosis, nanofiltration, and ultrafiltration. Membrane systems function ideally for small systems, require low energy for operation, and quickly attain a steady state (see Sections 9.1.1 and 9.2.2).

The principles of gas separation by membranes have been in industrial use for over a decade for gas and vapor separation in a variety of applications, including VOC recovery (see Section 8.2.5).

Pervaporation is a membrane separation process that extracts small quantities of materials from a larger matrix. Unlike size-dependent extraction (e.g., micro-filtration and ultra-filtration), pervaporation is a solubility-dependent separation method. Pervaporation is more desirable than traditional membrane separation in that it draws small quantities across the membrane instead of larger volumes. It is an excellent extraction method when attempting to maintain an appropriate living environment for cultured organisms (see Section 9.1.1).

## 2.8 CONSTRAINTS AND CHALLENGES

The implementation of Zero Emissions faces constraints and challenges, as well as new opportunities. For example, the use of dissipative materials poses a design challenge: If solvents and flocculants are no longer to be used, what would replace them? Chemical manufacturers need to work with design engineers to arrive

**TABLE 2.1. Key Separation Technologies**

| Liquid Phase | Solid Phase | Gas Phase |
|---|---|---|
| Ion exchange | Screening | Adsorption |
| Reverse osmosis | Magnetic separation | Membranes |
| Diffusion dialysis/electrodialysis | Air classification | Absorption |
| Electrolytic methods | | Condensation/cryogenics |
| Filtration/dewatering/evaporation | | |
| Distillation/extraction | | |

at an understanding of the constraints of separation technologies. Manufacturing any material without emissions is difficult, but working with chemicals is particularly challenging because of the need to develop nontoxic materials that are also bio-degradable. Two possible solutions are biological.

- Biopolymers are an outgrowth of chemurgy, the division of applied chemistry that deals with industrial utilization of organic raw materials, especially from agro-business. These substances, complex molecules formed in biological systems, can replace toxic, dissipative materials currently used as adhesives, absorbents, lubricants, soil conditioners, cosmetics, drug delivery vehicles, and textile dissipative. Substitutes for toxic materials and mechanical processes to substitute for dissipative materials are aspects of the same principle.
- Enzymes are natural catalysts that speed up chemical reactions without being consumed in the process. They function best in mild conditions, so their use requires up to one-third less energy than many synthetic chemicals, paradoxically, this lower need for energy can be an obstacle in a system that still rewards large-scale energy use with reduced rates. Enzymes are especially useful in systems designed to reduce or eliminate dissipative losses.

There is also a need for a taxonomy of environmental technologies that clarifies opportunities for fast developing, generic processes to address such recurring problems as treating large streams of contaminated water from various processes and oxidation in air. Chemical engineering and related professions ought to be able to make rapid advances in such areas.

Many of the industries in the investment recovery or "decomposer niche" are hard put to compete against large-scale facilities that produce materials from virgin materials. More recently, however, economies of scale for resource extractors and processors, along with cheap energy supplies, have been introduced almost everywhere in the world. For example, economies of scale have enabled chemical companies to produce plastics at a price that other manufacturers, as well as the individual consumer, can afford.

## REFERENCES

Allen, D. T., and Behmanish, N. (1994). *Wastes as Raw Materials*. In *The Greening of Ecosystems*, B. Allenby and D. Richards, Eds., Washington, D.C., National Academy Press.

Allen, D. T., and Rosselot, K. S. (1997). *Pollution Prevention for Chemical Processes*. Wiley, New York, NY, Chapter 2, pp. 19–32.

Allenby, B. R., and Richards, D. J. (1994). *The Greening of Industrial Ecosystems*. National Academy Press, Washington, D.C.

Ayres, R. U., Ayres, L. W., and Ayres, L. (1996). *Industrial Ecology: Towards Closing the Materials Cycle*. Edward Elgar, Cheltenham, UK.

Cote, R. P. (2003). *A Primer on Industrial Ecosystems: A Strategy for Sustainable Industrial Development*, Eco-Efficiency Center, School for Resources and Environmental Studies, Dalhousie University, Halifax, Nova Scotia, pp. 1–34.

Cote, R. P. (1995). Supporting pillars for industrial ecosystems. *Journal of Cleaner Production*, Vol. 5, Nos. 1–2, pp. 67–74.

Graedel, T. E., and Allenby, B. R. (1995). *Industrial Ecology*. Prentice Hall, Englewood Cliff, New Jersey.

Grann, H. (1994). The industrial symbiosis at Kalundborg, Denmark. Paper presented at the National Academy of Engineering's Conference on Industrial Ecology, Irvine, CA (May 9–13).

Herman, R. (1989). *Technology and Environment*. National Academy Press, Washington D.C.

Pauli, G. (1996). *Breakthroughs*. Epsilon Press Limited, Haslemere, Surrey, England.

World Commission on Environment and Development (1987). *Our Common Future*. Oxford University Press, Oxford, UK.

Zero Emissions Research Institute (ZERI), United Nations University (UNU), Tokyo, Japan. http://www.zeri.org; http://www.unep.or.jp/ietc/

# CHAPTER 3

# FUNDAMENTALS OF LIFE CYCLE ASSESSMENT

TAPAS K. DAS

Washington Department of Ecology, Olympia, Washington 98504, USA

## 3.1 WHAT IS LIFE CYCLE ASSESSMENT?

As environmental awareness increases, industries and businesses have started to assess how their activities affect the environment. Society has become concerned about the issues of natural resource depletion and environmental degradation. Many businesses have responded to this awareness by providing "greener" products and using "greener" processes. The environmental performance of products and processes has become a key issue, which is why some companies are investigating ways to minimize their effects on the environment. Many companies have found it advantageous to explore ways of moving *beyond* compliance using pollution prevention strategies and environmental management systems to improve their environmental performance. One such tool is called life cycle assessment (LCA). LCA was first defined in the way we know it today at the Vermont Conference of the Society of Environmental Toxicology and Chemistry (SETAC, 1991). The concept is holistic, promoting analysis, quantification, understanding of all the environmental impacts associated with an activity. The provision of such information aids decision making and helps in the formulation of environmental strategy and policy as such, LCA has been accepted into the mainstream of environmental thought and management (Azapagic, 2000; Bhander et al., 2003; Cooper, 2003; Curran, 1996, 2003; Das et al., 2001; ENDS, 1996; SETAC, 1993; Sharma and Das, 2002; Sharma et al., 2001; Tshudy, 1994; UNEP-SETAC, 2000).

In Chapter 2 we introduced the "cradle to grave" concept of materials. Life cycle assessment is such an approach for assessing industrial systems. It begins with the gathering of raw materials from the earth to create the product and ends when all materials are returned to the earth. LCA evaluates all stages of a product's life

*Toward Zero Discharge: Innovative Methodology and Technologies for Process Pollution Prevention,* edited by Tapas K. Das.
ISBN 0-471-46967-X   Copyright © 2005 John Wiley & Sons, Inc.

from the perspective of interdependence, focusing on how one operation leads to the next. LCA enables the estimation of the cumulative environmental impacts resulting from all stages in the product life cycle, often including the effects of activities not considered in more traditional analyses (e.g., raw material extraction, material transportation, ultimate product disposal). By including the impacts throughout the product life cycle, LCA provides a comprehensive view of the environmental aspects of the product or process and a more accurate picture of the true environmental trade-offs in product selection.

Most of the materials presented in the following sections are obtained from the U.S. Environmental Protection Agency's document, posted on the web at: http://www.epa.gov/ORD/NRMRL/lcaccess/lca101.htm, which provides an introductory overview of LCA.

The LCA process is a systematic, phased approach. Its components, discussed in detail in Section 3.2 can be listed briefly as follows:

1. **Goal Definition and Scoping**   Define and describe the product, process or activity. Establish the context in which the assessment is to be made and identify the boundaries and environmental effects to be reviewed for the assessment.

2. **Inventory Analysis**   Identify and quantify energy, water and materials usage and environmental releases (e.g., air emissions, solid waste disposal, wastewater discharge).

3. **Impact Assessment**   Assess the human and ecological effects of energy, water, and material usage and the environmental releases identified in the inventory analysis.

4. **Interpretation**   Evaluate the results of the inventory analysis and impact assessment in terms of the goal established at the outset. This makes it possible to select the preferred product, process or service with a clear understanding of the uncertainty and the assumptions used to generate the results.

### 3.1.1   Benefits of Conducting an LCA

An LCA will help decision-makers select the product or process that results in the least impact to the environment. This information can be used with other factors, such as cost and performance data to select a product or process. LCA data identify the transfer of environmental impacts from one medium to another (e.g., eliminating air emissions by creating a wastewater effluent instead) and/or from one life cycle stage to another (e.g., from use and reuse of the product to the raw material acquisition phase). Without an LCA, the transfer might be overlooked and excluded from the analysis because it is outside the typical scope or focus of product selection processes.

For example, when selecting between two rival products, it may appear that Option 1 is better for the environment because it generates less solid waste than Option 2. However, LCA might indicate that the first option actually creates larger cradle-to-grave environmental impacts when measured across air, water, and land. Perhaps, for example, it is seen to cause more chemical emissions during the manufacturing stage. Therefore, the second product although it produces solid waste may actually produce less cradle-to-grave environmental harm or impact than the first technology because its chemical emissions are lower.

This ability to track and document shifts in environmental impacts can help decision makers and managers fully characterize the environmental trade-offs associated with product or process alternatives.

By performing an LCA, researchers can:

Develop a systematic evaluation of the environmental consequences associated with a given product.

Analyze the environmental trade-offs associated with one or more specific products/processes to help gain stakeholder (state, community, etc.) acceptance for a planned action.

Quantify environmental releases to air, water, and land in relation to each life cycle stage and/or major contributing process.

Assist in identifying significant shifts in environmental impacts between life cycle stages and environmental media.

Assess the human and ecological effects of material consumption and environmental releases to the local community, region, and world.

Compare the health and ecological impacts between two or more rival products/ processes or identify the impacts of a specific product or process.

Identify impacts to one or more specific environmental areas of concern.

### 3.1.2 Limitations of LCA

Performing an LCA can be resource and time intensive. Depending upon how thorough an LCA the users wish to conduct, gathering the data can be problematic, and the unavailability of crucial data can greatly impact the accuracy of the final results. Therefore, it is important to weigh the availability of data, the time necessary to conduct the study, and the financial resources required against the projected benefits of the LCA.

LCA will not determine which product or process is the most cost effective or works the best. Nor does it take into account broader issues of acceptability. Therefore, the information developed in an LCA study should be used as one component of a more comprehensive process of assessing the trade-offs between performance and economic, geopolitical, and social costs.

## 3.2 CONDUCTING LCA

To see the interrelatedness of the four components of an LCA, it is useful to refer to the flowchart shown in Figure 3.1.

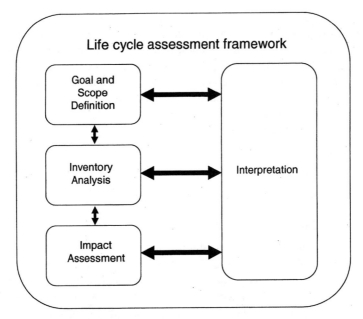

**Figure 3.1.** Life cycle stages, ISO, 1997.

### 3.2.1   Goal Definition and Scope

The LCA process can be used to determine the potential environmental impacts from any product, process, or service. The goal definition and scoping phase will determine the time and resources needed. The defined goal and scope will guide the entire process to ensure that the most meaningful results are obtained. Every decision made during goal definition and scoping impacts either how the study will be conducted, or the relevance of the final results. Goal definition and scoping will result in the determination of the following:

1. The goal(s) of the project
2. The type of information is needed to inform the decision makers
3. How the data should be organized and the results displayed
4. What will or will not be included in the LCA
5. The accuracy required of the data
6. Ground rules for performing the work

Each decision made in the fleshing out of these six areas has an impact on the LCA process, as explained in Section 3.2.1.1 through 3.2.1.6.

***3.2.1.1   Define the Goal(s) of the Project***   The primary goal of the LCA is to choose the best product, process, or service with the least effect on human health and the environment. There may also be secondary goals for performing an LCA, which would vary depending on the type of project. Some typical secondary goals are as follows:

- To prove one product is environmentally superior to a competitive product
- To identify stages within the life cycle of a product or process where a reduction in resource use and emissions might be achieved
- To determine the impacts to particular stakeholders or affected parties
- To establish a baseline of information on a system's overall resource use, energy consumption, and environmental loadings
- To help guide the development of new products, process, or activities toward a net reduction of resource requirements and emissions

### 3.2.1.2  Determine the Type of Information Needed to Inform the Decision-Makers

LCA can help answer a number of important questions. Identifying the questions that the decision-makers care about will help define the study parameters. Some examples include:

What is the impact to particular interested parties and stakeholders?

Which product or process causes the least environmental impact (quantifiably) overall or in each stage of its life cycle?

How will changes to the current product/process affect the environmental impacts across all life cycle stages?

Which technology or process causes the least amount of acid rain, smog formation, or damage to local trees (or any other impact category of concern)?

How can the process be changed to reduce a specific environmental impact of concern (e.g., global warming)?

Once the appropriate questions have been identified, the types of information needed to answer them will be apparent.

### 3.2.1.3  Determine How the Data Should Be Organized and the Results Displayed

LCA practitioners like to organize data in terms of a *functional unit* that appropriately describes the function of the product/process being studied. Comparisons between products and processes must be made on the basis of the same function, quantified by the same functional unit. This ensures that the activities being compared are true substitutes for each other. Careful selection of the functional unit to measure and display the LCA results will improve the accuracy of the study and the usefulness of the results.

An LCA study comparing two types of wall insulation to determine environmental preferability must be evaluated on the same function, the ability to decrease heat flow. Six square feet of 4-inch thick insulation Type A is not necessarily the same as six square feet of 4-inch thick insulation Type B. Insulation type A may have an R factor equal to 10, whereas insulation type B may have an R factor equal to 20. Therefore, type A and B do not provide the same amount of insulation and cannot be compared on an equal basis. If Type A decreases heat flow by 80 percent, you must determine how thick Type B must be to also decrease heat flow by 80 percent.

### 3.2.1.4 *What Will and Will Not Be Included*   Ideally, an LCA includes all four stages of a product or process life cycle: raw material acquisition, manufacturing, use/reuse/maintenance, and recycle/waste management. These product stages are explained in more detail below. To determine whether one or all of the stages should be included in the scope of the LCA, the following must be assessed: the goal of the study, the required accuracy of the results, and the available time and resources. Figure 3.2 presents a set of life cycle stages that could be included in a project related to treatment technologies. Note the "system boundary," which encompasses all aspects of the LCA. Additional examples of the four life cycle stages are explained in more detail below.

*Raw Materials Acquisition*   The life cycle of a product begins with the removal of raw materials and energy sources from the earth. For instance, the harvesting of trees or the mining of nonrenewable materials would be considered raw materials acquisition. Transportation of these materials from the point of acquisition to the point of processing is also included in this stage (USEPA, 1993).

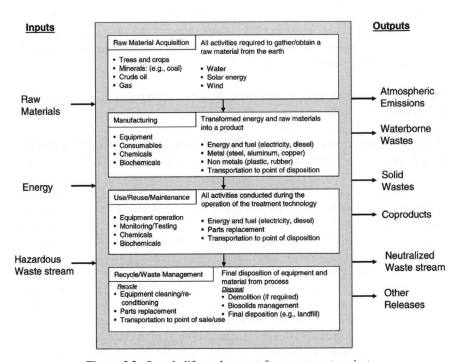

**Figure 3.2.** Sample life cycle stages for a treatment project.

*Manufacturing*   During the manufacturing stage, raw materials are transformed into a product or package, which is then delivered to the consumer. The manufacturing stage is broken into three parts: materials manufacture, product fabrication, and filling/packaging/distribution (USEPA, 1993). While those activities are self explanatory, it is noted that distribution, in which finished products are transported to retail outlets or directly to the consumer, entails environmental effects due to mode of transportation trucking, shipping, or others.

*Use/Reuse/Maintenance*   Once the product is in the consumer's hand, all activities associated with the useful life of the item must be identified: energy demands and environmental wastes from product storage and consumption, as well as any reconditioning, repaired or servicing may be required (USEPA, 1993). When the consumer no longer needs the product, it will be recycled or disposed of.

*Recycle/Waste Management*   Disposition of any product or material whether by recycling, incinerating, dumping, or other mode of waste management, requires energy and results in other environmental wastes (USEPA, 1993). These must be anticipated and listed.

### 3.2.1.5   *Accuracy Required of the Data*
The required level of data accuracy for the project depends on the use of the final results and the intended audience (Will the results be used to support decision-making in an internal process? In a public forum?). For example, if the intent is to use the results in a public forum to support product/process selection to a local community or regulator, then estimated data or best engineering judgment may not be accurate enough to justify basing policy decisions on them. In contrast, if the LCA is for internal decision-making purposes only, then estimates and best engineering judgment may be applied more frequently. This may reduce the overall cost and time required to perform the LCA, as well as enable completion of the study in the absence of precise, first-hand data. The criticality of the decision to be made and the amount of money at stake also come into play in determining the required level of data accuracy.

### 3.2.1.6   *Ground Rules for Performing the Work*
Prior to moving on to the inventory analysis phase it is important to define some of the logistical procedures for the project.

1. **Documenting Assumptions**   All assumptions or decisions made throughout the entire project must be reported along side the final results of the LCA project. If assumptions are omitted, the final results may be taken out of context or easily misinterpreted. As the LCA process advances from phase to phase, additional assumptions and limitations to the scope may be necessary to accomplish the project with the available resources.
2. **Quality Assurance Procedures**   Quality assurance procedures are important to ensure that the goal and purpose for performing the LCA will be met at the conclusion of the project. The level of quality assurance procedures employed

for the project depends on the available time and resources and how the results will be used. If the results are to be used in a public forum, a formal review process is recommended. Evaluators might include internal and external LCA experts and interested parties whose support of the final results is sought. If the results are to be used for internal decision-making purposes only, then an internal reviewer who is familiar with LCA practices and is not associated with the LCA study may effectively meet the quality assurance goals. A formal statement from each reviewer documenting his or her assessment of each phase of the LCA process should be included with the final project report.

3. **Reporting Requirements**   To ensure that the LCA meets appropriate expectations, participants should know from the outset how the final results are to be documented and exactly what is to be included in the final report. When reporting the final results, or results of a particular LCA phase, it is important to thoroughly describe the methodology used in the analysis. The report should explicitly define the systems analyzed and the boundaries that were set. The basis for comparison among systems and all assumptions made in performing the work should be clearly explained. The presentation of results should be consistent with the purpose of the study. The results should not be oversimplified solely for the purposes of presentation.

### 3.2.2   Life Cycle Inventory

A life cycle inventory (LCI) quantifies energy and raw material requirements, atmospheric emissions, waterborne emissions, solid wastes, and other releases for the entire life cycle of a product, process, or activity (USEPA, 1993). Such an inventory is in the form of a list of the quantities of pollutants released to the environment and the amounts of energy and materials consumed. The results can be segregated by life cycle stage, by media (air, water, land), by specific processes, or any combination thereof.

Without an LCI, no basis exists to evaluate comparative environmental impacts or potential improvements. The level of accuracy and detail of the data collected is reflected throughout the remainder of the LCA process. Life cycle inventory analyses can be used by industry for comparing products, processes, materials. Government policy makers, too, can use LCI analyses in the development of regulations targeting resource use and environmental emissions.

The U.S. Environmental Protection Agency published two guidance documents, *Life-Cycle Assessment: Inventory Guidelines and Principles* (USEPA, 1993) *and Guidelines for Assessing the Quality of Life-Cycle Inventory Analysis* (USEPA, 1995). These federal guidelines provide the framework for performing an inventory analysis and assessing the quality of the data used and the results. The two documents define the following steps of a life cycle inventory:

- Develop a flow diagram of the processes being evaluated
- Develop a data collection plan
- Collect data
- Evaluate and report results

***3.2.2.1 Step 1: Develop a Flow Diagram*** A flow diagram is a tool to map the inputs and outputs to a process or system. The system boundary varies for the LCA, as established in the goal definition and scoping phase and expanded to include process inputs and outputs serves as the system boundary for the flow diagram. Unit processes inside the system boundary link together to form a complete life cycle picture of the required inputs and outputs (material and energy) to the system. Figure 3.3 illustrates the components of a generic unit process within a flow diagram for a given system boundary. The more complex the flow diagram, the greater the accuracy and utility of the results. Unfortunately, increased complexity also means more time and resources must be devoted to this step, as well as the data collecting and analyzing steps.

Flow diagrams are used to model all alternatives under consideration (e.g., both a baseline system and alternative systems). For a comparative study, it is important that both the baseline and alternatives use the same system boundary and are modeled to the same level of detail. If not, the accuracy of the results may be skewed.

***3.2.2.2 Step 2: Develop an LCI Data Collection Plan*** An LCI data collection plan ensures that the quality and accuracy of data, characterized as part of the goal definition and scoping phase, meet the expectations of the decision-makers.

Key elements of a data collection plan include the following:

Defining data quality goals,

Identifying data sources and types,

Identifying data quality indicators, and

Developing a data collection worksheet and checklist.

**Define Data Quality Goals** Data quality goals provide a framework for balancing available time and resources against the quality of the data required to make a

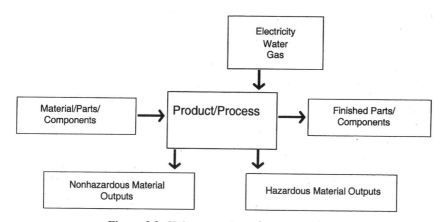

**Figure 3.3.** Unit process input/output template.

decision regarding overall environmental or human health impact (USEPA, 1989). Data quality goals, which are closely linked to overall study goals, both aid LCA practitioners in structuring an appropriate approach to data collection and serve as data quality performance criteria.

Although the number and nature of data quality goals necessarily depend on the level of accuracy required for a given LCA, the following list of hypothetical data quality goals is typical. Site-specific data are required for raw materials and energy inputs, water consumption, air emissions, water effluents, and solid waste generation. Approximate data values are adequate for the energy data category. Air emission data should be representative of similar sites in the U.S. A minimum of 95 percent of the material and energy inputs should be accounted for in the LCI.

**Identify Data Quality Indicators**    Data quality indicators are benchmarks against which the collected data can be measured to determine if data quality requirements have been met. Selection depends on which of the available indicators are most appropriate and applicable to the specific data sources being evaluated. Examples of indicators are precision, completeness, representativeness, consistency, and reproducibility (Diwekar, 2003).

**Identify Data Sources and Types**    For each life cycle stage, unit process, or type of environmental release, the data source and/or type that will provide sufficient accuracy and quality to meet the study's goal is specified. Doing this prior to data collection helps to reduce costs and the time required to collect the data. Data sources include:

- meter readings from equipment operating logs or journals; industry data reports, databases, or consultant's laboratory test results; government documents, reports, and databases; other publicly available databases or clearinghouses; journals, and papers, books; reference books; patents; and trade associations. Related LCI studies are also useful, as are equipment and process specifications best engineering judgment.

Examples of data types include full measured, modeled and sampled data; non-site-specific (i.e., surrogate) data; non-LCI data (i.e., not intended for use in an LCI); and vendor data.

The required level of aggregated data should also be specified. For example, the reader should be able to ascertain quickly whether data are representative of one process or several processes.

**Develop a Data Collection Worksheet and Checklist**    The LCI checklist should cover most of the decision areas in the performance of an inventory. This document can be prepared to guide data collection and validation and can enable construction of a database to store collected data electronically. The following general decision areas should be addressed on the inventory checklist:

Purpose of the inventory
System boundaries

Geographic scope

Types of data used

Data collection procedures

Data quality measures

Computational model construction

Presentation of results.

All inputs and outputs for each process modeled in the flow diagram should be recorded on an accompanying data worksheet.

The checklist and worksheet are valuable tools for ensuring completeness, accuracy, and consistency. They are especially important for large projects when several people collect data from multiple sources. The checklist and worksheet should be tailored to meet the needs of a specific LCI.

### 3.2.2.3  *Step 3: Collect Data*

The flow diagram(s) developed in step 1 provides the road map for data to be collected. Step 2 specifies the required data sources, types, quality, accuracy, and collection methods. Step 3 consists of finding and filling in the flow diagram and worksheets with numerical data. This may not be a simple task. If some data are difficult or impossible to obtain, and available data are difficult to convert to the appropriate functional unit, therefore, the system boundaries or data quality goals of the study will have to be refined to describe the results that can reliably be obtained from the data available. This iterative process is common for most LCAs.

Data collection efforts involve a combination of research, site-visits and direct contact with experts which generate large quantities of data. An electronic database or spreadsheet can be useful to hold and manipulate the data. Alternatively, it may be more cost effective to buy a commercially available LCA software package (see Section 3.3). Prior to purchasing an LCA software package the decision-makers or LCA practitioner should insure that it will provide the level of data analysis required.

A second method to reduce data collection time and resources is to obtain non-site specific inventory data. Several organizations have developed databases specifically for LCA that contain some of the basic data commonly needed in constructing a life cycle inventory. Some of the databases are sold in conjunction with LCI data collection software; others are stand-alone resources. Many companies with proprietary software also offer consulting services for LCA design.

### 3.2.2.4  *Step 4: Evaluate and Document the LCI Results*

When the data have been collected and organized, the accuracy of the results must be verified. In documenting the results of the life cycle inventory, it is important to thoroughly describe the methodology used in the analysis, define the systems analyzed and the boundaries that were set, and state all assumptions made in performing the inventory analysis. Use of the checklist and worksheet (see step 2) supports a clear process for documenting this information. The outcome of the inventory analysis is a list containing the quantities of pollutants released to the environment and the amount of energy and materials consumed. The information can be organized by life cycle stage, by media (air, water, land), by specific process, or any combination thereof that is consistent with the ground rules.

If the sensitivity of the LCI data collection efforts has not been properly determined before the next stage, life cycle impact assessment, is begun, the LCA itself may have to be repeated because the data are found to be insufficient to permit the drawing of the desired conclusions.

### 3.2.3   Life Cycle Impact Assessment

The LCIA phase is the evaluation of potential human health and environmental impacts of the environmental resources and releases identified during the life cycle inventory. Impact assessment should address ecological and human health effects; it can also address resource depletion. A life cycle impact assessment attempts to establish a linkage between the product or process and its potential environmental impacts. For example, an LCIA could determine whether one product or process causes more greenhouse gases than other, or could potentially kill more fish.

The key concept in this component is that of stressors. A stressor is a set of conditions that may lead to an impact. For example, if a product or process is emitting greenhouse gases, the increase of greenhouse gases in the atmosphere *may* contribute to global warming. Processes that result in the discharge of excess nutrients into bodies of water *may* lead to eutrophication. An LCIA provides a systematic procedure for classifying and characterizing these types of environmental effects.

#### *3.2.3.1   Why Conduct an LCIA?*   Although much can be learned about a process by considering life cycle inventory data, an LCIA provides a more precise basis to make comparisons. Thus we know that large releases of both carbon dioxide and methane are harmful, an LCIA can determine whether 9,000 tons of $CO_2$ or 5,000 tons of methane would have the greater potential impact. Using science-based characterization factors, an LCIA can calculate the impacts each environmental release has on problems such as smog or global warming. An impact assessment can also incorporate value judgments. In an air non-attainment zone, for example, air emissions could be of relatively higher concern than the same emission level in a region with better air quality.

#### *3.2.3.2   Key Steps of a Life Cycle Impact Assessment*   The following steps comprise a life cycle impact assessment.

1. **Selection and Definition of Impact Categories** identifying relevant environmental impact categories (e.g., global warming, acidification, terrestrial toxicity).

2. **Classification**  assigning LCI results to the impact categories (e.g., classifying $CO_2$ emissions to global warming).
3. **Characterization**  modeling LCI impacts within impact categories using science-based conversion factors. (e.g., modeling the potential impact of $CO_2$ and methane on global warming).
4. **Normalization**  expressing potential impacts in ways that can be compared (e.g. comparing the global warming impact of $CO_2$ and methane for the two options).
5. **Grouping**  sorting or ranking the indicators (e.g. sorting the indicators by location: local, regional, and global).
6. **Weighting**  emphasizing the most important potential impacts.
7. **Evaluating and Reporting LCIA Results**  gaining a better understanding of the reliability of the LCIA results.

The International Organization of Standardization standard for conducting an impact assessment states that impact category selection, classification, and characterization are mandatory steps for an LCIA and data evaluation (step 7) (ISO, 1998a). Whether the other steps are used will depend on the goal and scope of the study.

*Step 1: Select and Define Impact Categories*  The impact categories that will be considered as part of the overall LCA were selected as part of the initial goal and scope definition phase. To guide the LCI data collection process and requires reconsideration for an LCIA, impacts are defined as the consequences due to the input and output streams of a system on human health, on plants and animals (i.e., ecological health), or on the future availability of natural resources (i.e., resource depletion). Table 3.1 shows some of the more commonly used impact categories.

*Step 2: Classification*  For LCI items that contribute to only one impact category, the classification procedure is a straightforward assignment. For example, carbon dioxide emissions can be replaced in the global warming category.

For LCI items that contribute to two or more different impact categories, a rule must be established for classification. There are two ways of assigning LCI results to multiple impact categories (ISO, 1998a):

- Allocate a representative portion of the LCI results to the impact categories to which they contribute. This is typically allowed in cases when the effects are dependent on each other.
- Assign all LCI results to all impact categories to which they contribute. This is typically allowed when the effects are independent of each other.

For example, since one molecule of sulfur dioxide ($SO_2$) can stay at ground level or travel up into the atmosphere, it has the potential to affect either human health or acidification (but not both at the same time). Therefore, $SO_2$ emissions typically are divided between those two impact categories (e.g., 50 percent allocated to human

**TABLE 3.1. Commonly Used Life Cycle Impact Categories**

| Impact Category | Scale | Relevant LCI Data (i.e., Classification) | Common Characterization Factor | Description of Characterization Factor |
|---|---|---|---|---|
| Global warming | Global | Carbon dioxide ($CO_2$), nitrogen dioxide ($NO_2$), methane ($CH_4$), chlorofluorocarbons (CFCs), hydrochlorofluorocarbons (HCFCs), methyl bromide ($CH_3Br$) | Global warming potential | Converts LCI data to carbon dioxide ($CO_2$) equivalents. Note: global warming potentials can be 50, 100, or 500 year potentials. |
| Stratospheric ozone depletion | Global | Chlorofluorocarbons (CFCs), hydrochlorofluorocarbons (HCFCs), halons methyl bromide ($CH_3Br$) | Ozone depleting potential | Converts LCI data to trichlorofluoromethane (CFC-11) equivalents. |
| Acidification | Regional local | Sulfur oxides ($SO_x$), nitrogen oxides ($NO$), hydrochloric acid (HCl), hydroflouric acid (HF), ammonia ($NH_4$) | Acidification potential | Converts LCI data to hydrogen ($H_+$) ion equivalent |

| Impact category | Scale | Inventory data | Characterization factor | Description |
|---|---|---|---|---|
| Eutrophication | Local | Phosphate ($PO_4$), nitrogen oxide (NO), nitrogen dioxide ($NO_2$), nitrates ammonia ($NH_4$) | Eutrophication potential | Converts LCI data to phosphate ($PO_4$) equivalents |
| Photochemical smog | Local | Non-methane hydrocarbon (NMHC) | Photochemical oxidant creation potential | Converts LCI data to ethane ($C_2H_6$) equivalents |
| Terrestial toxicity | Local | Toxic chemicals with a reported lethal concentration to rodents | $LC_{50}$ | Convert $LC_{50}$ data to equivalents |
| Aquatic toxicity | Local | Toxic chemicals with a reported lethal concentration to fish | $LC_{50}$ | Convert $LC_{50}$ data to equivalents |
| Human health | Global regional local | Total releases to air, water, and soil | $LC_{50}$ | Convert $LC_{50}$ data to equivalents |
| Resource depletion | Global regional local | Quantity of minerals used Quantity of fossil fuels used | Resource depletion potential | Converts LCI data to a ratio of quantity resource used versus quantity of resource left in reserve |
| Land use | Global regional local | Quantity disposed of in a landfill | Solid waste | Converts mass of solid waste into volume using an estimated density |

health and 50 percent allocated to acidification). On the other hand, since nitrogen dioxide ($NO_2$) could potentially affect both ground level ozone formation and acidification simultaneously,, the entire quantity of $NO_2$ would be allocated to both impact categories (e.g., 100 percent to ground level ozone and 100 percent to acidification). The allocation procedure must be clearly documented.

*Step 3: Characterization*   Impact characterization uses science-based conversion factors, called characterization factors, to convert and combine the LCI results into representative indicators of impacts to human and ecological health. Characterization factors also are commonly referred to as equivalency factors. Characterization factors translate different LCI inputs — for example, the toxicity data for lead, chromium, and zinc — into directly comparable impact indicators. With such data in hand, estimates of the relative terrestrial toxicity of these metals could be made.

The impact categories listed in Table 3.1 have many possible endpoints, including the following:

**Global Impact**

| | |
|---|---|
| *Global Warming* | Polar melt, soil moisture loss, longer seasons, forest loss/ change, and change in wind and ocean patterns |
| *Ozone Depletion* | Increased ultraviolet radiation |
| *Resource Depletion* | Decreased resources for future generations |

**Regional Impacts**

| | |
|---|---|
| *Photochemical Smog* | "Smog," decreased visibility, eye irritation, respiratory tract and lung irritation, and vegetation damage |
| *Acidification* | Building corrosion, water body acidification, vegetation effects, and soil effects |

**Local Impacts**

| | |
|---|---|
| *Human Health* | Increased morbidity and mortality |
| *Terrestrial Toxicity* | Decreased production and biodiversity and decreased wildlife for hunting or viewing |
| *Aquatic Toxicity* | Decreased aquatic plant and insect production and biodiversity and decreased commercial or recreational fishing |
| *Land Use* | Loss of terrestrial habitat for wildlife and decreased landfill space |

Impact indicators are typically characterized using the following equation:

$$\text{Inventory Data} \times \text{Characterization Factor} = \text{Impact Indicators}$$

For example, all greenhouse gases can be expressed in terms of carbon dioxide equivalents by multiplying the relevant LCI results by a $CO_2$ characterization factor and then combining the resulting impact indicators to provide an overall indicator of global warming potential.

The intergovernmental Panel on Climate Change (IPCC) provides conversion factors for a number of industrial pollutants. In the following example, the global warming impacts of different amounts of chloroform ($CHCl_3$) and methane ($CH_4$) are characterized. The value of the conversion factor, called GWP, for global warming potential, is 9 for $CHCl_3$ and 21 for $CH_4$.

Thus we write:

$$GWP_{CHCl3} = 9 \times 20 = 180$$
$$GWP_{CH4} = 21 \times 10 = 210$$

Use of the conversion factors shows that 10 pounds of methane has a larger impact on global warming than 20 pounds of chloroform.

The key to impact characterization is using the appropriate characterization factor. For some impact categories, such as global warming and ozone depletion, there is a consensus on acceptable characterization factors. For other impact categories, such as resource depletion, a consensus is still being developed. Table 3.1 includes descriptions of possible characterization factors for some of the commonly used life cycle impact categories. A properly referenced LCIA will document the source of each characterization factor to ensure that they are relevant to the goal and scope of the study. For example, many characterization factors based on studies conducted in Europe, cannot be applied to American data unless it can be verified that they are appropriate to conditions in the United States.

*Step 4: Normalization*  Normalization is an LCIA tool used to express impact indicator data in a way that can be compared among impact categories. In this procedure the indicator results are divided by a reference value selected for the purpose. Reference value may be chosen from among numerous methods.

The following are representatives.

The total emissions or resource use for a given area that may be global, regional or local.

The total emissions or resource use for a given area on a per capita basis.

The ratio of one alternative to another (i.e., the baseline).

The highest value among all options.

The goal and scope of the LCA may influence the choice of an appropriate reference value. Note that normalized data can only be compared within an impact category. For example, the effects of acidification cannot be directly compared with those of aquatic toxicity because the characterization factors were calculated using different scientific methods.

*Step 5: Grouping*  Grouping assigns impact categories into one or more sets to facilitate the interpretation of the results into specific areas of concern. Typically grouping involves sorting or ranking indicators. The ISO (1998a) lists two possible ways to group LCIA data:

Sorting indicators by characteristics such as emissions (e.g., air and water emissions) or location (e.g., local, regional, or global)

Sorting indicators by a system of ranking based on value choices, such as high, low, or medium priority.

*Step 6: Weighting*    Weighting (also referred to as valuation) assigns relative values to the different impact categories based on their perceived importance or relevance. Weighting is important because the impact categories should also reflect study goals and stakeholder values. But since weighting is not a scientific process, its methodology must be clearly explained and documented. The weighting stage is the least developed of the impact assessment steps and also is the one most likely to be challenged for integrity. In general, weighting includes the following activities:

Identifying the underlying values of stakeholders.

Determining weights to place on impacts.

Applying weights to impact indicators.

Weighted data should not be combined across impact categories unless the weighting procedure is explicitly documented. The unweighted data should be shown together with the weighted results to ensure a clear understanding of the assigned weights.

In some cases, the impact assessment results are so straightforward that a decision can be made without the weighting step. For example, when the best-performing alternative is significantly and meaningfully better than the others in at least one impact category and equal to the alternatives in the remaining impact categories, then *one alternative is clearly better*.

Several issues make weighting a challenge. The first issue is subjectivity. According to ISO 14042, any judgment of preferability is a subjective statement of the relative importance of one impact category over another (ISO, 1998a). Additionally, these value judgments may change with location or time of year. For example, A resident of Los Angeles may place more importance on the values for photochemical smog than someone in Cheyenne, Wyoming. The second issue is derived from the first: how should users fairly and consistently make decisions based on environmental preferability, given the subjective nature of weighting? Trying to develop a truly objective (or universally agreeable) set of weights or weighting methods is not feasible.

However, several approaches to weighting do exist and are used successfully for decision-making.

*Step 7: Evaluate and Document the LCIA Results*    Now that the impact potential for each selected category has been calculated, the accuracy of the results must be verified. Documentation of the results of the life cycle impact assessment entails thoroughly describing the methodology used in the analysis, defining the systems analyzed and the boundaries that were set, and setting forth all assumptions on which the inventory analysis was passed.

The LCIA, like all other assessment tools, has inherent limitations, including the following:

> Lack of spatial resolution (e.g., a 4,000 gallon ammonia release is worse in a small stream than in a large river).
>
> Lack of temporal resolution (e.g., a 5 ton release of particulate matter during a one month period is worse than the same release spread through the whole year).
>
> Inventory speciation (e.g., broad inventory listing such as "VOC" or "metals" do not provide enough information to accurately assess environmental impacts).
>
> Threshold and nonthreshold impact (e.g., 10 tons of contamination is not necessarily 10 times worse than 1 ton of contamination).

The selection of more complex or site-specific impact models can help reduce the limitations of the impact assessment's accuracy. It is important to document these limitations and to include a comprehensive description of the LCIA methodology, as well as, a discussion of the underlying assumptions, value choices, and known uncertainties in the impact models with the numerical results of the LCIA to be used in interpreting the results of the LCA.

### 3.2.4 Life Cycle Interpretation

Life cycle interpretation is a systematic technique to identify, quantify, check, and evaluate information from the results of the life cycle inventory (LCI) and the life cycle impact assessment (LCIA), and communicate them effectively. Life cycle interpretation is the last phase of the LCA process. The ISO has defined the following objectives of life cycle interpretation:

1. Analyze results, reach conclusions, explain limitations and provide recommendations based on the findings of the preceding phases of the LCA and to report the results of the life cycle interpretation in a transparent manner.
2. Provide a readily understandable, complete, and consistent presentation of the results of an LCA study, in accordance with the goal and scope of the study (ISO, 1998b).

It is not always possible to use an LCA as the basis for stating that one alternative is better than the others. This does not imply that efforts have been wasted. Uncertainty in the final results notwithstanding the LCA process still provides decision-makers with a better understanding of the environmental and health impacts associated with each alternative, where they occur (locally, regionally, or globally), and the relative magnitude of each type of impact potentially attributed to each of the proposed alternatives investigated. This information more fully reveals the pros and cons of the alternatives.

The purpose of conducting an LCA is to better inform decision-makers by providing a particular type of information (often unconsidered), with a life cycle perspective of environmental and human health impacts associated with each product or process. However, LCA does not take into account technical performance, cost, or political and social acceptance. Therefore, it is recommended that LCA be used in conjunction with these other parameters.

### 3.2.4.1  *Key Steps to Interpreting the Results of the LCA*   The guidance provided thus far summarizes the information on life cycle interpretation in the ISO draft standard *Environmental Management — Life Cycle Assessment — Life Cycle Interpretation* (ISO, 1998b). The ISO draft standard covers the following steps in conducting a life cycle interpretation:

1. Identify significant issues
2. Evaluate the completeness, sensitivity, and consistency of the data
3. Draw conclusions and recommendations

Figure 3.4 illustrates the steps of the life cycle interpretation process in relation to the other phases of the LCA process.

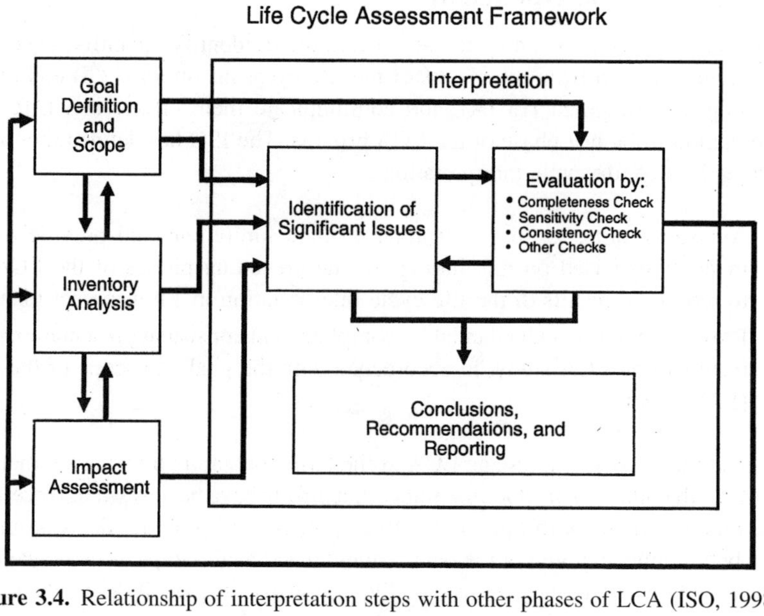

**Figure 3.4.** Relationship of interpretation steps with other phases of LCA (ISO, 1998b).

*Step 1: Identify Significant Issues*   Significant "issues" are the data elements that contribute most to the results of both the LCI and LCIA for each product, process, or service. Examples include:

   Inventory parameters (e.g., energy use, emissions, waste)

   Impact category indicators (e.g., resource use, emissions, waste)

   Essential contributions for life cycle stages to LCI or LCIA results such as individual unit Processes or groups of processes (e.g., transportation, energy production)

When these issues have been identified, the results are used to evaluate the completeness, sensitivity, and consistency of the LCA study (step 2). The identification of significant issues guides the evaluation step.

   Before determining which parts of the LCI and LCIA have the greatest influence on the results for each alternative, the previous phases of the LCA (e.g., study goals, ground rules, impact category weights, results, and external involvement, etc.) should be reviewed in a comprehensive manner.

   If a review of the information collected and the presentations of results developed indicates that the goal and scope of the LCA study have been met. If they have, the significance of the results can be determined. Several analytical approaches are possible.

   **Contribution Analysis**   the contribution of the life cycle stages or groups of processes are compared to the total result and examined for relevance.

   **Dominance Analysis**   statistical tools or other techniques, such as quantitative or qualitative ranking (e.g., ABC Analysis), are used to identify significant contributions to be examined for relevance.

   **Anomaly Assessment**   based on previous experience, unusual or surprising deviations from expected or normal results are observed and examined for relevance.

*Step 2: Evaluate the Completeness, Sensitivity, and Consistency of the Data*   The evaluation step of the interpretation phase establishes the confidence in and reliability of the results of the LCA. This is accomplished performing completeness, sensitivity, and consistency checks to ensure that products/processes are fairly compared.

   **Completeness Check**   The completeness check ensures that all relevant information and data needed for the interpretation are available and complete. A checklist should be developed to indicate each significant area represented in the results. Using the established checklist, it is possible to verify that the data comprising each area of the results are consistent with the system boundaries (e.g., all life cycle stages are included) and that the data is representative of the specified area (e.g., accounting for 90 percent of all raw materials and environmental releases).

   The result of this effort will be a checklist indicating that the results for each product/process are complete and reflective of the stated goals and scope of the LCA study. If deficiencies are noted, an attempt must be made to remedy them.

If this is not possible because data are not available, areas inadequately character-ized because of insufficient data must be highlighted in the final results and their impact on the comparison estimated either quantitatively (percent uncertainty) or qualitatively (alternative A's reported result may be higher because "X" is not included in its assessment).

**Sensitivity Check**   The objective of the sensitivity check is to evaluate the reliability of the results by determining whether the uncertainty in the significant issues identified in step 1 affect the decision maker's ability to confidently draw comparative conclusions. Three common techniques for data quality analysis can be used in performing sensitivity checks.

1. Gravity Analysis — Identifies the data that has the greatest contribution on the impact indicator results.
2. Uncertainty Analysis — Describes the variability of the LCIA data to deter-mine the significance of the impact indicator results.
3. Sensitivity Analysis — Measures the extent that changes in the LCI results and characterization models affect the impact indicator results.

Additional guidance on how to conduct a gravity, uncertainty, or sensitivity analysis can be found in the EPA document entitled *Guidelines for Assessing the Quality of Life Cycle Inventory Analysis*, (USEPA, 1995). If one of these analyses has been conducted as part of the LCI and LCIA phases, these results can be used. Then the sensitivity check will serve to verify that the goals for data quality and accuracy, defined early in have been met. If deficiencies exist, additional efforts are required to improve the accuracy of the LCI data collected and/or impact models used in the LCIA. If, better data or impact models cannot be obtained, the deficiencies for each relevant significant issue must be reported and its impact on the comparison estimated either quantitatively or qualitatively, as with the completeness check.

**Consistency Check**   The consistency check determines whether the assump-tions, methods and data used throughout the LCA process are consistent with the goal and scope of the study, and for each product/process evaluated. Verifying and documenting that the study was completed as intended at the conclusion increases confidence in the final results. A formal checklist should be developed to communicate the results of the consistency check. Table 3.2 lists seven categories and provides examples of inconsistencies that can creep into the data. The goal and scope of the LCA determines which categories should be used.

If, after completion of steps 1 and 2, it is determined that the results of the impact assessment and the underlying inventory data are complete, comparable, and accep-table as bases for drawing conclusions and making recommendations, stop! If any inconsistency is detected, document the role it played in the overall consistency evaluation. Although some inconsistency may be acceptable, depending upon the goal and scope of the LCA, the presence of inconsistencies usually means that it is necessary to repeat steps 1 and 2 until the results are able to support the original goals for performing the LCA.

**TABLE 3.2. Examples of Checklist Categories and Potential Inconsistencies**

| Category | Example of Inconsistency |
|---|---|
| Data source | Alternative A is based on literature and Alternative B is based on measured data. |
| Data accuracy | For Alternative A, a detailed process flow diagram is used to develop the LCI data. For Alternative B, limited process information was available and the LCI data developed was for a process that was not described or analyzed in detail. |
| Data age | Alternative A uses 1980's era raw materials manufacturing data. Alternative B used a one year old study. |
| Technological representation | Alternative A is bench scale laboratory model. Alternative B is a full-scale production plant operation. |
| Temporal representation | Data for Alternative A describe a recently developed technology. Alternate B describes a technology mix, including recently built and old plants. |
| Geographical representation | Data for Alternative A were data from technology employed under European environmental standards. Alternative B uses the data from technology employed under U.S. environmental standards. |
| System boundaries, assumptions, and models | Alternative A uses a Global Warming Potential model based on 500 year potential. Alternative B uses a Global Warming Potential model based on 100 year potential. |

*Step 3: Draw Conclusions and Recommendations*    The objective of this step is to interpret the results of the life cycle impact assessment (not the LCI) to determine which product/process has the overall least impact to human health and the environment, and/or to one or more specific areas of concern as defined by the goal and scope of the study. Depending upon the scope of the LCA, the results of the impact assessment will return either a list of unnormalized and unweighted impact indicators for each impact category for the alternatives or a single grouped, normalized, and weighted score for each alternative. In the latter case, the recommendation may simply be to accept the product/process with the lowest score. The assumptions underlying the analysis should be borne in mind, however. If an LCIA stops at the characterization stage, the LCIA interpretation is less clear-cut. The conclusions and recommendations rest on balancing the potential human health and environmental impacts in the light of study goals and stakeholder concerns.

It is essential to understand and communicate the uncertainties and limitations in the procedures that have produced the final recommendations. Perhaps no one product or process is better than another because of underlying uncertainties and limitations in the methods used to conduct the LCA. Perhaps insufficient good data were available, or restrictions on time or resources prevented analysts from thoroughly exploring certain aspects of the problem. Even so, the results of the LCA can be used to help inform decision-makers about the human health and environmental pros and cons and understand the significant impacts of each. Such

LCA results will reveal whether effects are occurring locally, regionally, or globally, and will provide at least a rough estimate of the magnitude of each type of impact in comparison to the proposed alternatives being investigated.

***3.2.4.2  Reporting the Results***   When the LCA has been completed, the materials must be assembled into a comprehensive report documenting the study in a clear and organized manner. This will help communicate the results of the assessment fairly, completely, and accurately to others interested in the results. The report presents the results, data, methods, assumptions and limitations in sufficient detail to allow the reader to comprehend the complexities and trade-offs inherent in the LCA study.

If the results will be communicated to parties who were not involved in the LCA study (e.g., stakeholders), the report will serve as a reference document, and which can help prevent any misrepresentation of the results.

***3.2.4.3  Conclusion***   Adding life cycle assessment to the decision-making process provides a level of understanding of human health and environmental impacts that traditionally has not been available to those responsible for selecting a product or process. This valuable information provides a way to account for the full impacts of decisions, especially those made off-site, that are directly influenced by the selection of a product or process. As emphasized earlier, LCA is a tool to better inform decision-makers; and other decision criteria such as cost and perform-ance must be weighed to reach a well-balanced decision.

As we have seen, LCA and LCI can be valuable tools in environmental analysis. To make sense of the many large data sets collected in any such study is clearly a task too complex to be undertaken without the assistance of computers. Section 3.3 describes some of the software tools that have been developed for this purpose.

## 3.3  LCA AND LCI SOFTWARE TOOLS

Table 3.3 lists the LCA and LCI tools we shall discuss in this section, along with the vendor or developer of each and that organization's internet address.

1. **ECO-it 1.0**
   ECO-it is a database tool used to assist an LCI and LCIA. ECO-it comes with over 100 indicator values for commonly used materials such as metals, plas-tics, paper, board and glass, as well as production, transport, energy and waste treatment processes.

2. **EcoManager**
   EcoManager is life cycle inventory tool designed to be used by persons who have or who gain a working knowledge of the LCI methodology for internal planning, screening, and evaluation. EcoManager uses a software program developed by Pira International of the U.K., in combination with U.S. life

**TABLE 3.3. LCA and LCI Software Tools**

| Name | Vendor | URL |
|---|---|---|
| 1. ECO-it 1.0 | PRé Consulting | http://www.pre.nl/eco-it.html |
| 2. EcoManager 1.0 | Franklin Associates, Ltd. | http://www.fal.com/software/ecoman.html |
| 3. EcoPro 1.5 | EcoPerformance Systems Switzerland | http://www.sinum.com/ |
| 4. GaBi 4.0 | Institute for Polymer Testing and Polymer Science, University of Stuttgart | http://www.pre-product.de/englisch/main/software.htm |
| 5. IDEMAT | Delft University of Technology | http://www.io.tudelft.nl/research/mpo/idemat/idemat.htm |
| 6. LCAD | Battelle Memorial Institute/Department of Energy | http://www.estd.battelle.org/sehsm/lca/LCAdvantage.html |
| 7. LCAiT 2.0 | Chalmers Industriteknik, Sweden | http://www.ekologik.cit.chalmers.se/lcait.htm |
| 8. REPAQ 2.0 | Franklin Associates, Ltd. | http://www.fal.com/software/repaq.html |
| 9. SimaPro 4 | PRé Consulting | http://www.pre.nl/simapro.html |
| 10. TEAM 2.0 | Ecobalance | http://www.ecobalance.com/software/team/team_ovr.htm |
| 11. TRACI | U.S. Environmental Protection Agency | http://www.epa.gov/ORD/NRMRL/std/sab/iam_traci.htm |
| 12. Umberto 3.0 | Institute for Energy and Environmental Research, Heidelberg | http://www.ifu.com/software/umberto-e/ |

cycle inventory data from the Franklin Associates, Ltd. (FAL) database. Developed for use with Microsoft Excel, EcoManager utilizes spreadsheets and program codes called "macros" to guide the user through the construction of a life cycle inventory, access data from the databases, perform file management functions, edit existing data and produce graphics and reports. The system provides on-line help at every stage of operation. The system also provides graphic and report features. And because the program has been created in Excel, all of the power of Excel for graphics, exporting, and file management is available for customized outputs.

3. **EcoPro1.5**

   EcoPro is a life-cycle-assessment (LCA) tool that models product life cycles with flow chart diagrams. This Windows-based application offers an online help, a toolbar, and icons; a large database is included.

4. **GaBi**

   GaBi is a Life Cycle Engineering or Life Cycle Assessment tool that uses many predefined data objects from industry and literature. Users can link data sets supplied with the GaBi database to their own data in order to calculate both Life Cycle Inventories and Impact Assessments. It allows clear weak point analyses of inventories and valuated balances. The structure is open to alterations and extensions.

5. **IDEMAT**

   Idemat is a computer database of over 365 materials from the Department of Environmental Product Development of the faculty of Industrial Design Engineering at the Delft University of Technology. It provides technical information about materials and processes in words, numbers and graphics, and puts emphasis on environmental information. The program was developed to be used by students of technically oriented academic disciplines like industrial design engineering, civil engineering, material science and aerospace engineering. Users must be quite familiar with the principles of life cycle assessment and the methods for characterization and evaluation of the environmental impacts as published by SETAC and the Center for Environmental Studies at the University of Leiden (CML), and used in SimaPro.

6. **LCAD**

   Life-Cycle Advantage, or LCAD5 is a life cycle modeling tool that has a graphical user interface and database structure. LCAD can model process flow diagrams with material and energy balances, and labor and revenue inputs. LCAD also can assess the data reliability. The LCAD system includes a basic commodity database for the U.S. covering fuels production and distribution, power generation, and cradle-to-gate operations for selected forest products, paper, metals, cement, and basic chemicals and plastics.

7. **LCAiT**

   LCAiT is a life cycle inventory tool that aids in generating an energy and materials balance. LCAiT also contains a cradle-to-gate information regarding certain materials.

8. **REPAQ**

   REPAQ is a life cycle inventory software program that permits users to examine energy and environmental emissions for the entire life cycle of a product, beginning with raw material extraction and continuing through refining and processing, material manufacture, product fabrication, and disposal. Products, processes, and packaging can be evaluated with REPAQ. Users may access the REPAQ database and enter their own data through the Custom Materials feature, which allows entry of data for any process for which LCI data can be gathered.

9. **SimaPro 4**

   SimaPro is a full-featured LCA software tool that facilitates the comparison and analysis of complex products with complex life cycles. The process databases and the impact assessment databases can be edited and expanded without limitation. SimaPro can trace the origin of any result that has been implemented. Special features include: multiple impact assessment methods, multiple process databases, automatic unit conversion. SimaPro comes with several well known impact assessment methods, including CML 1992 and 1996, Eco-points, and the Eco-indicator 95 method developed by PRé. All impact assessment data can be edited and expanded.

10. **TEAM**

    TEAM is a life cycle assessment software program that TEAM allows the user to build and use a database and model any system representing the operations associated with products, processes, and activities (waste management options, means of transportation, etc.). It is designed to describe and model complex industrial systems and to calculate the associated life cycle inventories, life cycle potential environmental impacts, and process-oriented life cycle costs.

11. **TRACI — A Model Developed by the US EPA**

    The tool for the reduction and assessment of chemical and other environmental impacts (TRACI) is described along with its history, the research and methodologies it incorporates, and the insights it provides within individual impact categories. TRACI, a stand-alone computer program developed by the U.S. Environmental Protection Agency (USEPA), facilitates the characterization of environmental stressors that have potential effects, including ozone depletion, global warming, acidification, eutrophication, tropospheric ozone (smog) formation, ecotoxicity, human health criteria-related effects, human health cancer effects, human health noncancer effects, fossil fuel depletion, and land-use effects. TRACI was originally designed for use with life-cycle assessment (LCA), but it is expected to find wider application in the future (Bare, 2003).

12. **Umberto 3.0**

    Umberto is a life cycle assessment tool that uses a graphical interface to model material flow networks to enter and track material and energy flows. Umberto also has a valuation system editor to assist with impact assessment.

## 3.4 EVALUATING THE LIFE CYCLE ENVIRONMENTAL PERFORMANCE OF TWO DISINFECTION TECHNOLOGIES

With increasing emphasis on promoting a sustainable ecological future and concern over introducing toxic chemicals in water, disinfection process design is leaning toward technologies that destroy pathogens while balancing the effects of the disinfected wastewater aquatic biota or on a drinking water supply. Since ultraviolet (UV) irradiation is not a chemical additive, it does not leave or produce toxic by-product in the wastewater, unlike traditional chlorination and dechlorination processes, therefore, the use of UV does not affect a drinking water supply or the aquatic biota in receiving waters.

This section discusses the use of life cycle assessment or analysis (LCA) to quantify the environmental and public health benefits of UV disinfection technology as opposed to chlorination and dechlorination methods. LCA tools are used to evaluate the short- and long-term environmental effects of both processes, and to select the best sustainable process. Our approach applies environmental LCA to these disinfection processes, incorporating economic criteria and all aspects of the environment: chemical use, electricity use, and releases to water, air and land.

### 3.4.1 The Challenge

With increasing emphasis on promoting sound ecological practices and concern over introducing toxic chemicals into water, designs for disinfection processes are increasingly leaning toward technologies that destroy pathogens while balancing the effects of this disinfected wastewater on aquatic biota or a drinking water supply.

In the USA and Canada the use of ultraviolet light irradiation for the disinfection of wastewater has become the accepted alternative to chlorination or chlorination/ dechlorination. There are several reasons for this move away from a proven technology. For example, because of current Uniform Fire Codes (UFC), containment requirements for volatile gases like chlorine, and public health concerns, municipalities are limited in the amounts of chlorine that can be stored in a water treatment plant. Moreover, the dechlorination process uses yet another chemical pollutant, sulfur dioxide, to remove chlorine from the effluent before discharged into the receiving water. Thus, and because chlorination/dechlorination of wastewater produces possible carcinogens in addition to destroying aquatic biota in the receiving waters, U.S. Environmental Protection Agencies (USEPA) started to look for an alternative wastewater disinfection system.

Various governments, municipalities, (Das, 2002, 2004; Das and Ekstrom, 1999; Ecology, 1998; LOTT, 1994; USEPA, 1986a, 1986b, 1992) and corporations have sponsored research (Loge et al., 1996; Scheible, 1987; Scheible and Forndran, 1986; White, 1999) that shows that the ultraviolet (UV) disinfection of wastewater was effective and economical. Another most important development was the parallel flow open channel modular UV system. This new design of the UV system for wastewater in the early 1980s opened up UV disinfection for both the retrofit market and new wastewater treatment plants. To promote a friendlier discharge to the marine environment, designers have begun to prefer alternative disinfection technologies, which emphasize sustainable and clean ecological disinfectants — such a clean technology is UV disinfection.

## 3.4.2 The Chlorination (Disinfection) Process

Despite the acknowledged advantages of disinfection by means of UV irradiation, however, chlorine continues to be the most widely used chemical for the disinfection of wastewater in the United States and elsewhere. The major advantages of chlorine over alternative disinfectants are cost-effectiveness, reliability, and efficacy against a host of pathogenic organisms. We turn now to a detailed examination of the chlorination process.

When chlorine ($Cl_2$) is dissolved in freshwater, a mixture of hypochlorous acid (HOCl) and hydrochlorite ion (OCl⁻) is formed. Chlorine exists predominantly as HOCl below pH 7.6 and as OCl⁻ above pH 7.6. HOCl and/or OCl⁻ is defined as free available chlorine, with the hypochlorous acid being the primary disinfectant.

Mono-, di-, and tri-chloramines ($NH_2Cl$, $NHCl_2$, and $NHCl_3$) are formed when chlorine reacts with nitrogen present in secondary effluent in the form of ammonia. Municipal effluents usually contain all these forms of chlorine in some proportion and taken together they are known as "total residual chlorine" (TRC). Because saltwater contains bromide, the addition of chlorine to saltwater will also form hypobromous acid (HOBr), hypobromous ion (OBr⁻), and bromamines.

Chlorine is typically supplied as liquefied gas in cylinders. Chlorinators apply gaseous chlorine to a feed stream which is then injected into a mixing zone in the chlorine contact chambers. Initial mixing and effective contact times are essential for good process performance. Generally, contact periods of 15–30 minutes are required at peak flow.

***3.4.2.1 Limitations*** Although chlorine disinfection is a largely reliable and effective process, it has certain limitations. For examples, chlorine reacts with certain chemicals in the wastewater, leaving only the residual for disinfection. Wastewater components that readily combine with chlorine include reduced iron and sulfur compounds, ammoniated-nitrogen, organic nitrogen, tannins, uric and humic acid, cyanides, phenols, and unsaturated organics. Cysts of *Entamoeba histolytica* and *Giardia lamblia*, *Mycobacterium tuberculosis*, some viruses, and eggs of parasitic worms show resistance to chlorine. Consistent disinfection in effluents containing organic nitrogen may pose problems, even when a measured free chlorine residual is present.

***3.4.2.2 Human Health and Environmental Impact*** Chlorine is toxic to aquatic, estuarine, and marine organisms. An additional hazard is the carcinogenic potential of chloro-organic compounds. Chlorine gas is potentially toxic when inhaled, and chlorine transport poses a risk. Special handling is required and emergency response plans are required under right-to-know regulations for on-site storage of gaseous chlorine. Chlorine gas and the hypochlorites are also highly corrosive. Chlorine gas concentrations of 15–20 ppm for 30–60 min are dangerous; higher concentrations for very brief periods can be fatal. Chlorination can result in the formation of carcinogenic chloro-organics.

The EPA has established toxicity criteria for total residual chlorine in receiving waters. In freshwaters the acute level is 19 mg/L (1-h average) and the chronic level is 11 mg/L (4-day average). The saltwater acute and chronic criteria are 13 mg/L (1-h average) and 7.5 mg/L (4-day average), respectively. Due to the toxicity of chlorine residuals at such low concentrations and the high limit of analytical detection (50–100 mg/L), chlorine induced toxicity in the receiving stream is difficult to control.

### 3.4.3   Dechlorination With Sulfur Dioxide

In the past decade, concerns over chlorine toxicity and protection of fish and wildlife have led to a dramatic growth in the practice of dechlorination to remove all or part of the chlorine residual and halogenated organics remaining after chlorination. Dechlorination also reduces or eliminates toxicity harmful to aquatic life in receiving waters.

Sulfur dioxide gas successively removes free chlorine, monochloramine, dichloramine, nitrogen trichloride, and poly-chlorinated compounds. When sulfur dioxide is added to wastewater, the following reactions occur:

$$SO_2 + H_2O \rightarrow HSO_3^- + H^+ \tag{3-1}$$

$$HOCl + HSO_3^- \rightarrow Cl^- + SO_4^{-2} + 2H^+ \tag{3-2}$$

$$SO_2 + HOCl + H_2O \rightarrow Cl^- + SO_4^{-2} + 3H^+ \text{ (free chlorine)} \tag{3-3}$$

$$SO_2 + H_2O \rightarrow HSO_3^- + H^+ \tag{3-4}$$

$$NH_2Cl + HSO_3^- + H_2O \rightarrow Cl - +SO_4^{-2} + NH_4^+ + H^+ \tag{3-5}$$

$$SO_2 + NH_2Cl + 2H_2O \rightarrow Cl^- + SO_4^{-2} + NH_4^+ + 2H^+ \text{ (combined chlorine)} \tag{3-6}$$

For the overall reaction between sulfur dioxide and chlorine (Eq. 3-3), the stoichiometric weight ratio of sulfur dioxide to chlorine is 0.9 : 1. In practice, it has been found that about 1.0 mg/L of sulfur dioxide will accomplish for the dechlorination of 1.0 mg/L of chlorine residue (expressed as $Cl_2$). Because the reactions of sulfur dioxide with chlorine and chloramines are nearly instantaneous, contact time is not usually a factor and contact chambers are not used; however, rapid and positive mixing at the point of application is an absolute requirement.

The ratio of free chlorine to the total combined chlorine residual before dechlorination determines whether the dechlorination process is partial or proceeds to completion. If the ratio is less than 85 percent, it can be assumed that significant organic nitrogen is present and that it will interfere with the free residual chlorine process (Metcalf and Eddy, 1991).

Since dechlorination removes most of the total residual chlorine from disinfected wastewaters, it reduces the toxicity of disinfected wastewater effluent to aquatic wildlife. In most situations, sulfur dioxide dechlorination is a very reliable unit process in wastewater treatment, provided the precision of the combined chlorine residual monitoring service is adequate. Excess sulfur dioxide dosages should be avoided not only because of the chemical wastage but also because of the oxygen demand (BOD and COD) exerted by the excess sulfur dioxide.

### 3.4.3.1 *Limitations*

Chlorination/dechlorination is more complex to operate and maintain than chlorination alone. Major difficulties are the inability to measure residual $SO_2$ and problems in the continuous measurement of a zero or low chlorine residual. Many halogenated organics are also rapidly formed upon chlorine addition, and are unaffected by application of $SO_2$. Heltz and Nweke who examined many plants, reported that the amount of residual chlorine in dechlorinated effluents considerably exceeded EPA criteria for receiving waters (Heltz and Nweke, 1995).

### 3.4.3.2 *Environmental Impact*

Sulfuric acid and hydrochloric acid are products of $SO_2$ dechlorination in small amounts but are generally neutralized in the wastewater. Based on laboratory experiments, residuals of sulfite dechlorination are at least three orders of magnitude less toxic than chlorine or ozone.

No cases of sulfur compounds affecting dissolved oxygen consumption or pH change in receiving waters or in dechlorinated effluents have been reported. In pilot studies, no significant oxygen depletion occurred until sulfur dioxide overdoses exceeded 50 mg/L. It is not uncommon, however, to find plants with post aeration after dechlorination to assure that dissolved oxygen requirements are met. Dechlorination with $SO_2$ would significantly reduce toxicity due to chlorination disinfection process.

Sulfur dioxide, which is used for dechlorination purposes, is stored on site in small quantities. The maximum anticipated storage inventory is quite small, and there need be no concerns about threats to the health of workers or of residents of nearby populated areas.

### 3.4.3.3 *Case Study: Accidental Chlorine Gas Release*

The Pierce County Chambers Creek Wastewater Treatment facility in Tacoma, Washington uses chlorine for disinfection; the gas is stored on site in a 1-ton container. The maximum anticipated storage inventory is ten containers (20,000 lb), which exceeds the threshold for applicability of the Risk Management Plan (RMP) requirements specified in the Accidental Release Prevention Program (USEPA, 1996). The threshold for chlorine is 2,500 lb. The potential off-site consequences of an accidental release of this poisonous gas are serious, and an evaluation of the situation is called for. The methodology used for this analysis follows the approach outlined in EPA's *Risk management program guidance for wastewater treatment plants* (USEPA, 1998) and in the guidance provided in the *Handbook of chlorination and alternative disinfectants* (White, 1999).

According to the Accidental Release Prevention Program (USEPA, 1996), the toxic gas worst case scenario must assume the release of the single largest vessel or container over a period of 10 min. The largest vessel at Chambers Creek is the 1-ton container noted earlier.

For worst-case scenarios, releases are assumed to take place without consideration for physical cause or likelihood of occurrence. Only passive mitigation measures (e.g., dikes, enclosures) can be considered, while active mitigation measures cannot be included. Furthermore, the rule states that toxic gases should be evaluated at ambient temperatures unless the gas is liquefied by refrigeration at ambient pressure. The chlorine containers at the treatment facility can be stored either inside or outside the building. For a worst case scenario, it was assumed that a container was ruptured outside the building.

The release rate, calculated for the characteristics of a vapor release, was 200 lb/min for 10 min. An alternative release scenario, assuming a slower rate of release owing to the use of active mitigation measures, was calculated as 31.3 lb/min over 20 min. For the remainder of this case study, we consider both the worst-case scenario and the alternative release possibly just described.

*Determination of Toxic Endpoint Distance*   The distance to the toxic endpoint was calculated for both the worst and alternate case release scenarios. The toxic endpoint for chlorine is 0.0087 mg/L (3 ppm). This value is based on the Emergency Response Planning Guidelines Level 2 (ERPG-2) value. ERPG-2 is defined as the maximum airborne concentration below which it is believed that nearly all individuals could be exposed for up to 1 hour without experiencing or developing irreversible or other serious health effects, or symptoms that could impair an individual's ability to take protective action (USEPA, 1996).

Highly specialized computer software was used to estimate the distance to the toxic endpoint for each scenario. The program, RMP*Comp, version 1.06, is a mathematical tool developed by EPA and NOAA specifically to determine the endpoint distances from various releases under the RMP rule. (for details, visit the U.S. government website http://response.restoration.noaa.gov/chemaids/rmp/rmp.html; USEPA, 1998).

Table 3.4 shows the input used in RMP*Comp. RMP*Comp assumes a meteorological condition of class F stability with a 1.5 m/s wind speed for worst case releases and, for alternate releases, a meteorological condition of class D stability and a 3 m/s wind speed. The model requires choosing between urban and rural dispersion patterns. Urban dispersion was chosen for this analysis because the site is surrounded by residential and industrial development.

*Determination of Exposed Population to this Scenario*   Population information was obtained using LandView III, an electronic mapping software system that includes database information from the EPA, the Bureau of Census, the U.S. Geological Survey, and other federal agencies. LandView III was developed at the EPA's Chemical Emergency Preparedness and Prevention Office in Washington, D.C. The software includes two different calculation methods for estimating the

**TABLE 3.4. RMP*Comp inputs**

|  | Worst Case Inputs | Alternative Case Inputs |
|---|---|---|
| Release type | Vapor | Vapor |
| Amount released (lb) | 2000 | 900 |
| Release rate (lb/min) | 200 | 31.3 |
| Release duration (min) | 10 | 20 |
| Wind speed (m/s) | 1.5 | 3.0 |
| Atmospheric stability class | F | D |
| Roughness type | URBAN | URBAN |
| Passive mitigation measures | Release outside building | Release inside building |
| Active mitigation measures | None | None |
| Tank parameters |  |  |
| Hole area | Not considered | $1.25 \times 10^{-4}$ square inches (0.15 inch diameter) |
| Tank temperature (°C) | 25 | 25 |
| Tank pressure (psi) | 120 | 120 |

population within a radius of a given point using an updated Census data. The methodology used for this analysis was the default method, called the Block Group Proration method, which sums the data for each census block group that has any portion falling within the radius of the circle and then prorates the result based on the ratio of the area of the circle to the total land area of the block groups. The results of the calculation are summarized for this scenario in Table 3.5.

### 3.4.3.4 Effects of Chlorine on Aquatic Life
The potential environmental and chemical effects of chlorine toxicity on 14 aquatic species are summarized in Table 3.6. Rainbow trout was the most sensitive of the species tested, followed by the golden shiner and three-spined stickleback. A calculated chlorine residual

**TABLE 3.5. Population Estimates Within Endpoint Circles**

|  | Worst Case | Alternate Case |
|---|---|---|
| Distance to endpoint (miles) | 1.3 | 0.1 |
| Estimated residential population within distance to endpoint | 13,322 | 5 |
| Public receptors within distance to endpoint | residences, commercial and industrial facilities, recreation areas, hospital | residences, recreation areas |
| Environmental receptors within distance to endpoint | Parks | None |

**TABLE 3.6. Select Summary of Acute and Chronic Toxic Effects of Residual Chlorine on Aquatic Life (Brungs, 1973)**

| Species | Effect Endpoint | Measured Residual Chlorine Concentration (mg/L) |
|---|---|---|
| Coho salmon | 7-day $TL_{50}^a$ | 0.083 |
| Pink salmon | 100 percent kill (1–2) | 0.08–0.10 |
| Coho salmon | 100 percent kill (1–2) | 0.13–0.20 |
| Pink salmon | Maximum nonlethal | 0.05 |
| Coho salmon | Maximum nonlethal | 0.05 |
| Brook trout | 7-day $TL_{50}$ | 0.083 |
| Brook trout | Absent in streams | 0.015 |
| Brown trout | Absent in streams | 0.015 |
| Brook trout | 67 percent lethality (4) | 0.01 |
| Brook trout | Depressed activity | 0.005 |
| Rainbow trout | 96-h $TL_{50}$ | 0.14–0.29 |
| Rainbow trout | 7-day $TL_{50}$ | 0.08 |
| Rainbow trout | Lethal (12) | 0.01 |
| Trout fry | Lethal (2) | 0.06 |
| Yellow perch | 7-day $TL_{50}$ | 0.205 |
| Largemouth bass | 7-day $TL_{50}$ | 0.261 |
| Smallmouth bass | Absent in streams | 0.1 |
| White sucker | 7-day $TL_{50}$ | 0.132 |
| Walleye | 7-day $TL_{50}$ | 0.15 |
| Black bullhead | 96-h $TL_{50}$ | 0.099 |
| Fathead minnow | 96-h $TL_{50}$ | 0.05–0.16 |
| Fathead minnow | 7-day $TL_{50}$ | 0.082–0.115 |
| Fathead minnow | Safe concentration | 0.0165 |
| Golden shiner | 96-h $TL_{50}$ | 0.19 |
| Fish species diversity | 50 percent reduction | 0.01 |
| Send | Safe concentration | 0.00.34 |
| Send | Safe concentration | 0.012 |
| *Daphnia magna* | Safe concentration | 0.003 |
| Protozoa | Lethal | 0.1 |

[a]$TL_{50}$ — median tolerance limit (50% survival).

of 0.03 mg/L, based on dilution of a measured concentration of 2.0 mg/L, reduced plankton photosynthesis by more than 20 percent of the value obtained with a dilution of effluent having no chlorine residual. Dechlorination with sodium bisulfite also eliminated chlorine-related toxicity. Dechlorination with sulfur dioxide also greatly reduced the acute and chronic toxicity to fish and invertebrate species.

In considering these data, it should be borne in mind that the toxicity of chlorine wastes in rivers depends not on the amount of chlorine added but on the concentration of chlorine remaining in solution (Merkens, 1958). The toxicity of this residual chlorine will depend on its composition (i.e., the relative proportions of free chlorine and chloramines). Arriving at this ratio, which in turn depends on at least five other variables, is a complex undertaking. Although free chlorine is more toxic than chloramines, research that has stood since the 1950s suggests that the difference between the toxicity to fish of free chlorine and chloramine is not very great (Dondoroff and Katz, 1950; Merkens, 1958).

*Results of Laboratory Bioassays*  Several studies (Dondoroff and Katz, 1950; Merkens, 1958; Thatcher, 1977) indicated that salmonids were the most sensitive fish species. Laboratory bioassays support this generalization. A residual chlorine concentration of 0.006 mg/L was lethal to trout fry in 2 days, and the 7-day median tolerance limits, or $TL_{50}$, for rainbow trout was 0.08 mg/L with an estimated median period of survival of 1 year at 0.004 mg/L (Merkens, 1958). The maximum nonlethal (in 23 days) concentration of residual chlorine for pink and coho salmon in seawater was 0.05 mg/L (Holland, 1960). Rainbow trout were killed at 0.01 mg/L in 12 days, and they avoided a concentration of 0.001 mg/L (Sprague and Drury, 1969). Brook trout had a mean survival time of 9 h at 0.35 mg/L, 18 h at 0.08 mg/L, and 48 h at 0.04 mg/L. Mortality was 67 percent after 4 days at 0.01 mg/L (Dandy, 1967, 1972). Fifty percent of brown trout were killed at 0.02 mg/L within 10.5 h and at 0.01 mg/L within 43.5 h (Pike, 1971). Trout, salmon, and some fish-food organisms are more sensitive than warm-water fish, snails, and crayfish (Tables 3.6 and 3.7).

**TABLE 3.7. Summary of Results of Brief Exposures of Fish to Residual Chlorine (Brungs, 1973)**

| Species | Effect Endpoint | Time | Measured Residual Chlorine Concentration (mg/L) |
|---|---|---|---|
| Chinook salmon | First death | 2.2 h | 0.25 |
| Brook trout | Median mortality | 90 min | 0.5 |
| Brook trout | Mean survival time | 9 h | 0.35 |
| Brook trout | Mean survival time | 18 h | 0.08 |
| Brook trout | Mean survival time | 48 h | 0.04 |
| Brown trout | Depressed activity | 24 h | 0.005 |
| Rainbow trout | Slight avoidance | 10 min | 0.001 |
| Rainbow trout | Lethal | 2 h | 0.3 |
| Fingerling Rainbow trout | Lethal | 4–5 h | 0.25 |
| Trout fry | Lethal | Instantly | 0.3 |
| Yellow perch | $TL_{50}^{a}$ | 1 h | <0.88 |
| Yellow perch | $TL_{50}$ | 12 h | 0.494 |
| Smallmouth bass | Median mortality | 15 h | 0.5 |
| White sucker | Lethal | 30–60 min | 1.0 |
| Largemouth bass | $TL_{50}$ | 1 h | <0.74 |
| Largemouth bass | $TL_{50}$ | 12 h | 0.365 |
| Fathead minnow | $TL_{50}$ | 1 h | <0.79 |
| Fathead minnow | $TL_{50}$ | 12 h | 0.26 |
| Miscellaneous | Initial kill | 15 min | 0.28 |
| Miscellaneous | Erratic swimming | 6 min | 0.09 |

[a]$TL_{50}$ — median tolerance limit (50 percent survival).

### *3.4.3.5  Relative Sensitivity of Pacific Northwest Fishes and Invertebrates to Chlorinated Seawater*   The introduction of chlorine into coastal

regions through numerous water and waste effluent treatment processes and for a wide variety of industrial applications necessitates an understanding of the potential impact of this powerful oxidant upon marine and estuarine organisms. Although chemical degradation and dilution reduces toxicity of chlorinated sea water by lowering the total residual oxidant (TRO) component, there are situations in which the TRO concentration could remain relatively high, or others in which low concentrations would remain over extended periods of time. In such cases, the TRO component of chlorinated seawater may be toxic to some aquatic organisms. TRO toxicity could occur from receiving chlorinated effluents in embayments and cooling water canals with low flushing or poor mixing areas, waters with low chlorine demand, and in areas where large-volume chlorination, such as "superchlorination" is practiced.

In this study, a series of 96-h continuous-flow bioassays was conducted to provide data on the median lethal concentration ($LC_{50}$); the goal was to enhance understanding of the potential impact of chlorination on Pacific Northwest estuarine and marine organisms. The relative sensitivity to chlorinated seawater was determined for 15 important species of estuarine and marine fishes and invertebrates. A thermal stress was included so that the experiments would simulate power plant cooling water effluents.

Eight fish and seven invertebrate species which occur in areas likely to be impacted by power plant cooling water effluents in the Pacific Northwest were selected as the experimental animals. The species selected are listed in Table 3.8. All the experimental animals, with the exception of the coho and chinook salmon, were collected in Sequim Bay, Washington. The cohos were from a stock raised from eggs at the Battelle Pacific Northwest Laboratories in Sequim. The chinook salmon came from a hatchery stock at Manchester, Washington (Thatcher, 1977).

Twelve of the species were acclimated for at least two weeks prior to testing. The other three, Pacific herring, pink salmon, mysids, were acclimated for only 2 or 3 days prior to testing, due to difficulties in maintaining them for several weeks under laboratory conditions. Each test chamber contained ten individuals for all species except mysids, in which case 40 were used.

We note here some of the highlights of the results summarized in Table 3.8. In general, the fishes (with the exception of the threespine stickleback) were more sensitive than the invertebrates. Among the fish species for which $LC_{50}$ values and 95 percent confidence interval (CI) limits were determined, the coho salmon (96-h $LC_{50}$: 0.32 mg/L TRO was more sensitive than all other species except the Pacific Herring (96-h $LC_{50}$: 0.065 mg/L TRO).

An examination of the 95 percent CI limits for each 96-h $LC_{50}$ concentration reveals that the species included in this study formed three rather distinct groups with differing sensitivity for chlorinated seawater. The most sensitive group included the three salmon species, Pacific herring, shiner perch, English sole, Pacific sand lance, and a shrimp, *Pandalus goniurus*. This group exhibited 96-h $LC_{50}$ values from 0.026 mg/L TRO up to 0.119 mg/L (taking into consideration

**TABLE 3.8. The Relative Sensitivity of 15 Fishes and Invertebrates to Chlorinated Sea Water as Indicated by 96-h LC$_{50}$ Values (Thatcher, 1977)**

| Name[a] | Number of Valid Tests | 96-h LC$_{50}$ (mg/L TRO) | 95 Percnt Fiducial Limits (mg/L TRO) | Probit Regression Line Slope |
|---|---|---|---|---|
| Coho salmon, j[b] | | | | |
| *Oncorhynchus kisutch* | 3 | 0.032 | 0.026–0.038 | 21.2 |
| Pink salmon, j | | | | |
| *Oncorhynchus gorbuscha* | 3 | >0.023 <0.052[c] | – | – |
| Chinook salmon, j | | | | |
| *Oncorhynchus tshawytscha* | 2 | >0.038 <0.065 | – | – |
| Pacific herring, j | | | | |
| *Clupea harengus* | 2 | 0.065 | 0.033–0.097 | 14.7 |
| Shiner perch, j & a | | | | |
| *Cymatogaster aggregata* | 5 | 0.071 | 0.045–0.098 | 23.5 |
| English sole, j | | | | |
| *Parophrys vetulus* | 3 | 0.73 | 0.044–0.103 | 10.3 |
| Pacific sand lance j & a | | | | |
| *Ammodytes hexapterus* | 6 | 0.082 | 0.062–0.102 | 21.9 |
| Shrimp, a | | | | |
| *Pandalus goniurus* | 3 | 0.090 | 0.063–0.119 | 27.8 |
| Shrimp, a[b] | | | | |
| *Crangon nigricauda* | 6 | 0.134 | 0.118–0.151 | 17.0 |
| Amphipod, a | | | | |
| *Anonyx* sp. | 4 | 0.145 | 0.118–0.173 | 12.8 |
| Mysid, a | | | | |
| *Neomysis* sp. | 4 | 0.162 | 0.150–0.175 | 14.1 |
| Threespine stickleback, j & a | | | | |
| *Gasterosteus aculeatus* | 3 | 0.167 | 0.141–0.193 | 21.6 |
| Coon stripe shrimp, j & a | | | | |
| *Pandalus danae* | 4 | 0.178 | 0.159–0.199 | 19.0 |
| Amphipod, j | | | | |
| *Pontogeneia* sp. | 5 | 0.687 | 0.583–0.864 | 27.0 |
| Shore crab, j & a | | | | |
| *Hemigrapsus nudus* and *H. oregonensis* | 5 | 1,418 | 1.240–1.530 | 14.2 |

[a]Identified using keys in Hart (1973) and Kozloff (1974).

[b]j = juvenile, a = adults.

[c]The low number of data points between 0 and 100% mcrtality prevented calculating LC$_{50}$ values for chinook and pink salmon. The mortality data obtained indicated sensitivity close to the coho and greater than the Pacific herring.

the 95 percent CI limits). The group representing intermediate sensitivity included the shrimp (*Crangon nigricauda*), the amphipod (*Anonyx* sp.), the mysid (*Neomysis* sp.), the threespine stickleback, and the coon stripe shrimp. Their 96-h $LC_{50}$ values ranged from 0.118 mg/L TRO to 0.199 mg/L. The most resistant group consisted of two invertebrates, the amphipod, *Pontogeneia* sp. and the shore crabs *Hemigrapsus nudus* and *H. oregonensis.* Their 96-h $LC_{50}$ values ranged from 0.583 to 1.530 mg/L TRO.

The test for parallelism of the probit regression lines determined that the slopes for all species were parallel at the 0.05 level of significance. Thus, the response (mortality) was similar for all species over the range of TRO concentrations employed (Thatcher, 1977).

### 3.4.4  UV Disinfection Process

A UV disinfection system transfers electromagnetic energy from a mercury arc lamp to an organism's genetic material, the chromosomes which contain DNA and RNA. When UV radiation penetrates the cell wall of an organism, it destroys the cell's ability to reproduce. In the disinfection process, UV radiation, generated by an electrical discharge through mercury vapor, penetrates the genetic material of microorganisms and causes molecular rearrangements that retard their ability to reproduce. Das gave a detailed review of the mechanisms of germicidal action, how UV light works, how UV-damaged DNA is repaired, and the effects of wastewater quality parameters on disinfection efficiency (Das, 2004).

UV disinfection systems are not mass-produced. Because their efficiency strongly depends on effluent characteristics that act to decrease the UV intensity in wastewater, each application must have a custom–designed system. Table 3.9 shows the major parameters that must be taken into consideration when a UV disinfection system is being designed for wastewater. In comparison to chlorination/dechlorination, UV disinfection offers a reduction of potential chlorinated hydrocarbons (including potential carcinogens) in the receiving waters, as well as considerably greater safety to wastewater treatment plant operators and to nearby populated areas.

**TABLE 3.9. Major Parameters Affecting the UV Disinfection of Wastewater**

| Parameters | Acceptable Values/Conditions |
| --- | --- |
| Percent transmittance (T) or absorbance | 35–65 |
| Total suspended solids (TSS) mg/L | 5–10 |
| Particle size distribution (PSD) μm | 10–40 |
| Flow rate or hydraulics design | ideal plug flow |
| Iron (mg/L) | less than 0.3 |
| Hardness (mg/L) | less than 300 |

### 3.4.4.1 UV Transmission or Absorbance

UV light's ability to penetrate wastewater is measured in a spectrophotometer at the same wavelength (254 nm) that is produced by germicidal lamps. This measurement is called the *percent transmission* or *absorbance* and it is a function of all the factors that absorb or reflect UV light. As the percent transmission gets lower (higher absorbance) the ability of the UV light to penetrate the wastewater and reach the target organisms decreases.

It cannot be estimated by simply looking at a sample of wastewater with the naked eye. The range of effective transmittances ($T$) will vary depending on the secondary treatment systems. In general, suspended growth-treatment processes produce effluent with $T$ varying from 60–65 percent. Fixed film processes range from 50–55 percent $T$ and lagoons 35–40 percent $T$. Industries that influence UV transmittance include textile, printing, pulp and paper, food processing, meat and poultry processing, photo developing, and chemical manufacturing. A discussion on other major parameters affecting the UV disinfection efficiency of wastewater is given by Das (2004).

### 3.4.4.2 Disinfection Standards

The level of disinfection required to obtain an EPA National Pollutant Discharge Elimination System NPDES permit is commonly less than 200 fecal coliform unit per 100 mL as a 30-day geometric mean. In general, a UV dose of 20–30 mW.s/cm$^2$ is required to achieve this level of disinfection in secondary-treated waste water with a 65 percent transmittance and total suspended solids (TSS) below 20 mg/L. The UV dose requirement to meet specific limits depends on the nature of the particle with respect to numbers, size, and composition. Therefore, UV dose requirements will vary (Table 3.9). A more stringent limit ($<2.2$ total coliform units per 100 mL) is required for water reuse in California and Hawaii. In such cases, filtered effluents with TSS of 2 mg/L or less and 65 percent transmittance may require UV dose as high as 120 mW.s/cm$^2$ to achieve this level of disinfection. The concentrations of solids, bacteria in the particles, and the PSD, are the main limiting factors in the design of systems that must meet stringent disinfection limits. It appears that the UV dose required to achieve the traditional coliform limits will achieve better virus inactivation than the comparable chlorine dose. Figure 3.5, which illustrates the relative doses of UV and chlorine required to inactivate selected organisms compared to fecal coliform indicator, also shows that the chlorine doses required to inactivate most organisms are much higher than comparable UV doses that would achieve the same level of disinfection (Trojan Technologies Inc., 2000).

### 3.4.4.3 Operation, Maintenance, and Worker Safety

A UV disinfection system must be run and maintained to ensure that unwanted is organisms receive enough radiation to render them sterile. All surfaces between the UV radiation and the target organisms must be cleaned, and the ballasts, lamps, and reactor must be functioning at peak efficiency. Inadequate cleaning is one of the most common causes of a UV system's ineffectiveness. In all cases, the quartz sleeves or Teflon tubes need to be cleaned regularly by mechanical wipers, ultrasound

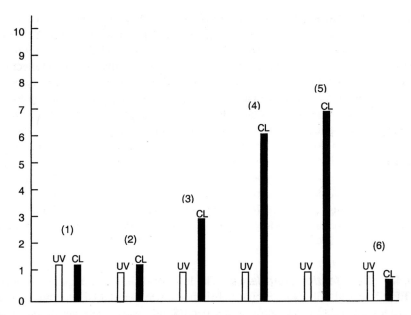

**Figure 3.5.** Comparison of the relative effectiveness of chlorine and UV on bacteria and viruses. (1) Escherichia coli, (2) Salmonella typhosa, (3) Staphylococcus aureus, (4) polio virus type 1, (5) Coxsackie AZ virus, (6) adenovirus type (Trojan Technologies Inc., 2000).

equipment, or chemicals. The cleaning frequency is very site-specific: some systems needing to be cleaned more often than others.

Chemical cleaning is most commonly done with citric acid. Other cleaning agents include mild vinegar solutions and sodium hydrosulfite. A combination of cleaning agents should be tested to find the agent most suitable for the wastewater characteristics without producing harmful or toxic by-products.

UV is generated on-site, and poses no significant safety concerns to surrounding communities. Worker safety requirements are directed to protection from exposure (primarily of the eyes to skin) from UV light, as well as to strict monitoring of electrical hazards and safe handling and disposal of expended lamps, quartz, and ballasts.

**3.4.4.4 Costs** Specifically, consideration must be given to how environmental matters affect our economic thinking and, conversely, how economic decisions affect the environment. Cost considerations are an integral part of the decision-making process at the stage of identifying potential improvements to a process, a product, or an activity.

A brief cost analysis for both chlorination/dechlorination and UV disinfection processes treating a wastewater treatment facility is presented in Figure 3.6, capable of processing 18 million gallons per day (mgd). These costs can vary considerably depending on the locality, demands, and supply. The annualized cost

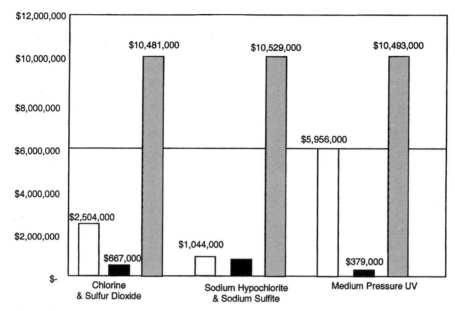

**Figure 3.6.** Disinfection system cost comparisons of an 18 mgd facility. Open bars, construction costs; solid bars, annual operations and maintenance; grey bars, total annualized costs.

values for alternatives were based on the design average plant flow of 18 mgd over a 20-year period and 5.5% interest rate (Temmer et al., 2000; Soroushian, 2001).

Figure 3.6 indicates that, although the construction cost for chlorination system is considerably less than the UV, annualized costs are about the same for both systems. There is no more real economic benefit for pursuing chlorination disinfection system.

### 3.4.4.5 Environmental Impacts of Energy Sources and Implications of Renewables
Like chlorination, UV disinfection systems demand energy supplies. One of the most important factors that can contribute to achieving sustainable development is the requirement for a supply of energy resources that is itself fully sustainable. Effective and efficient utilization of energy resources calls for such resources to be readily available at reasonable cost utilized for all required tasks without causing negative societal impact. Clearly, there is an intimate connection between renewable energy sources and sustainable development (Das, 2002).

Table 3.10 gives a brief summary of cost of generating powers, air quality, and environmental impacts of renewable and other forms of energy sources. It illustrates that renewable energy sources can make a significant contribution to reducing greenhouse and acid gas emissions. Renewable sources have their own environmental impacts but these are often small, site-specific, and local in nature.

**TABLE 3.10. Cost of Generating Power, Air Quality and Environmental Impacts (Energy Ideas Clearinghouse, 2001)**

| Source | Cost | Potential Environmental Impact and Pollutants of Concern |
|---|---|---|
| Solar photovoltaic (large scale) | ~18–20 U.S. cents/kWh | Corrosive acid, potentially toxic and hazardous substance, land use and loss of habitat, visual intrusion, uncontrolled dumping in landfills |
| Solar photovoltaic (small scale) | ~22–24 U.S. cents/kWh | Potentially toxic and hazardous substance |
| Wind | ~4.5–10 U.S. cents/kWh | Land use and habitat damage, noise, bird strike |
| Biomass | ~6–8 U.S. cents/kWh excluding ethanol based fuels | $CO_2$, $NO_x$, $PM_{10}$, CO, VOC, toxic air pollutants (TAPs) |
| Geothermal | ~5–8 U.S. cents/kWh | Emission of $H_2S$ during operation, groundwater and soil contamination, surface water, land use, visual impact |
| Solar thermal | ~5–25 U.S. cents/kWh | Thermal or chemical pollution of surface water |
| Large-scale hydro | ranges from 2.0–7.8 U.S. cents/kWh | Fish and other aquatic lives, terrestrial ecosystems, local climate, public health, water flow and supply, population displacement, loss of agricultural land |
| Existing hydro | ranges from 1.0–2.0 U.S. cents/kWh | Fish and other aquatic lives, water supply, irrigation, $SO_2$ |
| Nuclear | 5.5–6.0 U.S. cents per kWh | Radioactive substance, public health |
| Coal fired generators | 4.8–5.5 U.S. cents per kWh | $CO_2$, $NO_x$, $SO_2$, $PM_{10}$, CO, VOC, TAPs |
| Diesel generators | 10–15 U.S. cents/kWh | $CO_2$, $NO_x$, $SO_2$, $PM_{10}$, CO, VOC, TAPs |
| Natural gas Combined cycle Combustion Turbine | (at natural gas price of $5/MMBtu) = 5 U.S. cents/kWh | $CO_2$, $NO_x$, $PM_{10}$, CO, VOC, TAPs |
| Simple cycle Combustion Turbine | 6 U.S. cents/kWh at $5/MMBtu for natural gas | $CO_2$, $NO_x$, $PM_{10}$, CO, VOC, TAPs |

## 3.5 CASE STUDY: COMPARISON OF LCA OF ELECTRICITY FROM BIORENEWABLES AND FOSSIL FUELS

A series of life cycle assessments (LCA) was conducted on biomass, coal, and natural gas systems to quantify the environmental benefits and drawbacks of each. All those evaluations were conducted in a cradle-to-grave manner to cover all processes necessary for the operation of the power plant, including raw material extraction, feed preparation, transportation, waste disposal, and recycling. We summarize data on energy balance, global warming potential (GWP), air emissions, and resource consumption for each system.

The generation of electricity, and the consumption of energy in general, result in consequences to the environment. Using renewable resources and incorporating advanced technologies such as integrated gasification combined cycle (IGCC) may result in less environmental damage, but to what degree, and with what trade-offs? Life cycle assessment studies have been conducted on various power generating options in order to better understand the environmental benefits and drawbacks of each technology. Material and energy balances were used to quantify the emissions, energy use, and resource consumption of each process required for the power plant to operate. These include feedstock procurement (mining coal, extracting natural gas, growing dedicated biomass, collecting residue biomass), transportation, manufacture of equipment and intermediate materials (e.g., fertilizers, limestone), construction of the power plant, decommissioning, and any necessary waste disposal.

The following systems were studied:

- A biomass-fired integrated gasification combined cycle (IGCC) system using a biomass energy crop (hybrid poplar)
- A pulverized coal (pc) boiler with steam cycle, representing the average for coal-fired power plants in the U.S. today
- A system cofiring biomass residue with coal (15 percent by heat input will be presented here)
- A direct-fired biomass power plant using biomass residue (urban, primarily)
- A natural gas combined cycle (NGCC) power plant.

Each study was conducted independently and can therefore stand alone, giving a complete picture of each power generation technology. However, the resulting emissions, resource consumption, and energy requirements of each system can ultimately be compared, revealing the environmental benefits and drawbacks of the renewable and fossil based systems.

### 3.5.1 Results

***3.5.1.1 System Energy Balance*** The total energy consumed by each system includes the fuel energy consumed plus the energy contained in raw and intermediate materials that are consumed by the systems. Examples of the first type of energy use are the fuel spent in transportation, and fossil fuels consumed by the fossil-based power plants. The second type of energy is the sum of the energy that would be released during combustion of the material (if it is a fuel) and the total energy that is consumed in delivering the material to its point of use. Examples of this type of energy consumption are the use of natural gas in the manufacture of fertilizers and the use of limestone in flue-gas desulfurization. The combustion energy calculation is applied where non-renewable fuels are used, reflecting the fact that the fuel has a potential energy that is being consumed by the system. The combustion energy of renewable resources, those replenished at a rate equal to or greater than the rate of consumption, is not subtracted from the net energy of the system. This is because, on a life cycle basis, the resource is not being consumed. To determine the net energy balance of each system, the energy used in each process block is subtracted from the energy produced by the power plant. The total system energy consumption by each system is shown in Table 3.11.

To examine the process operations that consume the largest quantities of energy within each system, two energy measurement parameters were defined. First, the energy delivered to the grid divided by the total fossil-derived energy consumed by each system was calculated. This measure, known as the net energy ratio, is useful for assessing how much energy is generated for each unit of fossil fuel consumed. The other measure, the external energy ratio, is defined to be the energy delivered to the grid divided by the total non-feedstock energy to the power plant. That is, the energy contained in the coal and natural gas used at the fossil-based power plants is excluded. The external energy ratio assesses how much energy is generated for each unit of upstream energy consumed.

Because the energy in the biomass is considered to be both generated and consumed within the boundaries of the system, the net energy ratio and external energy ratio will be the same for the biomass-only cases (biomass-fired IGCC and direct-fired biomass). In calculating the external energy ratio, we are essentially treating the coal and natural gas fed to the fossil power plants as renewable fuels, so that upstream energy consumption can be compared. The energy results for each case studied are shown in Figure 3.7.

As expected, the biomass-only plants consume less energy overall, since the consumption of nonrenewable coal and natural gas at the fossil plants results in net

**TABLE 3.11. Total System Energy Consumption (Mann and Spath, 2002)**

| System | Total Energy Consumed (kJ/kWh) |
|---|---|
| Average coal | 12,575 |
| Natural gas IGCC | 8,377 |
| Biomass/coal cofiring (15 percent by heat input) | 10,118 |
| Biomass-fired IGCC using hybrid poplar | 231 |
| Direct-fired biomass power plant using biomass residue | 125 |

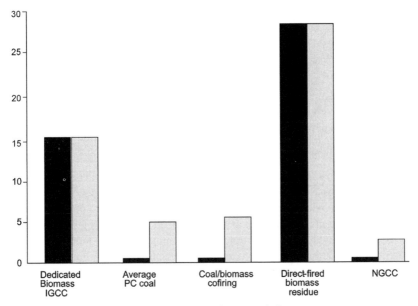

**Figure 3.7.** Life cycle energy balance.

energy balances of less than one. The direct-fired biomass residue case delivers the most amount of electricity per unit of energy consumed. This is because the energy used to provide a usable residue biomass to the plant is fairly low. Despite its higher plant efficiency, the biomass IGCC plant has a lower net energy balance than the direct-fired plant because of the energy required to grow the biomass as a dedicated crop. Residue resource limitations, however, may necessitate the use of energy crops in the future. Cofiring biomass with coal slightly increases the energy ratios over those for the coal-only case, even though the plant efficiency was derated by 0.9 percentage points.

In calculating the external energy ratios, the feedstocks to the power plants were excluded, essentially treating all feedstocks as renewable. Because of the perception that biomass fuels are of lower quality than fossil fuels, the external energy ratios for the fossil-based systems were expected to be substantially higher than those of the biomass-based systems. The opposite is true, however, due to the large amount of energy that is consumed in upstream operations in the fossil-based systems. The total non-feedstock energy consumed by the systems is shown in Table 3.12. In the case of coal, 35 percent of this energy is consumed in operations relating to flue-gas cleanup, including limestone procurement. Mining the coal consumes 25 percent, while transporting the coal is responsible for 32 percent. Greater than 97 percent of the upstream energy consumption related to the natural gas IGCC system is due to natural gas extraction and pipeline transport steps, including fugitive losses. Although upstream processes in the biomass systems also consume energy, shorter transportation distances and the fact that flue-gas desulfurization is not required, reduce the total energy burden.

**TABLE 3.12. Non-feedstock Energy Consumption (Mann and Spath, 2002)**

| System | Total Energy Consumed (kJ/kWh) |
| --- | --- |
| Average coal | 702 |
| Natural gas IGCC | 1,718 |
| Biomass/coal cofiring (15 percent by heat input) | 614 |
| Biomass-fired IGCC using hybrid poplar | 231 |
| Direct-fired biomass power plant using biomass residue | 125 |

*3.5.1.2 Global Warming Potential* Figure 3.8 shows the net emissions of the three greenhouse gases quantified for these studies: carbon dioxide ($CO_2$), methane ($CH_4$), and nitrous oxide ($N_2O$). The biomass IGCC system has a much lower GWP than the fossil systems because of the absorption of $CO_2$ during the biomass growth cycle. Sensitivity analyses demonstrated that even moderate amounts of soil carbon sequestration (1,900 kg/ha/seven-year rotation) would result in the biomass IGCC system having a zero-net greenhouse gas balance. Sequestration amounts greater

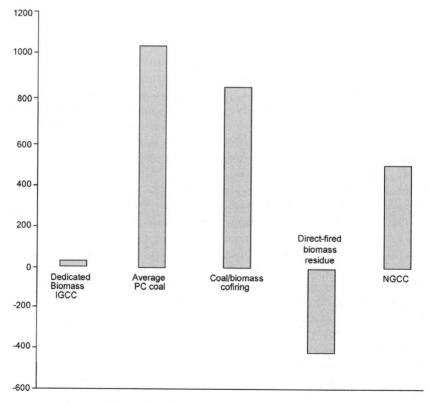

**Figure 3.8.** Net greenhouse gas emission.

than this would result in a negative release of greenhouse gases, and a system that removes carbon from the atmosphere overall.

The base case presented here assumes no net change in soil carbon, since actual gains and losses will be very site specific. The direct-fired biomass system in which the residue is used at the power plants, has a highly negative rate of greenhouse gas emissions because the generation of methane associated with biomass decomposition, is avoided. Based on current disposal practices, it was assumed that 46 percent of the residue biomass used in the direct-fired and cofiring cases would have been sent to a landfill and that the remainder would end up as mulch and other low-value products. Decomposition studies reported in the literature were used to determine that if the biomass residue had not been used at the power plant, approximately 9 percent of the carbon would have ended up as $CH_4$ and 61 percent as $CO_2$. The remaining carbon is resistant to decomposition in the landfill. Had all the residue biomass been decomposed aerobically, the $CO_2$ produced would have been 1.85 kg/kg biomass. If the biomass residue was not used at the power plant, the decomposition pathways described above would have resulted in total greenhouse gas emissions of 2.48 kg $CO_2$-equivalent/kg biomass (1.117 kg $CO_2 + 0.065$ kg $CH_4$). The net difference is the reason for the negative greenhouse gas emissions associated with the direct-fired system.

The natural gas combined cycle plant has the lowest GWP of all fossil systems because of its higher efficiency, despite natural gas losses that increase net $CH_4$ emissions. Natural gas losses during extraction and delivery were assumed to be 1.4 percent of the gross amount extracted. Because of the potency of methane as a greenhouse gas, nearly one-quarter of the total GWP of this system is due to these losses. Cofiring biomass with coal at 15 percent by heat input reduces the GWP of the average coal-fired power plant by 18 percent. The reduction in greenhouse gases is greater than the rate at which biomass is cofired because of the avoidance of methane emissions associated with decomposition that would have occurred had the biomass not been used at the power plant. Biomass disposal and decomposition emissions for this scenario are the same as those used in the direct-fired case.

### 3.5.1.3 Air Emissions
Figure 3.9 charts the following emissions: particulates, oxides of sulfur ($SO_x$), oxides of nitrogen ($NO_x$), $CH_4$, $CO_2$, and nonmethane hydrocarbons (NMHCs). Methane emissions are high for the natural gas case due to natural gas losses during extraction and delivery. The direct-fired biomass and coal/biomass cofiring cases have negative methane emissions, due to avoided decomposition processes (landfilling and mulching). CO and NMHCs are higher for the biomass case because of upstream diesel combustion during biomass growth and preparation. Cofiring reduces the coal system air emissions by approximately the rate of cofiring, with the exception of particulates, which are generated during biomass chipping and handling.

### 3.5.1.4 Resource Consumption
Figure 3.10 shows the total amount of non-renewable resources consumed by the systems investigated. Limestone is used in significant quantities by the coal-fired power plants for flue-gas desulfurization.

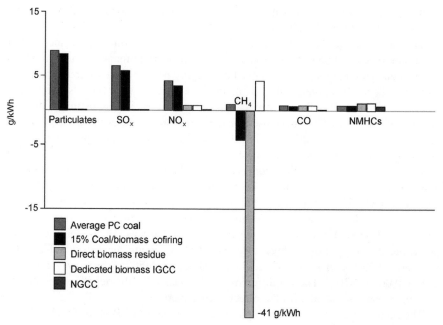

**Figure 3.9.** Emissions of PM, $SO_X$, $NO_X$, $CH_4$, CO, and NMHCs.

The natural gas IGCC plant consumes almost negligible quantities of resources, with the exception of the feedstock itself, including that lost during extraction and delivery.

### 3.5.2 Sensitivity Analysis

A sensitivity analysis was conducted on each system to determine which parameters had the most influence on the results and to pinpoint opportunities for reducing the environmental burden of the system. In general, parameters associated with increasing the system efficiency and reducing the fossil fuel usage had the largest effects. Additionally, for the biomass systems, variables associated with growing a dedicated feedstock and factors affecting how much $CO_2$ and $CH_4$ are avoided by using biomass residue significantly affected the GWP of the system. Overall, however, the sensitivity analyses demonstrated that the conclusions that can be drawn from these studies remain relatively constant as different parameters are varied.

### 3.5.3 Summary and Conclusions

It is evident that biomass power systems reduce the environmental burden associated with power generation. The key comparative results can be summarized as follows:

• The GWP of generating electricity using a dedicated energy crop in an IGCC system is 4.7% of that of an average U.S. coal power system.

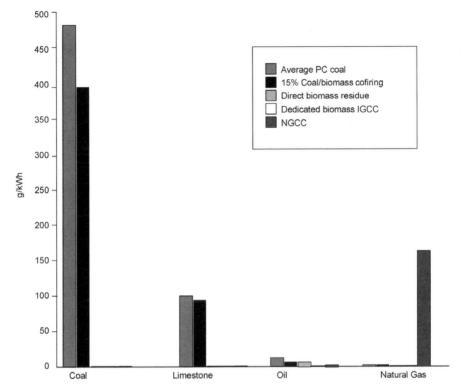

**Figure 3.10.** Total amount of non-renewable resources consumed by the systems.

- Cofiring residue biomass at 15 percent by heat input reduces the greenhouse gas emissions and net energy consumption of the average coal system by 18 percent and 12 percent, respectively.
- The life cycle energy consumption of the coal and natural gas systems are significantly higher than those of the biomass systems because of the consumption of non-renewable resources.
- Not counting the coal and natural gas consumed at the power plants in these systems, the net energy consumption is still higher than that of the biomass systems because of energy used in processes related to flue gas clean-up, transportation, and natural gas extraction and coal mining.
- The biomass systems produce very low levels of particulates, $NO_x$, and $SO_x$ compared to the fossil systems.
- System methane emissions are negative when residue biomass is used because of avoided decomposition emissions.
- The biomass systems consume very small quantities of natural resources compared to the fossil systems.
- Other than natural gas, the natural gas IGCC consumes small amounts of resources.

These results demonstrate that overall, biomass power provides significant environmental benefits over conventional fossil-based power systems. In particular, biomass systems can significantly reduce the amount of greenhouse gases that are produced, per kWh of electricity generated. Additionally, because the biomass systems use renewable energy instead of non-renewable fossil fuels, they consume very small quantities of natural resources and have a positive net energy balance. Cofiring biomass with coal offers an opportunity to reduce the environmental burdens associated with the coal-fired power systems that currently generate over half of the electricity in the United States. Finally, by reducing $NO_x$, $SO_x$, and particulates, biomass power can improve local air quality over coal-fired power generation.

## REFERENCES

Azapagic, A. (2000). Life Cycle Assessment: A Tool for Innovation and Improved Environmental Performance, Part VI35, pp. 519–530, In *Science, Technology and Innovation Policy: Opportunities and Challenges for the Knowledge Economy*. Quorum Books, Westport, Connecticut.

Bare, J. C. (2003). The tool for the reduction and assessment of chemical and other environmental impacts (TRACI), USEPA, 2003. TRACI is available on the Internet at no cost at http://epa.gov/ORD/NRMRL/std/sab/iam_traci.htm.

Bhander, G. S., Hauschild, M., and McAloone, T. (2003). Implementing life cycle assessment in product development. *Environmental Progress*, **22**(4), 255–267.

Brungs, W. (1973). Effects of residual chlorine on aquatic life. *J. Wat. Poll. Cont. Fed.* **45**(10), 2180–2193 (October).

Cooper, J. S. (2003). Life cycle assessment and sustainable development indicators. *J. Industrial Ecology*, **7**(1), 12–15.

Curran, M. A. (1996). *Environmental Life-Cycle Assessment*. McGraw-Hill, New York, NY.

Curran, M. A. (2003). Do bio-based products move us toward sustainability? A look at three USEPA case studies. *Environmental Progress*, **22**(4), 277–292.

Dandy J. W. (1967). The effects of chemical characteristics of the environment on the activity of an aquatic organism. Thesis, University of Toronto, Ontario, Canada.

Dandy, J. W. (1972). Activity response to chlorine in the Brook Trout, Salvelinus fontinalis (Mitchill). *Can. Jour. Zool.*, **50**, 405.

Das, T. K. (2002). Evaluating the life cycle environmental performance of chlorine disinfection and ultraviolet technologies. *Clean Technology and Environmental Policy*, Vol. 4, pp. 32–43.

Das, T. K. (2004). Disinfection. In the Kirk-Othmer *Encyclopedia of Chemical Technology*, 5th Ed., Vol. 8, pp. 605–672, Hoboken, NJ: Wiley.

Das, T. K., and Ekstrom, L. P. (1999). UV application to a major wastewater treatment plant on the west coast. Paper No. 100d, American Institute of Chemical Engineers Spring Annual Meeting, Houston (April).

Das, T. K., Sharma, M. P., and Butner, S. (2001). The concept of an eco-industrial park—a solution to industrial ecology. Presented at the Topical Conference: Industrial Ecology and Zero-Discharge Manufacturing, American Institute of Chemical Engineers Spring National Meeting, Paper No. 18c, Houston (April).

Diwekar, U. M. (2003). *Introduction to Applied Optimization*, Netherlands, Kluwer Academic Publishers.

Dondoroff, P., and Katz, M. (1950). Critical review of literature on the toxicity of industrial wastes and their components to fish. *Sew. & Ind. Wastes*, **22**, 1432.

Ecology, Washington Department of Ecology (1998). Ecology's criteria for sewage works design. Chapter T5 (December).

ENDS (1996). Dow Europe and the challenge of eco-efficiency. Company Report No. 170. ENDS Report, 252 (January 16–20).

Energy Ideas Clearinghouse (2001). Washington State University Cooperative Extension Energy Program, Olympia, Washington (http://www.energy.wsu.edu/nwalliance-eic).

Hart, J. L. (1973). Pacific fishes of Canada (Ottawa, Canada: Fisheries Research Board of Canada).

Heltz, G. R., and Nweke, A. C. (1995). Incompleteness of wastewater dechlorination. *Environmental Science & Technology*, **29**(4), 1018–1022.

Holland, G. A. (1960). Toxic effects of organic and inorganic pollutants on young salmon and trout. State of Washington Dept. of Fisheries, *Res. Bull.*, No 5, p. 198.

ISO, International Organization of Standardization (1997). Environmental Management — Life Cycle Assessment — Principles and Framework (ISO/FDIS 14040), ISO TC 207.

ISO, International Organization of Standardization (1998a). Life Cycle Assessment — Impact Assessment (ISO 14042), ISO TC 207/SC5/WG4.

ISO, International Organization of Standardization (1998b). Environmental Management — Life Cycle Assessment — Life Cycle Interpretation (ISO/DIS 14043), ISO TC 207.

Kozloff, E. N. (1974). Keys to the marine invertebrates of Puget Sound, the San Juan archipelago, and adjacent regional (Seattle, Washington), University of Washington Press.

Loge, F. J., Darby, J. L., and Tchobanoglous, G. (1996). UV disinfection of wastewater: probabilistic approach to design. *J. Env. Eng.*, p. 1078 (December).

LOTT (Lacey, Olympia, Tumwater, and Thurston) (1994). Pilot Study Final Report on UV Disinfection of Wastewater, The City of Olympia, Washington.

Mann, M. K., and Spath, P. L. (2002). Life cycle assessment comparison of electricity from biomass, coal, and natural gas. American Institute of Chemical Engineering Annual Meeting, Indianapolis, IN, USA, Paper #18d (November).

Merkens, J. C. (1958). Studies on the toxicity of chlorine and chloramines to the rainbow trout. *Water & Waste Treatment J.* (G.B.), 7, p. 150.

Metcalf & Eddy Inc. (1991). *Wastewater Engineering: Treatment, Disposal, and Reuse*, 3rd Ed., p. 344, McGraw-Hill International.

Pike, D. J. (1971). Toxicity of chlorine to brown trout, *New Zealand Wildlife*, No 33.

Scheible, O. K. (1987). Development of a rationally based design protocol for the ultraviolet light disinfection process. *J. Water Pollut. Control Fed.*, 59, p. 25.

Scheible, O. K., and Forndran, A. (1986). Ultraviolet disinfection of wastewaters from secondary effluent and combined sewer overflows, EPA-600/2-86/005, USEPA, Cincinnati, OH.

SETAC (1991). A technical framework for life cycle assessment, workshop report from the smugglers notch. Vermont, USA. The Society of Environmental Toxicology and Chemistry.

SETAC (1993). Guidelines for life-cycle assessment: a code of practice. Pensacola: The Society of Environmental Toxicology and Chemistry, Sesimbra, Portugal (31 March– 3 April).

Sharma, M. P., and Das, T. K. (2002). Green engineering in chemical engineering curriculum, Paper No. 219b, American Institute of Chemical Engineers' Annual National Meeting, Reno (November).

Sharma M. P., Das, T. K., and Butner, S. (2001). Green engineering, industrial ecology, and sustainability: the role of university and engineering educational institutions. Presented at the Topical Conference: Industrial Ecology and Zero-Discharge Manufacturing, AIChE Spring National Meeting, Houston (April).

Soroushian, F. (2001). Personal Communication, CH2M-Hill, Greenwood Village, CO (September 13).

Sprague, J. B., and Drury, D. E. (1969). Avoidance reactions of salmonid fish to representative pollutants. In *Advances in Water Pollution Research, Proc. 4th Intl. Conf. Water Poll. Res.* Pergamon Press, London, England, p. 169.

Temmer, B., Soroushian, F., and Noesen, M. (2000). UV emerges bright spot, *Pollution Engineering*, September, pp. 46–49.

Thatcher, T. O. (1977). The relative sensitivity of Pacific Northwest fishes and invertebrates to chlorinated sea water. Battelle, Pacific Northwest Laboratories, Sequim, Washington, Jolly, R. L., Gorchov, H., and, Hamilton, D. H (Ed.), Proc. Second Conf. Water Chlorination Environ, Impact and Health Effects, Vol. 2, Gatlinburg, TN, pp. 341–350, (Oct. 31–Nov. 4).

Trojan Technologies Inc. (2000). Overview of UV disinfection. 3020 Gore Road, London, Ontario, Canada N5V 4 TS.

Tshudy, J. A. (1994). Environmental life cycle analysis — the foundation for understanding environmental issues and concerns, *Proc. Int. Symp. Res. Converv. Environ. Technol. Metall. Ind.*, August 20–25, Lancaster, USA, pp. 229–237.

United Nations Environmental Program—Society of Environmental Toxicology and Chemistry (UNEP-SETAC) (2000), *Letter of Intent: Best Available Practice of Life Cycle Assessment with Generic Application Dependency*. Signed by the International Council of SETAC, and UNEP, Division Technology, Industry, and Economics (www.uneptie. org/pc/sustain/lca/letter-of-intent.htm). Accessed October 2002.

U.S. Environmental Protection Agency (1986a). *Design Manual: Municipal Wastewater Disinfection*. Office of Research and Development, Water Engineering Research Laboratory, EPA/625/1-86/021, Cincinnati, OH.

U.S. Environmental Protection Agency (1986b). *Design Manual, Odor and Corrosion Control in Sanitary Sewage Systems and Treatment Plants*. EPA/625/1-85/018 (October).

U.S. Environmental Protection Agency (1989). *EPA Quality Assurance Management Staff, Development of Data Quality Objectives: Description of Stages I and II*.

U.S. Environmental Protection Agency (1992). *Users Manual for UVDIS, Version 3.1 UV Disinfection Process Design Manual*. EPAG0703, Contract No. 68-C8-0023, USEPA, Cincinnati, OH.

U.S. Environmental Protection Agency (1993). *Office of Research and Development, Life Cycle Assessment: Inventory Guidelines and Principals.* EPA/600/R-92/245.

U.S. Environmental Protection Agency (1995). *Office of Solid Waste, Guidelines for Assessing the Quality of Life Cycle Inventory Analysis.* EPA/530/R-95/010.

U.S. Environmental Protection Agency (1996). *Chemical Accidental Prevention and Risk Management Programs.* 40 CFR 68 Subpart B (June). http://response.restoration.noaa.gov/chemaids/rmp/rmp.html

U.S. Environmental Protection Agency (1998). *EPA's Risk Management Program Guidance for Wastewater Treatment Plants.* 40CFR68, EPA 550-B-98-010, Washington D.C. (June).

White, G. C. (1999). *Handbook of Chlorination and Alternative Disinfectants*, 4th Ed., N.Y., Wiley, pp. 71–75.

# CHAPTER 4

# ASSESSMENT AND MANAGEMENT OF HEALTH AND ENVIRONMENTAL RISKS

TAPAS K. DAS

Washington Department of Ecology, Olympia, Washington 98504, USA

The concept of risk, the possibility of loss or injury, is familiar in every human endeavor. Chemical engineers routinely consider the risks of adverse health and environmental impacts from exposures to industrial byproducts released to air, water, and land. Risk assessment, thus, is an organized process for describing and estimating the likelihood of such events. Risk management is the decision-making process by which planners, who are not necessarily chemical engineers, attempt to minimize risks without undue harm to societal values.

This chapter explains the paradigms used in evaluating the impacts of chemical releases on human health and on the environment. Tools and procedures are described for monitoring exposure to chemicals and for communicating to the public the results of analysis of risk data. The coverage of risk management is briefer because management decisions should be made independently of risk analysis. However the important concepts in this field are explained and illustrated.

## 4.1  HEALTH RISK ASSESSMENT

Risks to human health may be assessed on the basis of information collected from environmental monitoring; alternatively, the data may be incorporated into models of human activity and exposure, in the workplace and elsewhere, to permit analysts to draw conclusions on the likelihood of adverse effects. In both

*Toward Zero Discharge: Innovative Methodology and Technologies for Process Pollution Prevention*, edited by Tapas K. Das.
ISBN 0-471-46967-X   Copyright © 2005 John Wiley & Sons, Inc.

cases, risk assessment is an essential tool used by those responsible for making decisions with environmental consequences.

The results of environmental risk assessment are generally intended to be incorporated into decisions that affect the public, along with the economic, social, technological, and political consequences of a proposed action. The need for such analytical data is indicated in Table 4.1, which lists selected U.S. laws that require or suggest human health risk assessment before regulations are promulgated. The list is enormous and probably will grow with time.

In contrast to the specific laws addressing concerns about environmental pollutants presented in Table 4.1, Figure 4.1 offers a generalize scheme of the routes by which these pollutants enter the environment, ending with human exposure.

The paradigm for assessing human health risk consists of formulation of the problem, assessment of the exposure, assessment of any toxic effects, and risk characterization.

### 4.1.1   Problem Formulation

Problem formulation comprises such activities as definition of the goals and spatial and temporal scale of the ERA, development of a site conceptual model, and selection of endpoint and nonhuman receptor species. The final step is to identify contaminants of potential concern and to determine whether, for example, threshold limit values (TLVs) have been exceeded. Examples 4-1 and 4-2 illustrate how the results of environmental monitoring of three common industrial pollutants can be used in calculations of the quantities present.

*Example 4-1 Monitoring*   An air sampler collects 16 mg of methyl chloride during six-hour work shift. The sampler has the following characteristics:

$$\text{Average flow rate} = 1.2 \, \text{L/min}$$
$$\text{Collection efficiency} = 0.9$$

What is the 8 hour TWA (total weighted average) of methyl chloride?

*Solution*   The total air collected was

$$6 \times 1.2 \, \text{L/min} \times 60 \, \text{min/h} = 432 \, \text{L}$$

$$\frac{432 \, \text{L}}{1000 \, \text{L/m}^3} = 0.432 \, \text{m}^3$$

Sampler content was 16 mg or $\frac{16}{0.9} = 17.8$ mg in the air sampled.

**TABLE 4.1.  United States Safety, Health, and Environmental Statutes that Imply Risk Assessment**

*Environmental Protection Agency*

| | |
|---|---|
| Atomic Energy Act (also NRC) | 42.U.S.C.2011 |
| Comprehensive Environmental Response, Compensation and Liability Act (CERCLA, or Superfund) | 42.U.S.C.9601 |
| Clean Air Act | 42.U.S.C.7401 |
| Clean Water Act | 33.U.S.C.1251 |
| Emergency Planning and Community Right to Know Act | 42.U.S.C.11001 |
| Federal Food and Drug, and Cosmetics Act (also HHS) | 21.U.S.C.301 |
| Federal Insecticide, Fungicide, and Rodenticide Act | 7.U.S.C.136 |
| Lead Contamination Control Act of 1988 | 42.U.S.C.300j-21 |
| Marine Protection, Research, and Sanctuaries Act (also DA) | 16.U.S.C.1431 |
| Nuclear Waste Policy Act | 42.U.S.C.10101 |
| Resource Conservation and Recovery Act | 42.U.S.C.6901 |
| Safe Drinking Water Act | 42.U.S.C.300f |
| Toxic Substances Control Act | 7.U.S.C.136 |
| Food Quality Protection Act of 1996 | $7.U.S.C. > 6$ |

*Consumer Product Safety Commission*

| | |
|---|---|
| Consumer Product Safety Act | 15.U.S.C.2051 |
| Federal Hazardous Substance Act | 15.U.S.C.1261 |
| Lead-Based Paint Poisoning Act (also HHS and HUD) | 42.U.S.C.4801 |
| Lead Contamination Control Act of 1988 | 42.U.S.C.300j-21 |
| Poison Prevention Packaging Act | 15.U.S.C.1471 |

*Department of Agriculture*

| | |
|---|---|
| Eggs Products Inspection Act | 21.U.S.C.1031 |
| Federal Meat Inspection Act | 21.U.S.C.601 |
| Poultry Products Inspection Act | 21.U.S.C.451 |

*Department of Labor*

| | |
|---|---|
| Federal Mine Safety and Health Act | $30.U.S.C. > 801$ |
| Occupational Safety and Health Act | 29.U.S.C.651 |

*Department of Transportation*

| | |
|---|---|
| Hazardous Liquid Pipeline Safety Act | 49.U.S.C.1671 |
| Hazardous Materials Transportation Act | $49.U.S.C. > 1801$ |
| Motor Carrier Safety Act | 49.U.S.C.2501 |
| National Traffic and Motor Vehicle Safety Act | 15.U.S.C.1381 |
| National Gas Pipeline Safety Act | 49.U.S.C.2001 |

*Sources*: Federal Focus, 1991; Roberts and Abernathy, 1997.

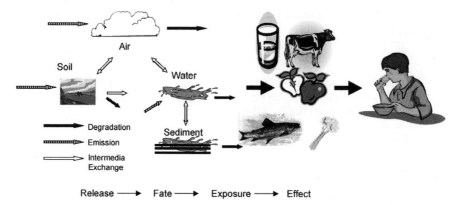

**Figure 4.1.** Generalized scenario for exposure by humans to environmental pollutants.


Average air concentration during the collection time

$$= 41.2 \, \text{mg/m}^3 \cdot 6\text{h}/8\text{h} = 31 \text{mg/m}^3$$

Was the TLV (threshold limit value) exceeded?
TWA for methyl chloride is 103 mg/m$^3$ and was not exceeded.


*Example 4-2 Continuous Monitoring*   Lead fumes are produced in a lead-burning operation. A continuous monitor in the work room is operated for 5 consecutive days. The following are known:

| | |
|---|---|
| Initial sample rate: | 1 cfm (cubic feet per minute) |
| Final sample rate: | 0.7 cfm |
| Filter efficiency: | 0.999 for particulates $\gg$ 20 μm AMD (average mean diameter) |
| | 0.83 for particulates <20 μm AMD |
| Mass of dust collected: | 1.7 g |
| % of lead in dust: | 23 |


*Solution*   Determine the TWA lead concentration in the work air.

Quality of air sampled = average sample rate · sampling time

Assume approximately linear reduction in sampling rate as filter loads. Hence, average sampling rate is

$$\frac{1 + 0.7}{2} = 0.85 \, \text{cfm}$$

and total air sampled is

$$0.85\,\text{cfm} \times 5\,\text{days} \times 1{,}440\,\text{min/day} - 6{,}120\,\text{ft}^3$$

Total dust collected $= 1.7$ g or $1.7 \times 0.23 = 0.39$ g of lead, since 23 percent of dust is lead. Since lead was present as a fume, AMD $< 20$ μm and total lead collected is

$$\frac{0.39}{0.83} = 0.47\,\text{g}$$

Average lead concentration is the total amount divided by sample volume or

$$\frac{0.47\,\text{g}}{6120\,\text{ft}^3} = \frac{470\,\text{mg}}{6120\,\text{ft}^3} \times \frac{1}{35.2\,\text{m}^3/\text{ft}^3} = 0.0022\,\text{mg/m}^3$$

Assuming lead generation occurs only during 8 hour work day, the TWA is

$$\text{TWA} = \frac{0.0022\,\text{mg}}{\text{m}^3} \times \frac{24}{8} = 0.007\,\text{mg/m}^3$$

Dusts pose a particular problem with respect to determining acceptable air concentrations. The particle size distribution needs to be taken into account, and the specific respirable fractions, as published in the literature, must be used to determine exposure. Dust particle sizing can be accomplished by size-selective samplers, typically multistage impactors.

These simple examples only scratch the surface of techniques used in problem formulation. They are, however, illustrative. Sometimes, as in Example 4-1, monitoring reveals that existing pollution prevention measures are working properly and that no problem exists. Example 4-2 shows that determining the TWA of a pollutant is not always sufficient to formulate a problem.

## 4.1.2 Exposure Assessment

Measures of ongoing exposure to health risks call for determination of concentration. In the next section, Example 4-3 shows how to determine bioconcentration, and Example 4-4 explains the use of slope factors, often used in quantifying risks of cancer.

### 4.1.2.1 Partition Coefficient and Bioconcentration Factor   While extensive research continues to refine existing methods and to develop new markers of exposure and effect, experience has identified several biomarkers of value to hazardous waste management decisions. Exposure to chemicals that bioaccumulate in tissue (see, e.g., Table 4.2) is easily accomplished by analyzing tissue residue. In the absence of tissue residues, empirical constants often derived and used by ecotoxicologists to predict bioaccumulation include the octanol–water partition

**TABLE 4.2. Selected Physicochemical Parameters and Bioconcentration Factors for Typical Compounds which Bioaccumulate**

| Compound | Log $K_{ow}$ | Log BCF (fish) | Log $K_{oc}$ | S (mg/L) |
|---|---|---|---|---|
| Chlordane | 5.15 | 4.05 | 4.32 | 0.0056 |
| DDT | 5.98 | 4.78 | 4.38 | 0.0017 |
| Dieldrin | 5.48 | 3.76 | 4.55 | 0.022 |
| Lindane | 4.82 | 2.51 | 2.96 | 0.150 |
| Chlorpyrifos | 4.99 | 2.65 | 4.13 | 0.3 |
| Cyhexatin | 5.38 | 2.79 | <3.64 | <1.0 |
| 2,4D | 1.57 | 1.30 | 1.30 | 900 |
| 2,4,5T | 0.60 | 1.63 | 1.72 | 2.38 |
| TCDD | 6.15 | 4.73 | 5.67 | 0.0002 |

$K_{ow}$ = octanol-water partition coefficient (unitless).
BCF = bioconcentration factor (L/kg).
$K_{oc}$ = organic carbon partition coefficient.
S = water solubility.

coefficient ($K_{ow}$), which approximates a chemical's behavior between the lipid water phase and the bioconcentration factor (BCF), which describes the relationship between the chemical concentration in an organism and the concentration in the exposure medium (typically water). The BCF is empirically expressed as:

$$C_{org} = (BCF)(C_w), \text{ or } BCF = C_{org}/C_w$$

Example 4-3 illustrates the use of this relationship.

*Example 4-3 Predicting Bioconcentration*    Routine monitoring of sediments and surface water downstream from a hazardous waste site indicates that reportable concentrations of the pesticide DDT average, respectively, 33 mg/kg and 0.25 μg/L. Considering that the U.S. Food and Drug Administration has identified levels of 5 ppm or higher of DDT as requiring action, is there a reason to be concerned about potential accumulation of DDT in fish tissue near this site?

*Solution*    Since the routine monitoring does not provide information regarding DDT residues in fish tissue, we refer to Table 4.2 and bioconcentration factor (BCF). Transforming the Log BCF of 4.78, we estimate a BCF of 60.256 (mg/kg DDT in tissue per mg/L DDT in water or L/kg).

Substituting our monitoring data and the empirical BCF for DDT in the following equation yields,

$$C_{org} = 60.256 \, (mg/kg)/(mg/L) \times (0.00025 \, mg/L) = 0.015 \, mg/kg$$

We estimate a DDT tissue residue of 0.015 mg/kg. Based on this estimate, it is probable that DDT residues in tissue are an issue that should be evaluated.

The calculation of carcinogenic risk involves the use of a carcinogen potency factor (CPF). A CPF is basically the slope or steepness of the dose–response curve at very low exposures, and is now referred to as slope factor. The dimensions of a slope factor are expressed as the reciprocal of the daily dose $(mg/kg/day)^{-1}$. Having derived a slope factor, the calculation of carcinogenic risk is straightforward. Quantification of carcinogenic risk of an exposure simply requires converting the dose to the appropriate terms (mg/kg/day) and multiplying it by the slope factor as demonstrated in Example 4-4.

*Example 4-4 Use of Slope Factors*    What are the maximal number of excess life-time cancer cases expected for a population of 100,000 adults with a daily intake of 0.14 mg of benzene?

*Solution*

1. The slope factor for benzene is 0.029 $(mg/kg/day)^{-1}$ taken.
2. Assuming an adult weight of 70 kg, the lifetime cancer risk is calculated as:

$$\text{Individual cancer risk} = (0.14 \text{ mg/day}) \frac{1}{70 \text{ kg}} 0.029 \text{ mg/kg/day}$$

$$= 0.000052 \text{ or } 0.0052\%$$
$$= \text{risk} \times \text{exposed population}$$
$$= 0.000052 \times 100,000$$
$$= \text{five excess lifetime cancer cases}$$

### 4.1.2.2  *Exposure Point Concentrations*

The risk analyst must next esti-mate the concentration of contaminants at the exposure points, which define the locations of the receptors, that is, the human or animal population whose exposure to a pollutant is of concern. Receptors are identified for each exposure scenario simply by overlaying the demographic information with the exposure pathway. An exposure point may be as close as the sources of waste at the site itself (e.g., the trespasser scenario) or at a considerable distance, particularly for pathways involving the food chain. It may take any pathway: air, ground and surface water, soils, sediments, and food (e.g., plants and fish). For present exposures, actual moni-toring data at the exposure point should be used wherever possible. For example, contaminant concentrations should be obtained for drinking water wells in the vicin-ity of a site. For on-site exposure points, representative concentrations in soil or ground water may be calculated as the arithmetic or geometric mean (depending on the statistical distribution of the site analytical concentration data).

Future conditions can differ starkly. A plume of gases may not have yet migrated to a potential exposure point. Remediation will, of course, drastically reduce migration. In a comprehensive risk assessment, the exposure concentrations for each for remedial alternative would be estimated including the present situation (baseline conditions of no remedial action). Determining the concentration of a contaminant at the exposure point for future conditions often requires the use of fate and transport modeling methods and standard references. The major effort with these models is calibration; once calibrated, successive runs can be made relatively easily to estimate concentrations for a range of conditions and assumptions.

For ground water contaminants, hydrogeologic models can be used to estimate the future concentration at a downstream well. For volatile organic compounds released to the atmosphere, a Gaussian diffusion model can be employed to estimate future downwind concentrations. In general, the level of effort employed in data collection and modeling will depend on the estimated severity of the risk. Nominal risks do not warrant the same level of analysis as the clearly significant risks.

All mathematical models require the making of assumptions. This is true, as well, of the mechanistic models on which many analyses are based.

### 4.1.2.3  *Mechanistic Models*

In mechanistic models, it is assumed that a certain number of reactions, events, or "hits" (concept derived from radiation biology), or transition stages, related to a critical target in the cell (DNA), are necessary to transform a normal cell to a cancer cell. These models are important because there seems to be general consensus that tolerance distribution models (e.g., Mantel-Bryan) should not be used unless the available data are accurate enough to exclude models involving a linear component in the low-dose region, or a genotoxic mechanism might be involved.

The multistage model, first proposed by Armitage and Doll (1954), is an extension of the one-hit model developed for ionizing radiations. One generalized version of this model where at least one of the stages is assumed to be dose-related takes the form:

$$P(D) = 1 - \exp[-(q_0 + q_1 D + q_2 D^2 + \cdots + q_k D^k)], \quad K > 1 \qquad (4\text{-}1)$$

where $P(D)$ = the probability of cancer at dose $D$
$q_k$ = coefficients to best fit to the data
$D^k$ = applied does raised to the $k$th power
$k$ = number of stages (usually set arbitrarily)
at the number of dose levels minus one.

The most likely estimate at very low doses becomes increasingly unstable with a small change in the response at experimental doses. Therefore, a further development in modeling was the replacement of the linear term in the polynomial function by its 95 percent confidence limit to achieve more stable estimates of risk above background than are obtained for the most likely estimates. This so-called linearized multi-stage model is used routinely by the EPA and has, therefore, become the most widely used model for estimation of cancer risk. At low doses the function becomes essentially linear. The multistage model is very flexible in fitting data sets because it is a polynomial function of dose. The so-called carcinogen potency factors provided by the EPA have been based on this model and are calculated from animal data using commercially available computer programs.

The multistage model is favored by the EPA because it generally gives conservative risk estimates for low exposures, but the model has several weaknesses.

As an example, pica is a real phenomenon and should be considered, but the assumption that toddlers who eat 200 mg/day of soil will continue this behavior for a 70-year lifetime is not realistic. Thus, the impact of assumptions on the exposure point concentration and, ultimately, the risk values should be examined to identify which of the myriad of input variables have the most significant impact on the resulting risk value. Such a refinement is called a sensitivity analysis.

### 4.1.2.4 Receptor Doses
The final step in the exposure assessment stage is to estimate the doses of the different classes of chemicals to which receptors are potentially exposed at the exposure points. There are three types of dose: the administered dose (the amount ingested, inhaled, or in contact with the skin), the intake dose (the amount absorbed by the body), and the target dose (the amount reaching the target organ).

For purposes of calculating risks, the dose should be in the same form as that of the dose-response relationship reported for the specific chemical and the exposure route under study. This will almost always be either administered dose or absorbed dose. Given the concentration of the contaminant at the exposure point, the calculation of administered dose is straightforward. In contrast, the calculation of absorbed dose based on administered dose requires consideration of some complex factors. The key factors influencing the uptake of contaminants by the body are simplified as follows:

| | |
|---|---|
| Ingestion | Contaminant concentration in the ingested medial |
| | Amount of ingested material |
| | Bioavailability to the gastrointestinal system |
| Inhalation | Concentration in air and dust |
| | Particle size distribution |
| | Bioavailability to the pulmonary system |
| | Rate of respiration |
| Dermal contact | Concentration in soil and dust |
| | Rate of deposition of dust from air |
| | Direct contact with soil |
| | Bioavailability |
| | Amount of skin exposed |

Other factors to be considered in determining the intake of contaminants include considerations of life style, frequency and duration of exposure (e.g., chronic, subchronic, or acute), and the body weight of the receptor. In the majority of hazardous waste sites, long-term (i.e., chronic) exposures are frequently of greatest concern.

The calculation of an administered dose is summarized in the following generic equation:

$$I = \frac{C \times CR \times EF \times ED}{BW \times AT} \tag{4-2}$$

where   $I$ = intake (mg/kg of body weight/day
   $C$ = concentration at exposure point (e.g., mg/L in water or $mg/m^3$ in air)
  $CR$ = contact rate (e.g., L/day or $mg/m^3/day$)
  $EF$ = frequency (day/year)
  $ED$ = exposure duration (yr)
  $BW$ = body weight (kg)
  $AT$ = averaging time (days)

Equation (4-2) is typical modified for specific exposure pathways. For example, the intake dose from the inhalation of fugitive dust may be calculated as

$$I = \frac{C \times CR \times EF \times ED \times RR \times ABS}{BW \times AT} \tag{4-3}$$

where     $RR$ = retention rate (decimal traction)
      $ABS$ = absorption into bloodstream (decimal fraction)

For fugitive dust, the concentration in the air is determined by:

$$C - C_s \times P_c$$

where     $C_s$ = concentration of chemical in fugitive dust (mg/mg)
      $P_c$ = concentration of fugitive dust in air ($mg/m^3$)

Often, appropriate parameters may be found in the literature. Considerable research has been done in recent years to define many basic parameters, such as skin surface areas, soil ingestion rates, and inhalation rates. Other parameters, such as exposure frequency and duration, are often based on site-specific information (if available) or professional judgment. For example, in the evaluation of a trespasser scenario, observations of trespassers during site investigation activities may dictate the values used for exposure frequency (e.g., number of days per year or number of events per year) and exposure duration (e.g., number of years that the activity occurred). Common sense also plays an important role in the selection of exposure parameters. For example, if one were evaluating residential use of two sites, one in North Dakota and one in Florida, it would not be reasonable to assume the exposure frequency (e.g., days per year that an activity occurred) would be the same in the two risk assessments for a child potentially exposed to surface soils as a result of playing in the backyard.

 Averaging time (AT) is another important parameter which must be defined in the intake equation. The averaging time selected will depend on the type of constituent being evaluated. For example, to assess long term or chronic effects associated with exposure to noncarcinogens, the intake is averaged over the exposure duration (expressed in days). Exposure to carcinogens, however, is averaged over a lifetime (assumed to be 70 years or 25,550 days), to be consistent with the approach used to develop slope factors.

Some parameters used for this type of calculation are shown in Table 4.3. It should be noted that these values can vary greatly depending on the assumed exposure conditions (i.e., the selected exposure scenario). As an example, the air breathing rate for adult males is 0.83 m³/h in Table 4.3. However, this rate can vary by an order of magnitude from 0.6 m³/h at rest to 7.1 m³/h for vigorous physical exercise.

Examples 4-5 and 4-6 illustrate the calculation of contaminant intake via the inhalation and dermal route, respectively.

*Example 4-5 Calculation of Contaminant Intake*  Determine the chronic daily inhalation intake, by adults, of a noncarcinogenic chemical as a function of concentration in fugitive dust at the ABC Landfill (Table 4-3).

*Solution*  For an adult exposed to a noncarcinogenic constituent, the intake ($I_N$) may be calculated from Eq. (4-2):

$$I_N = \frac{C \times CR \times EF \times ED \times RR \times ABS}{BW \times AT}$$

From Table 4.3, we take the air breathing rate for adults — 0.83 m³/h, and write.

$$CR = 0.83 \times 24 - 1.92 \text{ m}^3/\text{day}$$
$$EF = 365 \text{ days}$$
$$ED = 30 \text{ years for chronic exposure to noncarcinogens}$$

**TABLE 4.3. Standard Parameters for Calculation of Dosage and Intake Determined for a Landfill**

| Parameter | Adults (males) | Child Age 6–12 | Child Age 2–6 |
|---|---|---|---|
| Average body weight (kg) | 70 | 29 | 16 |
| Skin surface area (cm²) | 18,150 | 10,470 | 6,980 |
| Water ingested (L/day) | 2 | 2 | 1 |
| Air breathed (m³/h) | 0.83 | 0.46 | 0.25 |
| Retention rate (inhaled air) | 100% | 100% | 100% |
| Absorption rate (inhaled air) | 100% | 100% | 100% |
| Soil ingested (mg/day) | 100 | 100 | 200 |
| Bathing duration (minutes) | 30 | 30 | 30 |
| Exposure frequency (days) | 365 | 365 | 365 |
| Exposure duration (years) | 30 | 6 | 4 |

In the absence of better information, a conservative approach would assume the retention rate ($RR$) and the absorption into bloodstream would both equal 1.0.

$$BW = 70 \text{ kg (Table 4.3)}$$

$$AT = 365 \text{ days} \times 30 \text{ years}$$

$$I_N = \frac{C \times 19.92 \times (365 \times 30) \times 1.0 \times 1.0}{70(365 \times 30)}$$

$$I_N = 0.285 \text{ m}^3/\text{kg/day} \times /C$$

$$C = \text{exposure point concentration (mg/m3)}$$

*Example 4-6 Average Daily Intake From Dermal Contact with Soil*    Determine the average daily intake of chlorobenzene over one year of exposure for on-site workers from dermal contact of the soils:

Assume the following additional parameters

$A$ = skin exposed = 20% $-0.2 \times 18.150 \text{ cm}^2$ $- 3630 \text{ cm}^2$
$DA$ = dust adherence = 0.51 mg/cm$^2$
$ABS$ = skin absorption rate = 6%
$SM$ = effect of soil matrix = 15% (i.e., due to soil matrix only 15% of contamination is actually available for contact)
$EF$ = two exposure events per day; 156 exposure days per year
$ED$ = 1 year
$BW$ = 70 kg
$AT$ = 365 days

*Solution*

$$I_N = \frac{\left( C \times (\text{mg/kg}) \times A \text{ cm}^2 \times \dfrac{DA}{\text{exp.event}} (\text{mg/cm}^2) \times ABS \times SM \right.}{BW \times AT}$$
$$\left. \times \dfrac{2 \text{ exp.events}}{\text{day}} \times \dfrac{156 \text{ days}}{\text{year}} \times ED(\text{kg}/10^6\text{mg}) \right)$$

$$I_N = \frac{C \times 3630 \times 0.51 \times 0.06 \times 0.15 \times 2 \times 156 \times 10^{-6}}{70 \times 365}$$

$$I_N = 2.03 \times 10^{-7} \times 7 \text{ mg/kg/day}$$

The given average concentration of chlorobenzene in soil is 1.39 mg/kg. The daily intake of chlorobenzene is:

$$1.39 \text{ mg/kg} \times 2.03 \times 10^{-7}$$

Thus,

$$I_N = 2.82 \times 10^{-7} \times 7 \text{ (mg/kg/day)}$$

Again, the foregoing examples represent only two approaches to calculating contaminant intake in the course of an exposure assessment. The next factor to be analyzed is toxicity.

### 4.1.3  Toxicity Assessment

Broadly speaking toxicity is the degree to which a substance is poisonous. Most chemicals, toxic or otherwise, enter the body through eyes, respiratory tract, digestive tract, and skin. Two levels of toxicity are defined: acute "short-term" exposure that initiates poisoning and chronic "long-period" exposure that causes anemia, leukemia, and death.

In the context of the third stage of the risk assessment process, toxicity is defined by the dose–response relationship for each surrogate chemical. The output takes the form of mathematical constants for insertion into risk calculation equations. In addition to providing a set of mathematical constants for calculating risk, the toxicological assessment should also analyze the uncertainty inherent in these numbers, and describe how this uncertainty may affect the estimates of risk.

In a prudent approach to public health protection, the EPA has built several safety factors into its methods for establishing reference doses and carcinogenic slope factors. Virtually all errors of uncertainty are made in the direction of public health protection to insure that risks are over-estimated rather than underestimated. Four examples of this protective approach follow:

- For non-carcinogens, extrapolation of animal reference doses to humans utilizes two safety factors of ten-, one for animal-to-human extrapolation, a second for variation for toxic sensitivities within the human population.
- For carcinogens, the linearized multistage model assumes the upper-bound 95 percent confidence level of extrapolated data (i.e., it is likely to be conservative 19 out of 20 times).
- For carcinogens, the linearized multistage model extrapolates data from the 10–90% carcinogenesis range observed in experimental animals to the regulatory target of 0.0001% carcinogenesis, a step which could overstate risk by several orders of magnitude.
- Although evidence indicates, that, like non-carcinogens, nongenotoxic carcinogens have thresholds below which they fail to influence cellular differentiation or division, they are treated mathematically like genotoxic carcinogens according to the linearized multistage dose–response model.

### 4.1.4  Risk Characterization

The final stage of a four-stage human health risk assessment is to estimate risks. Cancer is not the only undesired health consequence of pollution: exposure to ionizing radiation and coal dust, for example, can lead to radiation sickness and birth defects, as well as cancer. To illustrate risk characterization, however, we shall focus on the calculations that yield quantitative estimates of both the carcinogenic and non-carcinogenic risks to receptors for all exposure scenarios considered. Estimates are typically calculated for all three exposure routes and for the maximally exposed individual (MEI) as well as the most probable exposed population. Such calculations are straightforward, and yield quantitative estimates of risk. The challenge lies not in making the calculations but in interpreting the results such that they are applied properly in decision making. The overall effort is referred to as risk characterization. Risks from toxic chemicals, depending on the context, may be defined, described, and calculated in different ways.

Risk is normally defined as the probability for an individual to suffer an adverse effect from an event. What is the probability that certain types of cancer will develop in people exposed to aflatoxin in peanut products or benzene from gasoline? What is the likelihood that workers exposed to lead will develop nervous system disorders? In the context of this text, a chemical release is an example of an event. As with any relationship expressing or using probability, there is no defined way of expressing (mathematically or with scientific rigor) a single deterministic value of a phenomenon that is probabilistic. A fairly simple conceptual way of expressing chemical risk is to write it as a function of hazard and exposure.

Hazard is the potential for a substance or situation to cause harm or to create adverse impacts on persons or the environment. The magnitude of the hazard reflects the potential adverse consequences, including mortality, shortened life-span, impairment of bodily function, sensitization to chemicals in the environment, or diminished ability to reproduce. Exposure denotes the magnitude and the length of time the organism is in contact with an environmental contaminant, including chemical, radiation, or biological contaminants.

When risk is in terms of probability, it is expressed as a fraction, without units. It has values from 0.0 (absolute certainty that there is no risk) to 1.0 (absolute certainty that an adverse outcome will occur).

For chemicals the term hazard is typically associated with the toxic properties of a chemical specific to the type of exposure. Similar chemicals would have similar innate hazards. However, one must examine the exposure to that hazard to determine the risk. For example, let us say you have three pumps that are all transporting the same chemical (same hazard), but one pump has a seal leak. Which pump poses the greatest risk to the worker? The pump with the seal leak has the greatest potential for exposure, while the hazards are equal (same chemical), so the seal leak pump poses the greatest risk. To expand, let's say we have three pumps that are transporting different chemicals; which one poses the greatest risk to the worker? In this case the engineer would need to examine the hazard, or innate inherent toxicity, of each of the chemicals, as well as the operation of the pumps to determine which poses the greatest risk.

### 4.1.4.1 *Risk for Average and Maximum Exposures*

In the exercises required for Superfund sites, both average and maximum exposure point concentrations are used to estimate risk. Performing a specific risk calculation using both values permits the estimation of a range of potential risks, which can frequently be useful in providing perspective regarding the potential hazards associated with a particular set of exposure conditions. The significance of either measurement depends, of course, on the amount of data and the associated confidence in that data. However, in general, calculation of potential risk using an average concentration permits a better estimate of risk associated with chronic exposures, since the average value represents a more likely estimate of the exposure point concentration to which a receptor would be exposed over time. Use of a maximum value is best in the estimation of shorter-term, subchronic risks, although its use can provide a useful upper bound estimate of potential risk.

It should be noted that current risk assessment guidance emphasizes the use of a single, upper bound estimate of exposure point concentration in the calculation of potential risks. This value is often taken to be 95% CI, the 95th percentile upper confidence limit of the arithmetic mean. Use of this number generally provides a worst-case estimate of risk, and can result in a significant overestimate of potential risk, especially when this value is used in combination with other worst-case assumptions to define a reasonable maximum exposure.

### 4.1.4.2 *Carcinogenic Risk*

*Carcinogenic Risk* Carcinogenic risk may be defined as the chronic daily intake dose (developed in the exposure assessment) multiplied by the carcinogenic slope (selected by the toxicity assessment). The product is a real term: the probability of excess lifetime cancer from exposure to this chemical. The computation is as follows:

$$\text{Risk} - CDI \times SF \qquad (4\text{-}4)$$

where
$$CDI = \text{chronic daily intake (mg/kg/day)}$$
$$SF = \text{carcinogen slope factor (mg/day/mg)}^{-1}$$

Characterization of carcinogenic risk first involves determining the intake for each chemical appropriate to the exposure route and pathway under study then multiplying each one by the proper exposure point concentration and slope factor. The classical methodology for risk assessment assumes additivity of risks from individual toxicants. For carcinogens this means that the total carcinogenic risk equals the sum by exposure route of carcinogenic risks from all individual substances. These calculations are repeated for each exposure scenario and exposed population. It must be emphasized that slope factors are specific to the exposure route (e.g., oral) and may be used only when the exposure data apply to the same route. With this caveat in mind, we can proceed to a consideration of assessing the risks associated with some familiar environmental pollutants.

## 4.2 ASSESSING THE RISKS OF SOME COMMON POLLUTANTS

Risk assessment is built on the principle that small exposures carry with them some risk of an untoward health effect such as development of a malignant tumor or leukemia at some time in the future. Such risks are generally considered to be stochastic or probabilistic in nature and are expressed in terms of a risk coefficient, an expression of the probability or chance of the specific health effect occurring per unit of exposure. Although response to ionizing radiation is generally considered to be linear with dose, the risk from chemical carcinogens may not be; thus, risk evaluation needs to consider such effects as the existence of a threshold dose below which the effect does not occur, as well as the shape of the dose-response curve.

Also, the latency period, the time between the exposure and the onset of disease, needs to be taken into account. For example, if the latency period is 25 years and exposure occurs at age 50, the individual may die from other causes prior to developing the specific health effect under consideration. Even if the individual lives long enough for the health effect to be manifested, the number of years of life lost, and productive years of life lost, will be less than if the exposure had occurred at an earlier age. Health and safety risks are sometimes expressed in terms of the number of years of productive (or health) life lost per thousand worker-years, as shown in Example 4-7.

*Example 4-7 Cancer Risk Assessment*    Estimate the total number of fatal leukemias in 50 years to a population of 100,000 exposed to an average annual dose of 500 $\mu$Gy if the risk coefficient is 0.03 $Gy^{-1}$.

*Solution*

$$\text{Fatal leukemias} = \frac{0.03}{Gy} \times \frac{500\,\mu Gy}{\text{person year}} \times 50\,\text{years} \times 100,000\,\text{persons}$$

$$= 75$$

### 4.2.1  NO$_x$, Hydrocarbons, and VOCs—Ground-Level Ozone

Ground-level ozone is one of the most pervasive and intractable air pollution problems in the United States. In more than 20 urban areas, its concentration exceeds one or more of the ambient air quality standards. In the language of regulators, these areas are said to be "nonattainment." (We emphasize difference between the "bad" ozone created at or near ground level from the "good" or stratospheric ozone that protects us from UV radiation.)

Ground-level ozone, a component of photochemical smog, is actually a secondary pollutant in that certain precursor contaminants are required to create it. The precursor contaminants are nitrogen oxides (NO$_x$, primarily NO and NO$_2$) and hydrocarbons. The oxides of nitrogen along with sunlight cause ozone formation, but the role of hydrocarbons is to accelerate and enhance the accumulation of ozone.

Oxides of nitrogen (NO$_x$) are formed in high-temperature industrial and transportation combustion processes. In 1997, transportation sources accounted for 49.2 percent and non-transportation fuel combustion contributed 45.4 percent of total NO$_x$ emissions. Health effects associated with short-term exposure to NO$_2$ (<3 h at high concentrations) are increased respiratory illness in children and impaired respiratory function in individuals with pre-existing respiratory problems.

Major sources of hydrocarbon emissions are the chemical and oil refining industries, and motor vehicles. In 1997, industrial processes accounted for 51.2 percent while the transportation sector contributed 39.9 percent of the total of manmade (non-biogenic) hydrocarbon sources. Solvents comprise 66 percent of the industrial emissions and 34 percent of total VOC emissions. It should be noted that there are natural (biogenic) sources of HCs/VOCs, such as isoprene and monoterpenes that can contribute significantly to regional hydrocarbon emissions and low-level ozone levels.

Ground-level ozone concentrations are exacerbated by certain physical and atmospheric factors. High-intensity solar radiation, low prevailing wind speed (dilution), atmospheric inversions, and proximity to mountain ranges or coastlines (stagnant air masses) all contribute to photochemical smog formation.

Human exposure to ozone can result in both acute (short-term) and chronic (long-term) health effects. The high reactivity of ozone makes it a strong lung irritant, even at low concentrations. Formaldehyde, peroxyacetylnitrate (PAN), and other smog-related oxygenated organics are eye irritants. Ground-level ozone also affects crops and vegetation adversely when it enters the stomata of leaves and destroys chlorophyll, thus disrupting photosynthesis. Finally, since ozone is an oxidant, materials with which it comes in contact, such as rubber and latex painted surfaces, undergo deteriorations.

### 4.2.2 Carbon Monoxide

Carbon monoxide (CO) is a colorless, odorless gas formed primarily as a by-product of incomplete combustion. The major health hazard posed by CO is its capacity to bind with hemoglobin in the blood stream and thereby reduce the oxygen-carrying ability of the blood. Transportation sources account for the bulk (76.6 percent) of total national CO emissions. Ambient CO concentrations have decreased significantly in the past two decades, primarily due to improved control technologies for vehicles. Areas with high traffic congestion generally will have high ambient CO concentrations: 80 urban areas in the United States are nonattainment for CO. High localized and indoor CO levels can come from cigarettes (second-hand smoke), wood-burning fireplaces, and kerosene space heaters. Example 4-8 considers exposure to CO in the workplace.

*Example 4-8 Total Weighted Average Concentration*

(a) In a plant, employees were exposed to ambient CO throughout the day at different levels as follows:

| Hour | Conc., ppm |
|---|---|
| 8–9 am | 0 |
| 9–11:30 am | 225 |
| 11:30–12:30 pm | 100 |
| 12:30–2:00 pm | 165 |
| 2:00–4:00 pm | 45 |

What is the total weighted average (TWA) in ppm? If national ambient air quality standard is 50 ppm for 8-h, did employees' exposure levels exceed the standard?

(b) CO concentration in a working environment is 150 ppm. Working area is 1,150,000 cu ft. How much air is needed to ventilate in the room to reduce the CO concentration to 50 ppm?

(c) For safety reasons, ventilation dilution is not recommended. Why not? (Note that the molecular weight of CO is similar to that of air.)

*Solutions*

(a) 8-h weight average

$$TWA = \frac{C_1 t_1 + C_2 t_2 + - - + C_n T_n}{\Sigma t_n}$$

$$= \frac{(0 \times 1) + (225 \times 2.5) + (100 \times 1) + (165 \times 1.5) + (45 \times 2)}{1.2.5 + 1 + 1.5 + 2}$$

$$= 125 \, \text{ppm}$$

Therefore, 125 ppm 8-h TWA concentration exceeded the national ambient air quality standard 50 ppm.

(b) Assure steady state with perfect mixing and exponential decay.

(c) $C = C_0 \times \exp\left(\dfrac{-t}{t_d}\right)$

$50 = 150 \exp\left(\dfrac{-1}{t_d}\right)$

$\therefore t_a = 0.91 \, \text{h} = 54.6 \, \text{min}$

$\therefore$ volume of air need to ventilate $= \dfrac{\text{area}}{\text{time}} = \dfrac{1,500,000}{54.6} = 27,472 \, \text{cfm}$

(d) Incomplete combustion in the heater unit may increase the concentration of CO and add other noxious contaminants to the air in the warehouse. It is true, as well, that fixed dilution ventilation rate may not be adequate for CO removal during peak generation period. In an isolated warehouse environment, moreover, dilution ventilation may not adequately remove pockets of air that are high in CO concentration. In any event, CO has an acute toxicity, and dilute ventilation is not recommended for acutely toxic gases.

## 4.2.3 Lead and Mercury

Lead in the atmosphere is primarily found in fine particulates, up to 10 $\mu$m in diameter, which can remain suspended in the atmosphere for significant periods of time. Tetraethyl lead was used as an octane booster and antiknock compound for many years before its full toxicological effects were understood. The Clean Air Act of 1970 banned all lead additives and the dramatic decline in lead concentrations and emissions has been one of the most important yet unheralded environmental improvements of the past 25 years. In 1997, industrial processes accounted for

74.2 percent of remaining lead emissions, with 13.3 percent resulting from transportation, and 12.6 percent from non-transportation fuel combustion (USEPA, 1998).

Lead also enters waterways in urban runoff and industrial effluents, and adheres to sediment particles in the receiving water body. Uptake by aquatic species can result in malformations, death, and aquatic ecosystem instability. There is a further concern that increased levels of lead can occur locally due to acid precipitation that increases lead's solubility in water and thus its bioavailability. Lead persists in the environment and is accumulated by aquatic organisms.

Lead enters the body by inhalation and ingestion of food (contaminated fish), water, soil, and airborne dust. It subsequently deposits in target organs and tissue, especially the brain. The primary human health effect of lead in the environment is its effect on brain development, especially in children. There is a direct correlation between elevated levels of lead in the blood and decreased IQ, especially in the urban areas of developing countries that have yet to ban lead as a gasoline additive.

According to EPA, coal-fired power generation companies currently emit 50 tons of mercury every year in the United States. There was no quibbling that these levels were high and a potential health concern to humans and wildlife. Eating mercury-tainted fish can trigger a variety of problems, ranging from hair loss and chronic fatigue in adults to nervous system impairment of fetuses and children.

Mercury taints the atmosphere worldwide, but there are large variations in how much of it drops onto land or water at any location. Recent experiments have begun identifying oxidizing gases, such as ozone and molecules containing the halogens bromine and chlorine, as triggers for that mercury fallout. Which oxidants dominate that process appears to depend on the environment, the season, the altitude of the airborne mercury, and even the amount of daylight.

Mercury enters the air easily. It is released when coal is burned, gold is mined, some chlorine is manufactured, and even when a fluorescent lightbulb breaks. Some 99 percent of the airborne metal is elemental. Fairly insoluble and unreactive in this form, it can circumnavigate the globe for up to two years. What's contaminating the Everglades, therefore may have originated in Miami, India, or Siberia.

However, atmospheric chemists have discovered that when elemental mercury encounters certain oxidants, it changes into so-called reactive gaseous mercury. Unlike the element, this form is both highly reactive and water soluble, so it remains airborne only hours to days and falls — in rain or snow or attached to dust — near where it's formed. In a lake or ocean, bacteria transform it into methylmercury, the harmful form of the metal that fish and, in turn, people and other predators accumulate in their tissues.

### 4.2.4 Particulate Matter

Particulate matter (PM) is the general term for microscopic solid or liquid phase (aerosol) particles suspended in air. PM exists in a variety of sizes, with diameters ranging from a few angstrom units to several hundred micrometers. Particles are either emitted directly from primary sources or are formed in the atmosphere by gas-phase reactions (secondary aerosols).

Since particle size determines how deep into the lung a particle is inhaled, there are two NAAQS for PM, $PM_{2.5}$, and $PM_{10}$, where the subscripts indicate particle diameter in micrometers. Particles smaller than 2.5 $\mu$m, called "fine," are composed largely of inorganic salts (primarily ammonium sulfate and nitrate), organic species, and trace metals. Fine PM can deposit deep in the lung where removal is difficult. Particles larger than 2.5 $\mu$m are called "coarse" particles, and are composed largely of suspended dust. Coarse PM tends to deposit in the upper respiratory tract, where removal is more easily accomplished. In 1997, industrial processes accounted for 42.0 percent of the emission rate for traditionally inventoried $PM_{10}$. Non-transportation fuel combustion and transportation sources accounted for 34.9 percent and 23.0 percent, respectively. As with the other criteria pollutants, $PM_{10}$ concentrations and emission rates have decreased modestly due to pollution control efforts.

Coarse particle inhalation frequently causes or exacerbates upper respiratory difficulties, including asthma. Fine particle inhalation can decrease lung functions and cause chronic bronchitis. Inhalation of specific toxic substances such as asbestos, coal mine dust, or textile fibers are now known to cause specific associated cancers (asbestosis, black lung cancer, and brown lung cancer, respectively).

An environmental effect of PM is limited visibility in many parts of the United States including some national parks. In addition, nitrogen and sulfur containing particles deposited on land increase soil acidity and alter nutrient balances. When deposited in water bodies, the acidic particles alter the pH of the water and lead to death of aquatic organisms. PM deposition also causes soiling and corrosion of cultural monuments and buildings, especially those that are made of limestone. More than 100 areas in the United States are nonattainment for $PM_{10}$.

### 4.2.5   $SO_2$, $NO_x$, and Acid Deposition

Sulfur dioxide ($SO_2$) is the most commonly encountered of the sulfur oxide ($SO_x$) gases, and is formed upon combustion of sulfur-containing solid and liquid fuels (primarily coal and oil). $SO_x$ are generated by electric utilities, metal smelting, and other industrial processes. Nitrogen oxides ($NO_x$) are also produced in combustion reactions; however, the origin of most $NO_x$ is the oxidation of nitrogen in the combustion air. After being emitted, $SO_x$ and $NO_x$ can be transported over long distances and are transformed in the atmosphere by gas phase and aqueous phase reactions to acid components ($H_2SO_4$ and $HNO_3$). The gas phase reactions produce microscopic aerosols of acid-containing components, while aqueous phase reactions occur inside existing particles. The acid is deposited to the earth's surface as either *dry deposition* of aerosols during periods of no precipitation or *wet deposition* of acid-containing rain or other precipitation. There are also natural emission sources for both sulfur and nitrogen-containing compounds that contribute to acid deposition. Water in equilibrium with $CO_2$ in the atmosphere at a concentration of 330 ppm has a pH of 5.6. When natural sources of sulfur and nitrogen acid rain precursors are considered, the "natural" background pH of rain is expected to be about 5.0. As a result of these considerations, "acid rain" is defined as having a pH less than 5.0. Figure 4.2 shows the major environmental cause and effect steps for acidification of surface water by acid rain.

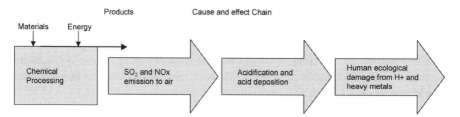

**Figure 4.2.** Environmental cause and effect for acid rain (Allen and Shonnard, 2002).

Major sources of $SO_2$ emissions are non-transportation fuel combustion (84.7 percent), industrial processes (8.4 percent), transportation (6.8 percent), and miscellaneous (0.1 percent) (USEPA 1998). Fifteen U.S. urban areas are nonattainment for $SO_2$. Emissions are expected to continue to decrease as a result of implementing the Acid Rain Program established in 1997 by EPA under Title IV of the Clean Air Act. The goal of this program is to decrease acid deposition significantly by controlling $SO_2$ and other emissions from utilities, smelters, and sulfuric acid manufacturing plants, and by reducing the average sulfur content of fuels for industrial, commercial, and residential boilers.

There are a number of health and environmental effects of $SO_2$, $NO_x$, and acid deposition. $SO_2$ is absorbed readily into the moist tissue lining the upper respiratory system, leading to irritation and swelling of this tissue and airway constriction. Long-term exposure to high concentrations can lead to lung disease and aggravate cardiovascular disease. Acid deposition causes acidification of surface water, especially in regions of high $SO_2$ concentrations and low buffering and ion exchange capacity of soil and surface water. Acidification of water can harm fish populations, by exposure to heavy metals, such as aluminum which is leached from soil. Excessive exposure of plants to $SO_2$ decreases plant growth and yield and has been shown to decrease the number and variety of plant species in a region (USEPA, 1998).

## 4.2.6  Air Toxics

Hazardous air pollutants (HAPs), or air toxics, are airborne pollutants that are known to have adverse human health effects, such as cancer. There are over 180 chemicals identified on the Clean Air Act list of HAPs (USEPA, 1998). Examples of air toxics include the heavy metals mercury and chromium, and organic chemicals such as benzene, hexane, perchloroethylene (perc), 1,3-butadiene, dioxins, and polycyclic aromatic hydrocarbons (PAHs).

The Clean Air Act defined a major source of HAPs as a stationary source that has the potential to emit 10 tons per year of any one HAP on the list or 25 tons per year of any combination of HAPs. Examples of major sources include chemical complexes and oil refineries. The Clean Air Act prescribes a very high level of pollution control technology for HAPs called MACT (maximum achievable control technology). Small area sources, such as dry cleaners, emit lower HAP tonnages but taken together are a significant source of HAPs. Emission reductions can be achieved by changes in work practices such as material substitution and other pollution prevention strategies.

HAPs affect human health via the typical inhalation or ingestion routes. HAPs can accumulate in the tissue of fish, and the concentration of the contaminant increases up the food chain to humans. Many of these persistent and bioaccumulative chemicals are known or suspected carcinogens.

## 4.3 ECOLOGICAL RISK ASSESSMENT

The four major components of the ecological risk assessment (ERA) paradigm are problem formulation, exposure assessment, effects assessment, and risk characterization (Anderson and Albert, 1998; Suter et al., 2000; USEPA, 1992, 1997, 1998). An ERA begins with problem formulation. Activities occurring during this phase include definition of the goals and spatial and temporal scale of the ERA, development of a site conceptual model, selection of endpoint and nonhuman receptor species selection, and preliminary identification of contaminants of potential concern. Exposure assessment and effects assessment follow and can be performed simultaneously. Exposure assessment evaluates the fate, transport, and transformation of chemicals in the environment, and quantitative uptake and intake of these substances in receptor organisms. Effects assessment establishes the relationship between exposure levels and toxic effects in receptors. Risk characterization is the last step in the ERA and is where exposure and toxic effect information are combined to describe the likelihood of adverse effects in receptors. Many of the evaluation criteria needed to evaluate an ERA are identical to those already presented for human health risk assessment (HHRA). This section focuses primarily on the unique aspects of ERAs and does not repeat material covered under HHRA that applies to both subjects.

### 4.3.1 Technical Aspects of Ecological Problem Formulation

Determining how many data are needed to address the ERA goals is part of the process of meeting a project's Data Quality Objective (DQO). All risk assessment stakeholders (e.g., the U.S. EPA, the State, the Fish and Wildlife Service, etc.) should be involved in this process. The DQO is ascertained at the beginning of an assessment, to define both the amount and quality of data required to complete the assessment. Scheduling time to complete DQOs at the beginning of the ERA may save the project time and money in the end. Once the goals and DQOs have been determined, the remainder of the problem formulation may be conducted. The ultimate goal of problem formulation is the site conceptual model.

A wide range of ecosystem characteristics may be considered during problem formulation. These include abiotic factors (e.g., climate, geology, soil/sediment properties) and ecosystem structure (e.g., abundance of species at different trophic levels, habitat size, and fragmentation). The environmental description may be documented using recent photographs and maps. Plant and animal species lists should be compiled.

The scale of the assessment is especially important if a large, complex site has been subdivided into several smaller sites. It also is not uncommon for Superfund sites to be located adjacent to each other. Hence the areal extent of the assessment must be defined. For example, if an off-site area is included in the assessment, the distance off-site must be specified. The development of the site conceptual model and the selection of assessment endpoints will be directly related to the spatial scale. For example, due to their large home ranges, effects of soil contamination on deer would not be assessed if the site encompasses only two acres; assessment of endpoint species with smaller home ranges, such as small mammals, would be more appropriate.

It is necessary to decide if the assessment must consider temporal changes. All historical information should be evaluated. Then, it may be determined how much new information is needed to adequately evaluate impacts and risks. Certain parts of the year may need to be included in the sampling season for the assessment. For example, environmental exposures may change over the course of a year, or over several years, due to various seasonal influences in either chemical form or organism behavior (e.g., salmon returning to a contaminated river to spawn; migrating birds making temporary use of a site).

**4.3.1.1** *Site Conceptual Models* The site conceptual model (SCM) describes a series of working hypotheses regarding how contaminants or other stressors may affect ecological receptors (ASTM, 2003). An SCM clearly illustrates the contaminated media, exposure routes, and receptors for the risk assessment. In addition to a written description, a diagrammatic SCM is easy to understand and is useful for ensuring that no relevant component is omitted from the assessment. The model ensures that all exposure scenarios are considered and allows for full documentation of the rationale behind selection and omission of pathways and receptors. The idea behind the SCM is that although many hypotheses may be developed during problem formulation, only those that are expected to contribute significantly to risks at the site are carried through the remainder of the ERA process.

During SCM development, all contaminant sources are identified (e.g., landfills, burial grounds, lagoons, air stacks, effluent pipes), and all contaminated media are represented (e.g., soil, water, sediment, air, biota). Groundwater usually is not considered an exposure medium until it becomes surface water, but contaminants in groundwater can migrate from soil to surface water and biota. An exception is shallow groundwater or seeps where plants may be exposed via their roots. All exposure pathways are represented, unless adequate rationale can be provided to exclude a pathway from the assessment. For example, an effluent pipe releasing metals into a stream would not need an air exposure pathway, and the only soils that would need to be considered are those of the floodplain. Thus, terrestrial receptors would be exposed by direct contact with or drinking from the stream, living in floodplain soils, or obtaining contaminated food from the stream and floodplain. An appropriate food web must be presented. A food web going from contaminated soil to earthworm to shrew may be appropriate for a 1 acre site, but a significantly larger site may require the food web to continue up to larger predators which have larger home ranges (see Figure 4.3).

For nonchemical stressors such as water level or temperature changes, or habitat disturbances, the SCM describes which ecological receptors are exposed to the physical disturbance, and the temporal and spatial scales of the alterations.

ERAs may have more than one SCM. In predictive ERAs, impacts on different components of the ecosystem from various activities may require several SCMS. In retrospective ERAs, a hypothetical future scenario often requires assessment. For example, an industrial area that now provides little habitat for wildlife (hence little exposure and little risk) may in future become covered in vegetation. It then would be more attractive as wildlife habitat, and hence the risk of exposure

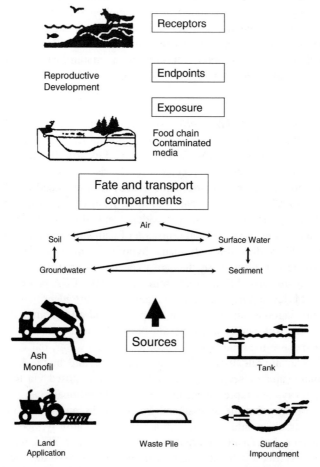

**Figure 4.3.** Environmental risk assessment multi pathways analysis. (Adapted from USEPA, 1995, Development of Human Health Based and Ecologically Based Emit Criteria from Hazardous Waste Identification Project.)

to contaminants would become greater. Similarly, a plume of contaminated ground-water that does not now pollute a pond may affect this water source in several years. The future risk must be evaluated.

### 4.3.1.2 *Identifying Endpoints*

Before the SCM can be completed, the assessment endpoints of the ERA must be defined and rationale given for their selection. An assessment endpoint is the actual environmental value that is to be protected (Suter et al., 1993, 2000). An example of an assessment endpoint would he "no less than a 20 percent decrease in the survival, growth, or reproduction in the large-mouth bass population in the creek." Desirable characteristics for assessment end-point species include the following (Suter et al., 1993, 2000):

- An assessment endpoint must be relevant to decision-making.
- The structure and function of components of the ecosystem must be understood in order to determine the ecological relevance or importance of the endpoint. Species that control the abundance and distribution of other species, and those that are involved in nutrient cycling and energy flow are generally considered to be ecologically relevant.
- Selection of endpoints may be influenced by societal involvement and concern.
- Only species that are present, or likely to be present at the site, should be used to cvaluate risks, regardless of the value or importance of the species.
- Since only some species at a site can be evaluated, endpoint species must be selected which are sensitive to the contaminants at the site, and are likely to receive high exposures. In this way, other species that may be less sensitive or receive lower exposures will also be protected. Other information necessary for each receptor species includes: diet composition; habitat preference/needs; home range size; intake rates of food, water, sediment, air, and soil; and body weight.
- Finally, an assessment endpoint must be able to be measured or modeled. If there is no method available to measure or model effects on an endpoint, evaluation of risk cannot be completed.

### 4.3.1.3 *Selecting Measurement Endpoints*

Because there are so many species and other ecosystem characteristics from which to choose assessment end-points, stakeholders must agree on the appropriate assessment endpoints early in the ERA process. The remainder of the assessment cannot he completed until these have been chosen. After assessment endpoints have been selected, ecological risk assessors can select appropriate measurement endpoints for each assessment endpoint. Measures of exposure and effect are measurable environmental characteristics related to the valued characteristic chosen as an assessment endpoint (Suter et al., 1993, 2000). There are three categories of measures (USEPA, 1989). "Measures of effect" are measurable changes in an attribute of an assessment end-point in response to a stressor to which it has been exposed (formerly referred to as "measurement endpoints"). "Measures of exposure" are measures of stressor existence and movement in the environment and this contact or co-occurrence with the assessment endpoint. "Measures of ecosystem and receptor characteristics" are measures of ecosystem characteristics that influence the behavior and location of assessment endpoints, the distribution of a stressor, and life history characteristics of the assessment endpoint that may affect exposure or response to the stressor. These three difference measures are especially important when completing a complex ERA.

ERAs that involve Superfund remedial actions must meet federal and state standards or criteria called ARARs, for "applicable or relevant and appropriate requirements" (USEPA, 1989). ARARs which may need to be considered at a site include those specified in the Clean Water Act, Clean Air Act, Endangered Species Act, Fish and Wildlife Conservation Act, Wild and Scenic Rivers Act, Migratory Bird Treaty Act, and many others. If numerical ARARs exist, modeled or measured chemical concentrations in site media cannot exceed these values.

During problem formulation, historical data and/or site investigation data are used to prepare a preliminary list of contaminants of potential ecological concern (COPEC). In order to obtain a meaningful ERA, selection of COPECs must ensure that all contaminants that may contribute significantly to risk are included. Reasoning must be provided for exclusion of chemicals from the COPEC list. In this initial screening of contaminants, valid reasons may include but not be limited to contaminant concentrations at or below background levels; concentrations below ARARS, other regulatory concentrations, or toxicity benchmarks; or chemicals infrequently detected. Exclusion of COPECs because the HHRA excluded them is not a valid reason. This is because protection of human health does not guarantee protection of nonhuman biota. Table 4.4 presents several aspects of this apparent paradox.

### 4.3.2   Ecological Exposure Assessment

ERA has several considerations that HHRA lacks. One of the most important factors affecting the exposure assessment is the spatial and temporal scale of the assessment. Spatially, exposure estimates must take into account the home range of, and the availability of, suitable habitat for the receptor species, relative to the areal extent of contamination. Temporal considerations include whether the receptor species is a resident or migrant species, and whether contaminant concentrations vary over the course of the year due to seasonal changes.

Another concept that is not often addressed in HHRA is the different level of protection afforded to different species. HHRAs are designed to protect individuals. In ERA, only threatened and endangered species, or other species of special legal (e.g., migratory birds) or public concern are evaluated for impacts at the individual level. For other species, protection is primarily afforded at the population level. For example, it is important to protect a population of deer at a site; individual deer will not be protected. Practically, this means that impacts on measures relevant to the population as a whole, such as survival and reproduction, are evaluated. Individual quality of life is not considered.

**TABLE 4.4.  Differences Between Human Health and Ecological Risk Assessments**

| Component | Human Health Risk Assessment (HHRA) | Ecological Risk Assessment (ERA) |
|---|---|---|
| Institutional controls | Institutional controls may be considered when selecting exposure parameters. | Nonhuman organisms are not excluded from waste sites by controls, such as fences or signs. |
| Standard exposure factors | The USEPA provides standard exposure parameters and toxicological benchmarks for humans. | Risk assessors must generate their own exposure parameters and toxicity data. |
| Receptor species | Humans only. | Nonhuman organisms (flora and fauna) and ecosystem properties (e.g., nutrient flow). |
| Exposure routes | Ingestion of food and water, incidental ingestion of soil, inhalation of contaminants from air, dermal contact, ingestion of fish fillets. | As well as the exposure routes common to HHRA, other routes exist, such as fish respiring water, benthic organisms consuming sediments, small mammals burrowing in soil leading to enhanced exposure, fish-eating wildlife consume the entire fish and chemicals accumulate to a different degree in different organs. |
| Chemical form | Total metals in water are assumed to be available to humans. | Dissolved metals are available to aquatic biota for gill uptake. |
| Spatial scale | Often assumes a residential scenario at the site, regardless of appropriateness. | Scale is important, since a small site (e.g., a few acres) cannot support a population of larger organisms (e.g., deer, hawks), but could support small animal populations (e.g., shrews). |
| Temporal scale | Often only considered when seasonality may change chemical concentrations. | Seasonality is more important in ERA, often because of habitat changes or changes in organism behavior. |

As in HHRA, for an exposure pathway to be complete, there must be a contaminated medium, a transport medium, receptor species, and an exposure route that enables the contaminant to enter the organism. However ERA has unique exposure routes, such as respiration of water by fish. In the exposure assessment, contaminant concentrations at an exposure point are determined, or intake rates calculated. In the risk characterization, these concentrations are related to toxicological benchmarks; which are contaminant concentrations that are assumed not to be hazardous to the receptor species.

The exposure scenario in an ERA may not be the same scenario as the HHRA. ERA does not have a default "residential scenario," or "industrial scenario." However, hazardous waste sites often are industrial in nature. Scenarios are developed which are appropriate to the current land use. Like the human health assessment, the ERA may make assumptions regarding future land use. This future scenario may assume the site is abandoned and undergoes natural succession. Therefore, it is unreasonable to assume that the same wildlife species will be present in the current and future scenarios, especially if the habitat changes. All assumptions regarding exposure scenarios must be documented early in the ERA process.

During characterization of the exposure environment, the relationship between the receptor species and the environment is detailed. Ecosystem characteristics can modify the nature and extent of contaminants. Chemicals may be transformed by microbial communities or through physical processes such as hydrolysis and photolysis. The environment also may affect bioavailability of contaminants. Physical stressors such as stream siltation and water temperature fluctuations may have considerable impact on ecological risks, and, therefore, must be described.

As part of the characterization of the exposure environment, it is also important to consider both the habitat requirements of receptor species and the amount of suitable habitat available at the site. Availability of habitat will determine the amount of use that a site receives. Because exposure cannot occur if receptor species are not present and receptor species will not be present if suitable habitat is not available, it is important to identify habitat requirements and availability early in the exposure assessment. Selecting exposure routes depends on the endpoints to be evaluated. Several endpoint categories and exposure routes are discussed in Sections 4.3.2.1 to 4.3.2.5.

### 4.3.2.1 Fish Community

Fish are exposed to contaminants in surface water through respiration and dermal absorption. They also may be exposed through the consumption of contaminated sediment or food. There are two important considerations for the fish community. The first is that for inorganic contaminants, it is the dissolved fraction of the contaminant in the surface water that the fish are exposed to by inhalation (i.e., gill uptake). Practically speaking, this involves filtering the water sample through a nanometer filter prior to analysis. HHRA calculates exposures using the total inorganic concentration in water. However, the particulate-bound fraction is not available to fish at the gill. Secondly, dermal absorption as a separate exposure route is not evaluated, because existing toxicity data for fish were generated either by feeding contaminated food to fish or exposing fish to contaminants in the water, without attempting separate evaluations of the various uptake routes.

### 4.3.2.2 Benthic Macroinvertebrate Community

Benthic macroinvertebrates live in or on contaminated sediments. They may be exposed through ingestion of the sediment or contaminated food. Also, benthic organisms may respire overlaying water or the sediment pore-water. Special considerations for this endpoint include the need for bulk sediment contaminant concentrations and pore water analyses, in order to compare these concentrations to benchmark concentrations. For nonionic/nonpolar organic contaminants, bulk sediment concentrations are used. The organic carbon content of the sediment is also required. For ionic/polar organic contaminants, the sediment pore water must be analyzed. For inorganic contaminants, either analysis is adequate.

### 4.3.2.3 Soil Invertebrate Species

Soil invertebrates, such as earthworms, are in direct contact with contaminated soil. Also, the earthworm ingests large amounts of soil during feeding. Contaminants are in contact with and may be absorbed by the gut of the worm.

### 4.3.2.4 Terrestrial Plants

Plants may take up contaminants from the soil at the root. Also, contaminants in shallow groundwater may be taken up by the plant roots. Airborne contaminants (e.g., ozone, acid gases) also may enter the plant through the leaf stomata and cause damage.

### 4.3.2.5 Terrestrial Wildlife

As terrestrial wildlife move through the environment, they may be exposed to contamination via three pathways: oral, dermal, or inhalation. Oral exposure occurs through the ingestion of contaminated food, water, or soil. Dermal exposure occurs when contaminants are absorbed directly through the skin. Inhalation exposure occurs when volatile compounds or fine particulates are respired into the lungs. While methods are available to assess dermal and inhalation exposure to humans, data necessary to estimate dermal and inhalation exposure are generally not available for wildlife However, these routes are generally considered to be negligible relative to other routes.

Ideally, seasonal data would provide the most complete evaluation of contaminants present in the environment. Wherever possible, site-specific data should be used, rather than modeled data. Where EPCs must be modeled, the same methods and considerations are applicable to ERA as in HHRA. EPCs are developed differently according to endpoint. For the fish community, the concentration of contaminant in water or sediment is used as the EPC. No exposure models are required. The upper 95 percent confidence limit on the mean water concentration may be used instead of the mean or maximum detected concentration. This is because chronic exposures of the maximally exposed aquatic organisms would be to spatially and temporally varying contaminant concentrations. For the benthic, soil invertebrate and plant communities, the concentration in the sediment or soil at each sample location is used as the EPC. Again, no exposure models are required. However, in each of these cases, the maximum concentration in the sediment or soil should be used as the EPC because these organisms are not particularly mobile. The entire community could be exposed to the maximum concentration present in the medium.

For wildlife species, contaminant concentrations in food, water and soil are used in exposure models to estimate dose. Because wildlife are mobile, use various portions of a site, and are exposed through multiple media, the upper 95 percent confidence limit on the mean best represents the spatial and temporal integration of contaminant exposure wildlife will experience. Exposure estimates for wildlife are usually expressed in terms of a body weight-normalized daily dose or mg contaminant per kg body weight per day (mg/kg/day). Exposure estimates expressed in this manner may then be compared to toxicological benchmarks for wildlife, or to doses reported in the toxicological literature. A very few wild animals consume diets that consist exclusively of one food type.

To account for varying contaminant concentrations in different food types, exposure estimates should be weighted by the relative proportion of daily food consumption attributable to each food type, and the contaminant concentration in each food type. Each parameter in a wildlife contaminant intake equation must be obtained from the literature because few site-specific values are likely to be available. The EPA *Wildlife Exposure Factors Handbook* (USEPA, 1993) contains a compilation of values for parameters such as diet composition, food intake rate, body weight, and home range for 15 birds, 11 mammals, and eight reptiles and amphibians. The primary and secondary literature must be consulted for any parameter values not contained in this document or if the values provided are not appropriate for the site or become outdated.

One advantage that ERA has over HHRA is the ability to sample the receptor species itself. Rather than introducing modeling uncertainties, fish, benthic macro-invertebrates, soil invertebrates, plants, and some wildlife species (e.g., small mammals) can be sampled directly to give an indication of the bioavailability of environmental contaminants. Of course, it is not acceptable to destructively sample many species, such as rare, threatened, and endangered species, or those with high societal value or low abundance. However, when possible the additional sampling and analytical costs will be worth the added certainty in the exposure assessment and risk characterization.

Ideally, contaminant analysis of whole fish are used when conducting an exposure assessment on piscivorous species. However, fish body burdens may be estimated using bioaccumulation factors. Professional judgment is required when selecting a parameter value for the exposure model. Full rationale for the selection of any parameter value must be provided in the exposure assessment. Exposure assessments will use a variety of data with varying degrees of uncertainty associated with them. Each assumption made will be a result of professional judgment, but will still have some uncertainty. It is important that the exposure assessment document and characterize each source of uncertainty, including those associated with analytical data, exposure model variables, contaminant distribution and bioavailability, receptor species presence and sensitivity, and other incomplete exposure information.

### 4.3.3   Ecological Effects Assessment

An ecological effects assessment includes a description of ecotoxicological benchmarks used in the assessment, toxicity profiles for contaminants of concern, and results of the field sampling efforts. The field data may include field survey information and toxicity test results.

Ecotoxicological benchmarks represent concentrations of chemicals in environmental media (i.e., water, soil, sediment, biota) that are presumed not to be hazardous to biota. There may be several benchmarks for each medium and each endpoint species, which allows for estimation of the magnitude of effects that may be expected based on the contaminant concentrations at the site. For example, there may be a benchmark for a "no-effect level," a "low-effect level," "chronic-effect level," a "population-effect level" and an "acute-effect level." Using all of these benchmarks will provide more information for decision makers than any one of the above.

There are few federal or state benchmarks currently available in the U.S. or elsewhere. Criteria that are used as benchmarks are the National Ambient Water Quality Criteria for the Protection of Aquatic Life (NAWQC) (USEPA, 1986). These are ARAR, and are used as benchmarks for the fish community and other water-column species (e.g., invertebrates such as daphnids). However, not all contaminants have these criteria. Therefore, other benchmarks are needed. Benchmarks for the fish, benthic, soil invertebrate, and plant communities, and wildlife are described briefly in Sections 4.3.3.1 to 4.3.3.4. The primary source of toxicity information used in the development of these benchmarks is the open literature (Risk Assessment Forum, 1996, 1999).

*4.3.3.1 Fish Community* The acute and chronic NAWQC or state water quality criteria are ARARs and must be used as benchmarks. However, these were developed as broadly-applicable values, and thus it may be more appropriate to determine benchmarks for the geographical location and species present at the site. The literature should be reviewed for chronic values in systems similar to that at the site, whether it be a freshwater, estuarine, marine, hard-water, or soft-water system. Laboratory toxicity tests have been conducted on many different aquatic species for many contaminants. In fact, the aquatic system currently has the largest readily-available data base of contaminant concentration/effects data.

*4.3.3.2 Benthic Community* There are several methods that may be used for calculating sediment benchmarks for the benthic community. For nonionic/nonpolar organic contaminants, the equilibrium partitioning approach is often employed. For inorganic contaminants, existing bulk sediment toxicity values from the literature may be used, or pore water concentrations of contaminant may be compared to existing NAWQC. Unfortunately, the database of single-contaminant exposure/effects data for sediments is limited. The majority of the data come from contaminated sites and, therefore, multiple contaminants were present. However, sediment contamination is receiving more attention, and risk assessors and managers must stay current with respect to advances in the areas of sediment toxicology and policy.

*4.3.3.3 Soil Invertebrate and Plant Communities* The plant community plays a dominant role in energy flow and nutrient cycling in ecosystems. Soil invertebrates and plants form the bases of many food webs. There is an extensive database for soil contaminants. However, the majority of endpoints used by researchers have been food crop species. While this information is crucial to human health risk assessors, it is not directly applicable to ecological risk issues. The primary literature will be the major source of toxicity information that must be used in the development of toxicity benchmarks. Soil contamination impacts on plant, invertebrate, and even microbial communities are recent important issues. Again, this is an area within ERA in which it is imperative to remain current.

*4.3.3.4 Wildlife*   Wildlife benchmarks are particularly complicated because wildlife may be exposed to contaminants in their drinking water, the soil around them, and in their diet (both from plant and animal sources). Therefore, wildlife benchmarks must account for these multiple exposure routes. Benchmarks may be derived for each exposure route separately (for cases where exposure is through only one route) and also for total exposure. In the case of exposures from multiple routes, a benchmark such as NOAEL or LOAEL (no/lowest observed adverse effects level) is selected and expressed as a dosage, such as milligrams of contaminant per kilogram of body weight (per day). Benchmarks for wildlife are species specific, in order to account for different species sensitivities, body weights, foraging habits, and diets. In the selection of appropriate benchmark values, the toxicological literature must be consulted, with emphasis on reproduction endpoints. Contaminant toxicity profiles assist risk assessment readers to clearly understand the toxic effects of contaminants in the environment. Toxicity profiles in a risk assessment can provide a concise summary of relevant toxicity information. It is worth repeating that the information must be relevant to the waste site and endpoints of concern. That is, the profile should not simply be a list of median lethal doses ($LD_{50}$s) for rats and mice. Dose/response information should be compiled for the contaminants that are found at the site, and for the receptor species of interest there.

Toxicity profiles and biological effects data for the species of interest also are useful for helping risk assessors and risk managers evaluate the extent and magnitude of risk. Contaminant concentrations at which lethal and sublethal effects (including behavioral modifications) are observed should be presented (i.e., dose/response information). Information such as the mobility of the chemical (e.g., water solubility, soil sorption, octanol/water partition coefficient), persistence in the environment, and interactions with other contaminants will help risk managers make an informed decision and educate the public.

### 4.3.4   Additional Components of Ecological Risk Assessments

*4.3.4.1 Sampling and Surveys*   Although general sampling issues will have necessarily been addressed before the ERA reached the effects assessment stage, it is worthwhile to note a few of them here. This will ensure that the risk assessor has mentioned and considered the potential impacts of these issues. Field surveys and ambient media chemical analyses are also addressed.

*Field Sampling*   Before sample locations are determined, sampling "reaches" must be defined. These are areas that may be impacted by specific contaminant sources. For example, one stream may have several contaminant sources along its length; a reach may be defined as that area between two sources. Sampling in reaches allows for the determination of the relative contribution of various sources to observed toxicity. It is important not to forget to sample an appropriate background (or reference) site. In fact, it is better to have a few reference sites, to account for natural variability in the environment. In the past, there was a distinction between background (meaning pristine) and reference (meaning not impacted by this particular site). Since, however, the distinction is losing currency, it is necessary to know which definition is being used.

One facet of field sampling that is often forgotten when schedules are set is the problem of seasonality in field parameters. For a large portion of the country, winter hinders sampling efforts. For example, it is difficult to sample worms or fish when the ground and creeks are frozen. Also, bats hibernate during the winter, birds migrate, and rare plants are more difficult to identify when they are not in bloom. It is better to delay completion of a risk assessment than to collect data at an inappropriate time.

A waste site investigation will necessarily involve the coordination of a variety of investigators covering the various sampling tasks. The coordination is important in order to obtain results useful for the ERA. For example water concentrations may change dramatically over a short period of time; thus water samples being tested for toxicity should be taken at the same time and from the same location as those taken for chemical analysis. It is less critical to coordinate other activities, such as collection of sediment samples, since sediment concentrations take longer period to show contamination.

In ERA, samples of ambient media do not consist exclusively of groundwater, surface water, sediment, soil, and air; the biota must be included as well, to permit evaluation of contaminant exposure and effects. This important source of information, available to ecological risk assessors, may allow greater certainty in the ERA results.

*Field Surveys*  Field surveys have the advantage of giving a real-world indication of effects. However, the cause of any observed effects is likely to be unknown. For example, a decrease in young fish during a certain year may be due to contaminants that impact fish eggs or larvae, or may be due to natural causes, such as a storm event which caused increased water flow that eroded the spawning beds. Another disadvantage is that small changes are unlikely to be detected. Usually a greater than 20 percent decrease in a field parameter (e.g., population size, number of species) is necessary for deleterious change to be detected, Field surveys may be further complicated because without appropriate and comparable reference sites, interpretation of effects observed at the site is extremely difficult.

In the case of predictive ERAS, field surveys provide information on the environment that may receive contaminants in the future. It is important to have this information in order to document any future adverse impacts. Surveys may include threatened and endangered species surveys, aquatic and terrestrial community surveys, and wetland survey. We mention the latter in passing before turning to a consideration of speciation.

In the United States, a wetland survey must be done for the site to identify and, if necessary, delineate wetlands. Note, it is easier (and less expensive) to identify than to delineate wetlands. It would only be necessary to delineate a wetland if remediation or other activities necessitated the destruction of all or part of the wetlands.

***4.3.4.2 Speciation***   Information on the speciation of the chemical in various media may be useful for contaminants, such as arsenic or chromium, that have species with very different relative toxicities. Before the samples are sent for analysis, it must be ascertained whether the analytical method used will have detection limits below the regulatory concentrations of interest (e.g., ARAR) and the concentration that would produce an unacceptable risk, unless this is not technically or economically feasible. If these detection limits cannot be met, there will be added uncertainty in the risk assessment, because it will not be known whether these contaminants are present or not, and hence whether they constitute a risk.

Chemical concentrations in media at a site, along with the abundant single chemical toxicity data available in the literature, may be used to determine the specific causes of the impacts observed in the field surveys or toxicity tests, and define the sources of the contamination. These data are used in predictive ERAs to model effects of contaminant exposures. However, the measured concentrations may not be indicative of the bioavailable fraction (e.g., chemicals may be bound to soil particles and hence not be available for uptake by organisms).

***4.3.4.3 Sources of Other Effects Information***   Supplementary information that may be useful in the interpretation of ecological data includes an analysis of biomarkers. Biomarkers serve as sensitive indicators in individual organisms of exposure to contaminants or other sublethal stressors. They are typically physiological or biochemical responses, such as enzyme concentrations, genetic abnormalities, histopathological abnormalities or body burdens of contaminants. While biomarkers give an indication of exposure to stressors, they cannot yield information on the impacts of this exposure on individuals. This is because ecological risk assessment is concerned primarily with the viability of organism populations, not physiological effects in a single fish or bird. However, some biomarkers are chemical-specific, and hence may provide valuable information on the potential cause of observed toxic effects. For example, increased blood levels of the enzyme $\delta$-amino-levulinic acid dehydratase (ALAD) indicates exposure to lead.

### *4.3.4.4 Additional Effects that Figure in Many ERAs*

*Global Warming*   The atmosphere allows solar radiation from the sun to pass through without significant absorption of energy. Some of the solar radiation reaching the surface of the earth is absorbed, heating the land and water. Infrared radiation is emitted from the earth's surface, but certain gases in the atmosphere absorb this infrared radiation, and redirect a portion back to the surface, thus warming the planet and making life, as we know it, possible. This process is often referred to as the *greenhouse effect*. The surface temperature of the earth will rise until a radiative equilibrium is achieved between the rate of solar radiation absorption and the rate of infrared radiation emission. Human activities, such as fossil fuel combustion, deforestation, agriculture and large-scale chemical production, have measurably altered the composition of gases in the atmosphere. Some believe that these alterations will lead to a warming of the earth-atmosphere system by enhancement of the greenhouse effect.

The primary greenhouse gases are water vapor, carbon dioxide, methane, nitrous oxide, chlorofluorocarbons, and tropospheric (ground-level) ozone. Water vapor is the most abundant greenhouse gas, but is omitted because it is generally not from anthropogenic sources. Carbon dioxide contributes significantly to global warming due to its high emission rate and concentration. The major factors contributing to global warming potential of a chemical are infrared absorptive capacity and residence time in the atmosphere.

*Visibility and Regional Haze*   The oxidizing agents, ozone ($O_3$), peroxyacetyl nitrate (PAN), peroxybenzoyl nitrate (PBN), and other trace substances react with nitrogen oxides and volatile organic compounds and produce smog. These components of photochemical smog which are most damaging to plants and detrimental to human health are these photochemical oxidants. The aerosols formed during the chemical reactions that create smog cause a marked reduction in visibility and regional haze, and give the atmosphere a brownish cast. Regional haze occurs at distances where the plume has become evenly dispersed into the atmosphere, such that there is no definable plume. The primary cause of regional haze are sulfates and nitrates (primarily as ammonium salts), which are formed from $SO_2$ and $NO_x$ through chemical reactions in the atmosphere.

*Eutrophication: A Widespread Ecological Effect*   *Eutrophication* is a natural process of aging of a body of water. It is a result of a very slow process of natural sedimentation of microscopic organisms which takes geologic times to complete. The completion of the process results in the extinction of the water body.

The process of eutrophication is propelled by increasing concentrations of nutrients necessary for biological activity. First, envision clean and clear water. In this condition, since the nutrients available are minimal, there is no significant biological activity in the water column that can support sedimentation. The water body is healthy and the condition is called *oligotrophic*. Over time, however, nutrients can build up. A water body with nutrient concentration supporting biological activity that is not objectionable but above that of oligotrophic conditions is considered mesotrophic. In the next stage of the life cycle the water becomes eutrophic-this is characterized by murky water with an accelerated rate of sedimentation of microorganisms. The final life stage before extinction is a pond, marsh, or swamp.

Although under natural conditions eutrophication occurs very slowly, over geologic times, human activity encourages the production of nutrients and shortens the cycle. People who have a green lawn are prolific producers of nutrients. Farmlands are an excellent source of nutrients. The Chesapeake Bay in Maryland, for example, is loaded with nutrients coming from farmlands as far away as New York.

To survive, microorganisms must be supplied with nutrients containing their components: carbon, hydrogen, oxygen, nitrogen, phosphorus, and trace elements. Since carbon, hydrogen, and oxygen are already in abundance in the environment, eutrophication control focuses on limiting the input of nitrogen and phosphorus. Nitrogen is utilized by organisms in the form of $NH_4^+$, $NO_2^-$, and $NO_3^-$; phosphorus is utilized in the form of orthophosphorus. Algae, the prime cause of eutrophication, can be controlled by limiting their access to nitrogen or phosphorus. Which should be chosen? It has been found that blue-greens algae, which can survive almost anywhere and can easily outgrow other algae under adverse conditions, can fix nitrogen from the air. For this and other reasons, the default choices, phosphorus control is recommended.

Eutrophication is our final example of the effects included in ecological risk analyses. There are, of course, many others; the literature is ample. Now, however, we are ready to consider another step in an ERA: toxicity testing.

### 4.3.5   Toxicity Testing

Ecological toxicity tests are classified according to duration (short-term, intermediate, and/or long-term), method of adding test solutions (static, recirculation, renewal, or flow-through), and (to satisfy permit requirements for a proposed pollutant discharge elimination system, and to determine mixing zones, etc). Detailed contemporary testing protocols are summarized in the regulatory literature (USEPA, 1985a, 1985b, 1988, 1989).

Toxicity testing has been widely validated in recent years. Even though organisms vary in sensitivity to effluent toxicity, the EPA has documented that effluent toxicity correlates well with toxicity found in the receiving waters when effluent dilution was measured and that predictions of impacts from both effluent and receiving water toxicity tests compare favorably with ecological community responses in the receiving waters. The EPA has conducted nationwide tests with freshwater, estuarine, and marine ecosystems. Methods include both acute as well as chronic exposures.

Current bioassay methods can assess several different phylogenic groups in 4–7 days (USEPA, 1988). The tests are based on species of nearly national distribution for which a large body of life history and toxicity sensitivity data are available. Proper testing protocol involves assessment of a range of sensitivities of test species to a particular effluent. Typically, two or three species are considered to eliminate uncertainty for this factor.

Common marine species include *Champia parvula*, the red alga; *Mysidopsis bahia*, the mysid shrimp; *Menidia beryllina*, the inland silversides; and *Cyrinidon variegatus*, the sheepshead minnow. Common freshwater species include *Pimephales promelas*, fathead minnow, and *Ceriodaphnia dubia*, the daphnid shrimp.

At the termination of the test, survival is determined and growth is measured (as an increase in dry weight) compared with control. The acute endpoint is the death of the fish.

#### 4.3.5.1   *Evaluation of Toxicity Test Results*   A number of terms are utilized in expressing toxicity test results. *Acute toxicity* is toxicity severe enough to produce a response rapidly (typically a response observed in 48 or 96 hours.) "Acute" does not necessarily imply mortality. In studies of marine organisms, the $LC_{50}$ is the concentration of effluent in dilution water that causes mortality to 50 percent of the test population; the $EC_{50}$ is the effluent concentration that causes a measurable negative effect on 50 percent of the test population. The NOAEL (no observed acute effect

level) is defined as the highest tested effluent concentration that causes 10 percent or less mortality.

*Chronic toxicity* is the toxicity impact that lingers or continues for a relatively long period of time, often $1/10$ or more of the target organism's life span. Chronic effects could include mortality, reduced growth, or reduced reproduction. The NOEC (no observable effect concentration) is the highest measured continuous concentration of an effluent or toxicant that causes no observable effect based on the results of chronic testing. The LOEC (lowest observed effect concentration) is defined as the lowest observed concentration having any effect. The LOEC is determined by an analysis of variance techniques.

Toxicity data are analyzed using the procedures developed by Stephan (1977). The $LC_{50}$ values are determined analytically using the Spearman–Kayber, moving average, binomial, and probit methods. Graphical methods, as illustrated in Example 4-9, can also be used to obtain estimated $LC_{50}$ values. Typically, $LC_{50}$ values are computed based on survival at both 48- and 96-hour exposures. Analysis of variance and Duncan's multiple comparison of means typically are utilized to compare chronic test results. Example 4-9 presents a simple comparison of results from acute toxicity testing.

*Example 4-9 Analysis of Toxicity Data* Use the following hypothetical data to determine the 48- and 96-hour $LC_{50}$ values in percent by volume.

| Concentration of Waste percent by volume | No. of Test Animals | No. of Test Animals Dead After[a] | |
|---|---|---|---|
| | | 48 h | 96 h |
| 40 | 20 | 17 (85) | 20 (100) |
| 20 | 20 | 12 (60) | 20 (100) |
| 10 | 20 | 6 (30) | 14 (70) |
| 5 | 20 | 0 (0) | 7 (35) |
| 3 | 20 | 0 (0) | 4 (20 |

[a]Percentage values are given in parentheses.

## Solution

1. Plot the concentration of wastewater in percent by volume (log scale) against test animals surviving in percent (probability scale), as shown in the accompanying diagram.

2. Fit a line to the data points by eye, giving most consideration to the points lying between 16 and 84 percent mortality.

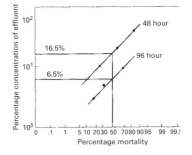

3. Find the wastewater concentration causing 50 percent mortality.

The estimated $LC_{50}$ values are seen to be 16.5 vol% for 48 h and 6.5 vol%.

*Comment* The estimated values of the $LC_{50}$ concentrations obtained graphically are usually quite close to the values obtained with formal probit analysis (Stephan, 1977). It should be noted that confidence limits cannot be obtained in a graphical analysis (APHA, 1989).

### 4.3.5.2 Toxic Units

The toxic units (TU) approach has become widely accepted for utilizing the toxicity test results. Both federal and state standards and/or criteria have been or are being formulated on the toxic unit basis. In the toxic units approach (USEPA, 1985c), a TU concentration is established for the protection of aquatic life.

A toxic unit acute ($TU_a$) is the reciprocal of the effluent dilution that caused the acute effect by the end of the acute exposure period.

$$TU_a = 100/LC_{50} \qquad (4\text{-}5)$$

A toxic unit chronic ($TU_c$) is the reciprocal of the effluent dilution that caused no unacceptable effect on the test organisms by the end of the chronic exposure period:

$$TU_c = 100/NOEC \qquad (4\text{-}6)$$

Where, as before, NOEC is the no observable effect concentration.

Formerly, acute to chronic ratios (ACR) were determined by the equation $ACR = LC_{50}/NOEC$. The chronic data were determined using extrapolation from acute data. Acute to chronic ratios have been found to vary tremendously between species and between different toxicants. Use of the whole effluent approach prevents implementation of overly stringent as well as overly lenient requirements.

### 4.3.5.3 Application of Toxicity Test Results

Water quality criteria ensure protection of designated uses by including magnitude (quantity of toxicant allowable), duration (period of time over which instream concentration is averaged), and frequency (how often criteria can be exceeded without unacceptable receiving water ecological community impacts). Contemporary water quality criteria are designed to protect against short-term (acute) effects through use of the criterion maximum concentration (CMC), and against long-term (chronic) effects through use of the criterion continuous concentration (CCC). These criteria generally apply after mixing. Typical water quality criteria contain a concentration limit, an averaging period, and a return frequency. The CMC is typically the four-day average concentration not to be exceeded more than once every three years, on the average, and the CCC is typically the one-hour average concentration not to be exceeded more than once every three years, on the average.

*Protection Against Acute Toxicity* For protection against acute toxicity, the CMC must not exceed 0.3 acute toxic unit, as measured by the most sensitive result of the tests conducted.

$$CMC = TU_a/CID \leq 0.3TU_a \qquad (4\text{-}7)$$

Where CID is the critical initial dilution. In ocean discharges, the CID is defined as the dilution achieved given "worst case" ambient conditions within the region close to the discharge, where mixing and dilution of the effluent plume are determined by the initial momentum and buoyancy of the discharge. In river discharges, the dilution achieved at the boundary of a mixing zone is usually taken as the CID. Based on the results of numerous 96-hour effluent toxicity tests, it was found that a factor of 0.3 accounted for 91 percent of the observed $LC_{50}$ to $LC_1$ ratios ($LC_1$ is equal to the concentration of effluent in the dilution water that causes mortality to 1 percent of the test population). Consequently, for acute protection, the CMC should not exceed $0.3\ TU_a$ based on the most sensitive species tested. The acute criterion thus equals the CMC that approximates the $LC_1$.

*Protection Against Chronic Toxicity*   To prevent chronic effects from occurring outside the initial mixing zone, or zone of influence, resulting from the discharge, the CCC must not exceed 1.0 chronic toxic unit, based on the results obtained with the most sensitive of at least three species tested.

$$CCC = TU_c/CID \leq 1.0Tu_c \qquad (4\text{-}8)$$

Compliance is based on comparison of the toxicity criteria (both TUa and TUc), expressed in toxicity units, in the effluent with critical initial dilution to determine if the EPA's recommended criteria will be met. The application of these criteria is illustrated in Example 4-10.

*Example 4-10 Application of Toxicity Test Results*   A critical initial dilution of $225:1$ is achieved for a treated effluent discharged to marine receiving waters. Toxicity tests were conducted with the wastewater treatment plant effluent using three marine species. The following acute and chronic toxicity test results indicate that *Champia parvula* exhibited the most sensitive species acute endpoint (2.59 percent effluent) in accordance with the $LC_{50}$, and also the most sensitive species chronic endpoint (1.0%) in accordance with the NOEC. Using the given toxicity data, determine the compliance with the CMC and CCC criteria.

*Results of Acute Toxicity Tests*

| Species | Exposure, h | Control Survival % | Percent Effluent | |
|---|---|---|---|---|
| | | | $LC_{50}$ | NOAEL |
| *Mysidopis bahia* | 96 | 100 | 18.66 | 10.0 |
| *Cyprinodon variegates* | 96 | 100 | >100 | 50.0 |
| *Champia parvula* | 48/168 | 100 | 2.59 | 12.25 |

*Results of Chronic Toxicity Tests*

| Species | Exposure, d | Control Survival % | Percent Effluent | |
| --- | --- | --- | --- | --- |
| | | | NOEC | LOEC |
| *Mysidopis bahia* | 7 | 82 | 6.0 | 10.0 |
| *Cyprinodon variegates* | 7 | 98.8 | 15.0 | >15.0 |
| *Champia parvula* | 7 | 100 | 1.0 | 2.25 |

*Solution*

1. Check compliance with the CMC criterion.
2. (a) Based on data for the most sensitive species tested, we use Eq. (4-5) to find the number of acute toxic units;

$$TUa = 100/LC_{50} = 100/2.59 = 38.6 \, units$$

(b) For acute protection, the criterion maximum concentration must not exceed 0.3 toxic unit, as shown in Eq. (4-7). Following an initial dilution of 225, the CMC is

$$CMC = TU_a/CID \leq 0.3TU_a$$
$$38.6/225 \leq 0.3(38.6)$$
$$0.17 \leq 11.58$$

The CMC (0.17) is considerably less than the value of 0.3 $TU_a$ (11.58) required for compliance with the CMC criterion.

3. Check compliance with the CCC criterion.
   (a) Based on data for the most sensitive species tested, we use Eq. (4-6) to find the number of chronic toxic units:

$$TU_c = 100/NOEC = 100/1.0 = 100 \, units$$

(b) For chronic protection, the criterion continuous concentration must not exceed 1.0 toxic unit as shown in Eq. (4-8). Following an initial dilution of 225, the CCC is

$$CCC = TU_c/CID \leq 1.0TU_c$$
$$100/225 \leq 1.0(100)$$
$$0.44 \leq 100$$

The CCC (0.44) is considerably less than the value of 1.0 $TU_c$ (100) required for compliance with the CCC criterion.

The use of whole effluent toxicity testing affords a number of advantages. In this approach, the bioavailability of the toxics is measured and the effects of any synergistic interactions are also considered. Because the aggregate toxicity of all components of a wastewater effluent is determined, the toxic effect can be limited by limiting only one parameter, the effluent toxicity. Because contemporary receiving water management strategies are based on site-specific water quality criteria, toxicity testing facilitates comparison of effluent toxicity with site-specific water quality criteria designed to protect representative, sensitive species and allow for establishment of discharge limitations that will protect aquatic environments.

### 4.3.6 Ecological Risk Characterization

Historically, the most common approach to risk characterization was the calculation of hazard quotients. This was adopted from the HHRA field, where this approach is still used. Simply, it compares chemical concentrations in ambient media to some toxicity benchmark. If the quotient exceeds 1, there is a potentially unacceptable risk. This approach has found use in predictive assessments and in screening-level (otherwise known as preliminary or tier I) retrospective ERAS, where it is used to refine the list of contaminants of concern and to focus a subsequent, more detailed assessment. However, for a baseline ERA, the quotient method should be used with caution. It is especially important to realize that the magnitude of the excess in the hazard quotient has no quantitative relation to the magnitude of potential toxic effects. Calculating several hazard quotients using different benchmarks (e.g., derived from different toxicity data, such as acute, chronic, or population level effects) has more direct applicability than using a single benchmark.

Because ecological effects can be measured in a retrospective ERA, an epidemiological, weight-of-evidence approach can be used. This approach depends upon weighing multiple lines of evidence, such as those provided by the field surveys, toxicity tests, and ambient media chemical analyses and literature toxicity data. Risk assessors, risk managers, and the public will have more confidence in a risk assessment that uses the weight-of-evidence approach, because it integrates all sources of information, attempts to reconcile conflicting data, and can account for the bioavailable fraction of chemicals in the environment, and the effects of multiple contaminants.

The primary line of evidence in the weight-of-evidence approach is the field survey data. Field surveys monitor actual ecological impacts, and therefore are the most credible line of evidence. However, as discussed in Section 4.3.4.1, field surveys have their limitations. Also, many ERAs will not have the budget necessary to conduct field surveys, and nocturnal, migratory, secretive, or wide-ranging species are not easily surveyed. Also, small impacts are not readily apparent in field surveys. Therefore, other lines of evidence are used as support.

Toxicity tests give an indication of whether ambient media are toxic. When several contaminants exceed benchmarks and there is an impact in the toxicity tests or field surveys, it is important and necessary to evaluate the magnitude of the effect caused by the contaminants which exceeded benchmarks. Using media contaminant analysis and the information provided in the toxicity profile, compiled in accordance with Section 4.3.5, an evaluation is conducted of which contaminants could be responsible for the observed toxicity. Combining all of these lines of evidence will present a picture of actual or potential impacts at the site and contaminants responsible for the impacts. In some cases, benchmarks may indicate unacceptable risk while field observations show no measurable impacts. Therefore, the weight of evidence suggests no unacceptable risks to a community, even though contaminant concentrations exceeded benchmarks. Reconciling multiple lines of evidence is difficult, and requires experience and understanding of the ecosystem being evaluated.

### 4.3.7  Uncertainties

Uncertainties are inherent in all risk assessments. The nature and magnitude of uncertainties depend on the amount and quality of data available, the degree of knowledge concerning site conditions, and the assumptions made to perform the assessment. In ERAs, there is uncertainty associated with the toxicity values selected as benchmarks. Because there is no one single benchmark for each contaminant, medium, and receptor, it is necessary to document any limitations in the use of a particular benchmark value.

Incomplete or absent toxicity information must be acknowledged. Obviously if no toxicity information can be found for certain contaminants, no toxicological benchmarks and profiles will be available for these substances and, therefore, risks cannot be assessed.

Also to be acknowledged and discussed are any uncertainties associated with the bioavailability of contaminants, especially if toxicity and field survey data are lacking for the assessment. Such data do provide an indication of contaminant bioavailability. Similarly, any specific uncertainties associated with field survey techniques must be documented.

Uncertainty in risk characterization often comes from insufficient lines of evidence: the fewer the lines of evidence, the less confidence in the risk characterization.

Uncertainties associated with the extrapolation of toxicity test results to effects on endpoint species must be addressed, as well. Toxicity tests typically use only a few common species that are easy to rear and maintain in the laboratory. Often, these are not the assessment endpoint species in the ERA. Species in the same genus may vary widely in their sensitivity to contaminants. For example, rainbow trout, brown trout, and brook trout have very different sensitivities. Quantitative uncertainty analysis may not be necessary if risk calculations indicate that the risk is clearly below a level of concern. However, if quantitative analysis is warranted, simple models or computer-assisted numerical approaches may be used. One common numerical approach is the Monte Carlo method.

## 4.4  RISK MANAGEMENT

Risk management, the process of decision-making that attempts to minimize risks without undue harm to other societal values, should be performed independently of risk analysis (NRC, 1983). Since, however, the effectiveness of a risk analysis depends on the successful interaction of the risk analyst and the risk manager, we discuss risk management briefly. The risk analysts do not perform the decision-making process for the risk manager, and the risk manager does not dictate what data, models, and assumptions the analysts use, but they must communicate for the risk analysis to be relevant and the decision to be well-founded.

As detailed earlier, this relationship should have begun with the definition of the source, environment, and endpoints of concern. Because these activities of the scientist/risk analyst require that value judgments be made that are not based on science, the only way to avoid improper injection of the analyst's values is by interjecting the values of the risk manager through either policy statements or ad hoc judgments. However, risk managers often are not environmental scientists, so the risk analysts are obliged to assist the risk managers in acquiring enough scientific background to be able to make informed judgments. In addition, the risk analyst must understand the needs and interests of the risk manager well enough to perform the risk characterization.

Figure 4.4 is a simple diagram of the factors involved in setting criteria for the allowable concentration of chemical in air, water, or other ambient medium. The ideal input from the risk analysis is a function relating the probability of unacceptable effects on the endpoint of concern to the criterion level, or more likely, functions for each of a number of endpoints. One former director of the EPA has said that the probability distributions are clearer to decision-makers than numeric values

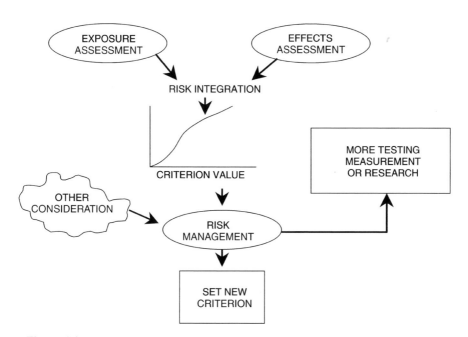

**Figure 4.4.** Diagrams of risk management for setting an environmental quality criteria.

(Ruckelshaus, 1984). From this function, the risk manager may choose a criterion that is unlikely to be overprotective from the upper end of the curve, one that is unlikely to be under protective from the lower end of the curve, or an intermediate criterion. Although not all risk managers are equally comfortable with probabilities, pressure on regulatory agencies to acknowledge and quantify uncertainty is increasing (Finkel, 1990).

In addition to the results of the human health and ecological risk assessments, the decision will be based on a variety of social, political, legal, engineering, and economic considerations. Legal considerations include the phrasing of the law being implemented and any judicial decisions that indicate the appropriate degree of protection. The principal engineering consideration is the availability of practical technologies to meet the criterion. The cost of meeting the criterion and the value of the resource at risk may be the subject of a formal risk-benefit or cost-benefit analysis. Other, more nebulous considerations include the degree of public concern and the political philosophy of the currently governing administration. The manager may set a criterion or may defer a decision and require that more testing, measurement, or research be performed to reduce the contribution of uncertainty to the slope of the function.

### 4.4.1  Valuation of Ecological Resources

In order to perform cost-benefit analyses or to set fines or compensation levels for environmental damage, it is necessary to predict, measure, or estimate the potential or actual loss of value of the system, and then convert that loss into monetary or equivalent units. Many ecologists and environmentalists have objected to the practice of placing monetary values on components of the natural environment, because they feel that economic analyses inevitably undervalue the environment relative to the value of a new chemical or development, or relative to the costs of treatment or cleanup. A counter argument is that decision-makers inevitably consider economics in their judgments, and, if environmental scientists do not develop methods for monetarizing the values of the environment, decision-makers are likely to underestimate the economic value of the environment. In the United States, for example, the Environmental Protection Agency is required to justify new environmental regulations by performing cost-benefit analyses.

A fundamental argument against valuation of environmental components is that they have intrinsic values that are independent of human values (Rolston, 1988). However, even if this premise is accepted, it is impossible for a human to determine what a forest's values are (i.e., would it rather be old growth or young and vigorously growing?), so a poetic description of the intrinsic worth of a forest is not utilitarian, but it is no less anthropocentric than a dollar value. All value systems devised by humans are inevitably based on human values.

In valuing ecological resources, it is helpful to consider that there are various classes of utilitarian and nonutilitarian values that can be assigned. Each class of environmental values must have its own methods of estimation and valuation.

***4.4.1.1 Utilitarian*** Utilitarian values are often classified in terms of commodities, potential commodities, recreation, and services of nature.

*Commodity Values* Trees, rangeland, commercial fish, and other species and communities are sold or leased, so they have market values that are easily determined by conventional market surveys.

*Potential Commodity Values* Plants and animals are sources of chemicals that may be marketed directly or be synthesized and marketed as drugs, pesticides, etc. In addition, all species constitute genetic resources that may be useful for genetic engineering of crops, livestock, industrial microbes, etc. Hence, all species have potential commodity values that are very difficult to estimate. However, this value is lost only in cases of extinction.

*Recreation* Fish and wildlife have recreational value for fishermen, hunters, bird watchers, and others who enjoy seeing animals in their natural habitats. Natural areas have recreational value for campers, hikers, and students of the environment.

*Services of Nature* Ecosystems moderate floods, control the local and global climate, degrade and sequester pollutants, reduce soil and nutrient loss, and perform other services that have palpable economic benefits. These can be estimated as costs of replacing the service (e.g., additional waste treatment) or costs resulting from the lack of the service (e.g., flood damage).

***4.4.1.2 Nonutilitarian*** Nonutilitarian values range from the existential and aesthetic to the scientific and cultured.

*Existence* Species and ecosystems have value to many people simply because they exist. Those people may not expect to see a golden lion marmoset or the Arctic National Wildlife Refuge, but they would experience a loss if the marmoset became extinct or the refuge was converted to an oil field.

*Aesthetic* The aesthetic experience provided by organisms and ecosystems has a value equivalent to that of paintings or dance performances, but is not so easily quantified.

*Scientific* Both species and ecosystems may have scientific value because they may serve as a particularly apt illustration of an unrecognized property of nature or because they are the subjects of long-term studies which may be disrupted. Of these, only the commodity values and recreational values are reasonably well characterized by existing methods. Research is needed to determine whether defensible methods can be developed for estimating losses of the other values, and monetarizing those losses. It can be argued that attempts to monetarize nonmarket values are inherently artificial and the balancing of apples and oranges should be performed by decision-makers rather than economists (NRC, 1975). Other issues relating to category errors, as well as to the appropriateness of discounting any values and to obligations to future generations are discussed by Costanza (1991), Desvousges and Skahen (1987), and many contributors to the journal *Ecological Economics.*

*Cultural and Ethical*   Certain ecosystems and species have cultural significance. This value is illustrated by the greater emotional impact in the United States of the decline of the bald eagle than the concurrent and more serious declines of the peregrine falcon and brown pelican.

Finally, it should be emphasized that cost-benefit analysis should not be considered to be the sole reliable basis for decision-making. It is one input that describes one particular criterion. In particular, it ignores moral and ethical considerations. Many, if not most, members of modern western cultures would argue that we have moral duties toward the environment that supersede considerations of utility. However, these duties must be balanced against duties toward the poor, the hungry, practitioners of traditional cultures, etc. (Randall, 1991). Therefore, environmental values are not absolute, even in moral terms, but must be balanced against competing values. Complexities like these put risk management beyond the scope of formal analysis.

### 4.4.2   Modeling Risk Management

Determining risk severity is one of a risk manager's first tasks. On this basis, he or she must decide whether a given risk is unacceptable, hence must be reduced or prevented with monitoring, or can be accepted, at least for the present. The risks associated with possible alternatives must be considered, as well. For example, the alternative to a pesticide that kills $x$ birds per year may be clean farming which destroys refugia for pests but also eliminates habitat for $y$ birds. Even if $y$ is much larger than $x$, clean farming may be more desirable because it is more readily embraced by environmental and animal rights groups. The integration of such considerations is nearly always an informal process.

Our single brief example is presented as if there is only one exposure model, one effects model, and one set of assumptions about how to parameterize them and combine them into a risk estimate. This approach is generally applicable to human health risk assessments where endpoints, models, and treatment of parameters are standardized. However, it is not appropriate for ecological risk assessments. One might estimate effects on fisheries with an ecosystem model, a population model, and a statistical model of an organism-level toxicity test. Ecosystem-level effects might be represented by results from a microcosm test, a lake ecosystem model, and a stream ecosystem model. Each risk estimate will have its own assumptions and associated uncertainties and those uncertainties may not be expressed equivalently. The separate lines of evidence must be evaluated, organized in some coherent fashion, and explained to the risk manager so that a weight-of-evidence evaluation can be made.

One approach to presenting alternate lines of evidence is to present them graphically using common axes. The axes should be chosen from the four dimensions of risk estimation so as to clarify the differences in the risk estimates that are most important to the decision. That is, does the decision depend primarily on the concentration at which an effect occurs, the time to recovery, the number of organisms dying, or some combination of axes? The output of the alternate models, assumptions, or data should be plotted on those axes (Diwekar, 2003).

### 4.4.3  Other Considerations for Risk Characterization

Those charged with risk management for human health are fully aware that some risks have consequences so dire that they must be avoided no matter what the cost. Risks so obviously unacceptable are called de manifestis. The companion concept, de minimis, which we shall explore shortly, points to the existence of risks with consequences so harmless that they are not worth treating, no matter how little the cost. Whipple (1987) has written about de minimis risks, and Travis et al. (1987) and Kocher and Hoffman (1991) have discussed both concepts in relation to cancer prevention.

Risk levels between the de manifestis and de minimis bounds are balanced against costs, technical feasibility of remediation, and other considerations to determine their acceptability. Because there have been no standard assessment endpoints for ecological effects, there have been no de minimis or de manifestis ecological risk levels. However, the protection provided to endangered species in the United States provides one basis for de manifestis risk levels because it ostensibly precludes economic and sociopolitical considerations. Establishment of de manifestis and de minimis risks would allow assessors to quickly dispense with clearly unacceptable or trivial risks after minimal assessments that do not establish the exact expected risk level, or balance costs and benefits. Consideration of what levels and types of ecological effects are clearly and invariably acceptable, and which are clearly and invariably unacceptable would provide risk assessors and risk managers with a good basis for working out the relation between societal values and ecological principles.

A final consideration in risk characterization and management is the need for audiences other than the risk manager to understand the risks and the bases for the decision. These include the public, environmental advocacy groups, and regulated parties. Risk communication to the nonprofessional has been a major topic of research and discussion with respect to health risks, but not ecological risks (NRC, 1989). This is understandable because the average citizen is more concerned with personal and family health than with ecological effects. However, when outside parties become concerned about potential ecological effects, the diversity of ecological endpoints and the uncertainty of their valuation make risk communication highly complex. Earlier we cited a hypothetical situation in which a nonpesticide approach to eliminating crop-destroying insects that would harm some wildlife was likely to be acceptable to groups opposed on principle to pesticides. Under different circumstances, there might be scientific support for replacing a pesticide with another substance that posed a higher risk to avian populations but a lower risk to zooplankton. Yet arguments for accepting more dead birds in exchange for fewer dead zooplankton would not be easily conveyed to a diverse audience.

### 4.4.4 Conceptual Bases for De Minimis Risks

The concept of de manifestis risk is not controversial because some effects are clearly unacceptable. However, the idea that some exposures to and effects of pollutants are acceptable (de minimis) is controversial. The use of the de minimis concept is based on the following considerations:

1. Some exposures and effects are manifestly trivial. For example, it is absurd to assess effects of exposure to one molecule per organism or to consider restricting a pollutant exposure because it causes a transient physiological response or causes one copepod species to replace another.

2. There are many genuinely significant risks so it is inappropriate to set them aside while assessing the trivial ones. As Weinberg (1987) remarked, one does not swat gnats while being charged by elephants.

3. Regulatory scientists and decision-makers do not have infinite time and resources. The benefits from their time and resources should be maximized by quickly setting aside trivial risks.

4. Very low-level exposures and effects are very difficult to accurately quantify. Therefore, any attempt to balance their costs against the costs of treatment or prevention, or the benefits of a product are likely to be confusing or misleading rather than enlightening.

5. Small increases in exposure to naturally occurring chemicals or other hazardous agents are likely to have trivial effects, because of adaptation to the background levels. This is particularly true if the increase is also small relative to variation in the naturally-occurring background levels.

6. Effects that are small relative to natural temporal or spatial variability of the biological parameter are likely to have trivial implications either at other levels of biological organization or for other components of the same level of organization. For example, an apparently large (e.g., 50 percent) temporary depression of the decomposition rate is unlikely to significantly affect primary production or, abundance of birds because they are adapted to much larger variance in the rate due to wetting and drying cycles and other natural variables.

7. Treating a trivial risk as if it were potentially significant unnecessarily raises public concerns that are not entirely extinguished by a subsequent finding that the risks are small.

Since we cannot prevent all human effects on the environment, and we cannot even carefully assess the benefits and costs of all human actions, it is desirable to develop criteria for eliminating clearly trivial risks from further consideration.

### 4.4.5 Ecological Risk Assessment of Chemicals

***4.4.5.1 One-Dimensional Models*** Most ecological assessments depend on one-dimensional models of toxicant/organism interaction, and in most cases that dimension is concentration. Because toxicology is historically the science of poisons, the fundamental paradigm of toxic effects is the single lethal dose. To the poisoner, duration of exposure, severities other than mortality and even the exact proportion responding are unimportant, so the only dimension of interest is concentration. This paradigm is appropriate in weed and pest control and, to a somewhat lesser extent, to unintentional acute poisoning, as when wildlife are sprayed with pesticides.

Use of only the concentration dimension simplifies assessment because the outcome is determined by the relative magnitudes of the exposure concentration and the effective concentration. However, most assessments in environmental toxicology involve some element of time and some concern for the severity and extent of effects, so use of only the concentration dimension requires that these other dimensions be collapsed. This collapsing of the other dimensions is most commonly done by using a standard test endpoint as the effective concentration. Standard test endpoints are used because (1) the assessor does not have the skill, time, latitude, or inclination to develop alternate models; (2) the assessor has a vaguely defined assessment endpoint and is willing to accept the toxicologists judgments as to what constitutes an appropriate duration of exposure, severity of effect, and frequency of effect; or (3) by luck the assessment endpoint corresponds to a standard test endpoint. The test endpoints most often used, median lethal concentrations and doses, $LC_{50}$ and $LD_{50}$, were introduced earlier. With these endpoints, proportion responding, severity, and time are collapsed by considering only 50 percent response, only mortality, and only the end of the test.

Occasionally, time is the basis of a one-dimensional assessment. If the concentration of a release is relatively constant, then the risk manager needs to know how long it can be released without unacceptable risk. For example, a waste treatment plant might be taken off-line for repairs, and the operator would like to know how long he can release untreated waste. The most common temporally defined test endpoint is the median lethal time ($LT_{50}$). In fact, many releases have relatively constant concentrations and time would be a more useful single dimension than concentration for assessment. However, because time is generally felt to be less important than concentration, it is more often considered in higher dimensional models.

The most common model for risk estimation, the quotient method (Barnhouse et al., 1982; Urban and Cook, 1986), relies on a one-dimensional scale. The method consists of dividing the expected environmental concentration by a test endpoint concentration (i.e., determining their relative position on the concentration scale); the risk is assumed to increase with the magnitude of the quotient.

### *4.4.5.2 Two-Dimensional Models*

*Concentration/Response* The most common two-dimensional model in ecological toxicology is the concentration/response function. This preeminence is explained by the desire to show that response increases with concentration of a chemical, thereby establishing a causal relationship between the chemical and the response. In this model, time is collapsed by only considering the end of the test, and either severity or proportion responding is eliminated so as to have a single response variable. Most commonly, proportion responding is preserved and severity is collapsed by considering only one type of response, usually mortality. Severity is most often used when a functional response of a population or community of organisms such as primary production is of interest, rather than the distribution of response among individuals.

A probit, logit, or other function is fit to the concentration-response data obtained in the test. Typically, that function is used to generate the standard $LC_{50}$ or $LD_{50}$ or median effective dose ($ED_{50}$) by inverse regression, thereby reverting to a one-dimensional model and throwing out much information. Because assessments are more often concerned with preventing mortality or at least preventing mortality to some significant proportion of the population, it would be more useful to calculate an $LC_1$, $LC_{10}$, or other threshold level, but information would still be thrown away. If both dimensions are preserved, the concentration-response function can be used in each new assessment to estimate the proportion responding at a predicted exposure concentration or to pick a site-specific threshold for significant effects. When only an $LC_{50}$ is available and the assessment requires estimation of the proportion responding, it is possible to approximate the concentration–response curve by assuming a standard slope for the function and using the $LC_{50}$ to position the line on the concentration scale. Concentration–response functions are an improvement over being stuck with an $LC_{50}$, but in many cases the proportion responding is less important than the time required for response to begin.

One improvement in concentration–response functions would be to match the test durations to the exposure durations, rather than using standard test durations and assuming that the exposure durations match reasonably well. Ideally, the temporal pattern of exposure would be anticipated and reproduced in the toxicity tests. For example, if a power plant will periodically "blowdown" chlorinated cooling water, on a regular schedule for a constant time period, that intermittent exposure can be reproduced in a laboratory toxicity test (Brooks and Seegert, 1977; Heath, 1977).

*Time–Response Functions*   Time–response functions, like concentration-response functions, are primarily generated in order to calculate a one-dimensional endpoint, in this case the $LT_{50}$. Time-response functions are useful when concentration is relatively constant but the duration of exposure is variable. With the time–response function defined, it is possible for the assessor to consider the influence of changes in exposure time on the severity of effects or the proportion responding.

*Time–Concentration*   The most useful two-dimensional model of toxicity that includes time is the time–concentration function or toxicity curve (Sprague, 1969). This is created from experimental data by recording data at multiple times during the test and, for each time, calculating the $LC_{50}$, $EC_{50}$, or the concentration causing some other response proportion or severity, as described above. These concentrations are then plotted against time and a function is fit (Figure 4.5). Such functions have been advocated by eminent scientists in prominent publications (Lloyd, 1979; Sprague, 1969), but are seldom used even though they require no more data than is already required by standard methods for flow-through $LC_{50}$s for fish (ASTM, 1991; USEPA, 1982). If the response used corresponds to a threshold for significant effects on the assessment endpoint, then this function can be used to determine whether any combination of exposure concentration and duration will exceed that threshold (i.e., the area above a line, as in Figure 4.6).

When time is a concern but temporal test data are not available, it is necessary to approximate temporal dynamics. The simplest approach is to treat the level of effects and the exposure concentration as constants within set categories of time. This approach is used in most environmental assessments. The assessor, faced with toxic effects expressed as standard test endpoints, tries to match the test durations to the temporal dynamics of the pollution in some reasonable manner. Time is divided into acute and chronic categories, and standard acute and chronic test endpoints are used as benchmarks separating acceptable and unacceptable concentrations. There are serious conceptual problems with this categorization, and we address these now.

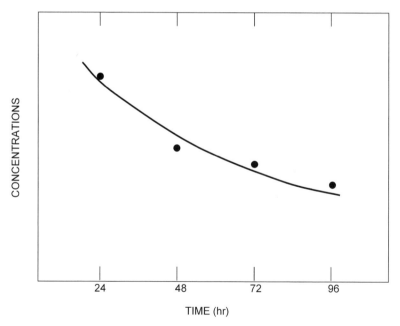

**Figure 4.5.** Toxic effects as a function of concentration and time. The function is derived by plotting $LC_{50}$s or other test end points against the times at which they were determined (i.e., 24, 38, 72 and 96 h).

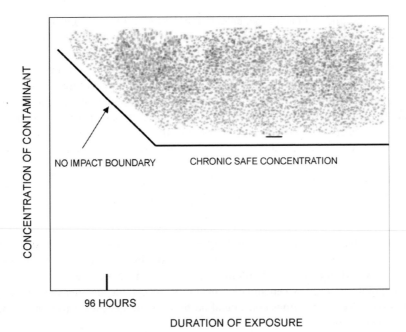

**Figure 4.6.** Toxic effects as a function of concentration and time, with time shown as a dichotomous variable, acute and chronic. Acute time ends at 96 hours, the end of a standard acute lethality test for fish.

*"Acute" and "Chronic" as Temporal Categories*   Ambiguities of terminology complicate the matching of test durations to exposure durations. Because "acute" and "chronic" refer to short and long periods of time, respectively, it is tempting to relate endpoints from acute tests to short exposures, and those from chronic tests to long exposures. However, the terms have acquired additional connotations, and using them to describe severity as well as duration leads to complications. Acute exposures and responses are assumed to be both of shorter duration and more severe than chronic exposures and toxicities. The implicit model behind this assumption is that chronic effects are sublethal responses that occur because of the accumulation of the toxicant or of toxicant-induced injuries over long exposures. Conversely, because of the cost of chronic toxicity tests, toxicologists have attempted to reduce testing costs by identifying responses that occur quickly and that are severe enough to be easily observed, but that occur at concentrations that are as low as those that cause effects in chronic tests (McKim, 1985). As a result, the relationship between the acute-chronic dichotomy and gradients of time and severity has become confused.

This confusion is illustrated by the standard test endpoints for fish. The standard acute endpoint is the 96-hour median lethal concentration ($LC_{50}$) for adult or juvenile fish (ASTM, 1991; OECD, 1981; USEPA, 1982). The standard chronic test endpoint has been the maximum acceptable toxicant concentration (MATC, also termed the "chronic value"), which is the threshold for statistically significant effects on survival, growth, or reproduction (ASTM, 1991; USEPA, 1982). Because this chronic endpoint is based on only the most sensitive response, life stages that appeared to be generally less sensitive have been dropped from chronic tests so that those tests have been reduced from life cycle (12–30 months) to early life stages (28–60 days) (McKim, 1985). Tests that expose larval fish for only 11 or 5–7 days have now been proposed as equivalent to the longer chronic tests. As a result, the chronic test endpoint for fish is now tied to events of short duration (the presence and response of larvae), whereas the acute endpoint is applicable to exposures or similar duration and to life stages that are continuously present.

Even the severity distinction between acute and chronic tests is not clear. Although the $LC_{50}$ clearly indicates a severe effect on a high proportion of the population, the fact that the MATC is tied to a statistical threshold rather than a specified magnitude of effect means that it too can correspond to severe effects on much of the population. For example, more than half of female brook trout exposed to chlordane failed to spawn at the MATC.

It would be advantageous to clarify the distinction between acute and chronic toxicity by restoring the original temporal distinction and expressing effects in common terms. Without that clear distinction, concentration–time functions like Figure 4.6 are uninformative.

*Concentration and Duration as Replacement for Concentration and Time*   In the absence of good time–concentration information from the test data, a concentration–duration function may be assumed. For example, the first version of the model of marine spill effects for type A damage assessments simply assumed a linear function (USDOI, 1986, 1987). Lee and Jones (1982) combined this linearity assumption for temporal effects in acute exposures with an assumption of time independence in chronic exposures to generate the concentration-duration function shown in Figure 4.6. This linearity assumption is chosen for its simplicity rather than any theoretical or empirical evidence. Parkhurst et al. (1981), assuming that $LC_{50}$ values are available for 24, 48, and 96 hours, linearly interpolated between these values and assumed time independence for exposures beyond 96 hours to generate the function shown in Figure 4.7.

These approaches depend on the assumption that temporal dynamics can be treated in terms of various durations of exposure to prescribed concentrations. In reality, organisms are subjected to a continuous spectrum of fluctuations in exposure concentrations due to variation in aqueous dilution volume or atmospheric dispersion, variation in effluent quality and quantity, intermittent release of effluents, and accidental spills or upset effluents. These can be treated conventionally if (1) the time between episodes is sufficient for recovery so they can be treated as independent, (2) the fluctuations are of sufficient frequency and of sufficiently low amplitude that the organisms effectively average them, or (3) certain frequencies predominate, so that temporal categories of exposure can be identified as discussed above. An example of the third possibility is Tebo's (1986) categorization of fluctuations in aqueous, point-source effluents as ponded and well mixed-wastes that

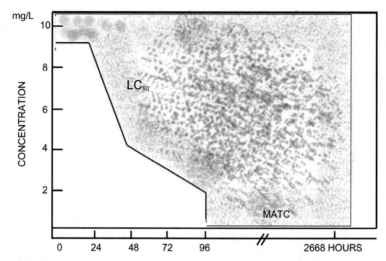

**Figure 4.7.** Toxic effects as a function of concentration and time, with acute time interpolated between measured values and extrapolated to zero and to the value for chronic time, which is expressed as a constant discrete variable (from Parkhurst et al., 1981).

are fairly uniform in character, wastes subject to short-term, daily fluctuations, and batch process wastes subject to severe fluctuations.

*Worst Case*   If it is not possible to characterize fluctuations in one of these ways, one can define a worst-case concentration and duration and assume that if that event does not cause unacceptable effects when it occurs in isolation, then it will also be acceptable when it occurs as part of a history of fluctuating exposures (Figure 4.8).

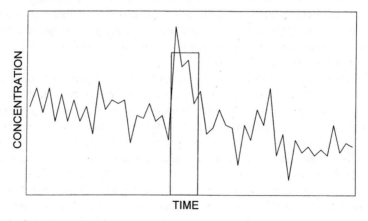

**Figure 4.8.** A fluctuating ambient exposure concentration can be represented in an assessment by a continuous exposure corresponding to the worst-case peak exposure, as highlighted by the vertical bar.

This assumption is commonly adopted in effluent regulation. For example, the worst-case dilution condition for aqueous effluents has traditionally been the minimum flow which occurs for seven days with an average recurrence frequency of 10 years, referred to as the 7Q10. The EPA recommends use of the lowest one-hour average and four-day average dilution flows that recur with an average frequency of three years (USEPA, 1985d). These correspond to highest one-hour and four-day exposures. The three-year recurrence frequency is assumed to allow for recovery of the system so that the peak exposures can be treated as independent events.

*Finding a Middle Ground*   The best solution would be to avoid the acute/chronic dichotomy and worst-case assumptions by identifying characteristic temporal patterns of exposures or biological responses, and either conducting toxicity tests to simulate those patterns or classifying test results into the environmentally-based temporal category in which their durations fall. That is, one can scale time to the characteristic temporal scales of the processes determining the risk. For example, exposures to gaseous pollutants from point sources might be classified as (1) plume strikes (an hour or less), (2) stagnation events (hours to a week), and (3) the growing season average exposure. Existing data on concentrations of an air pollutant causing phytotoxic effects might be plotted against time in terms of these categories (Figure 4.9), and then compared with estimated ground-level concentrations for each of the three categories of events. In any case, the matching of exposure

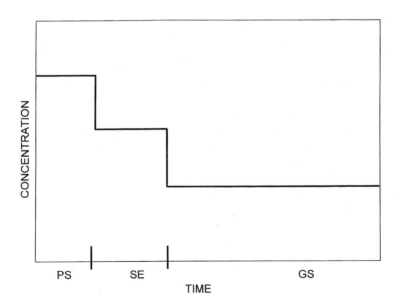

**Figure 4.9.** Toxic effects as a function of concentration and time with time expressed as a quartered variable defined in terms of categories of exposure PS, plume strike; SE, stagnation event; and GS, growing season.

durations with toxicological endpoints should be based on an analysis of the situation being assessed rather than on preconceptions about acute and chronic toxicity.

*Dose/Response Functions*  Although dose is defined in a variety of ways, all definitions concern the amount of material taken up by an organism. Traditionally, dose is simply the amount of material that is ingested, injected, or otherwise administered to the organism at one time. Exposure duration is not an issue for this definition (although time-to-response may be), and results are represented as dose-response functions. These functions, like the $LD_{50}$, values that are calculated from them, are useful to environmental assessors in cases like the application of pesticides in which very short duration exposures occur due to ingestion, inhalation, or surface exposure.

An alternate definition of dose is the product of concentration and time, as discussed earlier in connection with Figure 4.5, or, if concentration is not constant, the integral over time of concentration. This concept is applied to exposure concentrations as well as body burdens, allowing the calculation of dose-response functions for exposures to polluted media. It has been commonly used in the assessment of air pollution effects on plants, and is referred to as "exposure dose." For example, McLaughlin and Taylor (1985) compiled data on field fumigations with $SO_2$, of soy beans and snap beans, and plotted percent reduction in yield against dose in ppm-hours. Newcombe and MacDonald (1991) found that the severity of effects of suspended sediment on aquatic biota was a log linear function of the product of concentration and time.

The most strictly defined form of dose is delivered dose, the concentration or time integral of concentration of a chemical at its site of toxic action. This concept is applied empirically by measuring the concentration of the administered chemical in the target organ or tissue, in some easily sampled surrogate tissue such as blood, or in the whole body in the case of small organisms. By relating effects to internal concentrations, rather than (or in addition to) ambient exposure concentrations, this approach can provide a better understanding of the action of the chemical observed during tests, and it is useful as an adjunct to environmental monitoring. It allows body burdens of pollutants in dead, moribund, or apparently healthy organisms collected in the field to be compared to controlled test results to explain effects in the field. Toxicokinetic models provide a means of predicting dose to target organs from external exposure data.

Rather than simply being a function of peak body burden (i.e., mg/kg), effects may be a function of the product of body burden and time or, more generally, the time integral of body burden. This variable, termed "dose commitment," is used in estimating effects of exposures to radionuclides, and may be appropriate to some heavy metals and other environmental pollutants.

### 4.4.5.3  *Three-Dimensional Models*  Three-dimensional representations of toxic effects are rare. Only concentration-time-response models can be readily derived from conventional test data. Functions of this sort can be derived from the data that are required from flow-through 96-h $LC_{50}$ tests of fish by ASTM and EPA protocols (ASTM, 1991; USEPA, 1982). Such models have the obvious

advantage of allowing the assessor to estimate the level of effects from various combinations of exposure concentration and duration. The response surfaces developed by Richardson and Burton (1981) for two estuarine species exposed to ozonated water and for trees exposed to $SO_2$ are good examples.

Concentration–time–response relationships can also be a useful tool for dealing with diverse data. If test data are diverse with respect to both duration and severity of effects, then both of these dimensions must be related to concentration if regulatory criterion concentrations are to be calculated. Dourson (1986) explained how this is done to estimate allowable daily intake for humans. The various exposures are categorized as having no observed effect, no observed adverse effect, adverse effects, or frank effects. This severity scale is then translated into a series of symbols that are plotted on a dose and time surface. Lines are then fitted by eye separating the categories of severity and establishing the no adverse effect level, adverse effect level, and frank effect level. The same approach has been used by the EPA to set air quality criteria for phytotoxic effects of air pollutants (Figure 4.10).

Having characterized the risk of exposure to chemicals released to air, water and land, and the degree of uncertainty associated with the risk, the next step is to use the information to improve the basis for making decisions. This often involves the public, or at least embrace public concerns and attitude. It requires an examination of the question, "what is acceptable risk?" An essential part of this is risk communication — communication in terms of openness and transparency, understanding and engaging stakeholders, as well as providing balanced information to allow the public to make decisions on how to deal with risk.

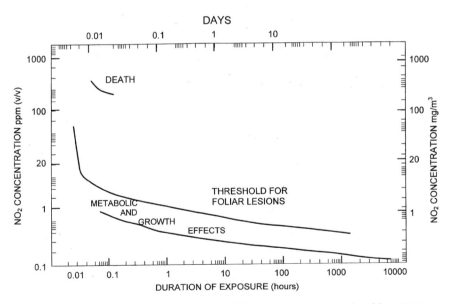

**Figure 4.10.** Categorization of effects of $NO_2$ on plants arrayed with respect to concentrations and duration of exposure (from USEPA, 1982).

## 4.5 COMMUNICATING INFORMATION ON ENVIRONMENTAL AND HEALTH RISKS

A key component of successful environmental and health risk management for any industry is an effective community outreach, or social responsibility program. This is even more important today than in the past, given the current explosion in public access and exposure to, and interest in, information regarding local health and environmental issues. The internet provides information on industrial facilities that is available to the public 24-hours a day; Web sites such as those provided in the United States by the Environmental Protection Agency (EPA) and the Environmental Defense Fund offer up-to-date information on the status of local industry to all users at the click of a mouse.

Facilities that have not taken a proactive approach to communication may find a public besieged with a wealth of negative information about a plant's operations and no positive data or reasons to trust the industry. In fact, in the absence of open communications or a public perception of information accessibility, a company may acquire a negative image even if there are no environmental or health risks associated with facility operations.

The chemical industry offers a good example of the potential divergence between real and perceived risk in the absence of communication. While major reductions in releases of both toxic chemicals and suspected carcinogens were documented between 1991 and 1997, and the accident rate for the chemical industry declined 32 percent, the public perception of the chemical industry did not improve. In fact, only 25 percent of people living near chemical plants have a favorable view of the industry. In general, the public does not hear about positive innovation or advances in risk reduction, and is skeptical of what they hear. These industries have to work at establishing trust. They actively must improve their public image if they want to mitigate and/or avoid future conflict with neighbors.

A commitment by industry to build and maintain an ongoing dialogue with local communities is essential. This dialogue needs to become an integral part of almost every industrial facility's day-to-day operations, with management providing access to information, in addition to learning and responding to the public's questions and concerns. As Section 4.5.1 demonstrates, facilities that have not opened the lines of communication with their neighbors may find themselves unwillingly drawn into the process — under circumstances that will leave them on the defensive and subject to public concern, or even outrage. The only effective solution is to build relationships over time with local communities and other key audiences.

### 4.5.1 From Concern to Outrage: Determinants of Public Response

What is likely to make people concerned or angry about a real or perceived risk? The answer depends on a number of risk perception and environmental vulnerability factors. A number of these environmental vulnerability factors are listed in Table 4.5. Without regular communication between industry and the public, people tend to use their general impressions to form an opinion regarding the

**TABLE 4.5. List of Environmental Vulnerability Factors**

Environmental Vulnerability Factors
- Things people can see, smell, hear (e.g., odors, clouds of steam, noise)
- High volumes of emissions or hazardous wastes
- History of community or employee health complaints
- History of unexplained odors or releases
- Presence of "dreaded" substances known to cause cancer
- Poor facility housekeeping or appearance
- Proximity to sensitive locations, including schools, nursing homes, and hospitals
- Reports of health problems among school children or staff
- Proximity to important scenic or cultural sites
- Active presence of organized environmental groups
- History of poor community outreach
- Lack of risk and crisis communication program

acceptability of a facility in their neighborhood. Impressions may be based on the "look" of the plant, for instance, whether the buildings and grounds are well cared for and maintained. If smoke or vapors are emitted, or if neighbors can smell or hear plant operations, a facility's environmental vulnerability is perceived to be greater. Environmental vulnerability will also be high if the plant has a history of community or employee health complaints or if high-profile substances such as radioactive materials are known to be on site. In addition, seeming appropriateness or inappropriateness of the facility in the community where it is located. People will be particularly sensitive to the risks imposed by the operations of facilities located near schools, hospitals, or important scenic or cultural sites.

The concept of environmental vulnerability helps to identify the factors that heighten public concern. However, environmental vulnerability does not clarify why people become outraged over certain kinds of risks and not others. In his classic discussion of "outrage factors," Sandman (1993) outlines the characteristics of a risk that help to determine the level of anger or fear in people's reaction to a risk. According to Sandman, the public's perception of risk goes beyond the narrow definition favored by scientists and engineers. The latter group uses a technical definition of risk, which can be established by multiplying the magnitude of the hazard by the probability of exposure, as follows:

$$Risk = Hazard \times Exposure \tag{4-9}$$

In this definition, risk increases with the hazardousness of the process or material (e.g., a chemical's toxicity) and with the amount and length of exposure by the person affected.

The public's understanding of risk, however, requires the incorporation into the equation of what Sandman has dubbed "outrage." Outrage accounts for all factors associated with a risk, not including the technical hazard. These factors, which are extremely important to public perception, include whether the risk is assumed voluntarily, as in smoking or not wearing a seatbelt, or involuntarily, as in dying in a plant explosion, whether the risk is controlled by the individual or by the system, and whether it is industrial or natural. The perceived risk associated with an industry or incident will be high if its effects are dreaded, concentrated in time and space, involuntary, memorable, and have no visible benefits, among other factors. With the incorporation of outrage, the layperson's view of risk can be expressed as follows:

$$Risk = Hazard + Outrage \qquad (4\text{-}10)$$

Where risk and hazard are the same as in Eq. (4-9) and outrage represents the incorporation of outrage factors. While the technical level of risk will also concern people, and they will expect something to be done, the outrage factors will help to determine their level of anger and will influence whether a risk is perceived as more or less dangerous. A qualitative breakdown of Sandman's outrage factors is given in Table 4.6.

Clearly, it is beneficial to industry to communicate to the public trustworthy information about potential or perceived problems before citizen response has reached the outrage state.

### 4.5.2 Sustainable Strategies for Communicating Environmental and Health Risks

While many industrial facilities may recognize the benefits of communication, many are reluctant to devote the necessary time and money to initiating community outreach programs. Some managers believe that if people are already getting information in ways over which industry has no control, there is no point in fighting the process. In fact, they are wise not to challenge the technical basis of environmental groups' statements: industry does not have the credibility that these

**TABLE 4.6. Outrage Factors\***

| More Risky | Less Risky |
| --- | --- |
| • Involuntary | • Voluntary |
| • Industrial/artificial | • Natural |
| • Exotic/unusual | • Familiar |
| • Memorable | • Not memorable |
| • Dreaded | • Not dreaded |
| • Concentrated in time and space | • Not concentrated in time and space |
| • Unfair | • Fair |
| • Morally relevant | • Morally neutral |
| • Untrustworthy | • Trustworthy |
| • Closed process | • Open process |
| • No visible benefits | • Visible benefits |

\*From Sandman, 1993.

groups have, and such challenges are likely to be seen as obfuscatory, further damaging the industry's image. However, the absence of communication espouses a negative perception. Communication should not be approached as a reaction to negative publicity or as combative. It needs to become an essential component of day-to-day operations in order to be effective.

### 4.5.2.1 *Breaking Down the Barriers to Communication*   Industry commonly uses the following rationalizations to defend a position of non communication.

*We're Too Small to do Outreach*   A low-profile or "no-profile" facility can actually find itself subject to increased public scrutiny because it lacks a history of outreach. Companies without a budget, activities or personnel allotted to community outreach may feel this is appropriate to their size and risk-potential. However, even a small chemical facility can be the object of the public's wrath, especially if people feel the risk has been hidden from them.

*Our Facilities and Operations are Environmentally Sound*   A solid environmental performance is certainly essential to a facility's efforts to be a good neighbor. However, a successful relationship can be achieved only when environmental performance is joined with regular communication and trust. Especially if there is no record of communication between industry and the public, people will base an opinion about whether a given operation is a good neighbor on the basis of general impressions of its facilities. Every facility is vulnerable in some way to public questions, concerns and even outrage (McDaniel et al., 1999).

*If We Talk to People, They Will Just Get Upset*   Sometimes after years or even decades of silence, an industrial facility will suddenly find itself engaged in an active and heated dialogue with its neighbors. The event precipitating the communication could be a minor chemical release, something that poses a small risk but requires a large response from the fire department and other emergency responders. While the technical level of risk will influence public response, the level of outrage will be in a large part determined by prior experiences with the company. Communication per se does not create outrage. Anger results when the appearance of unfairness leads to distrust and the perception that risks cannot or will not be controlled.

*Our People Are Already Overworked*   Risk communication and community outreach need to be part of each facility's day-to-day business plan and its employees' job descriptions. Section 4.5.2.2 discusses how these goals can be achieved. In the long run, time and money can be saved if public outrage is kept low and confrontation is avoided. A risk communication program should be thought of as a form of insurance against future problems with neighbors. By keeping in touch with community concerns, industry can largely avoid being blindsided by reactions to circumstances that should not be causes for alarm.

***4.5.2.2   Building a Communication Program***   The rationale for an effective risk communication program has been discussed, as have the theoretical underpinnings for determining public response or outrage to risk. On a practical level, industry needs to operationalize these concepts. While each facility will develop its own approach, three key steps to building sustainable strategies for risk communication are applicable to all operations.

- **STEP ONE: Begin with facility employees:**
  While the commitment to communication comes from top management, the success of an outreach program must involve employees at all levels. Employees are the best communication resource. They affect the way the facility is perceived, whether or not they have a specific responsibility for communication procedures. In fact, frontline employees, who themselves are exposed to whatever risk a facility might pose, possess greater credibility with the public than does management. If these employees do not believe the workplace is safe and environmentally sound, or that top management is responsive to their concerns, the community will have a similar outlook.

  Employees also need to be the first audience for communications. If the goal is consistent communication with the public at large, the employees need to be informed about risks. Frontline employees must be provided with risk and crisis communication training. They need the tools that will allow them to feel comfortable in fulfilling their communication responsibilities. Communication is a skill that can be learned, and if employees are expected to take an active role in the process, they need to be given a solid foundation for addressing public concern.

- **STEP TWO: Make communication a daily and ongoing commitment**
  Building trust and credibility with the public takes time. People reluctant to change long-held opinions of industry will do so only when they are satisfied that a new viewpoint is merited. To build trust, a company must make itself accountable to the public. This means establishing a track record of reaching out, following through, and hearing and responding to people's concerns.

  It is better to begin outreach with a series of small events rather than trying to address all audiences at one time. It is beneficial to seek out informal, as well as formal ways to interact, such as through meetings of civic organizations or homeowner's associations, and presentations at local schools or conferences. These interactions demonstrate an interest in the community and a willingness to participate on the public's terms and to relinquish control over communication infrastructure.

- **STEP THREE: Focus on listening and understanding**
  The key to communication is hearing what the public has to say. People want to receive information, but they also want to know that a company will listen to them and respond to answers. Never should conversation be limited to technical risk issues. A facility must be prepared to address a wide range of other issues and to take them seriously. Only then can industry construct a record of accountability.

### 4.5.3 Case Study: Environmental and Health Risk Communication Neglected Until After an Accident

To demonstrate the principles of risk communication, we present the simple case of a small oil-and-water separation plant that briefly released black smoke after a heavy rain caused oil to seep into a natural gas burner. Although the smoke was under control in less than ten minutes, in the aftermath of this seemingly small incident, the facility was nearly forced to shut down, and the company had to invest thousands of dollars and countless hours to retain its operating permit.

Management at the plant had been sure that the facility did not require a formal community-outreach program to explain its safe, unobtrusive oil/water separation process. A mixture of oil and water was piped in from offshore, heated, and transferred into settling tanks. The oil was then piped out through a pipeline, while the water was treated and discharged into the sewer system. The whole operation was so simple that for much of the time the plant was unstaffed.

Operations were quiet and odorless and produced no emissions. Likely due to the facility's unobtrusiveness, no one had ever asked about it, despite its location across the street from an elementary school. There was no signage on the perimeter fences, and most local residents could not identify the facility from the street.

#### 4.5.3.1 *The Accident and Aftermath*   Black smoke associated with burning oil brought the plant to the attention of its neighbors, who were concerned and had many questions regarding plant safety. One resident was particularly alarmed. He went to speak to the fire department and the facility manager, only to find that his questions were not taken seriously. Facility representatives refused to acknowledge that citizens' concerns had any validity, which served to make him more angry and suspicious. He began to go door-to-door throughout the neighborhood, compiling a list of anecdotal health concerns, including rashes, coughs, and problematic pregnancies. A number of residents teamed together with the objective of getting the facility shut down. The anger and level of activism among residents increased. Neighbors circulated a petition to the city council, requesting the plant's closure. News media picked up the story, which received considerable play in the papers and on television. The local air-pollution control district became involved, and the city came very close to withdrawing the plant's operating permit.

Why were the repercussions of a single and short-lived incident so extreme? The primary cause for the extended crisis was the fact that, in its 30 years of operation, the facility had not made an effort to talk to its neighbors or to learn about their potential concerns. In fact, most residents in the neighborhood were unaware that the plant existed. When people suddenly discovered the facility, via the black smoke emissions, they were angry. Residents felt that they had a right to know what kind of industrial operations existed in their neighborhood and what kinds of risks the facility posed to their environment and health.

#### 4.5.3.2 *The Costs of Noncommunication*   The oil/water separator facility was able to continue operations only after the company had funded two health risk

assessments, established health-protective levels for chemicals used during cleaning and turnaround, promised to provide extensive environmental monitoring during non-routine operations, and hosted four community meetings. Despite these efforts to gain the community's confidence, when the company attempted to reopen a similar facility 20 miles away, residents from the first neighborhood protested and succeeded in keeping the second facility from obtaining an operating permit.

Company representatives had discovered too late that fallout from the concerns of an ignored public could be very costly. Issues of concern to the local community include basic facts about the industries in the locality and the potential risks posed by their operations. In the United States, as a result of the Freedom of Information Act and many provisions of environmental statutes, people have a right to such information. If the oil/water separator facility had initiated a dialogue with the community early on, the meeting with the public would not have been associated with a developing crisis.

This case study illustrates how quickly a situation can escalate to crisis when there is no record of consistent, two-way communication between industry and the public. Several of the environmental vulnerability factors outlined in Table 4.5 came into play: people initially became concerned about an unexplained emission from an unexpected source; they wondered belatedly about the advisability of having the facility located so near to an elementary school and residential properties; those who had experienced unusual health problems began to suspect a connection to the oil/water separation facility.

In cases such as this one, where the public has little or no reason to trust the industry or facility that is creating the risk, the effect of the outrage factors is also likely to be intense (see Table 4.6). The unusual and memorable release of smoke led to concern that people had no control over the kinds of industry allowed in their own community. Management of the separation plant was not seen as trustworthy. There was resentment, as well, over the company's failure to communicate with its neighbors prior to the incident (McDaniel et al., 2000).

### 4.5.4 Lessons Learned

The key to communicating with the public lies not in simply explaining the technical hazards, but also in understanding and responding to the sources of people's outrage. Public communication does not end with the distribution of right-to-know information. To the contrary — as more and more community members acquire information, the tendency is to want more input into the entire process of risk management. Industry must be ready not only to communicate, but also to share decision-making power with local communities. Such efforts will succeed only if community involvement becomes an integral part of the facility's day-to-day routine and of the company's culture. Building good relationships with the public takes time. However, it is these relationships that will bring trust and credibility to the company. This does not mean that there will never be conflict, but it does mean that the facility will not be starting from ground zero when attempting to explain complex, technical information about health, environmental, and safety risks to the public.

## 4.6 ENVIRONMENTAL INFORMATION ON THE INTERNET

The Internet offers instant access to a wealth of resources for environmental (as well as chemical and other) information. The Internet makes available many materials that can be useful for environmental risk assessment and risk management. Not all sources are equally valuable, however. Section 4.6.2 gives some important caveats to those who use data from the following sources:

- Searchable databases of federal, state and local government regulations
- Technical information from universities and research organizations
- Data on chemical and environmental trends and specific events, as reported by hundreds of corporations and analyzed by trade associations, government agencies and non-governmental organizations (NGOs)
- Enforcement and inspection guidance, and penalty alternatives offered by the U.S. Occupational Safety and Health Administration (OSHA) and U.S. Environmental Protection Agency (EPA)
- Product listings and capability statements from providers of environmental and chemical engineering support services, and directories of such firms
- Links to material exchanges, as well as pollution prevention techniques and other resources for reducing the costs and liabilities associated with the handling, release and disposal of hazardous materials
- Online news services
- Publications, meeting announcements and other opportunities for professional exchange.

Following are some of the best environmental resources available on the Web, as well as lesser-known but incredibly valuable sites for data to support environmental and chemical decision-making, models and resources available from regulatory agencies, assistance with outreach and public communication, and environmental publications (Bolstridge, 2003).

### 4.6.1 Internet Sources

*EPA's Integrated Risk Information System (IRIS)* www.epa.gov/iriswebp/ iris/index.html. IRIS is EPA's database of dose-response data on human health effects from chemical exposure. The searchable database contains information on more than 500 substances, and is intended for use in risk assessments, decision-making and regulatory activities.

*EPA Chemical Accident Histories and Investigations*   http://yosemite.epa.-gov/oswer/ceppoweb.nsf/content/preventingaccident.htm. This page permits accessing documents and reports on recent chemical accidents, investigations by the EPA Chemical Accident Investigation Team, and Incident Summaries by the National Response Center, as well as related advisories and notices. EPA's Accidental Release Information Program (ARIP) database contains information on accidental releases of hazardous chemicals that have occurred in the U.S. at fixed facilities between 1986 and 2000. Facilities are asked by EPA Regional Offices to complete a questionnaire of 23 questions about the facility, the incident's circumstances and causes, any control and prevention practices and technologies that were in place prior to the event, and changes made as a result of the release. The questionnaire focuses on accident prevention, including hazards assessment, training, emergency response, public notification procedures, mitigation techniques, and prevention equipment and controls. The entire database (in compressed format) can be downloaded and searched using any database program that can handle dbase format files. A copy of the questionnaire and a data dictionary to explain the database fields are also included in the compressed file.

*International Toxicity Estimates for Risk (ITER) Database*   www.tera.org/iter. This database is a compilation of human health risk values for more than 500 chemicals that have been identified as environmental concerns by health organizations around the world. Values for actual risks (e.g., noncancer oral; cancer, oral; noncancer inhalation; cancer, inhalation) can be obtained for any listed chemical, as well as detailed information on the study or research that generated the risk value reported, including information on peer review and detailed bibliographic references.

*AIRNow's Links to Real-Time Air Pollution Data*   www.epa.gov/airnow. This site provides access to national air quality information, including air quality forecasts and real-time air quality for over U.S. 100 cities. Local air quality forecasts are available, as well as links to state and local websites that provide real-time air quality monitoring and forecasting data. Each state/local Internet site provides different types of information, but much is focused on the urban areas' ozone and air quality indices that are commonly reported in the news media.

*AWMA's State Agency Listing*   www.awma.org/resources/gad/state.htm. The Air and Waste Management Association (AWMA) site provides links to state agencies involved in environmental regulations and compliance enforcement.

*Acid Rain, Atmospheric Deposition and Precipitation Chemistry*   http://btdqs.usgs.gov/acidrain. The U.S. Geological Survey's National Atmospheric Deposition Program and National Trends Network collects data from 220 sites throughout the U.S. to determine if regulatory actions to reduce air pollution are improving the quality of precipitation. There is a five- to six-month time lag between data collection and availability of results, as would be expected when laboratory analysis is required. Online data access requires completing a form and providing the requestor's name and address and the intended use of the data. Data are collected for daily precipitation, field and laboratory pH, and weekly, monthly, and annual average concentrations and dry deposition data for sulfate ($SO_4$), nitrate ($NO_3$), ammonia ($NH_4$), calcium (Ca), magnesium (Mg), potassium (K), sodium (Na), chloride (Cl), and nitrogen (N) from $NO_3$ and $NH_4$.

*Predict Mixed Chemicals' Reactions*  http://response.restoration.noaa.gov/chemaids/react.html. The National Oceanic and Atmospheric Administration's (NOAA) Chemical Reactivity worksheet is a free software package that contains reactivity data on 4,000 chemicals and determines reactivity effects of accidental mixing of chemicals. Based on the classification of the chemical into one or more reactive groups, the software predicts the kind of reactivity likely to occur when members of the groups are mixed together. It does not limit the number of chemicals that can be selected to be mixed, but becomes less effective for more than ten chemicals.

*Model an Oil Spill*  http://response.restoration.noaa.gov/software/gnome/gnome.html. NOAA provides software that allows a scenario for an oil spill to be specified. The General NOAA Oil Modeling Environment (GNOME) model produces a predicted trajectory of the spill, using data on tides and currents.

*Hazards Analysis for Toxic Substances, Version 3 (HATS3)*  http://yosemite.epa.gov/oswer/ceppoweb/content/ds-epds.htm#hats. HATS is a menu-driven screen-show presentation for emergency planners and responders. It introduces EPA's Emergency Planning and Community Right-to-Know Act (EPCRA) program and the use of EPA's approach to analyzing the hazards of toxic substances to the community surrounding manufacturing plants where such materials are manufactured or used.

*EPA Center for Exposure Assessment Modeling (CEAM)*  www.epa.gov/ceampubl/index.htm. This site provides access to tabulated listings of exposure-assessment software packages to address modeling of organic chemicals and metals in groundwater, surface water, food chains and multimedia systems. The models are available for downloading along with descriptions of the packages' capabilities.

*Regulatory Air Models*  www.epa.gov/ttn/scram. EPA's Support Center for Regulatory Air Models provides access to the source code, data sets and user's guides for various air models, including regulatory models (ISCST3, UAM, BLP, CTDMP), receptor models (CMB7, CMB8), screening models (CAL3QHC, CTSCREEN, LONGZ, SHORTZ, VALLEY), other air models (DEGADIS, PLUVUE2, TOXST, MESOPUFF, RPM-IV, FDMI), related programs (CALMPRO, RAMMET, WRPLOT, MPRM, MIXHTS), and non-EPA models (SLAB, ADAM, AFTOX, SCREEN).

*EPA's Risk Management Consequence Analysis (RMP\* Comp)*   www.epa. gov/ceppo/tools/rmp-comp/rmp-comp.html. EPA provides Version 1.07 of this free software to perform the risk management consequence analysis that is required under Section 112(r) of the Clean Air Act.

*ATSDR's Primer on Health Risk Communication*   http://atsdr1.atsdr.cdc.gov/ HEC/primer.html. The Agency for Toxic Substances and Disease Registry (ATSDR) provides a framework of principles and approaches for communication of health risk information to diverse audiences. It is intended for spokespersons who must respond to public concerns about exposure to hazardous substances in the environment. The document contains a "do's and don'ts" list that describes approaches to use and pitfalls to avoid when communicating risk information.

*Disaster Communication From Indiana Law University*   www.law.india-na.edu/webinit/disaster/index.html. This site summarizes national laws and international conventions that affect global information and communication regarding natural and manmade disasters, including improving relief efforts and reducing overall impacts. The site provides access to documents that address the role of the news media in disaster reduction and links to other organizations involved in disaster communication.

*EPA's National Service Center for Environmental Publications (NSCEP)*   www.epa.gov/ncepihom/ordering.htm. This site taps into a central repository for all EPA documents that can either be downloaded or obtained by mail for free. Over 5,500 titles are available in paper and/or electronic format.

*The National Academies' Reports Online*   http://nap.edu/info/browse.htm. The National Academies Press publishes the reports issued by the National Academy of Sciences, National Academy of Engineering, the National Research Council and the Institute of Medicine. Many texts on a variety of environmental, health, natural resources and engineering topics can be reviewed or ordered online.

### 4.6.2   Implications and Limitations of Using the Internet

The Internet has become the ultimate data source for science and technology. However, caution must be used when applying Internet data to scientific efforts or regulatory compliance. Often, information is published on the Internet without any description of the study or research that was used to develop the data. When such documentation is lacking, consider sending an e-mail message to the organization or individual that maintains the site to request the reference information. Some online data are severely out-of-date or represent student research efforts that have not been reviewed for accuracy or quality. Finally, data provided on the Internet may have been developed for purposes other than environmental or chemical analysis, and as a result may contain hidden biases or inaccuracies that make them inappropriate for your intended use. Realize that you may not be getting exactly what you think you have when downloading free data from the Internet.

Reliance on official government sites can help you avoid being misled by the content of some Internet sites. However, university and non-profit organizations can offer more timely or detailed information on recent events and issues. Always review the organization behind any Internet location that is to be used as a resource. Focus on locating the types of funding sources, peer review and data quality evaluations, as well as the date of the last update, for additional insight into the true value represented by a particular site. Much of this information will be available online or can be obtained by sending an e-mail to the site's developer. Chemical and environmental engineering professionals require many types of information. Much of the required data, and even entire software packages, are now available to be downloaded from the Internet, free of charge. In fact, as the Internet matures, vast amounts of data and software are becoming available to everyone. However, converting data to information, and more importantly converting information into knowledge, has become the major challenge facing scientific professionals.

## Acknowledgment

Much of the work presented in the Section 4.6 was carried out and written by Dr. Richard D. Seigel and Dr. Mary F. McDaniel. Their contributions to this chapter and expert comments are greatly appreciated.

## REFERENCES

Allen, D. T., and Shonnard, D. R. (eds) (2002). *Green Engineering — Environmentally Conscious Design of Chemical Processes*. Prentice Hall PTR, Upper Saddle River, NJ.

Anderson, E. L., and Albert, R. E. (ed.) (1998). *Risk Assessment and Indoor Air Quality*. Lewis Publishers, Boca Raton, FL.

APHA, American Public Health Association (1989). *Standard Methods for the Examination of Water and Wastewater*, 17th Ed.

Armitage, P., and Doll, R. (1954). The age distribution of cancer and a multistage theory of carcinogenesis. *Brit. J. Cancer*, **8**, 1–12.

ASTM, American Society for Testing and Materials (1991). Annual Book of ASTM Standards, Sec. 11, Water and Environmental Technology, Philadelphia, Pennsylvania.

ASTM, American Society for Testing and Materials (2003). Standard Guide for Developing Conceptual Site Models for Contaminated Sites, Philadelphia, E1689–95.

Barnhouse, L. W. et al. (1982). *Methodology for Environmental Risk Analysis*. ORNL/ TM-8167, Oak Ridge National Laboratory, Oak Ridge, TN.

Bolstridge, J. (2003). Access environmental information on the internet. *Chemical Engineering Progress*, pp. 50–54 (January).

Brooks, A. S., and Seegert, S. R. (1977). The effect of intermittent chlorination on rainbow trout and yellow perch. *Trans. Am. Fish. Soc.*, **106**: 278–286.

Costanza, R. (1991). *Ecological Economics: The Science and Management of Sustainability.* Columbia University Press, New York.

Desvousges, W. H., and Skahen, V. A. (1987). *Techniques to Measure Damages to Natural Resources.* Final Report, PB88–100136, National Technical Information Service, Springfield, Virginia.

Diwekar, U. M. (2003). *Introduction to Applied Optimization*, Netherlands, Kluwer Academic Publishers.

Dourson, M. L. (1986). New approaches in the derivation of acceptable daily intake (ADI), *Comments Toxicology*, **1**: 35–48.

Federal Focus, Inc. (1991). *Towards Common Measures: Recommendations for a Presidential Executive Order on Environmental Risk Assessment and Risk Management Policy.* Federal Focus Inc., and The Institute for Regulatory Policy, Washington, D.C.

Finkel, A. (1990). *Confronting Uncertainaty in Risk Management, Resources for the Future.* Washington, D.C.

Heath, A. G. (1977). Toxicity of intermittent chlorination to freshwater fish: influence of temperature and chemical form, *Hydrobilogia*, **56**: 39–47.

Kocher, D. C., and Hoffman, F. O. (1991). Regulating environmental carcinogens: where do we draw the line? *Environ. Sci. Technol.*, **25**: 1986–1989.

Lee, G. F., and Jones, R. A. (1982). An Approach to Evaluating the Potential Significance of Chemical Contaminants in Aquatic Habitat Assessment. In *Acquisition and Utilization of Aquatic Habitat Inventory Information* American Fisheries Society, pp. 294–302, Washington, D.C.

Lloyd, R. (1979). Toxicity Tests with Aquatic Organisms In Lectures presented at the sixth ·FAO/SIDA workshop on aquatic pollution in relation to protecting living resources, U.N. Food and Agricultural Organization, Rome.

McDaniel, M. F., Siegel, R. D., and Leuschner, K. (1999). The time is right for right-to-know why your facilities need to be talking to their neighbors now. *Chemical Bond*, pp. 21–29.

McDaniel, M. F., Siegel, R. D., and Leuschner, K. (2000). Prepare for the next phase of public involvement. *Environmental Management*, pp. 35–40.

McKim, J. K. (1985). Early Life Toxicity Tests. In G. M. Rand and S. R. Petrocelli (eds), *Fundamentals of Aquatic Toxicology*, Hemisphere Publishing Corp., Washington, D.C.

McLaughlin, S. B., and Taylor, G. E. (1985). $SO_2$ Effects on Dicot Crops: Some Issues, Mechanisms, and Indicators. In W. E. Winner, H. A. Mooney, and R. A. Goldstein (eds), *Sulfur Dioxide and Vegetation*, pp. 227–249, Standford University Press, Standford, CA.

Newcombe, C. P., and MacDonald, D. D. (1991). Effects of suspended sediments on aquatic ecosystems. *N. Am. J. Fish. Manage.*, **11**: 72–82.

NRC, National Research Council (1975). *Decision Making for Regulating Chemicals in the Environment.* National Academy Press, Washington, D.C.

NRC, National Research Council (1983). *Risk Assessment in the Federal Government: Managing the Process.* National Academy Press, Washington, D.C.

NRC, National Research Council (1989). *Improving Risk Communication.* National Academy Press, Washington, D.C.

OECD, Organization for Economic Cooperation and Development (1981). *OECD Guidelines for Testing of Chemicals.* OECD, Paris.

Parkhurst, M. A., Onishi, Y., and Olsen, A. R. (1981). A Risk Assessment of Toxicants to Aquatic Life using Environmental Exposure Estimates and Laboratory Toxicity Data. In D. R. Branson and K. L. Dickson (eds). *Aquatic Toxicology and Hazard Assessment.* ASTM/STP 737, American Society for Testing and Materials, pp. 59–71, Philadelphia, Pennsylvania.

Randall, A. (1991). The value of diversity. *AMBIO* **20**: 64–68.

Richardson, L. B., and Burton, D. T. (1981). Toxicity of ozonated estuarine water to juvenile blue crabs (*Calinects sapidus*) and juvenile Atlantic menhaden (*Brevoortia tyrannus*). *Bull. Environ. Contam. Toxicol*, **26**: 171–178.

Risk Assessment Forum (1996). *Summary Report for the Workshop on Monte Carlo Analysis.* U.S. Environmental Protection Agency, Washington, DC.

Risk Assessment Forum (1999). Guiding Principles for Monte Carlo Analysis, Washington, 1997. Risk Assessment Forum, Report of the Workshop on Selecting Input Distributions for Probabilistic Assessments, Washington.

Roberts, W. C., and Abernathy, C. O. (1997). Risk Assessment: Principles and Methodologies. In *Toxicology and Risk Assessment*, Chapter 15, Fan, A. M. Ed. Academic Press.

Rolston, H. (1988). *Environmental ethics.* Temple University Press, Philadelphia, PA.

Ruckelshaus, W. D. (1984). Risk in a free society. *Risk Anal*, **4**: 157–162.

Sandman, P. (1993). *Responding to community outrage: strategies for effective risk communication.* Fairfax, VA: American Industrial Hygiene Association.

Sprage, J. B. (1969). Measurement of pollutant toxicity to fish I. bioassay methods for acute toxicity. *Water Research*, **3**: 793–821.

Stephan, C. E. (1977). Methods for Calculating an $LC_{50}$, F. L. Mayer and J. L. Hamelink, (eds): *Aquatic Toxicology and Hazard Evaluation.* ASTM STP 634, American Society for Testing and Materials, Philadelphia, pp. 65–84.

Suter, G. W. II et al. (eds) (1993). *Ecological Risk Assessment.* Lewis Publishers, Boca Raton, FL.

Suter, G. W. II et al. (eds) (2000). *Ecological Risk Assessment for Contaminated Sites.* Lewis Publishers, Boca Raton, FL.

Tebo, L. B. (1986). Effluent Monitoring: Historical Perspective. In H. L. Bergman, R. A. Kimerle, and A. W. Maki (eds), *Environmental Hazard Assessment of Effluents.* New York, Pergamon Press, pp. 13–31.

Travis, C. C. et al. (1987). Cancer risk management. *Environ. Sci. Technol.*, **21**: 415–420.

Urban, D. J., and Cook, N. J. (1986). *Hazard Evaluation, Standard Evaluation Procedure, Ecological Risk Assessment.* EPA-540/9–85–001, U.S. EPA, Washington, D.C.

USDOI, U.S. Department of the Interior (1986). *Natural Resource Damage Assessment.* Final Rule, *Fed. Regist.* **51**: 27674–27753.

USDOI, U.S. Department of the Interior (1987). *Natural Resource Damage Assessment.* Final Rule, *Fed. Regist.* **52**: 9042–9100.

U.S. Environmental Protection Agency (1982). *Guidelines and Support Documents for Environmental Effect Testing.* Washington, D.C.

U.S. Environmental Protection Agency (1985a). *Methods for Measuring the Acute Toxicity of 4 Freshwater and Marine Organisms.* U.S. EPA Environmental Monitoring and Support L EFIA-600/4–85/013, Cincinnati, OH.

U.S. Environmental Protection Agency (1985b). *User's Guide to the Conduct and Interpretation of Effluent Toxicity Tests at Estuarine/Marine Sites.* EPA-600/X-86/224, Washington, DC.

U.S. Environmental Protection Agency (1985c). *Short Term Methods for Estimating Chronic Effluents and Receiving Waters to Freshwater Organisms.* EPA-660/4–85/014.

U.S. Environmental Protection Agency (1985d). *Technical Support Document for Water Quality-Based Toxics Control.* U.S. EPA Office of Water, EPA-440/4–85/032, Washington, DC.

U.S. Environmental Protection Agency (1986). *Quality Criteria for Water, Office of Water.* Washington, D.C.

U.S. Environmental Protection Agency (1988). *Short Term Methods for Estimating the Chronic Effluents and Receiving Waters to Marine and Estuarine Organisms.* EPA-600/4–88/028.

U.S. Environmental Protection Agency (1989). *Short Term Methods for Estimating Chronic Effluents and Receiving Waters to Freshwater Organisms.* EPA-660/2nd.

U.S. Environmental Protection Agency (1992). *Framework for Ecological Risk Assessment, Risk Assessment Forum.* Washington, D.C.

U.S. Environmental Protection Agency (1993). *Wildlife Exposure Factors Handbook.* Vol. 1, Office of Research and Development, Washington, D.C.

U.S. Environmental Protection Agency (1995). *Development of Human Health Based and Ecologically Based Exit Criteria for the Hazardous Waste Identification Project.* Washington, DC.

U.S. Environmental Protection Agency (1997). *Ecological Risk Assessment Guidance for Superfund Process for Designing and Conducting Ecological Risk Assessments.* Interim Final, Emergency Response Team, Washington, DC.

U.S. Environmental Protection Agency (1998). *Guidelines for Ecological Risk Assessment, Risk Assessment Forum.* Washington, DC.

Weinberg, A. (1987). Science and Its Limits: The Regular's Dilemma, pp. 27–38. In C. Whipple (ed.), *De Minimus Risk.* New York, Plenum Press.

Whipple, C., (Ed.) (1987) *De Minimus Risk.* New York, Plenum Press.

# CHAPTER 5

# ECONOMICS OF POLLUTION PREVENTION: TOWARD AN ENVIRONMENTALLY SUSTAINABLE ECONOMY

TAPAS K. DAS

Washington Department of Ecology, Olympia, Washington 98504, USA

The failure of many pollution-prevention programs can be traced to the inability of the engineers and scientists to convince business leadership to change manufacturing processes unfavorable to the environment. Often, this reluctance to change is not because the recommended process improvements were technically unsound, but because the engineering team failed to speak the language of business, that is, dollars and cents. The role of economics in pollution prevention is very important, even as important as the ability to identify technology changes to the process, new and emerging technologies, zero-discharge technologies, technologies for biobased engineered chemicals, products, renewable energy sources and its associated costs.

Market acceptance of new technologies, products, processes, and services depends upon the complex interplay of cost, physical properties, environmental performance, public policy, cultural prejudices, and other factors. Accurate forecasting is a difficult and time consuming activity best left to the experts. However, cost estimating is a valuable skill that allows an engineer to obtain "ballpark" approximation of project costs. The goal is to obtain an estimate that is within $\pm 30$ percent of the actual cost if the enterprise were pursued. Such estimates are relatively easy to develop. Engineers use a relatively small set of financial tools to assess alternative capital investments and to justify their selections to upper management. Most often, the engineer's purpose is to show how a recommended investment will improve the company's profitability. Spending on capital improvements is largely voluntary. Adding an assembly line or changing from one type of gasket material to another can be postponed, or rejected outright. This is not the case with

*Toward Zero Discharge: Innovative Methodology and Technologies for Process Pollution Prevention,*
edited by Tapas K. Das.
ISBN 0-471-46967-X   Copyright © 2005 John Wiley & Sons, Inc.

**161**

pollution control devices, which are necessary for compliance with State and Federal pollution standards and generally must be installed by or before a mandated deadline. Consequently, a decision to install device X may not originate with the engineer. Instead, the need for pollution control equipment might be identified by a company's environmental manager who then passes the decision to acquire device X down to the engineer. This chapter gives some cost-estimating methods that can be used to assess the costs of implementing pollution prevention technologies; these methods are useful, as well, in cost comparisons, and in considerations of cost-effectiveness, best available technologies, and emerging technologies. Also provided is some information about the cost of biorenewable resources and the cost of manufacturing biobased products from such feedstocks. We begin, however, with a few remarks about the economic matrix in which pollution control technology is implemented.

## 5.1 ECONOMICS OF POLLUTION CONTROL TECHNOLOGY

In the present economic system the goal of a sustainable development process is to maintain intergenerational equity by ensuring quality of life for future generations, which essentially requires the stopping of further damage to the environment. From this point of view the necessity of an integrated system consisting of modern production and controlling technologies is being gradually recognized. To this end, various regulatory measures are being implemented, mainly in the industrial sector, which is the major contributor of pollution, for adopting such technologies.

### 5.1.1 Cost Estimates

The costs and estimating methodology in this section are directed toward the "study" estimate with a nominal accuracy of ± 30 percent. According to Perry's Chemical Engineer's Handbook, a study estimate is "used to estimate the economic feasibility of a project before expending significant funds for piloting, marketing, land surveys, and acquisition. However, it can be prepared at relatively low cost with minimum data" (Perry and Green, 1997). Specifically, the development of a study estimate calls for knowledge of the following:

- Location of the source within the plant
- Rough sketch of the process flow sheet (i.e., the relative locations of the equipment in the system)
- Preliminary sizes of, and material specifications for, the system equipment items
- Approximate sizes and types of construction of any buildings required to house the control system
- Rough estimates of utility requirements (e.g., electricity)

- Preliminary flow sheet and specifications for ducts and piping
- Approximate sizes of motors required

In addition, since the accuracy of an estimate (study or otherwise) depends on the amount of engineering work expended on the project, the user will need an estimate of the labor hours required for engineering and drafting activities. There are four other types of estimate.

The order-of-magnitude estimate a rule-of-thumb procedure, and can be used only for plant installations of the repetitive type for which there exists good cost history. Its error bounds are greater than ±30 percent. (However, Perry states that "no limits of accuracy can safely be applied to [an order-of-magnitude estimate].") The sole input required for making an order-of-magnitude estimate is the control system's capacity (often measured by the maximum volumetric flow rate of the gas passing through the system).

The other three types of estimate, listed next, are preferable.

- Scope, Budget Authorization, or Preliminary. This estimate, nominally of ±20 percent accuracy, requires more detailed knowledge than the study estimate regarding the site, flow sheet, equipment, buildings, etc. In addition, rough specifications for the insulation and instrumentation are also needed.
- Project Control or Definitive. These estimates, accurate to within ±10 percent, require yet more information than the scope estimates, especially concerning the site, equipment, and electrical requirements.
- Firm or Contractor's or Detailed. This is the most accurate (±5 percent) of the estimate types, requiring complete drawings, specifications, and site surveys. Consequently, detailed cost estimates are typically not available until right before construction. Common sense suggests that there is seldom enough time to prepare such estimate before approval to proceed with the project has been obtained.

## 5.1.2 Elements of Total Capital Investment

Total capital investment (TCI) includes all costs required to purchase equipment needed for the control system (purchased equipment costs), the costs of labor and materials for installing that equipment (direct installation costs), costs for site preparation and buildings, and certain other costs (indirect installation costs). TCI also includes costs for land, working capital, and off-site facilities.

Direct installation costs include costs for foundations and supports, erecting and handling the equipment, electrical work, piping, insulation, and painting. Typical indirect installation costs are engineering costs, construction and field expenses (i.e., costs for construction supervisory personnel, office personnel, rental of temporary offices, etc.), contractor fees (for construction and engineering firms involved in the project), start-up and performance test costs (to get the control system running and to verify that it meets performance guarantees), and associated costs with contingencies. "Contingencies" is a catch-all category that covers unforeseen costs that may arise if, for example, it becomes necessary to redesign or modify equipment, to pay higher costs of equipment or field labor costs, or to compensate for delays in start-up. Contingency costs are not the same as uncertainty and retrofit factor costs.

The elements of total capital investment are displayed in Figure 5.1. Note item "Battery Limits" cost, which is the sum of the purchased equipment cost, direct and indirect installation costs, and costs of site preparation and buildings. By definition, this is the total estimate for a specific job; any support facilities that may be needed (e.g., control systems) are assumed to already exist at the plant and are not included in the battery limits. For systems installed in new plants, off-site facilities (special facilities for supporting the control system) also might be excluded from the battery limits. Off-site facilities are exemplified by units to produce steam, electricity, and treated water; laboratory buildings; and railroad spurs, roads, and other transportation infrastructure items. Pollution control devices rarely consume enough energy to warrant dedicated off-site capital units. However, it may be necessary—especially in the case of control systems installed in new or "grass roots" plants to build extra capacity into the site's generating plant to service the system. (A venturi scrubber, which often requires large amounts of electricity, is a good example). Nevertheless, that the capital cost of a device does not include utility costs, even if the device were to require an offsite facility. Utility costs are

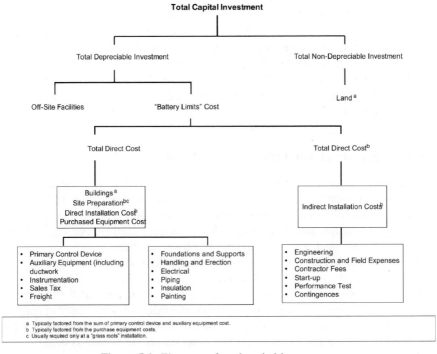

**Figure 5.1.** Elements of total capital investment.

charged to the project as operating costs at a rate that covers both the investment and operating and maintenance costs for the utility.

As Figure 5.1 shows, the installation of pollution control equipment may also require land, but since most add-on control systems take up very little space (a quarter-acre or less) this cost tends to be relatively small. Certain control systems, such as those used for flue gas desulfurization (FGD) or selective catalytic reduction (SCR), require larger quantities of land for the equipment, chemicals storage, and waste disposal. In these cases, especially when a retrofit installation must be performed, space constraints can significantly influence the cost of installation, and the purchase of additional land may be a significant factor in the development of the project's capital costs. Land retains its value over time, however, land is not treated the same as other capital investments because it retains its value over time. The purchase price of new land needed for sitting a pollution control device can be added to the TCI, but it must not be depreciated. Instead, if the firm plans to dismantle the device at some future time, either the land should be excluded from the analysis or its value included at the disposal point as an "income" to the project, to net it out of the cash flow analysis.

One might expect to include initial operational costs (the initial costs of fuel, chemicals, and other materials, as well as labor and maintenance related to start-up) in the operating cost section of the cost analysis instead of in the capital component, but such an allocation would be inappropriate. Routine operation of the control does not begin until the system has been tested, balanced, and adjusted to work within its design parameters. Until then, all utilities consumed, all labor expended, and all maintenance and repairs performed are a part of the construction phase of the project and are included in the TCI in the "Start-Up" component of the indirect installation costs.

### 5.1.3 Elements of Total Annual Cost

Total annual cost (TAC) has three elements: direct costs (DC), indirect costs (IC), and recovery credits (RC), which are related by the following equation:

$$TAC = DC + IC - RC \qquad (5\text{-}1)$$

Clearly, the basis of these costs is one year, a period that allows for seasonal variations in production (and emissions generation) and is directly usable in financial analyses. The various annual costs and their interrelationships are displayed in Figure 5.2.

Direct costs are those that tend to be directly proportional (variable costs) or partially proportional (semi-variable costs) to some measure of productivity-generally the company's productive output, but for purposes of pollution control, the proper metric may be the quantity of exhaust gas processed by the control system per unit time. Conceptually, a variable cost can be graphed in cost/output space as a positive sloped straight line that passes through the origin. The slope of the line is the factor by which output is multiplied to derive the total variable cost of the system. Semivariable costs can be graphed as a positive sloped straight line that

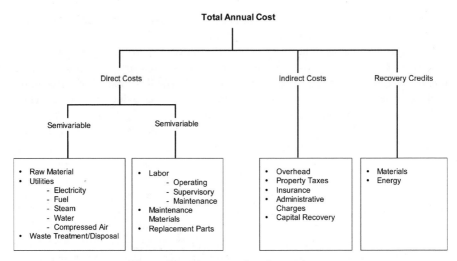

**Figure 5.2.** Elements of total annual cost.

passes through the cost axis at a value greater than zero - that value being the "fixed" portion of the semi-variable cost and the slope of the line being analogous to that of the variable cost line discussed above.

### 5.1.4 Economic Criteria for Technology Comparisons

Technology evaluation encompasses not only technical feasibility of a particular technology but also the economics of its implementation. While there are many measures of economic merit, two measures — investment and net present value (NPV) — provide a complete set of information on which to base an informed, economic decision. "Investment" refers to the money that must be spent initially to design and build the new facilities. Net present value is the after-tax worth in today's dollars of all the future cash that an alternative will either consume or generate; it includes the effects of new investment, working capital, operating costs, revenues, and income taxes. NPV is the most popular singular measure of the economic merit of an alternative, and many spreadsheet programs can easily calculate it.

To evaluate alternative pollution control devices, the analyst must be able to compare them in a meaningful manner. Since different controls have different expected useful lives and will result in different cash flows, the first step is to use the principle of the time value of money to normalize the returns of the alternatives being compared. The process through which future cash flows are translated into current dollars is called present value analysis. When the cash flows involve income and expenses, it is also commonly referred to as net present value analysis. In either case, the calculation is the same: adjust the value of future money to values based on the same (generally year zero of the project), employing an appropriate interest (discount) rate and then add them together. The decision rule for NPV analysis is that projects with negative NPVs should not be undertaken; and for projects with positive NPVs, the larger the net present value, the more attractive the project. Derivation of a cash flow's net present value involves the following steps:

- Identification of alternatives — for example, the choice between a fabric filter baghouse and an electrostatic precipitator for removing particulate matter from a flue gas stream.
- Determination of costs and cash flows over the life of each alternative.

The relationship of the future value of money to the present value of a sum of money compounded annually at an interest rate, $i$, for a total of $n$ years is shown by the following equation:

$$FV = PV(1 + i)^n$$

where

$$FV = \text{future value}$$
$$PV = \text{present value}$$
$$i = \text{annual earnings rate}$$
$$n = \text{years in the future}$$

This equation is used to determine how many future dollars will be realized by investing present dollars at the interest rate $i$ for $n$ years. For example, $100 invested today at 5 percent for 12 years will yield the following return:

$$FV = \$100(1 + 0.05)^{12} = \$180$$

Similarly, inflation at 5 percent per year makes $100 worth of merchandise today cost $180 in 12 years. Discounting is the opposite of compounding: the present value of a sum of money to be spent or received in the future is reduced according to:

$$PV = \frac{FV}{(1 + i)^n}$$

where $i$ is now referred to as the discount rate. Thus, if the discount rate is 5 percent per year, we calculate the present value of $100 to be spent in 12 years as follows:

$$PV = \frac{\$100}{(1 + 0.05)^{12}} = \$56$$

Since the present value of money to be spent in the future declines with increasing interest rate and number of years into the future, the importance of accounting for the time at which money is spent or received is evident.

### 5.1.5   Calculating Cash Flow

A complete financial analysis requires the inclusion of cash flow. On a year-by-year basis, the cash flow from a manufacturing investment is determined through calculations of cash costs, total costs, and cost-effectiveness.

Cash costs, those associated with operating and overhead, include raw materials (chemicals, catalysts, solvent, etc) utilities (steam, electricity, natural gas, water, etc.), maintenance materials and labor, operations labor, technical support, startup costs, taxes and insurance, and administrative costs.

Total costs include the sum of cash costs as just defined and depreciation. Note that depreciation is not a cash cost and is used only to calculate income taxes. If the investment generates revenues or savings from which the total costs may be subtracted, the resulting amount is known as the "pretax earnings." Income taxes (calculated on pretax earnings) are next subtracted to yield the after tax earnings. A year-by-year cash flow can then be determined by summing the four real cash flows: investment, cash costs, revenues, and income taxes.

Net present value is the sum of the present values of a series of cash flows (CF):

$$NPV = CF_0 + (CF_1 \times D_1) + (CF_2 \times D_2) + (CF_3 \times D_3) + \cdots$$

where, $CF_0$, $CF_1$, $CF_2$, represent cash flow in year 0, 1, 2, and so on, and the $D$ terms are discount factor for the respective years.

In the context of pollution control, cost-effectiveness is defined as the annualized cost of the control option divided by the annual emission reductions resulting from the control option. The following information is required to calculate the cost-effectiveness of a proposed control option: (1) the capital cost of purchasing and installing the control equipment or making a process modification; (2) the annual operating costs of the control option, and (3) an estimate of the emissions before and after application of the control option.

The capital costs of purchasing the control option should be determined from actual vendor price quotes for each proposed control option. Installation costs should also be based on vendor price quotes. If vendor price quotes are unavailable, elements of the installation cost may be estimated (Vatavuk, 2002). To illustrate cost-effectiveness calculations, let us consider a facility that is a major source of $NO_x$ emissions.

*Example 5-1: Cost-Effectiveness Calculation*   For a boiler operating at 100 percent of its operating capacity ($50$ Btu $\times 10^6$), the following information was determined for the purposes of calculating the cost-effectiveness of installing selective catalytic reduction as a control option. The facility is a major source of $NO_x$ emissions.
Given:

Capital cost of control option $= \$1,500,000$
Capital recovery factor (CRF) $= 0.1627$ (assuming 10 percent interest for 10 years)
Annual operating and maintenance costs $= \$98,000$

Uncontrolled emissions = 21 tons of $NO_x$ per year

Capture efficiency = 100 percent

Control efficiency = 80 percent

Solution: We insert the data into the following equations:

Cost-effectiveness ($/ton) = annualized costs/tons of pollutant reduced

Annualized costs ($) = (capital cost × CRF) + annual operating costs

Pollutant reduced (tons) = (uncontrolled emissions × capture efficiency × control efficiency)

The result is:

$$\text{Cost-effectiveness} = (\$150{,}0000 \times 0.1627 + \$98{,}000)/(21 \text{ tons} \times 1.0 \times 0.80)$$
$$= \$24{,}000/\text{ton}$$

Inducing industries to adopt effective new production methods and pollution control technologies in timely fashion is a challenge for regulators everywhere. Environmental protection costs money, and the charges that will be incurred by a given enterprise depend on the nature and extent of the pollution it generates. These factors, in turn, are functions of the type and scale of production, the plant's input, the technology used, and the size of the plant.

Generators of pollution range from private individuals and small business to municipalities, large corporations, and the government itself. Each polluting entity must decide how to invest in the socially desirable goal of a cleaner environment; influencing such decisions, inevitably, will be the benefit the entity expects to obtain from the investment.

When business owners consider how to implement the installation of modern pollution control technology, their decisions incorporate the results of comparative analyses of the costs and benefits to the investments contemplated. Later we shall present a case study focused largely on financial costs and benefits associated with biobased energy and products (Section 5.4). To provide the necessary background for this material, we turn now to some basics of cost estimation and capital investments.

## 5.1.6  Achieving a Responsible Balance

In the real world, taking socially desirable action for pollution prevention may not be in the best interest of an industry in the short run, since environmental protection costs money. The type of production, input used, technology employed, space used as well as plant size and the scale of production determine the nature and extent of pollution which, in turn, influence the abatement cost. Hence, the decisions of the profit maximizing private investor for implementation of pollution abatement measure are influenced by the benefit expected as a result of the investment required. On the other hand, the task of government planners and policy makers is to encourage industry to adopt modern production and pollution control technologies to provide a better environment for the society. In upgrading their facilities to meet the near antipollution regulations, industries turn for guidance to the principles of best available control technology (BACT).

## 5.2   BEST AVAILABLE CONTROL TECHNOLOGY: APPLICATION TO GAS TURBINE POWER PLANTS

In the United States, best available control technology (BACT) comprises the equipment and methods needed to achieve the maximum degree of reduction of pollutants subject to federal regulations that are emitted from any proposed major stationary source or major modification of such a source. The pollutants subject to review under the PSD regulations, and for which a BACT analysis is required, include nitrogen oxides ($NO_x$), carbon monoxide (CO), particulates less than or equal to 10 microns in diameter ($PM_{10}$), and volatile organic compounds (VOC). All PM is assumed to be $PM_{10}$. The BACT review follows the "top-down" approach recommended by the EPA.

There is no single BACT for any industry; it is determined on a case-by-case basis, taking into account cost-effectiveness, economic, energy, environmental, and other effects of proposed solutions. Three important terms in BACT analysis are "Prevention of significant deterioration" (PSD), "significant emission rate" (SER), and ambient air quality analysis and national ambient air quality standards (NAAQS). The purpose of the PSD program is to implement the Federal Clean Air Act requirements for prevention of "significant" deterioration of air quality. These insure that the permitting of proposed new industrial facilities and the associated economic growth will occur in a manner consistent with the preservation of clean air resources. The program provides for special emphasis on implementation of BACT, protecting of scenic areas such as national parks, and informed public participation. Significant emission rate refers to net emissions increase or the potential of a source to emit pollutants, a significant emission rate equal or greater than the rates listed in Title 40 CFR 51. An Ambient air quality analysis must be carried out for each regulated pollutant and must demonstrate that the source will not cause nor contribute to a violation of any applicable NAAQS.

As industries age and expand, they acquire new emissions units for pollutants and modify old ones. Federal law calls for a BACT analysis of each such new or altered unit of "major stationary sources". Major stationary source, as defined in Title 40 CFR Part 51, Subpart 1, review of new sources and modification — any stationary source that (1) emits, or has the potential to emit (PTE), 250 tons per year (tpy) or more of any pollutant under the CAA; or (2) emits, or has the PTE, 100 tpy or more of a regulated pollutant within one of the 28 sources categories (Title 40 CFR).

The USEPA has consistently interpreted the statutory and regulatory BACT definitions as containing two core requirements that the agency believes must be met by any BACT determination. First, the BACT analysis must *consider* the most stringent available technologies (i.e., those which provide the "maximum degree of emissions reduction"). Second, any decision to require a lesser degree of emissions reduction must be justified by an objective analysis of "energy, environmental, and economic impacts."

BACT must be at least as stringent as any New Source Performance Standard (NSPS) applicable to the emissions source. With this set of data as a baseline, a BACT analysis often proceeds by way of the so-called top-down approach recommended by the EPA. The first step is to determine for the emission unit in question the most stringent control available for a similar or identical source or source category. If it can be shown that this level of control is technically infeasible for the unit in question, the next most stringent level of control is determined and similarly evaluated. This process continues until the BACT level under consideration cannot be eliminated by any substantial or unique technical or environmental concerns. The remaining technologies are evaluated on the basis of operational and economic effectiveness (New Source Review Workshop Manual: Prevention of Significant Deterioration and Nonattainment Area Permitting USEPA, 1990).

The emission units of a gas turbine power plant for which a BACT analysis is required include the combustion turbines, duct burners, the auxiliary boiler, and the cooling towers. Due to their status as emergency/backup units and/or very limited run time, the emergency diesel generator and the diesel fire water pump are not included in the BACT analysis. In the top-down BACT approach, analysts must look not only at the most stringent emission control technology previously approved but also must evaluate all demonstrated and potentially applicable technologies, including innovative controls, lower polluting processes, etc. For a gas turbine power plant, these technologies and emissions data can be identified through a review of sources made available by the EPA. Foremost among these is RBLC, an umbrella term standing for RACT/BACT/LAER clearinghouse, where RACT is reasonably available control technology and LAER is lowest achievable emission rates. Other sources are EPA's New Source Review (NSR) and Clean Air Technology Center (CTC) websites, the EPA Technology Transfer Network, STAPPA/ALAPCO (the State and Territorial Air Pollution Program Administrators and the Association of Local Air Pollution Control Officials), and Clean Air World.

Table 5.1 presents the technologies we shall consider in the sections that follow and their approximate control efficiencies.

## 5.2.1  $NO_x$ BACT Review

### 5.2.1.1  *Combustion Turbines and Duct Burners*  $NO_x$ is produced through two mechanisms: high temperature processes, which create thermal $NO_x$ (products of the reaction of nitrogen and oxygen gases in the air) and combustion of nitrogen-containing materials, which produces fuel $NO_x$. Table 5.2 lists the technologies that were identified for controlling $NO_x$ emissions from gas turbines and their effective emission levels.

### 5.2.1.2  *SCONO$_x$*  SCONO$_x$ is an emerging proprietary catalytic and absorption technology that has shown some promise for turbine applications. Unlike selective catalytic reductions, which requires ammonia injection, this system does not require

**TABLE 5.1. Technologies and Their Approximate Control Efficiencies**

| Pollutant | Technology | Potential Control Efficiency, % |
|---|---|---|
| $NO_x$ | $SCONO_x$ | 70–95 |
| | **Selective Catalytic Reduction (SCR)** | 50–95 |
| | Dry Low $NO_x$ Combustors | 40–60 |
| | **Selective Non-Catalytic Reduction (SNCR)** | 40–60 |
| | **Water/steam Injection** | 30–50 |
| | **Good Combustion Practices** | Base case |
| CO | Catalytic Oxidation | 60–80 |
| | **Good Combustion Practices** | Base case |
| $PM_{10}$ | **Good Combustion Practices** | 10–30 |
| | Fuel Specification: Clean-Burning Fuels | Base case |
| VOC | Catalytic Oxidation | 60–80 |
| | **Good Combustion Practices** | Base case |

(Source: USEPA, STAPPA/ALAPCO Clean Air World).

ammonia as a reagent; its parallel catalyst beds are alternately taken off line through means of mechanical dampers for regeneration.

Despite its advantages, however, the process $SCONO_x$ catalyst is subject to the same fouling or masking degradation that is experienced by any catalyst operating in a turbine exhaust stream. There is also a small energy loss from the performance loss due to the pressure drop across the catalyst. Process details are provided in Chapter 7 in connection with the case study of $NO_x$ emission from gas turbine engines of General Electric.

**TABLE 5.2. $NO_x$ Control Technologies and Effective Emission Levels**

| Technology | Typical Control Range (Percent Removal) | Typical Emission Level |
|---|---|---|
| $SCONO_x$ | 90–95 | 2–2.5 ppm |
| $XONON_{TM}$ Flameless Combustion | 80–90 | 3–5 ppm |
| Selective Catalytic Reduction (SCR) with low-$NO_x$ combustor or SCR with water injection | 50–95 | 2–6 ppm |
| SCR with water/steam injection or advanced low-$NO_x$ combustor | 50–95 | 6–9 ppm |
| Dry Low-$NO_x$ combustor and/or aggressive water injection | 30–70 | 9–25 ppm |
| Water/steam injection or low-$NO_x$ burners | 30–70 | 25–35 ppm |

*Sources*: USEPA, STAPPA/ALAPCO Clean Air World; Catalytica Energy Systems, Mountain View, CA, USA June 2004; "Best Available Control Technology Guidelines", South Coast Air Quality Management District, August 17, 2000 (http://www.arb.ca.gov/bact/bact.htm).

The vendor of SCONO$_x$ guarantees performance to all owners and operators of natural gas-fired combustion turbines, regardless of size or gas turbine supplier. The system is designed to reduce both CO and NO$_x$ emissions from natural gas-fired power plants to levels below ambient concentrations. Indeed, the EPA considers SCONO$_x$ a technically feasible and commercially available air pollution control technology and expects its emission levels for criteria pollutants such as NO$_x$, CO and VOC to be comparable or superior to previously applied technologies for large combined cycle turbine applications.

SCONO$_x$ has been demonstrated successfully on smaller power plants, including a 32 MW combined-cycle General Electric LM2500 gas turbine in Los Angeles, as discussed in Chapter 7. This facility uses water injection in conjunction with SCONO$_x$ to achieve a NO$_x$ emissions rate of 0.75 ppm on a 15-minute rolling average. The SCONO$_x$ technology has also been successfully demonstrated on a 5 MW Solar Turbine Model Taurus 50 at the Genetics Institute in Andover, Massachusetts. The system is reducing NO$_x$ down to 0.5 ppm NO$_x$, on a one-hour rolling average. The permit for the power plant was originally issued for 2.5 ppm NO$_x$.

The manufacturer guarantees CO emissions of 1 ppm and NO$_x$ emissions of 2 ppm. According to one set of figures, when NO$_x$ is reduced from 12.18 ppm (gas turbine with duct burner firing) to 2 ppm, the cost effectiveness is \$13,627 per ton of NO$_x$ removed (Catalytica Energy Systems Inc., 2004).

### 5.2.1.3 *XONON*

Several companies are reported to be working on a second technology for the control of NO$_x$. Introduced commercially by Catalytica Combustion Systems, it is being marketed under the name XONON. This technology replaces traditional flame combustion with flameless catalytic combustion. NO$_x$ control is accomplished through the combustion process using a catalyst to limit the temperature in the combustor below the temperature where NO$_x$ is formed. The XONON has been demonstrated to achieve near-zero emissions. The XONON combustion system consists of four sections: (1) the preburner, for start-up, acceleration of the turbine engine, and adjusting catalyst inlet temperature if needed; (2) the fuel injection and fuel-air mixing system, which achieves a uniform fuel-air mixture to the catalyst; (3) the flameless catalyst module, where a portion of the fuel is combusted flamelessly; and (4) the burnout zone, where the remainder of the fuel is combusted.

The single field installation of the XONON technology at a municipal power company, is being used to perform engineering studies of the technology at Silicon Valley Power, in Santa Clara, California. NO$_x$ emissions are well below 2.5 ppm on the 1.5 MW Kawasaki M1A-13A gas turbine. Catalytica has a collaborative commercialization agreement with General Electric Power Systems, committing to the development of XONON. In conjunction with General Electric Power systems, the XONON system was specified for use with the GE 7FA turbines at the proposed 750 MW natural gas-fired Pastoria Energy Facility, near Bakersfield, California. The project entered commercial operations in 2003. Because the NO$_x$ emissions limitations of 2.5 ppm have been demonstrated in practice by a commercial facility, this technology is considered commercially available at this time (Catalytic Energy Systems Inc., 2004).

### 5.2.1.4  *Selective Catalytic Reduction*   SCR systems selectively reduce $NO_x$ by injecting ammonia ($NH_3$) into the exhaust gas stream upstream of a catalyst. $NO_x$, ammonia, and oxygen react on the surface to form molecular nitrogen ($N_2$) and water. The overall chemical reaction can be expressed as:

$$4NO + 4NH_3 + O_2 \rightarrow 4N_2 + 6H_2O$$

Parallel plates or honeycomb structures, permeated with the catalyst, are installed in the form of rectangular modules, downstream of the gas turbine in simple-cycle configurations, and into the heat recovery steam generator (HRSG) portion of the gas turbine downstream of the superheater in combined-cycle and cogeneration configurations.

The turbine exhaust gas must contain a minimum amount of oxygen and be within a somewhat narrow temperature range in order for the selective catalytic reduction system to operate properly. The temperature range is dictated by the catalyst: if it is too low, the reaction efficiency drops and increased amounts of $NO_x$ and ammonia are released from the stack; it becomes too high, the catalyst may begin to decompose. Turbine exhaust gas is generally too hot to be passed through the catalyst, so is cooled by the HRSG, which extracts energy from the hot turbine exhaust gases and creates steam for use in other industrial processes or to turn a steam turbine. In simple-cycle power plants where no heat recovery is accomplished, high temperature catalysts (e.g., zeolite) are an option. Selective catalytic reduction can typically achieve $NO_x$ emission reductions in the range of about 80 to 95 percent.

SCR is the most widely applied post-combustion control technology in turbine applications and is currently accepted as LAER for new facilities located in ozone non-attainment regions. It can reduce $NO_x$ emissions to as low as 4.5 ppmvd for standard combustion turbines without duct burner firing, and as low as 2.2.5 ppmvd when combined with lean-premix combustion (again without duct burner firing). SCR uses ammonia as a reducing agent in controlling $NO_x$ emissions from gas turbines. The portion of the unreacted ammonia passing through the catalyst and emitted from the stack is called ammonia slip. The ammonia is injected into the exhaust gases prior to passage through the catalyst bed. There is also a potential for increased particulate emissions from formation of ammonia salts. SCR may also result in the generation of spent vanadium pentoxide catalyst, which is classified as a hazardous waste. In addition, there is an energy loss from the performance loss due to the pressure drop across the SCR catalyst.

Gas turbines using SCR typically have been limited to 10 ppmvd ammonia slip (emissions of ammonia that has not reacted with nitrogen) at 15 percent oxygen. However, levels as low as 2 ppmvd at 15 percent oxygen have been proposed and guaranteed by control equipment vendors. In addition, Massachusetts and Rhode Island have established ammonia slip LAER levels of 2 ppmvd. Massachusetts has permitted at least two large gas turbine power plants using SCR reduction with 2 ppmvd ammonia slip limits. California recommended that the establishment of ammonia slip levels below 5 ppmvd at 15% oxygen on the basis of guarantees from control equipment vendors of single-digit levels for ammonia slip.

Data supplied by the vendor show that when $NO_x$ is reduced from 15 ppm (gas turbine with duct burner firing) to 3.5 ppm, the cost effectiveness is \$9,473 per ton of $NO_x$ removed.

### 5.2.1.5  Lean-Premix Technology or Dry-Low $NO_X$

Processes that use air as a diluent to reduce combustion flame temperatures achieved reduce $NO_x$ by premixing the fuel and air before they enter the combustor. This type of process is called lean-premix combustion, and goes by a variety of names, including the Dry-Low $NO_x$ (DLN) process of General Electric, the Dry-Low Emissions (DLE) process of Rolls-Royce, and the $SoLoNO_x$ process of Solar Turbines.

Lean premixed designs reduce combustion temperatures, thereby reducing thermal $NO_x$. In a conventional turbine combustor, the air and fuel are introduced at an approximately stoichiometric ratio and air/fuel mixing occurs at the flame front where diffusion of fuel and air reaches the combustible limit. A lean premixed combustor design premixes the fuel and air prior to combustion. Premixing results in a homogeneous air/fuel mixture, which minimizes localized fuel-rich pockets that produce elevated combustion temperatures and higher $NO_x$ emissions. A lean air-to-fuel ratio approaching the lean flammability limit is maintained, and the excess air serves as a heat sink to lower combustion temperatures, which lowers thermal $NO_x$ formation. A pilot flame is used to maintain combustion stability in this fuel-lean environment. Lean-premix combustors can achieve emissions of about 9 ppmvd $NO_x$ at 15 percent oxygen (approximately 94 percent control).

To achieve low $NO_x$ emission levels, the mixture of fuel and air introduced into the combustor must be maintained near the lean flammability limit of the mixture. Lean-premix combustors are designed to maintain this air/fuel ratio at rated load. At reduced load conditions, the fuel input requirement decreases. To avoid combustion instability and excessive CO emissions that occur as the air/fuel ratio reaches the lean flammability limit, lean-premix combustors switch to diffusion combustion mode at reduced load conditions. This switch to diffusion mode means that the $NO_x$ emissions in this mode are essentially uncontrolled.

Lean-premix technology is the most widely applied pre-combustion control technology in natural gas turbine applications. It has been demonstrated to achieve emissions of approximately 9 ppmvd $NO_x$ at 15 percent oxygen (Catalytica Energy Systems Inc., 2004).

### 5.2.1.6  Steam/Water Injection

In steam/water injection, the technology commonly chosen to reduce the $NO_x$ emissions in natural gas turbine, higher combustion temperatures are used to achieve greater thermodynamic efficiency. In turn, more work is generated by the gas turbine at a lower cost. However, the higher the gas turbine inlet temperature, the more $NO_x$ that is produced. Diluent injection, or wet controls, can be used to reduce $NO_x$ emissions from gas turbines. Diluent injection involves the injection of a small amount of water or steam via a nozzle into the immediate vicinity of the combustor burner flame. $NO_x$ emissions are reduced by instantaneous cooling of combustion temperatures from the injection of water or steam into the combustion zone. The effect of the water or steam injection is to increase the thermal mass by mass dilution and thereby reduce the peak flame temperature in the $NO_x$ forming regions of the combustor. Water injection typically results in a $NO_x$ reduction efficiency of about 70 percent, with emissions below 42 ppmvd $NO_x$ at 15 percent oxygen. Steam injection has generally been more successful in reducing $NO_x$ emissions and can achieve emissions less than 25 ppmvd $NO_x$ at 15 percent oxygen (approximately 82 percent control).

### 5.2.1.7  Summary of NO<sub>X</sub> BACT for Turbines and Duct Burners

*5.2.1.7  Summary of NO$_X$ BACT for Turbines and Duct Burners*  Table 5.3 provides information on the emissions, control effectiveness, economics, and environmental impacts of measures for the control of $NO_x$ discussed in Sections 5.2.1.2, 5.2.1.4, and 5.2.1.5. The analysis was performed on a unit (turbine and duct burner) basis.

$SCONO_x$ provides the highest level of $NO_x$ reduction. However, this very new technology has yet to prove itself for long-term commercial operation on large scale combined-cycle plants. It is the relatively high cost per emission reduction of this control technology ($13,627 per ton of $NO_x$ removed) that rules out $SCONO_x$ as a control option. The next most effective control technology for $NO_x$ is a combination of DLN combustors and SCR. The adverse environmental impact of SCR is primarily from the emissions of ammonia which is on the EPA's list of extremely Hazardous Substances. Although, these adverse impacts of the SCR process can be minimized with proper system design and operation, it is ruled out as a control operation because the cost is prohibitive ($10,191 per ton of $NO_x$ removed).

The next most effective control technology is DLN combustors. At reduced loads, combustion instability requires that a switch to diffusion combustion mode, in which $NO_x$ emissions are essentially uncontrolled. However, this can be minimized with proper system design and operation.

## 5.2.2  CO BACT Review: Combustion Turbines and Duct Burners

$SCONO_x$ reduces CO emissions by oxidizing CO to $CO_2$. When CO is reduced from 11.51 ppm (gas turbine with duct burner firing) to 1 ppm, the cost effectiveness is $21,706 per ton of CO removed.

**TABLE 5.3. NO$_x$ Emissions, Control Effectiveness, Economics, and Environmental Impacts**

| Technology Effectiveness | NO$_x$ Emissions Reduction (TPY) | Capital Cost (rep) | Annualized Cost ($) | Cost Effectiveness ($/tons) | Adverse Environmental Impacts |
|---|---|---|---|---|---|
| SCONO$_x$ (2 ppm) | 399.5 | 14,922,733 | 5,444,139 | 13,627 | Yes |
| DLN + SCR (3.5 ppm) | 366.5 | 3,476,578 | 3,471,362 | 9,473 | Yes |

In the United States, combustion turbines and duct burners are subject to the federal New Source Performance Standard (NSPS), but the regulations provide no CO emission limits. The following sections assess the control strategies that are potentially feasible for decreasing CO emissions from the facility.

### 5.2.2.1  *Catalytic Oxidation*

The rate of formation of CO during natural gas combustion depends primarily on the efficiency of combustion. The formation of CO occurs in small, localized areas around the burner where oxygen levels cannot support the complete oxidation of carbon to $CO_2$. Efficient burners can minimize the formation of CO by providing excess oxygen or by mixing the fuel thoroughly with air. CO emissions resulting from natural gas combustion can be decreased via catalytic oxidation. The oxidation is carried out by the well-known overall reaction:

$$CO + \tfrac{1}{2}O_2 \rightarrow CO_2$$

Several noble metal-enriched catalysts at high temperatures promote this reaction. Under ideal operating conditions, this technology can achieve an 80 percent reduction in CO emissions. Prior to entering the catalyst bed where the oxidation reaction occurs, the exhaust gas must be pre-heated to about 400–800°F. Below this temperature range, the reaction rate drops sharply, and effective oxidation of CO is no longer feasible. Moreover, there is an energy loss because of the reduction in performance due to the pressure drop across the CO catalyst.

Sulfur and other compounds in the exhaust may foul the catalyst, leading to decreased activity. Catalyst fouling occurs slowly under normal operating conditions and may be accelerated by even moderate sulfur concentrations in the exhaust gas. The catalyst can be chemically washed to restore its effectiveness, but eventually, irreversible degradation occurs. Catalyst replacement is usually necessary every five to ten years depending on type and operating conditions.

An economic analysis for the catalytic oxidation of CO emissions based on vendor information estimates the cost at $5,084 per ton of CO removed. This cost level is considered to be economically infeasible for BACT. In addition to cost, catalytic oxidation would lead to increased downtime for catalyst washing and would present hazardous waste concerns during catalyst disposal. Due to the high cost and concerns with downtime and hazardous material disposal, catalytic oxidation is not selected as BACT for control of CO emissions from the turbines and duct burners.

### 5.2.2.2  *Good Combustion Practices*

Clearinghouse (RBLC) data show that the majority of BACT determinations for CO relied on the use of good combustion practices. Since add-on controls for CO were shown to be economically infeasible, the proposed BACT for CO emissions is the use of good combustion practices.

### 5.2.2.3  *Combustion Control*

Because CO results from the incomplete combustion of fuel, combustion control is an inherent design feature of combustion turbines and duct burner. Control is accomplished by providing adequate fuel residence time and high temperature in the combustion zone to ensure complete combustion. These control methods, however, also result in increased emissions of $NO_x$. Conversely, a low $NO_x$ emission rate achieved through flame temperature can result in higher levels of CO emissions. Thus, a compromise is needed to set the flame temperature at the level that will achieve the lowest $NO_x$ emission rate possible while keeping CO emissions rates at acceptable levels.

### 5.2.2.4   *Summary of CO BACT for Turbines and Duct Burners*   Table 5.4 provides information on the emissions, control effectiveness, economics, and environmental impacts associated with control of CO. The analysis was performed on a unit (turbine and duct burner) basis.

SCONO$_x$ provides a higher level of CO reduction than a combination of DLN combustors and CO oxidation. The adverse impacts include an energy loss from the performance loss due to the pressure drop across the CO catalyst and emissions of sulfates condense as additional $PM_{10}$ or $PM_{2.5}$. However, at a cost of \$5,084 per ton of CO removed, and the second option is more than four times as cost-effective as a control option.

Combustion control is selected as BACT for CO control with the limit of 9 ppm at 15 percent $O_2$ (annual average) without duct burners firing and 11.5 ppm at 15 percent $O_2$ with duct burner firing (annual average).

### 5.2.3   PM$_{10}$ Control Technologies

There are no specific particulate emission limits established in the NSPS for combustion turbines, but a particulate emission limit of 0.03 lb per million Btu has been set for duct burners.

Properly tuned burners firing natural gas inherently emit low levels of particulate matter (<0.007 lb per million Btu). The RBLC database indicates that good combustion practices are widely accepted as BACT for turbines and duct burners firing natural gas. Thus, it is reasonable to propose good combustion practices as BACT for PM$_{10}$ emissions. Proposed PM$_{10}$ BACT limits for the turbines and duct burners combined are 0.0136 lb per million Btu.

**TABLE 5.4. CO Emissions, Control Effectiveness, Economics, and Environmental Impacts**

| Technology Effectiveness | CO Emissions Reduction (TPY) | Capital Cost ($) | Annualized Cost ($) | Cost Effectiveness ($/tons) | Adverse Environmental Impacts |
|---|---|---|---|---|---|
| SCONO$_x$ (1 ppm) | 251 | 14,922,733 | 5,444,139 | 21,706 | Yes |
| DLN + CO Catalyst (3.4 ppm) | 194 | 1,302,514 | 2,208,461 | 5,084 | Yes |

## 5.2.4 VOC Control Technologies

The NSPS regulations for combustion turbines and duct burners provide no VOC emission limits. The following sections assess the control strategies that are potentially feasible for decreasing VOC emissions from the facility.

### 5.2.4.1 *Catalytic Oxidation*
The formation of VOC in combustion units depends primarily on the efficiency of combustion. Inefficient combustion leads to the formation of aldehydes, aromatic carbon compounds, and various other organic compounds by several mechanisms. Catalytic oxidation decreases VOC emissions by facilitating the complete combustion of organic compounds to water and carbon dioxide. Prior to entering the catalyst bed where the oxidation reaction occurs, the exhaust gas must be pre-heated to about $400-800°F$.

The RBLC database shows few instances of catalytic oxidation being selected as BACT for VOC at any gas-fired turbine power plant nationwide. An economic analysis for the catalytic oxidation of VOC emissions based on vendor information estimates the cost at \$30,811 per ton of VOC removed. This cost level is economically infeasible for VOC removal. Thus, catalytic oxidation would not be selected as BACT for VOC emissions from turbines and duct burners proposed in the present time frame.

### 5.2.4.2 *Good Combustion Practices*
All the RBLC database BACT determinations for VOC outside California and New York show the use of combustion control or good combustion practices. Thus in most areas good combustion practices are acceptable as BACT for VOC, with limits of 7.25 ppmvd at 15 percent $O_2$, as methane.

### 5.2.4.3 *$SO_2$ and $H_2SO_4$ BACT Review: Gas Turbines and Duct Burners*
Control techniques available to reduce $SO_2$ and $H_2SO_4$ emissions include flue gas desulfurization (FGD) systems and the use of low sulfur fuels. Although FGD systems are common in boiler application, the RBLC database shows no known FGD systems on combustion turbines. Thus, the use of an FGD system is not warranted and an FGD system should be rejected as a BACT control alternative.

Another available technique is the use of low sulfur fuels. There are no adverse environmental or energy impacts associated with the properly specified use of pipeline natural gas with low sulfur content, and this control alternative should be acceptable as BACT.

## 5.3 CASE STUDY: THE NORTH CAROLINA CLEAN SMOKESTACKS PLAN

From the mountains to the sea, North Carolina has long been recognized for its natural beauty and high quality of life, a reputation that makes it difficult to accept that the state's air quality is now on the decline. The North Carolina Clean

Smokestacks Plan documents the serious public health and environmental problems resulting from coal-fired power plants and other sources of air pollution. It then offers a policy framework for executive and legislative action to reduce emissions from coal-fired power plants and clean up North Carolina's air. This plan illustrates how a timely and effective air pollution prevention measure can be a win–win situation in terms of future economic and environmental benefits.

The sources of air pollution of greatest concern in North Carolina are coal-fired power plants, followed by mobile sources. Power plants emit 82 percent of all sulfur dioxide air emissions, 45 percent of nitrogen oxides, and 65 percent of mercury. Automobiles and other mobile sources emit 48 percent of the nitrogen oxides.

### 5.3.1  Consequences of Dirty Air

Emissions of nitrogen oxides ($NO_x$), sulfur dioxide ($SO_2$), and mercury, as well as carbon dioxide ($CO_2$) undermine public health, the environment, and the economy of North Carolina in the following ways:

○ **Public Health.** North Carolina's air quality consistently ranks among the least healthy in the U.S.; for example, in 1999, the state had the fifth highest number of unhealthy air days. The American Lung Association found the Charlotte metropolitan area to have the eighth smoggiest air in the nation, and the Raleigh–Durham area ranked seventeenth. It is estimated that $NO_x$ pollution from power plants triggers more than 200,000 asthma attacks across the state each year and $SO_2$ pollution causes more than 1,800 premature deaths, ranking North Carolina as the fourth worst state in the nation for power-plant related deaths. Airborne mercury falls into the state's rivers and estuaries, contaminating freshwater and saltwater fish populations. Mercury compounds bioaccumulate in the food chain, making king mackerel, bass (in some areas), and bowfin unfit for human consumption by children and women of childbearing age.

○ **Visibility.** Visibility in the southeast has declined by 75 percent from natural levels. One should be able to see out 93 miles on an average day in the Smoky Mountains, but now air pollution has reduced this to an average of 22 miles. On any given summer day in the mountains, there is a good chance that views may be entirely obscured by pollution.

○ **Ecosystems.** Air pollution causes acid rain and nitrogen deposition, which make vegetation more susceptible to disease and pests, contributing to stunted growth and significant declines in populations of dogwood, spruce, fir, beech, and other tree species. Rainfall in the Great Smoky Mountains National Park is five to 10 times more acidic than normal rainfall. In the east, airborne nitrogen adds to nutrient pollution in sensitive coastal watersheds, contributing to algal blooms and fish kills.

○ **Economic Consequences.** North Carolina's dirty air threatens the vitality of the state's economy. Dirty air is estimated to cost the state over $3.0 billion annually in morbidity and mortality costs. Ozone's effect on plants is "significant stress factor in agricultural production". Air pollution reduces crop yields, which causes North Carolina farmers to lose more than $175 million each year. Frequent smog alerts atop the Great Smoky Mountains discourage hiking and other outdoor activities. The impairment of visibility undermines North Carolina's $12 billion tourism industry. The loss in economic activity in the area around the Great Smoky Mountains National Park is estimated to cost more than $200 million each year.

○ **Global Climate Change.** Carbon dioxide pollution from power plants and other sources is one of the primary pollutants that contributes to global warming, which over the next thirty years is expected to raise sea levels off the North Carolina coast by 7.5 inches, a rise that could completely inundate or change the coastline at Wrightsville Beach, Topsail Beach, and the Outer Banks.

○ **Quality of Life.** A loss of air quality clearly diminishes the quality of life for all North Carolinians, but putting a price tag on it is difficult. Nonetheless, some indirect measures are possible. For example, quality of life has long been a major factor in persuading new businesses to locate in the state, which implies that as air quality declines, it will be harder for the state to attract new investment and jobs.

The statistics just given are not comprehensive: the economic effects of acid rain, eutrophication, mercury exposure, and other environmental problems are not included. Thus, the actual costs of dirty air are much higher than those reflected by the dollar amounts cited. Nor can any economic calculation quantify the pain experienced by a child who suffers from asthma or a grandparent with cardiopulmonary disease whose death is hastened by exposure to high levels of fine particulate matter.

## 5.3.2 Technologies and Clean Energy

Technologies to control both $NO_x$ and $SO_2$ emissions from North Carolina's old, coal-fired power plants are both affordable and available. For example, selective catalytic reduction, discussed in Section 5.2, can reduce $NO_x$ emissions by more than 80 percent from uncontrolled levels. In addition, though not deemed useful in gas turbine application, flue gas desulfurization (scrubbers) can reduce sulfur dioxide emissions by more than 90 percent from uncontrolled levels.

Clean energy sources offer another attraction alternative. Renewable energy sources such as wind power essentially eliminate all air pollution from power generation. Efficiency, conservation, and switching to cleaner fuels also substantially reduce air pollution.

As of 2002, only three of the fourteen major power plants in the state had committed to installing SCR systems, and no facilities use scrubbers. Installation of scrubbers combined with SCR systems could reduce mercury emissions by up to 95 percent from uncontrolled levels.

Carbon injection and other cutting-edge technologies also are being used to control mercury from other combustion sources, such as municipal waste and medical waste incinerators. These technologies show promise in substantially reducing mercury emissions from power plants, although the widespread implementation of such technologies at coal-fired power plants may require further work. Switching fuels to natural gas or using other cleaner sources of energy such as wind power achieve even higher $NO_x$, $SO_2$, and mercury reduction levels and reduce carbon dioxide as well.

### 5.3.3   Costs and Benefits of Reducing Emissions

Table 5.5 outlines some specific reduction targets that are both feasible and vital to public health and environmental protection in North Carolina and elsewhere in the United States. North Carolina, however, has yet to make the dramatic strides that would result in cleaner air statewide. One reason for this is that the state's coal-fired power plants are in the so-called grandfathered power sector, which means that they are largely uncontrolled. Thus there is a need to phase in, over a reasonable length of time (say until 2007), new laws and regulations that impose stiffer requirements on grandfathered facilities. This would enable utility companies to plan and retrofit their plants as cost-effectively as possible.

It has been found that if the costs for reducing emissions of $NO_x$, $SO_2$, and mercury to the levels recommended by the North Carolina clean Smokestacks Plan were passed on to consumers, the average household's power bill would go up by $4.09 a month. (Benefits and costs of reducing $CO_2$ were not included in this analysis because climate change issues are slated for further treatment in a later report.

Overall, the benefits of improved air quality are projected to far outweigh the costs. The developers of the Clean Smokestacks Plan have calculated that the cost savings resulting from reducing coal-fired power plant emissions by the amounts called for in the plan would be $3.5 billion annually, compared with a cost of compliance to the utilities of $450 million.

---

**TABLE 5.5.  Reduction Targets for 2007, Except as Noted, from the Power Plant Sector**

---

○ **Summertime Nitrogen Oxides.** Cap summertime $NO_x$ emissions at 23,000 tons, an 80 percent reduction over 1998 levels.

○ **Year-Round Nitrogen Oxides.** Cap year-round $NO_x$ emissions at 50,000 tons, an 80 percent reduction over 1998 levels.

○ **Sulfur Dioxide.** Cap emissions of $SO_2$ at 85,000 tons annually, an 82 percent reduction over 1998 levels.

○ **Mercury.** Reduce year-round emissions of mercury by 90 percent from 1998 levels.

○ **Carbon Dioxide.** Cap new $CO_2$ emissions at 1990 levels, as called for by the United Nations Framework Convention on Climate Change, which has been ratified by the United States.

---

## 5.4  SUSTAINABLE ECONOMY AND THE EARTH

Evidence that the economy is in conflict with earth's natural systems can be seen in the daily news reports of collapsing fisheries, shrinking forests, eroding soils, deteriorating rangelands, expanding deserts, rising carbon dioxide levels, falling water tables, rising temperatures, more destructive storms, melting glaciers, rising sea level, dying coral reefs, and disappearing species. These trends, which mark an increasingly stressed relationship between the economy and earth's ecosystem, are taking a growing economic toll. At some point, this could overwhelm the world-wide forces of progress, leading to economic decline. The challenge for our generation is to reverse these trends before environmental deterioration leads to long-term economic decline, as it did for so many earlier civilizations.

### 5.4.1  What is a Sustainable Economy?

An environmentally sustainable economy — an eco-economy — requires that the principles of ecology be observed in establishing the framework for formulating economic policy and that economists understand that all economic activity, indeed all life, depends on the earth's ecosystems-the complex of individual species living together, interacting with each other and with their physical habitat. Millions of plant and animal species exist in an intricate balance, woven together along the food chain by nutrient cycles, the hydrological cycle, and global climate systems. Economists know how to translate goals into policy. Economists and ecologists, along with engineers, scientists and policy makers, working together are being challenged to design and build an eco-economy, one that can sustain progress (L. Brown, 2001; R. Brown, 2001; Western Governors' Association December 3, 2003).

Through industrial ecology, eco-industrial parks, eco-efficiency, zero-discharge manufacturing, and engineering new products from agricultural materials (biorenewable resources), an eco-economy that facilitates sustainable economic progress offers the potential for the achievements, over the long-term, of a sustainable economy and a healthy planet. Humankind has gone outside the biotic environment for the majority of its material needs only recently. Plant-based resources were the predominant source of energy, organic chemicals, and fibres in the West as recently as 150 years ago, and they continue to play important roles in many developing countries.

The transition to non-biological (or non-renewable) sources of energy and materials was relatively swift and recent in the history of the world. Biorenewable resources are by definition sustainable natural resources; that is, they are self-renewing at a rate that ensures their availability for use by future generations. Biorenewable resources can be converted into either bioenergy or biobased engineering products. Bioenergy, also known as biomass energy, is the result of the conversion of the chemical energy of a biorenewable resource into heat and stationary power. Biobased products include transportation fuels, chemicals, and natural fibers derived from biorenewable resources. Transportation fuels are generally liquid fuels, such as ethanol or biodiesel, but compressed hydrogen and methane have also been proposed and evaluated for use in vehicular propulsion.

Chemicals may include pharmaceuticals, nutraceuticals, and other fine chemicals, but the emphasis here is on commodity chemicals, which produce high demands for biorenewable resources. An example is polylactic acid, which is derived from the fermentation of sugars hydrolyzed from cornstarch and can be converted into biobased polymers used in a variety of consumer products such as carpets.

The cost of a biorenewable resource is related to the demand for the resource by a supply curve. Figure 5.3 is a generalized representation of a supply curve for the three kinds of biorenewable resources. The cost of biorenewable resources is highly variable and dependent on local conditions of supply and demand. This is particularly true for the costs involved in processing residue and wastes. Table 5.6 provides an estimate of the availability and cost of several kinds of residues and wastes along with a comparison of the costs of a few crops. The cultivation and harvesting of dedicated energy crops, on the other hand, is amenable to standardized cost estimating since information on "unit operation" such as planting, fertilizing, and harvesting, can be readily obtained from knowledgeable sources (R. Brown, 2001).

### 5.4.2   Costs of Manufacturing Various Biobased Products

Over the next decades, a much larger fraction of fuels, chemicals and materials will be produced from renewable plant materials. These biobased industrial products offer the potential for a much more sustainable economy based on

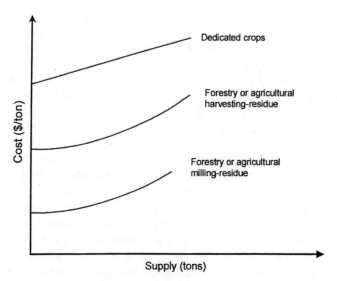

**Figure 5.3.** Example of supply curves to different kinds of biorenewable resources.

**TABLE 5.6.  Availability and Cost of Potential Feedstocks**

| Feedstocks | Production (106 tons/year) | Price (194 $/kg) |
|---|---|---|
| Corn | 191 | 0.09 |
| Potato | 17 | 0.16 |
| Sorghum | 16 | 0.09 |
| Beet molasses | 1 | 0.09 |
| Cane molasses | 1 | 0.03 |
| Sugar cane | 25 | 0.03 |
| Agricultural residues | | |
|   Low cost | 4 | 12.9 |
|   Mid cost | 36 | 38.8 |
|   High cost | 50 | 47.4 |
| Forest residue-logging | | |
|   Low cost | 3 | 12.9 |
|   Mid cost | 3 | 25.9 |
|   High cost | 3 | 43.1 |
| Forest residue mill | 3 | 17.2 |
| Municipal solid waste | | |
|   Mixed paper | 26 | 0–19 |
|   Packaging | 14 | 0–5.2 |
|   Urban wood | 3.5 | 12.9–25.9 |
|   Yard waste | 11 | 0–12.9 |

*Sources*: Crop data from K. Polman (1994); Waste and residue data from L. R. Lynd (1996).

environmentally-superior products. This Section briefly describes the associated costs of producing electricity, fuels and chemicals from various feedstocks using biorenewable resources.

### 5.4.2.1  *Electricity from Combustion of Biomass*

The capital and operating costs for steam power plants fired with biomass are relatively well known because of significant operating experience with these systems. The capital cost for a new plant ranges between $1,400 and $1,800 per kilowatt capacity. Accordingly, a 50-MW biomass plant based on direct-combustion would cost approximately $80 million. On the basis of a target price of $41.90/GJ for biomass, the cost of production for direct-fired biomass power is about $0.06/kWh (Environmental Law & Policy Center, 2001).

### 5.4.2.2  *Electricity from Gasification of Biomass*

The capital cost for a gasification plant, including fuel feeding and gas cleanup, is dependent on both the size and the operating pressure of the system. In the United States, an atmospheric-pressure gasifier producing 50 MW thermal energy would cost about $15 million. A gasification/gas-turbine power plant producing 50 MW of electricity would have total capital cost of between \$75 million and \$138 million (between \$1500 and \$2750/kW), the smaller number reflecting improved technical know-how after building at least ten plants. Electricity production costs would range from \$0.05 to \$0.09/kWh if fuel is available at an optimistic price range of \$1.00–\$1.50/GJ.

Capital costs for high temperature fuel cells suitable for integrated gasification/fuel cell power plants currently cost \$3000/kW. Molten carbonate fuel cells are expected to be \$1500/kW at the time of market entry, decreasing to about \$1000/kW for a commercially mature unit. The cost of electricity from a mature unit operating on natural gas is projected to be between \$0.049 and \$0.085 per kWh. More attractive economics result if less expensive fuel is available. The cost of electricity generated from landfill gas using mature fuel-cell technology is expected to be comparable to that for an internal combustion engine/electric genera-tor set, i.e., about \$0.05/kWh (Bridgewater, 1995; Hirschenhoofer et al., 1994; William and Larson, 1993).

### 5.4.2.3  *Biogas from Anaerobic Digestion*    Anaerobic digestion is commer-cially developed for the purpose of treating wastewater. Power production from anaerobic digestion is in its infancy. The methane generated is often fired in internal combustion engines to produce electricity in an effort to help offset costs of waste treatment, but there are no immediate prospects for it to replace natural gas. Capital costs for anaerobic digestion facilities processing more than 200 ton/day of volatile solids are estimated to be between \$44,000 and \$132,000 for each ton/day of volatile solids processed. Methane yields will be approximately 0.38 $m^3$/kg of volatile solids. Thus, a 200 ton/day anaerobic digestion plant could produce 28 million cubic meters of methane per year, representing almost 2900 GJ/day of chemical energy. Projected operating costs for producing methane from dedicated energy crops were in the range of \$5–\$6/GJ in 1986 dollars. In comparison, the cost of natural gas in the United States, which shows large seasonal and geographical variations, ranges between \$1.90 and \$4/GJ. In niche markets, where the feedstock is inexpensive and natural gas is not available, biogas can be a viable alternative energy source (Benson et al., 1986). In a similar biogas manufacturing using poultry waste-to-energy by catalytic steam gasification process is described in Chapter 10.

### 5.4.2.4  *Ethanol from Biomass*    The cost of producing ethanol from biomass varies tremendously depending on the feedstock employed, the size and manage-ment of the facility, and the market value of co-products generated as part of some conversion processes. Cost information for ethanol plants to be built in the United States is most reliable for those using cornstarch, the basis of the U.S. ethanol industry. A 5,000 barrel per day plant (about 265 million liters per day) built from the ground up will have a capital cost of about \$140 million in 1987 dollars, or \$0.53/L of annual capacity. Smaller facilities can have capital costs as high as \$0.79/L of annual capacity, and poorly designed facilities of any size may cost \$1.06/L of annual capacity. On the other hand, ethanol plants that are

built from existing facilities such as refineries or chemical plants, or ethanol plants integrated into a larger industrial facility, can have substantially lower capital costs, often in the range of $0.26–$0.40/L of annual capacity.

Low-end production costs are about $0.26/L. However, the volumetric heating value of ethanol is only 66 percent that of gasoline. This production cost, therefore, is equivalent to gasoline selling for $0.39/L before tax, transportation, or profit. In contrast, refinery price for gasoline in 1990 dollars was $0.20/L. Currently, the economics of fermentation are such that the commercial viability of ethanol is entirely dependent on government incentives in the form of tax credit, currently $0.16 for each liter of ethanol used for fuel blending. Also, a strong market for fermentation by-products is key factor in the economic viability of ethanol-from-corn.

Technology to convert lignocellulose to sugar is expected to reduce the cost of fuel ethanol, although dedicated economic information is not currently available. Capital cost for a 5,000 barrel per day plant to produce ethanol from lignocellulose using simultaneous saccharification and fermentation (SSF) is estimated to be $175 million (1994 dollars). Assuming wood costs $42/dry ton, ethanol can be produced for about $0.31/L. Combining economies of scale with advances in processing technology are projected to decrease production costs to $40.31/L. However, some reports have suggested that ethanol from cellulose will have to cost as little as $0.08–$0.11/L to be competitive with the gasoline prices anticipated early in the twenty-first century (Lynd et al., 1996; National Advisory Panel, 1987).

### 5.4.2.5  *Methanol from Biomass-derived Syngas*

Capital investment for a 7,500 barrel per day plant to produce methanol from biomass would be about $280 million in 1991 dollars. The cost of methanol from $40/dry ton of wood is projected to be about $0.27/L. Since the volumetric heating value of methanol is only 49 percent that of gasoline, the production cost from this plant is equivalent to gasoline selling for $0.55/L. Methanol from natural gas can be produced at significantly lower cost, but this assumes much larger plant capacities to capture economies of scale. Such large plants are not feasible for widely dispersed biomass feedstocks. New methanol-synthesis technologies may be able to significantly reduce this price. The U.S. Department of Energy's methanol from biomass program has a goal of $0.15/L ($7.90/GJ) based on feedstock cost of $1.90/GJ (Klass, 1998).

### 5.4.2.6  *Bio-oil from Fast Pyrolysis*

Capital investment for a 5,000 barrel per day plant to produce pyrolysis liquids would be $63 million in 1987 dollars. Assuming biomass feedstock was available at $1.70/GJ, this size plant could produce pyrolysis liquids for $0.18/L, which has an energy value of $6.70/GJ (Elliott et al., 1990).

### 5.4.2.7  *Biodiesel from Vegetable Oils*

Capital costs for a biodiesel facility are relatively modest, costing about $250,000 for a 50 barrel per day (3.2 million liters per year). However, feedstock costs for production of biodiesel are higher than feedstocks for production of other kinds of fuel, ranging from $0.16–$0.26/L for waste fats to $0.53–$0.79/L for vegetable oils. Under the best scenarios, a biodiesel plant might produce fuel for $0.44/L. Diesel fuel produced from petroleum typically sells for less than $0.25/L (Gavett et al., 1993). The biodiesel manufacturing process, emission inventories, and costs are described in detail later (see Chapter 8).

**5.4.2.8 Succinic Acid** Succinic acid is used in producing food and pharmaceutical products, surfactants and detergents, biodegradable solvents and plastics, and ingredients to stimulate animal and plant growth. Although it is a common metabolite formed by plants, animals, and microorganisms, its current commercial production of 15,000 tons per year is from petroleum. However, the recently discovered rumen organism *Actinobacillus succinogenes* produces succinic acid with yields as high as 110 g/L, offering prospects for producing this chemical from biorenewable resources. In contrast to most other commercial fermentations, the process consumes $CO_2$ and, integrated with a process like ethanol fermentation, succinic acid production could contribute to reduction in greenhouse-gas emissions.

Optimum yields occur under pH conditions where succinate salt rather than free acid is produced. Thus recovery entails concentration of the salt, conversion back to free acid, and polishing of the acid to the desired purity. Downstream purification can account for 60–70 percent of the product cost.

**5.4.2.9 Lactic Acid** Lactic acid, a three-carbon molecule, is used in the production of *polylactide* (PLA) resin, a biodegradable polymer expected to compete with polyethelene and polystyrene in the synthetic fibers and plastic markets. Lactic acid is currently produced by milling corn, separating the starch, hydrolyzing the starch to glucose, and anaerobically fermenting the glucose to lactic acid with *Bacillus dextrolactius* or *Lactobacillus delbrueckii*. Esterification with ethanol produces ethyl lactate, which can be polymerized to polylactate resin.

## REFERENCES

Benson, P. H., Hayes, T. D., and Isaacson, R. (1986). Regional and Community Approaches to Methane from Biomass and Waste: An Industry Perspective. In Proceedings Energy from Biomass and Waste X, Orlando, Florida.

Best Available Control Technology Guidelines (2000). South Coast Air Quality Management District (August 17) (http://www.arb.ca.gov/bact/bact.htm).

Bridgewater, A. V. (1995). The technical and economic feasibility of biomass gasification for power generation. *Fuel*, **74**: 631–653.

Brown, L. R. (2001). *Eco-Economy: Building an Economy for the Earth*. W. W. Norton & Company, New York, NY.

Brown, R. C. (2001). *Biorenewable Resources: Engineering New Products from Agriculture*. Iowa State Press.

Catalytica Energy Systems Inc. (2004). Mountain View, CA. (http://www.cataly ticaenergy.com/xonon/technology.html).

Elliott, D. C., Osman, A., Borje Gevett, S., Beckman, D., Solanantausta, Y., Hornell, C., and Kjellstrom, B. (1990). In Energy from Biomass and Wastes XIII, D. L. Klass (ed.), Chicago IL: Institute of Gas Technology.

Environmental Law & Policy Center (2001). Repowering the Midwest: The Clean Energy Development Plan for the Heartland, (available on the World Wide Web: www.repower midwest.org/).

Gavett, E. E., van Dyne, D., and Blasé, M. (1993). In Energy from Biomass and Waste XIII.: D. L. Klass (ed.), Chicago IL: Institute of Gas Technology.

Hirschenhoofer, J. H., Stauffer, D. B., and Engleman, R. R. (1994). *Fuel Cells: A Handbook.* Department of Energy Technical Report DOE/METC-94/1006 (DE94004072).

Klass, D. L. (1998). Thermal Conversion, Chap.10 in *Biomass for Renewable Energy, Fuels, and Chemicals.* San Diego, Academic Press.

Lynd, L. R. (1996). Overview and evaluation of fuel ethanol from cellulosic biomass: technology, economics, the environment, and policy. *Annual Rev. Energy Environ*, **21**: 403–465.

Lynd, L. R., Elander, R. T., and Wyman, C. E. (1996). Likely features and costs of manure biomass ethanol technology. *Applied Biochemistry and Biotechnology*, **57/58**: 741–761.

National Advisory Panel on Cost Effectiveness of Fuel Ethanol Production (1987). Fuel Ethanol Cost-Effectiveness Study Final Report.

New Source Review Workshop Manual: Prevention of Significant Deterioration and Nonattainment Area Permitting (1990). USEPA, Office of Air Quality Planning and Standards, Research Triangle Park, NC 27711, Draft Report (October).

Perry, R. H., and Green, D. W. (1997). *Perry's Chemical Engineers' Handbook*, 7th Ed., McGraw-Hill, New York, NY, pp. 9–10 and 9–16.

Polman, K. (1994). Review and analysis of renewable feedstocks for the production of commodity chemicals. *Appl. Biochem & Biotech.*, **45/46**: 709–722.

Title 40 Code of Federal Regulation (CFR): Part 51, Subpart 1, Review of New Sources and Modifications.

U.S. EPA Technology Transfer Network, Clean Air Technology Center, RACT/BACT/LAER Clearinghouse (http://cfpub1.epa.gov/rblc/htm/b102.cfm).

U.S. EPA, STAPPA/ALAPCO Clean Air World, (http://www.cleanairworld.org/scripts/us_temp.asp?id = 307).

Vatavuk, W. M. (2002). Control Cost Manual, Emissions Standards Division of the Office of Air Quality, Planning and Standards, U.S. EPA, EPA/452/B-02-001 (January).

Western Governors' Association (2003). Principles for Environmental Management in the West. (December 3, 2003) (http://www.westgov.org/wga/policy/accessed on February 22, 2004).

William, R. H., and Larson, E. D. (1993). *Advances in Gasification-based Biomass Generation, Renewable Energy: Source for Fuels and Electricity.* T. B. Johnson, H. Kelly, A. K. Reddy and R. H. Williams (eds.), Island Press, Washington, D.C.

# CHAPTER 6

# SUSTAINABILITY AND SUSTAINABLE DEVELOPMENT

TAPAS K. DAS

Washington Department of Ecology, Olympia, Washington 98504, USA

## 6.1 INTRODUCTION

This chapter provides a framework within which to measure how well a company or organization is doing in terms of resource consumption and pollution emissions and mitigation while extracting more value from its processes. This framework supports the decision-making process by providing mechanisms for benchmarking performance, tracking improvement over time, evaluating the products and processes involved, and developing strategies for improvement.

The idea of sustainability and sustainable development emerged from the World Commission on Environment and Development (WCED), the so-called Brundtland commission, which defined sustainable development in "Our Common Future" (1987) as that which meets the needs of the present without compromising the ability of future generations to meet their own needs. Within a few years after publication of the Brundtland report the international business community come up with a more concrete definition of this term, envisioning a sustainable development that "combines environmental protection with economic growth and development," as the International Chamber of Commerce stated (Welford, 1997, p. 69). Similarly, the Business Council for Sustainable Development agreed that "sustainable development combines the objectives of growth with environmental protection for a better future" (Welford, 1997, p. 75). The President's Council on Sustainable Development, in the United States, also believes that "it is essential to seek economic prosperity, environmental protection, and social equity together (PCSD, 1996)." Former President Clinton asked the Council to recommend a national action strategy for sustainable development at a time when Americans are confronted with new challenges that have global ramifications. The Council concluded that in order to meet the needs

*Toward Zero Discharge: Innovative Methodology and Technologies for Process Pollution Prevention*, edited by Tapas K. Das.
ISBN 0-471-46967-X   Copyright © 2005 John Wiley & Sons, Inc.

of the present while ensuring that future generations have the same opportunities, the United States must change by moving from conflict to collaboration and adopting stewardship and individual responsibility as tenets by which to live. Similarly, the European Organization for Economic Cooperation and Development (OECD) issued a major report on eco-efficiency in an effort to promote "sustainable" economic growth (OECD, 1998).

Sustainability further provides a framework for integrating environmental, social, and economic interests into effective business strategies. Earlier, the economist, Repetto (1986) wrote in this context, "Current decisions should not impair the prospects for maintaining or improving future living standards." Another definition, recently arrived at by a group of experts from various disciplines at the Environmental Protection Administration, is that sustainability occurs when we maintain or improve the material and social conditions for human health and the environment over time without exceeding the ecological capability that support them. A sustainable industry is one that is capable of operating indefinitely and providing a good standard of living in a healthy environment. Sustainability has three interlinked strands portrayed in Figure 6.1 (Sikdar, 2002, 2003).

The attainment of sustainability requires the incorporation of three interlinked endeavors:

**Economic**   The industry should earn sufficient profit to pay its costs for materials and personnel, while making a return that allows the replacement of obsolescent processes and products with new ones. An economic analysis would be based on value-added per unit value of sales or per direct employee, gross margin per direct employee, return on average capital per employee, percent increase or decrease in capital per employee, and R&D expenditure as a percent of sales.

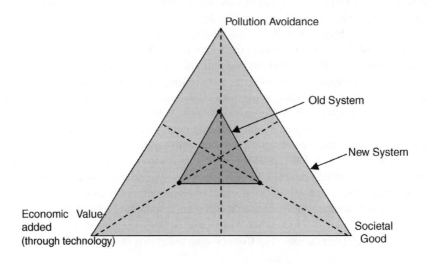

## TOWARDS SUSTAINABILITY

**Figure 6.1.** Sustainability trajectories (Sikdar, 2002, 2003).

**Environmental** Industry should only discharge to the environment quantities and types of materials that can be absorbed indefinitely without ill effect. Industry should use only quantities and types of raw materials that can be renewed indefinitely by natural processes.

**Social** Industry should provide a safe and rewarding environment for its workers. Industry should not pose hazards to people not working in its employ, and should provide products and services that enrich their lives (directly or indirectly) (IChemE, UK, 2000).

In order to make progress towards sustainability, all three aspects of economic development, environmental protection and societal good have to be improved. If, for instance, we can identify pollution avoidance, economic value-added, and societal good as the apexes of a triangle, as shown in Figure 6.1, progress towards sustainability would imply an increase in the surface area of the triangle (Sikdar, 2002, 2003).

To meet these three requirements for sustainability, the nature of the human enterprise will need to change. With respect to chemical manufactures and the use of chemicals, the output of production facilities must not be persistent, toxic, bioaccumulative in the environment. It is the task of waste minimization and green chemistry to address these problems of waste production and unwanted products of manufacture. In a related vein, safer processing addresses the need to prevent catastrophes that may not only pose immediate danger but also result in environmental contamination. The challenge to scientists and engineers is to develop sustainable processes that not only meet our definition but are also profitable, meet or exceed regulations, satisfy customers' needs, engender support from the community, and continue to provide jobs. The common goals are to produce zero manufacturing waste; develop molecules that are not persistent, toxic, and bioaccumulative yet meet society's need; and have a nonhazardous manufacturing process. These goals typically result in the rise of common practices among the technologies, highlighting the need to better coordinate the application of these technologies when designing a new process or upgrading an existing one (Mulholland et al., 2000).

## 6.1.1 Green Chemistry and Green Engineering

Green chemistry is the process of designing of chemical products and chemical processes that reduces or eliminate the use and/or generation of hazardous substances. This process involves definition of the feed materials, reaction pathways, products, and reactor conditions to minimize the impact on the environment, including workers.

Green Chemistry uses all the chemical principles and techniques at its disposal to prevent pollution at its source. Green Chemical Engineering does likewise but also includes the design and operation of processes as well as the design and manufacture of products to minimize pollution and the risks to human health and the environment. Progress has been made in the search for processes using fewer toxic chemicals and producing less waste while requiring less energy (Matlack, 2003).

Green chemistry and green chemical engineering promote the goal of achieving the so-called triple bottom line of economic prosperity and continuity for corporations, social well-being for communities and employees, and environmental protection and resource conservation. Economic, financial, and environmental accounting and risk management tools are being developed to assess green technologies include life cycle assessment (LCA: see Chapter 3), total cost assessment (TCA: see Chapter 5), and ecological risk assessment (ERA: see Chapter 4). The goal of achieving green and sustainable chemical manufacturing has been embraced by a number of government and non-governmental organizations (Mathew, 2003).

**6.1.1.1 *The Twelve Principles of Green Chemistry*** The design and implementation of completely green products and processes is an enormous challenge. Because the number of chemical synthesis pathways is enormous, there is no one systematic, fail-safe method for ensuring that the chemistry being implemented is green. Indeed, it is more nearly correct to inquire if a proposed chemical manufacturing process is simply "greener" than other alternatives. Thus, green chemistry recognizes the importance of incremental improvements. Anastas and Warner (1998) have formulated a set of 12 principles to define and guide the scope of green chemistry.

*1 Prevent Waste* It is better to prevent waste than to treat or clean up waste after it has been created. In the past, the earth and its rivers, oceans, atmosphere, and soil have been considered infinite sinks for the discharge of pollution and waste. The command and control philosophy implies that it is acceptable to generate waste, as long as the waste is safely treated and disposed or stored. Green chemistry's highest principle is that waste should not be created at all. From an economic standpoint, waste is a cost that requires paying twice: the first cost is for purchase of raw materials, and the second is for management or disposal of the waste.

*2 Employ the Atom Economy When Practical* Synthetic methods should be designed to maximize the incorporation of all materials used in the process within the final product. Thus, it is not sufficient to achieve even a 100 percent yield of the desired product if the synthesis also produces by-products. Rather, it is desirable to incorporate all of the atoms of the raw materials into the products. The atom economy, as discussed by Trost (1991), and atom utilization, a term coined by Sheldon (2000), embody the same concept with slightly different definitions. The atom economy is defined as the formula weight of the desired product divided by the stoichiometrically weighted formula weights of all reactants. The degree of atom utilization is the quotient of the formula weight of the desired product and the formula weight of all products and by-products. In situations where the exact composition of all by-products is difficult to determine, the atom economy is more practical one to use.

Sheldon (2000) also defines the E factor, which is the mass ratio of waste to desired product. E factors for bulk commodity chemicals are typically less than five, and often less than one. By contrast, fine chemicals and pharmaceuticals typically generate large amounts of waste per unit of product, with E factors perhaps exceeding 100. Neither the atom efficiency nor the E factor provides any quantitative measure of the potential for human harm of the waste products.

*3 Use the Least Hazardous Chemical Syntheses* Wherever practical, synthetic methods should be designed to use and generate substances that possess little or no toxicity to human health and the environment. This principle promotes the use of less toxic reagents and intermediates, and the generation of less toxic by-products. The use of less toxic feedstocks drives the development of, for instance, renewable raw materials. Better catalysts and chemical reactor designs are also critical to developing less hazardous syntheses.

*4 Design Safer Chemicals* Chemical products should be designed to achieve their desired function while minimizing their toxicity. This principle requires matching the desired function of a given chemical compound with its chemical structure. It also requires the ability to predict beforehand the possible toxic effects of a given chemical substance. Thus, the research and development of methods for establishing structure–property relationships promote this principle. In making the decision to develop and market a particular product, industry should go beyond considerations of the traditional performance properties (vapor pressure, color, stability, etc.) and add as performance metric such favorable environmental properties as reduction in toxicity and carcinogenicity.

*5 Use Safer Solvents and Auxiliaries* The use of auxiliary substances, such as solvents, separation agents, and other additives, should be made unnecessary wherever possible and innocuous if they must be used. Safer solvents such as water, alcohols, supercritical carbon dioxide, or biodegradable surfactant solutions are preferable to chlorinated solvents. Separation processes like liquid extraction require the use of mass separation agents, but such alternatives as reactive separations through membrane reactors (see Chapter 8) may provide a more environmentally acceptable alternative.

*6 Design for Energy Efficiency* The energy requirements of the various chemical processes should be recognized for their environmental and economic impacts, which should be minimized. If possible, synthetic methods should be conducted at ambient temperature and pressure. The chemical industry has substantially reduced its use of energy in manufacturing, starting with the oil embargo and energy shortages of the 1970s. Heat integration and other engineering tools are well developed in the chemical industry and can be applied within a given process. However, it is still true that a substantial portion of energy usage in manufacturing can be attributed to the chemical industry. Thus, advances that can fundamentally alter the required energy usage in manufacturing a product are a key aspect of green chemistry.

*7  Use Renewable Feedstocks*   Raw material, or feedstock, should be renewable rather than depletable, whenever technically and economically practical. The process of manufacturing chemicals from agricultural feedstocks is known as Chemurgy. Fuels derived from biomass may include such cultivated ones as ethanol from corn or fuels from waste biomass. Natural polymers such as chitin and cellulose are receiving attention for their use in manufacturing engineered materials. Of particular interest is the use of $CO_2$ as a raw material for $C_1$ chemistry.

*8  Reduce or Avoid the Use of Derivatives*   The unnecessary use of derivatives in blocking groups, for protection–deprotection, and the temporary modification of physical–chemical processes should be minimized or avoided when possible, because such steps require additional materials, increased solvent usage, and greater energy requirements. This principle is related to principles 2 and 5 above.

*9  Use Catalytic Reagents*   Catalytic reagents, employed as selectively as possible, are superior to stoichiometric reagents. The use of catalysts, whether homogeneous or heterogeneous, is essential to realizing principles 2 and 3. Anastas and co-workers (Anastas et al., 2001) referred to the use of catalysis as a foundational pillar of green chemistry because catalysts can help achieve a number of green chemistry's goals, including decreased material usage, increased atom efficiency, and reduced energy demands.

*10  Design for Degradation*   Chemical products should be designed so that at the end of their function they are broken down into innocuous products that degrade and do not persist in the environment. This principle clearly intersects with the concept of design for the environment (DFE: see Chapter 2). Product design involves communication along a diverse chain of stakeholders: customers, sales representatives, engineers and materials scientists, and synthetic chemists and others. Green chemistry therefore requires close collaboration of chemists with these other stakeholders.

*11  Use Real-Time Analysis to Pollution Prevention*   Analytical methodologies need to be developed further to allow for real-time, in-process monitoring and control prior to the formation of hazardous substances. Waste production is not always the result of synthetic chemistry; sometimes it is simply the result of poor process monitoring and control. The analytical methods currently in use frequently rely on sampling the product at the end of the production line, which may be accomplished in minutes but also may last for hours. In such cases, by the time the product analysis has been completed it is typically too late to adjust the production process for optimum performance. For this reason real-time process monitoring and control of temperature, pressure, feed rate, feed composition, and product composition are vital to green chemical engineering.

*12  Choose Inherently Safer Chemistry for Accident Prevention*   Substances and the forms of them used in a chemical process should be chosen to minimize the potential for chemical accidents, including releases, explosions, and fires. There is some overlap of this principle with principle 5 regarding safer solvents and auxiliaries: for instance, supercritical carbon dioxide is an inherently nontoxic substitute for chlorinated solvents. Substitutes for lead compounds in paint and gasoline are also examples of principles 5 and 12. This principle also applies to strategies such as eliminating the bulk transport and storage of hazardous intermediates in favor of generating these intermediates on-site. While the ultimate goal of green chemistry would be to eliminate the hazardous material entirely, the possibility of a massive release of the hazardous compound is reduced or eliminated by generating it on-site instead of relying on bulk storage.

These principles address an important criterion necessary for achieving the goal of a green technology-preventing pollution on all fronts. This includes conserving materials (feedstocks, reagents, and solvent) and energy (decreased byproduct formation and increased conversion, which, in turn, minimizes the number of process steps), using improved catalysts or catalytic processes in place of non-catalytic ones, and designing safer chemicals and chemical reactions.

As a new technology surpasses initial bench-level development, the researcher should address the 12 additional principles of green chemistry, introduced by Winterton (2001), as follows:

1. Identify byproducts; quantify if possible
2. Report conversions, selectivities, and productivities
3. Establish a full mass balance for the process
4. Quantify catalysis and solvent losses
5. Investigate basic thermochemistry to identify exotherms (safety)
6. Anticipate other potential mass and energy transfer limitations
7. Consult a chemical or process engineer
8. Consider the effect of the overall process on choice of chemistry
9. Help develop and apply sustainable measures
10. Quantify and minimize use of utilities and other inputs
11. Recognize where operator safety and waste minimization may be incompatible
12. Monitor, report and minimize wastes emitted to air, water, and solids from experiments or process.

### 6.1.1.2  *Green Engineering*

Green engineering is the design, commercialization, and use of processes and products that are feasible and economical while minimizing both the generation of pollution at the source and the risk to human health and the environment. The discipline embraces the concept that decisions to protect human health and the environment can have the greatest impact and cost effectiveness when applied early in the "design and development phase of a process or product" (AIChE CWRT, 2000; AIChE IFS, 2004). Green Engineering transforms existing engineering disciplines and practices to those that promote sustainability. This new discipline incorporates the development and implementation of technologically and economically viable products, processes, and systems that promote human welfare while protecting human health and elevating the protection of the biosphere as a criterion in engineering solutions. To fully implement green engineering solutions, engineers use the following principles:

1. Engineer processes and products holistically, use systems analysis, and integrate environmental impact assessment tools.
2. Conserve and improve natural ecosystems while protecting human health and well-being.
3. Use life-cycle thinking in all engineering activities.
4. Ensure that all material and energy inputs and outputs are as inherently safe and benign as possible.
5. Minimize depletion of natural resources.
6. Strive to prevent waste.
7. Develop and apply engineering solutions that take into consideration local geography, aspirations, and cultures.
8. Create engineering solutions that go beyond current or dominant technologies; improve, innovate and invent technologies to achieve sustainability.
9. Actively engage communities and stakeholders in development of engineering solutions, recognizing green engineering's duty to inform society about its practices.

## 6.2 SUSTAINABLE PRODUCTION AND GROWTH FOR THE CHEMICAL PROCESS INDUSTRIES

There are two distinct schools of thought regarding the road to sustainability, one involving incremental improvement, the other radical change (Sikdar, 2003). How to reach sustainability in the production and growth of the chemical process industries includes, but is not limited to, using the following tools:

Impact assessment, risk assessment, and risk management
Waste minimization
Computer modeling
Green chemistry and green engineering
Life cycle assessment and life cycle process and product design
Clean processes and clean products
Bio-based engineered chemicals and materials
Renewable and clean energy sources

### 6.2.1 Impact Assessment, Risk Assessment, and Management

All chemical processes are in some ways hazardous, ranging from nonambient temperature and pressures to dangerous solvents and chemical reactions. To protect workers, the public, and the environment, potential impact assessments need to be evaluated, and typically layers of protection are put in place to meet the conditions described in the impact assessments.

Risk per se is a concept used in the chemical industry and by practicing chemical engineers. The multifaceted term *risk* is used in many respects: raw materials supply (e.g., single source, back integration), plant design and process change (new design, impact on the bottom line), and so on. As we have seen, risk assessment, a systematic, analytical method used to determine the probability of the occurrence of adverse effects, is commonly applied in the evaluation of the human health and ecological impacts of chemical releases to the environment. Information collected from environmental monitoring or modeling is incorporated into portrayals of human or worker activity and exposure, and conclusions on the likelihood of adverse effects are formulated.

Risk assessment is an important tool for making decisions having environmental consequences. Almost always, when the results from environmental risk assessment are used–they are incorporated into the decision-making process along with economic, societal, technological, and political consequences of a proposed action. The basic concept of environmental risk and risk assessment as applied to a chemical's manufacturing, processing, or use, and the impact of exposure to these chemicals on human health and the environment are given in Chapter 4.

## 6.2.2 Green by Design Process

In addition to developing a new or improved chemistry for a reaction, the researcher must be cognizant of its engineering aspects. The additional 12 principles of green engineering for chemical reactions have been introduced by Anastas and Zimmerman (2003) to govern design activity as follows:

1. Inherent rather than circumstantial
2. Prevention instead of treatment
3. Design for separation
4. Maximize mass, energy, space, and time
5. Output-pulled versus input-pushed
6. Conserve complexity
7. Durability rather than immortality
8. Meet need, minimize excess
9. Minimize material diversity
10. Integrate local material and energy flows
11. Design for commercial "afterlife"
12. Renewable rather than depleting

There is a duty to inform society of the practice of green engineering.

## 6.3 INTEGRATING LIFE CYCLE ASSESSMENT IN SUSTAINABLE PRODUCT DEVELOPMENT

One of the internal applications of LCA is in process or product design and development. Recently, an LCA-related tool called Life Cycle Product/Process Design (LCPPD) has started to emerge as an extension of life cycle thinking and as an aid in sustainable product and process design and development. The LCPPD procedure is outlined in Figure 6.2.

LCA is used throughout the development procedure, initially with reference to an existing process or product. This holistic approach is dynamic, involving a continuous exchange of information within and outside the design team to explore systematically the possibilities for improvement. LCA, green chemistry, green engineering, design-for-environment, industrial ecology, and other emerging areas of chemical engineering and environmental engineering provide greater opportunities for developing sustainable and innovative processes and products (Azapagic, 2000; Bhander et al., 2003; Cooper, 2003; Curran, 2003; Das, 2002; Das et al., 2001, 2003; Gonzalez and Smith, 2003; Linke et al., 2003; Sharma and Das, 2002; Sharma et al., 2001).

Once the main environmental impacts have been quantified, potential improvements are identified and the subsequent design focuses on these. The improvements are achieved through the selection of materials and technologies so as to minimize the environmental impacts but still satisfy such other parameters as efficiency, technical performance, costs, legislation, and the needs of customers and suppliers. When all of these requirements have been met, LCA is performed again, to identify and quantify the improvements that have been made.

Clearly the LCA process is iterative, with a continuous exchange of information between the stakeholders, and yields a number of possibilities for improvements. Thus LCPPD offers the potential for technological innovation in the product or

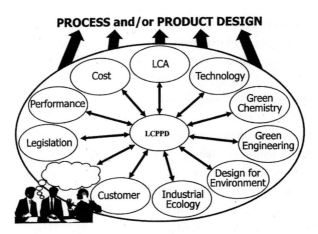

**Figure 6.2.** Methodological framework for life cycle process or product design as a tool for innovation.

process concept and structure through selecting the best material and process alternatives over the whole cycle. This can be of particular importance in the context of International Standards Organization (ISO) 14000, which require companies to maintain full awareness of the environmental consequences of their actions, both on and off site (Azapagic, 2000).

Other applications are related to identifying possibilities for improvements in the environmental performance of an existing process or product or for the design of new ones. Studies made pursuant to ISO 14000, which are usually industry-specific, are mainly used for product or process innovation as well as to demonstrate the environmental progress of a company. Another use of LCA is for public policy making by governments

Because of its holistic approach to system analysis, LCA is becoming an increasingly important decision-making tool in environmental system management. Its main advantage over other, site-specific methods for environmental analysis, such as environmental impact assessment or environmental audits, lies in broadening of system boundaries to include all burdens and impacts in the life cycle of a product or a process, not merely focusing on the emissions and wastes generated by a plant or manufacturing site.

LCA can be used both internally by a company or externally by industry, policy makers, planners, educators, and others with an interest in the outcome. If the results of LCA are to be used internally by a company, the possible areas where it can be useful include, but are not limited to, the following:

- Strategic planning or environmental strategy development
- Problem solving in the system
- Environmental reporting, auditing, and marketing
- Identification of opportunities for, and the tracking of, environmental improvements
- Process and product design, innovation, improvement, and optimization

The external applications of LCA include its use as a marketing tool, to support environmental labeling or claims, for educational or informational purposes, and to support policy decisions. LCA, like design for environment, industrial ecology, and other emerging areas of chemical engineering and environmental engineering provides greater opportunities for developing sustainable and innovative processes and products (Azapagic, 2000, Bhander et al., 2003; Cooper, 2003; Curran, 2003; Das et al., 2001, 2003; Gonzalez and Smith, 2003; Linke et al., 2003; Sharma and Das, 2002; Sharma et al., 2001).

## 6.4  BIORENEWABLE ENERGY RESOURCES

### 6.4.1  Biomass Fuel

Concern over possible global warming has been linked with steady increases in greenhouse gases such as carbon dioxide. This gas, $CO_2$, is emitted whenever fossil fuels or biomass materials, such as wood and agricultural wastes, are burned. When used as a renewable fuel, however, the $CO_2$ released by biomass during combustion ideally equals that which was used as the fuel was grown. Thus, biomass should not contribute to $CO_2$-related global climate change (Golob and Brus, 1994; Lamarre, 1994, 1995).

Global demand for electricity can be expected to increase substantially. Once that occurs, only massive expansion in the use of biomass and other nonfossil fuel sources can slow the annual increase in global $CO_2$ emissions.

Wood fuel is also low in sulfur, ash, and trace toxic metals. Wood-fired power plants emit about 45 percent less nitrogen oxides, $NO_x$, than coal-fired units. Legislation intended to reduce sulfur oxides, $SO_x$, and $NO_x$ emissions may therefore result in encouraging the use of wood-burning or cofiring wood with coal.

In the United States, up to about $4 \times 10^{15}$ Btu/year of biofuels are consumed for electricity generation, raising process heat, and domestic heat. Indeed, many nations meet much of their energy need with biofuels, including wood and wood waste, spent pulping liquors, bagasse, and municipal waste. Some use is also made of dried corn-cobs, rice hulls, and a wide variety of agricultural wastes used in niche applications.

About 6,000 MW of electricity generating capacity in the United States is based on the operation of several hundred wood-fired plants. Most such plants are owned by paper companies and saw mills, which burn their own scrap wood to generate heat and electricity, primarily for on-site use. Any excess electricity is commonly sold to local utilities. Fewer than 10 of the country's wood-fired plants, generating less than 300 MW, are actually operated by utilities, the rest having been built and operated by independent producers. The largest units range between 50–60 MW.

Extensive efforts are now under way to use wood and agricultural wastes as fuel, including increasing emphasis on capturing these materials. Also, the traditional direct combustion technologies used for electricity generation have been joined by developments in the use of circulating fluidized-bed technology for biofuels and cofiring as a technology for using biofuels in pulverized coal boilers. Existing and emerging gasification technologies are also under study for use in supporting conventional-combustion turbine technology or for systems using integrated gasification-combined cycle (IGCC). Experiments continue with direct-fixed gas turbines.

The higher usable energy cost of a wood-fired plant, compared with the corresponding cost of a coal-fired plant, arises from the higher costs entailed in collecting and transporting wood and the lower energy heating value of wood relative to coal.

Short-rotation woody crop (SRWC) practices offer one means to reduce wood-collection costs. SRWC programs seek to develop a steadily replenishable, predictable supply of wood for energy production. For example, SRWC plantations, focusing mainly on fast-growing, short-lived hybrid willow trees, use many typical agricultural practices, including fertilizers and pesticides, to maximize yields of genetically improved trees. The fastest-growing hybrids have produced growth rates exceeding 24.7 dry tons/ha/year. The first harvests are expected three to four years after planting. Sponsors for such efforts include the U.S. Department of Energy, the Electrical Power Research Institute (EPRI), Empire State Electric Energy Research Corporation, New York State Electric & Gas Corporation, and Niagara Mohawk Power Corporation (Torrero, 2000).

### 6.4.2 Solar-Thermal Power

The first solar-electric technology to arouse industry interest was solar-thermal energy (Winter et al., 1991; Golob and Brus, 1994; Mancini et al., 1994; Markvart, 1994; Lamarre, 1995). Under favorable circumstances it can be cost-effective, as evidenced by the fact that solar-thermal gas-hybrid plants are now producing more than 350 MW of commercial power in southern California. This power is used during peak demand to supplement that available from conventional generation.

The level of benefits from tax credits and favorable terms for buying power that helped give these installations their start has recently diminished. Nevertheless, with the remaining tax credits available and given (ca. 1995) natural gas prices, solar-thermal technology can deliver power at 8–12¢/kW.h and an installed cost of $2,500–$3,000/kW.

Solar-thermal technology uses tracking mirrors to concentrate sunlight onto a receiver. The receiver absorbs solar energy as heat, warming a fluid that then drives a turbine generator. Most solar-thermal plants require cooling water. The efficiency of a solar-thermal power plant is the product of the collector efficiency, field efficiency, and steam-cycle efficiency. Overall, solar-thermal power plants can reach annual efficiencies about 15–18 percent.

### 6.4.3 Wind Power

Like solar-thermal technology, wind is providing utility customers with electricity (Golob and Brus, 1994; Lamarre, 1995, 1994, 1992; Jayadev, 1995). The technology is advancing despite the expiration of favorable tax credits in the mid-1980s. About 17,000 mid-size wind turbines, nearly all in California, are producing approximately 1,500 MW of electricity and more are planned. Many of these turbines are operated by Kenetech Corporation, the world's largest developer of wind power. The electricity produced costs 7–9¢/kW.h, and state-of-the-art technology can reduce this amount to 5¢/kW.h (constant dollars). However, even this amount exceeds the present cost of wholesale gas of about 2–3¢/kW.h. Total installed costs for state-of-the-art wind systems range from $900–$1,200/kW.

Wind turbines have had a varied history. Once widely used for electric power generation in remote areas in the United States, they fell into disuse by the 1940s as a result of rural electrification. In the 1970s, shortages stimulated a revival of interest in wind energy. Government and industry combined efforts to develop, build, and test more than a dozen new turbines, the largest of which was capable of generating several megawatts. However, their imposing size and complexity as well as the high costs associated with their operation and maintenance discouraged potential commercial developers. Even the turbines that were producing something on the order of hundreds of kilowatts were not yet cost competitive with conventional forms of electric generation. In the early 1980s, favorable tax credits and energy rates for independent power producers encouraged the development of wind farms in California that used 50–100-kW turbines. Simpler and relatively easy to design, build, install, and repair compared to the earlier, larger turbines, this new generation of wind machines was also relatively reliable and provided lower cost electricity than the previous ones. Nevertheless, as numerous manufacturers were drawn into the business and wind farms grew in California, machines of widely varying effectiveness were deployed. However, enough of the wind farms pulled through to renew utilities' interest in them.

Typical wind turbines consist of rotor blades mounted atop a tower and connected by gears to a drive shaft that spins a generator. Another common design is the vertical axis turbine, which has an eggbeater-shaped rotor attached directly to a vertical shaft. The rotor's blade length and wind speed determine the amount of electric power that can be delivered. In general, wind speeds of at least 6.7 m/s are sought for power generation.

Most of the best locations for wind projects lie outside California. In the northern Rocky Mountain states and northern Great Plains, several states possess substantial wind, as do the Northeast and Texas. In all, about 14 states possess potential wind energy equal to or greater than that of California.

A cooperative alliance of Kenetech Corporation, Pacific Gas and Electric Company, Niagara Mohawk Power Corporation, and EPRI has developed an advanced, utility-grade variable speed wind turbine that can deliver 300 kW. Commercially available and in widespread use, the turbine uses advanced power electronics to increase turbine efficiency, improve power quality, and lengthen turbine life. Future efforts are directed at developing turbines capable of individually producing 500–1,000 kW for eventual deployment on a large scale.

Traditionally, wind turbines have operated at constant rpm to produce 60 Hz alternating current. Because the extra torque generated by wind gusts must be absorbed by the drive trains of constant-speed wind turbines, these require heavier designs than comparable variable speed models. By contrast, variable-speed turbines employ an electronic power converter between the generator and the utility power line. The converter allows the rotor and generator to speed up in response to gusts winds, but without increasing drive-train torque. The increased rotational energy is converted into additional electricity. The amount of energy captured increases by 10 percent or more, and stresses on the turbine are reduced. The variable-speed design produces wind power at a cost of 5¢/kW.h, a level that is also achieved in moderate wind resources by other, fixed-speed turbines, e.g., the AWT-26, the Z, and the Z-46.

In Europe, government policies in the mid 1990s called for a steadily increasing commitment to wind power. Combined, the European programs call for the installation of at least 4,000 MW of such power by the twenty-first century, a level that would have dominated world production. Environmental concerns are the incentives behind Europe's wind targets. With over 2,000 MW of wind power already installed, Europe is well on its way to achieving its goal.

### 6.4.4  Hydropower

Falling water has been used to generate electricity for over 100 years (Golob and Brus, 1994; Department of Energy, 1993). The first hydroelectric or hydro power plant was built at Niagara Falls in 1879. In succeeding years, hydro plants were installed at other natural waterfalls, as well as artificial ones created by dams, at most of the nation's most favorable sites for such facilities. In the mid 1990s, hydropower energy accounted for some 93,000 MW of power, or about 12 percent of the electric generating capacity in the United States bulk power market. The United States also imports some hydropower from Canada. The average generating cost of hydropower is less than 3¢/ kW.h. Hydropower provides an essential contribution to the national power grid: its capability to respond in seconds to large and rapidly varying loads, which other baseload plants with steam systems powered by combustion or nuclear processes cannot accommodate. Also, the ownership of hydropower facilities is spread over a broad base. The owners comprise federal and state agencies, cities, metropolitan water districts, irrigation companies, and public and independent utilities. Individual persons also own small plants at remote sites for their own energy needs and for sale to utilities.

The amount of electricity that can be generated at a hydro plant is determined by two factors: head and flow. Head is the distance in elevation between the highest level of the damned water to the point where it goes through the power-producing turbine. Conventional hydro plants must have a head of water that is at least 3 m high to provide sufficient water pressure to operate the turbine. Flow is the rate of water moving through the system. In general, a high head plant needs less water flow than a low head plant to produce the same amount of electricity.

Some hydro plants use pumped storage systems, which have proven to be among the most reliable energy storage systems available. Pumped storage systems use recycled water instead of tapping free-flowing water. After flowing through the turbine, the water resource is pumped, usually through a reversible turbine, from a lower reservoir back to an upper one. Whereas pumped-storage facilities are net energy consumers, in that they require more energy in total is for pumping than is generated by the plant, they can nevertheless be valuable to a utility, because they operate in a peak-power production mode, when electricity is most costly to produce. The pumping to replenish the upper reservoir is performed during off-peak hours using the utility's least costly resources.

Because the best sites for hydropower dams have already been developed, further construction of additional large, conventional plants is unlikely. However, existing projects could be modified to provide additional generating capacity. Also, many existing dams not equipped for electricity production—only about 3 percent of the nation's 80,000 dams are used to generate power—could be outfitted with generating capacity. Other opportunities for development are offered by small-low-head (from 3–9 m) hydro plants at new sites. These concepts have helped to tap vast hydropower resources.

### 6.4.5  Wave Energy

Ocean waves are formed by the wind driving water toward shore. The wave energy thus created depends strongly on wind speed, this energy being a fifth-power function of speed. Most methods to convert this irregular and oscillating low frequency energy source to grid power employ pneumatic, hydraulic, or hydropower technology (Seymour, 1992).

Pneumatic systems use the wave motion to pressurize air in an oscillating water column (OWC). The pressurized air is then passed through an air turbine to generate electricity. In hydraulic systems, wave motion is used to pressurize water or other fluids, which are subsequently passed through a turbine or motor that drives a generator. Hydropower systems concentrate wave peaks and store the water delivered in the waves in an elevated basin. The potential energy thus supplied with the stored seawater then runs a low-head hydro plant.

The world's total capacity of grid-connected electric power derived from wave energy is less than half a megawatt, distributed among several demonstration plants. The largest unit, the 350-kWe Tapered Channel plant in Norway, uses the hydropower approach. The plant developed by Norwave AS, has operated continuously since 1986. Encouraged by this durability, orders have been placed by firms for two systems in other parts of the world. Other experimental wave-energy systems have been investigated in California and Hawaii, including a 30-MWe heaving-buoy design.

### 6.4.6   Tidal Power

Tidal power is caused by the gravitational pull of the sun and especially the moon, as they pull at the earth. Reacting to this pull, the ocean's waters rise, causing a high tide where the moon is closest. The difference between low and high tide can range from a few cm to several meters. Harnessing tidal power for electricity production by the use of dams requires a tidal difference of at least 4.5 m, a requirement met at few locations in the United States. Thus, the principal demonstration sites of tidal power are in Canada, China, and France.

## 6.5   APPLYING THE METRICS OF SUSTAINABILITY TO TRANSFORM BUSINESS PRACTICES AND PUBLIC POLICY

In recent years, many companies have adopted the concept of sustainable development as a core business value. Although sustainability can be defined in many ways, its underlying premise is that economic well-being is inextricably linked to the health of the environment and the success of the world's communities and citizens (OECD, 1998; PCSD, 1996; WCED, 1987).

For those businesses that have recognized the need to embrace sustainable development, the next step is to understand how to implement it. Putting this concept into operation requires identifying the practical indicators of sustainability and understanding how they can be measured over time to determine if progress is being made. The metrics of sustainability are designed to consolidate key measures of environmental, economic and social performance. The development of metrics that relate environmental and economic performance for production processes is an excellent way for many companies to begin to incorporate the goal of sustainability into management decision-making. Linking the business concept of creating value with environmental performance is termed eco-efficiency. A management strategy that incorporates eco-efficiency strives to create more value with less impact. It enables more-efficient production processes and the creation of better products and services while reducing resource use, waste and pollution along the entire value chain. Often, metrics are designed to meet the following criteria:

- Simplicity: not requiring large amounts of time or manpower to develop
- Useful to management decision-making and relevant to business
- Understandable to a variety of audiences, from people in operations to finance to strategic planning
- Cost-effective in terms of data collection
- Reproducible: incorporating decision rules that produce consistent and comparable results
- Robust and nonperverse, indicating that progress has in fact been made toward sustainability when improvement has indeed occurred
- Stackable along the supply chain so that metrics are useable beyond the particular fenceline for which the calculation was performed
- Protective of proprietary information: preventing the back calculation of confidential information

The five basic indicators of the degree of sustainability are its

- Material intensity
- Energy intensity
- Water consumption
- Toxic emissions
- Pollutant emissions

Putting the concept of sustainability into operation requires practical, cost-effective ways to assess performance and measure progress. These metrics of sustainability give managers simple yardsticks to calibrate how well their company is doing in terms of resource consumption and pollution emissions while extracting more value from their processes. These metrics support decision making by providing a mechanism for benchmarking performance, tracking improvement over time, evaluating products and processes, and developing strategies for improvement. (For instance, an excellent case study using two manufacturing processes for hexamethylene-diamine is presented by Schwarz et al., 2002.)

## 6.6  TOWARD A HYDROGEN-BASED SUSTAINABLE ECONOMY

A long-term hydrogen-based scenario of the global energy system has been developed (Figure 6.3). The scenario illustrates the key role of hydrogen in a long-term transition

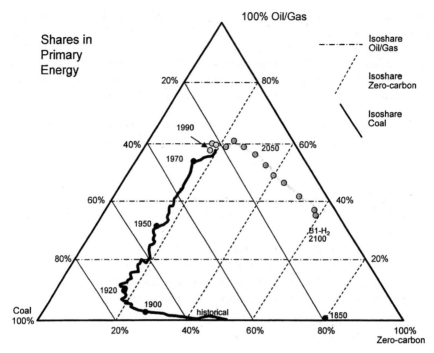

**Figure 6.3.** Global shares in primary energy use, coal, oil/gas, and non-fossil energy, illustrated with an "energy triangle" (in percent). Constant market shares of coal, oil/gas, and non-fossil (0-carbon) energies are denoted by their respective isoshare lines. Historical data from 1850 to 1990 (black) are based on Nakicenovic et al., 1998. The development in the B1-H$_2$ scenario is shown for the years 1990 to 2100 (ten year time steps).

towards a clean and sustainable energy future. In an affluent, low-population-growth, equity and sustainability-oriented B1-H$_2$ (population trajectory shows in the B1 scenario and H$_2$ denotes the key role of hydrogen energy use for the years 1990 to 2100) world, hydrogen technologies experience substantial but plausible performance and cost improvements and are able to diffuse extensively. Corresponding production and distribution infrastructures emerge. The global hydrogen production system, initially fossil-based, progressively shifts towards renewable sources. Fuel cells and other hydrogen-using technologies play a major role in a substantial transformation towards a more flexible, less vulnerable, distributed energy system which meets energy needs in a cleaner, more efficient and cost-effective way. This profound structural transformation of the global energy system brings substantial improvements in energy intensity and an accelerated decarbonization of the energy mix, resulting in relatively low climate impacts (Barreto et al., 2002).

The B1-H$_2$ scenario is based upon the IIASA-B1 scenario developed for the IPCC Special Report on Emission Scenarios (SRES, 2000), with updated information on technology characteristics for hydrogen technologies gathered from a technology assessment. The B1 world has been chosen because it provides a good context to outline the role that hydrogen could play in the global energy system if ideal conditions for its penetration were in place.

The B1-H$_2$ scenario illustrates a relatively smooth transition towards a post-fossil global energy system. Fossil fuels still dominate the primary energy supply until 2050, but during this period, the system shifts away from coal and oil, which reduce their shares substantially, towards natural gas. In turn, natural gas operates as the main transitional fuel to the post-fossil era, which unfolds in the second half of the 21st century. During this period, remarkable structural changes become evident. Renewable energy sources, in particular biomass, increase their shares substantially. A transition to a decentralized energy system takes place.

Likewise, the amount of final energy per unit of GDP decreases at an accelerated rate, as the economy shifts towards less energy, and material-intensive activities, and more efficient end-use technologies improve and diffuse. At the global level, final-energy intensity declines at an average rate of 2 percent per year between 1990 and 2100. Although different world regions follow different paths, fast improvements are evident in all of them, particularly in the developing regions where economic growth and capital turnover are faster.

Global shares in primary energy use, coal, oil/gas, and non-fossil energy are illustrated with an "energy triangle" (in percent). Constant market shares of coal, oil/gas, and non-fossil (0-carbon) energies are denoted by their respective isoshare lines. Historical data from 1850 to 1990 (black) are obtained from an earlier work by Nakicenovic et al. (1998). The development in the B1-H2 scenario is shown for the years 1990 to 2100 (ten year time steps).

Globally, hydrogen is produced with a diversified mix of technologies (see Figure 6.4). Steam reforming of natural gas and gasification of biomass play the leading roles. Along the majority of the time horizon, steam reforming holds the largest share of supply. In the last decades of the 21st century, however, the rapidly increasing production from biomass becomes the most important supply source at the global scale. Significant contributions are also made by the solar thermal technology and, to a lower extent, by coal gasification. In the regions where coal gasification is introduced, it operates as a transition technology towards a renewable-based hydrogen supply structure. Nuclear high-temperature reactors and electrolysis play marginal roles. Still, they constitute valuable complementary options in particular niche markets. Following considerable economic and technological structural changes and substantial energy efficient improvements that reduce the demand for final energy carriers in the long-term, global hydrogen production peaks at 330 EJ/year around the year 2080 and declines afterwards.

In the B1-H$_2$ scenario, electricity production strongly shifts away from traditional centralized fossil-based technologies towards post-fossil and zero-carbon generation systems. Such transition contributes substantially to achieve sustainable-development goals in the electricity system. By the end of the 21st century, hydrogen-based fuel cells, renewables and nuclear power plants become the leading suppliers, while coal and oil power plants are completely phased out. The only fossil fuel that remains is natural gas and its share is small compared to other options. However, natural gas power plants, more specifically the gas-fired combined cycle, play an

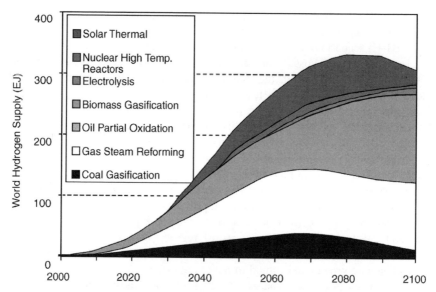

**Figure 6.4.** Global hydrogen supply mix in the B1-H$_2$ scenario. Steam reforming of natural gas and gasification of biomass are the dominant technologies.

important role "bridging" the long-term transition to advanced post-fossil systems. Figure 6-5 presents the market shares of generation technologies in the global electricity mix for the years 2020, 2050 and 2100.

The transformation of the global electricity sector is substantial, not only regarding primary fuels, but also regarding its very nature. Large-scale centralized power plants give way to small-scale distributed generation systems that operate nearer the point of use. A substantial number of highly efficient, cost-effective and less vulnerable micropower systems penetrates the global electricity markets at a quick pace, driven by technological breakthroughs and accompanied by a favorable institutional and regulatory revolution (Dunn, 2000, 2001).

By the end of the 21st century, decentralized systems, mainly hydrogen-based fuel cells and on-site solar photovoltaic installations, hold almost a 50 percent share of the global electricity market. Fuel cells, in particular, experience a dramatic growth. Electricity co-generation in industrial and residential stationary fuel-cell applications and generation from mobile hydrogen-based fuel cells in the transportation sector (e.g., fuel cell-powered cars generating electricity while parked) become major contributors to the generation mix, accounting for approximately 38 percent of the global generation market in 2100.

During the course of the 21st century, the final energy mix of the B1-H$_2$ scenario changes considerably, as the trend towards cleaner, more flexible and convenient energy carriers continues (see Figure 6.6). Solid fuels, such as coal and biomass, are gradually phased out of the final energy market. Oil products, today's prevailing fuels, reduce their share drastically. Grid-delivered energy carriers such as

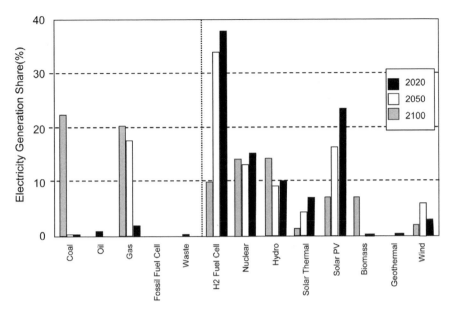

**Figure 6.5.** World market shares for aggregate electricity generation technologies in the year 2020, 2050 and 2100 in the B1-H$_2$ scenario. Although hydrogen fuel cells (H$_2$FC) are shown here in the group of zero-carbon technologies, they become a true zero-carbon option only when hydrogen production system ceases to be fossil based.

electricity and hydrogen increasingly dominate the final-energy mix. Hydrogen, in particular, driven by the penetration of efficient end-use technologies, increases its share dramatically, accounting for approximately 49 percent of the global final consumption by the end of the 21st century and becomes the main final energy carrier.

Fuel cells and related technologies drive forward this transformation. They play a key role in the transportation sector, residential/commercial stationary applications and in key industrial niches. In the transportation markets, for instance, fuel cells penetrate extensively, displacing currently prevailing technologies such as the internal combustion engine. Figure 6.7 presents the evolution of the market share of fuel cells versus the aggregate of other technologies in the global transportation sector during the 21st century in our scenario. The aggregate share of fuel cells is already 51 percent in the year 2050 and rises to 71 percent in 2100. The main role is played by hydrogen based fuel cells, but alcohol-based fuel cells operate as important complementary options.

From today's perspective, if penetration of hydrogen and fuel cells is to be successful, opportunities in different sectors must be tapped. Transportation constitutes indeed a primary target market for both fuel cells and hydrogen. Market potential is huge and benefits can be very significant. However, barriers for the penetration, ranging from supply infrastructures to onboard storage problems and perceived safety risks, appear to be high. Other market segments, where barriers are less

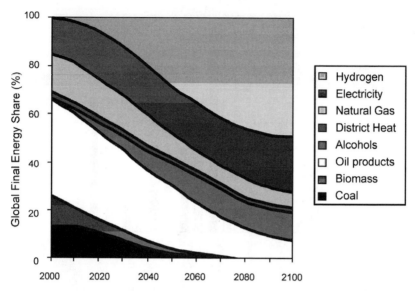

**Figure 6.6.** Evolution of global market shares of different final-energy carriers for the period 1990–2100 in the B1-$H_2$ scenario. The alcohols category includes methanol and ethanol.

severe, may offer attractive opportunities to stimulate early introduction. Potential synergies between, for instance, the buildings and vehicles markets could be used to make hydrogen a more attractive alternative and help to overcome the initial infrastructure barrier (Lovins and Williams, 1999).

This calls for a coordinated strategy for the deployment of hydrogen technologies in different market segments, in order to profit from potential synergies and benefit from the cost reductions as the volume of manufacturing and sales builds up. Research and development, demonstration and commercialization strategies must be targeted at overcoming the barriers specific to each market segment while exploiting the advantages of hydrogen (NHA, 2000).

The $CO_2$ emissions resulting in the B1-$H_2$ scenario are presented in Figure 6.8. In order to provide an adequate perspective of the effects of this hydrogen-based energy path we compare this emissions path to a "dynamics-as-usual case." Generally, "dynamics-as usual" scenarios assume that rates of change for the main scenario drivers, such as technological enhancement, demographic changes, and economic development follow historical experience. Hence, they tend to result in relatively high levels of GHG emissions and climate impacts. As the dynamics-as-usual scenario, we selected the IIASA B2 scenario developed for the IPCC Special Report on Emission Scenarios (SRES, 2000; Riahi and Roehrl, 2000; Marita and Lee, 1998). In B2, global carbon emissions from energy use and industrial sources rise from 6.2 GtC in 1990 to 14.2 GtC in 2100. In contrast, $CO_2$ emissions in B1-$H_2$ peak at about 10.5 gigatons of carbon (GtC) in 2040 and reach 5.7 GtC in 2100, a lower value than in 1990.

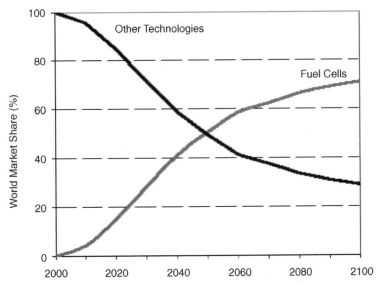

**Figure 6.7.** Evolution of the market share of fuel cells vs the aggregate of other technologies in the global transportation sector in the B1-H$_2$ scenario.

The unfolding of a sustainable hydrogen-based energy system such as the one portrayed here could bring profound changes to the current energy markets and standard business practices. In particular, the emergence of fuel cells and other distributed electricity generation alternatives could alter fundamentally the structure of the power generation and transportation business, driving the creation of new

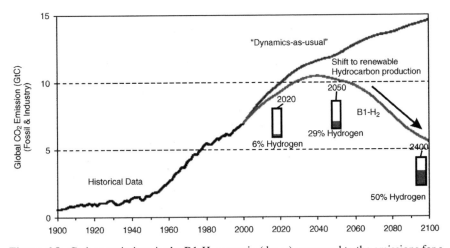

**Figure 6.8.** Carbon emissions in the B1-H$_2$ scenario (down) compared to the emissions for a dynamics-as-usual development (up), and actual development from 1900–2000.

products and values, service standards, innovative business partnerships etc. Fuel cells have significant potential to become an important element of the portfolio of options to meet ever-increasing demands for energy services while responding to more stringent reliability and power quality standards, mounting environmental constraints, cost-effectiveness pressures and other challenges that energy systems will face in the future. Achieving the large-scale transformations of the global energy system that result in a clean and sustainable hydrogen future requires substantial efforts in a number of fields and the involvement of many different social actors. In particular, the combination of government measures and business actions is necessary to stimulate the growth of a sustainable hydrogen energy industry.

## 6.7 BIO-BASED CHEMICALS AND ENGINEERED MATERIALS

Since the 1980s and specially in the 1990s, bio-based chemicals, polymers, plastics, natural fibers, fuels, and some engineered materials have been developed, and they have been playing an increasingly important role in various applications in our daily use. Bio-based raw materials are being used to produce both biodegradable materials and materials that do not easily biodegrade. Superiority in environmental benefits and part of sustainable chemical manufacturing has been an important driver for the increased use of bio-based raw materials and final products. Section 5.4 presents a number of case studies on bio-based chemicals, polymers, and fuels and its economic and environmental benefits (Brown, 2001; Patel, 2002).

## 6.8 ECO-EFFICIENCY AND ECO-INDUSTRIAL PARKS

To an increasing extent, the environmental aspects of economic activity are now being ranked alongside the financial issues. It is against this background that many chemical manufacturing companies have come to develop an in-house tool for eco-efficiency analysis. The use of this instrument provides for early recognition and the systematic detection of economic and environmental opportunities and risks in existing and future business activities.

The specific benefit to the customer always lies at the heart of eco-efficiency analysis. In the majority of cases, a customer having particular needs and requirements is able to choose between a number of alternative products or processes. In the context of this choice, eco-efficiency analysis compares the economic and environmental pros and cons of each solution. In doing so, the method goes beyond the isolated assessment of the company-specific product and weighs up the relevant decision options. The eco-efficient solutions emerge as those that provide the specific customer benefit more effectively than others from the standpoint of both financial cost and the environment.

Using data and analyses on the relative costs and environmental impact, an eco-efficiency portfolio is drawn up that makes it easy to see the strengths and weaknesses of a particular product at a given point. Eco-efficiency analysis is therefore a tool for in-house as well as for external decision makers. Hitherto highly subjective and perhaps vague ideas about the costs and environmental aspects of a product from the viewpoint of the end user are captured in the form of different "screen shots" and so more amenable to discussion. The aim is to get results in a short time, with a maximum of information and aspects for discussion during the decision-making process. As we have seen, substantial environment and economic benefits are possible when the waste or by-products from one facility can be used directly as feedstocks for another facility. The facility generating the waste is thus spared the burden of treatment or disposal, the facility accepting the waste can reduce the costs of its own feedstock materials, and the environment burdens typically associated with waste disposal and raw material extraction are minimized or eliminated.

Industrial ecology-the practice of applying the principles of ecology to industrial and regional economic development-offers considerable promise for improving economic performance while reducing industry's environmental and ecological footprint. However, it also faces substantial technical and entrepreneurial challenges, some of which may be intrinsic to the current eco-industrial development model itself, to realizing that promise. Overcoming these challenges is especially important, given our current state of resource depletion and waste, with its effects on the environment, eco-system, health, equity, and the productivity of society and the planet.

Industrial ecology applies design principles drawn from nature to industry and engineered systems. In so doing it transforms environmental issues from a financial burden to a source of strategic business advantage — and demands environmental policy changes far more profound than the shift from "pollution control" to "pollution prevention," all the way to "zero discharge." Facilities at Kalunborg, Denmark, provide examples of materials exchanges between industries within an eco-industrial park, one key element in industrial ecology and zero discharge (Ehrenfeld and Chertow, 2002). Several examples of process networking and materials exchanges in the eco-industrial parks are given in Chapters 2 and 7.

The current industrial practices have massive economic and environmental impacts on businesses and communities. Industrial ecology offers a design and management framework with clear directives for resource use:

- Balance resource use with natural system regeneration
- Improve the efficiency of resource use
- Find productive uses for industrial by-products
- Eliminate waste altogether

These principles raise important challenges for government policies and scientific understanding. Managers, policymakers and researchers must work together to create dynamic systems characterized by elegant cycles of materials and energy.

The environmental performance of chemical processes is governed not only by the design of the process but also by how the process integrates with other processes and materials and energy flows. Consider a classic example, the manufacture of vinyl chloride, billions of pounds of which are produced annually. Approximately half of this production occurs through the direct chlorination of ethylene. The ethylene then reacts with molecular chlorine to produce ethylene dichloride (EDC). The EDC is pyrolyzed, producing vinyl chloride and hydrochloric acid (Allen and Shonnard, 2002).

$$Cl_2 + H_2C = CH_2 \rightarrow ClH_2C - CH_2Cl$$
$$ClH_2C - CH_2Cl \rightarrow H_2C = CHCl + HCl$$

In this synthesis, one mole of hydrochloric acid is produced for every mole of vinyl chloride. Considered in isolation, this process might be thought wasteful, because half the original chlorine winds up not in the desired product but in a waste acid. However, the process is not operated in isolation. The waste hydrochloric acid from the direct chlorination of ethylene can be used as a raw material in the oxychlorination of ethylene. In this process, hydrochloric acid, ethylene, and oxygen are used to manufacture vinyl chloride (Allen and Shonnard, 2002).

$$HCl + H_2C = CH_2 + 0.5O_2 \rightarrow H_2C = CHCl + H_2O$$

Section 2.3 of Chapter 2 highlighted more on the design principles of industrial ecology, its implications for the operations of our industrial society, and the key policies and management practices that can enable the transformation of industrial society.

## 6.9  PROCESS INTENSIFICATION AND MICROCHANNEL REACTION

There are a series of technologies that enable equipment sizes to be radically reduced. These include process intensification and microreaction technology, as well as spinning disc reactors, high g distillation, high specific surface heat exchanger, microchannel reaction systems, the catalytic plate reactor, and a chemical microsystem for pervaporation. Such technologies enable to plant sizes to be correspondingly reduced. The very low inventories have environmental benefits and there are also claimed cost benefits (Ramshaw and Rogers, 2003; Tsouris et al., 2003). An incidental benefit is that the processes may be economic at a smaller scale (mostly batch processes), and that partly contributes to economic sustainability.

Putting into operation the concept of sustainability-incorporating economics, the environment, and social performances-into operation requires practical, cost-effective ways to assess performance and measure progress. In sort, sustainable development is a journey, not an endpoint.

# APPENDIX

*The Hannover Principles*
Developed by William McDonough and Michael Braungart (2002), the Hannover Principles were among the first to comprehensively address the fundamental ideas of sustainability and the built environment, recognizing our interdependence with nature and proposing a new relationship that includes our responsibilities to protect it. The Principles encourage all of us — you, your organization, your suppliers and customers — to link long term sustainable considerations with ethical responsibility, and to re-establish the integral relationship between natural processes and human activity. When you make decisions in your organization, remember these essential principles:

1. Insist on rights of humanity and nature to co-exist in a healthy, supportive, diverse and sustainable condition.
2. Recognize interdependence. The elements of human design interact with and depend upon the natural world, with broad and diverse implications at every scale. Expand design considerations to recognizing even distant effects.
3. Respect relationships between spirit and matter. Consider all aspects of human settlement including community, dwelling, industry and trade in terms of existing and evolving connections between spiritual and material consciousness.
4. Accept responsibility for the consequences of design decisions upon human well-being, the viability of natural systems and their right to co-exist.
5. Create safe objects of long-term value. Do not burden future generations with requirements for maintenance or vigilant administration of potential danger due to the careless creation of poor products, processes or standards.
6. Eliminate the concept of waste. Evaluate and optimize the full life-cycle of products and processes, to approach the state of natural systems, in which there is no waste.
7. Rely on natural energy flows. Human designs should, like the living world, derive their creative forces from perpetual solar income. Incorporate this energy efficiently and safely for responsible use.
8. Understand the limitations of design. No human creation lasts forever and design does not solve all problems. Those who create and plan should practice humility in the face of nature. Treat nature as a model and mentor, not as an inconvenience to be evaded or controlled.
9. Seek constant improvement by the sharing of knowledge. Encourage direct and open communication between colleagues, patrons, manufacturers and users to link long term sustainable considerations with ethical responsibility, and re-establish the integral relationship between natural processes and human activity.

The Hannover Principles should be seen as a living document committed to the transformation and growth in the understanding of our interdependence with nature, so that they may adapt as our knowledge of the world evolves.

*Source*: *http://www.virginia.edu/arch/pub/hannover_list.html;   http://policyworks.gov/org/main/ mp/gsa/sd2.html*

## REFERENCES

AIChE, CWRT (2000). American Institute of Chemical Engineers, Center for Waste Reduction Technology, *Collaborative Projects, Focus Area: Sustainable Development, Development of Baseline Metrics*. http://www/aiche.org/cwrt/projects/sustain.htm

AIChE, IFS (2004). American Institute of Chemical Engineers, Institute for Sustainable Development, http://www.aiche.org/sustainability/accessed April 15, 2004.

Allen, D. T., and Shonnard, D. R. (eds) (2002). *Green Engineering — Environmentally Conscious Design of Chemical Processes*. Prentice Hall PTR, Upper Saddle River, NJ.

Anastas, P. T., and Warner, J. C. (1998). *Green Chemistry: Theory and Practice*. New York: Oxford University Press.

Anastas, P. T., and Zimmerman, J. B. (March, 2003). Design through the 12 principles of green engineering. *Environ. Sci. Technol.*, **37**(5), 94A–101A.

Azapagic A. (2000). Life cycle assessment: a tool for innovation and improved environmental performance. Part VI35, p 519–530, In *Science, Technology and Innovation Policy: Opportunities and Challenges for the Knowledge Economy*. Quorum Books, Westport, Connecticut.

Barreto, L., Makihiro, A., and Riahi, K. (December 12-13, 2002). The hydrogen economy in the 21st century: a sustainable development scenario. Sustainable Technologies for the 21st Century European Climate Form Workshop, Oldenburg, Germany.

Bhander, G. S., Hauschild, M., and McAloone, T. (2003). Implementing life cycle assessment in product development. *Environmental Progress*, **22**(4), 255–267.

Cooper, J. S. (2003). Life cycle assessment and sustainable development indicators. *J. Industrial Ecology*, **7**(1), 12–15.

Curran, M. A. (2003). Do bio-based products move us toward sustainability? A look at three USEPA case studies. *Environmental Progress*, **22**(4), 277–292.

Das, T. K. (2002). Evaluating the life cycle environmental performance of chlorine disinfection and ultraviolet technologies. *Clean Technology and Environmental Policy*, **4**, 32–43.

Das, T. K., Tsoka, C., Johns, W., and Kokossis, A. (2003). Sustainability metrics and practice: an evaluation of industrial practice. Presented at the AIChE National Meeting Topical Conference: Sustainability and Life Cycle, San Francisco, CA (November).

Das, T. K., Sharma, M., and Butner, S. (2001). The concept of an eco-industrial park — a solution to industrial ecology. Presented at the AIChE National Meeting Topical Conference: Industrial Ecology and Zero-Discharge Manufacturing, Houston, TX (April).

DOE, Department of Energy (1993). Hydropower program biennial report 1992–1993, DOE/ ID-10424, Idaho National Engineering Laboratory, Idaho Falls.

Dunn, S. (2000). Micropower: the next electrical era. Worldwatch Paper 151, Worldwatch Institute, Washington D.C., USA.

Dunn, S. (2001). Hydrogen Futures: Towards a Sustainable Energy System. Worldwatch Paper 157, Worldwatch Institute, Washington D.C., USA.

Ehrenfeld, J. R., and Chertow, M. R. (2002). Industrial Symbiosis: the Legacy of Kalundborg, Ayres, R. and Ayres, L. (eds). In *Handbook of Industrial Ecology*.

Golob, R., and Brus, E. (1994). The Almanac of Renewable Energy. New York: Henry Holt and Co.

Gonzalez, M. A., and Smith, R. L. (Dec. 2003). A methodology to evaluate process sustainability. *Environmental Progress*, **22**(4), 269–276.

IChemE. U.K., Institution of Chemical Engineers, UK (2000). The sustainability metrics. *Inst. Chem. Eng.*, UK.

Jayadev, J. (Nov. 1995). Harnessing the wind. *IEEE Spectrum*, 78–83.

Lamarre, L. (Nov.–Dec. 1995). Renewables in a competitive world. *EPRI J.*, 16–25.

Lamarre, L. (Dec. 1992). A growth market in wind power. *EPRI J.*, 4–15.

Lamarre, L. (Jan.–Feb. 1994). Electricity from whole tree. *EPRI J.*, 16–24.

Linke, P., Kokossis, A., Das, T. K., and Tsoka, C. (Nov. 16-21, 2003). Environmental informatics: tools and technologies to monitor trends and changes. Paper Presented at the AIChE Topical Conference: Sustainability and Life Cycle, San Francisco, CA.

Lovins, A., and Williams, B. (1999). A strategy for the hydrogen transition. Paper presented at the 10th Annual US Hydrogen Meeting, National Hydrogen Association, Vienna, Virginia, US.

Mancini, R., Chavez, J. M., and Kolb, G. J. (Aug. 1994). Solar thermal power today and tomorrow. *Mech. Eng.*, 74–79.

Markvart, T. (ed.) (1994). *Solar Electricity*. Wiley, New York, NY.

McDonough, W., and Braungart, M. (2002). *Cradle to Cradle*. North Point Press, New York.

Morita, T., and Lee, H.-C. (1998). IPCC SRES Database, Version 0.1, Emission Scenario Database Prepared for IPCC Special Report on Emission Scenarios. http://www-cger. nies.go.jp/cger-e/db/ipcc.html.

Mathew, M. A. (2003). Green Chemistry. Kirk-Othmer Encyclopedia of Chemical Technology. On-line Issue, Feb. 14, Wiley: New York, NY.

Matlack, A. (2003). Some recent trends and problems in green chemistry. *Journal of the Royal Society of Chemistry*, G7–G12.

Mulholland, K. L., Sylvester, R. W., and Dyer, J. A. (2000). Sustainability: waste minimization, green chemistry and inherently safer processing. *Environmental Progress*, **19**(4), 260–268.

Nakicenovic, N., Grübler, A., and McDonald, A. (eds). (1998). Global Energy Perspectives. Cambridge, Cambridge University Press, UK.

NHA, National Hydrogen Association, (2000). Strategic plan for the hydrogen economy: the hydrogen commercialization plan. http://www.hydrogenus.com/commpln.htm

Organization for Economic Cooperation and Development (OECD) (1998). *Eco-efficiency*. OECD, Paris.

PCSD, President's Council on Sustainable Development (1996). Sustainable America: a new consensus for prosperity, opportunity and a healthy environment for the future. http://www.whitehouse.gov/PCSD/Publications/political life. University Press of America, Lanham, MD.

Ramshaw, C., and Rogers, J. E. L. (2003). *Process Intensification*. Topical Conference Proceedings, AIChE Spring National Meeting, March 30–April 3, AIChE Pub. No. 181, American Institute of Chemical Engineers, New York, NY.

Repetto, R. (1986). *The Global Possible: Resources, Development, and the New Century*. Yale University Press, New Haven, CT.

Riahi, K., and Roehrl, R. A. (2000). Greenhouse gas emissions in a dynamics-as-usual scenario of economic and energy development. *Technological Forecasting and Social Change*, **63**: 175–205.

Schwarz, J., Beloff, B., and Beaver, E. (July 2002). Use sustainability metrics to guide decision-making. *Chemical Engineering Progress*, 58–63.

Seymour, R. J. (ed.) (1992), *Ocean Energy Recovery: The State of the Art*. American Society of Civil Engineers, New York.

Sheldon, R. A. (2000). Atom efficiency and catalysis in organic synthesis. *Pure Appl. Chem.*, **72**(7), 1233.

Sharma, M. P, Das, T. K, and Butner, S. (2001). Green engineering, industrial ecology, and sustainability: the role of university and engineering educational institutions. Presented at the AIChE National Meeting Topical Conference: Industrial Ecology and Zero-Discharge Manufacturing, Houston, TX (April).

Sharma, M. P. and Das, T. K (2002). Green engineering in chemical engineering curriculum. Paper No. 219b, AIChE Annual National Meeting, Reno, NV (November).

Sikdar, S. (Nov. 2002). American Institute of Chemical Engineers (AIChE) Environmental Division Cecil Award Lecture, AIChE Annual Meeting, Indianapolis, IN.

Sikdar, S. (2003). Journey towards sustainable development: a role for chemical engineers. *Environmental Progress*, **22**(4), 227–232.

SRES, Special Report on Emission Scenarios (2000). A Special Report on Emissions Scenarios for Working Group III of the Intergovernmental Panel on Climate Change (IPCC). Cambridge University Press, Cambridge, UK.

Torrero, E. A. (Dec. 2000). *Renewable Energy Resources. Kirk-Othmer Encyclopedia of Chemical Technology*, On-Line Ed., Wiley, New York, NY.

Trost, B. M. (1991). The atom economy — a search for synthetic efficiency. *Science*, **219**, 245.

Tsouris, C., Porcelli, J., and Rogers, J. E. L. (April, 2003). Process intensification — has its time finally come?" Final Report on Topical Conference and Workshop Sponsored by AIChE's Sustainable Engineering Forum and Institute for Sustainability. AIChE Spring Meeting, New Orleans,

Welford, R. (1997). *Hijacking Environmentalism: Corporate Responses to Sustainable Development*. Earthscan, London.

Winter, C. J., Sizman, R. L., and Vant-Hull, L. I. (eds.) (1991). *Solar Power Plants: Fundamentals, Technology, Systems, Economics*. Springer-Verlag, New York.

Winterton, N. (2001). Twelve more green chemistry principles. *Green Chemistry*, **6**(1), G73–G75.

WCED, World Commission on Environment and Development (1987). *Our Common Future.* Oxford University Press, Oxford, UK.

**PART II**

---

TECHNOLOGIES AND APPLICATIONS

---

# CHAPTER 7

# ZERO DISCHARGE TECHNOLOGY

TAPAS K. DAS

Washington Department of Ecology, Olympia, Washington 98504, USA

Zero discharge (ZD) is applied industrial ecology at the manufacturing level: a practical approach with a concrete methodology to redesign industrial processes so they have no discharges as described in Chapter 2. This chapter describes some specific ZD processes and technologies that have been successfully operating.

## 7.1 ZERO WATER DISCHARGE SYSTEMS

### 7.1.1 Why Aim for Zero Water Discharge?

The increasing scarcity of water coupled with escalating cost of fresh water and its treatment has prompted industry to think of water conservation and recycle in most industries, the case for ZD is neither compelling nor farfetched, though many a time regulatory authorities dictate the implementation of a ZD system. It is prudent for existing facilities to develop systematic approaches to effective and efficient plant-wide water management rather than to implement full-fledged ZD systems. Later in the chapter (Section 7.1.5), we present a case study of a cement plant utilizing captive power plant effluents.

Industrial operations use water for processing, conveying, heating, cooling, steam production and housekeeping. The bulk of the water consumed (85–90 percent) by industry is discharged as process wastewater. The rising price of fresh water and stringent environmental regulations with respect to effluent discharge are now compelling industries to consider reduction in water consumption, as well as recovery and recycling of water.

While there are several practical definitions of ZD, a ZD system is most commonly defined as one from which no water effluent stream is discharged by the processing site. All the wastewater after secondary or tertiary treatment is converted to a

*Toward Zero Discharge: Innovative Methodology and Technologies for Process Pollution Prevention*, edited by Tapas K. Das.
ISBN 0-471-46967-X   Copyright © 2005 John Wiley & Sons, Inc.

solid waste by evaporation processes, such as brine concentration followed by crystallization or drying. The solid waste may then be landfilled. It is, however, important to be clear in one's definition of ZD, as and when mentioned by any regulatory body (Byers, 1995; Dalan, 2000; Goldblatt et al., 1993; Kiranmayee and Manian, 2000; Rosain, 1993).

### 7.1.2 Establishing a Database

Before starting an evaluation of water reuse with a view to moving toward ZD, it is essential to define the current baseline of information on plant water use and wastewater generation. Developing a database for water utilization for an existing plant will require:

- Looking at unit process and instrument designs
- Field verifying piping connections
- Recording documented water uses
- Estimating raw water and process water quality requirements
- Determining current water treatment capability and costs
- Accessing process wastewater flow and composition
- Developing current wastewater treatment capability and cost
- Flow and mass balances for the water/wastewater management systems
- Looking for non-documented water uses
- Documenting the water quality requirements for each plant water use

Although water can be treated to remove the contaminants, if a particular process can use such water with little or no treatment, the overall economics of water reuse will improve. When an appropriate database is available, it is time to think about the pros and cons of a ZD system.

### 7.1.3 Advantages and Disadvantages

Zero-discharge systems offer many advantages. The principal ones are as follows:

- Minimize consumption of fresh water
- Allow the recovery of valuable resources
- Reduce volumes for sludge handling
- Improve product quality by yielding water of better quality than raw water
- Facilitate site selection (since site location is less limited if a receiving waterway is not needed for wastewater effluent)

The principal disadvantages can be described briefly as follows: maintenance problems, reduced plant reliability, and trace chemicals.

Scaling, especially on heat transfer surfaces is quite common as the salt concentration of the water increases. It is also quite common for the resulting water quality to be incompatible with metallurgy selected for different conditions. Efforts made to combat scale and corrosion through local pH adjustments or changed flow configurations can be handled only temporarily.

With respect to reduced plant reliability, it must be borne in mind that a failure or shutdown in the plant could curtail water availability or change water quality in a way that affects operation in another part of the plant.

Finally, water reuse may cause a buildup of trace metals and organic solvents in the water system, again necessitating increased maintenance.

### 7.1.4  Design Principles

To implement a successful zero-discharge system, the design must accomplish the following:

**Minimization of raw water consumption** This can be accomplished by reusing plant effluents or using secondary treated effluent from a nearby municipal wastewater treatment plant etc. resulting in reduced fresh water make-up for the plant.

**Source reduction** Vary the process design parameters and raw materials (wherever applicable) to minimize the wastewater generation. A careful planning in the selection of process, equipment, raw materials, and operating conditions can reduce the wastewater flows.

**Segregation and reuse of wastewater streams** Segregated treatment may be particularly effective if the removal of only one or two contaminants will allow the wastewater stream to be reused directly or will reduce the size or complexity of the end-of-the-pipe treatment system. Usually, an integrated water reuse system will likely employ a combination of segregated and end-of pipe treatment systems to achieve cost-effective water reuse.

**Advanced treatment and processing** Advanced processing that removes suspended solids as well as dissolved solids may produce boiler quality water from the wastewater. These treatments may include precipitation softening, multimedia filtration, carbon adsorption, de-ionization, reverse osmosis or distillation.

**Disposal** Once all steps to minimize and reuse wastewater streams are taken, the remaining wastewater is normally treated and then disposed of. However, if an ideal zero liquid discharge system is to be implemented, the treated water is completely reused.

With these general principles in mind, we are ready to look at three case studies involving wastewater recovery in three very different industrial areas: the manufacture of cement, automobiles, and chemicals.

Also essential to good design is consideration of the economics of water reuse. This includes the availability and cost of supply water, restrictions on and costs of discharge water, recycle stream characteristics and effects on production or product quality, and purchase and operating costs of a water purification system. In the case of an existing plant, complete implementation of true zero-liquid discharge entails extensive re-piping or costly unit operations.

### 7.1.5   Case Study: A Cement Plant in India

Industrial wastewater recovery and reuse systems are cropping up all over the world. This chapter's first case study explores reuse of effluents from a captive power plant (CPP) used in the making of cement in India.

The unit studied is a $3 \times 23$ MW coal-based CPP. The environmental clearance document for the project stated that all effluents generated in the plant activities were to be collected in the central effluent treatment plant and treated to ensure adherence to specified standards. It was expected that the concept of ZD would be adopted.

The detailed engineering specifications revealed that the quantity of effluent, mainly the cooling tower blowdown, was too large to be used entirely in the cement plant and the CPP together; therefore adopting a ZD system would be extremely difficult. Other alternatives, such as using the effluent in neighborhood industries, were evaluated, but were not practical. The installation of a complete ZD system would have involved adopting technologies like reverse osmosis and brine concentrators. While the reverse osmosis process recovers about 70–85 percent water based on feed water characteristics, it leaves behind a very high total dissolved solids (TDS) effluent, a briny substance that can be disposed of only by means of brine concentrators, expensive devices that would have exceeded the project's budget.

#### 7.1.5.1 *Review of Existing Water Balance at the Plant*   An extensive review was conducted on the base-line information on existing cement plant water use, the wastewater generation, the existing wastewater treatment facilities and the water quality required for various plant processes and equipment. A large water tank receives makeup water from a nearby irrigation canal for the cement plant. The captive power plant receives water from a nearby river. There are mainly two areas in the cement plant where large quantities of water are required. These are the gas conditioning towers (GCTs) and the cement mills. The plant owner provided the following rates of water consumption and loss:

- Gas conditioning tower in phase 1—1000 $\text{m}^3/\text{day}$
- Gas conditioning tower in phase 2—1100 $\text{m}^3/\text{day}$
- Flow rate for four cement mills—576 $\text{m}^3/\text{day}$

All the other water used by the facility (e.g., for kiln bearing cooling, packing plant compressors etc.) is under circulation. Also to be considered are domestic, horticultural, and nondocumented water uses (service waters, washing etc.) and evaporative losses from the large open water tank and cooling towers. Since the wastewaters were intended to be reused after appropriate treatment in critical process operations such as in GCTs, the effluents were restricted to cooling tower blowdown, the boiler blowdown, and the neutralized effluent from demineralizer regeneration, strict monitoring and control over the quality of combined effluents was possible. The use of service wastewater and any other effluent for horticulture or dust suppression was possible because these discharges were small and well within the norms laid out for discharge onto land. The first approach was to segregate all the streams so that recycling could be accomplished with as little treatment as possible and the effluents reused with a minimum of operational complexities. Combining all the effluents in a single effluent treatment plant would have the following disadvantages:

- The effluent treatment plant would have to be large. This is because a large amount of effluents (2350 $m^3$/day) was to be handled with only three cycles of concentration (COC).
- Operational complexities would be severe.
- Maintaining and monitoring water quality for a particular process unit would be difficult.

Although the cooling tower blowdown quality was sufficient for use directly in the cement mills, it would have to be treated in accordance with the process licensor's specifications before it could be used in the GCTs. Analysis of the licensor's data indicated that while the TDS of cooling tower blowdown, the chlorides as well as the sulfates, were well within specifications, the hardness and alkalinity in the stream were of concern. To validate the consequences of this variation, water samples from gas conditioning towers of various cement plants were analyzed. Although GCT water analysis results from some cement plants were off spec, no cement plant had serious operating problems.

### 7.1.5.2 Effluent Reuse Options
Two approaches to the reuse of effluents were considered: the use of scaling/corrosion inhibitors and the use of a softener. Both measures are intended to reduce the hardness and alkalinity of the cooling tower blowdown. Blowdown was an area that could not be overlooked because the values developed in the process licensor's analysis were too high to meet the specifications.

The idea of using scaling and corrosion inhibitors was discarded because the vendors lacked experience with this kind of service. Thus bids were obtained for softening equipment, which would allow the blowdown to be treated by both ion-exchange and lime-soda softening. However, the ion-exchange system would produce a regeneration effluent of about 180 $m^3$/day with a very high TDS (11,000 mg/L), introducing, in turn, a very difficult disposal problem. The lime-soda system, on the other hand, results in huge quantities of sludge, which would have to be removed to a landfill.

In addition to hardness and alkalinity, frequent problems with excessive chlorides ($>$ 13 mg/L as $Cl^-$) had led cement plant process personnel to suggest that chloride reduction be incorporated into the design. A reverse osmosis (RO) system could have served this purpose, but it would have been prohibitively expensive. Moreover, the creation of another waste stream was undesirable because the environmental clearance mandated a common effluent collection point from which all the power plant effluents could be recycled or reused.

Ultimately, the softener option was selected, but without RO, to accommodate the requirement to use a single treatment plant for all the effluents. In addition, the COC for the cooling tower, originally envisaged as three, was set at six. As a result, the amount of effluent was greatly reduced, but the concentration of pollutants increased accordingly.

### 7.1.5.3 Effluent Treatment Plant Design

In view of the intermittent nature of blowdowns, it was necessary to ensure that all the effluents had sufficient retention time to attain uniform concentration and therefore, two RCC effluent treatment pits of 500 m$^3$ each were provided. The combined effluent quality is well within specified norms for discharge into inland surface water or for irrigation. Only a pH adjustment was found to be necessary. The final expected quantity and quality of effluents is given in Table 7.1. The final effluent quality is maintained such that 2,100 mg/L TDS is never exceeded for discharge onto land for irrigation, especially for purposes of gardening. The effluents from the CPP could be utilized as summarized in Table 7.2. Seasonal variations were also considered.

Thus, for a 3 × 23 MW CPP, the effluent quantity would be about 1,307 m$^3$/day with the cement plant water requirement being the same at 1,372 m$^3$/day in summer. In summer in this part of India, all power plant effluents from cement mills can be used; in winter, there is an excess of effluent, which may be diverted for use in GCTs with additional dilution with fresh water. Thus, the cement plant at Mithapur is an ideal example of an economically viable ZD system.

### 7.1.6 Case Study: DaimlerChrysler's Zero-Discharge Wastewater Treatment Plant in Mexico

DaimlerChrysler's production complex in Toluca, Mexico, home of the Chrysler PT Cruiser, has received much attention not only because of its in-demand product but also because of its state-of-the-art ZD wastewater treatment plant (WWTP). Located 37 miles north of Mexico City, Toluca suffered for years from a worsening water shortage due to urban sprawl, regional drought and increased industrial activity. The city is one of the leading producers of beverages, textiles, and automobiles in Mexico, as well as a center for food processing.

DaimlerChrysler, one of Mexico's largest manufacturers, mindful of the mounting strain on the world's natural resources, has consistently sought ways to decrease operational waste, reduce costs and increase process efficiencies. Upon locating in Mexico, the automaker began to study the region's rapidly dropping aquifer, hoping to be able to minimize the stress on this valuable resource, yet keep its operations in compliance with the federal government's water quality standards.

In 1999, the company hit upon a solution. It would build its own $17 million WWTP that would treat sanitary and manufacturing-process water generated by

**TABLE 7.1. Water Balance for 3 × 23 MW Coal-Based Captive Power Plant (Kiranmayee and Manian, 2000)**

| Effluent | Flow m³/day | pH | Temp °C | TSS mg/L | TDS mg/L | Hardness mg/L | Alkalinity mg/L | Cl mg/L |
|---|---|---|---|---|---|---|---|---|
| Typical raw water quality | | 8 | Ambient (30–35°C) | 5 | 286 | 146 | 194 | 19 |
| Cooling tower blowdown | 1080 | 9 | 40 | 3 | 2100 | 1686 | 1164 | 115 |
| Boiler blowdown | 206 | 10.8–11.8 | 100 max | – | 200 | 0 | 20 | 0 |
| DM regeneration | 21 | 7 | 30 | 7 | 4700 | 1800 | 1950 | 2272 |
| Total effluents (mixed) | 1307 | 9.0–11.0 | 47.00 | 2.59 | 1842.31 | 1422.10 | 996.32 | 131.53 |

**TABLE 7.2. Water Requirements in Various Places (Kiranmayee and Manian, 2000)**

| Effluent | Flow m³/day | | pH | Temp °C | TSS mg/L | TDS mg/L | Hardness mg/L | Alkalinity mg/L | Cl mg/L |
|---|---|---|---|---|---|---|---|---|---|
| | Summer | Winter | | | | | | | |
| Gas conditioning tower | 2100 | | Neutral | Amb. | NIL | 1000 | 130 | 200 | 55 |
| Cement mills | 560 | | Neutral | Amb. | No limits | | NA | NA | NA |
| Dust suppr. in mines | 120 | | | | | | | | |
| Coal quench | | | | | | | | | |
| In cement plant | 90 | 45 | Neutral | Amb. | No limits | No limits | NA | NA | |
| In power plant | 10 | 5 | Neutral | Amb. | No limits | No limits | NA | NA | |
| Total — coal quench | 100 | 50 | | | | | | | |
| Horticulture — total | 592 | 295 | 5.5–9 | Amb. | 200 | 2100 | No limits | | 600 |
| Total, max | 3472 | 3125 | | | | | | | |
| Water req. without GCT | 1372 | 1025 | | | | | | | |
| Power plant effluents | 1307 | 1307 | | | | | | | |

the facility's four separate plants—engine, transmission, stamping and assembly. And to make this WWTP truly state of the art, a comprehensive zero liquid discharge (ZLD) system would be installed. By using a ZLD system, the Toluca complex would avoid further depleting the local aquifer, the environmentally friendly and cost-efficient system would discharge no process water, but rather would recycle it to use throughout the facility. It was projected that implementing a ZLD solution and thus reusing water could extend the facility's life without disrupting production and causing costly overhauls.

### 7.1.6.1 *System Design*

DaimlerChrysler put in operation ZLD systems of two kinds. The first uses reverse osmosis (RO) to produce a concentrate of total dissolved solids (TDS), which is sent to a large evaporator and eventually, on to a lagoon or solar evaporator pond. Used in dry, arid areas of low elevation, this system is frequently found in the WWTPs of Northern Mexico's automotive facilities. The other system, used at the Toluca facility, softens and removes silica from the RO concentrate through microfiltration before sending the water on to another RO unit where it is further concentrated. Water is then returned and blended with the water from the first stage water, where the concentrate is sent on to either an evaporator or a crystallizer to dry TDS to powder and eliminate the need to dispose of liquid.

In essence, industrial plants with ZLD installations can expect to recover nearly 100 percent of water that would otherwise be discharged to the environment as wastewater. At the Toluca facility, the WWTP recovers 95 percent or more of the water used for processing, with a recovery rate of up to 237,500 gallons per day (gpd). In actuality, the ZLD installation at the Toluca facility is two separate systems: a sanitary water system that biologically treats wastewater from the complex's restrooms, showers, cafeterias and other domestic areas, and a manufacturing-process water system that chemically treats wastewater mixed with heavy metals and paint from the assembly plant. The latter also treats wastewater containing emulsified and soluble oils from the facility's stamping, transmission and engine plants.

In the sanitary water system, domestic water is collected and sent through a screening mechanism before moving on to the biological treatment system's equalization tank, ensuring a constant, even flow of water through the system. This water is then passed through jet aeration sequential batch reactors that treat the water with microorganisms and air to reduce the biological and chemical oxygen demands (BOD and COD), as well as suspended solids. The complex uses the 150,000 to 200,000 gpd of disinfected water to irrigate its landscape. The microorganisms and solids recovered from the batch reactors are then sent through a sludge digester and eventually a filter press that eliminates the water. While the dewatered sludge is used as fertilizer, the filtered water re-enters the system.

Wastewater from the Toluca facility's three machining plants is directed through the manufacturing-process system where it is first chemically treated, passing through a filtering screen. In a separate tank, chemicals are used to de-emulsify the free-floating oils that comprise most of the waste. Afterwards, the oils are removed and stored in another tank before disposal. The process water from the machining plants is then mixed with water from the assembly plant that contains residue from the spray painting, phosphating, E-coating and body-wash operations. Upon being mixed with a combination of ferric chloride, lime and magnesium oxide, metal pollutants and silica are rendered insoluble and turned into sludge that is removed and sent to a landfill. Then, to further lower the proportion of unwanted organic compounds, the water is pumped to a biological system that reduces the BOD to 20–30 ppm.

### 7.1.6.2 Results

Since installing the wastewater recovery system, the Toluca facility has noted several benefits, including decreased production and operation costs, reduced aquifer use, better environmental friendliness, and greater employee safety (Zacerkowny, 2002). Moreover, the integrated system helps preserve the environment, is safe for employees to work with and provides almost 7,000 jobs to local residents. The Toluca industrial complex uses approximately 250,000 gpd of water, recovering more than 95 percent of its processing water. The ZLD system allows the facility to treat more than 550,000 gpd, significantly reducing the amount of water that must be drawn from the local aquifer. Using treated water might also extend the life of the facility's equipment, as the salt content of the processed industrial water is much lower than that of the aquifer.

### 7.1.7   Case Study: Zero Effluent Systems at Formosa Plastic Manufacturing, Texas

Formosa Plastic Complex is a wholly owned subsidiary of Formosa Plastics Corporation, USA. with operations at Point Comfort, Texas. The company is a vertically integrated plastics manufacturer whose core business is the production and processing of commonly chemicals and plastic resins.

### 7.1.7.1 Complying with ISO 14001

Formosa Plastic's complex in Point Comfort was the first major chemical plant in the United States to be certified as complying with ISO 14001, the series of international standards developed for managing environmental impacts. The standards address six distinct, but related, components that together form the basis of a comprehensive environmental management system. To be in compliance, the company's environmental management program must include a specific plan that describes actions proposed to meet each objective and target, the person(s) responsible for meeting each objective, and the time schedule for attaining each target (Delaney and Schiffman, 1997).

In addition to meeting the ISO requirements, the company entered into a pact that has become a model good industry-community relations, the "Wilson-Formosa Zero Discharge Agreement" (Ford et al., 1994a,b). The parties to this agreement are Formosa Plastics-Texas, community activist Diane Wilson, the U.S. Environmental Protection Agency (EPA), the Texas Natural Resource Conservation Commission (TNRCC), and the Formosa Technical Review Commission (TRC). By this agreement, the company made a commitment to studying and implementing alternative methods to reduce and/or recycle and/or remove the wastewater generated at its Point Comfort facility. The goal was to create a process to resolve disagreements between the parties regarding the feasibility of wastewater recycling and/or reduction and/or removal programs to be studied and/or implemented at its Point Comfort facility with a goal of zero discharge to Lavaca Bay.

### 7.1.7.2 The Quest for Zero Discharge

The Wilson-Formosa Agreement launched the company on a comprehensive analysis of the possibility of a "zero discharge" system for the facility. Although complete recycling in a complex chemical facility is virtually impossible, serious attempts to reduce pollution from hazardous wastes must be made. Thus a list of candidate solutions was developed, all based on water quality realities and cost-effectiveness. According to the agreement, the system selected was to be "economically beneficial, environmentally superior, and technically proven to be effective in similar industrial applications" (Ford et al., 1994a,b). It is noted that a successful "zero discharge" scenario eliminates much of the costly monitoring as well as offering other potential cost-savings.

Meetings were held between the parties to the Agreement to identify potential alternatives. A summary of these candidate systems and the associated estimated costs (1999) is presented in Table 7.3 (Blackburn and Ford, 1998). These alternatives concentrated on comprehensive concepts for removal of the wastewater from Lavaca Bay. While the work on the zero discharge alternatives was proceeding, studies were made of the various contributing waste streams and the ability to recycle or reuse them.

At the initiation of the Wilson-Formosa Agreement, the Point Comfort facility was recycling its treated sanitary wastewater to the cooling towers. Additional analysis of the water use patterns the plant identified three waste streams that could be recycled and reused to further reduce water use and wastewater generation: (1) low strength organic wastewater from the polyvinyl chloride process; (2) cooling tower discharges; and (3) evaporator process condensate recycle.

The low strength organic waste stream can be segregated from the other organic wastes and biologically treated to be suitable for reuse. A step that will reduce water use and effluent production by approximately one million gallons per day. By increasing the cycles on the cooling tower, both water consumption and the wastewater it generates can be reduced by another million gallons per day. The third stream, the recycled evaporator process condensate can itself be recycled, thus eliminating a waste stream of approximately 600,000 gpd. Together, these three alternatives represent a reduction in the volume of effluent from the plant of 2.6 million gpd, approximately 32 percent, with a comparable reduction in water use. These alternatives were recommended by the TRC and adopted by both Wilson and Formosa.

Until late 1990s, however, a total zero discharge option has not been adopted, largely because complications associated with the concentrated brine system could not be quickly resolved.

Nevertheless, water conservation by recycle and reuse has been successful at Formosa Plastics. The plant only uses 17 mgd of the 26.6 contract allowance and

**TABLE 7.3. Summary of Costs and Environmental Issues for Viable Zero Discharge Alternatives**

| System No. | Description | Capital $MM | O&M $/1,000 Gal Removed from Discharge | Economic and Environmental Issues | |
|---|---|---|---|---|---|
| | | | | Advantages | Disadvantages |
| 1 | No additional action | 0 | $4.15 | Meets permit requirements<br>No additional capital costs | Not zero discharge<br>Continuous monitoring costs<br>Continued permitting costs<br>Local public opposition<br>Potential damage to Lavaca Bay |
| 2 | Reverse osmosis with deep well injection | $37.1 | $7.12 | Low energy cost<br>No additional solids generated or atmospheric emissions<br>Proven technology | Potential negative public opinion<br>Holding capacity or spare wells<br>Water lost to well<br>Significant capital investment |
| 3 | Reverse osmosis with deep well injection | $164.9<br>($72.9) | $18.92 | Meets zero discharge<br>Low energy consumption<br>Potential for $CaSO_4$ by-product recovery | Significant land sacrifices to dead salt lake<br>Technology unproven in full-scale<br>Large quantities of chemicals input to system<br>Technology unproven in full-scale<br>Large quantities of chemicals input to system<br>Variable brine quality may impact production |

| | | | | | |
|---|---|---|---|---|---|
| 4 | Reverse osmosis with wastewater cooling towers with sulfate removal | $29.6 | $12.27 | Meets zero discharge<br>Potential for $CaSO_4$<br>by-product recovery | Large quantities of chemicals input to system<br>Technology unproven at 26 percent salt concentration |
| 4A | Reverse osmosis with wastewater cooling towers and vapor compression concentrator with sulfate removal. | $35.5 | $11.05 | Meets zero discharge<br>Wastewater recycled not lost to atmosphere or in well<br>Proven technology<br>Potential for $CaSO_4$<br>by-product recovery | High energy requirement<br>Large quantities of chemicals input to system<br>Large quantities of sludge generated requiring disposal |
| 5 | Reverse osmosis with vapor compression concentrate and sulfate removal | $33.2 | $13.27 | Meets zero discharge<br>Wastewater recycled not lost to atmosphere or in wall<br>Proven technology<br>Potential for CaSO4<br>by-product recovery | High energy requirement<br>Large quantities of chemicals input to system<br>Large quantities of sludge generated requiring disposal<br>Significant capital investment |

discharges less than 8 mgd to Lavaca Bay, despite permitted allowance of 9.7 mgd average and 15.1 mgd maximum. The path to this significant reduction in effluent flow has been extensively documented in the literature (Ford, 1996, 1997; Formosa Plastic Corp., 1991; Ford et al., 1994a,b; Ford and Blackburn, 1997; Morris, 1993; Parsons Engineering-Science, 1993; The University of Texas at Austin, 1993). Indeed, the goal of "zero discharge" to Lavaca Bay could well be realized in the future.

The TRC has considered two other options for the disposal of the brine stream. One of these options is to return the brine to the salt dome. This option is technically feasible but expensive. The second option is to chemically treat the brine stream and reuse the brine stream as a feedstock. This alternative is technically the best solution. However, it presents difficult chemical engineering challenges. Designers of the chlorine plant have concurred that this recycle concept is technically feasible and its feasibility is receiving more detailed scrutiny.

**7.1.7.3 Summary** The attempt to approach ZD at Formosa Plastics, a chemical manufacturing facility, has been a complex, multidisciplinary, regulation-sensitive, and technically challenging decades-long project. By virtually all yardsticks, progress is being made, demonstrating how industrial expansion and economic growth can coexist with environmental control and enhancement; how constructive and creative agreements allow a process to move forward concomitant with oversight and controls; how allocation of resources can be better applied for scientific evaluation and evolution as compared with litigation; and how an overall higher probability of improving both human health and environment can be attained.

## 7.2 CASE STUDY: GAS TURBINE NO$_x$ EMISSIONS AT GENERAL ELECTRIC

The requirement for gas turbines to meet ever lower NO$_x$ emission levels results from a regulatory approach that preceded the development of combustion systems capable of achieving single digit NO$_x$ without add-on controls (such as selective catalytic reduction). Yet dry low NO$_x$ (DLN) combustors are now demonstrating 9 ppm NO$_x$ at General Electric for its gas turbines identified as Frames 7FA, 7EA, and 6B. This case study compares the energy, environmental, and economic impacts of requiring add-on emission controls to achieve a lower level of NO$_x$ for a gas turbine combustion system that is already capable of achieving single digit NO$_x$. The conclusion reached is that ratcheting NO$_x$ down to lower and lower levels through the use of add-on emission controls reaches the point of diminishing returns when the gas turbine combustion system is capable of achieving single digit NO$_x$. The cost of add-on emission controls to achieve a lower NO$_x$ level becomes excessive, the heat rate increases and the overall environmental impacts are actually worsened. One way to eliminate this unintended consequence would be for the USEPA to amend the regulatory process to allow permit authorities to consider conflicting environmental, energy and economic impacts in nonattainment areas, as they now can in attainment areas, in cases where add-on emission controls will result in only a small reduction in emissions (Schorr and Chalfin, 1999).

## 7.2.1 Regulatory Background

The decade of the 1980s was one of rapid change for both gas turbine emission control regulations and the technologies used to meet those regulations. The primary pollutant of concern from gas turbines has been, and continues to be, oxides of nitrogen. The Gas Turbine New Source Performance Standards (NSPS), issued in 1979, did not regulate the emissions of carbon monoxide or unburned hydrocarbons from gas turbines because the levels are very low at base load. However, in December 1987, EPA began to require industry to adopt "top-down approach" for determining the best available control technology (BACT). This pushed allowable gas turbine NO$_x$ emission levels to levels significantly lower than the NSPS. As the allowable NO$_x$ levels decreased, with steam or water injection the primary technology used for NO$_x$ control, carbon monoxide emissions started to become more of a concern. Increases in carbon monoxide levels resulted from massive amounts of steam or water being injected to reduce NO$_x$ to the lower levels and part load operation in cogeneration applications. Subsequent advances in dry low NO$_x$ combustion technology and new add-on emission controls allowed gas turbine operators to achieve very low levels of NO$_x$ without injection. The Clean Air Act Amendments of 1990 have resulted in new emission control requirements, not only for NO$_x$, but also for carbon monoxide and volatile organic compounds in ozone nonattainment areas.

## 7.2.2 Gas Turbine Emissions

Potential pollutant emissions from gas turbines include oxides of nitrogen (NO and NO$_2$, collectively referred to as NO$_x$), carbon monoxide (CO), unburned hydrocarbons (UHC, usually expressed as equivalent methane), oxides of sulfur (SO$_2$ and SO$_3$), and particulate matter (PM). Unburned hydrocarbons are made up of volatile organic compounds (VOCs), which contribute to the formation of ground level atmospheric ozone, and compounds such as methane and ethane, which do not contribute to ozone formation. SO$_2$, UHC and PM are generally considered negligible when burning natural gas. Thus, NO$_x$ and possibly CO are the only emissions of significance when combusting natural gas in combustion turbines. The NO$_x$ production rate falls sharply either as the combustion temperature decreases or as the fuel−air ratio decreases, due to an exponential temperature effect. Therefore, the introduction of a small amount of any diluent into the combustion zone will decrease the rate of thermal NO$_x$ production. This is the physics behind the injection of water or steam and of lean combustors. Because the diluent effect is a thermal one, the higher specific heat of steam means that less steam needs to be introduced than air and less water than steam to achieve the equivalent NO$_x$ reduction. However, the introduction of steam or water to the gas turbine combustor is a thermodynamic loss, whereas redistributing combustor airflow splits (combustion vs dilution/cooling) has no impact on the cycle efficiency. As a result, the use of very lean combustors to achieve the lower $NO_x$ levels is more desirable than steam/water injection.

### 7.2.3 $NO_x$ Control Technologies

The technologies available for the control of $NO_x$ emissions from gas turbine engines include: (1) injection of water or steam into the combustion zone, a control technology that lowers flame temperature; (2) selective catalytic reduction; (3) the new proprietary post-combustion catalytic technology $SCONO_x$; (4) dry low $NO_x$ combustion (DLN), a technology that uses staged combustion and lean-premixed fuel-air ratios; and (5) catalytic combustion, a new technology that holds promise of achieving extremely low emissions levels. Injection, DLN, and catalytic combustion are so-called front-end technologies. SCR and $SCONO_x$ are referred to as "back-end" exhaust gas cleanup systems.

#### 7.2.3.1 *Water-Steam Injection*    Most of the experience base with gas turbine $NO_x$ emission control prior to 1990 was with diluent injection into the combustion zone. The injected diluent provides a heat sink that lowers the combustion zone temperature, which is the primary parameter affecting $NO_x$ formation. As the combustion zone temperature decreases, $NO_x$ production decreases exponentially.

Manufacturers continue to develop machines having higher firing temperatures as a way to increase the overall thermodynamic efficiency. However, higher firing temperatures mean higher combustion temperatures, which produce more $NO_x$, resulting in the need for more diluent injection to achieve the same emission levels of $NO_x$. There has also been a reduction of allowable $NO_x$ emissions and lower $NO_x$ levels require even more injection. The increased injection rate lowers the thermodynamic efficiency, seen as an increase in heat rate (fuel use), due to taking some of the energy from combustion gases to heat the water or steam. Furthermore, as injection increases, dynamic pressure oscillation activity (i.e., noise) in the combustor also increases, resulting in increased wear of internal parts. Carbon monoxide, which may be viewed as a measure of the inefficiency of the combustion process, also increases as the injection rate increases. Basically, as more and more water or steam is injected into the combustor to lower the combustion temperature, flame stability is affected. If increased sufficiently, the water will literally put out the flame. Thus, a design dichotomy exists whereby increasing firing temperature to increase the efficiency of the combustion process produces more $NO_x$, requiring more injection, which lowers the thermodynamic efficiency, producing more CO and also decreasing parts life. Increased injection to meet lower $NO_x$ emission limits simply exacerbates the problems associated with increased injection. The lowest practical $NO_x$ levels achieved with injection are generally 25 ppm when firing natural gas and 42 ppm when firing oil.

#### 7.2.3.2 *Selective Catalytic Reduction (SCR)*    In the SCR process, ammonia injected into the gas turbine exhaust gas stream as it passes through the heat recovery steam generator (HRSG), reacts with nitrogen oxides in the presence of a catalyst to

form molecular nitrogen and water. SCR has been found to work best in base-loaded combined-cycle gas turbine applications where the fuel is natural gas. This is because of the dependence on temperature of the catalytic NO$_x$-ammonia reaction and the catalyst life, and because of major problems associated with the use of sulfur bearing (liquid) fuels. The reaction takes place over a limited temperature range, 600–750°F: above approximately 850°F the catalyst is damaged irreversibly, and below 600°F, reaction efficiency drops unacceptably. In addition, because of the temperature dependency of the chemical reaction and catalyst life, SCR cannot be used in simple cycle configurations, except possibly in lower exhaust temperature systems. Other issues associated with SCR include exhaust emissions of ammonia (known as ammonia slip), concerns about accidental release of stored ammonia to the atmosphere, environmental concerns and costs of disposal of spent catalyst.

**Ammonia Release**   The use of ammonia in the SCR chemical process for NO$_x$ control presents several problems, since as noted earlier, ammonia is on EPA's list of Extremely Hazardous Substances. Releases of ammonia to the atmosphere may occur when ammonia that has not reacted with nitrogen exits the stack (ammonia "slip"), or as a result of accidents during transport, transfer, or storage. In addition, ammonia is a PM$_{10}$ precursor emission (i.e., it gives rise to emissions of particulate matter smaller in diameter than 10 μm). Some ammonia slip is unavoidable with SCR because the reacting gases are not uniformly distributed. Thus, some ammonia and unreacted NO$_x$ will pass through the catalyst, and in fact some catalyst manufacturers recommend operating with excess ammonia to compensate for imperfect distribution. An ammonia slip of 10–20 ppm is generally permitted in a new system (although higher slip has been noted) and will increase with catalyst age. In the past, ammonia slip was not considered to be a problem by regulatory agencies: since the release was from an elevated stack, the ground level concentration was relatively low. However, it has never appeared to be good environmental policy to allow ammonia to be released to the atmosphere in place of NO$_x$, and ammonia emissions are now of concern because emissions of particulate matter as small as 2.5 μm in diameter (PM$_{2.5}$) has attracted the attention of regulators.

**The Use of Sulfur-Bearing Fuels**   SCR has never succeeded as a measure for NO$_x$ control when a gas turbine was fired with sulfur-bearing oil. However, some regulatory agencies require the use of SCR, even when distillate oil, which contains sulfur, is used as a backup fuel. In most cases regulators have simply pointed to the many combined cycle plants with SCR permitted with oil as the backup fuel, even though most of those plants actually operate almost exclusively on gas and use little or no oil fuel. Those that have used oil have experienced significant problems.

The problems associated with the use of sulfur bearing fuels are due to the formation of the ammonium salts ammonium bisulfate, $NH_4HSO_4$, and ammonium sulfate, $(NH_4)_2SO_4$. These compounds are formed by the chemical reaction between the sulfur oxides in the exhaust gas and the ammonia injected for NO$_x$ control. Ammonium bisulfate causes rapid corrosion of boiler tube materials; and both ammonium compounds cause fouling and plugging of the boiler and an increase of PM$_{10}$ emissions. Ammonium bisulfate forms in the lower temperature section of the HRSG where it deposits on the walls and heat transfer surfaces.

These surface deposits can lead to rapid corrosion in the HRSG economizer and downstream metal surfaces resulting in increased pressure drop and reduced heat transfer (lower power output and cycle efficiency).

While ammonium sulfate is not corrosive, its formation also contributes to plugging and fouling of the heat transfer surfaces (leading to reduced heat transfer efficiency) and higher particulate emissions. The increase in emissions of particulates due to the ammonium salts can be as high as a factor 5 due to conversion of $SO_2$ to $SO_3$. Some of the $SO_2$ formed from the fuel sulfur is converted to $SO_3$ and it is the $SO_3$ that reacts with water and ammonia to form ammonium bisulfate and ammonium sulfate. The increase is a function of the amount of sulfur in the fuel, the ammonia slip (ammonia that does not react with $NO_x$) and the temperature. It can also be increased by supplementary firing of the HRSG and by the use of a CO oxidizing catalyst (which significantly increases the conversion of $SO_2$ to $SO_3$). The only effective way to inhibit the formation of ammonium salts appears to be to limit the sulfur content of the fuel to very low levels (or switch to a sulfur free fuel such as butane) and/or limit the excess ammonia available to react with the sulfur oxides.

Pipeline quality natural gas usually has a sulfur content so low that ammonium salt formation, though present, has not been a significant problem with natural gas-fired units. However, the sulfur content of even very low sulfur distillate oil (e.g., 0.05 percent) or liquid aviation fuel (Jet–A) may not be low enough to curtail the formation of ammonium bisulfate sufficiently to avoid the problems discussed above (ambient sulfates may also contribute). This potential is usually handled by a requirement to limit the operating time on the low sulfur distillate oil to a few hundred hours between shutdowns and then clean the HRSG internals (although disposal of the deposits may be a problem due to the presence of hazardous materials).

Lowering the ammonia slip or the sulfur concentration may lengthen the time between cleanings. Limiting the ammonia that is available to react with the sulfur oxides to negligible levels does not appear practical at $NO_x$ removal efficiencies above 80 percent because higher excess ammonia levels are required to achieve the higher $NO_x$ removal efficiencies. Limiting the excess ammonia may work at lower $NO_x$ removal efficiencies because the lower $NH_3/NO_x$ ratios required ensure that all the ammonia is consumed. However, when oil is to be used as the primary fuel, experience indicates that SCR should not be used, as there appears to be significant risk of equipment damage or failure, performance degradation and increased emissions of fine PM.

**Disposal of Spent Catalyst** SCR materials typically contain heavy metal oxides such as vanadium and/or titanium, thus creating a human health and environmental risk related to the handling and disposal of spent catalyst. Vanadium pentoxide, the most commonly used SCR catalyst, is on the EPA's list of Extremely Hazardous Materials. The quantity of waste associated with SCR is quite large, although the actual amount of active material in the catalyst bed is relatively small.

**7.2.3.3 SCONO$_x$** SCONO$_x$ is a postcombustion catalytic system that removes both $NO_x$ and CO from the gas turbine exhaust, but without ammonia injection.

The catalyst is platinum, and the active NO$_x$ removal reagent is potassium carbonate. At present, the only operating SCONO$_x$ system is being used with GE gas turbine engine, model LM2500, injected with steam to 25 ppm NO$_x$ at a facility in Vernon, California (Greater Los Angeles). Stack NO$_x$ is maintained at 2 ppm or less and CO at less then 1 ppm.

**How SCONO$_x$ Works**  The exhaust gases from a gas turbine flow into the reactor and react with potassium carbonate which is coated on the platinum catalyst surface. The CO is oxidized to CO$_2$ by the platinum catalyst and the CO$_2$ is exhausted up the stack. NO is oxidized to NO$_2$ and then reacts with the potassium carbonate absorber coating on the catalyst to form potassium nitrites and nitrates at the surface of the catalyst. When the carbonate becomes saturated with NO$_x$ it must be regenerated. The effective operating temperature range is 280–750°F, with 500–700°F the optimum range for NO$_x$ removal. The optimum temperature range is approximately the same as that of SCR.

Regeneration is accomplished by passing a dilute hydrogen reducing gas (diluted to less than four percent hydrogen using steam) across the surface of the catalyst in the absence of oxygen. The sections of reactor catalyst undergoing regeneration are isolated from exhaust gases using sets of louvers on the upstream and downstream side of each reactor box. The Vernon LM2500 facility has 12 vertically stacked catalyst reactor boxes, nine of which are in the oxidation/absorption cycle at any given time, while three are in the regeneration cycle. When regeneration has been completed in the three reactor boxes, the louvers open on those reactors and the louvers on three other reactors close and those reactors go into the regeneration cycle. Motor drives outside each box drive the shaft that opens and closes the louvers on each side of the box (inlet and outlet sides).

**SCONO$_x$ Issues**  There are several issues associated with the use of SCONO$_x$. First, it is very sensitive to the presence of sulfur, even the small amount in pipeline natural gas. Second, the initial capital cost is about three times the cost of SCR, although the price may come down when there are more units in operation. Third, the reliability of moving parts and performance degradation due to leakage may pose significant problems, especially on scale-up to bigger gas turbines (a 7FA would require 20 modules of four reactor boxes each vs the LM2500, which uses three modules of four reactor boxes). Last, use of any exhaust gas treatment technology (SCR or SCONO$_x$) results in a pressure drop that reduces gas turbine efficiency (SCONO$_x$ produces about twice the pressure drop of SCR). Thus, the requirement for a back-end cleanup system means that more fuel must be burned to reduce NO$_x$.

### 7.2.3.4 The GE Dry Low NO$_x$ Combustor

GE began development of a dry low NO$_x$ combustor in 1973, primarily in response to increasingly stringent emission control requirements in California. The initial goal was a NO$_x$ level of 75 ppmvd at 15 percent oxygen, the NSPS requirement for utility gas turbines. An oil-fired combustor designed for a Frame 7 gas turbine achieved this goal in the laboratory in 1978. Field testing of the prototype dry low NO$_x$ combustor design demonstrated that the combustor was capable of meeting the NSPS. The design, tested in Texas at Houston Lighting and Power in 1980, has evolved into a system that is achieving a $NO_x$ level of 9 ppmvd at 15 percent oxygen in GE Frame 7EA, FA, and 6B gas turbines fired on natural gas.

### 7.2.4  Discussion

#### 7.2.4.1  Cost in $/ton of NO$_x$ Removed/Energy Output Reduction   The annual cost of using SCR to reduce $NO_x$ emissions from 9 ppm to 3.5 ppm for a GE Frame 7FA, 170 MW class gas turbine operating 8,000 h/year is $8,000 to $12,000 per ton of $NO_x$ removed when a non-sulfur bearing fuel is used and $15,000 to $30,000 when the fuel contains sulfur. The cost will be the same or more than that with $SCONO_x$, which cannot be used with sulfur-bearing fuels without incurring additional costs for sulfur removal.

Most gas turbine combined-cycle or cogeneration systems today operate with natural gas as the primary fuel and fuel oil as the backup fuel. SCR operating and maintenance costs include continuous ammonia injection, periodic catalyst replacement, and the cost associated with a small decrease in power output ($>650$ kW for a Frame 7FA turbine). The output drop is due to power for auxiliaries associated with ammonia injection, catalyst pressure drop in the new and clean condition, which increases as ammonia-sulfur salts build up, and decrease in heat transfer as the salt buildup increases over time. This cost is considered too high for BACT in ozone attainment areas by most states. The decrease in output efficiency results in an increase in $CO_2$ emissions due to the need to burn more fuel to make up for the output reduction. It is often argued that economics should not be considered at all in determining lowest achievable emission rates (LAERs). There is, however, an implicit "reasonableness test" in all LAER determinations. Thus, no regulator has required that trains of multiple SCR be utilized to reduce $NO_x$ to zero (although this is technically possible) because the cost would be unreasonably high. This same rationale should apply to adding any emission control if the cost is unreasonably high, as is the case for adding SCR or $SCONO_x$ to a combustion system achieving 9 ppm $NO_x$ in a combined cycle.

#### 7.2.4.2  Ammonia Slip/Ammonium-Sulfur Salts   The impact of slip on the environment may be at least as detrimental as that of $NO_x$. Where an ammonia emission limit is imposed, and there is often no such emission limit, slip is generally targeted at 10–20 ppm, although there are units operating with ammonia slip well below and well above that level. Most recent SCRs operate with 5 ppm slip or less, but slip is expected to be on the high side when the $NO_x$ level entering the catalyst bed is already very low. Unless there is perfect mixing, the ammonia molecules must "find" the fewer $NO_x$ molecules in order to react and this will require adding more excess ammonia. Thus, 20 ppm or more ammonia slip would be released in place of the reduction in $NO_x$ in going from 9 to 3.5 ppm. Table 7.4 shows that a Frame 7FA turbine with 20 ppm ammonia slip (base load, 8,000 h/year, 45°F ambient, natural gas) emits 24 tons per year (tpy) more ammonia than results from $NO_x$ reduction by lowering $NO_x$ from 9 to 3.5 ppm with SCR. There also is an increase of 5.6 tpy in particulate matter emitted, or 33.6 tpy if a CO catalyst is

**TABLE 7.4. Estimated Tons/Year Change in Emissions for 7FA Turbine (with SCR and COC, Based Load, 8,000 h/year, 20 ppm Ammonia Slip, 45°F Ambient)**

| | 9 ppm NO$_x$ w/o SCR | 3.5 ppm NO$_x$ w/SCR | TPY | 3.5 ppm NO$_x$ w/SCR and COC | TPY |
|---|---|---|---|---|---|
| Natural gas only | | | | | |
| NO$_x$ | 240 | 92 | −148 | 92 | −148 |
| PM$_{2.5}$ | 36 | 41.6 | +5.6 | 69.6 | +33.6 |
| NH$_4$ | 0 | 172 | +172 | 164 | +164 |
| SO$_x$ | 40 | 39 | −1 | 25 | −15 |
| Gas + 400 h/year oil | | | | | |
| NO$_x$ | 294 | 116 | −178 | 116 | −178 |
| PM | 37.6 | 45.8 | +8.2 | 86 | +48.4 |
| NH$_4$ | 0 | 172 | +172 | 161 | +161 |
| SO$_x$ | 57 | 56 | −1 | 36 | −21 |

SCR — selective catalytic reduction.
COC — CO oxidizing catalyst.

also used. Note also that as the catalyst ages, ammonia slip increases as the efficiency of conversion decreases, until at the end of catalyst life the ammonia slip may be much higher than a new and clean catalyst. In fact that is one way that catalyst replacement is indicated. Some ammonia released to the atmosphere will be converted to NO$_x$ and ultimately to ozone.

Finally, accident studies of transport and on-site storage of ammonia for use with SCR, performed for the Massachusetts Department of Environmental Protection and California's South Coast Air Quality Management District, resulted in a change from anhydrous ammonia to aqueous ammonia. Aqueous ammonia has a lower ammonia concentration and lower storage pressure (resulting in a slower release rate) than anhydrous. Anhydrous ammonia was used until these studies revealed the potential public hazard in the event of catastrophic release. The hazard was reduced, but not eliminated. GE Power Systems analysis of measurements of ammonia emissions on six plants with SCR showed a great deal of inconsistency (<1–30 ppm). All of the tests were performed using different ammonia sampling methodologies. EPA Method 206 for ammonia is applicable to coal-fired plants. There is no specific method for gas turbine plants. The conclusion drawn from this study is that the ammonia slip on plants with SCR is not known with any accuracy.

### 7.2.4.3 Spent Catalyst
From a policy standpoint, the disposal of spent catalyst as hazardous waste simply converts an air problem (NO$_x$) into a long-term solid waste disposal problem. This is not a good environmental tradeoff.

### 7.2.4.4 Use of Sulfur Bearing Fuels
It has been a long-standing policy at GE that SCR should not be used in gas turbine applications that use a sulfur-bearing fuel,

such as distillate oil. In response to EPA's promulgation of the National Ambient Air Quality Standards for fine particulate matter ($PM_{2.5}$), the company increased its determination to avoid the use of SCR in conjunction with sulfur-bearing fuels. Unreacted ammonia from the SCR, and sulfur from the fuel react to form ammonium salts that are released as particulate matter, as previously discussed. EPA is very concerned about emissions of $PM_{2.5}$ because these very fine, inhalable particulates would increase significantly.

Aside from the important health risks that EPA has indicated are posed by $PM_{2.5}$, the impact of the increase in fine particulates on regional haze should also be considered. A CO oxidizing catalyst, supplementary firing catalysts, and noble metal catalysts will all result in much higher $SO_2$ to $SO_3$ conversion and greater sulfur salt formation. Note that particulate emission controls have never been used on gas turbines. Although many gas turbine combined-cycle plants using SCR are permitted to use distillate oil as the backup fuel, successful large-scale operation with this combination has not been achieved. Indeed, operating experience indicates that ammonium- sulfur salt formation and boiler damage occur without exception, when ANY sulfur bearing fuel is fired in the gas turbine and SCR is used for $NO_x$ control. Seldom accounted for in BACT determinations, such damage adds significant cost and should be considered. Besides absorbing the downtime associated with periodic cleaning, it is necessary to pay for either periodic replacement of the low pressure tube sections of the HRSG damaged by ammonium bisulfate corrosion or the development and installation of an alternative design HRSG. Schorr documented the damage done to the HRSGs on several representative plants (Schorr, 1995).

## 7.3   QUESTIONS OF REGULATORY POLICY

Based on the performance of $SCONO_x$ at the single facility in California, $NO_x$ permit levels as low as 2 ppm are being required in some states. Yet a study by the American Society of Mechanical Engineers calculated that the need to work around errors in monitoring to verify compliance would force industry to meet a standard even more stringent than 2 ppm $NO_x$ (ASME, 1998). If, then, it is known that gas turbines can achieve an uncontrolled $NO_x$ level of 9 ppm, and that 2 ppm is not a practical emission limit for these engines, a question is raised about the feasibility of laws requiring manufacturers of these engines to obtain at any cost $NO_x$ levels nearer to 2 ppm than 9 ppm. The cost of reducing $NO_x$ from 9–3.5 ppm with SCR is $15,000 to $30,000 a ton. Is this reasonable for a BACT or LAER determination? If the cost of an add-on control is $100,000/ton, should this major expenditure be required even to achieve LAER in nonattainment areas? A million dollars a ton? Although state agencies can impose requirements more stringent than those of the EPA, are states justified in requiring businesses to use the top-down approach in determining BACT, to ignore cost-effectiveness, or to set an arbitrary effectiveness threshold that is much higher for some gas turbines than for other sources of emissions? Since BACT, by definition, is supposed to be site/project specific, a one-size-fits-all approach to regulating gas turbines is hard to rationalize.

Some state environmental agencies require the use of add-on controls for gas turbines able to achieve otherwise satisfactory $NO_x$ emission levels (i.e., 9 ppm) to bring emissions down to 2 or 3 ppm in attainment and nonattainment areas, ignoring all other factors, simply because the technology to reach the lower levels exists. In the extreme case, a gas turbine manufacturer who achieves an uncontrolled $NO_x$ level of 3 ppm might be required to use add-on controls to reduce the $NO_x$ another 0.5 ppm, no matter what the cost. It is not certain that the Clean Air Act anticipated this kind of situation.

Regulators often state that economics cannot be considered in determining LAER. It should also be noted that the controls required to reduce emissions of a nonattainment pollutant sometimes have negative environmental impacts. It seems at best unwise to ignore these in determining LAER. For example, one might ask whether emitting ammonia in place of $NO_x$ is a good environmental trade-off, or whether it makes sense to reduce atmospheric loadings of $NO_x$ while permitting emissions of ammonia of the same order of magnitude.

The ASME study found that 2 ppm $NO_x$ emission levels probably can be measured, but these engineers did not thoroughly investigate whether the present generation of gas turbine engines can be controlled at such low levels under all operating conditions. The one unit operating with $SCONO_x$ that appears to be achieving the 2 ppm level runs only at full load, with no load following. In any event, the foregoing practical questions would seem to be mooted by the limitations of monitoring equipment. Specifically, 10 ppm is the lowest scale certified for a process $NO_x$ analyzer.

Gas turbine manufacturers will continue to develop lower $NO_x$ combustion systems only as long as economic incentives to do so exist. If it becomes apparent that add-on controls such as SCR will be required even for equipment able to achieve uncontrolled $NO_x$ levels well below 9 ppm, the development of lower $NO_x$ systems will be discouraged.

## 7.4 SMOKESTACK EMISSIONS

In the United States, the contribution of coal to electricity generation is more than half, while more than a third of electricity generated worldwide comes from coal. The dominance of coal in electricity production is expected to continue well into the 21st century (EIA, 2002). Several facts render coal essential to current and future world energy needs. Coal represents over 80 percent of the world's proven recoverable fossil fuel (by heat content), and the United States has one-fourth of the world's supply. Heavy reliance on coal for energy production, however, poses serious environmental problems.

Ever since the issues of acid rain (wet and dry acid deposition), and lower atmospheric ozone and smog formation, began to raise important environmental and public health concerns, utilities have been under pressure to rid smokestack emissions of sulfur dioxide ($SO_2$) and nitrogen oxides ($NO_x$), a matter made more pressing by the ongoing fuel switch from oil and gas to less expensive coal. In fact, despite some significant reductions of $SO_2$ emissions since 1970, the major cause of such emissions (coal combustion) has almost tripled.

### 7.4.1   Clear Skies Initiative

The Clear Skies Initiative (CSI), designed to help protect the environment and public health, and to restore affordable energy supply, was introduced by the US Environmental Protection Agency in 2002 (CSI, 2002). The intention of the proposal is to achieve significant emission reductions of $SO_2$, $NO_x$, $CO_2$, and Hg from the coal-fired electric power generation industry between 2012 and 2020. The industry will be required to comply with existing and contemplated regulations mandated by the 1990 Clean Air Act (CAA). Under the CSI's regulatory framework, such reductions would mitigate the health and environmental effects of fine particles ($PM_{2.5}$), ozone, regional haze, acid rain, and eutrophication. According to EPA, coal-fired power generation companies currently emit about 11 million tons (Mt) of $SO_2$ and 5 Mt of $NO_x$ every year. The scale of pollution from these coal-fired smokestacks is immense, and so is the damage to public health. Data supplied by EPA in late 1990s estimate the annual health bill: more than 10,800 premature deaths; at least 5,300 chronic bronchitis incidents; more than 5,100 hospital emergency visits; and over 1.5 million lost work days. To these must be added the severe damage to natural resources due to acid rain that attacks soils and plants, and deposits of nitrogen and sulfur in many critical bodies of water (USEPA, 1997). The CSI offers a framework for addressing these crucial issues and taking action in a coordinated and phased manner. Under the CSI, the $SO_2$ cap will be lowered to 4.5 Mt by 2010, and to 3 Mt in 2018, for $NO_x$ will be 2.1 Mt in 2008 and 1.7 Mt in 2018, and for Hg will be 26 tons in 2010 and 15 tons in 2018, respectively. The CSI also set a national goal to reduce the greenhouse gases by 18 percent (equivalent of 32 metric tons carbon) by 2012 (CSI, 2002).

The CSI proposal, which has yet to be adopted by Congress, would mark a major shift in regulating pollutant emissions from the current "command-and-control" regulatory program to a market-based, flexible, "cap-and-trade" scheme. Such a shift in implementing environmental laws, especially if expanded to other industrial sectors, would have significant impacts not only on the stakeholders in the regulatory process but also on the environmental profession as a whole. Even so, total $NO_x$ removal under the CSI proposal might not be enough to comply with the revised National Ambient Air Quality Standards (NAAQS) for ozone, and likewise, overall $SO_2$ reduction might not suffice to comply with the proposed regional haze rule (Regional Haze Regulations, 1999).

A technology developed in Japan in the late twentieth century holds promise for boosting industry compliance with present and anticipated rules. The Japanese system represents a refinement of the electron beam ammonia method of converting $SO_2$ and $NO_x$ to powdered fertilizer feedstock.

### 7.4.2 The Ebara Process

Electron beam irradiation of a flue gas containing both $SO_2$ and $NO_x$ (about 95 percent NO and 5 percent $NO_2$) acts on these pollutants by oxidizing them to nitrogen dioxide and sulfur trioxide anions. Addition of water and ammonia to these ionized gases yields solid products, ammonium nitrate ($NH_4NO_3$) and ammonium sulfate (($NH_4)_2SO_4$) that can be separated and sold as fertilizers. The simplified chemical reactions are as follows:

$$NO + O \rightarrow NO_2$$
$$NO_2 + OH \rightarrow HNO_3$$
$$SO_2 + O \rightarrow SO_3$$
$$NH_3 + HNO_3 \rightarrow (NH_4NO_3)$$
$$2NH_3 + H_2O + SO_3 \rightarrow (NH_4)_2SO_4$$

A similar process has been developed and marketed by the Tokyo-based Ebara International Corporation (Ebara, 2000; Chemical Week, 1983). The main attraction of the Ebara process (Figure 7.1) is the simultaneous removal of both $SO_2$ and $NO_x$. The complex reaction mechanism entails ionization, the formation of excited electronic states, the transfer of excitation energy between molecules, molecular dissociation, electron capture, neutralization, and radical reactions. The flue gas is first cooled and humidified, then irradiated by means of a high-intensity electron beam. The flue gas is routed first through electrostatic precipitators (ESPs) or other mechanical collectors to remove particulates, such as fly ash. The gas is

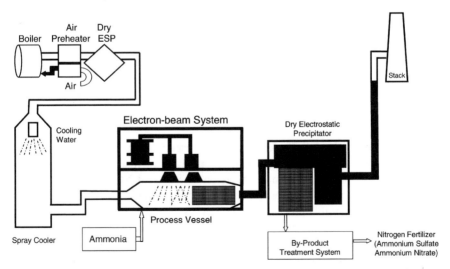

**Figure 7.1.** Schematic diagram of the electron beam-ammonia conversion of $SO_2$ and $NO_x$ to fertilizer (Ebara Corp., 2000).

next piped to a spray cooler and humidified and cooled to a temperature in the 160–200°F range, the optimal temperature. Anhydrous ammonia is introduced in stoichiometric amounts at this point. The ammoniated flue gas is passed to an electron-beam reactor chamber, where it is irradiated (1.5 Mrad dose), producing nitrate and sulfate anions. These species react with ammonia to yield a dry, particulate mixture of $NH_4NO_3$ and $(NH_4)_2SO_4$.

### 7.4.2.1 *Advantages and Disadvantages*  There are at least five advantages of the electron beam process:

1. $SO_2$ and $NO_x$ are removed together in the same piece of equipment by the addition of one chemical. The method can meet all New Source Performance Standards (NSPS) for $SO_2$ removal because it has the potential to remove 98 percent of all $SO_2$ and 70–90 percent $NO_x$ depending on the process condition (CFR-40, 1999). High $SO_2$ removal can be achieved with very low power consumption (1 percent of the block). Most of the energy is consumed for $NO_x$ reduction. NSPS standards are complex, but in the case of 3–6 percent sulfur coal, a 90 percent reduction is required.
2. Volatile organic compounds are removed in the process as well.
3. Since no waste is produced, there is no need to devise a means of disposal.
4. Since the process produces a dry product, high maintenance costs associated with abrasive and corrosive slurries are avoided.
5. A salable product results.

Among the corresponding disadvantages and challenges associated with this process is the need to keep ammonia injection close to the stoichiometric point to alleviate any ammonia slip. In addition, large electron beam generators are needed for a large-scale plant because the current electron beam has a limited range. Technical concerns about this process include questions about the ability to scale up the electron accelerators for a full-scale application and the associated power requirements for the technology. To accelerate radiation, a combined microwave-electron-beam process has been developed and patented, and this hybrid radiation system will be suitable for a large scale plant (Ebara, 2000).

### 7.4.2.2 *Cost Analysis*  Table 7.5, which presents an annualized cost analysis, including a projected overall return on investment for the installation and operation of a system of E-beam-ammonia conversion of $SO_2$ and $NO_x$. The cost analysis shows that the coal-fired power generating industry can earn about $11 million dollars per year from this process. The system itself is much cheaper than conventional flue gas treatments — with 25 percent less construction costs and 20 percent less running costs — and requires considerably less space.

But there are caveats in all such calculations. For starters, the Ebara process would be sensitive to the cost of anhydrous ammonia. To control this variable, research is being conducted to discover a way to extract ammonia from municipal

**TABLE 7.5. Return on Investment Calculations: Electron Beam-Ammonia SO$_2$ and NO$_x$ Conversion (Ebara Corp., 2000)**

| Sample Calculation | Total Profit/Cost ($million) |
|---|---|
| Profit from product sales | |
| Fertilizer: 110,000 tons/year (400 tons/day) × $100/ton = | $11/year profit |
| Operating cost | |
| Personnel expenses, etc. | −$0.9/year |
| Ammonia: 29,000 tons/year (105 tons/day) × $130/ton = | −$3.8/year |
| | −$4.7/year cost |
| Facility construction cost = | −$38.1 cost |

*Note:* At an annual interest rate is 12 percent, facility construction cost would have been paid back within 15 years.

wastewater (Ebara, 2000). If ammonia were available from a nearby municipal wastewater treatment plant, it would be even more economical to run this process. In any event, the fertilizer-producing variant of the Ebara process would be economically beneficial only in locations well sited with respect to markets for the fertilizer materials produced as byproducts, and variability in capital and electricity costs.

*Mini-Case Study 7-1: Early Ebara Facilities in China and Japan*  The first fertilizer plant using the Ebara process and these gases from a coal-fired power station was inaugurated in Chengda, China, in 1997. This city, in Szechuan Province has a high dependence on coal-fired power generation, as well as strong demand for fertilizer, and the Ebara facility treats 300,000 N-m$^3$/h of flue gas emitted from a coal-fired thermal power plant.

With the plant in China running successfully, Ebara continued to expand the capacity of their electron-beam facilities. In 1999, on the Japanese island of Housha, the company started construction of a 620,000 N-m$^3$/h facility for Chubu Electric Power Company. Ebara is also refining new technologies that will result in even higher capacity.

***7.4.2.3 The Ebara Mebius System***  Another important product from Ebara's is the company's Mebius system, which allows several forms of waste to be mixed and simultaneously treated with fermentation technologies to create methane gas, a useful energy source, without producing dioxins. Ebara initially became involved in this field in 1996, introducing technologies developed by CITEC of Finland. In the following year, they sublicensed this technology to six Japanese water treatment companies, with which Ebara is conducting joint research at a demonstration plant in Kanagawa Prefecture.

In addition, a local government in Niigata Prefecture commissioned Ebara to build a waste treatment facility based on its Mebius system and said to be the first in Japan to simultaneously treat night soil, septic tank sludge and kitchen waste. The plant has a daily throughout of 70 kL of night soil, 70 kL of sludge and 8 tons of kitchen waste. Each ton of waste produces around $100 \, m^3$ of methane gas, which is used to power the facility.

Ebara's Zero Emission activities have generated considerable commercial benefits, and the company continues to refine new technologies.

***7.4.2.4 Future Prospects***   The Ebara process exemplifies how pollution prevention, industrial ecology, and zero emission concepts are being implemented into sustainable business strategies by integrating environmental protection, societal, and economic interests. Efforts are under way around the world to promote the concept and actualization of zero emissions and zero discharge industries. These industries focus on zero waste as a means of improving the profitability of manufacturing while mitigating its environmental impact. The zero emission or discharge industry seeks to develop clusters of industries that use each other's waste as inputs to their own processes (Ebara, 2000; Das et al., 2001).

The Ebara process offers industry a way of reducing environmental pollution generated by individual facilities. This is the so-called end-of-the-pipe approach. Another approach to pollution control has a more holistic focus. Figure 7.2 provides a context for the discussion that follows by comparing the end-of-the-pipe and zero discharge approaches.

## 7.5   ECO-INDUSTRIAL PARKS: MODEL ZERO DISCHARGE COMMUNITIES

This section reviews several key publications based on research and initiatives in the United States, Japan, Denmark, England, and other countries in the growing fields of industrial ecology, zero-discharge manufacturing, and establishment of eco-industrial parks (Hawken, 1993; Graedel, 1994; Mehrota, 1995; Salvesen, 1996; Cornell University Work and Environment Initiative, 1995; Ebara Corporation in Japan, 1997, 1998, 1999).

Presently, resources in many parts of the world are so plentiful that the flow of materials from one stage in an industrial ecology to another is independent of all other flows. Anticipating future needs, a cyclic ecology in an eco-industrial park setup would allow all resources to be conserved. This means that all waste will be reused as raw materials; energy of course must be put into the system, but nothing classifiable as waste should come out. Research and development efforts are underway around the world to promote the concept and actualization of eco-industrial parks and ZD industries.

### 7.5.1   Industrial Ecology

There are several practical definitions of industrial ecology (IE) as follows. It is said, for example, to be a means of deliberately and rationally approaching and

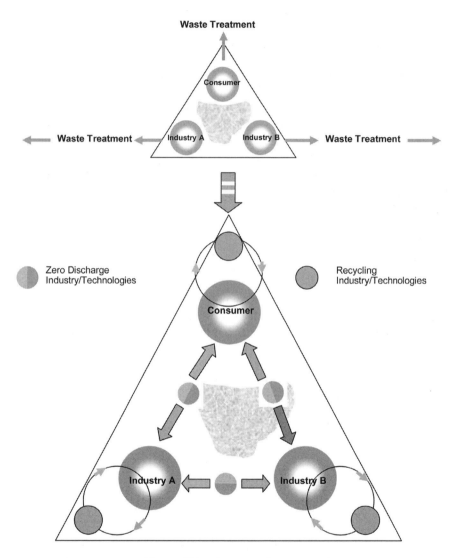

**Figure 7.2.** Schematic diagram of the end-of-pipe and zero-emission approaches (Ebara Corp., 2000).

maintaining a desirable carrying capacity, given continued economic, cultural, and technological development. The concept requires that an industrial system be viewed not in isolation from its surrounding systems, but in concert with them. It is a view of a system in which one seeks to optimize the total materials cycle from virgin material to finished material to component to product to obsolete product and to ultimate disposal. Factors to be optimized include resources, energy, and capital.

Industrial ecology is a dynamic, systems-based framework that enables management of human activity on a sustainable basis by minimizing energy and materials usage, ensuring acceptable quality of life for people, minimizing ecological impacts of human activity to levels natural systems can sustain, and maintaining economic viability of systems for industry, trade and commerce. The industrial ecology approach involves (1) application of science to industrial systems; (2) defining the system boundary to incorporate the natural world; and (3) seeking to optimize that system. In this context, industrial systems apply not just to private sector manufacturing and service but also to government operations, including provision of infrastructure (Diwekar).

Industrial ecology is a framework for designing and operating industries as living systems interdependent with natural systems, and therefore it is essential to grasp the concept of industrial metabolism. Whereas the biosphere's metabolism is a near-perfect recycling system because so many of its components are capable of biological regeneration, in industry, metabolism depends largely on the combustion of fossil fuels that are not regenerated within the system.

Thus students of industrial metabolism investigate the mass flows of key industrial materials of environmental significance and their associated emissions. With this information in hand, industrial ecologists are better able to balance environmental concerns and economic performance, factoring in the real value of nonrenewable resources and the real costs of environmental pollution. These efforts are enhanced by improvements in understanding of local and global ecological constraints.

Industrial ecology supports coordination of design over the life cycle of products and processes. IE enables creation of short-term innovations in a long-term context. While much of the initial work in IE has focused on manufacturing, a full definition of industrial systems includes service, agricultural, manufacturing, military and other public operations, as well as infrastructure such as landfills, water and sewage systems, and transportation systems.

In 1989, the concept of an "industrial ecosystem" received wide attention when Scientific American published an article by two General Motors researchers who suggested that the days of finding an "open space beyond the village gates" for the by-products of industrial activity were quickly fading (Frosch and Gallopoulos, 1989). Since that time, the concept of industrial ecology has spawned an ever increasing amount of research and activities. At the most basic level, industrial ecology describes a system where one industry's wastes (outputs) become another's raw materials (inputs). Within this "closed loop," illustrated earlier in Figure 7.2, fewer materials would be wasted. Thus, if businesses were able to turn waste into feedstocks, they could sharply reduce pollution and the need for raw materials (Van Der Ryn and Cowan, 1996).

Many industries have long had symbiotic relationships where wastes and materials are transformed internally or by others. For example, metal industries use scrap materials in the production process; the advent of the electric arc furnace increased the ability of steel manufacturers to use scrap materials. Petrochemical and chemical companies are adept at finding new production uses or markets for waste materials (Richards et al., 1994). The growth in rubber, plastics, paper, and glass recycling has generated new uses for previously discarded materials. As Ernest Lowe suggests, industrial ecology is a broad holistic framework for guiding the transformation of the industrial systems. The shift from the linear model (mine pit to producer to consumer to dump) to a closed-loop model more closely resembling the cyclical flows of ecosystems has stimulated new ways of thinking in forward-looking companies, in a number of universities, and in governmental agencies like the Environmental Protection Agency and the Department of Energy in the United States (Lowe, 1995).

### 7.5.2 What Is an Eco-Industrial Park?

According to Ernest Lowe, John Warren, and others, an Eco-industrial park is a community of manufacturing and service businesses seeking enhanced environmental and economic performance through collaboration in managing environmental and resource issues including energy, water, and materials (Lowe and Warren, 1996). By working together, the community of businesses seeks a collective benefit that is greater than the sum of the individual benefits each company would realize if it optimized its individual performance only. The goal of an EIP is to improve economic performance of the participating companies while minimizing their environmental impact.

Industrial activity releases wastes into local, regional, and eventually global ecosystems. Since the 1970s, myriad laws and regulations have been promulgated to limit emissions to air and water, and to regulate solid and hazardous waste disposal. Further, industrial activity results in non-point-source pollution caused by general runoff, spills, or illegal dumping. Business and the environment have traditionally been considered natural enemies. The assumption has long been "more environmental protection corresponds to higher costs for business"; however, new developments in research and business operations are challenging this assumption. Many companies such as 3 M have long realized the economic benefits of applying environmental principles to business operations. The goal of an eco-industrial park is to minimize the ecological impact of industrial activity and to improve business performance. In the United States and Canada several projects are testing this idea, which has already demonstrated its effectiveness in Europe, most spectacularly perhaps, in Kalundborg, Denmark, as discussed in Chapter 2.

### 7.5.3 Eco-Industrial Park Development

Eco-industrial park development is a new paradigm for achieving excellence in business and environmental performance. It opens up innovative new avenues for managing businesses and conducting economic development. By creating linkages among local "resources," including businesses, nonprofit groups, governments, unions, and educational institutions, communities can creatively foster dynamic and responsible growth. Antiquated business strategies, based on isolated enterprises, are no longer responsive enough to market, environmental, and community requirements.

Economic development is a never-ending challenge for communities. As the global marketplace has become increasingly competitive, municipalities, counties, states, and regions seek new strategies for attracting good investments with good jobs. Moreover, communities everywhere are demanding improvements in local ecosystems. In the past, economic development and environmental protection were seen as mutually exclusive. However, new practices and activities are challenging that assumption. One broad category of activities falls under the umbrella of eco-industrial park development (EIPD).

Sustainable EIPD looks systematically at development, business, and the environment attempting to stretch the boundaries of current practice. On one level, it is as directly practical as making the right connections between wastes and resources needed for production. At another level, it is a whole new way of thinking about doing business and interacting with communities. The eco-industrial approach has many ways of being applied. At a most basic level, each organization seeks higher performance internally. However, most eco-industrial activity is moving ahead by increasing interconnections between companies.

Just as in nature interconnected systems work together to ensure survivability and efficient use of resources and energy, in the business world, strategic partnerships, networked manufacturing and preferred supplier arrangements assist companies to grow, to contain costs, and to reach for new opportunities. Eco-industrial development can help to achieve these goals by offering businesses access to cost-effective, quality resources for producing products or delivering services.

The vital task of securing community support for eco-industrial projects calls for multi-stakeholder engagement in setting a vision for local development. Companies that offer attractive models in their community outreach will have an edge in obtaining popular support. Attention to getting wide backing for a project in the early planning stages can reduce delays due to the reflexive community opposition that often crops up when a major change is proposed. A positive and proactive stance in anticipating and overcoming residents' concerns can be most helpful.

The three mini-case studies presented in Section 7.5.4 illustrate the range of results that are possible when an attempt is made to husband the resources and improve the environment of communities dominated by a single industry.

### 7.5.4 Eco-Industrial Parks — Mini Case Studies in India

Ramesh Ramaswamy and Suren Erkman have played a central role in introducing industrial ecology into India through field research, conferences, and workshops. Their organization, Institute for Communication and Analysis of Science and Technology (ICAST), organized a major workshop on Industry and Environment in

Ahmendabad in 1999, working in collaboration with the Confederation of Indian Industries and the Indo-Dutch Project on Alternatives in Development.

ICAST has conducted industrial metabolism studies on several industrial systems in India, including a paper mill/sugar mill complex, foundries, and leather processing. Mini-Case Studies 7-2 to 7-4 describe these projects briefly. For more details, consult the paper the industrial ecologists presented to their colleagues at a UN conference (Erkman and Ramaswamy, 2001).

*Mini-Case Study 7-2: Seshasayee Paper and Board Ltd.* In this success story, a corporate group devised and adopted a growth strategy whereby most of the wastes from one activity were converted to feedstock for another activity.

Seshasayee's original enterprise was a paper mill. To ensure a regular supply of raw material, the company then set up a sugar mill. Bagasse, a waste from the sugar mill, was used as a raw material for paper-making. Another waste product, molasses, was used in a distillery for the production of ethyl alcohol. To guarantee a regular supply of sugarcane, the company took interest in the cultivation of this crop by organizing the farmers in the region. Seshasayee struck long-term agreements with the farmers to buy back their produce and, in turn, took the responsibility of supplying them with water. Part of the water for cultivation was treated wastewater from the paper manufacturing operations. The company also used bagasse pith (a waste after the paper making) and other combustible agricultural wastes in the region, as energy sources. This example could be viewed as an agro industrial eco-complex (Figure 7.3).

## ECO SUGAR COMPLEX

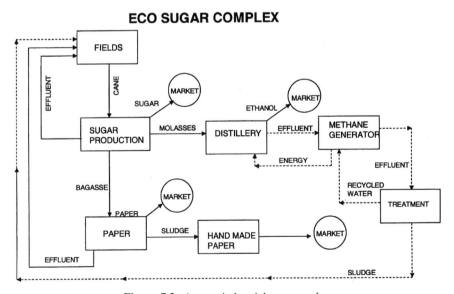

**Figure 7.3.** An agroindustrial eco-complex.

*Mini-Case Study 7-3: The Foundries in Howrah*   There are nearly 500 foundries in Howrah, a suburb of Kolkata (formerly Calcutta), in eastern India, and authorities have been insisting that these facilities install pollution control systems to improve the quality of the air, which is heavily polluted by coke by-products.

A new technology for the use of natural gas instead of coke in the foundries has been developed, and the authorities are likely to mandate conversion to natural gas to eliminate the pollution problem. Since natural gas is not available in the region, however, this conversion could substantially increase the cost of production. Because the engineering industry in the eastern region has been depressed, the foundries now subsist on manufacturing very low value-added products like manhole covers. A requirement to adopt a much pricier fuel would put them out of business.

A detailed study of the foundries and the region showed that the industry could adapt the new technology to use coke oven gases instead of natural gas. As the eastern region is a major coal-producing region and as there are many independent coke ovens, coke oven gas is easily available in the Kolkata region and is often wasted. Depending on the economics and logistics of individual situations, either a foundry could be relocated near coke ovens or the coke oven gases could be transported to a foundry.

*Mini-Case Study 7-4: The Leather Industry in Tamil Nadu*   Here we highlight the option of strategic relocation of an industry segment to ensure its long-term survival. Tamil Nadu, a state in the south of India, is the country's premier center for the processing of leather. The leather industry, which has been flourishing in the region for decades, comprises thousands of small businesses, many of which date back to the period of British rule in India, when Madras (now Chennai), the first city of Tamil Nadu, was a major trading center. The industry remains a major foreign exchange earner and important to the economy of the state. Paradoxically, compliance with strict environment regulations has rendered the processing of leather very expensive in the developed world, and the industry in Tamil Nadu has grown as buyers in the developed countries began to look to India for their tanned products.

In a country where the average per capita water availability for human settlements is estimated at around 30 L/day, the leather industry uses 30,000–50,000 L for each ton of hide. Water is often trucked in to plants, at high cost. Moreover, the wastewater is very saline and unusable for most human needs. The sludge from water used in tanning, estimated at 250 kg/1000 kg of hide processed, is carelessly dumped, and the pollutants leach into the groundwater. Unsurprisingly, the industry has been under pressure from pollution control authorities, and civil suits have been brought by citizens' groups.

A detailed study in the context of industrial ecology created a new perspective on the problem, which had heretofore been viewed merely as failure to meet specifications laid down by the pollution control authorities. Now the other dimension of the situation — the monopolization by one industry in Tamil Nadu, which has survived thus far because the slow judicial process in India favors the status quo. The possibility of social pressure bringing matters to a halt, however, is no longer remote.

In the long term, an alternate solution will need to be found. One of the options that may be considered in the context of industrial ecology would be to relocate the entire industry to the coast, where tanners can draw seawater, desalinate it for use, treat the wastewater and discharge the still-saline wastewater back into the sea. Desalination is expensive, but it may be possible to reduce the cost by setting up a thermal power plant and using the waste heat to run the desalinators. The sludge from the process could also incinerated and the energy used in the desalination process.

### 7.5.5 Conclusions

At its core, an eco-industrial park is very simple: it strives to increase business success while reducing pollution and waste.

Reuse of water offers a substantial benefit to participants in an EIP. Because many firms use large amounts of water in manufacturing, collaborative efforts can reduce the need for water and minimize the amount of effluents entering water treatment systems and the ecosystem.

Much of the attention to environmental technologies for EIPs swirls around the development of new processes for reusing wastes and byproducts. Central to an EIPs ability to exchange wastes and materials will be conversion or separation technologies capable of preparing former wastes for other uses.

It is certain that in the future concern for the environment will grow, resources will become more scarce and less reliable, and demands for quality and cost competitiveness will increase. To meet these challenges a new paradigm of industrial organization asks individual communities and companies to consider how they might work together to use resources more wisely.

## REFERENCES

ASME Codes and Standards Committee B133, Subcommittee 2, Environmental Standards for Gas Turbines, Report 9855-3, *Low NO_x Measurement: Gas Turbine Plants*, (Dec. 4, 1998).

Blackburn, J., and Ford, D. (1998). Wilson-Formosa discharge agreement — summary.

Byers, W. (1995). Zero discharge: a systematic approach to water reuse. *Chemical Engineering*, **102**, 96.

CFR-40 Part 60, Chapter 1, Subpart Db, Page 100–121, Revised as of July (1999).

*Chemical Week* (1983). Electron guns clean up $SO_2$ and $NO_x$. pp. 27–28 (Nov. 23).

Clear Skies Initiative (CSI), The White House: Washington, DC, February 14 (2002), available online at http://www.whitehouse.gov/news/releases/2002/02/clearskies.html (accessed April 2002).

Cornell University Work and Environmental Initiative (1995). Cornell University's Center for the Environment, *Fairfield Ecological Industrial Park Baseline Study*, Ithaca, New York.

Dalan, J. (2000). 9 things to know about zero liquid discharge. *Chem. Eng. Progress*, **98**(11), 71–76.

Das, T. K., Sharma, M. P., and Butner, S. (2001). *The concept of an eco-industrial park — a solution to industrial ecology*. Presented at the Topical Conference: Industrial Ecology and Zero-Discharge Manufacturing, American Institute of Chemical Engineers Spring National Meeting, Paper No. 18c, Houston, TX.

Delaney, T., and Schiffman, R. I. (1997). Organizational issues associated with the implementation of ISO 14000. *Environmental Engineer*, **33**(1).

Diwekar, U. M. (unpublished). Process analysis approach to industrial ecology.

Ebara Corporation, Tokyo, Japan (1997–2000). Progress on our zero emission challenge. Summary of Annual Reports.

Energy Information Administration (EIA), "International Energy Outlook-2002," Office of Integrated Analysis and Forecasting, U.S. Department of Energy, Washington, DC 20585, DOE/EIA-0484 (2002).

Erkman, S., and Ramaswamy, R. (2001) *Industrial Ecology as a Tool for Development Planning. Case Studies in India*. Sterling Publishers, New Delhi and Paris.

Ford, D. (1996). Zero discharge and environmental regulations, the toxic release inventory, and natural laws. *Environmental Engineer*, **32**(4).

Ford, D., and Blackburn, J. (1997). Proceedings, Bio 1997, Ninth Annual Council of Biotechnology Centers Meeting, Houston, Texas, June B-12.

Ford, D. (1997). *Formosa plastics — an industrial case study*. International Conference on Global Development, Rice University, Houston Advanced Research Center, National Academy of Sciences, James A. Baker III Institute for Public Policy, The Woodlands, TX (March).

Ford, D., Blackburn, J., and Mounger, K. (1994a). Project summary prepared by the technical review commission. Wilson-Formosa agreement.

Ford, D., Blackburn, J., and Mounger, K. (1994b). Wilson-formosa zero discharge agreement (July).

Formosa Plastics Corporation (1991). Application for Texas water commission wastewater discharge permit for the proposed Formosa Plastics Corporation expansion facility. Point Comfort, Texas.

Frosch, R., and Gallopoulos, N. (1989). Strategies for manufacturing. *Scientific American*, 144–152 (September).

Goldblatt, M. E., Eble, K. S., and Feathers, J. E. (1993). Zero discharge: what, why, and how? *Chem. Eng. Progress*, **89**, 22.

Graedel, T. (1994). *Industrial Ecology and Global Change*, In R. Socolow, C. Andrews, F. Berkhout, and Thomas, V. (eds), Cambridge University Press, Great Britain.

Hawken, P. (1993). *The Ecology of Commerce*. Harper Business, New York, NY.

Kiranmayee, V., and Manian, C. V. (2000). Zero discharge systems — a practical approach. Indian Chemical Engineering Congress, Science City, Kolkata, India, Paper # WTR21, (Dec. 18–21).

Lowe, E., and Warren, J. (1996). *The Source of Value: An Executive Briefing and Sourcebook on Industrial Ecology*. Pacific Northwest National Laboratory, Richland, Washington.

Lowe, E. (1995). The eco-industrial park: a business environment for a sustainable future. Presented at Designing, Financing and Building the Industrial Park of the Future Workshop, San Diego CA.

Mehrotra, P. (1995). *Industrial Ecology.* Pentap Press, Stanford, CA.

Morris, G. D. (1993). Formosa plastics labors to clean up its image. *Chemical Week*, June 23, pp. 18–19.

Parsons Engineering-Science (1993). Outfall diffuser modeling studies for Formosa Plastics, Inc.

Regional Haze Regulations; Final Rule. Fed. Regist. 64, 35714 (1999).

Richards, D. J., Allenby, B. R., and Frosch, R. A. (1994). *The Greening of Industrial Ecosystems: Overview and Perspective*, In B. R. Allenby, and D. J. Richards (eds), *The Greening of Industrial Ecosystems*, pp. 1–19, Washington, D.C., National Academy Press.

Rosain, R. M. (1993). Reusing water in CPI plants. *Chemical Engineering Progress*, **89**, 28.

Salvesen, D. (1996). Making industrial park sustainable. *Urban Land*, **55**(2), 29.

Schorr, M. M., and Chalfin, J. (1999). Gas turbine NOx emissions approaching zero — is it worth the price? general electric power generation. Internal Report #GER 4172, Schenectady, NY (September).

Schorr, M. M. (1995). NOx emission control for gas turbines: a 1995 update on regulations and technology. CIBO NOx Control Conference (March).

Texas Department of Water Resources Permit No. 02436 (1983).

The University of Texas at Austin (1993). Operating training course for Formosa Plastics, conducted by D. L. Ford and E. E. Gloyna, Jan. 19–21.

U. S. EPA (1997). *The Benefits and Costs of the Clean Air Act, 1970 to 1990.* (October 15). www.epa.gov/oar/sect812/design.html.

Van der Ryn, S., and Cowan, S. (1996). *Ecological Design*, Island Press, Washington, D.C. (1996).

Zacerkowny, O. (2002). Not a drop leaves the plant. *Pollution Engineering*, 19–22 (Feb.).

# CHAPTER 8

# TECHNOLOGIES FOR POLLUTION PREVENTION: AIR

Air pollution control has become an essential part of operations for many industries, particularly the chemical process industries. In developed countries, air quality problems are attributable to the by-products of combustion processes used in the private and public transportation sectors of the economy, as well.

Frequently, however, planners fail to acknowledge that control systems themselves are industrial processes that consume energy and can emit significant amounts of pollutants into the atmosphere. Regulators have tended to pursue the control of each target pollutant independently, with little consideration of secondary pollutants.

In the United States, for example, regulation of air pollution started with the largest sources because they had the most potential for immediate environmental improvement and because the major corporations responsible for these sources could reasonably be asked to assimilate the costs of added controls. The next regulatory phase saw a progressive tightening of standards and application of limits to more and smaller sources, with priority pollutants targeted as separate and distinct entities to be controlled.

Nevertheless, there are increasing opportunities to reduce energy consumption and secondary pollutants simply by selecting sustainable pollution control and prevention technologies in accordance with the principles introduced in Chapter 6. In this chapter we present a sampling of such technologies for the prevention of air pollution. The control processes described in Sections 8.1.1 to 8.1.3

*Toward Zero Discharge: Innovative Methodology and Technologies for Process Pollution Prevention*, edited by Tapas K. Das.
ISBN 0-471-46967-X   Copyright © 2005 John Wiley & Sons, Inc.

accomplish the destruction of low levels of volatile organic compounds by electron-beam-generated plasma, the biofiltration of organic vapors, and the use of sorbents and carbon injection to remove dioxins, furans, and toxic metals from municipal and medical wastes. In Section 8.2 we shall move on to a consideration of emerging technologies that are even more innovative.

## 8.1   SOME ON-GOING POLLUTION PREVENTION TECHNOLOGIES

Industrial air pollution prevention (P2) efforts have focused on both source and waste reduction, and on reuse and recycling. A key approach is preventing air pollution within a company's manufacturing processes, particularly in chemical process industries. Frequently we fail to consider that control systems themselves are industrial processes that consume energy and can emit significant pollutants into the atmosphere. There has been a tendency to pursue the control of each target pollutant independently while ignoring secondary pollutants. Opportunities to reduce energy consumption and secondary pollutants simply by selecting sustainable pollution control technologies are increasing.

Air pollution arises from a number of sources: stationary such as factories and other manufacturing processes; mobile such as automobiles, recreational vehicles, snowmobiles and watercraft; and area sources which are all other emissions associated with human activities. Air quality problems are closely associated with combustion processes occurring in the industrial and transportation sectors of the economy.

Air pollution control is an area where P2 and sustainability concepts are relatively new and just beginning to be applied, and the promise of significant environmental and economic benefits is strong. However, the regulatory development processes and framework have not fully embraced them.

Since large industries could reasonably assimilate the cost of controls into their cost of doing business, the United States' air quality regulatory approach began with these sources having the most potential for immediate environmental improvement. The regulatory requirements have progressively tightened standards and applied limits to smaller sources too. The regulations began targeting priority pollutants as separate and distinct entities to be controlled. These tendencies are part of the problem Congress was addressing through the *Pollution Prevention Act of 1990*, as stated in finding number 3:

> "The opportunities for source reduction are often not realized because existing regulations, and the industrial resources they require for compliance, focus upon treatment and disposal, rather than source reduction; existing regulations do not emphasize multi-media management of pollution; and businesses need information and technical assistance to overcome institutional barriers to the adoption of source reduction practices"
>
> United State Congress, Pollution Prevention Act of 1990;
> Pollution Prevention, 1997; State of California, January 2001.

However, some P2 concepts and the lack of cost-effective treatment technologies for smaller sources have also guided the regulatory approach. For example, many paint and coatings industry regulations have set limits on volatile organic compounds (VOCs) in the product formulations as the primary compliance mode, with end-of-pipe treatment as an alternative compliance option (State of California, 2001). On the other hand, current regulations also have requirements such as Best Available Control Technology (BACT), requiring large sources to achieve a high level of treatment for each target pollutant with only limited consideration of the cost, in either money or secondary environmental impacts.

An area of particular opportunity is the control of VOCs and odors. Traditional control technologies, such as thermal and catalytic oxidation, are effective in reducing VOC emissions by more than 95 percent. They have become the standard by which other VOC control technologies are measured for acceptability through regulatory concepts such as BACT. However, these technologies frequently demand significant energy and produce secondary wastes such as $NO_x$, $SO_x$, and greenhouse gases. It is not uncommon for a facility to have challenges meeting $NO_x$ emission standards due to the secondary emissions from their VOC removal systems. VOC control technologies frequently combust large quantities of natural gas to maintain the required temperature for destruction of the VOCs, resulting in significant emissions of greenhouse gases.

The following sub-sections of Chapter 8 present technologies for air pollution prevention. Some selected innovative air pollution control processes for criteria and toxic air pollutants are described. The control processes are as follows: (1) destruction of low level of VOC by e-beam generated plasma; (2) biofiltration of organic vapor; and (3) removal of dioxin/furans and toxic metals using sorbents and carbon injection.

## REFERENCES

*Pollution Prevention 1997: A national progress report*, available at http://www.epa.gov/opptintr/p2 97/foreword.pdf, accessed April 2004.

State of California, South Coast Air Quality Management District Rule 1151 (amended December 1998) and Rule 1132 (adopted January 2001).

United States Congress, Pollution Prevention Act of 1990, available at http://www.epa.gov/opptintr/p2home/p2policy/act1990.htm, accessed April 2004.

### 8.1.1 Destruction of Low Level of VOCs by Electron-Beam-Generated Plasma

ATUL C. SHETH, PROFESSOR, CHEMICAL ENGINEERING
The University of Tennessee Space Institute, Tullahoma, TN 37388, USA

The traditional approach of detoxification of airstreams polluted with toxic and hazardous organic chemicals by thermal incineration is still in use in many industries all over the world. Other alternatives include adsorption by granular activated carbon bed, absorption, catalytic oxidation, and ultraviolet photo thermal detoxification.

Detoxification by exposure to plasma generated by an electron beam is an effective way to convert hazardous volatile organic compounds (VOCs) in host gas streams, such as air, into innocuous chemical compounds without incurring the costs of high ignition temperatures needed for the thermal incineration process. Also, there is less chance of producing $SO_x$, $NO_x$ and other types of organic/inorganic pollutants using electron beam plasma.

The destruction of parts per-million (ppm) levels of volatile organic compounds in a dry air stream by high-energy electron-beam irradiation has been investigated in a pilot plant at the University of Tennessee Space Institute (UTSI), Tullahoma, Tennessee. Based on economic analyses carried out at UTSI and the Massachusetts Institute of Technology (MIT), this e-beam based approach to VOC destruction has been identified as an attractive option.

### 8.1.1.1 Background

Electron-beam bombardment has long been known to break down complex molecular structures. Its ability to destroy VOCs in an air stream was first appreciated by a team of scientists from Zapit Technology, Inc., and Lawrence Livermore National Laboratory (LLNL) (Zapit Technology Inc., 1994). Results of a series of beam irradiation tests using trichloroethylene (TCE) at 300 ppmv in dry air in one-liter Tedlar bags showed that the irradiation could destroy not only TCE, but also intermediate organic compounds formed by the irradiation, such as dichloroacetyl chloride (DCAC) and phosgene ($COCl_2$), reducing them to $CO_2$, HCl, and water.

In related work Slater and Douglas-Hamilton of Avco Everett Research Laboratory (Slater and Douglas-Hamilton, 1981) have reported the decomposition of low concentrations of vinyl chloride ($C_2H_3Cl$) in nitrogen, argon, and dry air. The major reaction mechanism was thought to be via charge transfer from the host gas to the constituent with the lowest ionization potential. These investigators developed a reaction kinetics model relating the destruction of the vinyl chloride to the radiation dose. This model was fitted to experimental data on the destruction of vinyl chloride at very low doses (3 kGy in dry air, 4 kGy in nitrogen, and 6 kGy in argon) by assuming a linear relationship between the $V/V_o$ (where V is the concentration of vinyl chloride ratioed to the initial concentration of vinyl chloride $V_o$) and $D/V_o$ (where the radiation dose D is defined as energy absorbed per unit mass of the irradiated material is ratioed to $V_o$). However, the resulting model was not able to predict results of experiments carried out at higher doses ($\leq 60$ kGy).

Researchers at MIT's Plasma Fusion Center (Bromberg and Cohn, 1992) have reported tests conducted on $CCl_4$ concentrations of 10–250 ppm and host gas flow rates of 0.3–0.7 cfm in a laboratory-scale electron beam reactor with a 150-keV energy beam. The results suggested dissociative electron attachment as the decomposition mechanism. The energy requirements for humid air in those tests were found to be approximately twice as large as those required for dry air. This group reported similar results for electron-beam destruction of dilute concentrations of trichloromethane ($CHCl_3$) in air (Koch and Cohn, 1993).

Scientists from Zapit Technology, Inc. and LLNL also made efforts to ascertain the effect of the presence of water in an air stream on destruction and removal efficiency (DRE), but found that the destruction of TCE in dry air and TCE in wet air (50 percent relative humidity) required similar doses. Energy requirements were drastically decreased, however, by the addition of a promoter such as hydrogen peroxide (Zapit Technology Inc., 1994)

Very little information is available regarding the mechanisms of destruction of VOCs for low and high doses in air. Although considerable work has been done on the detoxification of hazardous chemicals in aqueous solution by an electron beam, the detoxification of hazardous chemicals in an air stream is largely an unexplored area.

### 8.1.1.2 Process Concept

Ionization of VOCs in air by an electron beam induces chemical reactions. Slater and Douglas-Hamilton (1981) have reported that a pollutant compound present in low concentrations in a host gas is indirectly ionized by an electron beam via a charge transfer process, provided the pollutant compound has an ionization potential sufficiently lower than that of the carrier host gas. In the charge transfer reaction, instead of being ionized directly, a neutral molecule (B) can be ionized by losing an electron to an adjacent positive ion ($A^+$), thus: $A^+ + B \rightarrow A + B^+$.

The charge transfer process was also observed in experiments with mixtures of methane and an inert gas in which methane was the minor component (Spinks and Woods, 1990). Based on these observations, Anshumali and co-workers at UTSI assumed that VOCs are ionized by a charge transfer process (Anshumali, 1996). This process can be rapid, provided the VOC has an ionization potential sufficiently lower than that of the donor ions, which is frequently true for VOCs in air. Table 8.1, which gives the literature-reported ionization potential (IP) values for nitrogen, oxygen, and other candidate VOCs, shows that the IP of nitrogen is considerably higher than the IPs of the various commonly observed VOCs.

Since the experimental value of the energy required to form an ion pair in dry air is 33.85 eV, which is higher than the IPs for both nitrogen and oxygen, ion pairs and excited molecules are formed. Because of the substantially higher IP of nitrogen

**TABLE 8.1. Ionization Potential for Various Species**

| Species | Ionization Potential (eV) |
| --- | --- |
| Nitrogen | 15.60 |
| Oxygen | 12.06 |
| 1,1,2-Freon | 11.99 |
| Propane | 10.95 |
| 1,1,1-Trichloroethane | 11.04 |
| Vinyl chloride | 9.99 |
| Trichloroethylene | 9.47 |
| Perchloroethylene | 9.32 |

*Source: CRC Handbook of Chemistry and Physics*, 74th Ed., CRC Press, 1993–1994, pp. 10–208 to 10–226.

relative to the IPs of VOCs, the ion pairs undergo change transfer reactions with the VOCs, yielding free radicals. Anshumali suggested that these free radicals react with the excited molecules and form VOC-removal agents, $R$, which then react with the VOCs and destroy them (Anshumali, 1996). He proposed the following overall destruction scheme:

Initiation:

$$\text{host gas (air} + \text{VOC)} \rightarrow R \tag{8-1}$$

Termination:

$$R + \text{VOC} + O_2 \xrightarrow{k_1} Y + \text{products (CO}_2, \text{H}_2\text{O, etc.)} \tag{8-2}$$

Competing reactions:

$$Y + R + O_2 \xrightarrow{k_2} Z + \text{products} \tag{8-3}$$

$$R + O_2 \xrightarrow{k_3} \text{NO}_x \tag{8-4}$$

where $k_1$, $k_2$, and $k_3$ are the respective rate constants for reactions (8-2), (8-3), and (8-4). Because the energy required to form an ion pair in air is constant, the production of $R$ in reaction (8-1) would depend upon the formation of free radicals and, therefore, on the ionization potential difference ($\delta$IP) between nitrogen and the target VOC.

Experimental measurements made at UTSI showed destruction levels of VOCs levelling off at high doses, indicating the existence of other reactions competing with reaction (8-2) as scavenging steps for R. Hence, Anshumali posted the production of an unknown inhibitor species $Y$ which would react with $R$ to form $Z$, and thus compete with the destruction of VOC via reaction (8-3). Because Anshumali had used air as a carrier gas, NO$_x$ species were observed in the effluents during pilot-scale tests. Thus he suggested reaction (8-4) to account for the potential removal of $R$ from the system by reaction with $O_2$, forming NO$_x$. In this reaction, which competes with reactions (8-2) and (8-3), $R$ was believed to be the source of nitrogen.

Invoking a quasi-steady-state assumption for $R$ and modeling the electron-beam reactor as an ideal plug-flow reactor, Anshumali developed the following performance equation (Anshumali, 1996):

$$\frac{GD\rho}{C_{V_0} + a_1 D} = \ln \frac{C_{V_0}}{C_V} \tag{8-5}$$

where

$G$ = moles of $R$ formed/joules of energy absorbed

$D$ = dose (kGy) = $\dfrac{\text{joules of energy absorbed}}{\text{mass } (g) \text{ of irradiated material}}$ ;

$\rho$ = density of host gas at operating conditions (g/L)

$C_v$ = concentration of VOCs (mole/L of host gas)

$C_{v_0}$ = initial concentration of VOC (moles/L of host gas)

$a_1$ = proportionality constant relating D to the ratio of the reaction rates of Eqs. (8-2) and (8-4).

Equation (8-5) suggests that a high value of $G$ will result in a high destruction. It also shows that higher values of $C_{v_0}$ will result in relatively less destruction. The appearance of $D$ in both the numerator and denominator of Eq. (8-5) suggests that the destruction will level off at very high doses. Equation (8-5) can be linearized in the following form:

$$\frac{D}{\ln(C_{V_0}/C_V)} = \frac{a_1}{G\rho} D + \frac{C_{V_0}}{G\rho} \tag{8-6}$$

The plot of $\lfloor D/\ln(C_{V_0}/C_V t)\rfloor$ vs $D$ gives a straight line with a slope of $a_1/G\rho$ and intercept of $C_{V_0}/G\rho$. This can then be used to verify the goodness of the fit to the experimental data.

### 8.1.1.3 UTSI's Pilot Facility: Description and Typical Results  The pilot facility used at UTSI is shown schematically in Figure 8.1. A simplified sketch of the electron-beam reactor is shown in Figure 8.2. The major components of this pilot plant were the electron beam reactor, a caustic soda scrubber, and an activated carbon bed adsorber (used as a final polishing step). The scrubber used a caustic soda (NaOH) solution to neutralize any inorganic acids formed during the destruction process. The pH value in the scrubber was maintained by means of a control loop that automatically injected caustic soda. The carbon bed adsorber was installed to capture any organics remaining in the flow after treatment.

A total hydrocarbon analyzer was available for on-line detection of hydrocarbons at the reactor inlet and outlet, scrubber outlet, and adsorber outlet. All the diagnostic sensors and control mechanisms were connected to a control and data acquisition computer to facilitate testing and data collection.

In the UTSI's pilot tests, the following chemicals were tested to determine their destruction and removal efficiency (DRE). (1) vinyl chloride ($C_2H_3Cl$); (2) 1,1,1-trichloroethane ($C_2H_3Cl_3$); (3) 1,1,2-trichlorotrifluoroethane ($C_2Cl_3F_3$, 1,1,2-freon); (4) perchloroethylene ($C_2Cl_4$); (5) trichloroethylene ($C_2HCl_3$); and (6) propane ($C_3H_8$).

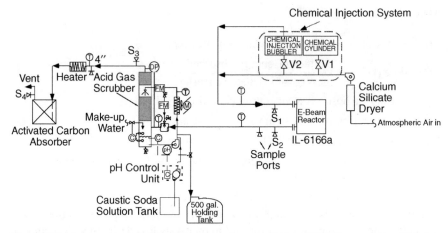

**Figure 8.1.** Schematic of electron-beam-based test facility at UTSI: C, controller; FM, flow meter; P, pressure; S, sample points; T, temperature gauge; V, valves.

UTSI tests were carried out using a mixture of dry air and target VOC simulating a contaminated air stream. For gas species (e.g., propane) mixing was done simply by injecting the gaseous chemicals into the air stream at an appropriate flow rate. Mixing for chemicals that are liquids at ambient conditions was accomplished by bubbling a portion of the desired air flow through a bubbler containing the target chemical, which had been placed in a temperature-controlled water bath, then combining the vapor-containing flow with the main air stream.

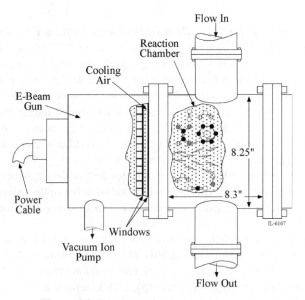

**Figure 8.2.** Blow-up of electron beam reactor.

Beam voltage was kept constant during testing, while beam current was allowed to rise slowly from its initial value near zero to its maximum. In this manner, the dose received by the flow was gradually increased from low to high levels, allowing the measurement of DRE as a function of dose. To determine the effect of applied dose and inlet VOC concentration on the extent of destruction at least two tests were carried out for each VOC using the same carrier gas flow rate but different VOC inlet concentrations. Typical results for TCE in a single pass through the electron beam reactor are shown in Figure 8.3. Generally, destruction increased rapidly with dose in the lower dose ranges, but tended to level off at the higher doses. Also, higher initial concentrations of VOC resulted in lower destruction efficiencies.

Anshumali (1996) had also tested the fitness of Eq. (8-5) to his pilot test data. he calculated the necessary parameters from regression techniques using one test and then used these parameters to predict the performance during other tests that were not included in the parameter determination. Typical comparisons between the experimental and predicted data as well as the residual (i.e., difference between the experimental and predicted value) values for three TCE tests are shown in Figures 8.4, 8.5 and 8.6. At higher doses (>200 kGy), Anshumali's model deviates appreciably from the data and may need modification. However, up to 200 kGy dose, the model fits very well. Similar results were obtained for the other VOCs tested.

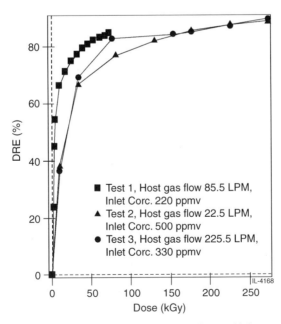

**Figure 8.3.** Dose vs DRE data for TCE. These DRE values, which are typical for all the VOCs tested, are for the electron-beam reactor only; removal by other portions of the flow train is not represented.

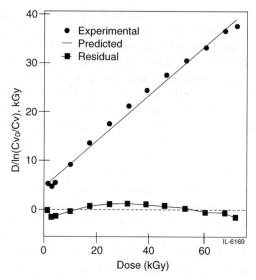

**Figure 8.4.** Comparison between experimental and predicted data for test 1 for TCE.

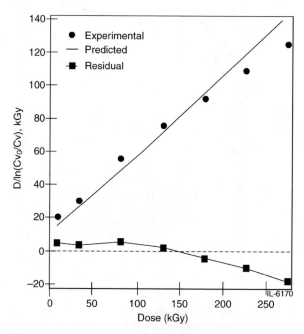

**Figure 8.5.** Comparison between experimental and predicted data for test 2 for TCE.

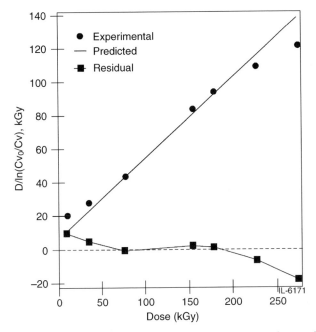

**Figure 8.6.** Comparison between experimental and predicted data for test 3 for TCE.

Anshumali also attempted to correlate particular VOCs' DRE values at a dose of approximately 20 kGy and the respective ionization potential difference relative to nitrogen. This correlation (Table 8.2) supports the assumption of a change transfer process, as the VOCs with the largest $\delta$IP provided the highest DRE value. The DRE values reported by Anshumali were not obtained under optimum conditions: much

**TABLE 8.2. Relationship Between Destruction and Removal Efficiency (DRE) and Ionization Potential**

| Species | Inlet VOC Concentration (ppmv) | DRE[a] (%) | $\delta$IP [b] (eV) |
|---|---|---|---|
| 1,1,2-Freon | 392 | 14.03 | 3.61 |
| 1,1,1-TCA | 795 | 16.23 | 4.56 |
| Propane | 525 | 20.95 | 4.65 |
| Vinyl chloride | 252 | 62.30 | 5.61 |
| TCE | 221 | 75.10 | 6.13 |
| PCE | 166 | 78.92 | 6.28 |

[a]All DRE data are estimated at the dose level of 20–25 kGy and correspond to a single pass through the electron beam reactor.
[b]Values are calculated from the literature reported in Table 8.1 for the VOCs tested by Anshumali (1996).

higher values can be obtained by recycling the partially treated reactor effluent. Also, all DRE values given by him are species non-specific, since he had calculated DRE values based on the total hydrocarbon (THC) analyzer measurements. While measured THC inlet concentrations were clearly due to the chemical under test, measured outlet concentrations could potentially reflect the presence of a mixture of the test chemical and other unknown but detectable species created by the irradiation process. This was not confirmed. In establishing the DREs the outlet concentrations were attributed entirely to the chemical being tested and, therefore, represent the low limits of species-specific DREs.

### 8.1.1.4 Preliminary Process Economics

Besides carrying out pilot plant tests, Anshumali had developed a preliminary cost estimate of an electron-beam based detoxification plant capable of processing an air stream contaminated with 500 ppmv of TCE (Anshumali, 1996). This design presented schematically in Figure 8.7, followed the guidelines provided by the Electric Power Research Institute (EPRI, 1986) and had a capacity to process 100 million cubic feet of contaminated air stream per year. Anshumali's estimate assumed negligible hydrocarbon present at the outlet of the carbon bed. The efficiency of the electron gun (defined as percentage of the input power to the gun delivered to the host gas) was assumed to be 17 percent and a detailed economic cost analysis was prepared. Since, more recent gun designs have much greater efficiencies (40–60 percent), additional cost data were developed for the same plant with 50 percent gun efficiency. In Figure 8.8 which shows how the cost of TCE detoxification plant would change with the dose level at these two gun efficiency levels; at higher gun efficiency the optimum (minimum) point in the cost plainly shifts toward the higher dose level.

The minimum points seen in the cost versus dose curves in Figure 8.8 result from the hybrid nature of the treatment plant. Anshumali's design utilized both electron beam and activated carbon bed detoxification. The relationship between electron beam based destruction and dose level is also nonlinear. Thus, the higher destruction via the electron beam, while requiring greater power consumption, is achieved with lower consumption of carbon. At low-dose levels, the cost for increased destruction via the electron beam is less than the cost for removal by carbon; however, the decreasing effectiveness of the electron beam as the dose is increased eventually results in carbon adsorption becoming the more economically attractive of the two processes.

### 8.1.1.5 Comparison With Other VOC Destruction Options

The electron beam process of converting hazardous contaminants into innocuous compounds

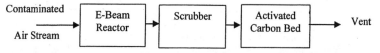

**Figure 8.7.** Block diagram of the hypothetical plant for TCE detoxification.

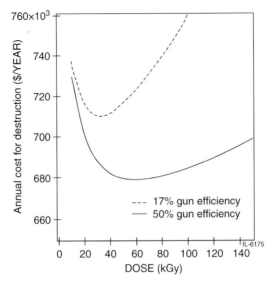

**Figure 8.8.** Cost of TCE destruction at two different electron gun efficiencies.

and generating minimal secondary waste has numerous attractive features which include the following:

- Simple but versatile option
- Uses commercially available and proven electron beam technology
- Can be operated at ambient conditions
- Capable of handling contaminants in air or suitable carrier gas at a wide range of levels (few ppm to thousands)
- Can treat liquid/gaseous waste stream at acceptable throughput rate
- Capable of providing very high DRE, which can be achieved by appropriately tuning power to gun
- Needs little/no supervision and can be easily automated
- Low capital, labor, and maintenance costs
- Acceptable from environmental and community standpoints because it does not produce undesirable by-products
- Can be designed as a "transportable" or "small/modular" system depending upon needs
- Can be used in conjunction with other options such as carbon adsorption

Among the alternatives for destroying and removing VOCs are thermal incineration, carbon bed absorption, and solvent extraction. Other options, some relatively new, include ultraviolet photo thermal irradiation, ozonolysis, polymerization, dechlorination, microbial degradation, wet air oxidation, and catalytic oxidation.

Anshumali (1996) began his comparisons by calculating the percentage reduction in carbon consumption for different VOCs using the electron-beam process with carbon adsorption as a polishing step and using carbon adsorption alone. Figure 8.9 shows the results. For this comparison, the host-gas flow rate and VOC concentration were assumed to be same as those for the hypothetical plant discussed previously, (see Figure 8.7), but an electron gun efficiency of 50 percent was assumed. Data required to determine the amount of carbon needed were taken from Yaws and Nijhawan (1995). The reduction in carbon consumption shown is for the optimal dose conditions. As can be seen from Figure 8.9, the reduction in carbon consumption is significant for all the VOCs except 1,1,2-freon, suggesting that the electron-beam-based detoxification process can be economically attractive, except possibly for contaminants, such as 1,1,2-freon, that have IPs quite close to that of nitrogen.

Anshumali (1996) also compared the cost of electron-beam detoxification with the ultraviolet irradiation based destruction option. Based on US government data (USEPA, 1995), the operational cost for UV detoxification of TCE at a plant of capacity similar to that considered by Anshumali was estimated to be about $29/h. The expected operational costs for the electron gun, electricity, and activated carbon bed (as a polishing step) in a scheme considered by Anshumali would be only about $15/h.

Later, Hadidi and co-workers at MIT (Hadidi et al., 1999) calculated the energy required for destruction of halogenated hydrocarbons for different levels of DRE.

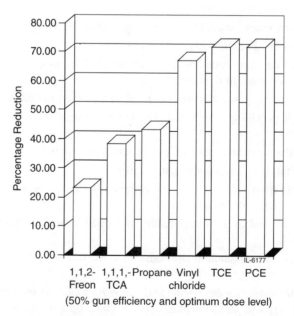

**Figure 8.9.** Percent reduction in activated carbon consumption using e-beam process with carbon adsorption instead of carbon adsorption alone.

To calculate energy required in electron-volts per molecule of TCE and TCA (trichlovoethane), they used the following equations (Vitale et al., 1997):

$$\varepsilon_{TCA} = 166.7 \frac{\ln[1/(1-\eta)]}{\eta T_0^{0.22}} \tag{8-7}$$

$$\varepsilon_{TCE} = 70.9 \frac{\ln[1/(1-\eta)]}{\eta T_0^{0.22}} \tag{8-8}$$

where $\varepsilon$ is the energy per molecule destroyed in electron-volts, $\eta$ is the fractional removal, and $T_0$ is the initial concentration of the VOC. Hadidi reported that TCE had the lowest energy expense of various compounds studied, and TCA had one of the highest energy expenses (Hadidi et al., 1999). $CCl_4$ was reported to have a somewhat lower energy expense (by a factor of about two) than TCA. Figure 8.10, provides the calculated energy in electron-volts per molecule destroyed for TCE and TCA for 95, 99 and 99.99 percent DRE as a function of inlet concentration.

Hadidi and co-workers (1999) also compared the overall cost of VOC destruction by the e-beam process and the scrubbing of the final byproducts with costs of thermal incineration, catalytic oxidation and granulated activated carbon bed adsorption, calculated from the EPA handbook (USEPA, 1991). They assumed capital costs amortized over a 10-year period with an interest rate of 10 percent per year. Plant availability was assumed to be 90 percent. They also used a standard industrial multiplier index to cover the research and development costs, marketing costs, and a profit margin for thermal incineration, catalytic oxidation, and carbon bed adsorption. The cost estimate for thermal incineration was estimated for 75 percent heat recovery. For catalytic oxidation it was calculated based on 50 percent heat recovery and 95 percent DRE. For e-beam based option (which they described as THP or tunable hybrid plasma technology), they assumed negligible

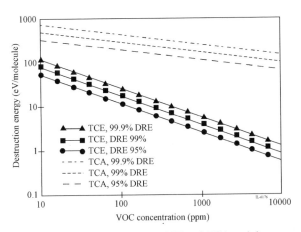

**Figure 8.10.** Energy expense per molecule for TCE and TCA vs inlet concentration for 95, 99, and 99.9 percent DRE. (Taken from Hadidi et al., 1999.)

**TABLE 8.3. Flow Rates for a Typical Concentration of 100, and 200 ppm for which the Cost Ratio is Equal to 1**

| VOC | Flow Rates (cfm) | | Technology Compared With |
|-----|----------|----------|--------------------------|
| | 100 ppm | 200 ppm | |
| | 0–10,000 | 0–10,000 | Thermal incineration |
| TCE | 0–10,000 | 0–10,000 | Catalytic oxidation |
| | >100 | >100 | Granulate activated carbon bed adsorption |
| | <5,000 | <2,500 | Thermal incineration |
| $CCl_4$ | <5,000 | <2,500 | Catalytic oxidation |
| | >100 | >100 | Granulate activated carbon bed adsorption |
| | <2,500 | >500 | Thermal incineration |
| TCA | <2,500 | <500 | Catalytic oxidation |
| | >100 | >100 | Granulate activated carbon bed adsorption |

*Source:* Hadidi et al. (1999).

labor operating costs. The e-beam technology was competitive for flow rates smaller than 2500 cfm for air streams with $CCl_4$ and smaller than 1000 cfm in the case of TCA. For TCE, the e-beam option was competitive at high as well as low flow rates.

To get a better idea of influence of contaminant concentration on the attractiveness of E-beam option, Hadidi defined a cost ratio ($C_r$) as the ratio of cost per pound of VOC processed by candidate technology over the cost per pound of same VOC processed by the e-beam option (Hadidi et al., 1999). Table 8.3 provides the flow rates for which the e-beam option was found to be less expensive than the other candidate options for typical concentrations of 100 and 200 ppm of TCE, $CCl_4$ and TCA. For TCE, the cost ratio between the two thermal technologies and e-beam was higher than 1 for the 0 to 10,000 cfm range. Hence, the upper limit on flow rates given in Table 8.3 for TCE can be considered undetermined.

Clearly, the electron-beam-based detoxification process has the potential to be an attractive alternative to other technologies for the destruction of VOC in air streams.

# REFERENCES

Anshumali (1996). Destruction of low levels of volatile organic components in air by an electron beam generated plasma. M.S. Thesis, University of Tennessee, Knoxville, TN.

Bromberg, L., and Cohn, D. R. (1992). *Decomposition of dilute concentration of carbon tetrachloride in air by an electron beam generated plasma*, Plasma Fusion Center, MIT, Cambridge, MA, PFC/JA-92-021.

EPRI, Electric Power Research Institute (1986). Technical Assessment Guide, EPRI P-4463-SR, V. 1.

Hadidi, K., Cohn, D. R., Vitale, S., and Bromberg, L. (1999). Economic study of the tunable electron beam plasma reactor for volatile organic compound treatment. *J. AWMA*, **49**, 225–228.

Koch, M., and Cohn, D. R. (1993). Electric field effects on the decomposition of dilute concentrations of CHCl$_3$ and CCl$_4$ in electron beam generated plasma. *Physics Letters A*, **184**(1), 109–113.

Slater, R. C., and Douglas-Hamilton, D. H. (1981). Electron beam initiated destruction of low concentration of vinyl chloride in carrier gas. *J. Appl. Physics*, **52**(9), 5820–5828.

Spinks, J. W., and Woods, R. J. (1990). *An Introduction to Radiation Chemistry*. New York, Wiley.

U.S. Environmental Protection Agency (1991). *Control Technologies for Hazardous Air Pollutants*, U.S. Environmental Protection Agency Technology Transfer Handbook, Washington, DC, EPA/625/6-91/014.

U.S. Environmental Protection Agency (1995). *Development of a Photothermal Detoxification Unit*, EPA/540/SR-95/526.

Vitale, S. A., Hadidi, K., Cohn, D. R., Bromberg, L., and Falkos, P. (1997). Electron beam generated plasma decomposition of 1,1,1-trichloroethane, and effect of a carbon-carbon double bond on 1,1,1-TCA and TCE decomposition in an electron beam generated plasma reactor. *Plasma Chemistry and Plasma Processing*, **17**(1), 508.

Yaws, C. L., and Nijhawan, S. (1995). Determining VOC adsorption capacity. *Pollution Engineering*, **27**(2), 34–37.

Zapit Technology, Inc. (1994). Personal communication.

## 8.1.2 Biofiltration of Organic Vapors

NIRUPAM PAL, PROFESSOR

Civil and Environmental Engineering, California Polytechnic State University, San Luis Obispo, CA 93407, USA

The vapors from volatile organic chemicals (VOCs) are present almost everywhere in the environment because of the widespread use in industry of such substances as benzene, toluene, xylenes, ethers and esters; methanol, ethanol, butanol, and other alcohols; and chlorinated compounds such as methylene chloride, chloroform, tri- and tetrachloromethane. In addition to health hazards posed by vapors from these chemicals, the processing of fruit and vegetable products generate large quantities of low molecular weight organic alcohols, ketones, and other VOCs, resulting in severe odor problems.

There are a number of existing technologies for controlling the VOC emissions. Incineration, catalytic oxidation, adsorption on charcoal or other strong adsorbents and absorption of contaminants in water or other soluble liquid, mentioned in Section 8.1.1, are popular, well-developed technologies that can be used in most situations. However, there are a number of shortcomings. Both incineration and catalytic oxidation are costly and generate gases that are themselves pollutants. Burning of TCE has been linked with generation of toxic chemicals similar to DDT. The adsorption and absorption technologies do not provide a complete solution because they simply transfer contaminants from one medium to another. The biggest drawsback of all these technologies is very low efficiency at lower concentration of the contaminant. These shortcomings combined with stricter environmental regulation have motivated engineers and researchers to develop new and environmentally friendly technologies. Among the new technologies are several discussed in later sections (ultraviolet oxidation, membrane separation, destruction of VOC by corona plasma), and biofiltration.

### 8.1.2.1 *What is Biofiltration?*    Biofiltration is a simple process in which contaminated air is passed over an active microbial population immobilized on a suitable support material (Figure 8.11). The microbes use these chemicals in their metabolism, converting them to innocuous products such as carbon dioxide and water. The objective is to maximize the removal of VOC at a minimum cost.

To maintain maximum microbial activity, the incoming contaminated air may be humidified before it enters the biofilter bed. Intermittent spraying of water mixed with suitable nutrients is another means of maximizing microbial activity. A simple flow chart of the process appears in Figure 8.12.

Biofiltration is not a new idea. In India people have traditionally covered their cow dung waste and compost piles with dried hay, which was then moistened with water sprays. Microbes grew on the moist hay and worked as a biofilter to reduce odors. The principle has been used for years in Europe for the removal of odors from airsteams, wastewater treatment plants, and sewer mains. However, the application of biofilters to remove industrial VOCs is comparatively new. The first scientific model for an engineering biofilter was published only a generation

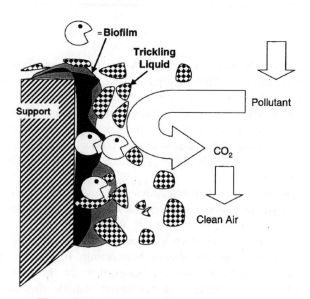

**Figure 8.11.** Schematics of biofiltration process.

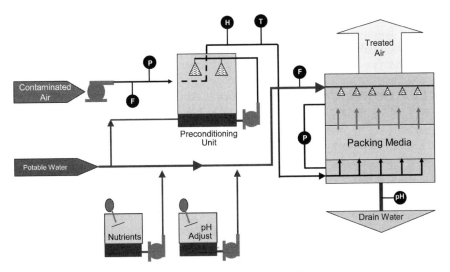

**Figure 8.12.** Schematic diagram of a biofilter setup.

ago (Ottengraf and Van Den Oever, 1983). Primitive systems, however, were in place in Germany and in the United States in 1960s, becoming widespread in Germany in the next decade. The 1980s saw the introduction of biofiltration for the treatment of toxic emissions and VOCs from industry. In Europe in the 1990s, the process gained popularity and began to spread to the United States, where early in the twenty first century there were more than 4,000 biofilters installations.

Biofilters have been well accepted by regulators and industry, as well as by the public. Among the reasons for the popularity of this technology are the following:

- **VOCs and HAPs (Hazardous Air Pollutants) are oxidized at ambient temperatures** The low temperature oxidation eliminates the high costs associated with combustion and no production of any other combustion products.
- **Intrinsically Safe** The low temperature oxidation and high moisture content eliminate the fears associated with combustion.
- **Low Annual Operating Cost** There are only two major power consumers in a biofiltration system: a recirculation pump for humidification and a fan to pull the gas stream through the equipment.
- **Low Pressure Drop** Much lower pressure drop than catalytic or regenerative thermal oxidizers resulting in fan power consumption savings.
- **Proven Effective Technology** Biofiltration systems are in place with historical operating experience demonstrating regulatory compliance.

- **Low Maintenance**   Very few moving parts result in lower maintenance cost.
- **Environmentally Friendly**   Zero $NO_x$ emissions, Zero $SO_x$ emissions and substantially lower Carbon Dioxide emissions, since part of carbon becomes biomass.

### 8.1.2.2 Factors Affecting Biofiltration   Although, in principle biofiltration is a very simple, its engineering design is quite complex, and process optimization is challenging. The factors that affect biofiltration are organics, elemental, environmental and mechanical, as described in the subsections that follow.

**Microorganisms**   The concept of biofiltration is based on the ability of microorganisms to oxidize (known as degradation) VOCs. Primarily, prokaryotes are used as the microorganisms. However molds and fungi have been seen to aid in the process (Diks and Ottengraf, 1991). Naturally occurring packing materials such as pearlite and peat, composted wood chips, contain the organisms capable of degrading some VOCs. In most cases, the biofilter must be inoculated with microbial cultures specifically suitable for the VOC to be treated.

**Oxygen and Other Nutrients**   All biofiltration process is aerobic in nature. Proper distribution and availability of oxygen in the biolayer is a major design consideration as is the availability of oxygen itself. Often the oxygen concentration in the water layer (Figure 8.13) is insufficient necessitating the use of forced air. For microbes to work properly, the biofilter must contain sources of nitrogen and phosphorus as well as certain micronutrients. Peat, compost, and bark contain

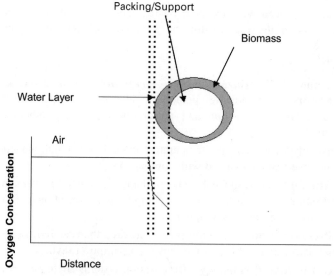

**Figure 8.13.** Typical biofilter oxygen profile (not to scale).

these nutrients, but when synthetic polymeric materials are used as packing materials, micro and macronutrients must be provided.

**Moisture and Temperature Effects**   Water is required in biofilters for proper biological activities. It has been reported (Zilli et al., 2000) that operation of biofilters is optimal when 50 percent of the pore volume is filled with water. In commercial biofilters, moisture sensors that trigger automatic addition of water are beneficial. Due to high air volume rates used in commercial biofilters, beds can dry out very quickly even though the ambient temperature remains unchanged. To alleviate the problem of bed drying, contaminated air streams are humidified before entering the bed. When the streams fail to compensate for the evaporation of water induced by biodegradation, an exothermic process, water must be added periodically. Almost all commercial biofilters are equipped with automated sprinkler systems. Sprinkling of water has been reported to create two problems: formation of an anaerobic zone due to non homogeneous water distribution (Ottengraf and Van Den Oever, 1983) and creation of lumps of materials leading to reduction of contact surface between gas and biofilter.

Another approach to maintaining proper moisture is use to steam to supersaturate the incoming gas. This, however increases bed temperature, and decreases the amount of VOCs absorbed on the packing materials, and leads to reduced biological activity (Van Lith, 1989; Van Lith et al., 1997). Drying of bed can also initiate channeling of the gas stream, resulting in severe loss of removal efficiency.

Operating temperatures for biofilter units should be between 5°C and 50°C (Bohn, 1992). Above certain temperature ranges, one can use an Arrhenius expression to describe the effect of temperature on biodegradation and consequently on biofiltration (Zilli et al., 2000). The efficiency of styrene vapor removal can be increased by a factor of 2 by raising the temperature by 7°C.

**Pressure Drop**   Experimental studies (Brenner et al., 1993) have shown that pressure drop in an optimally run biofilter is very low. Typical values are around 1–2 in. of water/meter filter bed.

$$\frac{\Delta P}{H} = \frac{150\,\mu v_0\,(1 - \epsilon)^2}{D_p^2}\,\frac{}{\epsilon^3} + \frac{1.75\,\rho v_0^2\,(1 - \epsilon)^2}{D_p}\,\frac{}{\epsilon^3}$$

Estimation of pressure drop can be made using the Ergun equation (Ottengraf and Van Den Oever, 1983), typically expressed as follows: Where $D_p$ is the effective particle diameter in the bed, $V_0$ is the superficial gas velocity, $\mu$ is the viscosity of air, $\varepsilon$ is the bed porosity, $\rho$ is the gas density and $\Delta P/H$ is the pressure drop per unit height of the bed.

### 8.1.2.3 Packing Materials Used in Biofilters   A wide variety of materials have been used in commercial biofilters and media selection is a critical factor in biofilter design. For a biofilter to operate efficiently, the media must provide a

suitable environment in which microorganisms can live and reproduce, have good moisture holding capacity, and have a high porosity for minimal back pressure. The media should be locally available for minimal construction costs. Four critical material properties, porosity, moisture holding capacity, nutrient content, and useful life before the materials decompose, affect the following operating parameters: pressure drop, airflow rate, moisture addition, and pH buffering.

For smaller scale filters, peat, pearlite, and compost materials have been used with great success. Results from a typical industry that uses agricultural packing media are shown in Figure 8.14. Recently, however, a number of synthetic biofilter media have been developed. These products, characterized by high surface area, high porosity, and longevity, require neither maintenance nor replacement. Therefore, they are the choice of most commercial biofilter companies in their permanent installations. Synthetic media come in four basic forms. Two inexpensive US-made synthetic filtering media are Bio Strata and Bio-Fill.

Bio Strata consists of black sheets of polyvinyl chloride (PVC) glued in a block form suitable for insertion into filtration unit. This product has negative buoyance and is preferred where loose media is not acceptable. It is well suited for submerged and trickling biofilters. Bio-Fill is a white PVC ribbon with a high surface area. Like Bio Strata, it has negative buoyancy.

Another inexpensive option is small woven polypropylene pads, like the scrubbers sold for home kitchen use. These products offer much more surface area for the price, are positively buoyant, come in assorted colors, and contain no harmful chemicals. Finally, Siporax, developed by prestigious Schott Group of Germany, is a biofilter medium made of sintered glass. In addition to having an extremely

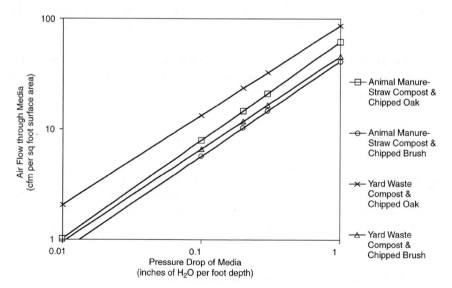

**Figure 8.14.** The relationship airflow rate and pressure drop for various packing materials. (From University of Minnesota Agricultural Engineering: www.bae.umn.edu.)

high surface area ($82,000$ ft$^2$/ft$^3$), it has the unique ability to perform both nitrification and denitrification functions.

**Kinetics and Design**   In their study of the kinetics and design of biofilters (Ottengraf and Van den Oever, 1983), the following basic assumptions made:

1. In the biolayer, the nutrients are transported by diffusion, which can be described by an effective diffusion coefficient $D$.
2. The biolayer thickness $\delta$ is small compared to the support material particles.
3. The substrate elimination reaction in the biolayer obeys zero-order microkinetics with respect to the nutrient concentration. This is true for Monod's model in case of high substrate concentration.
4. The gases pass through the filter bed in plug flow mode.

With the above assumptions, the differential equation describing the concentration of nutrient component $C_l$ inside the biolayer is given by

$$D' \frac{d^2 C_l}{dx^2} - k = 0 \tag{8-9}$$

where $k$ is the zeroth reaction rate constant.

Upon stating the following boundary conditions

$$\text{At } X = 0, \qquad C_l = C_g/m \tag{8-10a}$$

$$\text{At } X = \delta, \qquad \frac{dC_l}{dx} = 0 \tag{8-10b}$$

We can solve Eq. (8-9) as

$$\frac{C_l}{C_g/C_m} = 1 + 0.5 \frac{\phi^2}{C_g/C_{go}} (\sigma^2 - 2\sigma) \tag{8-11}$$

where $\phi = \sqrt{Km/D'C_{g0}}$ is the Thiele modulus or Thiele number, $\sigma = x/\delta$, the dimensionless length coordinate, and m $= C_g/C_l$, the distribution coefficient.

In the biofiltration, two separate scenarios can happen and one must distinguish these situations.

- **Diffusion limitation of the wet biobarrier**   When the biolayer pores are full with water or when the concentration of the contaminants in the inflowing gas is too low, the biolayer is not fully active and we say that diffusion limitation has occurred. We may also say that the depth of penetration in the biolayer is smaller than the actual layer thickness. In that case the rate of conversion is controlled by diffusion of the contaminant in the biolayer, referred as diffusion limitation.

**TABLE 8.4. Range of Design and Process Parameters**

| Parameter | Parameter Value Range |
|---|---|
| Bed depth | 3–6 feet |
| Loading rates | 3–5 scfm/ft$^2$ |
| Residence time | 25–60 seconds |
| Oxygen: VOC | 100 : 1 ppm |
| Temperature | 25–37°C |
| Inlet VOC concentration | <500 mg/ft$^3$ |
| Moisture | Depends on packing materials 12–25 percent by weight |
| Pressure drop | 0.1 to 0.5 inch of water/ft |

• **Reaction Limitation**   When the biolayer is fully active, the rate of conversion is controlled by the rate of reaction in the biolayer. Thus the efficiency of a bio-filter can vary significantly depending on the conditions of operation.

***8.1.2.5***   Biofiltration process parameters vary significantly from one applic-ation to another, and are proprietary. Available data (Whittle and Quimby, 1997), however, allow us to list some general design and process parameters (Table 8.4).

***8.1.2.6***   Industrial Applications of Biofilters VOCs and many other chemicals are subject to biofiltration all over the world. Among the well-known generators of VOCs are the petroleum industry, including manufacturers of polymers; the makers of electrical and electronic equipment, including intregrated circuits; and the pharmaceutical, paints and coatings, pulp and paper, wood products, and printing

**TABLE 8.5. Some Compounds Treated in Biofilters by US Companies and Industries**

| Company Uses Biofilters | Compounds Bioremediated |
|---|---|
| Amoco | Styrene |
| Corn Products | Methanol, ethanol, SO$_2$ |
| Tamko Asphalt Products | Formaldehyde |
| Dow Elanco | Pesticide emissions |
| Weyerhaeuser | Terpene, alcohol, and formaldehyde |
| General Tire | Styrene |
| 3M | Alcohols, ketone, toluene |
| Union Camp | Flavor odors |
| Brown Foreman Whiskey | Ethanol |
| Tobacco Industry | Odors |
| Gentex Polycarbonate Lenses | Alcohols and ketone |
| Weyerhaeuser | Zimpro waste water odors |

**TABLE 8.6. Removal Rates for Various Odorous Compounds in a Biofilter**

| Odorous Compounds | Removal Rates (%) |
|---|---|
| Aldehydes | 92–99.9 |
| Amines, amides | 92–99.9 |
| Ammonia | 92–95 |
| Benzene | 90–92 |
| Dimethylsulfide | 91 |
| Polyaromatic hydrocarbons | 96–100 |
| Ethanol, diacetyl | 96 |

industries. Wastewater treatment plants, facilities for the remediation of groumdwater and soils, and enterprises that uses viscose processes also produce VOCs.

Food processing is another source of environmental contamination from VOCs. These substances are released by plants that make food for humans, for pets, and for fish, by operations for the processing of meat and poultry, and by rendering plants and slaughterhouses. Table 8.5 lists a few representative companies and the compounds they generate and then treat with biofilters.

Process efficiency depends not only on biofilter design but also on the contaminant itself. Table 8.6 shows approximate removal efficiency for various chemical groups, based on the results obtained from industrial applications (Zeager Bros. Inc., 1992).

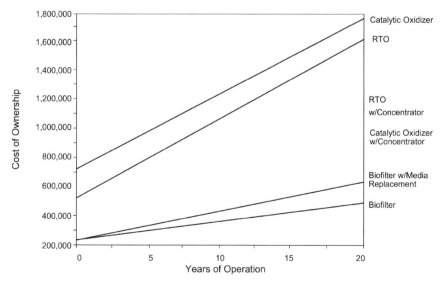

**Figure 8.15.** Comparative cost of various VOC treatment technologies; RTO — regenerative thermal oxidizers. (Chart courtesy of PPC Biofilters.)

***8.1.2.7 Cost*** The cost of biofiltration systems include both capital and operating costs. Capital cost may be broken down into cost of pretreatment, quantity of air being treated, type of media used, ductwork, and removal efficiency requirements. Operating costs included costs of electricity, media replacement (if necessary), periodic inspection and testing, and any sidestream treatment required. Figure 8-15 shows that even after 20 years of operation, biofilters are less expensive to own than regenerative thermal oxidizers and catalytic oxidizers, with and without concentrators.

Shareifdeen (1994) found that treating 100 ft³ of VOC-contaminated air by biofiltration costs roughly 9 times less than using combustion and half as much as activated carbon adsorption.

For biofilters for agricultural use, which function more to control odors than to remove toxic substances, capital expenditures are significantly lower. Installation costs are limited to materials (fans, media, ductwork, and plenum) and labor. Annual operation and maintenance costs of such biofilters include increased cost of electricity needed to push the air through the biofilters and the cost of replacing the media every five years. Both capital costs and operation and maintenance costs are quite variable.

## REFERENCES

Bohn, H. L. (1992). Consider biofiltration for decontaminating gases. *Chemical Engineering Progress*, **88**: 34–40.

Brenner, R. C., Sorial, G. A., Smith, F. L., Suidan, M. T., Smith, P. J., and Biswas, P. (1993). Evaluation of biofilter media for treatment of air streams containing VOCs. Proceedings of the 66th Annual Conference and Exposition of the Water Environment Federation, pp. 429–439 (Oct. 3–7).

Diks, R. M., and Ottengraf, S. P. (1991). Verification studies of a simplified model for the removal of dichloromethane from waste gases using a biological trickling filter: part I and part II. *Bioproc. Eng.*, **6**: 93–99, 131–140.

Ottengraf, S. P., and Van Den Oever, A. H. (1983). Kinetics of organic compound removal from waste gases with a biological filter. *Biotechnology and Bioengineering*, **25**, 3089–3103.

Shareifdeen, Z. (1994). Engineering analysis of a packed-bed biofilter for removal of volatile organic compounds (VOC) emissions. Ph.D. Thesis, New Jersey Institute of Technology.

Whittle, T., and Quimby, J. (1997). Biological filters for odor control. 27th Annual Conference, Environmental Engineering in the Food Processing Industry.

Van Lith, C. (1989). Design criteria for biofilters. Presented at the 82nd Annual Meeting and Exhibition of the Air and Waste Management Association, Anaheim, CA (June 25–30).

Van Lith, C., Leson, G., and Michelse, R. (1997). Evaluating design options for biofilters. *Journal of the Air & Waste Management Association*, **47**, 37–47.

Zeager Bros. Inc. (1992). Odor remediation using biofiltration.

Zilli, M, Del Borghi, A., and Converti, A. (2000). Toluene vapour removal in a laboratory-scale biofilter, *Appl. Microbiol Biotechnol.*, **54**(2), 248–254.

### 8.1.3 Using Sorbents and Carbon Injection to Remove Dioxin/Furans and Toxic Metals

TAPAS K. DAS[1] and ARINDAM GHOSH[2]
[1]Washington Department of Ecology, Olympia, Washington 98504, USA
[2]National Environmental Engineering Research Institute (NEERI), Nagpur, India 440020

Emissions of dioxins, furans, and toxic metals, particularly mercury, generated in municipal and medical waste incineration processes can be controlled by means of sorbents and carbon injection. Polychlorinated dibenzo-*p*-dioxins (PCDDs) and polychlorinated dibenzofurans (PCDFs) are a group of tricyclic compounds with one to eight substituted chlorine atoms. This results in 210 different compounds: 75 PCDDs and 135 PCDFs. Seventeen cogeners are more toxic than the others. These toxic cogeners all have chlorine atoms at the 2, 3, 7, and 8 positions. PCDDs and PCDFs are very stable compounds and have very long residence times in the environment and in organisms, including humans. Their hydrophobicity promotes accumulations in sediments and organisms, resulting in concentration in both sediment and organisms.

#### 8.1.3.1 *Sources of Dioxins and Furans*

PCDDs and PCDFs are found in fly ash and flue gas from waste incineration processes. Municipal and medical waste incinerators seem to be the most important source for emissions into the air of PCDDs and PCDFs. These pollutants, which do not exist at temperatures of more than 1,800°F, are formed in the postcombustion zone as the gases cool (between 500 and 850°F). Some specific metals (particularly copper) are important components of the fly ash that can catalyze the formation of PCDD and PCDFs. Metals enter an incinerator as trace species in the fuel matrix from oil and coal, as well as from solid wastes from municipal and medical sources.

Dioxin and Furans are formed as unwanted byproducts in certain industrial processes (e.g., associated with the manufacture of pulp, paper, and pesticides), and as products of incomplete combustion when chlorine and complex mixtures containing carbon are present. These pollutants are emitted from incinerators that burn medical waste, municipal solid waste, hazardous waste, sewage sludge, wood, and recycled oil; from motor vehicles; and from metal recovery processes. Among them, medical and municipal solid waste incinerators have the potential for posing the greatest risk of dioxin/furans sources currently identified. It has been determined that dioxins pose a risk of cancer from exposure through inhalation, ingestion, and skin (dermal) absorption. Figure 8.16 shows the structure and numbering system for dioxin and cogeners. The most toxic of all dioxins is 2,3,7,8-tetrachloro dibenzo-*p*-dioxin (2,3,7,8-TCDD). The 16 other dioxins and furans with chlorines at the 2,

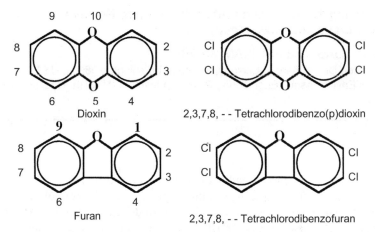

**Figure 8.16.** Molecular structure of dioxin and furan.

3, 7 and 8 positions have been assigned toxicity values, called toxicity equivalency factors (TEFs), relative to 2,3,7,8-TCDD.

2,3,7,8-TCDD has a TEF of 1, and the other cogeners are assigned values of less than 1. TEFs are used to express the total toxicity of dioxins when the concentration of each cogener is multiplied by its TEF and all the products are added up (called dioxin equivalents or TEQs).

**8.1.3.2 Formation of Dioxins and Furans** Since the discovery of dioxins and furans in municipal waste incinerator (MWI) fly ash and flue gas in 1977, much attention has been paid to the formation of these compounds. Theoretical calculations by Shaub and Tsang (1983, 1985) showed that formation is unlikely to take place through homogeneous gas phase reactions. Instead they proposed that in the heterogeneous environment of the postcombustion zone of the incinerator, the fly ash itself catalyzez the formation of PCDDs and PCDFs.

Two mechanisms of formation are proposed: De-novo synthesis and the Deacon reaction.

*De-Novo Synthesis* PCDD and PCDF are formed from all organic carbon sources and chlorine in the presence of metal catalysts and in more direct synthesis from chlorinated organic precursors like benzenes, phenols, naphthalenes and biphenyls. The copper ion $Cu^{2+}$ has a strong catalytic effect on the formation of PCDDs/PCDFs at 300-325°C (500–550°F), while $Fe^{3+}$, $Pb^{2+}$ and $Zn^{2+}$ ions have minor effects. Additional significant effects are attributed to carbon sources (morphology), percentage of oxygen, temperature, and aerosol and fly ash size distributions.

*Deacon Reaction* The Deacon process involves the oxidation of HCl in the presence of a metal catalyst to produce $Cl_2$ and $H_2O$ (Edwards et al., 2001). Chlorinated plastics, principally PVC (polyvinyl chloride), are major source of chlorine and HCl, according to the following reaction:

$$2HCl + \tfrac{1}{2}O_2 \rightarrow Cl_2 + H_2O \,(\text{below } 900^\circ C) \tag{8-12}$$

Chlorination of aromatic ring structures with HCl is a thermodynamically unfavored reaction: chlorination of benzene (Bz) with HCl, for example, proceeds with a positive Gibbs free energy of formation (Griffin, 1986).

$$Bz(g) + 2HCl(g) \rightarrow Cl_2Bz(g) + H_2(g) \quad \Delta G = 34.3 \,\text{kcal} \tag{8-13}$$

When HCl is first converted to $Cl_2$ a favored reaction is obtained

$$2HCl(g) + \tfrac{1}{2}O_2(g) \rightarrow Cl_2(g) + H_2O(g) \quad \Delta G = -9.07 \,\text{kcal} \tag{8-14}$$

$$Bz(g) + Cl_2(g) \rightarrow Cl_2Bz(g) \Delta G = -11.23 \,\text{kcal} \tag{8-15}$$

$$CuCl_2 + \tfrac{1}{2}O_2 \rightarrow CuO + Cl_2 \tag{8-16a}$$

$$CuO + 2HCl \rightarrow CuCl_2 + H_2O \tag{8-16b}$$

$$2HCl + \tfrac{1}{2}O_2 \rightarrow Cl_2 + H_2O \tag{8-16c}$$

Reaction (8-14) can be catalyzed by copper, according to reactions (8-16a, 8-16b, and 8-16c) (Hagenmaier et al., 1987).

This reaction is called the Deacon reaction and could play a role in the chlorination of carbon (Hagenmaier et al., 1987; Vogg et al., 1987; Stieglitz et al., 1990a). The chlorine formed in the Deacon reaction can react back to hydrogen chloride with sulfur dioxide as follows:

$$SO_2 + Cl_2 + H_2O \rightarrow SO_3 + 2HCl \tag{8-17}$$

Reaction (8-17) will inhibit the chlorination of carbon and, hence, the formation of PCDDs and PCDFs (Griffin, 1986; Stieglitz et al., 1990a).

Chlorinated aromatic compounds can be intermediates in the formation of dioxins and furans. The correlation between some chlorinated compounds and these pollutants present in municipal waste incinerators supports this route of formation. A strong correlation between the presence of chlorinated benzene and the formation of PCDD and PCDFs has been reported (Schooneboom, 1995). In a reaction catalyzed by fly ash that results in the formation of PCDD and PCDFs, the amount of pollutants increased when the fly ash was heated at temperature around $500^\circ F$ in an excess oxygen atmosphere ($\leq 1.0\%$). In this case PCDD and PCDFs are probably formed from residual carbon present on the fly ash surface.

### 8.1.3.3 *Carbon Source of Dioxin and Furans*   The origin of carbon that results the formation of PCDDs and PCDFs is important. For example, soot and

sugar coal are less reactive than charcoal (Vogg et al., 1987). Graphite produces only minor amounts of PCDD and PCDFs (Stieglitz et al., 1990b). The crystal lattice of graphite is probably more resistant to an attack of chlorine/oxygen than the already disturbed graphite structure of the other carbon samples. There is no correlation between the surface area of the carbon and the formation of PCDD and PCDFs, so the variance of their formation from different kinds of carbon is not caused by variance in surface area (Stieglitz et al., 1990b). In another study, Eichberger and Stieglitz (1994) found that formation of PCDDs, PCDFs, and PCBs is correlated with the activity of the carbon.

Chlorinated aromatic compounds other than PCDDs and PCDFs are formed during the municipal waste incineration process. In laboratory studies, many of those compounds are found to be formed from carbon. The most important group of compounds are chlorobenzenes, which are also dominant in municipal waste incineration. Besides chlorobenzenes, chlorinated naphthalenes, biphenyls, and phenols, and a large number of trace compounds are formed from carbon (Stieglitz et al., 1989b; Jay and Stieglitz, 1991; Milligan and Altwicker, 1993). It is suggested that those compounds are formed when the carbon is degraded similarly to the formation of PCDD and PCDFs. Some of these products, together with the compounds present or formed in the gas stream, could act as precursor compounds for the formation of PCDDs and PCDFs (Stieglitz et al., 1989a, 1989b, 1993; Olie et al., 1998). A separate study on the role of inorganic chlorine in the formation of PCDD/F from residual carbon on incinerator fly ash reported that if fly ashes were doped/impregnated with KCl and HCl and with copper as catalytic material, a positive correlation with the amount of PCDD/F was observed (Addink et al., 1998).

**8.1.3.4 Role of Metals**  As noted earlier ions of heavy metals or the transition metal group are essential for the formation of dioxins and furans. Copper has been identified as the strongest catalyst. The addition of only 0.08 percent $Cu^{2+}$ gives rise to a significant amount of PCDD and PCDFs (Stieglitz et al., 1989b). The catalytic action of $CuCl_2$ is poisoned by $NH_3$, which results in lower PCDD/F amounts in the presence of $NH_3$. It is therefore important to understand the fate of metallic species in combustion environments. An understanding of the fate of the metallic species at different temperatures is also essential in the process of cutting down dioxin formation, as well as in devising appropriate control strategies for preventing their emissions (Olie et al., 1998).

**8.1.3.5 Inhibiting the Formation of Dioxins and Furans**  Raghunathan and Gullett (1996) suggested that the presence of sulfur and chlorine and the S/Cl ratio in the feedstocks can decrease the downstream formation of chlorinated organic compounds, particularly PCDD and PCDFs. Cofiring municipal solid waste or combination wood waste with sulfur-bearing coal may reduce these emissions. Thus they report that the depletion of molecular chlorine ($Cl_2$), an active chlorinating agent, by $SO_2$ through a gas-phase reaction appears to be significant inhibition mechanism in addition to previously reported $SO_2$ deactivation of copper catalysts. In other work, Lindbauer (1994) proposed the influence of added sulfur compounds on PCDD/F is governed by $SO_3$, not $SO_2$. The PCDD/F formation inhibition is attributed to masking of the surface of catalytic dust particles as a result of sulfatization by $SO_3$.

Inhibition of the formation of PCDD and PCDFs from carbon was also studied by Addink (1995). Ethylenediamine-tetra-acetic acid (EDTA), nitrilotriacetic acid (NTA), and sodium sulfide ($N_2S$) were found to inhibit the formation of PCDD and PCDFs from carbon. Sodium thiosulfate ($Na_2S_2O_3$) also inhibits formation of these compounds, but less effectively. The inhibition was proposed to proceed by formation of stable complexes with transition metals, especially copper.

### 8.1.3.6 *Role of Scavenging on Ash Particles*

The number density and size distribution of ash particles resulting from combustion or incineration is an important parameter that establishes the scavenging of metallic species by the entering particles, as the aerosol stream travels through various stages of post-combustion operation. Das and Mahalingam (2000) calculated the scavenging of both refractory mineral and trace metal species contained in the entering and exiting ashes, using a previously developed code AEROSOL (Nadkarni and Mahalingam, 1985). This AEROSOL code, which calculates the existing particle size distribution and containing as well a selected single condensable specie distributed between the gas phase and the particulate phase, has been verified through detailed laboratory experimentation on coal gasification (Tumati and Rees, 1992; Das and Mahalingam, 2000). For municipal waste combustion, the code can be used to predict the concentration variation of $Cu^{2+}$ adsorbed on fly ashes as the aerosol stream travels through temperature gradients over various stages of post-combustion operations. Hence this analysis can track the fate of toxic compounds in the air, including dioxin and furans in the entrained, deposited, and emitted fly ash. Such an analysis should ultimately facilitate the improved design of the thermal recovery equipment and emission control equipment for the minimization of air toxics release. Depending upon the type and origin of coal, the ash constitute could be as high as 26 percent. In typical combustion operations, about 60 percent of this ash is entrained as fly ash while the rest is discharged as bottom ash (Das, 2003).

### 8.1.3.7 *Control Technologies*

Figure 8.17 illustrates possible pathways for metallic species in the gas phase of the combustor: transfer of the metallic species to the gas phase, transformation of metallic species by chemical reactions in the gas phase, and aerosol formation and growth dynamics. The submicrometer aerosol is primarily formed from the gas phase by nucleation of the supersaturated vapors as the temperatures decrease downstream from the combustor, followed by growth by condensation, then coagulation. A comprehensive description of aerosol mechanisms supported by some mathematical treatment is given by Biswas and Wu (1998).

Two commercially available technologies can be applied to reduce emissions of dioxin and metal vapor (e.g., mercury) from incinerators: sorbent injection and carbon injection. A description of each technology as a means of controlling dioxin and metals is summarized in this article. The subsections that follow describe

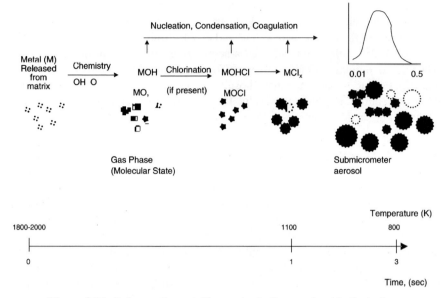

**Figure 8.17.** Pathways for metallic species in the gas phase in the boiler.

these technologies and, briefly, a way of adsorbing elemental mercury by means of activated carbon treated with sulfuric acid.

*Sorbent Injection*  Sorbent injection into a combustor chamber is effective in reactively scavenging the vapors of metallic species, thereby suppressing nucleation at the combustor exit. This process, encapsulates the metal compounds in larger sized particulate matter, which can be more effectively removed by conventional particulate-control devices. Moreover, these sorbents may also help immobilize the metallic species by forming glasslike complexes. The pathways and a mechanistic description of the various transformations are shown in Figure 8.18. The metals transformation pathways are important because effective sorbent methodologies cannot be developed unless these pathways are well understood. Three aspects of metals transformation are critical: the release into the gas phase in the combustor (this depends on the form of the metal entering the combustion system), the transformations (function of the environment and chemical reactions that take place), and the subsequent aerosol dynamics (formation, growth by condensation, and coagulation).

Typical mineral sorbents have an affinity for the metal oxide; the sorbent should therefore be injected prior to the formation of the aerosol mode or before other chemical transformations take place (such as halogenation and sulfation). Biswas and Wu (1998) have developed a gas-phase precursor injection method of producing directly in the combustor an agglomerate of sorbent oxide having a high surface area and high mass. This sorbent oxide is stable at the high temperatures encountered in

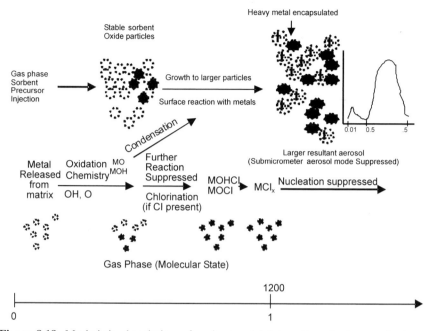

**Figure 8.18.** Mechnistic description of sorbent-metal interactions in the combustor, at 1,200 K.

the combustor, and provides a surface for condensation and reaction of metal species vapors. Because sorbent metal oxide or metal hydroxide radical species are preferred over the halogenation reactions (as the sorbent encounters these intermediate species in the high-temperature zones), the gas-phase sorbent precursor is effective in chlorine environments (Owens et al., 1995).

Researchers have proposed the use of several sorbent materials (e.g., silica, alumina, alumino-silicates, bauxite, limestone, hydrated lime, titania) for metals control. Owens et al. (1995) performed equilibrium calculations for a number of toxic metal species reacting with sorbents to establish their relative efficiencies. For a specific sorbent and toxic metal, there is a temperature range over which it is most effective. For example, more than 80 percent of lead was predicted to be captured by silica between 600 and 1,600°K. In the presence of chlorine, this effective temperature range was reduced to 700–1,200°K. Alumina was predicted to be more effective for capturing beryllium and nickel than silica. Silica-based sorbent was most effective in removing submicrometer particles of mercury (500–600°F) and lead (600–1,200°F). Table 8.7 presents a summary of these findings reported by Biswas and Wu (1998).

*Carbon Injection for Dioxin and Mercury Removal*   Studies in the United States and Europe have demonstrated that carbon injection (CI) systems are effective in removing dioxins with powdered activated carbon produced from reactivated

**TABLE 8.7. Effectiveness of Different Sorbents for Metals Removal**

| Metal | Alumina | Silica | Titania |
|-------|---------|--------|---------|
| As | 100 (<1400) | N/A | <1 |
| Ba | 100 (900–1600) | 100 (600–1600) N/A | |
| Be | 97 (400–800) | <10 | N/A |
| Cd | 37 | 100 (500–900) | 100 (700–1000) |
| Ni | 63 | <1 | N/A |
| Pb | N/A | 97 (600–1000) | 100 (600–1200) |
| Hg | N/A | 92 (500–600) | N/A |

N/A — Data not available.
The numbers are maximum capture efficiencies (%) and the temperature range ($^\circ$K) in brackets below is for capture greater than 80 percent.

granular coal-based carbons. The powdered carbon is injected into the system and when it has adsorbed dioxins and other contaminants, it is removed along with other particulate matters in electrostatic precipitators or fabric filters. The material is then disposed of with other plant solid wastes in an off-site landfill.

The use of activated carbon injection has been shown to be an extremely effective means of controlling dioxin emissions from municipal, medical, and hazardous waste incinerators fed with chlorinated waste materials. However, the technology will be effective only in system configurations in which a dry flue gas stream is held at relatively low temperatures (<400$^\circ$F).

The CI system is more than 95 percent effective in reducing total dioxin emissions as shown in Table 8.8. For 2,3,7,8-TCDD, emissions have been reduced by more than 99 percent through CI. The use of CI can also improve mercury-emission control, depending on concentrations of the dioxins and mercury present in the gas stream. Tests conducted with the CI system showed 97 percent Hg removal efficiency. Mass carbon-to-mercury ratios of > 100,000 : 1 may be required at residence times of a second or less upstream of an electrostatic precipitator or a baghouse at 300$^\circ$F (149$^\circ$C) to achieve 90–97 percent Hg removal efficiency (Brown et al., 1999). In another process, used in Europe, the flue gas is heated to

**TABLE 8.8. Selected Emission Results Before and After Carbon Injection Installation**

| Emissions Parameters | Pre-Control Data | Post-Control Data | Relative Efficiency[a] |
|----------------------|------------------|-------------------|------------------------|
| Total PCDD/PCDFs ng/m$^3$ @ 7% O$_2$ | 133 | 6.0 | 95.5% |
| Total TEQs ng/m$^3$ @ 7% O$_2$ | 2.4 | 0.083 | 96.5% |
| 2,3,7,8-TCDD, pg/train | 107 | 0.4 | 99.6% |
| 2,3,7,8-TCDF, pg/train | 904 | 24 | 97.4% |
| 1,2,3,7,8-PCDD, pg/train | 483 | 10 | 97.9% |
| 2,3,4,7,8-PCDD, pg/train | 3,645 | 138 | 96.2% |

[a]The efficiency is based on average data prior to and after CI installation.

about 250°F in a fixed-bed or entrainment process and carbon is injected to coat the bags in a fabric filter. This process, which requires an investment of several million dollars is much more expensive than the CI system discussed earlier (Durham, 2001; Kraus, 2002; Roeck and Sigg, 1996; Rossler, 2002).

In addition to heavy metals and chlorinated organics (particularly PCDD/Fs), several other major pollutants are of concern, including acid gases, particulate matter ($PM_{10}$), and nitrogen oxides. Several best available control technologies (BACT) are applied in removing these pollutants from incineration sources to meet the state and federal emission standards. Table 8.9 is a summary of the removal efficiencies of major air pollutants using combinations of selected control systems. Some systems are capable of removing pollutants at a rate as high as 99 percent.

*Elemental Hg Adsorption by Activated Carbon Treated with Sulfuric Acid*
Unlike most other heavy metals, which are emitted as particles, mercury has been reported to be released in its elemental form in the gas phase. This is primarily because of its closed, outer-shell electronic structure ($5\,d^{10}\,6\,s^2$). Activated carbons treated with $H_2SO_4$, however exhibit high $Hg^0$ sorption capacity. Granular activated carbon is immersed in an aqueous 70% $H_2SO_4$ solution and then dried it at 100–200°C for several hours; at 110–120°C the treated activated carbon achieves a very high $Hg^0$ removal. Since activated carbon is used to remove $H_2SO_4$ in industrial applications, it would be mutually beneficial if activated carbon units already impregnated with sulfuric acid from another process could be used for the control of $Hg^0$ emissions. Furthermore, studies of $Hg^0$ adsorption by the $H_2SO_4$ treated carbon could lead to a better understanding of the $Hg^0$ adsorption mechanism and, hence, to the development of a more effective carbon sorbent. Li et al. (2001) reported that more than 70 percent $Hg^0$ was physically captured by activated carbons treated with $H_2SO_4$.

**TABLE 8.9. Estimated Collection Efficiencies for Combined Control Systems**

| Control System(s) | Collection Efficiency in Percent | | | | | |
|---|---|---|---|---|---|---|
| | Particulates | $SO_2$ | HCl | Hg | Other Metals | PCDD/Fs |
| ESP | 95.5–99.9 | -0- | -0- | 20–30 | 95–98 | 25–50 |
| SD/ESP | 98.5–99.9 | 60–75 | 95–98 | 50–80+ | 95–98 | 70–80 |
| SD/FF | 99.0–99.9 | 65–80 | 95–98 | 80+ | 99+ | 90–99+ |
| DI/ESP | 98.5–99.9 | 60–70 | (70–80) | — | 95–98 | (60–70) |
| DI/FF | 99.0–99.9 | 70–80 | 80–90 | — | 99+ | 90–99+ |
| SD/DI/FF | 99.0–99.9 | 80–90 | 95–98 | (80+) | 99+ | 90–99+ |
| ESP/WS (1) | 98.5–99.9 | 50–60 | 95 + | (85+) | 95–98 | (80–90) |
| ESP/WS(2) | 98.5–99.9 | 90–95 | (95+) | (85+) | 95–98 | (90–99+) |
| SD/CI/FF | 99.0–99.9 | 85–95 | (98+) | (98+) | (99+) | (90–99+) |

SD — spray dryer; ESP — electrostatic precipitator; FF — fabric filter or bag house, DI — dry injection; WS — wet scrubber (stage 1, 2); CI—carbon injection.

***8.1.3.8 Summary*** Formation of dioxins and furans during incineration of munici-pal and medical wastes is prompted by the catalytic properties of metals, which are typically found in the waste itself. All types of carbon can serve as a carbon source. Chlorine can also be used in the different forms in which it is present. PCDDs and PCDFs can be prevented from forming by poisoning the catalysts with complexing compounds. The pollutants are formed in a complex set of reactions in which the chlorination of carbon and dechlorination of the initially formed compounds are the most important. Copper has been identified as the strongest catalyst. Sorbent methods are often used for capturing metals in high-temperature systems. For the sorbent to be most effective, not only must it be chemically reactive but it must also suppress submicrometer particles of metals. Most heavy metals will react with sorbents such as silica, titania, alumina, and calcium-based compounds.

Carbon injection in the incinerators has been demonstrated to be more than 95 percent effective in removing PCDDs and PCDFs, as well as mercury emissions, from the flue gas stream. Mass carbon-to-mercury ratios of exceeding $100,000:1$ may be required residence times of a second or less to achieve 90–97 percent Hg removal efficiency. A greater portion of this removal is achieved through scavenging of the trace elements and dioxin/furans by the entering activated carbon and ash par-ticles, as the aerosol stream travels through various stages of postcombustion operations.

## REFERENCES

Addink, R. (1995). Ph.D. thesis, University of Amsterdam.

Addink, R., Espourteille, F., and Altwicker, E. (1998). Role of inorganic chlorine in the for-mation of polychlorinated dibenzo-p-dioxins/dibenzofurans from residual carbon on incinerator fly ash. *Env. Sci. Tech*, **32**(21), 3356–3359.

Biswas, P., and Wu, C. Y. (1998). Control of toxic metal emissions from combustors using sorbents: a review. *Journal of the Air & Waste Manag. Assoc.*, **48**, 113–127.

Brown, T. D., Smith, S. N., Hargis, R. A., and O'Dowd, W. J. (1999). Mercury measurement and its control: what we know, have learned, and need to further investigate. Critical review series. *J Air & Waste Manag. Assoc.* 1–97 (June).

Das, T. (2003). Hog fuel boiler RACT (reasonably available control technology) determi-nation. Washington Department of Ecology, Olympia, WA, Report #03-2-009. (http://www.ecy.wa.gov/biblio/0302009.html).

Das, T., and Mahalingam, R. (2000). Investigation of ash deposition and air toxics pathways in boiler downstream operations. Proceedings of the International Conference on Incinera-tion and Thermal Treatment Technologies (IT3), Portland, OR, (May).

Durham, M. D. (2001). Controlling mercury emissions from coal-fired utility boilers, *Maga-zine for Environmental Managers*, 27–33 (July).

Edwards, J. R., Srivastava, R. K., Lee, C. W., and Kilgroe, J. D. (2001). A Computational and experimental study of mercury speciation as facilitated by the deacon process. *Air & Wast. Manag. Assoc. Specialty Conf. on Mercury Emission: Fate, Effects, and Control*, Chicago, IL (August 20–23).

Eichberger, M., and Stieglitz, L. (1994). In *DIOXIN'94, Organohalogen compounds*. Kyoto University, Kyoto, Vol. 20, pp. 385–390.

Griffin, R. D. (1986). A new theory of dioxin formation in municipal solid waste combustor. *Chemosphere*, **15**(9–12), 1987–1990.

Hagenmaier, H., Kraft, M., Brunner, H., and Haag, R. (1987). Catalytic effects of fly ash from waste incineration facilities on the formation and decomposition of polychlorinated dibenzo-p-dioxins and polychlorinated dibenzofurans. *Environ. Sci. Technol.*, **21**, 1080–1084.

Jay, K., and Stieglitz, L. (1991). On the mechanism of formation of polychlorinated aromatic compounds with copper(II) chloride. *Chemosphere*, **22**, 987–996.

Kraus, K. (2002). Mercury control: find solutions that balance environmental risk and energy impacts. *Air & Wast. Manag. Assoc. Magazine for Environmental Managers*, 22–28 (April).

Li, H. Y., Serre, S. D., Lee, C. W., and Gullett, B. K. (2001). Elemental mercury adsorption by activated carbon treated with sulfuric acid. *Air & Wast. Manag. Assoc. Specialty Conf. on Mercury Emission: Fate, Effects, and Control*, Chicago, IL (August 20–23).

Lindbauer, R. L. (1994). PCDD/F emission control for municipal solid waste incineration by $SO_3$ —addition. *Organohalogen Compd.*, **19**, 355–361.

Milligan, M. S., and Altwicker, E. R. (1993). The Relationship between de novo synthesis of polychlorinated dibenzo-p-dioxins and dibenzofurans and low-temperature carbon gasification in fly ash. *Environ. Sci. Technol.*, **27**, 1595–1601.

Nadkarni, A. R., and Mahalingam, R. (1985). Aerosol behavior in temperature and concentration gradient fields in nonisothermal tube flow. *Amer. Ins. Chem. E. J.*, **3**, 603–614.

Olie, K, Addink, R., and Schoonenboom, M. (1998). Metals as catalysts during the formation and decomposition of chlorinated dioxins and furans in incineration processes. *Journal of the Air & Waste Manag. Assoc.*, **48**, 101–105.

Owens, T. M., Wu, C. Y., and Biswas, P. (1995). An equilibrium analysis for reaction of metal compounds with sorbents in high temperature systems, *Chem. Eng. Commun.*, **133**, 31–52.

Raghunathan, R, and Gullett, B. (1996). Role of sulfur in reducing PCDD and PCDF formation, *Env. Sci. Tech.*, **30**(6), 1827–1834.

Roeck, D., and Sigg, A. (1996). Carbon injection proves effective in removing dioxins — a case study, *Environmental Protection*, 34–34 (April).

Rossler, M. T. (2002). The electric power industry and mercury regulation: protective, cost-effective, and market-based solutions. *Air & Wast. Manag. Assoc. Magazine for Environmental Managers*, 16–21 (April).

Schooneboom, M. (1995). Ph.D. Thesis, University of Amsterdam.

Shaub, W. M., and Tsang, W. (1983). Dioxin formation in incinerators. *Environ. Sci. Technol.*, **17**, 721–730.

Shaub, W. M., and Tsang, W. (1985). In *Chlorinated dibenzodioxins and dibenzofurans in the total environment II*. G. Choudhary; L. H. Keith; C. Rappe, (eds). MA, Butterworth, Stoneham, pp. 469–487.

Stieglitz, L., Zwick, G., Beck, J., Bautz, H., and Roth, W. (1989a). Carbonaceous particles in fly ash a source for the de nova synthesis of organochloro compounds. *Chemosphere*, **19**, 283–290.

Stieglitz, L., Zwick, G., Beck, J., Roth, W., and Vogg, H. (1989b). On the de-novo synthesis of PCDD/PCDF on fly ash of municipal waste incinerators. *Chemosphere*, **18**, 1219–1226.

Stieglitz, L., Vogg, H., Zwick, G., Beck, J., and Bautz, H. (1990a). *Chemosphere*, **23**, 1255–1264.

Stieglitz, L., Zwick, G., Beck, J., Bautz, H., and Roth, W. (1990b). *Chemosphere*, **20**, pp. 1953–1958.

Stieglitz, L., Eichberger, M., Schleihauf, J., Beck, J., Zwick, G., and Will, R. (1993). The oxidative degradation of carbon and its role in the de-novo synthesis of organohalogen compounds in flyash. *Chemosphere*, **27**, 343–350.

Tumati, P., and Rees, D. (1992). ESP performance in controlling trace elements. CONSOL, Pittsburgh, PA.

Vogg, H., Metzer, M., and Stieglitz, L. (1987). Recent findings on the formation and decomposition of PCDD/F in municipal solid waste incineration. *Waste Manag. Res.* **5**, 285–294.

## 8.2  SOME EMERGING AND INNOVATIVE PROCESSES

In the latter days of the twentieth century, certain technologies emerged that showed promise of bringing the advantages of pollution prevention and sustainability into the air pollution control marketplace. We discuss a few of these in this section: the use of nonthermal plasma to control multiple pollutants, electron-beam-based oxidation treatment for volatile organic compounds (VOCs), biodiesel fuels, membrane-based removal and recovery of VOCs from air, and carbon sequestration.

These emerging and innovative technologies have the potential to provide economical control of VOCs, oxides of nitrogen, sulfur dioxide, mercury vapors, carbon dioxide, and other greenhouse gases. As sustainable technologies, they offer secondary emission and energy reduction benefits that need to be recognized during the technology selection and permitting stages of industrial development.

As the policies and practices of pollution prevention and sustainability continue to gain currency, the advantages these technologies will become increasingly apparent. Looking at air pollution control from the more holistic point of view of pollution prevention will allow tremendous progress to be achieved in this field, paralleling advances in the areas in which those principles have been accepted and applied.

### 8.2.1  Nonthermal Plasma for Control of Multiple Pollutants from Coal-Fired Utility Boilers

VIRENDRA K. MATHUR, PROFESSOR
Department of Chemical Engineering, University of New Hampshire, Durham, NH 03824, USA

To achieve continued economic growth, environmental protection, and energy security, nations require an adequate supply of affordable energy produced in an environmentally friendly way. In the United States, there are coal reserves that should last for several hundred years. A major criticism of coal as a fuel source for power generation is that it produces large amounts of pollutants. During the past several years, control of emissions of oxides of nitrogen and sulphur ($NO_x$, $SO_x$) and mercury vapors has become a national priority. Emissions of $NO_x$ and $SO_x$ are a leading contributor to acid rain as well as to photochemical smog (through the formation of ozone). State-of-the-art technologies for controlling $NO_x$ either attempt to control the temperature, residence time and stoichiometry in the combustion zone, thereby lowering the $NO_x$ formation, or provide a postcombustion reducing agent that consumes the oxygen in the $NO_x$ molecules, producing nitrogen and water. $SO_2$ is mostly removed during combustion or by flue gas wet scrubbers. Most of the studies for mercury control in utility flue gas are still under development (Huang et al., 1996; Galbereath and Zygarkicke, 1996; Carey et al., 1998; Dunham et al., 1998; Granite et al., 2000; Serre and Silcox, 2000). Mercury speciation analysis results suggest that the proportions of the different Hg forms in flue gases vary widely.

Conventional adsorption techniques, such as sorbent injection (Section 8.1.3), have long been studied as potential methods for mercury control in utility flue gases. Most studies focus on screening activated carbons and other commercially available materials as sorbents. Activated carbons, the most useful sorbents, have essentially reached their maximum potential. One of the most promising classes of new technologies for the treatment of gaseous oxide and mercury pollutants is the use of nonthermal plasma to overcome some of the difficulties associated with current technologies.

The production and acceleration of free electrons under the influence of an electric field is an electrical discharge. Through collisions with molecules in gases, electrons cause excitation, ionization, electron multiplication and the formation of atoms, and metastable compounds. It is the formation of atoms and compounds that gives an electrical discharge its unique chemical environment and makes it useful for chemical processing. Dielectric barrier discharges (DBD) are one of the electrical breakdown products that can be classified as nonequilibrium discharges, also called nonthermal plasmas.

Investigation of the electrical discharge-initiated reaction of nitrogen oxides was first undertaken by S. S. Joshi at the University College, London in 1927 (Joshi, 1927). Since then, particularly in 1980s and 1990s there were many investigations into the use of coronas for controlling $NO_x$ and $SO_2$ emissions (Chang, 1997; Matteson et al., 1972), pulse (Clements et al., 1989; Veldhuizen et al., 1998), e-beam (Maezawa and Izutsu, 1993; Frank and Hirano, 1993) or dielectric barrier discharge (Kogelschatz et al., 1997; Penetrante et al., 1996; Urashima et al., 1997). Research and development of the DBD technique has been conducted at the University of New Hampshire for several years (Mathur et al., 1992; Breault et al., 1993; McLarnon, 1996; McLarnon and Mathur, 2000; Mathur et al., 2000; Chen and Mathur, 2002). Based on this work, several efforts are in progress to commercialize this technique for the simultaneous removal of $NO_x$, $SO_2$ and Hg vapour emissions from utility boiler flue gas by means of a nonthermal plasma technique.

### *8.2.1.1 Experimental Apparatus and Procedure*

A schematic diagram of the benchtop reactor system for $NO_x$, $SO_2$, and Hg removal is shown in Figure 8-19. It includes the barrier discharge reactor, high voltage power supply and measurement, gas supply and gas analytical instrumentation. Details can be found in earlier publications (McLarnon, 1996; McLarnon and Mathur, 2000; Chen and Mathur, 2002).

*Dielectric Barrier Discharge Reactor*   The reactor geometry used to initiate a barrier discharge in a gas space containing oxides of nitrogen is two concentric stainless steel tubes. Each cylinder serves as an electrode, but the outer cylinder also serves as the pressure boundary. A high voltage is applied between the electrodes, with the outer electrode (tube) at ground potential for safety reasons. The distance between the two electrodes is 10 mm. Two quartz tubes are installed adjacent to the electrodes for producing barrier discharge and the gas spacing between the two quartz tubes is 6 mm. The effective length of the reaction is 305 mm; both electrodes cover this space and generate microdischarges along it.

A simulated gas mixture enters the inlet plenum of the reactor and exits through the outlet plenum. The outer stainless steel electrode for the benchtop reactor is

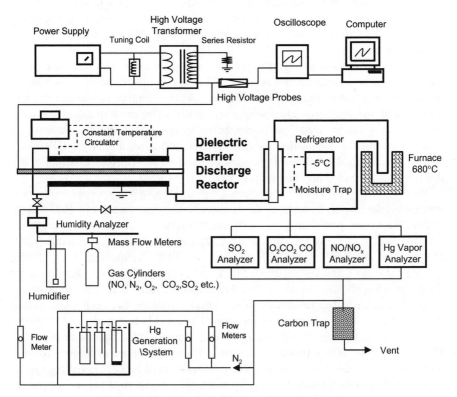

**Figure 8.19.** Schematic diagram of benchtop system.

wrapped with copper tubing fed by a constant temperature bath, cooling or heating the reactor as necessary for temperature control.

*HV Power Measurement*   Power consumption in the DBD reactor is calculated from the voltage and current traces recorded and stored by the oscilloscope. A curve of point power obtained by multiplying the corresponding current and voltage values at the same phase angle is averaged to give the discharge power. An on-line computer reads the values recorded by the oscilloscope and calculates the discharge power. The voltage and current signals are taken from the second circuit of the transformer, where some loss of energy occurs; however, inefficiencies due, for example, to inductors and resistances, are negligible compared to discharge power consumption.

*Steady Generation of Mercury Vapor*   The mercury generation system illustrated in Figure 8.19 is used to provide a steady stream of mercury vapour and nitrogen gas. At the start, 2 cc of liquid mercury (26 g) is added to the first evaporator to maintain a relatively large surface area for better mercury evaporation. The water bath is maintained at 30°C and nitrogen gas is passed through the first vessel with mercury at a rate of 2 cubic centimetres per minute (CCM). This stream containing mercury vapor is then mixed with the pure nitrogen stream, diluting the mercury concentration by 100 times to a level that can be used for the preparation of simulated gas for the plasma reactor.

*Gas Analytical Instrumentation*   The flow system shown in Figure 8.19 allows sampling of the inlet and outlet gas streams from the DBD reactor so that concentrations of pollutants can be measured. All outlet gas samples are passed through a moisture trap (glass tube packed with glass wool and cooled with chilled water to catch $HgO$ and minimize the water content before the gas streams are fed to analytical equipment). A furnace is installed downstream of the moisture trap to decompose ozone and $NO_2$ and minimize their effect on mercury analysis. Several commercial gas analyzers are used on-line.

Sampling and analysis of mercury need considerable attention. A Jerome 431-X Mercury Analyzer (Arizona Instruments, Tempe, AZ) is installed on line. This instrument detects only mercury and is unaffected by interference common to UV analyzers. With its Gold Film Sensor, it can detect mercury in the range of 0.003 to 0.999 $mg/m^3$ (0.3–10 ppbv) in just 13 seconds.

**8.2.1.2 Results and Discussion**   Destruction of pollutants is measured in terms of inlet and outlet concentrations from the DBD reactor as follows:

$$\text{Pollutant Conversion } (\%) = \frac{C_{\text{inlet}} - C_{\text{outlet}}}{C_{\text{inlet}}} \times 100 \qquad (8\text{-}18)$$

where $C_{\text{inlet}}$ is $C_{NO} + C_{NO_2}$ in the inlet stream and $C_{\text{outlet}}$ represents the same gases in the outlet stream. Accordingly, NO conversion can be referred to percentage of inlet

NO converted to $NO_2$, $HNO_3$, $N_2O$, and/or $N_2$. $NO_x$ conversion, the measurement of removal of both NO and $NO_2$, is also calculated from Eq. (8-18). $NO_x$ conversion can be used to estimate inlet NO converted to $N_2$ when $HNO_3$ and $N_2O$ produced are considered to be negligible. Energy density is the power deposited into a liter of gas mixture at standard conditions (J/L), that is, discharge power divided by total gas flow rate.

The electrical variables in a barrier discharge affect the operation of the system in two major ways. First, the effective use of these parameters can help achieve an optimal design by enhancing the efficiency of the discharge process as measured by the pollutant conversion, voltage, and energy requirements. Secondly, the reaction path sequence and consequently the product slate can be controlled.

All experimental results are obtained from a mixture of a pollutant and other gases. The reactor is operated at atmospheric pressure and room temperature except under conditions of high moisture, when it is maintained at 65°C. Chemical concentrations and discharge power are experimentally determined. The plasma start-up usually needs 5–10 minutes to reach steady state. A concentration reading is taken after the plasma has been within an acceptable range for 5 minutes. Discharge power is calculated by the computer and usually 3–5 readings of discharge power are taken to get one averaged data point.

*Electric Frequency Selection*   In an ac-powered barrier discharge system, the reactions are a function of the electric energy deposited into the gases passing through. An increase in voltage or frequency will result in an increase in energy input. Total energy consumption is the sum of discharge power and other circuit consumption (transformer, inductors, etc.). In investigating the effects of frequency on total energy consumption and discharge power, it was found that the ratio of discharge power to total energy consumption reached its maximum at a frequency of 150 Hz for this barrier discharge system. Therefore, most of the experiments were conducted at 150 Hz.

*Effects of Chemical Compounds Present in the Inlet Gas on NO/NO$_x$ Conversions*   The make-up of the gas entering the barrier discharge reactor can significantly affect the performance of the reactor and thus the products of the discharge-initiated reactions. Exhaust gas from fossil fuel combustion generally contains several compounds including $N_2$, $H_2O$, $CO_2$, $O_2$, $NO_x$, $SO_2$ and Hg, among which $N_2$, $H_2O$, $CO_2$ and $O_2$ are present in large concentrations of 1–2 digital percentage while $NO_x$ and $SO_2$ are present at the level of hundreds of parts per million. The amount of each compound varies depending on the fuel and the way it is burned as well as the amount of excess air used in the combustor.

*Effect of Oxygen*   Figure 8.20 shows the effect of oxygen concentration in a mixture of NO, $O_2$, and $N_2$ on NO/NO$_x$ conversions. At low $O_2$ concentrations (<1 percent), a small increase in $O_2$ content results in a significant drop in NO/NO$_x$ conversions at 150 J/L. An increase in energy density to 320 J/L greatly enhances NO/NO$_x$ conversions. At higher $O_2$ concentration (>1 percent), NO con-

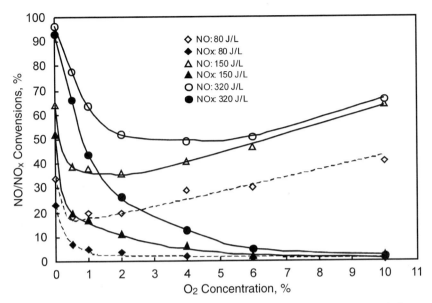

**Figure 8.20.** Effect of oxygen concentration on NO/NO$_x$ conversions at total flowrate 1,000 CCM, 250 ppm NO with N$_2$ balance, room temperature and at a frequency of 150 Hz. (From Chen and Mathur, 2002.)

version increases as O$_2$ concentration is increased while NO$_x$ conversion drops nearly to zero. An increase in O$_2$ concentration from 1 percent to 1–10 percent leads to a significant increase in NO$_2$ formation. McLarnon and Penetrante (1998) proposed that the following reactions occur in the plasma.

$$e + N_2 \rightarrow e + N + N \tag{8-19}$$

$$N + N \rightarrow N_2 + O \tag{8-20}$$

$$e + O_2 \rightarrow e + O(^3P) + O(^3P) \tag{8-21}$$

$$e + O_2 \rightarrow e + O(^3P) + O(^1D) \tag{8-22}$$

$$O(^3P) + NO + M \rightarrow NO_2 + M \tag{8-23}$$

$$O + O_2 + M \rightarrow O_3 + M \tag{8-24}$$

$$O_3 + NO \rightarrow NO_2 + O_2 \tag{8-25}$$

where O($^3P$) and O($^1D$) are ground-state and metastable excited-state oxygen atoms, respectively and M is either N$_2$ or O$_2$.

Variations in NO conversions at different energy densities are due to the competition between reactions (8-19) and (8-23) or (8-25). At O$_2$ concentration less than 1%, the dominant reaction (8-20) mainly contributes to NO conversion into N$_2$ although the presence of O$_2$ competes for consuming electrons via reactions

(8-21) and (8-22). At $O_2$ concentration greater than 1 percent, however, reaction (8-23) or (8-25) becomes dominant and more oxygen molecules compete for electrons, resulting in increasing $NO_2$ formation while NO conversion into $N_2$ via reaction (8-20) decreases. Both reactions (8-23) at low energy density ($<100$ J/L) and reaction (8-25) at high energy density ($>100$ J/L) demonstrate dominance. In comparing NO conversions, it is observed that when $O_2$ concentration is greater than 5 percent, an increase in energy density from 150 J/L to 320 J/L does not help NO conversion because reaction (8-25) needs three oxygen atoms to convert one NO molecule while reaction (8-23) only needs one oxygen atom. This counterbalances the effect of reactions (8-21) and (8-22) at high energy density. In the presence of a third body (M), O combines with $O_2$ to form ozone via reaction (8-24).

*Effect of Water Vapor*  In the absence of $O_2$ and $H_2O$, the dominant reaction is the reduction of NO to $N_2$ as shown in reaction (8-20) and Figure 8.21. However, both NO and $NO_x$ conversions significantly decrease with the addition of 1 percent $H_2O$. Electro-impact dissociation of $H_2O$ competes for electrons and accounts for the conversion drop as follows:

$$e + H_2O \rightarrow e + H + OH \tag{8-26}$$

$$OH + NO + M \rightarrow HNO_2 + M \tag{8-27}$$

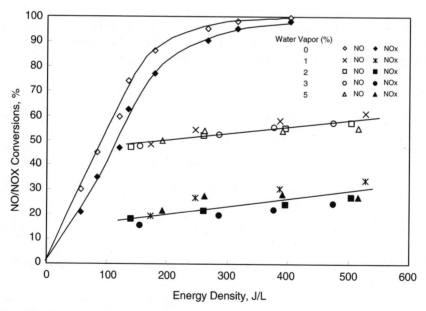

**Figure 8.21.** Effect of water vapor concentration on $NO/NO_x$ conversions at 65°C and a frequency of 150 Hz: total flowrate 1,000 CCM, 250 ppm NO with $N_2$ balance. (From Chen and Mathur, 2002.)

As $H_2O$ concentration is increased from 1–5 percent, there is little change in conversions of NO and $NO_x$. Fewer electrons are available for reaction (8-19) and thus less nitrogen atoms are available for reaction (8-20), possibly leading to lower NO and $NO_x$ conversions. However, reactions (8-26) and (8-27) enhance NO conversion, which counterbalances the effect of reaction (8-20), resulting in little overall change in $NO/NO_x$ conversions. An increase in energy density helps N and OH formation but does not promote NO and $NO_x$ conversions significantly, as shown in Figure 8.21.

*Effect of Water Vapor and Oxygen*   The experiments were conducted in the presence of 1 percent water vapor and 8.3 percent $O_2$ in the inlet $NO/N_2$ gas stream. Higher $NO/NO_x$ conversions were observed in the presence of combined $O_2$ and $H_2O$ than in the presence of either $O_2$ or $H_2O$ alone (Figure 8.22).

*Effect of Carbon Dioxide*   Addition of $CO_2$ decreased NO and $NO_x$ conversions slightly as shown in Figure 8.23. The large CO formation is attributed to the electro-impact dissociation of $CO_2$ in a mixture of NO, $O_2$, $N_2$, and $CO_2$ in a plasma system.

*Effects of Carbon Monoxide, Methane, and Ethylene in the Inlet Gas Mixture on $NO/NO_x$ Conversions*   Chemical compounds such as CO, $CH_4$ and $C_2H_4$ have

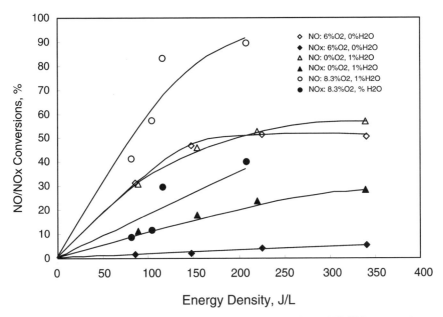

**Figure 8.22.** Effect of 1 percent water vapor and 8.3 percent $O_2$ on $NO/NO_x$ conversions at 65°C and a frequency of 150 Hz: total flowrate 1,000 CCM, 250 ppm NO with $N_2$ balance. (From Chen and Mathur, 2002.)

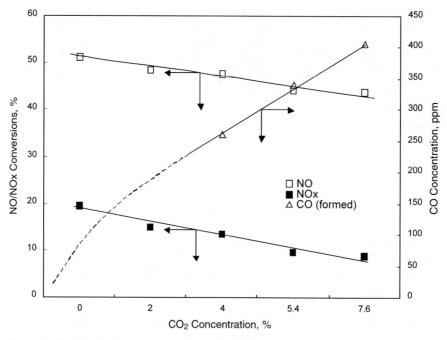

**Figure 8.23.** Effect of $CO_2$ concentration on $NO/NO_x$ conversions at room temperature and a plasma energy density of 320 J/L; total flowrate 1,000 CCM, 250 ppm NO, 3 percent $O_2$, $N_2$ balance. (From Chen and Mathur, 2002.)

played an important role in the catalytic reduction of $NO_x$. Now we describe some experiments conducted to study their effects in the presence of nonthermal plasma.

Figure 8.24 shows only a small increase in NO conversion when energy density was increased from 150 J/L to about 400 J/L. However, there was a significant increase in CO formation.

Methane was added to the mixture of NO, $O_2$ $CO_2$, and $N_2$ to investigate its effect on $NO/NO_x$ conversions. Figure 8.25 shows NO conversion slowly increasing as methane concentration was increased, while $NO_x$ conversion decreased a little.

Figure 8.26 shows that ethylene ($C_2H_4$) significantly enhanced NO conversion to $NO_2$. The following two reactions are proposed to occur in the plasma process:

$$C_2H_4 + O \rightarrow CH_2CO + H + H \qquad (8\text{-}28)$$

$$C_2H_4 + O + O \rightarrow HCO + HCO + H + H \qquad (8\text{-}29)$$

Unsaturated hydrocarbon $C_2H_4$ consumes most of the O atoms via reactions (8-28) and (8-29) that might otherwise react with NO to form $NO_2$. As $C_2H_4$ concentration

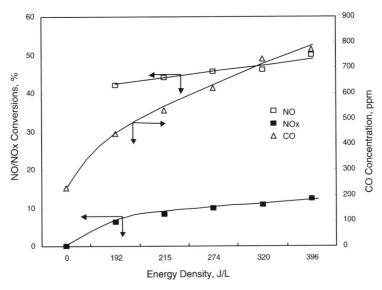

**Figure 8.24.** NO/NO$_x$ conversions and CO concentration vs. energy density: total flowrate 1,000 CCM; inlet, 250 ppm NO, 230 ppm CO, 3 percent O$_2$, 7.6 percent CO$_2$, N$_2$ balance. (From Chen and Mathur, 2002.)

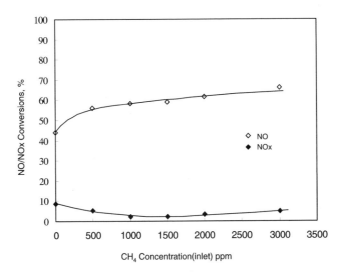

**Figure 8.25.** Effect of CH$_4$ concentration on NO/NO$_x$ conversions at room temperature and a plasma energy density of 320 J/L; total flowrate 1,000 CCM, 250 ppm NO, 3 percent O$_2$, 7.6 percent CO$_2$, N$_2$ balance. (From Chen and Mathur, 2002.)

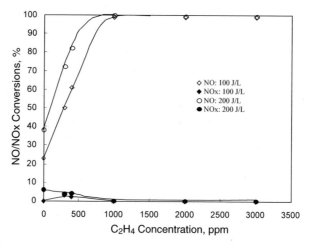

**Figure 8.26.** $C_2H_4$ concentration effect on $NO/NO_x$ conversions at 60°C and a frequency of 150 Hz: total flowrate 1,000 CCM, 250 ppm NO, 3 percent $O_2$, 7.6 percent $CO_2$, $N_2$ balance. (From Chen and Mathur, 2002.)

is increased, more $HO_2$ is produced via reaction (8-30) (Penetrante et al., 1998):

$$HCO + O_2 \rightarrow CO + HO_2 \qquad (8\text{-}30)$$

$$HO_2 + NO \rightarrow NO_2 + OH \qquad (8\text{-}31)$$

Reaction (8-31) becomes dominant, leading to almost 100 percent conversion of NO to $NO_2$.

*By-Product $N_2O$ Formation*  Nitrous oxide ($N_2O$) is one of the six hazardous gases listed on the Kyoto Protocol as hazardous with respect to global warming. A literature review reveals that $N_2O$ may be formed during the plasma process. The reaction

$$NO_2 + N \rightarrow N_2O + O \qquad (8\text{-}32)$$

takes place and produces $N_2O$ as a stable product (Matzing, 1991). In the presence of $O_2$ and $H_2O$, up to 10 ppm $N_2O$ is formed in a positive-pulsed corona discharge reactor (Mok et al., 2000). In a series of experiments conducted to investigate possible $N_2O$ formation in the DBD reactor, no $N_2O$ was detected in the exit gas stream when 2 percent water vapor was present in the inlet gas.

*Sulphur Dioxide Oxidation in Barrier Discharge*  The nonthermal plasma experimental system described earlier was used to investigate $SO_2$ oxidation as well. The effects of cylindrical, square, and rectangular reactor geometries on the conversion of $SO_2$ to $SO_3$ were studied (Mathur et al., 1999). A rectangular reactor was used to compare effects of different barrier (dielectric) materials and their thickness on $SO_2$ conversion. Most of the experiments were conducted with an inlet gas mixture of 770 ppm $SO_2$, 4000 ppm $O_2$, and the balance $N_2$ at 400 Hz. The $SO_2$ concentration measurements were made by an infrared analyzer.

Barrier (dielectric) materials such as composite mica, quartz and soft glass with dielectric constants of 2.3, 3.7 and 6.4, respectively, were evaluated as barrier materials. A lower dielectric constant material was found to provide higher $SO_2$ conversion.

*Effect of Geometry on $SO_2$ Conversion*   Figure 8.27 shows that the energy requirements of cylindrical, square and rectangular reactors for the same conversion were in increasing order under identical experimental conditions. For an energy density of 3000 J/L, the $SO_2$ conversions were about 80, 33, and 22 percent for cylindrical, square, and rectangular reactors, respectively.

## MERCURY OXIDATION IN BARRIER DISCHARGE

*Hg Oxidation with Low $O_2$ Concentration*   To minimize ozone formation in the plasma reactor which would interfere with the mercury analyzer, a low concentration (0.1 percent) of $O_2$ was used. The concentration of Hg vapor in the outlet stream of the plasma reactor, presented in Figure 8.28, was measured according to the following procedure. First a stream of $N_2$–$O_2$ with 0.073 mg (Hg)/$m^3$ was passed through the plasma reactor without power. When a voltage of 21.2 kV at 150 Hz was applied to the reactor, the outlet Hg concentration dropped to about 0 mg/$m^3$, it increased to the feed level of 0.073 mg/$m^3$ when power was turned

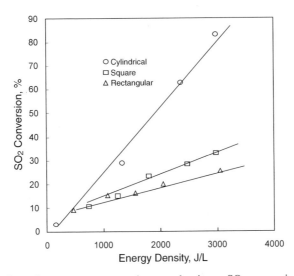

**Figure 8.27.** Effect of reactor geometry and energy density on $SO_2$ conversions. (From Chen and Mathur, 2002.)

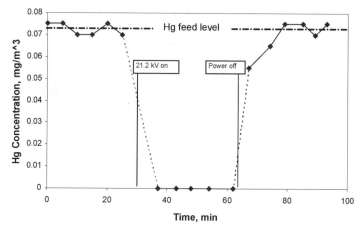

**Figure 8.28.** Effect of 0.1 percent $O_2$ on Hg oxidation.

off. Thus the 0.1 percent $O_2$ in the inlet $N_2$ gas stream was found to be adequate to oxidize almost all mercury vapors.

The following is proposed to be the main oxidation reaction in the plasma reactor:

$$Hg^0 + O_3 \rightarrow HgO + O_2 \qquad (8\text{-}33)$$

The effect of $O_2$ on mercury oxidation is further investigated by increasing $O_2$ concentration from less than 1 ppm to 0.1 percent. The Hg concentration was kept constant at 0.068 mg/m$^3$ in the inlet $N_2$ stream, and a voltage of 21.6 kV at 150 Hz was applied to the plasma reactor. The total gas flow rate was 1,000 CCM. Figure 8.29 shows the outlet Hg concentrations and conversions as a function of four $O_2$ concentrations: 0.1, 0.06, 0.04 percent, and less than 1 ppm (present in the $N_2$ cylinder). Conversion was about 80 percent at $O_2$ concentrations between 0.03 and 0.04 percent. As the $O_2$ concentration was increased to 0.06 percent, Hg conversion reached 100 percent.

*Effect of Energy Density on Mercury Conversion*   Energy input is a key parameter for a barrier discharge reactor because it has a strong effect on reactions inside the reactor. When a gas stream consisting of 0.061 mg/m$^3$ Hg, 0.1 percent $O_2$, and $N_2$ balance was passed through the plasma reactor, the initial discharge provided energy input of 86 J/L, resulting in 45 percent Hg conversion. The Hg conversion was 100 percent at an energy density of 114 J/L. In general, a higher energy density results in a higher Hg conversion into HgO (Figure 8.30).

### 8.2.1.3 *Commercialization Efforts*   Work on the removal of NO$_x$, SO$_x$, and Hg was extended from laboratory scale to commercialization at several companies. Powerspan, Inc., in New Durham, New Hampshire, is currently conducting field tests of a nonthermal plasma barrier discharge system for the NO$_x$, SO$_x$, and Hg

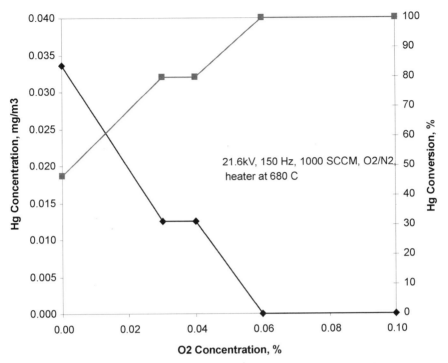

**Figure 8.29.** Effect of different $O_2$ concentrations on Hg conversion.

removal at a pilot test unit, located at First Energy Corp.'s R. E. Burger Plant near Shadyside, Ohio. The test system consists of one dry electrostatic precipitator (ESP), two parallel barrier discharge reactors, and one wet ESP, processing up to 4,000 cfm of flue gas drawn from the exhaust of a 125 MW coal-fired power plant. The inlet flue gas contains $SO_2$ and $NO_x$, about 1,700–1,800 and 240–400 ppm, respectively. The gas is at 300°F. Overall removal efficiencies of about 90, 97, and 80 percent have been reported for $NO_x$, and $SO_2$, and Hg, respectively (McLarnon and Jones, 2000; Powerspan Corp., 2000). Powerspan is installing a commercial demonstration unit of this technology at the same plant, designed to process about 110,000 cfm slipstream using about 6,000 reactor tubes.

***8.2.1.4 Summary*** A barrier discharge nonthermal plasma technique was investigated on a laboratory scale for the simultaneous removal of $NO_x$, $SO_2$, and Hg vapors. Study of the conversion of $NO_x$ in the presence of other gases ($O_2$, $H_2O$, $CO_2$, CO, $CH_4$ and $C_2H_4$) revealed that NO undergoes chemical oxidation into $NO_2$ or reduction into $N_2$ depending upon the inlet gas composition. High $O_2$ content and energy density promote NO oxidation. The presence of $H_2O$ has a strong effect on NO conversion. Addition of ethylene significantly improves NO oxidation into $NO_2$: as the concentration of $C_2H_4$ is increased from 0 to 3,000 ppm, almost 100 percent NO conversion to $NO_2$ is observed.

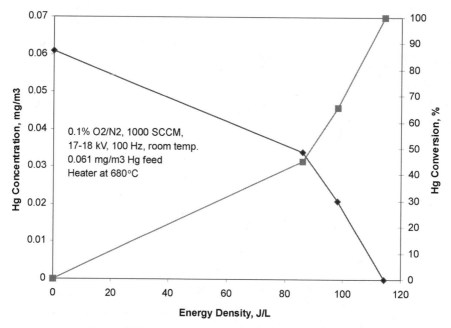

**Figure 8.30.**  Effect of energy density on Hg conversions.

Plasma reactor performance was also studied with respect to geometry and dielectric materials for $SO_2$ oxidation. A cylindrical reactor offers a significant advantage over square and rectangular rectors in terms of $SO_2$ conversion. Evaluation of barrier (dielectric) materials such as composite mica, quartz, and soft glass showed that a lower dielectric constant material provide higher $SO_2$ conversion. Barrier discharge is equally effective in the oxidation of Hg to HgO, which can be removed in the downstream processing by scrubbing (Boyle and Bernier, 2003).

## ACKNOWLEDGMENT

The author is thankful to his graduate students Zongyuan Chen, Claire A. Golden and Christopher R. McLarnon who worked on this project.

## REFERENCES

Boyle, P., and Bernier, D. (2003). Commercial demonstration of eco multi-pollutant control technology, Presented at Coal-Gen, Columbus, OH, (August).

Breault, R., McLarnon, C., and Mathur, V. (1993). Reaction Kinetics for Flue Gas Treatment of $NO_x$, *Nonthermal Plasma Techniques for Pollution Control.* B. M. Penetrante, and S. E.

Schultheis (eds.) NATO ASI Series, Volume 34, Part B: Electron Beam and Electrical Discharge Processing, Springer-Verlag, Berlin, 239–256.

Carey, T., Hargrove, O., Richardson, C., Chang, R., and Meserole, F. (1998). Factors affecting mercury control in utility flue gas using activated carbon. *Journal of the Air & Waste Management Association*, **48**, 1166.

Chang, J. (1997). Simultaneous removal of $NO_x$ and $SO_x$ from a combustion flue gas by corona discharge radical shower systems. *Applications of Electrostatics Workshop*, University of Cincinnati, (August 22).

Chen, Z., and Mathur, V. (2002). Nonthermal plasma for gaseous pollution control. *Ind. Eng. Chem. Res.*, **41**(9), 2082–2089.

Clements, J., Mizuno, A., Finney, W., and Davis, R. (1989). Combined removal of $SO_2$, $NO_x$, and fly ash from simulated flue gas using pulsed streamer corona. *IEEE Transactions on Industry Applications*, **25**(1), 62.

Dunham, G., Miller S., Chang R., and Bergman, P. (1998). Mercury capture by an activated carbon in a fixed-bed bench-scale system, *Environmental Progress*, **17**(3), 203.

Frank, N. W., and Hirano, S. (1993). The history of electron beam processing for environmental pollution control and work performed in the United States. *Nonthermal Plasma Techniques for Pollution Control*, B. M. Penetrante, and S. E. Schultheis (eds.) NATO ASI Series, Volume 34, Part B: Electron beam and electrical discharge processing, Springer-Verlag, Berlin, 1–26.

Galbereath, K., and Zygarlicke, C. (1996). Mercury speciation in coal combustion and gasification flue gases. *Environmental Science & Technology*, **30**(8), 2421.

Granite, E., Pennline, H., and Hargis, R. (2000). Novel sorbents for mercury removal flue gas. *Industrial & Engineering Chemistry Research*, **39**(4), 1020–1029.

Huang, H., Wu, J., and Livengood, C. (1996). Development of dry control technology for emissions of mercury in flue gas. *Hazardous Waste & Hazardous Materials*, **13**(1), 107.

Joshi, S. (1927). The decomposition of nitrous oxide in the silent electric discharge. *Trans. Faraday Soc.*, **23**, 227.

Kogelschatz, U., Eliasson, B., and Egli, W. (1997). Dielectric-barrier discharges principle and applications. Printed plenary lecture. *International Conference on Phenomena in Ionized Gases (ICPIG XXXIII)*, Toulouse, France, July 17–22.

Maezawa, A., and Izutsu, M. (1993). Application of e-beam treatment to flue gas cleanup in Japan. *Nonthermal Plasma Techniques for Pollution Control*, B. M. Penetrante, and S. E. Schultheis (eds.) NATO ASI Series, Volume 34, Part B: Electron beam and electrical discharge processing, Springer-Verlag, Berlin, 47–54.

Mathur, V., Breault, R., McLarnon, C., and Medros, F. (1992). $NO_x$ reduction by sulfur tolerant corona-catalytic apparatus and method. U.S. Patent 5,147,516.

Mathur, V., Chen, Z., Carlton, J., McLarnon, C., and Golden, C. (2000). *Barrier discharge for gaseous pollution control*, Proceedings of the 3rd Joint China/USA Chemical Engineering Conference, Beijing, China (September 25–28).

Mathur, V., Chen, Z., and Carlton, J. (1999). Electrically induced chemical reactions in a barrier discharge reactor. Technical Report, Zero Emissions Technology, Inc. (August).

Matteson, M., Stringer, H., and Busbee, W. (1972). Corona discharge oxidation of sulphur dioxide. *Environmental Science & Technology*, **6**(10), 895.

Matzing, H. (1991). Chemical kinetics of flue gas Cleaning by irradiation with electrons. In *Advances in Chemical Physics Volume LXXX*, L. Prigorgine, and A. R Stuart (eds). Wiley.

McLarnon, C., and Jones. M. (2000). Electro-catalytic oxidation process for multi-pollutant control at first energy's R.E. Burger generating station. *Proceedings of Electric Power 2000 Conference in Cincinnati.*

McLarnon, C., and Mathur, V. (2000). Nitrogen oxide decomposition by barrier discharge. *Ind. Eng. Chem. Res.*, **39**(8), 2779.

McLarnon, C., and Penetrante, B. (1998). Effect of gas composition on the NO$_x$ conversion chemistry in a plasma. *SAE Technical Paper Series* 982433, San Francisco, CA (October).

McLarnon, C. R. (1996). Nitrogen oxide decomposition by barrier discharge. Ph.D. dissertation, University of New Hampshire, Durham, NH.

Mok, Y., Kim, J., Nam, I., and Ham, S. (2000). Removal of NO and formation of byproducts in a positive-pulsed corona discharge reactor. *Ind. Eng. Chem. Res.*, **39**, 3938.

Penetrante, B., Brusasco, R., Merritt, B., Pitz, W., Vogtlin, G., Kung M., Kung, H., Wan, C., and Voss, K. (1998). NO$_x$ conversion chemistry in plasma-assisted catalysis. *Proceedings of the 1998 Diesel Engine Emissions Reduction Workshop*, Maine Maritime Academy, Maine (July 6–7).

Penetrante, B., Hsiao M., Merritt B., Vogtlin G., and Wallman, P. (1996). Pulsed corona and dielectric-barrier discharge processing of NO in N$_2$. *Applied Physics Letter*, **68**(26), 3719.

Powerspan Corp. (2002) http://www.powerspan.com, Powerspan Corp. News (March 15).

Serre, S., and Silcox, G. (2000). Adsorption of elemental mercury on the residual carbon in coal. *Industrial & Engineering Chemistry Research*, **39**(6), 1723–1730.

Urashima, K., Chang, J., and Ito, T. (1997). Reduction of NO$_x$ from combustion flue gases by superimposed barrier discharge plasma reactors. *IEEE Transactions on Industry Applications*, **33**(4).

Veldhuizen, E., Zhou, L., and Rutgers, W. (1998). Combined effects of pulsed discharge removal of NO, SO$_2$ and NH$_3$ from flue gas. *Plasma Chemistry and Plasma Processing*, **18**(1), 91.

## 8.2.2 E-Beam Based Oxidation Treatment for VOC Contaminated Aqueous Phase Waste Streams

ATUL C. SHETH, PROFESSOR, CHEMICAL ENGINEERING
The University of Tennessee Space Institute, Tullahoma, TN 37388, USA

A technology based on irradiation by a high-voltage electron beam that removes volatile organic contaminants (VOCs) from liquid waste streams has substantial advantages over most competing technologies. Not only can it be directed specifically to the total destruction of organic contaminants with little effect on the carrier stream, but residual contamination is significantly reduced.

There are numerous options available to treat the VOCs in aqueous streams such as drinking water, groundwater, surface run-offs, and landfill leachates. However, most of these methods simply concentrate the pollutants, leaving a waste stream, which requires additional treatment. Only options that are based on the oxidation principle do not need additional treatment. The oxidation technologies can be classified into four basic categories: biological, thermal, chemical, and radiation or radiant energy based systems (Materi, 1998). We precede our discussion of electron-beam-based systems by noting the disadvantages of the four technologies just named.

### 8.2.2.1 Drawbacks of Oxidation Technologies

Biological treatment methods usually operate at low temperatures, are cost effective and tend to work well on low level organics. These methods are however, slow and show a susceptibility of microorganisms to poisoning. In the presence of high organic and inorganic contamination, the micro organisms used in the system are generally rendered ineffective, and therefore, these types of options are not used for higher concentrations of organic and inorganic contamination.

Thermal oxidation technologies, some of which are highly controversial, are currently considered the most viable method for the treatment of organics in water. These techniques are generally expensive and have been found, in many cases, to be ineffective and to create undesirable by-products. Incineration and catalytic oxidation, although fast and generally effective, have been found to create products of incomplete combustion (PICs: e.g., dioxins and furans) and acid gases ($NO_x$, $SO_2$) that are less desirable than the original pollutants. In addition, catalytic systems have shown a propensity for fouling and poisoning in the presence of inorganic compounds, which renders them ineffective. Wet air and supercritical oxidation are two additional thermal oxidation techniques. Both of these systems operate at temperatures below 750°F, but the pressure is generally on the order of 3000 psi. This makes these systems not only costly and difficult to maintain, but also somewhat unsafe. In addition, these techniques have been shown not to work on a number of compounds, yielding either incomplete oxidation products or limited effectiveness.

Ozone and hydrogen peroxide systems are the most widely used chemical oxidation systems. Although effective treatment techniques, they have been found to be ineffective for a variety of compounds. In many instances they require the use of catalysts and long reaction times in order to be effective. In addition, the reactions must be carefully controlled to ensure that explosive conditions are minimized.

Radiant energy treatment systems using gamma rays, x-rays, electron beams, radio frequency (RF) waves, microwaves, and ultraviolet (UV) radiation are all cited in the literature. Gamma ray and x-ray systems have the inherent liability of radiation exposure to the operators of the equipment and as such generally require extremely thick safety shields. In addition, these techniques have been found to have limited effectiveness in the treatment of many organic compounds.

Electron beam and ultraviolet oxidation systems are currently limited to the treatment of water with very low levels of organics. Not unlike other radiant energy systems, they have been found to be ineffective on a number of compounds and show evidence of undesirable by-products. In the past, electron beams have been used to treat waste streams either in solely gas phase or liquid phase. In the flue gas desulfurization (FGD) applications, in which e-beam is used to treat gas phase inorganic contaminants such as $SO_2$ and $NO_x$, the electron beam oxidizes $SO_2$ and $NO_x$, whereupon the contaminants react with ammonia to form the respective ammonium salts. These salts are then captured in downstream baghouse filler (BH) or dry electrostatic precipitator (ESP).

As discussed in Section 8.1.1, electron beams can be used to treat gas phase VOCs such as trichloroethylene, vinyl chloride, and methylene chloride (Anshumali, 1996; Anshumali et al., 1997). The primary limitation of these processes is that they are not capable of treating organically contaminated liquids and consequently are not suitable for the treatment of most semivolatile and water soluble organic contaminants. In addition, these processes are known to form water-soluble organic by-products, which must be captured and treated by other means.

Electron beam processes have also been used in the treatment of liquid phase, volatile, semi-volatile and water soluble organics. In one such process, wastewater is passed over a weir and allowed to flow down a wall forming a thin water film. A rectangular electron beam is focused along this wall such that it completely covers the width of the water but only a small portion of its length. This and other processes, which are described in the patent literature, have been effective only on wastewater with fairly dilute organic concentrations, however (US Patents, 1975, 1980, 1991, 1995).

To handle large volumes of contaminated liquids, thermal and/or catalytic combustion (to increase the rate of oxidation), adsorption using granulated activated carbon, absorption using suitable solvents or neutralizing agents, and condensation (when VOC concentrations are high) have been considered. The typical field application methods are usually smaller in scale and designed for site-specific conditions. Usually the following parameters determine the choice of appropriate destruction system (Patrick and Winkleman, 2001):

- Type of VOC
- Physical state and concentration of VOC
- Physical state of the medium in which the VOC is present
- Level of clean-up required
- Allowable cost of clean-up and operation time, and
- Potential of producing undesirable by-products, intermediates and end-products

However, at present, there does not exist a single option capable of handling gas and liquid phase contaminants that is broadly applicable in all concentration ranges to wide variety of VOCs and be cost-effective (Patrick and Winkleman, 2001).

### 8.2.2.2 Electron-Beam Chemistry

To induce chemical reactions, a highly energetic electron beam, like other forms of ionizing radiation, induces the formation of ion pairs, free radicals, and excited molecules in all phases in which it comes into contact. In contrast to liquid-phase systems, however, gas-phase reactions are not limited to those that are generally concentrated along the particle track. Electron charge transfer in the bulk gas plays a significant role in the formation of excited species and free radicals.

In addition, the radicals that are formed readily diffuse and react throughout the system. The transient species formed tend to have longer lifetimes and lose energy more slowly because the rate of collision is slower. Thus, mechanisms of radiolysis are generally much more complex in gaseous systems than in the liquid counterparts. Because such systems usually are operated in an oxidizing mode, the primary characteristics of the gas-phase composition are considerably different than in the liquid. The bulk of the gas, $N_2$ and $O_2$, is in far greater supply than the organic contaminants and water vapor. Thus, the transient reaction products and pathways will prominently include much higher levels of activated oxygen species than found in the liquid phase. In other words, excited oxygen, ozone, and the ionizing species $O^-$ and $O^{2-}$ will participate in far more reactions than in the liquid phase (Hager, 1990).

For an electron beam facility of the type pioneered in Florida, discussed in Section 8.2.2.3 the high-energy electron irradiation of water solutions and sludges produces a large number of very reactive chemical species, including the aqueous electron ($e_{aq}^-$), the hydrogen radical (H·), and the hydroxyl radical (OH·) (Florida International University and University of Miami, 1994). These short-lived intermediates react with organic contaminants, transforming them to nontoxic by-products. In the principal reaction, the $e_{aq}^-$ ion transfers to halogen-containing compounds, breaking the halogen-carbon bond and liberating halogen anions, such as chloride ($Cl^-$) or bromide ($Br^-$). The hydroxyl radical can undergo addition or hydrogen abstraction reactions, producing organic free radicals that decompose in the presence of other hydroxyl radicals and water. In most cases, organics are converted to carbon dioxide, water, and salts. Lower molecular weight aldehydes and carboxylic acids form at low concentrations in some cases, however these compounds are biodegradable end products (Florida International University and University of Miami, 1994).

Hager (1990) has proposed a similar oxidation mechanism for reaction of formic acid with UV-catalyzed hydrogen peroxide. This photochemical oxidation process is described as:

$$H_2O_2 \xrightarrow{\text{UV}<400\,\text{nm}} 2 \cdot OH$$
$$HCOOH + \cdot OH \rightarrow H_2O + HCOO\cdot$$
$$HCOO \cdot + \cdot OH \rightarrow H_2O + CO_2$$

A glance at oxidation potentials for common oxidants (Table 8.10) shows that the hydroxyl radical is second only to fluorine in oxidative power. Ozone also forms hydroxyl radicals under UV light catalysis. Neither hydrogen peroxide nor ozone contains metals or halogens that can lead to undesirable by-products during the organic oxidation process. However, hydrogen peroxide is a more cost-effective reactant because each molecule of hydrogen peroxide forms two hydroxyl radicals.

TABLE 8.10. Oxidation Potential of Oxidants[*,†]

| Relative Oxidation Power | Species | Oxidative Potential (Volts) |
|---|---|---|
| 2.23 | Fluorine | 3.03 |
| 2.06 | Hydroxyl radical | 2.80 |
| 1.78 | Atomic oxygen | 2.42 |
| 1.52 | Ozone | 2.07 |
| 1.31 | Hydrogen peroxide | 1.78 |
| 1.25 | Perhydroxyl radical | 1.70 |
| 1.24 | Permanganate | 1.68 |
| 1.17 | Hypobromous acid | 1.59 |
| 1.15 | Chlorine dioxide | 1.57 |
| 1.10 | Hypochlorous acid | 1.49 |
| 1.07 | Hypoiodous acid | 1.45 |
| 1.00 | Chlorine | 1.36 |
| 0.80 | Bromine | 1.09 |
| 0.39 | Iodine | 0.54 |

[*]Taken from Hager (1990).
[†]Used with permission from the Journal of the Air and Waste Management Association, permission granted through the Copyright Clearance Center.

Furthermore, hydrogen peroxide has other inherent advantages as the preferred oxidant over ozone. Hydrogen peroxide is supplied commercially as an easily handled liquid (30–50 percent) that is infinitely soluble in water. Ozone is a toxic gas with limited water solubility. The water solubility of hydrogen peroxide greatly simplifies the reactor design, in terms of oxidant addition, mixing of the reactants, and elimination of fugitive toxic gases. In addition, hydrogen peroxide storage and feed systems are relatively inexpensive compared to ozone generation and feed equipment.

In comparison, the important factor in the effective reaction of organics in air is ionization potential. The material to be destroyed must have an ionization potential lower than that of the carrier gas. The ionization potential is 15.60 eV for nitrogen, 12.06 eV for oxygen, and 12.61 eV for water. Organics with the greatest differences in ionization potentials compared to these are generally destroyed most rapidly; organics with ionization potentials in the 12–16 eV range can still be destroyed; however, recirculation may be necessary to achieve a reasonable destruction efficiency. Ionization potentials for selected substances are shown in Table 8.11.

### 8.2.2.3 Electron Beam Systems

*AOS System*   In the e-beam facility patented by Advanced Oxidation Systems (AOS), VOC-contaminated liquid is introduced into a reaction chamber that is positioned to expose the liquid to the electron beam (Figure 8.31) (Materi, 1998). A gaseous phase is generated first by exposing the liquid to e-beams, and then by recirculating the gas around the reactor loop. This virtually simultaneous exposure of gas

**TABLE 8.11. Ionization Potentials of Typical Organic Contaminants***

| Species | Ionization Potential (eV) | Species | Ionization Potential (eV) |
|---|---|---|---|
| Nitrogen | 15.60 | Methanol | 10.85 |
| Oxygen | 12.06 | Acetonirile | 12.19 |
| Water | 12.61 | Propane | 10.95 |
| Perchloroethylene | 9.32 | Butane | 10.53 |
| Trichlorethylene | 9.47 | C5 and above alkanes | 9–11 |
| Vinyl chloride | 9.99 | 1,1.1-Trichloroethane | 11.04 |
| Benzene | 9.25 | Methyl-t-butyl ether | 9.54 |
| Toluene | 8.82 | Dichloromethane | 10.36 |
| Xylene | 8.56 | Chlorotrifluoromethane | 12.39 |
| Acetone | 9.71 | Dichlorodifluoromethane | 11.75 |
| Ethyl acetate | 10.01 | Trichlorofluoromethane | 11.77 |
| Tetrahydrofuran | 9.41 | 1,1,2-Trichlorotrifluoroethane | 11.99 |
| Dimethyl formamide | 9.13 | Carbon tetrachloride | 11.47 |

*Taken from Patrick and Winkleman (2001).

and liquid phase to the e-beam results in complete oxidation of the VOC species. Also, the liquid containing VOCs can be recirculated in the reaction chamber, further facilitating the oxidation process (Materi, 1998).

According to the patent for the AOS system, the active/intermediate species formed in the gas phase reactions further react with the VOCs in the aqueous phase, leading to more breakdown of organics in the liquid phase (Materi, 1998). In addition, the breakdown of VOCs in the liquid phase coupled with the stripping

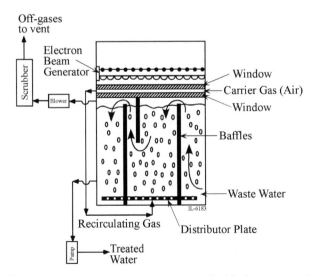

**Figure 8.31.** Schematic of e-beam apparatus to treat liquid phase waste. (Adopted from Materi, 1998.)

action of the circulating gas allows further oxidation of VOCs in the gas phase. Such beneficial effects are not available in the systems relying on earlier technologies.

In the novel reaction chamber designed and patented by AOS, an electron beam is generated and then focused and directed through two titanium (primary and secondary) windows into the main section of the reactor chamber, where it interacts with the gases and water (Materi, 1998). The electron beam configuration matches the gas-liquid interface so accurately that it contacts the entire cross-sectional surface area of the water it is treating. The space between the two windows forms a plenum and allows for the passage of carrier gas or air used for removing VOC gases. Wastewater to be treated enters the reaction chamber and passes through various compartments by going over the baffles. This ensures the complete interaction of the electron beam with the water and also prevents backmixing and short-circuiting. Gas from the plenum between two electrodes is dispersed through the liquid by suitable means. Bubbles generated in the liquid provide agitation as well as help in vaporizing the VOCs. The gases collected over the liquid surface are also exposed to the e-beam, where some gas-phase VOC destruction is also going on. The portion of this gas is bled to the off-gas treatment for acid gas removal and any required polishing step (such as guard column of activated carbon) to remove final traces of VOCs before it is released to air. The treated water is taken out of the chamber and undergoes any post treatments or disposed off. Because of their special contacting configuration, AOS claims that complete oxidation of the VOCs is possible.

*Florida International University/ University of Miami System*    In a joint project, Florida International University and the University of Miami built an electron beam research facility designed to handle tank trucks carrying up to 6000 gallons of waste (Florida International University and University of Miami, 1994). Initially, researchers tested six organic compounds (TCE, PCE, chloroform, benzene, toluene, and phenol) to determine removal efficiencies at three solute concentrations and three pHs, and in the presence and absence of three percent clay. Trace quantities of formaldehyde and other low molecular weight aldehydes were detected, as well as formic acid, at low VOC concentrations. Subsequently, the treatment process was demonstrated at the US Department of Energy's Savannah River site in Aiken, South Carolina, where chlorinated solvents from fuel and target manufacturing operations had been discharged to an unlined basin for almost 20 years (High Voltage Environmental Applications Inc., 1994a).

Later a mobile trailer-mounted electron beam treatment facility and delivery system was built (High Voltage Environmental Applications Inc., 1994a). Its computer-automated 500 kV e-beam accelerator produced a continuously variable beam current from 0 to 40 mA. At full power, the system was rated at 20 kW with a maximum flow rate of up to 50 gallons per minute. The flow rate and beam current could be varied to obtain doses up to 2,000 kilorads in a one-pass, flow through mode.

This self-contained system included a 100 kW generator, but the contaminated site had to provide a mixing tank to slurry the treatable solids. The initial studies showed that electron beam irradiation removed from aqueous streams more than 99 percent of the compounds initially tested in Florida (TCE, PCE, chloroform, benzene, toluene and phenol). They also demonstrated effective removal of 2,4,6-trinitrotoluene from soil slurries. Later on in a bench-scale study, a multisource hazardous waste leachate containing 1 percent dense nonaqueous phase liquid was successfully treated (High Voltage Environmental Applications Inc., 1994b).

*The University of Tennessee Space Institute System*    A contacting configuration developed at the University of Tennessee Space Institute (UTSI) uses flashing (or sudden reduction in pressure) to release vapors of VOCs from water. A schematic of the UTSI concept, which is based on the AOS design (Figure 8.31), is shown in Figure 8.32. In the reaction chamber, a pool of wastewater is maintained at the bottom by circulating contaminated ground water. High-pressure air (or carrier gas) is released through this pool of water, which then expands above the liquid surface. This expansion of air (or carrier gas) causes dissolved VOCs to be released from the water and also provides agitation in the form of rising air bubbles. The air containing VOCs in the gas phase is introduced in the electron beam chamber where it is destroyed. The exposed gas is then recirculated by mixing with the make-up gas and recompression. A slip stream of recycle gas is taken out and treated to remove

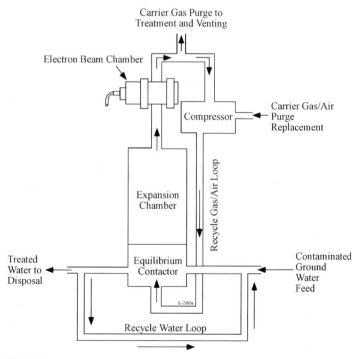

**Figure 8.32.** Schematic of electron beam-based ground water treatment unit. Used with permission from the Journal of the Air and Waste Management Association, permission granted through the Copyright Clearance Center.

acid gases before it is released in atmosphere. In contrast to the AOS system, in this concept the electron beam does not need to cover entire liquid surface, and therefore, can be smaller, concentrated and more efficient. Such a system can be less costly to build and if necessary can be moved around on a truck.

In general, the e-beam radiation can be applied to contaminated wastewater in similar contacting configurations to accommodate site specific needs. The efficiency of VOC removal in each design will depend upon the contacting pattern and interface area between the e-beam and VOC-containing gas or liquid stream.

*UTSI Tests and Results*   In a test program carried out at UTSI, three surrogate samples were tested (Materi, 1998; Patrick and Winkleman, 2001). The test system contained the following components:

- E-beam generator
- Reaction chamber for holding liquid sample and off-gases to be treated during the test
- Plenum used to supply an oxidizing gas
- Liquid recirculating system
- Computer monitoring system and gas analysis system

The following procedure was used on each waste stream until a total dose of 125 kGy was achieved:

- Approximately 2.5 L of VOC waste containing liquid was added to the reactor. The pump was activated to start recirculation of the liquid sample.
- The oxidant was turned on to begin the introduction of air and ozone to the system.
- Samples of waster were taken initially and at accumulated does of 5, 10, 25, 50, 100, and 125 kGy. Samples were stored in ice until analyzed.
- After each sample was collected, the e-beam was turned on to allow the liquid waste to achieve the accumulated dose required for the next sample point.
- The steps were repeated until a dose of 100 kGy was attained at which time 2.5% of the stoichiometrically required $H_2O_2$ was added to the reactor and the final accumulated dose was achieved.

The results of the UTSI tests are given in Tables 8.12 and 8.13. Table 8.12 shows the overall reduction in the chemical oxygen demand (COD) and the total cyanide (CN) and total ammonia as nitrogen ($NH_3$-N) species remaining in the treated water for a dose of 125 kGy. Table 8.13 shows the overall reduction in the individual components contained in the three surrogates tested after a dose of 125 kGy. Figure 8.33 depicts the overall reduction in COD values for surrogates #2 and #3. Figure 8.34 depicts the data displayed in Table 8.13 for surrogates #2 and #3. Two waste streams tested employed initial component concentrations varying from 81 ppm to 35,220 ppm. As can be seen from Figure 8.34, destruction

TABLE 8.12. Waste Stream Treatment Results at 125 kGy*

| Waste Stream | %COD Reduction | Total CN ppm | Total NH3-N ppm |
|---|---|---|---|
| Surrogate 1 | 84.19 | NA | NA |
| Surrogate 2 | 72.57 | 0.21 | 107 |
| Surrogate 3 | 43.70 | 0.07 | 156 |

*Taken from Materi (1998).

efficiencies varied from a low of about 13–100 percent. The lower destruction efficiencies were not considered to be related to ionization potential but were attributed to failure to optimize the waste materials and the process. Evaluation of the data did indicate that the use of ozone as part of the oxidizing carrier gas and the introduction of hydrogen peroxide could assist in speeding the reduction process. According to the patent, however, in a fully optimized system utilizing the proprietary AOS' reaction chamber, oxidants such as ozone and hydrogen peroxide will not be needed except for the recalcitrant wastes (Materi, 1998).

TABLE 8.13. Individual Contaminant Reductions at 125 kGy*

| Contaminant | Starting ppm | Ending ppm | % Reduction |
|---|---|---|---|
| Surrogate 1 | | | |
| Methylene chloride | 34,668 | NA | NA |
| Surrogate 2 | | | |
| Acetonitrile | 21,819 | 5,844 | 73.22 |
| Acetone | 5,236 | 938 | 82.09 |
| Dimethylformamide | 26,227 | 18,845 | 28.15 |
| Methanol | 10,429 | 6,691 | 35.84 |
| Ethylacetate | 421 | 4 | 99.12 |
| Toluene | 81 | 0 | 100.00 |
| Sodium chloride | 10,393 | | |
| Calcium chloride | 10,393 | | |
| Surrogate 3 | | | |
| Methylene chloride | 35,220 | 2,428 | 93.11 |
| Acetonitrile | 20,985 | 9,211 | 56.11 |
| Acetone | 5,036 | 2,077 | 58.76 |
| Dimethylformamide | 25,225 | 22,008 | 12.75 |
| Methanol | 10,031 | 7,640 | 23.83 |
| Ethylacetate | 405 | 56 | 86.10 |
| Toluene | 78 | 7 | 90.91 |
| Sodium chloride | 9,996 | | |
| Calcium chloride | 9,996 | | |
| Zinc hydroxide | 2,999 | | |

*Taken from Materi (1998).

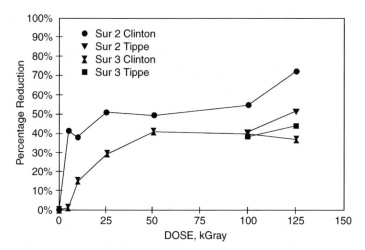

**Figure 8.33.** Overall COD reduction for surrogates 2 and 3 as a function of dose level. (From Materi, 1998.)

***8.2.2.4 Comparison of Oxidation Options*** Concentrated aqueous hydro-carbon waste solutions result from a variety of chemical processes. The disposal of these concentrated solutions without air emissions or without the use of landfills is a formidable challenge for many industries. Radiation based chemical oxidation offers an alternative solution for the on-site destruction of numerous concentrated organic wastes.

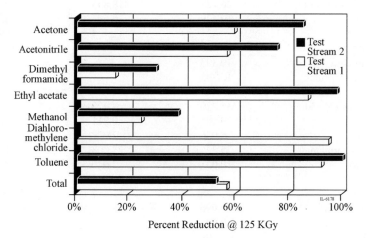

**Figure 8.34.** Percent VOC reduction in two test waste streams. (From Patrick and Winkleman, 2001.)

Both thermal and chemical oxidation converts hydrocarbons to carbon dioxide and water. However, there are a number of factors, to be considered in addition to the basic economics of each process.

An important advantage of the chemical oxidation process is that it can be used as a pre-treatment to detoxify aqueous wastes. With thermal oxidation it is necessary to operate at very high temperatures (1,200°C) to minimize formation of toxic compounds. In so doing, much fuel is consumed evaporating the water.

On the contrary, radiation based chemical oxidation may be easily operated at less than full oxidation by control of the amount of oxidant and power. If intermediate reaction products are not objectionable, considerable savings in oxidant and power can be attained by this procedure. Chemical oxidation is particularly applicable to a detoxification step prior to biological treatment. Phenolic aqueous wastes, for example, can be oxidized to biodegradable intermediate chemicals at a fraction of the cost of thermal oxidation.

Halogenated or sulfonated hydrocarbons are not desirable for combustion in incinerators. These compounds generally have poor fuel value and more importantly, produce acid gas discharges that are corrosive and illegal to discharge to the atmosphere. Expensive alloy construction and gas scrubbers are required to burn these compounds. There are no gaseous emissions with chemical oxidation at economical concentrations, and the lower temperature liquid (less than 83°C) discharge stream is less corrosive and more easily neutralized than the exhaust gases from incineration.

In radiation based chemical oxidation, the oxygen requirement is supplied by the oxidizer and not air. Hager has estimated that about 3.2 kg of $H_2O_2$ is required for chemical oxidation of 1 kg of methanol (Hager, 1990). In the e-beam option tested at UTSI, no $H_2O_2$ or $O_3$ type oxidizers are needed, except for very recalcitrant wastes. Also, Hager's estimate shows that about 0.6 kg of $CO_2$ will be produced per kilogram of methanol plus water by chemical oxidation method (Hager, 1990). However, the solubility of $CO_2$ at standard temperature and pressure in water is about 168 kg per cubic meter of water, which is much higher than 0.6 kg of $CO_2$ produced. As a result, chemical oxidation options such as e-beam will have no noticeable emissions of $CO_2$ from oxidation of 0.12 kg of methanol in 1 cubic meter of water. Compared to this, almost 311 cubic meters of combustion products will be produced by thermal incineration of the same solution.

An additional factor in favor of chemical oxidation of these compounds is that the halogen exerts no demand for oxidant while greatly increasing the weight of the contaminating molecule. The net result is that chemical oxidation tends to be less expensive than thermal oxidation when applied to chlorinated compounds. Because of these factors, incineration of halogenated or sulfonated hydrocarbons will not usually be an economical choice.

Chemical oxidation equipment comes in many sizes, designs, and capacities. A given unit can vary over a 30:1 range in capacity depending upon the resistance of the VOCs to chemical oxidation. As a result, the capital cost in terms of dollars per cubic meter and waste to be handled will also vary widely. Chemical oxidation equipment will also vary in cost per unit of capacity, with larger units costing less. A balance will need to be struck between many units scattered at point of use to minimize collection and operating costs for smaller units versus equipment costs for the larger units.

***8.2.2.5 Economic Considerations*** A thorough economic analysis of a VOC destruction system based on e-beam radiation for aqueous phase waste has yet to be performed. However, a prototype e-beam plant capable of processing 100,000,000 cubic feet of air per year contaminated with about 500-ppmv trichloroethylene has been proposed (Anshumali, 1996). The plant includes an e-beam reactor, an acid gas scrubber to remove acidic by-products, and a final carbon adsorber to remove any residual organics. The plant was intended to operate continuously (i.e., 330 days per year) with an e-beam dose of 100 kGy. The inlet temperature was 35°C and inlet pressure 1.1 atm. The e-beam gun efficiency was estimated at 17 percent, which is highly conservative given that current e-beam designs can achieve substantially higher efficiencies. At these conditions and assuming once-through operation, the e-beam destruction efficiency was estimated to be about 82 percent. The scrubber was sized to remove the estimated quantities of hydrochloric acid and the activated carbon bed was sized to remove the residual organics. Using typical chemical engineering cost estimating techniques, the combined annualized fixed cost and annual operating cost was estimated to be $757,285, or about $7.57 per thousand cubic feet of gas treated. In terms of operating cost alone, the annual operating cost for this plant was estimated to be about $6.51 per thousand cubic feet of gas or $0.23 per cubic meter of air containing 500 ppm of TCE.

Although for reasons of space, this section has not given detailed coverage of ultraviolet/hydrogen peroxide systems, for completeness we mention some data provided by Hager on the annual operating costs of the destruction of toluene, TCE, methylene chloride, phenol and methanol using a $UV/H_2O_2$ Perox-pure[TM] system. The costs of decontaminating VOC concentrations ranging from 300 to 10,000 mg/L varied between $0.79 and $95.11 per cubic meter of wastewater (Hager, 1990). The smaller units had far lower operating costs. Thus, the operating cost of $0.23 per cubic meter for the e-beam-based option, even though calculated at present only for a gas phase system, is certainly attractive and comparable to the $UV/H_2O_2$ based chemical oxidation system.

The vapor pressure of TCE is 60 mmHg at 20°C and 100 mmHg at 31.4°C. Thus, at ambient conditions, a concentration of 500 ppm in TCE in air could be achieved by just bubbling of air through a pool of TCE contaminated wastewater. Even assuming that an e-beam system would destroy only gas phase TCE, the cost of $0.23/m^3$ of air seems reasonable.

These operating cost data can be compared to those of thermal incineration. Hager (1990) assumed a fuel cost of $0.50/10^5$ J and an efficiency of 75 percent to arrive at an operating cost of $15.85 per cubic meter for thermal incineration of VOCs such as toluene, TCE, methylene chloride phenol and methanol. In comparing the values for handling emissions of TCE wastes containing concentrations below 1000 mg/L, the value of $0.23 per cubic meter calculated for e-beam destruction is certainly more attractive than the \$0.79–1.85 per cubic meter values estimated for $UV/H_2O_2$ systems by Hager (1990). In the AOS patent (Materi, 1998), it is further claimed that the process and apparatus can be set up to produce usable by-products such as concentrated hydrochloric acid (HCl) solution and concentrated streams of $CO_2$.

Thus an e-beam based detoxification process has the potential to be an attractive alternative to technologies currently used or considered for the destruction of VOCs in air and wastewater streams. Preliminary economic data look comparable to those for a commercial $UV/H_2O_2$ system, and far better than thermal incineration and adsorption on activated carbon.

## REFERENCES

Anshumali (1996). Destruction of low levels of volatile organic compounds in air by an electron beam generated plasma. M.S. thesis, University of Tennessee, Knoxville, TN.

Anshumali, Winkleman B. C., and Sheth A. (1997). Destruction of low levels of volatile organic compounds in dry air streams by an electron-beam generated plasma. *JAWMA*, **47**, 1276–1283.

Florida International University and University of Miami (1994). *Electron Beam Research Facility*. EPA's Superfund Innovative Technology evaluation Program, Technology Profiles, EPA/540/R-94/526, 7th Ed., 264–265.

Hager, D. G. (1990). UV-catalyzed hydrogen peroxide chemical oxidation of organic contaminants in water, published in *Physical/Chemical Processes, Innovative Hazardous Waste Treatment Technology Series*, 2, 143–153, Ed. by H. M. Freeman.

High Voltage Environmental Applications, Inc. (1994a). *High-energy electron irradiation, EPA's superfund innovative technology evaluation program, technology profiles.* EPA/540/R-94/526, 7th Ed., 82–83.

High Voltage Environmental Applications, Inc. (1994b). *High-energy electron beam irradiation, EPA's superfund innovative technology evaluation program, technology profiles.* EPA/540/R-94/526, 7th Ed., 342–343.

Materi, G. E. (1998). Electron beam process and apparatus for the treatment of an organically contaminated inorganic liquid or gas. U. S. Patent No. 5,807,491, assigned to Advanced Oxidation Systems, Inc., Gaithersburg, Maryland.

Patrick, D. R., and Winkleman B. C. (2001). Electron beam destruction of VOC contaminated waste streams, paper #23, Session no. AE-2b. Presented at the Air and Waste Management Association's 94th Annual Conference and Exposition, Orlando, Florida.

US Pat. No. (1975). 3,901,807 issued to Trump on August 26.

US Pat. No. (1980). 4,230,947 issued to Cram on October 28.

US Pat. No. (1991). 5,072,124 issued to Kondo et al. on December 10.

US Pat. No. (1995). 5,451,790 issued to Enge on September 19.

## 8.2.3 Biodiesel

TAPAS K. DAS

Washington Department of Ecology, Olympia, Washington 98504, USA

Biodiesel is a mono-alkyl ester based oxygenated fuel made from vegetable or animals fats. It is commonly produced from oilseed plants such as soybean or canola, or from recycled vegetable oils (Brown, 2001). Biodiesel has similar properties to petroleum diesel fuel and can be blended with petroleum diesel fuel at any ratio. The most common blend, referred to as "B20," is 20 percent biodiesel and 80 percent petroleum diesel. Pure or neat biodiesel is termed B100.

Biodiesel is a domestically produced, renewable motor fuel which is non-toxic and biodegradable. Biodiesel is registered as a fuel and fuel additive with the US Environmental Protection Administration (USEPA) and has passed the EPA's Tier 1 Health Effects Testing under the Clean Air Act section 211(b). Neat biodiesel, B100, has also been classified as an alternative fuel by the US Department of Energy, and meets California Air Resources Board (CARB) clean diesel standards. The American Society of Testing and Materials, the US fuel standard-setting body, recently issued a new specification (D-6751) for biodiesel fuel. This specification applies to all biodiesel bought and sold in the US (NBB, http://nbb.org/).

***8.2.3.1 Emissions from Biodiesel*** There is a growing body of emission data for biodiesel. Compared with conventional diesel fuel, the use of B100 significantly reduces particulate matter (PM), carbon monoxide (CO), and hydrocarbons or volatile organic compounds (VOCs) but increases nitrogen oxides ($NO_x$) (NBB; CARB, 2000; USEPA, 2002). When B20 is compared with conventional diesel fuel, the changes in emissions are directionally the same, but smaller. CARB reports that B100 and B20 reduce PM emissions by 30 percent and 22 percent, respectively, in comparison to conventional diesel fuel (CARB, 2000). The National Biodiesel Board indicates similar emissions benefits, and reports PM reductions of 40 percent for B100 and 8 percent for B20 (NBB). In a draft report on biodiesel emissions, EPA reported an average PM reduction of 10.1 percent for soybean based B20 fuel, and a 2 percent increase in $NO_x$ emissions (USEPA 2002). These emissions varied with the source of biodiesel used (soybean, rapeseed, animal fats), and emissions benefits appeared to be consistent across engine model years.

In a study on life-cycle emissions from biodiesel and petroleum diesel, the US Department of Energy (DOE) concluded that tailpipe $PM_{10}$ emissions are 68 percent lower for biodiesel, while biodiesel life-cycle particulate emissions are 32 percent lower than conventional diesel fuel (Sheehan, 2001). A summary of biodiesel emissions is presented in Table 8.14. Actual emission reductions will vary with application.

**TABLE 8.14. Biodiesel Emissions Compared to Petroleum Diesel Fuel**

| | CARB | | NBB | |
|---|---|---|---|---|
| Pollutants | B100 (%) | B20 (%) | B100 (%) | B20 (%) |
| Particulate matter ($PM_{10}$) | −30 | −22 | −40 | −8 |
| Oxides of nitrogen | +13 | +2 | +6 | +1 |
| Polyaromatic hydrocarbon | −80 | −13 | −80 | −13 |
| Carbon dioxide | −100 | −20 | −100 | −20 |

*Source:* California Air Resources Board (CARB); National Biodiesel Board (NBB).

Biodiesel can be blended with conventional diesel to reduce the sulfur content of petroleum diesel fuels. Because biodiesel contains 0–1 ppmw of sulfur, exhaust emissions of sulfur oxides and sulfates are eliminated (NBB). Further, the absence of fuel sulfur suggests that after-treatment technologies such as diesel oxidation catalysts and particulate traps would perform well with biodiesel. In fact, a study conducted by Southwest Research Institute showed that catalyst conversion efficiency of total particulates improved with increased biodiesel content (Sharp, 1998). PM reductions for B20 versus conventional diesel went from 5–15 percent, to 10–22 percent when an oxidation catalyst was used. Similarly, PM reductions for B100 as compared to conventional diesel fuel averaged 30%–50%, while PM reductions increased to 50–60 percent with the addition of a catalyst.

Biodiesel can also be blended with ultralow sulfur diesel (ULSD). The Washington State Metropolitan Area Transit Authority recently investigated bus emissions resulting from the use of conventional diesel fuel, ULSD, and BD20, a blend of 20 percent biodiesel and 80 percent ULSD (National Research Center for Alternative Fuels, Engines and Emission, 2002). During the ULSD and BD20 tests the transit bus was equipped with a catalyzed particulate trap. The BD20 fuel showed virtually similar PM reduction efficiencies as the ULSD fuel and reduced PM emissions by greater than 98 percent as compared to the baseline diesel fuel. While showing a slight increase in $NO_x$ emissions, the BD20 blend also reduced both CO and HC emissions by 90 percent over the ULSD fuel.

### 8.2.3.2 Cost of Using Biodiesel

The cost of biodiesel depends primarily on the market price for vegetable oils or other feedstock. At a feedstock price of $0.10 per pound, a production cost of about $1 per gallon is projected for a 10 million gallon per year (mgy) facility. If the price of feedstock doubles, production costs for a 10 mgy plant may double as well, although lower increases have been reported (Tyson, National Renewable Energy Lab, 2001). Transportation costs will also impact the sale price for biodiesel. Currently biodiesel sold into markets in the state of Washington is shipped from the midwest or east coast. Lilyblad Petroleum, a biodiesel distributor in Tacoma, estimates that transportation or freight charges add about $0.20 per gallon to the price of B100 sold in Washington (Tegan, 2002).

Nationally, B20 costs about \$0.15–\$0.30 per gallon more than diesel (Sheehan, 2001). B100 costs about \$0.50–\$1.00 more than a gallon of conventional diesel fuel. Price will vary locally due to production, transportation and distribution costs, and on the volume of fuel purchased. The city of Seattle received price quotes from World Energy for B20 fuel ranging from \$0.259 per gallon over diesel for 500–2500 gallon lots to a price premium of \$0.199 per gallon for 5,001 gallons or more (Graham, 2002). Elsewhere in Washington, the city of Tacoma is having B20 delivered by mobile refueling for \$1.20 per gallon (Hennesey, 2002). About 13 cents per gallon of this amount is a delivery charge. In the same year that the city made this purchase of BD20, Lilyblad quoted an ex-tax bulk purchase price of \$1.60 per gallon of B100 fuel (Tegan, 2002).

Biodiesel prices are expected to increase if the current US Department of Agriculture Commodity Credit Corporation (CCC) US Bioenergy Program is stopped. The program provides reimbursements for bioenergy producers who convert targeted commodities into bioenergy. These direct payments to producers were passed on to consumers and reduced the price of biodiesel by over \$1.00 per gallon. This price cut has been the single biggest contributor to making biodiesel market acceptance possible.

Although biodiesel demands a price premium, it does not require engine modifications, nor does it require any infrastructure changes. To offset biodiesel's higher price, many states have reduced the state fuel tax paid for biodiesel. Similar federal tax incentives for biodiesel blends may be written into agricultural legislation as a matter of national energy policy.

### 8.2.3.3 Advantages and Other Properties of Biodiesel

Biodiesel fuel is biodegradable, non-toxic, and has a higher flash-point than petroleum diesel fuel. Biodiesel is also a renewable, domestically produced fuel that can provide local economic benefits. According to an energy lifecycle study completed by the US Department of Energy, biodiesel yields 3.2 units of fuel energy for every unit of fossil fuel consumed. By comparison, petroleum diesel yields 0.83 unit of fuel energy per unit of fossil energy consumed. Because biodiesel is derived from vegetable oils, carbon is also recycled. As a result, biodiesel can reduce $CO_2$ emissions by as much as 78 percent over petroleum diesel (Sheehan, 2001).

*Availability and Use* The US dedicated production capacity of biodiesel is estimated at around 80 mgy, with more than 12 companies actively producing and marketing the fuel (NBB). However, new plants are being proposed throughout the country, including Washington State, with a typical facility taking approximately one year to come on-line. Additional production capacity may be available within the oleo-chemical industry, where it is estimated that as much as 200 million gallons of capacity may be available for biodiesel production.

Biodiesel is commercially available in Washington State for delivery or pickup from Alternative Fuel Works in Seattle, Lilyblad Petroleum in Tacoma, Acme Energy Service in Olympia, Albina Fuels in Vancouver, and Soundoil.com, Oak Harbor. Arrangements can also be made for bulk purchase deliveries from national suppliers including World Energy, Chelsea, Massachusetts, and US Pacific Northwest Biodiesel, located in Aloha, Oregon. A list of biodiesel suppliers is available at the National Biodiesel Board website.

*Handling* Biodiesel is handled similarly to petroleum diesel fuels, with some notable differences. For example, handling neat biodiesel may lead to material compatibility issues because of the fuel's inherent solvent properties. Rubber seals and hoses will degrade after prolonged exposure to biodiesel and should be scheduled for replacement accordingly. Fuel filters should be checked when first using biodiesel as they may become plugged with accumulated sediments. For the same reasons, spills need to be cleaned up quickly, as biodiesel is an effective paint remover.

Biodiesel should not be stored for more than one year to avoid fuel quality problems. In addition, operators should be aware of biodiesel's cold flow properties and take any necessary precautions, including adding pour point depressants in colder climates. Finally, to ensure fuel quality, biodiesel should meet ASTM specification D-6751, the fuel provider should guarantee quality in case of engine related problems. A guidebook entitled *Biodiesel and Handling and Use Guidelines* is available from the US Department of Energy.

*Fuel Economy* Because of its lower BTU content, engine fuel economy and power are about 10 percent lower when running on neat biodiesel, and about 2 percent for a B20 blend. Biodiesel also has excellent lubricity characteristics, and can be added to petroleum diesel fuel in quantities as low as $1-2$ percent to provide lubricity improvements that meet or exceed the specifications of the original equipment manufacturer.

*Warranty* Although engine manufacturers warrant their engines and not the fuel, most major engine companies have stated formally that the use of biodiesel blends up to 20 percent will not void their parts and workmanship warranties. Some engine companies have already specified that the biodiesel must meet ASTM D-6751 as a condition, while others are still in the process of adopting D-6751 within their company (NBB). A list of engine manufacturer comments on biodiesel use in their engines is maintained at the National Biodiesel Board website.

## REFERENCES

Brown, R. C. (2001). *Biorenewable Resources: Engineering New Products from Agriculture.* Iowa State Press.

CARB, California Air Resources Board (2000). Fuels Report: Appendix to the Diesel Risk Reduction Plan, Appendix IV (October).

Graham, L. (2002). Personal communication, City of Seattle, WA.

Hennesey, S. (2002). Personal communication, City of Tacoma, WA.

National Research Center for Alternative Fuels, Engine and Emissions (2002). Biodiesel fuel comparison final data report—Washington Metropolitan Transit Authority, West Virginia University (August 15).

NBB (National Biodiesel Board), website at http://nbb.org/

Sharp, C. (1998). The effects of biodiesel on diesel engine exhaust emissions and performance, Southwest Research Institute (June).

Sheehan, J. (2001). A life-cycle inventory analysis, biodiesel vs petroleum diesel, National Renewable Energy Lab, presented at the Renewable Diesel Workshop, Seattle, WA (Sept. 27).

Tegan, M. (2002). Personal communication, Lilyblad Petroleum Inc., Tacoma, WA.

Tyson, S. (2001). Personal communication, National Renewable Energy Lab.

U. S. Environmental Protection Agency (2002). A comprehensive analysis of biodiesel impacts on exhaust emissions. EPA-Draft Technical Report, EPA420-P-02-001.

## 8.2.4 Fuel Cells for Clean and Efficient Energy

NIKUNJ GUPTA, SUBBARAO VARIGONDA, AND SHUBHRO GHOSH
United Technologies Research Center, MS 129-15, 411 Silver Lane, East Hartford, CT 06108, USA

A fuel cell is a device that generates electrical power by reacting hydrogen fuel with oxygen. The British inventor Sir William Grove discovered the principle of the fuel cell in 1839. As is the case with most new technologies, practical realization of fuel cell technology took some time, and the first significant applications were in the US space program starting in the 1960s. Fuel cells have long been unviable from a cost standpoint. In the stationary and distributed power generation markets, fuel cells face stiff competition from cheaper technologies like micro-turbines and diesel engines. In the automotive market, fuel cells still cannot compete in cost with the internal combustion engine.

The cost of commercializing fuel cell technology remains high even today and the market for fuel cells is confined to niche areas like space programs and high reliability uninterrupted power segment. However, over the last few decades, rapid technology development has resulted in significant cost reduction, generating much optimism for large-scale adoption of fuel cells in stationary, automotive and portable applications. Fuel cells are seen as one of the most promising technologies for environmentally friendly and highly efficient energy generation.

*8.2.4.1 An Overview of Fuel Cell Technology*  A fuel cell produces electrical energy, water and heat by electrochemically combining hydrogen and oxygen. Typically, using pure hydrogen and oxygen is not viable and they are obtained from other sources. Hydrogen is produced by reforming hydrocarbon fuels like natural gas, gasoline, or methanol, whereas oxygen is taken from air. An excellent introduction to various kinds of fuel cells and the associated fuel processing can be found in (Larminie and Dicks, 2000).

The fuel cell itself is a small part of the power generation system. Several individual fuel cells are connected in series to produce the required output voltage. This assembly is called a "fuel cell stack". The lack of widespread hydrogen supply infrastructure has promoted the development of fuel processing technology in order to generate hydrogen for the cell stack from readily available hydrocarbon fuels. The fuel processor can be based on one of several different reforming technologies depending on the fuel and detailed performance requirements. The fuel cell produces electric power, with water and heat as by-products. A "power conditioning system" converts the DC power from the cell stack into AC power, which is fed to the electric grid or used to power a load directly. Proper continuous operation of the cell stack requires efficient thermal and water management. Controls and monitoring functions must be added to ensure that the complete system meets safety, performance and availability requirements.

Figure 8-35 is a schematic of a fuel cell. The basic fuel cell consists of two electrode plates (anode and cathode) separated by a solid or liquid electrolyte and chambers on the anode and cathode sides where reactant gases come into contact with the electrodes. Hydrogen rich fuel is fed to the anode and oxygen or air is fed to the cathode. The electrolyte layer between the electrodes facilitates the

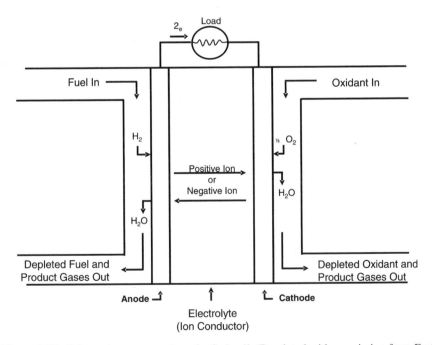

**Figure 8.35.** Schematic representation of a fuel cell. (Reprinted with permission from Fuel Cell Handbook, 1998.)

migration of ions but prevents direct contact between the reactant gases. The electrolyte also blocks the electrons produced at the anode from reaching the cathode. When an external electrical load is connected to the fuel cell, the electrons travel via the load to the cathode, creating a current. The heat produced in the reaction is removed by a circulating coolant fluid or by the product gases themselves. The exact nature of reactions, the ions involved and the efficiency in converting chemical energy to electrical energy depends on the electrodes, electrolyte, composition of the reactant gases and the operating conditions.

The voltage obtained from a single fuel cell is of the order 0.5–1.2 V. Many such cells, typically hundreds are stacked together to raise the voltage to the level needed for power generation applications. The most important characteristic of a fuel cell is how the voltage varies with current density. Figure 8.36 shows the typical fuel cell *V-I* (voltage vs current density) characteristic curve — commonly referred to as the "polarization curve" — at given temperature and reactant concentrations. The ideal voltage is the value obtained when the entire free energy available from the chemical reaction is converted into electrical energy and there are no losses of any kind. However, as current is drawn from the cell, there are energy losses due to kinetic, electric, and transport resistance in the electrodes, membrane, and the gas phases, respectively (Larminie and Dicks, 2000). The voltage drops nonlinearly due to these factors and the drop becomes severe at high current densities. The *V-I* curve also depends on other parameters like temperature, hydrogen and oxygen partial pressures and poisoning of the catalyst on the electrode. The desirable operating point is in the intermediate current range where the voltage curve is mostly linear and has a small slope.

Fuel cells are distinguished mainly on the basis of the electrolyte material used. The electrolyte serves to conduct the active ions from the fuel to the oxidant or vice

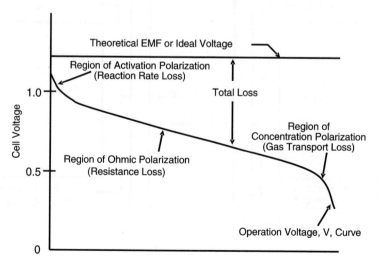

**Figure 8.36.** *V-I* polarization curve. (Reprinted with permission from the Fuel Cell Handbook, 1998.)

versa. This section presents a brief overview of fuel cells of various types and their respective applications. The predominant fuel cell types and their characteristics are shown in Table 8.15.

One of the key issues in fuel cell systems is the tolerance of the fuel cell electrode to carbon monoxide (CO) in the fuel stream. When hydrogen is obtained by reforming other hydrocarbon fuels, CO is produced as a byproduct. Some fuel cells, notably the polymer electrolyte membrane (PEM) type, are highly intolerant to CO and require an elaborate CO removal mechanism. Table 8.15 also compares the CO tolerance levels of various fuel cells.

The efficiency of the fuel cell can be defined as the fraction of total chemical energy available from the reactants that is converted to electrical energy. In Sections 8.2.4.2 and 8.2.4.3 we discuss, respectively, the PEM fuel cell (PEMFC) and the solid oxide fuel cell (SOFC) technology, which is rapidly emerging as a promising long-term solution for stationary power generation as it provides for high-quality waste heat for combined heat and power generation (co-generation), leading to very high overall system efficiency. Sections 8.2.4.4 to 8.2.4.7 briefly review the characteristics of other fuel cell types.

### 8.2.4.2 Polymer Electrolyte Membrane Fuel Cells (PEMFC) Polymer electrolyte membrane fuel cells, which use a thin solid polymer membrane as electrolyte, and were developed at General Electric in the early 1960s. These devices, also known as proton exchange membrane fuel cells, were considered for space applications by NASA (see the Fuel Cell History Project of the Smithsonian Institution) but was largely unsuccessful due to technology limitations. Today, with the availability of improved materials and designs, PEMFC is one of the most promising options for commercial production of fuel cells for stationary as well as transportation applications.

Pure hydrogen or a reformate fuel rich in hydrogen is fed to the anode, where hydrogen reacts to produce electrons and $H^+$ ions (protons).

$$\text{Anode reaction: } H_2 \rightarrow 2\,H^+ + 2\,e^-$$

The polymer electrolyte membrane allows proton migration to the cathode. The electrons travel to the cathode via the external electrical load. At the cathode, pure oxygen or oxygen from air reacts with protons and electrons to produce water and energy:

$$\text{Cathode reaction: } O_2 + 4H^+ + 4e^- \rightarrow 2\,H_2O$$

The operating temperature is in the range $50-100°C$, and water is used as coolant to remove the heat generated in the reaction.

PEMFC has an advantage over other fuel cells that use liquid electrolytes because the electrolyte is a noncorrosive solid, and does not need a circulation system, leading to simpler system designs relative to alternate technologies. The low operating temperature makes PEMFC attractive in small scale and automotive applications by increasing reliability and reducing startup time. A key challenge

**TABLE 8.15. Principal Types of Fuel Cells**

| Fuel Cell Type | Mobile Ion | Electrolyte | Operating Temperature (°C) | Common Fuels | Oxidant | Type of Reforming | Poisons | Main applications |
|---|---|---|---|---|---|---|---|---|
| PEMFC | $H^+$ | Polymer membrane | 50–100 | $H_2$, Natural gas | $O_2$, Air | External | CO intolerant | Very low to high power stationary and automotive applications |
| SOFC | $O^{2-}$ | Ceramic solid | 500–1000 | Natural gas | Air | Internal or external | N/A | Large scale power generation |
| PAFC | $H^+$ | Immobilized liquid phosphoric acid | 160–220 | $H_2$, Natural gas | $O_2$, Air | External | Moderately CO tolerant | Medium to large stationary power |
| MCFC | $CO_3^{2-}$ | Immobilized, molten carbonate salt | 630–650 | Natural gas | Air ($CO_2$ free) | Internal or external | N/A | Medium to large scale stationary power applications |
| AFC | $OH^-$ | 35–50 percent KOH | 50–200 | $H_2$ | $O_2$ | N/A | $CO_2$ intolerant | Used in space missions Apollo and the Shuttle |

in operating PEMFC systems is the intolerance of the PEM cell to carbon monoxide, necessitating an elaborate process for CO removal from the reformate fuel. PEMFC technology is currently the closest to being commercially viable. However several challenges still lie ahead, as outlined below.

## Major Challenges in PEMFC Technology Development

MEMBRANE DRY-OUT   One of the critical performance indices for PEMFC is water saturation in the membrane. To ensure desirable performance (e.g., proton conductivity) in the PEM membrane, adequate water saturation levels are required. During transient operation, it is usually seen that at times the membrane gets dry leading to a huge drop in its performance. In recent years, new technologies have been developed to avoid the membrane dry-out. The principal idea in most of these technologies is to maintain saturated anode gas during dynamic operation.

STARTUP FROM FREEZE CONDITIONS   While normal startup of PEMFC systems is already challenging, startup from subzero conditions, especially for automotive applications, is an order of magnitude more difficult. It has been observed that sub-optimal startup from freeze conditions could result in membrane dry-out, flooding of the porous components, formation of hot spots, thermal runaway, etc., many of which result in catastrophic failure of the fuel cell.

CELL STARVATION   Although good operating efficiency requires maintaining high hydrogen utilization in the cell, high utilization may result in hydrogen starvation during dynamic load transients. The starvation can potentially damage the cell, which often forces operation at lower utilizations and lower over-all efficiencies. While increasing hydrogen utilization is a key technical challenge, it is also the gateway to higher system efficiencies. Thus there is a problem of trade-off between the efficiency of a cell and its desirable performance.

CO POISONING   Almost all reforming processes generate a lot of carbon monoxide (CO) along with hydrogen. However, CO is a poison for the electrode because of its strong reducing properties. The presence of CO in the fuel cell anode feed stream, even at levels of a few parts per million, can be detrimental to a PEM cell because the effects are cumulative, resulting in continuous decay in the electrode performance. CO tolerant electrodes are a challenging new area in the development of PEMFC technology.

### 8.2.4.3 Solid Oxide Fuel Cells   Research in the technology of solid oxide fuel cells (SOFC) goes back to the 1930s (Fuel Cell History Project of the Smithsonian Institute). Technology advancements have made the SOFC a strong candidate for large-scale power generation applications. Tubular and planar cell designs are currently available, but the tubular technology is more mature (Wien) for commercial deployment. Planar designs can save material costs by reducing the operating temperature (Cambridge Energy Associates, 1998) but face several challenges such as

material degradation during thermal cycling, low power density, and lack of electrode stability.

The SOFC employs an oxide ion-conducting ceramic material as the electrolyte. The reactions are as follows:

$$\text{Anode reactions: } H_2 + O^{2-} \rightarrow H_2O + 2e^-$$

$$CO + O^{2-} \rightarrow CO_2 + 2e^-$$

$$\text{Cathode reaction: } O_2 + 4e^- \rightarrow 2O^{2-}$$

The SOFC operates at high temperatures, typically around 1,000°C, providing additional opportunities for energy recovery from waste heat, which in turn can potentially increase the overall system efficiency from about 50–70 percent. Since the electrolyte is solid as with PEMFC, the design is simplified. A significant advantage of SOFC over PEMFC is that SOFC is completely tolerant to CO in the fuel and in fact, CO is used to generate more energy. High temperature also allows internal reforming of fuels, eliminates the need for an external fuel processor and can lead to a compact design.

#### 8.2.4.4 Phosphoric Acid Fuel Cells

Phosphoric acid fuel cells (PAFC) gained attention in the late 1960s due to advances in electrode materials that made them attractive relative to other types of fuel cells (Fuel Cell History Project). Many commercial PAFC units of up to 200 kW capacity are in operation today.

Concentrated phosphoric acid held in a porous matrix structure is used as the electrolyte in PAFC. Porous electrodes coated with platinum are used. The electrolyte conducts protons and the reactions are the same as in PEMFC.

$$\text{Anode reaction: } H_2 \rightarrow 2\,H^+ + 2\,e^-$$

$$\text{Cathode reaction: } O_2 + 4H^+ + 4e^- \rightarrow 2H_2O$$

With an operating temperature in the range of 160–220°C, PAFC is attractive for small-scale power generation. The higher CO tolerance (about 2 percent CO) and full tolerance to $CO_2$ offer benefits over PEMFC and AFC. The drawbacks of PAFC are the corrosive nature of the electrolyte and low power density. Although there are few hundred commercial installations of PAFC stationary power plants worldwide, the cost remains prohibitively high for greater market penetration.

#### 8.2.4.5 Alkaline Fuel Cells

Alkaline fuel cells (AFC) were demonstrated for several applications in the 1960s. NASA employed them for the Apollo missions as well as for the Space Shuttle program. However, they are very expensive and require pure reactants. Configurations with either circulating or static electrolyte are available.

AFC uses potassium hydroxide (KOH) as the electrolyte. The performance of KOH matches the performance of acid electrolytes with less corrosiveness.

The electrolyte conducts hydroxyl ions ($OH^-$). The reactions at the electrodes are as follows:

$$\text{Anode reaction: } 2H_2 + 4OH^- \rightarrow 4\,H_2O + 4\,e^-$$

$$\text{Cathode reaction: } O_2 + 2H_2O + 4e^- \rightarrow 4OH^-$$

The operating temperature of AFC is in the range 50–200°C. Despite the many success stories of AFC in the past, research in AFC has slowed down due to technological advances in other kinds of fuel cells. The biggest limitation of AFC is that the alkaline electrolyte cannot tolerate $CO_2$ in the reactants. Both anode and cathode reactant gases must be free of $CO_2$. The need to circulate the liquid electrolyte adds to the complexity of the system.

### 8.2.4.6 Molten Carbonate Fuel Cells (MCFC)

Molten carbonate fuel cells (MCFC) were developed mainly as an alternative to SOFC in the 1950s to circumvent the problems of poor conductivity and adverse chemical reactions associated with the SOFC technology while retaining the benefits of high operating temperature.

MCFCs use a mixture of lithium and sodium/potassium carbonate in the molten state as the electrolyte. Conduction is provided by the carbonate ($CO_3^{2-}$) ions. Unlike the other fuel cells, $CO_2$ needs to be supplied at the cathode along with $O_2$. Carbonate ions form at the cathode, migrate through the electrolyte and react at the anode with hydrogen to produce water and $CO_2$. The reactions are as follows:

$$\text{Anode reaction: } H_2 + CO_3^{2-} \rightarrow H_2O + CO_2 + 2e^-$$

$$\text{Cathode reaction: } O_2 + 2CO_2 + 4e^- \rightarrow 2CO_3^{2-}$$

The high operating temperature (600–700°C) makes MCFCs good candidate for stationary power generation applications. Waste heat can be recovered in a turbine to increase efficiency. Since the temperature is well below that used for SOFCs, material degradation issues are far less severe. High temperature allows internal reforming as well. Since, however, the electrolyte is liquid and $CO_2$ injection is required at the cathode, the MCFC is less attractive than the SOFC. The corrosiveness of the electrolyte and leakage over time are also key environmental concerns.

### 8.2.4.7 Direct Methanol Fuel Cells

Direct methanol fuel cells (DMFCs) are akin to PEMFC because they use a polymer electrolyte. However, they do not require external reforming and generate protons at the anode directly from methanol. Because DMFCs are seen to be very promising in powering portable electronic devices like mobile phones and laptops, interest in the technology has grown over the last decade. Advances in the catalysts, electrode and cell design have led to increased efficiency (up to 40 percent). DMFC-powered electronic devices are already in use by the military.

The reactions in a DMFC are as follows:

$$\text{Anode reaction: } CH_3OH + H_2O \rightarrow {} + CO_2 + 6H^+ + 6\ e^-$$
$$\text{Cathode reaction: } 3/2O_2 + 6H^+ + 6e^- \rightarrow 3H_2O$$

DMFCs operate in the temperature range 50–100°C, which is particularly attractive for low power applications. The technology is attractive for automotive applications because the liquid fuel can be easily stored and delivered. For similar reasons, ethanol fuel cells (DEFCs) are also being developed for automotive applications, although, at present they are less efficient than DMFCs. The main disadvantage of DMFCs is the crossover of methanol fuel from anode to cathode by diffusion, which results in a loss of efficiency. Several companies are working on overcoming this problem (Meyer et al., 2001).

### 8.2.4.8 Fuel Processing System

As the preceding sections have indicated, it is the lack of hydrogen supply infrastructure that has motivated design and development of fuel processing systems able to generate hydrogen from more widely available fuels for fuel cell power plants. A fuel processing system converts hydrocarbon or other organic fuels through reforming into hydrogen-rich reformate of composition and purity suitable for fuel cell operation. The fuels include petroleum-derivative liquids, such as naphtha and gasoline, petroleum-derivative gases, such as methane and propane, and other fuels such as methanol and ethanol. The product, reformate gas must be supplied at temperature, humidity and purity level determined by the type of fuel cell.

The conversion process is carried out in a fuel processing "train," in which chemical reactors in a locomotive-like series sequentially effect changes to the fuel to bring it to fuel cell requirements. One of the most critical steps in the process train is the step to remove sulfur from the source fuel. Reducing the fuel sulfur content to very low (parts per billion, ppb) levels makes the subsequent processing steps more reliable and ensures high product purity.

The second step in the train is the primary conversion step. Here the source fuel is broken down and the reformate product is primarily hydrogen, carbon dioxide and carbon monoxide. The composition at this point depends on the primary reforming process, which is usually either catalytic steam reforming (CSR), auto-thermal reforming (ATR), or catalytic partial oxidation (CPO). In reforming processes, steam is supplied to the reactor to facilitate the equilibrium limited, endothermic steam reforming reaction. Addition of steam reduces coke formation on the catalyst surface as well and enhances the quantity of hydrogen produced and the life of the catalyst.

The next process step, shift conversion, reduces the carbon monoxide level of the product stream by reacting it with additional steam. This shift process is equilibrium limited, slightly exothermic and operates at lower temperatures relative to the upstream reforming process, favoring higher CO conversions. The shift converter also adds to the hydrogen composition of the stream.

The final purification and conditioning steps remove impurities such as ammonia and CO and adjust stream temperature and humidity to fuel cell inlet conditions. These terminal fuel processing steps are not always necessary depending upon the completeness of reaction in the earlier conversion steps and/or the type of fuel cell to which fuel is being provided. For example, PAFCs can tolerate much higher CO levels in the reformate fuel than PEMFCs and shift conversion is sufficient for CO reduction; the PEMFC require further CO cleanup using catalytic preferential oxidation, which oxidizes CO preferentially with respect to oxidizing hydrogen.

A fuel processing system based on CSR or ATR incorporates a steam supply system since the primary reforming process as well as the shift converter requires steam. Typically, the steam supply system consists of a condenser through which all the exhaust gases of the fuel cell power plant have to pass. The condenser removes water from the exhaust gases, making the power plant independent of extraneous water supply. The water is purified by a de-ionization bed and is then stored in a tank ready for evaporation to steam. Water is evaporated in the steam generator on demand in response to process needs.

### 8.2.4.9 Components of the Fuel Processing System

*De-Sulfurizer (DS)*    Almost all the reactors in the fuel processing train are catalytic reactors and sulfur is a known poison for most of the catalysts. Sulfur blocks the active sites of the catalysts and progressively renders the catalyst unusable for subsequent use. The fuel processing system, like the power plant, is designed for a given life, and thus sulfur must be eliminated from the fuel stream before it is fed to the reformer and the rest of the fuel processing train. The tolerance criteria established for sulfur in the fuels is very stringent (5–10 ppb) and requires de-sulfurization of even the low sulfur (few ppm sulfur) fuels.

The prevalent de-sulfurization technology in fuel processing is hydro-desulfurization. This is a hydrogen-assisted process; hydrogen is readily available (through a recycle) in the power plant. The hydrogen is used to convert organic sulfur (sulfur attached to hydrocarbon) to inorganic sulfur (hydrogen sulfide, $H_2S$), which is readily adsorbed by an adsorption bed. Zinc Oxide (ZnO) is one of the commonly used adsorbents for adsorption of inorganic sulfur. ZnO adsorbs sulfur through chemical adsorption by reacting with sulfur to form Zinc Sulfide. The reaction occurring in a ZnO hydro-desulfurizer can thus be summarized as:

$$C_xH_yS + H_2 \rightarrow C_xH_y + H_2S$$
$$H_2S + ZnO \Leftrightarrow H_2O + ZnS$$

Equilibrium favors the (forward) chemical adsorption reaction in the second reaction above, which makes it possible to attain very low sulfur levels. Other adsorbents include manganese oxide and zeolites.

*Reforming and Catalytic Partial Oxidation*    The primary process in the fuel processing train is reforming the conversion of the fuel to a hydrogen-rich gas mixture.

It is the single most important process in the system and it determines the performance as well as the architecture of the entire fuel processing system.

Most fuel processing systems use CSR, ATR, or CPO as the process for converting hydrocarbon fuels. In all cases, the primary process is followed downstream by the shift converter and/or other subsequent purification steps.

The CSR involves reforming of the hydrocarbon fuel using steam and is an endothermic (requiring heat supply) process. The main reactions that take place in a CSR are steam reforming and water gas shift. The reactions can be expressed as follows:

$$C_xH_y + xH_2O \Leftrightarrow xCO + \left(\frac{y}{2} + x\right)H_2 \quad \text{(steam reforming)}$$

$$CO + H_2O \Leftrightarrow H_2 + CO_2 \quad \text{(water gas shift)}$$

where the steam reforming reaction is expressed in generic form for hydrocarbons. It is easily seen that great amounts of carbon monoxide are also produced along with hydrogen. This is typical of all the reforming processes, as we will see later. The double-sided arrows indicate equilibrium-limited reactions, and the maximum conversion, which is always less than complete conversion, is determined by thermodynamics at the operating conditions.

The ATR differs from the CSR in that it supplies some of its own heat requirements by burning a little amount of fuel using oxygen. The generic ATR reaction can be expressed as:

$$C_{x+z}H_y + xH_2O + \left(\frac{x}{2} + z\right)O_2 \Leftrightarrow xCO + zCO_2 + \left(\frac{y}{2} + x\right)H_2$$

In addition to oxygen the ATR also uses steam for reforming. Thus, it is not completely self-sustaining and does require steam generation.

Catalytic partial oxidation, although a well-known process, is relatively new in its application to fuel processing systems. The CPO reaction, as the name suggests, is partly oxidizing the fuel and is an intermediate reaction, which takes place during (complete) oxidation. The reaction is expressed as:

$$C_xH_y + \frac{x}{2}O_2 \Leftrightarrow xCO + \left(\frac{y}{2}\right)H_2$$

CPO must be a carefully controlled process to ensure that the reaction stops at the intermediate step and does not completely burn out (oxidize) the fuel. This is because the operating temperature, which is dictated by the ratios of oxygen to carbon ratio and steam to carbon ratio, and by the catalyst chosen, is very high ($600-1{,}000°C$).

Fuel cells operating at higher temperatures (e.g., SOFC and MCFC) can use *in-situ* (internal) reforming in the fuel cell itself. However, performance of the SOFC can be enhanced with an external reformer that provides $H_2$ and $CH_4$ by converting a more complex fuel. CSR is preferred for high temperature fuel cells because it provides for better heat integration, allows a simpler system design, and results in higher system efficiency relative to ATR- or CPO-based systems. SOFCs and MCFCs have CO-tolerant electrodes and hence there is no need for CO removal — which makes internal reforming viable.

The selection of the primary process step depends on its integration in the power plant and more importantly on the customer needs for a specific application. Table 8.16 compares CSR, ATR and CPO processes based on the most common customer needs.

In stationary commercial power system applications, high overall system efficiencies favor fuel cell economics and CSR-based power plants are preferred since they typically have superior efficiency relative to ATR or CPO based systems. In applications like stand-alone residential power systems or portable power, weight or volume of the power plant become dominant considerations, in which case the CPO or ATR based systems can be favorable. In transportation applications, when transient response for rapid start/stop is a key consideration, the CPO systems can be superior to the alternatives. In almost all applications, system cost is a major driver for selection of fuel processing system technology, and efficient integration of the power plant is always critical.

*Water Gas Shift*   Carbon monoxide acts as a poison for platinum electrodes in PEM fuel cells. Fuel processing for PEM-based fuel cells therefore require a CO cleanup subsystem that is designed to reduce CO in the anode feed (reformate) to levels which can be tolerated by the PEM stack. The typical operating requirements are that CO in the anode feed stream should be reduced to less than 10–50 ppm, from 10–20 percent CO mole fraction at the inlet of the CO cleanup sub-system. The CO cleanup sub-system in prototypical fuel processors consists of a train of reactors.

**TABLE 8.16. Comparison of Primary Step Hydrogen Generation Processes and Their Integration into Fuel Cell Systems (Adapted from Meyer et al., 2001)**

| Primary Step Process | Integrated Power Plant Efficiency | Integrated Weight/Volume | Range of Fuels | Integrated Startup Time |
|---|---|---|---|---|
| Catalytic steam reforming (endothermic) | Best | Good | Gases through light paraffin liquids | Good |
| Auto-thermal reforming (adiabatic) | Better | Better | Gases through low boiling point liquids | Better |
| Catalytic partial oxidation (exothermic) | Good | Best | Gases through high boiling point liquids | Best |

Two major classes of reactors constitute the CO cleanup train, the water gas shift (WGS) and the preferential oxidation (PrOx) reactors. The WGS reactors essentially mitigate CO and at the same time enrich the hydrogen content of the reformate gas. There is only one predominant reaction (water gas shift) taking place in the WGS reactor; the reaction stoichiometry is expressed as:

$$CO + H_2O \leftrightarrow CO_2 + H_2$$

The WGS reaction is an equilibrium-limited reaction, and the maximum achievable conversion (or equivalently, equilibrium conversion) is solely determined by the thermodynamics of the process. However, the choice of the catalyst is mainly determined by the (kinetic) rate at which the thermodynamic equilibrium is approached, much like the reforming process. Under typical operating conditions, the reformate gas is injected with water (or steam) before being fed to the WGS reactor to drive the reaction. The equilibrium limitations in the WGS reaction require that prototypical CO cleanup trains consist of a two-stage WGS system with an intermediate heat exchanger/vaporizer to favorably shift the equilibrium.

*Preferential Oxidation*    Equilibrium limitations inside the WGS reactor preclude its economical use past certain threshold CO levels (typically 0.5–1 percent). The PEM fuel cells have very low CO tolerance ($< 50$ ppm), however, and this calls for an additional process to reduce CO concentration in the reformate stream to within the tolerance level of the stack. This is usually achieved by using a preferential (selective) oxidation process. The process is called selective oxidation because it favors CO oxidation over $H_2$ oxidation. Obviously, since hydrogen is the fuel for the fuel cell, oxidation of hydrogen to water would have an adverse impact on the efficiency of the whole process.

Selectivity is one of the most important parameters in designing a PrOx reactor for fuel cell applications as there are two global reactions that can adversely impact system efficiency: $H_2$ oxidation and the reverse water gas shift (RWGS) reaction. As the name suggests, RWGS is the water gas shift reaction going in the opposite direction, a highly detrimental reaction that not only consumes hydrogen but also produces carbon monoxide. The global reactions occurring in the PrOx reactor can be represented as follows:

$$CO + \tfrac{1}{2}O_2 \rightarrow CO_2$$
$$H_2 + \tfrac{1}{2}O_2 \rightarrow H_2O$$
$$CO_2 + H_2 \leftrightarrow CO + H_2O$$

Like all the other reactors in the fuel processing system, the PrOx reactor is a catalytic device, and choice of the catalyst depends primarily on its selectivity and operating range. The PrOx system, much like the WGS, is also usually a two-stage reactor system with intermediate heat exchangers for cooling.

*Catalytic Burner*   Unlike the other components of the fuel processing system, catalytic burner (CB) is downstream from the fuel cell stack in a typical power plant configuration. The CB is used to burn the effluents of the fuel cell power plant. In a fuel cell being operated with reformate feed, only a certain fraction ($\sim$80–90 percent) of the hydrogen can be utilized on the fuel cell anode. The rejected hydrogen is combusted in the catalytic burner to bring it down to within flammable thresholds set by regulatory bodies, such as the US Environmental Protection Agency.

Unlike other catalytic components, the catalytic burner has multiple functions. The hot exhaust gases coming out of the CB are used to preheat the fuel entering the reformer. Moreover, during startup some of the fuel itself is burnt inside the catalytic burner to provide additional heat for rapid startup of a CSR. Due to these additional applications, the CB is considered as a part of the fuel processing system, not the balance of plant (BOP).

### 8.2.4.10 Fuel Processing System Efficiency and Performance   Fuel cell
system efficiencies and performance depend on the performance of the individual components, as well as on the total system integration. Unsurprisingly, losses in the efficiency of a fuel processing system lead to significant reductions in the overall efficiency of fuel-cell-based power plants. The system efficiency losses are caused by combustion instead of partial oxidation, production of CO, thermal losses, and many other factors.

To define overall fuel processor efficiency ($FP_{eff}$) adequately, one must account for heat generated by the rejected hydrogen. With this in mind, $FP_{eff}$ can be defined as the lower heating value of the hydrogen ($LHV_{H_2}$) converted on the fuel cell anode, divided by the lower heating value of the fuel ($LHV_{fuel}$) fed to the reformer. The hydrogen converted on the fuel cell anode is the product of the total hydrogen generated in the fuel processor and the hydrogen utilization on the anode (Carpenter et al., 2001).

$$
\begin{aligned}
FP_{eff} &= \frac{LHV_{H_2} \times H_2 \text{ used on anode}}{LHV_{fuel} \times FP \text{ fuel feed}} \\
&= \frac{LHV_{H_2} \times \text{ total } H_2 \text{ produced} \times H_2 \text{ utilization fraction}}{LHV_{fuel} \times FP \text{ fuel feed}}
\end{aligned}
$$

The overall FPS efficiency ranges from 70–78 percent. The efficiency critically depends on the choice of reformer (CSR, ATR, or CPO) and also the choice of fuel.

Commercialization of fuel processing technology is impeded by several challenges with respect to performance, cost and durability. Some of the key technological challenges in fuel processing are described in Section 8.4.2.11.

### 8.2.4.11 Major Technical Challenges in Fuel Processing

*Part-Load and Transient CO Control*   One of the major technical challenges currently faced in fuel processing is the control of CO levels during part load and dynamic operation. Quite often it is observed that during dynamic switching between power levels, the CO concentration at the exit of the fuel processor exceeds the tolerance level of the cell stack. One of the major reasons for transient CO spikes is the high sensitivity of the PrOx reactor to its operating conditions and a suitable control scheme is required to ensure acceptable dynamic PrOx performance.

*CPO Control*   In many applications, especially in the field of transportation, CPO is the prevalent choice for reforming because of its compactness, fast transient response, and ease of system integration (e.g., does not require steam cogeneration). However, CPO controllability remains a challenging issue in fuel processing technology. The CPO (as a choice of reformer) is not only the most critical component of the system, it is also among the most challenging components from a controls perspective due to its high sensitivity to inlet flow, composition and temperature disturbances. As mentioned earlier, the CPO typically operates at very high temperatures (600–1000°C), and its performance mainly depends on the temperature of operation and the inlet gas composition. Uncontrolled fluctuations in the inlet composition (fuel-to-air or fuel-to-steam ratio) can drastically undermine performance, which therefore presents a disturbance-rejection challenge. Sustained drift in inlet composition from design values could result in thermal runaway of the reformer, leading to catastrophic failure from catalyst melting and/or sintering. Any controls implementation has to also address sensor limitations in the high-temperature CPO environment.

*Cost and Volume Reduction*   Reductions in cost and volume, which are essential if fuel processing is to be commercially viable, will have to come through changes in the whole process sequence and not just changes in the components. Achieving such reductions is particularly challenging because both cost and volume are linearly dependent on the throughput (load).

*Durability/Life*   Another vexing issue in fuel processing is the durability of the individual components. The catalytic reactors face a serious problem of progressive deactivation with use. This deactivation can be caused by sintering of the catalyst, impurities in the feed, sulfur in the feed, and other common industrial conditions. In anticipation of catalyst deactivation, the reactors must be sized *a priori* according to a specified life. This requirement results in an enormously complicated design.

*De-sulfurization*   At the component level, de-sulfurization is currently the most difficult step in fuel processing. The challenges here are severe because of the low sulfur tolerance level. No existing proven technology can steadily (over the life of the unit) mitigate the sulfur level to a range of few ppbs. Since de-sulfurization is the first step in the fuel processing system, poor performance here can spell doom for rest of the components as sulfur quickly poisons the catalyst beds in the components downstream, making them ineffective for any further use.

**TABLE 8.17. Comparison of Various Fuel Choices**

| Fuel | Advantages | Disadvantages |
|------|-----------|---------------|
| Hydrogen | • Low cost, highly integrated packaged fuel processing & vending apparatus (for automotive application) and hydrogen generation infrastructure (for all other applications) will be required and needs extensive development<br>• Eliminates the need for onboard fuel processing, increasing the power plant efficiency. | • High public safety risks associated with storage and commercial usage of hydrogen<br>• Lack of hydrogen supply infrastructure |
| Gasoline/naphtha | • Attractive for the Automotive application because of available distribution systems | • Sulfur contamination of the fuel cell and FPS components will be a key issue<br>• Gasoline Reformer Technology challenges |
| Natural gas (methane) | • Easy availability and existing distribution systems for stationary applications | • Sulfur contamination of the fuel cell and FPS components will be a key issue<br>• Natural gas reformer technology adequately developed. |
| Methanol | • Methanol has a relatively high energy density | • Acute toxicity of methanol in human contact will require a variety of mitigations as well as strenuous public education<br>• The methanol reformer's success is crucial, both in performance and cost. |
| Ethanol | • Regenerative fuel | • Ethanol cannot be a standalone fuel, due both to availability and price; must be paired with naphtha or gasoline. |

*Source*: Bevilacque Knight, Inc. (2001).

### 8.2.4.12. Choice of Fuel for Fuel Cells

There are several choices of fuels for powering fuel cells. Table 8.17 summarizes advantages and disadvantages of the different fuels in the context of different applications of fuel cell systems. The fuel cell research community continues to devote much effort to eliminating or compensating for the disadvantages listed. Even now, however, there are significant social and environmental benefits, as Section 8.4.2.13 shows.

**8.2.4.13 Social and Environmental Benefits**  Fuel cell technology offers many benefits over other existing power generation technologies. Fuel cells can reduce local air pollution, groundwater contamination, and greenhouse gases due to direct conversion of stored chemical energy into electrical energy as well as increased overall energy efficiency. Conventional power plants using fossil fuels convert the chemical energy first into heat and then to useful mechanical energy. The efficiency in converting heat to work is fundamentally limited by thermodynamics. Fuel cells produce electrical energy directly and are not constrained by this limitation. The heat produced in the fuel cells can also be used to augment the overall efficiency in cogeneration applications.

When operating on hydrogen, fuel cells have zero emissions and only produce water. When using hydrogen from a reformate fuel, the efficiency of the reformer is also a major factor in determining the overall efficiency. Higher overall efficiency means that less fuel is used by a fuel cell power plant than a conventional one to produce the same amount of energy. This translates into significantly lower carbon dioxide ($CO_2$) emissions. By using regenerative fuels, that is, fuels produced from plants or organic waste (like methanol and ethanol), the balance of $CO_2$ can be preserved in the environment. The lower operating temperatures of fuel cells result in low or no $NO_x$ productions. Argonne National Laboratories' GREET model (Argonne National Laboratory, GREET) provides for a relatively comprehensive study of the green house gas emissions from automotive vehicles using various feedstock and technologies including fuel cells. Unlike turbines or combustion engines, fuel cells have no moving parts and this contributes to increased reliability and reduced noise levels as well.

As noted throughout the preceding sections, fuel cells can run on a variety of fuels and flexible designs that can operate with many different fuels are possible (Thomas et al., 2002). External as well as internal reforming technologies that can work with multiple fuels are under development. The possibility of using natural gas and regenerative fuels can reduce the dependence on crude oil and provide increased national energy security.

**8.2.4.14 Future Prospects**  Fuel cell power systems hold unprecedented promise to deliver energy at high efficiencies while producing very low levels of harmful emissions. Fuel cell technologies and the attendant systems integration, control and optimization have advanced rapidly over the past couple of decades and several corporations around the globe are involved in a high-stakes race towards commercialization. Prototypical fuel cell power plants developed by a number of corporations have already demonstrated technological feasibility, although several challenges remain for mass-market applications. The advent of PEM technology has made fuel cell technology particularly attractive for automotive applications. PEM technology is also well on its way towards displacing previous generation of fuel cell technologies for stationary applications. Meanwhile, continued advancements in SOFC technology and highly efficient cogeneration power systems is making SOFCs look increasingly attractive for stationary power generation.

Fuel cell power systems constitute a revolutionary technology that is on a collision course with two highly established and mature technologies: the internal-combustion engine in the transportation sector and grid electricity in the stationary power generation sector. These technologies represent decades-long cost reduction and engineering refinements, highly developed support infrastructure, entrenched political support, and a regulatory framework that supports their dominance. With such cheap and widely available substitutes, fuel cells face an uphill battle to gain market share away from competing technologies. However, if fuel cells do fulfill their present promise and continue the march towards displacing their less-efficient and pollution-causing substitutes in the transportation and power generation sectors, the implications for the environment and the world's growing energy crisis will be profound and staggering. The recent pace of technological advancements and cost reductions for fuel cell power systems bodes well for this revolutionary technology and we hope technologists and governments will continue to work together to help achieve the full promise of fuel cells as an alternative source of clean and efficient energy.

## REFERENCES

Argonne National Laboratories, *The greenhouse gases, regulated emissions, and energy use in transportation (GREET) model*, http://greet.anl.gov/.

Bevilacqua Knight Inc. (2001). *Bringing fuel cell vehicles to market: scenario and challenges with fuel alternatives*, Consultant Study Report prepared for California Fuel Cell Partnership.

Cambridge Energy Associates (1998). *The next generation: fuel cells and micro-turbines.*

Carpenter, I., Edwards, N., Ellis, S., Frost, J., Golunski, S., van Kuelen, N., Petch, M., Pignon, J., and Reinkingh, J. (2001). *On-board hydrogen generation for PEM fuel cells in automotive applications*, fuel cell technology for vehicles, Richard Stobart (ed.), Progress in Technology Series.

*Fuel Cell Handbook* (1998). 4th Ed., US Department of Energy, DOE/FETC-99/1076, (November).

*Fuel Cell History Project.* National Museum of American History, http://fuelcells.si.edu/.

Larminie, J., and Dicks, A. (2000). *Fuel Cell Systems Explained.* Wiley, NY.

Meyer, A. P., Schroll, C. R., and Lesieur, R. (2001). *Development and evaluation of multi-fuel fuel cell power plant for transportation applications.* Fuel Cell Technology for Vehicles, ed. by Richard Stobart, Progress in Technology Series.

Thomas, C. E., Lomax, Jr, F. D., and Bue, E. (2002). *Fuel infrastructure options for fuel cell vehicles: gasoline, methanol or hydrogen?* Proceedings of the Topical Conference on Fuel Cell Technology, AIChE Spring National Meeting, New Orleans, LA.

Wien, B. *Future of Fuel Cells*, http://www.benwiens.com/energy4.html.

## 8.2.5   Membrane-Based Removal and Recovery of VOCs from Air

SUDIPTO MAJUMDAR

Compact Membrane Systems, Wilmington, Delaware, USA

and

KAMALESH K. SIRKAR

New Jersey Institute of Technology, Newark, New Jersey, USA

Streams of air and nitrogen gas released or vented into the atmosphere from a variety of sources often contain volatile organic compounds (VOCs). The sources of these emissions are various: the chemical, food, petrochemical, and pharmaceutical industries, fuel storage tanks and fuel transfer operations, coating/finishing and air-stripping operations, water treatment plants, and so on. Removal and recovery of the VOCs emitted in such streams will provide an opportunity to recoup the cost of antipollution equipment by selling the products recovered. A variety of separation techniques have been investigated to this end; among those being practiced in the commercial marketplace are absorption, adsorption, condensation, and membrane vapor permeation.

The optimum scale of operation of each of these technologies has been identified for the emission stream vis-à-vis the gas stream flow rate and its VOC concentration level (Humphrey and Keller, 1997). Of the four techniques just mentioned, membrane vapor permeation, the latest entrant into the marketplace, is generally recommended for medium to low volume air streams having somewhat higher VOC levels. Pressure swing adsorption (PSA) processes are recommended for lower VOC concentrations and lower gas stream flow rates, whereas absorption processes are recommended for high gas stream flow rates and lower VOC levels (Humphrey and Keller, 1997). Membrane-based processes for VOC removals are of interest here, specifically, the two techniques of vapor permeation and membrane-based absorption and stripping.

***8.2.5.1 Vapor Permeation***   In membrane vapor permeation processes, the membranes are generally polymeric in nature; further they are nonporous and rubbery. For example polydimethylsiloxane (PDMS) (silicone) is the most common rubbery polymer used. An effluent air stream containing the VOC is exposed to one side of such a polymer membrane while conditions on the other side are maintained such that the partial pressure of the VOC is less than that on the effluent air side (Figure 8.37). We can use an integrated form of Fick's law to describe $J_i$, the permeation flux of VOC species $i$ through such a nonporous rubbery membrane:

$$J_i = \frac{Q_{im}}{\delta_m}(p_{if} - p_{ip}) \tag{8-34}$$

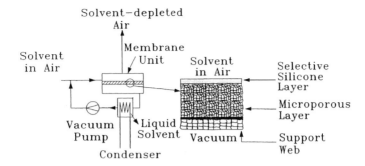

**Figure 8.37.** Vapor permeation process using a spiral wound membrane: vacuum on the permeate side.

where $p_{if}$ and $p_{ip}$ are, respectively, the partial pressures of VOC species $i$ in the feed air and the permeated gas stream and $Q_{im}$ is the permeability of $i$ through the nonporous rubbery polymer membrane of thickness $\delta_m$. If the feed air stream (the effluent) is at atmospheric pressure, the VOC partial pressure difference ($p_{if} - p_{ip}$) can be made positive by maintaining a vacuum on the permeate side (Figure 8.37). Alternately, if the permeate side gas stream is at atmospheric pressure, the effluent feed air stream can be compressed, to raise the value of $p_{if}$ with respect to $p_{ip}$ (Figure 8.38) (Baker et al., 1996).

During such a VOC permeation process, $N_2$, $O_2$ and other effluent gases will also permeate the nonporous polymeric membrane. The permeation flux of, say, $N_2$, $J_{N_2}$, can also be described by an equation similar to Eq. (8-34):

$$J_{N_2} = \frac{Q_{N_2 m}}{\delta_m}(p_{N_2 f} - p_{N_2 p}) \qquad (8\text{-}35)$$

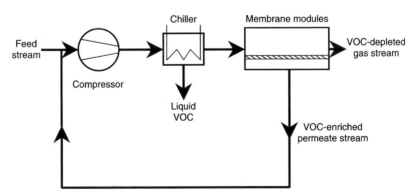

**Figure 8.38.** Flow diagram showing the application of a membrane vapor permeation system to effluent gas treatment and VOC recovery: condenser on the compressed fee gas side. (Adopted from Lokhandwala et al., 1999.)

If $Q_{im}$ greatly exceeds $Q_{N_2}m$, the gas stream that permeates the membrane from the effluent gas side to the other side will be highly enriched in VOC species $i$. In fact, the ideal selectivity of the membrane, $\alpha^*_{i-N_2}$ is expressed as follows:

$$\alpha^*_{i-N_2} = \frac{Q_{im}}{Q_{N_2 m}} \tag{8-36}$$

For the silicone membranes conventionally used, this ratio can be anywhere from 5-10 to 200 and up, depending on the VOC species $i$. The VOCs in the VOC-enriched gas stream on the permeate side are recovered in a condenser by condensation when the vacuum is on the permeate side (Figure 8.37). When the feed gas is compressed, the condenser is on the feed gas side, with the result that after condensation, the compressed gas is exposed to the membrane unit (Figure 8.38) and the permeated gas enriched in the VOC is recycled to the compressor (Baker et al., 1996).

Equations (8-34) and (8-35) may also be written in terms of feed and permeate pressures, $P_f$ and $P_p$, respectively, and the mole fractions of VOC and $N_2$ in the permeate stream, $y$ and $1 - y$, respectively, as well as those in the feed stream, x and $1 - x$, respectively:

$$J_i = \frac{Q_{im}}{\delta_m}(P_f x - P_p y) \tag{8-37a}$$

$$J_{N_2} = \frac{Q_{N_2 m}}{\delta_m}\{P_f(1 - x) - P_p(1 - y)\} \tag{8-37b}$$

The selectivity $\alpha_{i-N_2}$ achieved in practice is defined by

$$\alpha_{i-N_2} = \frac{y(1 - x)}{(1 - y)x} \tag{8-38}$$

This selectivity approaches $\alpha^*_{i-N_2}$ when the pressure ratio $(P_p/P_f)$ tends to zero. As Baker and Wijmans (1991) have shown, the permeate concentration of VOC approaches high values when $\alpha^*_{i-N_2}$ is high and $P_p/P_f$ is low. For example, for a feed containing 0.5 percent VOC with $\alpha^*_{i-N_2} = 50$, $y$ would be around 0.2 for a very low pressure ratio of $10^{-3}$. The pressure ratio employed in practice is usually higher. The value of $Q_{N_2 m}/\delta_m$ observed in practice is around $100 \times 10^{-6} \, cm^3$ (STP)/$cm^2.s.cm$ Hg for silicone membranes on an appropriate porous substrate (Baker et al., 1998).

According to Eq. (8-34), the flux of a VOC depends on $Q_{im}$ (should be high), $\delta_{im}$ (should be low), and $(p_{if} - p_{ip})$ (should be as large as possible). Handling large volumetric flow rates of gas streams, however, calls for the membrane surface area $A_m$ to be large in a small equipment volume. Of the two conventional physical forms of membranes, flat in spiral-wound form and hollow fibers, the latter provide much larger surface area per unit equipment volume (by as much as seven times) leading to highly compact membrane devices for processing a given effluent air flow rate. Although a large fraction of the commercial membrane vapor permeation units employ spiral-wound devices (Baker et al., 1996), a significant number of units have a flat leaf structure (Nitsche et al., 1998; Ohlrogge et al., 1995). Highly compact hollow-fiber-based devices are under development (Majumdar et al., 2003).

As indicated earlier, the higher the value of $Q_{im}$, the lower the membrane area required for removing a given amount of VOC from a gas stream. However, VOCs tend to plasticize rubbery membranes, especially silicone. As a result, the value of the VOC permeability, $Q_{im}$, increases strongly with the concentration level of the VOC in the effluent gas stream. This is unlike conventional gas permeation membranes, where $Q_{im}$ is generally independent of the concentration of species $i$ in the gas stream. The need to accommodate such changes in $Q_{im}$ makes VOC permeator design calculations quite complicated.

The nonporous VOC-selective rubbery membrane is applied generally as a coating over a porous more rigid polymeric support membrane (e.g., polysulfone, polyetherimide). The permeability $Q_{im}$ and the selectivity $\alpha_{i-N_2}$ depend on the properties of the rubbery coating membrane, which is exposed to the effluent gas stream. In practice, there is gas pressure drop in the permeate flowing through the porous substrate. As a result, the applied VOC partial pressure driving force $p_{if}-p_{ip}$ is reduced significantly. Therefore, the support should be highly permeable, as much as 10–100 times that of the coating. In conventional VOC-selective rubbery membranes, the feed gas stream (at a higher pressure) cannot be applied to the substrate side because it would damage the rubbery polymeric coating easily. On the other hand, when plasma-polymerized silicone membrane is used on the outside surface of porous hollow fibers, the feed gas stream can be introduced into the fiber bore at a higher pressure and a vacuum pulled on the shell side (Cha et al., 1997). This mode of operation avoids the pressure drop in the permeating gas stream as it passes through the porous substrate in the membrane in the conventional operational mode.

### 8.2.5.2 Membrane-Based VOC Absorption and Stripping

The VOC level in an effluent air stream may vary from as low as 10 ppm or less to a few percent or even 30 or 40 percent. Obviously when the VOC level in the feed is low, the driving force $(p_{if} - p_{ip})$ will be very low; further creation and maintenance of a high vacuum is very costly in a large scale. Under such circumstances, in conventional chemical engineering, the VOC would be scrubbed with a nonvolatile solvent in an absorber and then removed from the nonvolatile solvent at a higher temperature in a stripper. For low VOC levels in an effluent gas stream and smaller gas flow rates, where traditional large gas absorbers are not useful, however, nondispersive absorbers and strippers based on compact hollow fiber membranes have been developed. These devices allow the absorbed VOC to be scrubbed and stripped from a nonvolatile absorbent at a higher temperature (Poddar et al., 1996a,1996b; Xia et al., 1999).

To see how a membrane-based nondispersive absorption and stripping process works, consider a porous hydrophobic relatively solvent resistant membrane (Figure 8.39a). An essentially nonvolatile organic solvent (for example, a suitable silicone oil) acting as an absorbent is allowed to flow on one side of the membrane,

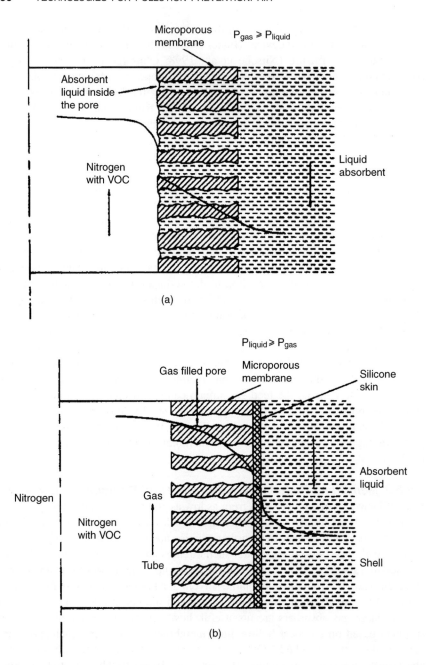

**Figure 8.39.** Local partial pressure and concentration profiles of VOCs (a) being absorbed in a microporous/porous hollow fiber module, (b) being absorbed in microporous/porous hollow fibers having a nonporous silicone skin on the outer surface, and (c) being stripped from absorbent in a hollow fiber module. (Adopted from Poddar et al., 1996b.)

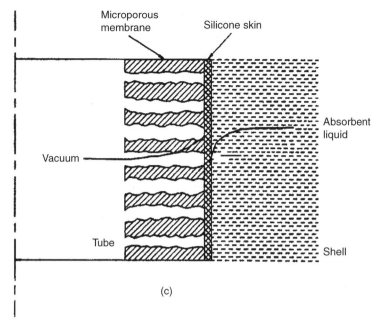

**Figure 8.39.** (*Continued*).

and the gas stream flows on the other side, thus conveniently implementing VOC absorption into the oil. The solvent wets the pores. To prevent the solvent from breaching the pores to the other side (gas side), the gas pressure is maintained higher than the liquid pressure. These design features assure nondispersive operation (Majumdar et al., 2001; Poddar et al., 1996a; Xia et al., 1999).

A related development makes use of a porous hydrophobic membrane having a nonporous rubbery silicone skin on the fiber outside diameter (OD) (Figure 8.39b). The nonvolatile oil is made to flow on the shell side (fiber OD) and the gas flows through the fiber bore. Although this configuration avoids the considerable diffusional resistance of a VOC through the stagnant liquid in the pores encountered in noncoated fibers, to prevent bubbling of the gas in the oil phase, the liquid phase pressure must be maintained higher than that of the gas phase (Poddar et al., 1996a). The use of hollow fiber membranes allows nondispersive countercurrent flow of the gas and the liquid (oil) phases.

Next, the nonvolatile absorbent (organic solvent) oil containing the VOCs is regenerated in a membrane stripper that uses the same hollow fiber membrane, and the VOC-loaded absorbent oil flows on the shell side as in Figure 8.39b. On the tube side, namely, the fiber bore, a vacuum is pulled to strip the oil of its VOC (Figure 8.39c).

In the absorption-stripping setup, the loaded absorbent oil is heated as it leaves the absorber and is passed through the stripper. The oil is cooled upon leaving the stripper, before it is allowed into the absorber for contact with the VOCs

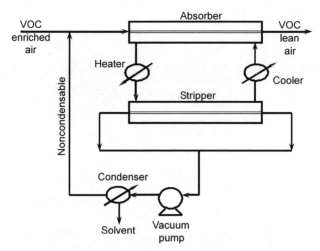

**Figure 8.40.** Schematic process diagram for VOC removal from air by absorption and stripping in hollow fiber modules. (Adopted from Podder et al., 1996b.)

(Figure 8.40). Simultaneous absorption and stripping can be achieved by inserting between the two sets of hollow fibers a stationary oil layer acting as a liquid membrane without circulating the absorbent (Obuskovic et al., 1998). Feed gas will flow through the bore of one set of fibers, and a vacuum may be applied to the bore of the other set of fibers. The liquid is not heated, however, and this may reduce stripping efficiency.

The VOC mass transfer coefficient for the overall process may be described as the sum of three resistances: the liquid film resistance, the membrane resistance and the gas film resistance. Although the rate of mass transfer may be determined much as is done for conventional dispersive absorbers/strippers, the membrane resistance must be added to the calculations. If the VOC-containing gas phase is flowing through the hollow fiber bore (radius $r_i$), the liquid phase is on the shell side and the pores are filled with the absorbent liquid, the overall gas-phase-based mass transfer coefficient $K_{oG}$ based on the fiber inside radius may be expressed via the resistances-in-series model (Poddar et al., 1996a):

$$\frac{1}{2r_i K_{oG}} = \frac{1}{2r_i k_g} + \frac{1}{d_{lm} k_m H_i} + \frac{1}{2r_o k_l H_i} \tag{8-39}$$

where $H_i$ is the Henry's law constant of the VOC between the gas phase partial pressure $p_{ig}$ and the absorbent phase concentration $C_{il}$ ($=H_i\,p_{ig}$), $k_l$ is the liquid-phase VOC transfer coefficient, $k_m$ is the membrane phase mass transfer coefficient describing VOC diffusion in the oil in the pore, $k_g$ is the gas phase mass transfer coefficient, and $d_{lm}$ is the logarithmic mean diameter of the fiber. The membrane mass transfer coefficient in an oil-filled porous membrane (Figure 8.39a) may be estimated from

$$k_m = \frac{\varepsilon_m D_{il}}{\tau_m (r_o - r_i)}$$ (8-40)

for a membrane having a porosity $\varepsilon_m$ and tortuosity $\tau_m$; $D_{il}$ is the diffusion coefficient of the VOC in the absorbent liquid. If the membrane pores are not filled with the liquid and the porous membrane has a VOC permeable skin on the fiber outside diameter (Figure 8.39b), then the overall mass transfer resistance may be described by

$$\frac{1}{2r_i K_{oG}} = \frac{1}{2r_i k_g} + \frac{1}{d_{lm} k_m} + \frac{1}{2r_o k_{cm}} + \frac{1}{2r_o k_l H_i}$$ (8-41)

where $k_m$ is the mass transfer coefficient of the gas-filled pore and $k_{cm}$ is the mass transfer coefficient for VOC permeation through the coating on the fiber OD which is exposed to the absorbent on the shell side.

Depending on the membrane or the configuration being utilized, the membrane pores may or may not be filled with the absorbent oil; correspondingly $k_m$ will vary. To reduce the overall resistance, it is desirable to operate without the absorbent in the pores of the membrane; this is achieved in two of the configurations shown earlier (Figures 8.39b, 8.39c). The coating membrane here should be highly permeable to VOCs but not to hot silicone oil or other absorbents being employed.

The overall gas scrubbing efficiency of such an absorption-stripping process is dependent on the extent of regeneration of the spent absorbent in the stripper. By heating the spent absorbent, the level of VOC in the stripped absorbent can be reduced drastically. Such a regenerated absorbent when cooled (Figure 8.40) will have a very high VOC sorption capacity and therefore a high gas-cleaning capacity when it contacts the VOC-containing gas stream in the absorber. Removal of the VOCs to levels as low as 1–3 ppm can be achieved when the inlet VOC concentration is around 5–30 ppm. (Majumdar et al., 2001). Although the membrane vapor permeation process may be employed to treat a gas stream containing VOC at 25–50 ppm level and achieve 90%+ VOC removal, the selectivity is quite low, around 4–5 (Majumdar et al., 2003); therefore a large vacuum pump would be required, increasing the process cost. On the other hand, because the solubility of $O_2/N_2$ in the absorbent oil is quite low in the absorption-stripping process, the vacuum pump for the stripper (Figure 8.40) can be correspondingly low in capacity. Further, conventional dispersive strippers may be used instead of a membrane-based stripper. However, two devices are needed in the absorption-stripping process in addition to a pump for circulating the absorbent; additionally heating and cooling devices for the absorbent and a vacuum pump for the VOC-stripper are needed.

### 8.2.5.3 Design Equations, Principles and Strategies    At steady state, the differential equation governing the VOC permeation in a hollow fiber permeator

is given by

$$\frac{d(Lx)}{dl} = \pi d_{lm} \left( \frac{Q_{im}(P_f x, P_p y)}{\delta_m} \right) (P_f x - P_p y) \tag{8-42}$$

where $Q_{im}$ is the VOC permeability, $L$ is the gas flow rate per fiber, $x$ is the feed side VOC mole fraction, $P_f$ is the feed side total gas pressure, $P_p$ is the permeate side gas pressure, and $y$ is the permeate side VOC mole fraction for a membrane of thickness $\delta_m$ on the fiber outside diameter $d_o$, and $d_{lm}$ is the logarithmic mean diameter of the fiber. For an exact analysis of permeation, many other equations are needed which are solved numerically (Cha et al., 1997). However an approximate approach leading to an analytical solution has been developed based on the following assumptions (Cha et al., 1997):

1. $P_p y \ll P_f x$ corresponding to high vacuum on the permeate side.
2. Pressure $P_f$ is essentially constant on the feed side.
3. $Q_{im}/\delta_m$ varies only with $P_f x$ in the following manner.

$$\frac{Q_{im}}{\delta_m} = a \exp(b P_f x) \tag{8-43}$$

where $a$ and $b$ are constants for the specific VOC and the membrane.
4. Change in $L$ with $l$, the length coordinate of the permeator, is negligible.

The analytical equation to predict the membrane performance is given by the equation

$$\left( \frac{\pi d_{lm} a P_f}{L} \right) l_f = \left[ \ln x - \frac{(b P_f x)}{1.1!} + \frac{(b P_f x)^2}{2.2!} - \cdots \right]_{x_w}^{x_f} \tag{8-44}$$

Permeances and their dependence on VOC concentrations [e.g., Eq. (8-43)] are determined experimentally. Both $x_f$ and $x_w$, the feed and residue VOC mole fraction, respectively, are fixed for design purposes. Since $d_{lm}$, $a$, $b$, $P_f$ and $L$ are also known, one can easily obtain $l_f$, the length of hollow fiber required for the given separation from Eq. (8-44).

Along the permeator length, it is also possible to incorporate a small variation of $L$ due to $N_2$ permeation. The analytical solution for such a situation is given by:

$$\frac{a}{(Q_{N_2 m}/\delta_m)} \ln \left\{ 1 - \frac{l_f}{l_f - L_f/[\pi d_{lm} P_f(Q_{N_2 m}/\delta_m)]} \right\} = \left[ \ln x - \frac{(b P_f x)}{1.1!} + \frac{(b P_f x)^2}{2.2!} - \cdots \right]_{x_w}^{x_f} \tag{8-45}$$

Although the design equations are developed for a hollow fiber module, a similar approach can be applied for a spiral wound module when high vacuum exists in the permeate side.

For a given VOC level in the feed stream (the rest being generally $N_2$ with some $O_2$ or both in the form of air), the process design considerations are:

(1) How much of the VOC is to be recovered? Are there any considerations for VOC recycling within the facility? What would be the VOC level in the treated stream to be vented or discharged?

(2) Will the membrane selectivity be sufficient for one-stage operation? Is two-stage operation needed?

(3) Is the vapor permeation process sufficient to meet the specifications? Or, is a hybrid of this process with adsorption or absorption or condensation necessary?

(4) How does the gas stream flow rate magnitude influence such decisions?

Answers to these questions require detailed design calculations for the given VOC-containing stream and the membrane. The membrane selectivity, the feed stream pressure and the permeate vacuum level achievable in the facility for atmospheric pressure feed streams go into the design process. Additional items to be considered during the design phase include requirements for recycling the VOC-stripped gas stream (say, $N_2$) in the facility and reuse of the VOC-enriched permeate stream. Baker et al. (1998) offer useful guide to such design issues for recovery of ethylene or propylene monomer from degassing vent streams. For example, to recycle $N_2$ from the vent gas stream in a polyolefin polymerization plant containing 10–20 vol.% valuable hydrocarbon in nitrogen, the nitrogen should have 99 percent purity for recycle within the plant; in a polypropylene plant, propylene should have 95% purity for recycle. A single-stage vapor permeation process with a membrane selectivity between 10 and 50 (the latter is difficult to achieve) will not suffice. A multi-stage multistep membrane separation system was devised to produce a 99 percent $N_2$ stream and a 95 percent propylene stream with a membrane selectivity of 10 (between $C_3H_6$ and $N_2$) and a $N_2$ permeance of $100 \times 10^{-6} \, \text{cm}^3$ (STP)/ $\text{cm}^2.\text{s.cm Hg}$ (Figure 8.41).

A step in the direction of increased selectivity of the VOC vis-à-vis $N_2/O_2$ for ultrathin silicone rubber membranes was taken by Obuskovic et al. (2003). These investigators used porous hollow fibers with an ultrathin silicone rubber skin on the fiber outside surface, depositing an ultrathin layer of silicone oil in the pores next to the coating. Following the studies of Cha et al. (1997), the feed gas was introduced through the fiber bore. The liquid layer thickness was around 5 μm, about one-fifth the thickness of the porous polypropylene fiber wall. This silicone oil layer reduced the $N_2$ flux drastically and the VOC-$N_2$ separation factor was increased 5–20 times depending on the VOC and the feed gas flow rate. The VOCs studied were acetone, methanol, and toluene, with toluene showing the highest selectivity increase. Such VOC-$N_2$ selectivity enhancement will provide

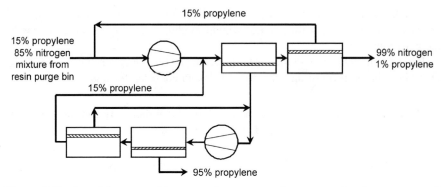

**Figure 8.41.** A multistage, multistep membrane separation system for the recovery of propylene monomer and the recycling of nitrogen from resin degassing vent streams. (Adopted from Baker et al., 1998.)

additional flexibility in the design of vapor permeation membrane separation systems; the design of multistep multi-stage membrane separation systems of the type shown in Figure 8.41 may be potentially simplified via these higher selectivity membranes.

A major difference in the design of a membrane-based absorption-stripping process from that of a membrane vapor permeation process is that in the former, the VOC concentration in the stream obtained by vacuum stripping the loaded absorbent liquid at a higher temperature is bound to be high. From a VOC-recovery point of view, the multiple steps in Figure 8.41 may not be necessary. However, the number of steps still depends on the nature of the nonvolatile absorbent and the solubility of $O_2$ and $N_2$ in such an absorbent. Simulation and design procedures for a membrane-based absorption-stripping process are provided in Poddar et al. (1996b).

The membrane vapor permeation process is known to be efficient for higher feed VOC concentrations. The treatment of the higher VOC level feed gas in a membrane vapor permeation unit will produce a retentate having a low VOC level. Such a low VOC-level feed will be ideal as a feed for the membrane-based absorption-stripping process. Poddar and Sirkar (1997) have experimentally demonstrated such a hybrid process (Figure 8.42) for a 6,000 ppmv methylene chloride containing $N_2$ stream. Overall VOC removals of around 99.5 percent were achieved with a hybrid process; the membrane vapor permeation process alone gave much lower removals. Other hybrid processes investigated in literature include pressure swing adsorption instead of absorption-stripping.

### 8.2.5.4 Commercial Membrane Modules
The membranes that are used for vapor separation applications are thin film composite membranes with a dense rubbery polymer as a top layer. Silicone rubber (polydimethylsiloxane) is the most common VOC-selective membrane material. A silicone rubber layer only 0.2–5 μm thick is coated onto an open, highly permeable microporous polymeric layer (Figure 8-37), which provides the mechanical strength required to support

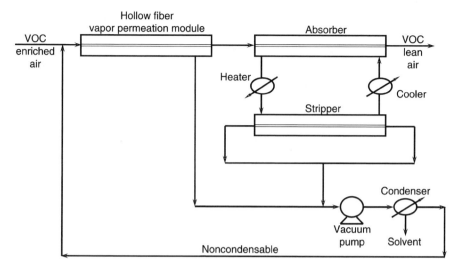

**Figure 8.42.** Schematic diagram of VOC removal by a hybrid process of vapor permeation and membrane-based absorption-stripping. (Adopted from Poddar and Sirkar, 1997.)

the selective layer. The support web is an industrially produced nonwoven fabric with a very porous structure.

The commercial membrane modules commonly used for vapor separation comprise flat sheets of membrane packaged either in spiral-wound modules or in an envelope format. In a spiral-wound module, a membrane is rolled around a perforated central collection tube like a jelly roll, with appropriate spacers in between the individual membrane layers (Wijmans and Helm, 1989). The feed air passes parallel to the membrane surface and the vapors permeate through the membrane perpendicular to the flow (Lahiere et al., 1993). The packing density (the ratio of membrane area over device volume) of a spirally wound module varies from $300-1,000 \text{ m}^2/\text{m}^3$. These modules are available from Membrane Technology and Research (MTR), in Menlo Park, California.

Envelope modules for vapor separation are manufactured by GKSS, Geesthacht, Germany. The design, a hybrid of the plate-and-frame module and spiral-wound modules, features a stack of membrane envelopes each consisting of two flat sheet membranes. Spacers and fleeces in between two membranes provide an open space for unhindered permeate flow. A sandwich structure is formed as the membrane sheets are sealed at the edges by thermal welding. The membrane envelopes are mounted on a perforated central permeate collection pipe. They are divided into groups and are separated in asymmetric compartments by means of baffle plates (Ohlrogge et al., 2003). The feed flow is parallel over all sandwiches in a group but serial for the groups separated by the plates. The packing density varies from $270-450 \text{ m}^2/\text{m}^3$.

Commercial-size hollow fiber membrane modules are now being employed for vapor separation (Majumdar et al., 2003). These modules consist of thousands of self-supporting hollow fine fibers in a simple shell-and-tube parallel flow design from Applied Membrane Technology, in Minnetonka, Minnesota. The hollow fibers are made out of microporous hydrophobic polypropylene having an ultrathin plasma-polymerized silicone skin $1-2$ μm thick. The process employs a lumen side air feed essentially at atmospheric pressure and vacuum on the shell side (Bhaumik et al., 2000). The membrane cartridges have a very high value of around 4500 m$^2$/ m$^3$ surface area per unit device volume. These modules are also suitable for stripping as well as absorption operations in membrane-based absorption-stripping process.

The most well-known commercial module for gas liquid contacting (for VOC absorption in oil) is the Liqui-Cel$^®$ Extra-Flow system manufactured by Membrana, in Charlotte, North Carolina. The module uses microporous hydrophobic polypropylene fibers that are knitted into an array and wound around a central liquid distribution tube. The design incorporates a baffle in the middle of the module which directs the liquid radially across the hollow fiber array. The baffle minimizes shell side bypassing and creates a liquid flow which is normal to the membrane surface. This results in a higher mass transfer coefficient than that achieved with parallel flow modules.

### 8.2.5.5 Applications

The membrane vapor separation process has been applied in chemical, petrochemical and pharmaceutical industries to separate and recover a wide range of common organic compounds from gas streams. Many systems have been installed worldwide to recover hydrocarbons from petroleum transfer operations in gasoline tank farms, to control emissions at car refueling gasoline stations (Ohlrogge et al., 1995), and to recover chlorofluorocarbons (Baker et al., 1993), vinyl chloride monomer (Lahiere et al., 1993), olefin and other high value chemicals (Baker et al., 2000).

### Gasoline Vapor Recovery at Tank Farms

Off gases containing gasoline vapor are generated during loading operations of trucks, tankers and barges. The volumetric flow rate of the off gases varies and depends on the loading schedule. A buffer tank is normally used to shave off the loading peaks. To achieve high recovery rates and to keep investment and operating costs low, a combination of different unit operations is used. The vapor goes first through an absorption unit, where most of the recovery takes place, and then through a membrane vapor permeation unit.

A post treatment is carried out with a pressure swing adsorption (PSA) unit (Ohlrogge et al., 2003; Nitsche et al., 1998). The vapor from the buffer tank is fed to a scrubber via a compressor. Lean gasoline liquid is used as a scrubbing fluid to absorb the recovered product. Enriched gasoline is returned to the storage tank. The gas leaving the top of the scrubber is fed through the membrane unit. A vacuum pump is used in the permeate side of the membrane to draw gas stream with enriched hydrocarbon vapor. The permeate is recycled and mixed with the incoming vapor from the buffer tank before it is introduced in the absorption unit. The retentate of the membrane stage is fed to the PSA unit for further treatment.

Sometime a second membrane stage is used instead (Ohlrogge et al., 1990). The capacities of the vapor separation units range from $100-4,000 \text{ m}^3/\text{h}$.

*Vapor Recovery at Gasoline Stations*   Gasoline dispensing stations often utilize vacuum-assisted vapor return systems to reduce hydrocarbon emissions to the atmosphere. These systems use a small pump to draw saturated vapor displaced from the vehicle fuel tank during refueling and transport these vapors through the nozzle and vapor return lines to the underground storage tank. To capture the entire vapor stream, a surplus of air-vapor volume must be returned to the storage tank. This causes the pressure in the tank to increase, and the vapor is vented through a stack. As new vehicles equipped with mandatory onboard refueling vapor recovery (ORVR) systems are introduced in the market, even more vapor is being vented into the atmosphere. This is because fresh air, rather than saturated gasoline vapor, is introduced into the storage tank during fueling of such vehicles, resulting in significant additional pressurization of the tank as liquid gasoline evaporates. Two modes of operation have been considered for this particular situation.

In one approach, when the pressure in the tank reaches a preset value, a vacuum pump on the downstream side of a membrane system is activated. The excess vapor volume from the storage tank is introduced through the membrane stack at a pressure slightly higher than atmospheric. The gasoline vapors are separated from the off-gas and are concentrated in the permeate stream, which is recycled into the storage tank. The clean air is discharged to the atmosphere (Ohlrogge and Sturken, 2001).

In the second approach, a compressor in the feed line is used instead of a vacuum pump in the permeate side. Part of the hydrocarbon vapors condenses and is returned to the storage tank as a liquid; the remaining hydrocarbons permeate the membrane and are returned to the tank as concentrated vapor (MTR, 2004). Clean air is vented. Hydrocarbon emissions are reduced by $95-99$ percent. It has been suggested that in addition to virtually eliminating hydrocarbon emissions, the unit may pay for itself with the value of the recovered gasoline.

Vapor Systems Technologies, Dayton, OH has introduced a gasoline vapor processor utilizing membrane module from Compact Membrane Systems, in Wilmington, Delaware. The membrane operates differently from the ones discussed so far; it preferentially permeates air rather than hydrocarbons. Cleaned air (permeate stream) is vented to the atmosphere, while supersaturated gasoline vapors are returned to the underground storage tank where they partially condense and remain as saleable product (Bowser et al., 2004). Since the membrane permeates air it is not necessary to remove all of the air from the gasoline vapor. The advantage here is that the system does not need to permeate a large volume of air to be an effective vapor processor. Removal of any fraction of air brings the pressure in the storage tank below atmosphere, thus preventing outward vapor leaks and meeting emission requirements (Koch, 2001).

*Monomer Recovery From Olefin Polymerization Vents*   Olefin polymerization takes place at high pressure in a reactor, where monomer is introduced along with catalyst, various comonomers, solvents and stabilizers. After polymerization the resin is introduced into a bin where nitrogen purge is used to remove absorbed unreacted monomer and process solvents. The composition of the off gas varies considerably depending on the type of product polymer, polymerization process and degree of purification. A single stage membrane unit with a compressor can separate the gas mixture into a monomer rich permeate that can be recycled to the reactor and a nitrogen rich retentate stream that can be recycled to the purge bin (Baker et al., 2000). However, a single stage process may not produce a monomer pure enough to be recycled to the reactor. With the multistage, multistep process illustrated earlier (Figure 8.41), polymerization plant vent gases yield 500–1,000 lb/h of recoverable monomer and twice as much recoverable nitrogen. There is no appropriate recovery technology other than membrane separation technology.

The preceding description of vapor permeation modules and membrane configurations suggests that applications and process/vent streams of certain types will require membranes and operations/process configurations of particular types. For example, lack of space in an existing plant, or the need to avoid the high cost of compressing the feed stream may preclude the use of a compressor (Figure 8.37). The possibility of a vacuum pump-based configuration is of interest here provided the vacuum pump size is not too large. The role of a condenser on the vacuum pump side is of some importance here especially if the VOCs have to be recovered.

Bhaumik et al. (2000) and Majumdar et al. (2003) have made a number of experimental studies using pilot-scale and commercial-size hollow fiber membrane modules. These were primarily carried out for two different applications. The *first application* involved removal and recovery of VOCs at high concentrations from a $N_2$ stream leaving a batch reactor in a pharmaceutical plant and exiting through a water-cooled condenser (Figure 8.43). The VOC concentrations at the condenser exhaust were as high as 14 percent (for methanol); other VOCs studied include toluene and ethyl acetate, with or without water vapor. The VOC-containing stream at a pressure slightly above or below atmospheric was passed through the bore of the hollow fibers in the cartridges; a vacuum was maintained on the shell side to remove the highly-enriched VOC stream which was sent to a condenser. A number of such cartridges were employed in parallel to treat higher gas stream flow rates. Under carefully chosen conditions, VOC recoveries of up to at least 98 percent could be achieved.

The *second application* involved removal of low levels of VOCs (5–40 ppmv) from a paint-booth exhaust stream (Majumdar et al., 2003). Even though the inlet VOC concentration levels were quite low, removals of 90–95 percent or better were achieved from the dilute gas streams. However, the VOC-$N_2$ selectivities were significantly lower (5–10), suggesting a need to use a much smaller membrane area to treat the permeated stream to achieve sufficiently high VOC levels in the permeate suitable for condensation-based recovery. Bhaumik et al. (2000) had carried out a study using a much smaller module on the exhaust stream from a pharmaceutical research facility. This exhaust stream having somewhat higher levels of VOC could be successfully treated also. An aspect of some importance was that even though the VOC levels and types in this exhaust were fluctuating, the membrane device was responding essentially instantaneously with the desired selectivity to the fluctuations.

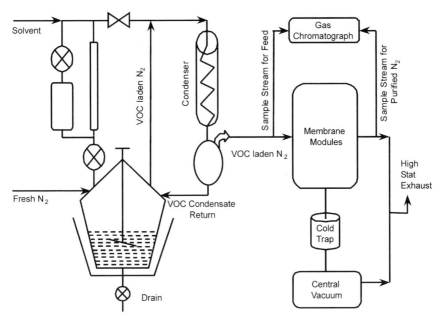

**Figure 8.43.** Removal and recovery of solvent vapor from a pharmaceutical batch reactor exhaust utilizing a vapor permeation system. (Reprinted from Majumdar et al., 2003.)

There is no known commercial installation for VOC removal and recovery using a membrane based absorption-stripping process. However, a pilot scale demonstration of the technology at a paint spray booth has been reported (Majumdar et al., 2001). Tests were performed on slipstreams of real time air emissions from scheduled intermittent painting operations, so the VOC concentration in the vent air fluctuated with time. Volatile organic compounds were efficiently removed from the contaminated air stream by an inert organic nonvolatile absorbent (silicone oil) in a hollow fiber membrane based process. The air stream was cleaned by absorbing the VOCs in a module containing microporous fibers. Spent absorbent liquid was continuously and effectively regenerated under vacuum in a module having silicone coated hollow fibers. The concentrations of the incoming VOCs in the air were in the range of 0–350 ppmv. The process successfully removed as much as 95+% of the VOC present.

### 8.2.5.6 Concluding Remarks
Membrane vapor permeation processes are likely to be increasingly utilized to remove and recover a variety of VOCs from various vent streams. The applications, which will involve moderate gas flow rates and higher VOC levels, are unlikely to compete with wheel-based adsorption technology (Humphrey and Keller, 1997) for very large gas flow rates and low VOC concentration levels. More compact devices, higher effective VOC-$N_2$ selectivities to cut down on the vacuum pump capacity, and lower cost will potentially expand

this technology's reach into existing plants and operations where a small footprint is required. Hollow fiber devices are particularly useful in this regard.

For dilute VOC-containing streams, membrane-based absorption-stripping process as well as the membrane vapor permeation process having higher VOC-$N_2$ selectivity membranes show significant potential. Further larger-scale studies are needed to establish the role of these technologies vis-à-vis other technologies in the marketplace. Similarly, the novel technology in which nitrogen/oxygen is selectively removed through a membrane from a mixture containing VOCs may find significant applications. Combinations of these disparate technologies will provide new process design tools for control and recovery of VOCs from emissions from a variety of sources.

## REFERENCES

Baker, R. W., and Wijmans, J. G. (1991). Membrane fractionation process. US Patent 5,032,148 (July 16).

Baker, R. W., Simmons, V., and Wijmans, H. (1993). Membrane recovery of volatile organic compounds (VOCs) in the pharmaceutical industry. *Pharmaceutical Engineering*, **13**(5), 44–48.

Baker, R. W., Kaschemekat, J., and Wijmans, J. G. (1996). Membrane systems for profitable VOC recovery. CHEMTECH, **26**(7), 37–43.

Baker, R. W., J. G. Wijmans, and J. H. Kaschemekat (1998). The design of membrane vapor-gas systems. *J. Membr. Sci.*, **151**, 55–62.

Baker, R. W., Lokhandwala, K. A., Jacobs, M. L., and Gottschlich, D. E. (2000). Recover feedstock and product from reactor vent streams. *Chem. Eng. Prog.*, **96**(12), 51–57.

Bhaumik, D., Majumdar, S., and Sirkar, K. K. (2000). Pilot-plant and laboratory studies on vapor permeation removal of VOCs from waste gas using silicone-coated hollow fibers. *J. Membr. Sci.*, **167**, 107–122.

Bowser, J., Grantham, R., Stutman, M., and Nemser, S. (2004). Novel air/VOC separation process. Paper presented at the AIChE Spring National Meeting, New Orleans, LA (April 25–29).

Cha, J. S., Malik, V., Bhaumik, D., Li, R., and Sirkar, K. K. (1997). Removal of VOCs from waste gas streams by permeation in a hollow fiber permeator. *J. Membr. Sci.*, **128**, 195–211.

Humphrey, J. L., and Keller, G. E. II (1997). *Separation Process Technology*. McGraw-Hill, NY, Chapter 7.

Koch, W. H. (2001). Developing technology for enhanced vapor recovery: Part 1 – vent processors, *Petroleum Equipment & Technology*, **6**(2), 16–22.

Lahiere, R. J., Hellums, M. W., Wijmans, J. G., and Kaschemekat, J. (1993). Membrane vapor separation: recovery of vinyl chloride monomer from PVC reactor vents. *Ind. Eng. Chem. Res.*, **32**, 2236–2241.

Lokhandwala, K. A., Segelke, S., Nguyen, P., Baker, R. W., Su, T. T., and Pinnau, I. (1999). A membrane process to recover chlorine from chloralkali plant tail gas. *Ind. Eng. Chem. Res.*, **38**, 3606–3613.

Majumdar, S., Bhaumik, D., Sirkar, K. K., and Simes, G. (2001). A pilot scale demonstration of membrane-based absorption-stripping process for removal and recovery of volatile organic compounds. *Env. Prog.*, **20**(1), 27–35.

Majumdar, S., Bhaumik, D., and Sirkar, K. K. (2003). Performance of commercial-size plasma-polymerized PDMS-coated hollow fiber modules in removing VOCs from $N_2$/air, *J. Membr. Sci.*, **214**, 323–330.

MTR, Gasoline Vapor Recovery from Fuel Pumps Using MTR Membrane Process. http://www.mtrinc.com/Pages/Petrochemical/OtherApplications/petro_other.html, retrieved April 15, 2004.

Nitsche,V., Ohlrogge, K., and Sturken, K. (1998). Separation of organic vapors by means of membranes. *Chem. Eng. Technol.*, **21**, 925–935.

Obuskovic, G., Majumdar, S., and Sirkar, K. K. (2003). Highly VOC-selective hollow fiber membranes for separation by vapor permeation. *J. Membr. Sci.*, **217**, 99–116.

Obuskovic, G., Poddar, T. K., and Sirkar, K. K. (1998). Flow swing membrane absorption-permeation. *I&EC Res.*, **37**(1), 212–220.

Ohlrogge, K., Peinemann, K. V., Wind, J., and Behling, R. D. (1990). The separation of hydrocarbon vapors with membranes. *Sep. Sci. Tech.*, **25**(13–15), 1375–1386.

Ohlrogge, K., Wind, J., and Behling, R. D. (1995). Off-gas purification by means of membrane vapor separation systems. *Sep. Sci. Tech.*, **30**(7–9), 1625–1638.

Ohlrogge, K., and Sturken, K. (2001). The separation of organic vapors from gas streams by means of membranes, In *Membrane Technology in the Chemical Industry*, S.P Nunes, and K.V. Peinemann (eds), WILEY-VCH, 71-94.

Ohlrogge, K., Wind, J., and Brinkmann, T. (2003). Membrane based hybrid systems to treat organic vapor loaded gas streams. Paper presented at the 6th Italian Conference on Chemical and Process Engineering (ICheaP-6), Pisa, Italy, June 8–11.

Poddar, T. K., Majumdar, S., and Sirkar, K. K. (1996a). Membrane-based absorption of VOCs from a gas stream. *AIChE J.*, **42**, 3267–3282.

Poddar, T. K., Majumdar, S., and Sirkar, K. K. (1996b). Removal of VOCs from air by membrane-based absorption and stripping. *J. Membr. Sci.*, **120**, 221–237.

Poddar, T. K., and Sirkar, K. K. (1997). A hybrid of vapor permeation and membrane-based absorption-stripping for VOC removal and recovery from gaseous emissions. *J. Membr. Sci.*, **132**, 229–233.

Wijmans, J. G., and Helm, V. D. (1989). A membrane system for the separation and recovery of organic vapors from gas streams. *AIChE Symp. Ser.*, **85**(272), 74–79.

Xia, B., Majumdar, S., and Sirkar, K. K. (1999). Regenerative oil scrubbing of volatile organic compounds from a gas stream in hollow fiber membrane devices. *I &EC Res.*, **38**(9), 3462–3472.

## 8.2.6 Carbon Sequestration

ANTONIS C. KOKOSSIS[§], PATRICK LINKE[§] AND TAPAS K. DAS[§§]

[§]Centre for Process & Information Systems Engineering, Department of Chemical and Process Engineering, University of Surrey, Guildford, Surrey, UK
[§§]Washington Department of Ecology, Olympia, Washington, USA

The past four decades have seen rising concern over the environmental impact of emissions and the extreme natural phenomena they appear to have caused. In particular, much effort has gone into investigating the relationship between emissions and climate change (Keeling and Whorf, 1997). The finding that emissions and extreme natural phenomena are related launched numerous studies into the consequences of failing to prevent further damage to the environment. In 1997, the representative of 178 of the world's nations signed the agreement about global emissions known as the Kyoto Protocol. It was decided to determine separate national standards for the reduction of emissions based on the level of technological advancement of each country. The pollutants to be monitored and reduced are mainly the greenhouse gases (GHGs), which are:

- Carbon dioxide ($CO_2$)
- Methane ($CH_4$)
- Nitrous oxide ($N_2O$)
- Chlorofluorocarbons (CFCs) and ozone ($O_3$)

Carbon dioxide, the most significant GHG, has been linked to acid rain as well as the global warming. This news has resulted in the development of new technologies and materials for the capture of carbon dioxide emitted from industrial processes. Carbon sequestration holds great potential to reduce GHG emissions at cost and impacts that are economically and environmentally acceptable. There are five technology pathways to the reduction of GHG emissions,

Separation and capture
Geological sequestration
Terrestrial sequestration
Oceanic sequestration
Novel sequestration systems

encompassing a broad set of opportunities for both technology development and partnership formation for national and international cooperation. This section deals mainly with separation and capture, although some novel sequestration systems are presented in Section 8.2.6.3. Fossil fuel power plants, the prime targets for $CO_2$ capture, produce about a third of all $CO_2$ emissions worldwide. The remainder come from the manufacture of iron, cement, steel and chemicals, oil and natural gas operations, and the production of hydrogen gas.

The power generation schemes that have been considered for $CO_2$ capture from fossil fuel power plants include the following:

- PC + FGD (pulverized coal fired plant equipped with flue gas desulfurisation facilities) and operating with a sub-critical or supercritical high temperature steam cycle. This is established technology with emerging improvements in the development of higher efficiency through advanced ultra-supercritical steam conditions. PC + FGD is expensive to retrofit with capture technology (Segalstad, 1998).

- NGCC (natural gas combined cycle) power plants. NGCC is established technology that is difficult to retrofit with capture technology (Brown and Heim, 1996; Houghton et al., 1995).
- IGCC (integrated gasification combined cycle) in which coal is gasified. This is advanced and emerging technology that represents more efficient and potentially the cleanest of the available coal technologies. There are low additional costs for carbon dioxide capture, particularly if the plant is oxygen driven.
- $O_2/CO_2$ recycle for pulverized coal-fired plants that use $O_2/CO_2$ flue gas recycle combustion. This technology has not been demonstrated yet and while the costs of oxygen are high, it could provide an immediate retrofit opportunity (Kyoto, 1997).

**8.2.6.1 Flue GAS Capture**   In flue gas, or postcombustion, $CO_2$ is first separated from the other byproducts of the burning of fossil fuels in electricity generation or other industrial applications. Generally it is then compressed to a high pressure (80–120 bars) to liquefy the gas prior to transport and storage. In many cases, pre-treatment to remove contaminants such as particulates, sulfur dioxide, and $NO_x$ will be required. Flue gas separation and capture methods include absorption after contact with solvents, adsorption on activated carbon or other materials, cryogenic separation and membrane separation (Kimball et al., 1995; Bolin, 1998). The characteristics of flue gas (composition, pressure, temperature, and volume) vary according to industrial process, fuel type, and power generation scheme.

*Chemical and Physical Absorption*   Carbon dioxide can be removed from gas streams by physical or chemical absorption. Physical absorption processes are governed by Henry's law; that is, they are temperature and pressure dependent, with absorption occurring at high pressures and low temperatures. Typically, these processes are used when the concentration (i.e., partial pressure of $CO_2$) exceeds 525 kPa. Natural gas production wells in remote locations can conveniently use amines for chemical absorption, achieving removal of 0.1–6 percent $CO_2$. This approach is the most widely deployed commercial technology for capture. However, in other commercial applications, the typical solvents for physically absorbing $CO_2$ include glycol-based compounds (e.g., the dimethyl ether of polyethylene glycol) and cold methanol.

Chemical absorption is preferred for low to moderate $CO_2$ partial pressures. Because $CO_2$ is an acid gas, chemical absorption of $CO_2$ from gaseous streams such as flue gases depends on acid- base neutralization reactions using basic solvents. Most common among the solvents in commercial use for neutralizing $CO_2$ are alkanolamines such as monoethanolamine (MEA), diethanolamine (DEA), and methyldiethanolamine (MDEA). Hot potassium carbonate, discussed in Section 8.2.6.2, and ammonia are also used. Flue gases are typically at atmospheric pressure. Depending on the $CO_2$ content of the flue gas, the partial pressure of $CO_2$ can vary from 3.5–21.0 kPa. At such low partial pressures, alkanolamines are the best chemical solvents to enable good $CO_2$ recovery levels; however, use of these solvents must be balanced against the high energy penalty of using steam stripping to regenerate them.

As noted elsewhere in connection with other processes, flue gases typically contain contaminants such as $SO_x$, $NO_x$, $O_3$, hydrocarbons and particulates, which can reduce the absorption capacity of amines as well as create operational problems such as corrosion. To avoid such conditions, these contaminants are often pretreated to reduce them to acceptable levels. Some commercial processes use pretreatment and/or chemical inhibitors to facilitate absorption. However, these processes tend to be more expensive than the conventional alkanolamine-based absorption processes.

*Physical and Chemical Adsorption* Selective separation of $CO_2$ may be achieved by the physical adsorption of the gas on high-surface-area solids. The large surface areas result from the creation of very fine surface porosity through surface activation methods using, for example, steam, oxygen, or $CO_2$. Some naturally occurring materials (e.g., zeolites) have high surface areas and efficiently adsorb some gases. Adsorption capacities and kinetics are governed by numerous factors including adsorbent pore size, pore volume, surface area, and affinity for the adsorbent.

The International Energy Agency (IEA, 1998) evaluated physical adsorption systems based on zeolites operated in pressure swing adsorption (PSA) and thermal or temperature swing adsorption (TSA) modes. In PSA operation, gases are absorbed at high pressures, isolated, and then desorbed by reducing the pressure. A variant of PSA, is called vacuum swing adsorption, uses a vacuum desorption cycle. In TSA operation, gases are adsorbed at lower temperatures, isolated, and then desorbed by heating. These processes are somewhat energy-intensive and expensive. The IEA report concludes that PSA and TSA technologies are not attractive options for the gas and the coal-fueled power systems investigated.

Nevertheless, PSA and TSA are commercially practiced methods of gas separation and capture and are used to some extent in hydrogen production and in removal of $CO_2$ from sub-quality natural gas. Therefore, these methods clearly are applicable for separation and capture of $CO_2$ from some relatively large point sources (Khesgi et al., 1997; Lehtila et al., 1997; Marland et al., 1994).

### 8.2.6.2 Additional CO₂ Capture Approaches

*Low-Temperature Distillation* Low-temperature distillation is widely used commercially for the liquefaction and purification of $CO_2$ from high purity sources (typically a stream with >90 percent $CO_2$). However, such processes are not used for separating $CO_2$ from significantly cleaner $CO_2$ streams. The application of distillation to the purification of lean $CO_2$ streams necessitates low-temperature refrigeration ($<0°C$) and solids pressing below the triple point of $CO_2$ ($-57°C$). A patented process to separate $CO_2$ from natural gas, providing liquid $CO_2$, is an example of such a low temperature approach (Marland et al., 1995; OECD, 1996b).

Distillation generally has good economies of scale, as it is cost-effective for large-scale plants, and it can generally produce a relative pure product. Distillation is the most cost-effective when feed-gases contain components with widely different boiling points, and when the feed gas is available at high pressure and most of the products are also required at high pressure. Low-temperature distillation enables direct production of liquid $CO_2$ that can be restored at high pressure via liquid pumping. The major disadvantage of this process is that any components having their freezing points above normal operating temperatures must be removed before the gas stream is cooled to avoid crystallization and eventual blockage of process equipment. Another disadvantage is the amount of energy required to provide the refrigeration necessary for the process (OECD, 1996a,b; Wigley, 1997).

Most $CO_2$ emissions being considered for $CO_2$ capture are produced in combustion processes. Such streams contain water and other trace combustion by-products such as $NO_x$ and $SO_x$, several of which must be removed before the stream is introduced into the low-temperature process. The by-products are usually generated at near atmospheric pressure. These attributes coupled with the energy intensity of low-temperature refrigeration, tend to make distillation less economical than other routes. The application of low-temperature distillation, therefore, is expected to be confined to feed sources at high pressure with high $CO_2$ concentrations such as gas wells (Adger and Brown, 1994; Arnell and Reynard, 1994).

*Gas Separation Membranes*   Gas separation membranes are of many different types, and although only a few membranes for separating and capturing $CO_2$ have demonstrated efficiency, their potential is generally viewed as very good. Diffusion mechanisms in membranes differ greatly depending on the type of membrane used. Generally, gas separation is accomplished via interaction between the membrane and the gas separated (Figure 8.44). For example, polymeric membranes transport

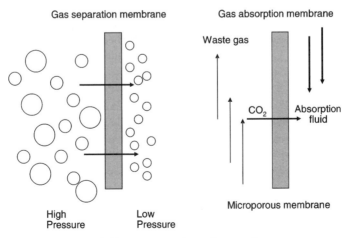

**Figure 8.44.** Gas separation using membranes.

gases by a solution-diffusion mechanism (i.e., the gas is dissolved in the membrane and transported through the membrane by a diffusion process). Polymeric membranes, although effective, typically achieve low gas transport flux and are subject to degradation. However, polymer membranes are inexpensive and can achieve large ratios of membrane area to module volume (DTI, 1998; Howe and Hendersen-Seller, 1997; IPCC, 1996; Kapoor and Sirkar, 1993).

Palladium membranes are effective in separating $H_2$ from $CO_2$, but gas fluxes are typically very low, and palladium is subject to degradation in sulfur-containing environments. Porous inorganic membranes, metallic or ceramic, are particularly attractive because of different transport mechanisms that can be used to maximize the separation factor for various gas separations. Permeability is measured by the volume of gas transported through a membrane per unit of surface per unit of differential pressure. Porous inorganic membranes, which can be 100 to 10,000 times more permeable than the polymeric type, are very expensive, however, and the ratio of membrane area to module volume is 100 to 1,000 times smaller than that for polymer membranes. These factors tend to equalize the cost per membrane module. The inorganic membrane cycle is generally expected to be much longer. Inorganic membranes can be operated at high pressures and temperatures and in corrosive environments with long life cycles. They are also less prone to fouling and can be used in applications to which polymer membranes are not suited.

*Membranes with Small Pore Diameters*   Inorganic membranes with effective pore diameters as small as 0.5 nm can be made from a wide range of materials, and many pore sizes, and materials can be changed to improve permeability and separation factors. Large separation factors are essential to achieve the desired performance in a single stage. Inorganic membranes can be made to separate large molecules from larger molecules (molecular sieves) or to separate certain large molecules from smaller molecules (enhanced surface flow). The latter effect is important because it allows separation that will keep the desired gas either on the high-pressure or the low-pressure side of the membrane. Operating conditions play an important role in determining the change in mole fraction across a membrane and the amount of the desired gas that can be recovered or captured. There must a partial pressure gradient of the desired gas across the membrane to achieve a flow of that gas through the membrane.

With all the design parameters available it is likely that an inorganic membrane can be made that will be useful for separating $CO_2$ from almost any other gas if appropriate operating conditions can be achieved. However, for multiple gas mixtures, several membranes with different characteristics may be required to separate and capture high-purity $CO_2$ (Croiset and Thambimuthu, 2001; Miyame et al., 1994; IEA, 1998, 1999).

*Sodium Carbonate Process*   The sodium carbonate process is used in a number of dry ice plants in the United States although its operating efficiency is generally not as high as that of the process using other solutions. These plants obtain the carbon dioxide from flue gases as well as limekiln gases (Arnell et al., 1994).

Carbon dioxide can be recovered from gases containing other diluents by taking advantage of the reversibility of the following reaction:

$$Na_2CO_3 + H_2O + CO_2 \leftrightarrow 2NaHCO_3$$

The reaction proceeds to the right at low temperatures and takes place in the absorber where the carbon dioxide-bearing gases are passed counter-current to the carbonate solution. The amount of carbon dioxide absorbed in the solution varies with temperature, pressure, partial pressure of carbon dioxide in the gas and solution strength.

The reaction is reversed when heat is supplied in a lye boiler. A heat exchanger preheats the strong lye approaching the boiler and cools the weak lye returning to the absorber. The lye cooler further cools weak lye to permit the reaction to proceed further to the right in the absorber. The carbon dioxide gas and vapour released from the solution in the boiler pass through a steam condenser where the water condenses and returns to the system. The cool carbon dioxide proceeds to the gasholder and compressors. The absorber is generally a carbon-steel tower filled with coke, Raschig rings, or steel turnings. The weak solution is distributed evenly over the top of the bed and contacts the gas on the way down. Some plants operate with the tower full of sodium carbonate solution and allow the gas to bubble up through the liquid. Although this may afford a better gas-to-liquid contact, an appreciable amount power is required to force the gas through the tower. The lye boiler is usually steam heated but may be direct-fired. Adding a tower section with bubble-cap trays may increase separation efficiency. To permit the bicarbonate content of the solution to build up, many plants are designed to re-circulate the lye over the absorber tower with only 20–25 percent of the solution flowing over this tower passing through the lye boiler. Several absorbers may also be used in series to increase absorption efficiencies.

*Potassium Carbonate Process*  The potassium carbonate process is similar to the sodium carbonate process but more efficient since potassium bicarbonate is more soluble than the corresponding sodium salt. The equipment layout is the same and the operation technique is similar to the sodium carbonate process.

There are several variations of the potassium carbonate process. The hot potassium carbonate process does not involve cooling of the solution flowing from the boiler to the absorber. Absorption takes place at essentially the same temperature as solution regeneration. In its simplest form this process uses an absorption column, a regeneration column, a heated boiler or re-boiler, and a circulating pump. This arrangement minimizes energy requirements and capital costs, but because the absorbing temperature is higher, the carbon dioxide removal is less complete than it is with lower temperatures.

One modification, which improves removal, is the use of split stream to the absorber. Part of the solution is cooled and used at the top of the absorber. The balance of the solution, un-cooled, is added part way down the absorber. The combined solution from the absorber then flows to the regenerator. In another modification, featuring two stage absorption and regeneration, the carbonate solution from the absorber first flows to the regeneration column. Then part of the solution is withdrawn from the column at an intermediate point and pumped, un-cooled, to an intermediate point in the absorber. The remaining solution undergoes a more complete regeneration and pumped to the top of the absorber.

Three commercial processes that use these various hot carbonate flow arrangements are the enhanced Benfield process, the Catacarb process, and the Giamarco-Vetrocoke process. Each uses an additive described as a promoter, activator or catalyst, which increases the rates of absorption and desorption, improves removal efficiency, and reduces the energy requirements. The processes use corrosion inhibitors, which allow use of carbon steel equipment. The Benfield and Catacarb processes do not specify additives. Vetrocoke uses boric acid, glycine or arsenic trioxide, which is the most effective.

These processes have been used in plants to remove carbon dioxide from natural gas or ammonia synthesis (Figure 8.45). They are most effective in the gas stream being treated at elevated pressure (17 atm or higher). This increases the carbon dioxide partial pressure so that the hot potassium carbonate solution absorbs a substantial amount of carbon dioxide. The stripping tower or regenerator operates at or near atmospheric pressure (Arnell et al., 1994).

**8.2.6.3 Novel Concepts**  Novel concepts, an evolving category, merit much further discussion in order to broaden thinking on innovative approaches to carbon dioxide removal. Unlike the other capture systems, which are established approaches, it is important to emphasize that these concepts have not yet been proven. However, they provide 'out of the box' ideas that may offer breakthroughs. Following is a small sample of the many novel concepts that have been proposed.

*New Power Cycles*  New power cycles are variations of the oxygen combustion approach. Carbon dioxide is used as the working fluid of the power cycle and the

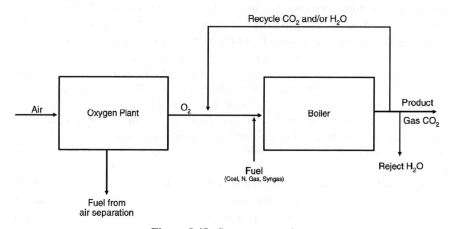

**Figure 8.45.**  Syngas approach.

fuel is combusted in relatively pure $O_2$ in an $O_2/CO_2$ atmosphere. The major advantage of these new cycles is that $CO_2$ could be obtained in liquid form as the output stream from combustion without using an energy-consuming separation process. Cycles based on this concept include the MATIANT and COPERATE cycles.

*Chemical Looping*   In chemical looping combustion, direct contact between the fuel and the combustion air is avoided by using a metal oxide to transfer $O_2$ to the fuel. Overall combustion takes place in two reaction steps in two separate reactors. In the reduction reactor, the fuel is oxidized by the metal oxide and the metal oxide is reduced to a lower oxidation state in the reaction with the fuel (Figure 8.46). The metal oxide is then transported to the second reactor (the oxidation reactor) where it is re-oxidized by $O_2$ in air. In order to achieve a high thermal efficiency, chemical looping combustion should be integrated in a combined gas turbine steam power process (OECD, 1996a,b).

*Dry Ice Co-generation*   By expanding highly compressed $CO_2$ from a quasi-combined cycle, solid $CO_2$ dry ice can be produced. Sublimation of $CO_2$ due to heat through thermal insulation is slow enough to delay emissions of an ordinary power plant over about a millennium. The dry ice can be used in the food industry or stored in large thermally insulated repositories.

*Biological $CO_2$ Fixation with Algae*   Algae are marine plants with rapid growth rates that divide into micro-algae and macro-algae (seaweeds). Micro-algae are currently used commercially for production of high value products such as food

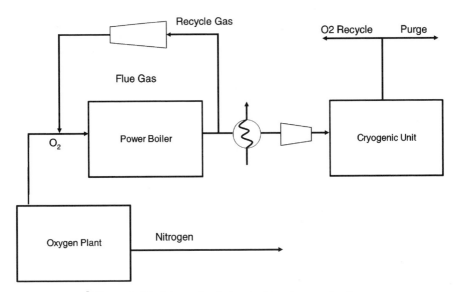

**Figure 8.46.** Schematic of chemical looping combustion.

supplements and for wastewater treatment. Biological fixation utilizes algae that are optimized for high $CO_2$ feeding and photosynthesis of organic matter. $CO_2$ produced by ordinary fossil-fuel power plants is used to cultivate algae in large open ponds (a pond area of about 50–100 km$^2$ would be needed for a 500 MW power station). The algae can be converted into either methane-gas or bio-diesel, or can alternatively be incorporated into building materials. Some algae might also be used as a bio-fuel at power plants (Figure 8.47).

Micro-algae require a combinaton of land, water and climatic resources which are not often found at power plant locations. There is one commercial micro-algae production plant in Hawaii that uses flue gas from a power plant to grow algae. However, it is a relatively small industry and in order to be cost competitive with other fuels or make an impact as a carbon management strategy, there would need to be large productivity increases and markets to support them.

*Direct Capture of $CO_2$ from Air*   Ambient $CO_2$ could be removed from the air which would involve processing of up to 1 percent of the earth's atmosphere each

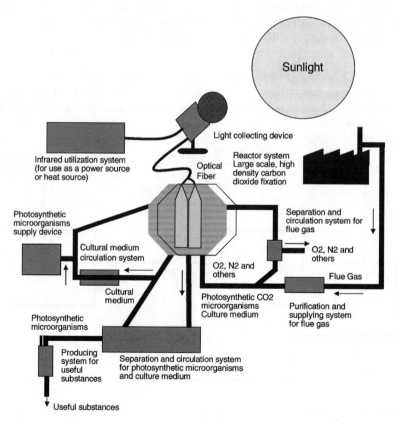

**Figure 8.47.**  Schematic of biological $CO_2$ fixation with algae.

year. The low $CO_2$ concentration in air limits the choice of collection methods and chemical or physical absorption appear to be the only viable options. As the air blows over an absorber surface, lime can be used as the capture agent. A slight pressure gradient must be maintained in order to keep the gas flowing through an absorption system and the rate of absorption depends on the wind speed and the efficiency of the absorber. One example from a number of possible design options is to pump water to the top of a convection tower to cool the air, which causes a downdraft inside the tower. The air leaving at the bottom could drive wind turbines or flow over $CO_2$ absorbers. Based on the volumes of airflow and the potential energy of the cold air generated on the top of the tower, the tower could generate electricity after pumping water to the top. The same airflow would carry $CO_2$ through the tower for disposal. Large amounts of energy would be required for this process and many issues surrounding its viability require examination.

*Zero Emissions Coal*   The zero emission coal (ZEC) power plant utilizes anaerobic $H_2$ production to generate electricity. This process integrates coal gasification, $H_2$ production via calcium oxide carbonation, and limestone calcinations, while incorporating a fuel cell system. The anaerobic $H_2$ production is an industrial, elevated-temperature process that involves no combustion, and requires no heat input. Aside from coal, the process requires water and CaO as inputs, which are continuously recycled. ZEC technology features a mineral carbonation concept but could employ other techniques to dispose of carbon dioxide (Croiset and Thambimuthi, 2001).

## ACKNOWLEDGMENT

The information presented here has been collected during a master thesis prepared by George Yiannou at the University of Surrey, UK. Dr. Antonis Kokossis and Dr. Patrick Linke were the project lead, and Dr. Tapas K. Das supervised this research project as an extramural advisor.

## REFERENCES

Adger, N., and Brown, K. (1994). *Land Use and the Causes of Global Warming.* Wiley, Chichester.

Arnell, N., Jenkins, A., Herrington, B. W., and Dearnely, M. (1994). Impact of Climate Change on Water Resources in the United Kingdom: Summary of Project Results. Climatic Research Unit, University of East Anglia: Norwich.

Arnell, N., and Reynard, N. (1994). Impact of Climate Change on River Flow Regimes in the United Kingdom, Climatic Research Unit, University of East Anglia: Norwich.

Bolin, B. (1998). The Kyoto Negotiations on Climate Change: A Science Perspective, *Science,* **279**, 330–331.

Brown, W. O., and Heim, R. R. (1996). National Climate Data Centre, Climate Variation Bulletin 8, Historical Climatology Series 4–7, December [http//www.ncdc.noaa.gov/ol/documentlibrary/cvb.html/].

Croiset, E., and Thambimuthu, K. V. (2001). $NO_x$ and $SO_2$ emissions from $O_2/CO_2$ recycle plant combustion, *Fuel*, **80**, 2117–2121.

DTI, Department of Trade and Industry (1998). Digest of UK Energy Statistics 1998, The Stationery Office, London.

Houghton, J. T. et al. (1995). Report of the Intergovernmental Panel on Climate Change. Cambridge University Press.

Howe, W., and Henderson-Sellers, A. (eds) (1997). Assessing Climate Change. Gordon & Breach Science Publishers, UK.

IEA, International Energy Agency (1999). Greenhouse Gas R&D Programme and Stork Engineering Consultancy B. V. Assessment of Leading Technology Options for Abatement of $CO_2$ Emissions (December).

IEA, International Energy Agency (1998). Greenhouse Gas R&D Programme, Carbon Dioxide Capture from Power Stations.

IPCC, Intergovernmental Panel on Climate Change (1996). Houghton, J. T., Meira Filho, L. G., Callander, B. A., Harris, N., Kattenberg, A. and Maskell, K. (eds). Climate Change 1995: The Science of Climate Change, Cambridge University Press, Cambridge, pp. 572.

Kapoor, S., and Sirkar, K. K. (1993). Gas absorption studies in micro-porous hollow fiber membrane modules, *Ind. Eng. Chem. Res.* **32**, 674–684.

Keeling, C. D., and Whorf, T. P. (1997). *Trends Online: A Compendium of Data on Global Change*. Carbon Dioxide Information Analysis Centre, Oak Ridge National Laboratory. [http://cdiac.asd.ornl.gov/ftp/ndp/001r7].

Khesgi, H. S., Jain, A. K., and Wuebbles, D. J. (1997). *Analysis of proposed $CO_2$ emissions reductions in the context of stabilization $CO_2$ concentration*. In Proceedings of the Air and Waste Management Association's 90th Annual Meeting and Exhibition, Toronto, Ontario, Canada, 97-TA5302.

Kimball, et al. (1995). Productivity and water use of wheat under free-air $CO_2$ enrichment, *Global Change Biology*, **1**, 429–442.

Kyoto Protocol to the United Nations Framework Convention on Climate Change (1997). Adoption of this protocol would sharply limit GHG release for one-fifth of the world's people and nations, including the United States.

Lehtila. A., Savolainen, I., and Sinisalo, J. (1997). Indicators describing the development of national greenhouse gas emissions, EPOO: VTT Energy, p. 18.

Marland, G., Anters, R. J., and Boden, T. A. (1995). Global, regional and national $CO_2$ emissions. Data available online at the anonymous FTP server ftp//cdiac.esd.ornl.gov

Marland, G., Anters, R. J., and Boden, T. A. (1994) *Global, regional and national $CO_2$ emissions*. In Boden, T. A., Kaiser, D. P., Sepanski, R. J., Stross, F. W. (eds), Trends 1993: A Compendium of Data on Global Change. Oak Ridge: Carbon Dioxide Analysis Centre, Oak Ridge National Laboratory, pp. 505–585.

Miyame, S., Kiga, T., Tkano, S., Omata, S., and Kimura, N. (1994). Bench-scale testing on $O_2/CO_2$ combustion for $CO_2$ recovery. Paper presented at the JFRC/AFRC Symposium on Combustion, Hawaii, USA.

OECD (1996a). Organisation for Economic Co-Operation and Development, Paris. In *Main Economic Indicators*, p. 200 (February).

OECD (1996b). Organisation for Economic Co-Operation and Development. In *Short term economic indicators: transition economies*. Centre for Co-Operation with the Economies in Transition, Paris, p. 169 + Appendix 35 pp.

Segalstad, T. V. (1998). Global warming: the continuing debate. Cambridge UK: *Europ. Sci. and Environ.*, R. Bate (ed.), 184–218.

Wigley, T. M. (1997). Implications of recent $CO_2$ emission-limitation proposal for stabilisation of atmospheric concentrations, *Nature*, **390**, 267–270.

# CHAPTER 9

# TECHNOLOGIES FOR POLLUTION PREVENTION: WATER

## 9.1 ADVANCES IN WASTEWATER TREATMENT TECHNOLOGIES

Advances in the effectiveness and reliability of wastewater treatment technologies have improved the capacity to produce reclaimed wastewater that can serve as a supplemental water source in addition to meeting water quality protection and pollution abatement requirements. In developing countries, particularly those in arid parts of the world, reliable low-cost technologies are needed for acquiring new water supplies and protecting existing water resources from pollution. The implementation of wastewater reclamation, recycling, reuse, and zero discharge systems in conjunction with water conservation and watershed protection programs, promotes the preservation of limited water resources. In the planning and implementation of water reclamation and reuse, the intended water reuse applications dictate the extent of wastewater treatment required, the quality of the finished water, and the method of distribution and application. Thus, the goal of the designer of current and future wastewater treatment plants is to devise the facilities that provide much higher levels of treatment.

Some of the treatment methods being developed and utilized in new and upgraded treatment facilities include vortex separators, high rate clarification, membrane bioreactors, pressure-driven membrane filtration (ultrafiltration and reverse osmosis, see Section 9.1.1), technologies for zero discharge based on supercritical $CO_2$ and $H_2O$ (see Section 9.1.2), and ultraviolet disinfection (see Section 9.1.3). Some of the new technologies, especially those developed in Europe, are more

*Toward Zero Discharge: Innovative Methodology and Technologies for Process Pollution Prevention*, edited by Tapas K. Das.
ISBN 0-471-46967-X  Copyright © 2005 John Wiley & Sons, Inc.

compact and are particularly well suited for plants where space for expansion is limited. In addition, recent years have seen the emergence of numerous proprietary wastewater treatment processes that offer potential savings in construction and operation.

### 9.1.1 Membranes for Environmental and Pharmaceutical Applications

D. BHATTACHARYYA AND D. SHAH

Department of Chemical & Materials Engineering, University of Kentucky, Lexington, Kentucky 40506-0046, USA

Membrane processes provide a highly flexible separation technique for water, solvent and solute recovery. The special features for membrane processes that make them attractive for environmental applications are their compactness, ease of fabrication, operation, and modular design. The development of thin-film reverse osmosis and nanofiltration membranes has enhanced the environmental applications of pressure driven membranes from traditional desalination to organics and toxic metals separation at high water flux values. The uses of membranes range from new biofunctional approaches to water purification with functionalized membranes and membrane contactors, gas separation, water reuse and material recovery, removal of suspended impurities and dialysis (Bhattacharyya and Butterfield, 2003; Koros and Mahajan, 2000; Williams et al., 1999; Klassen et al., 2003; Ritchie et al., 2001; Zydney, 1998; Hestekin et al., 1998). Already being used in water and wastewater treatment and material recovery, membranes will likely find greater use, given advances in membrane materials, module designs, techniques in fouling control, and integration of membrane into hybrid systems with conventional technology.

*9.1.1.1 Membrane Types and Properties* Depending on the pore structure, membranes may be classified into various types such as microfiltration (MF), ultrafiltration (UF), nanofiltration (NF), reverse osmosis (RO), and pervaporation (PV). The typical pore size of a microfiltration membrane, which is mainly used to remove colloidal particles and suspended impurities from process streams, lies in the range of $0.1-1$ μm. Ultrafiltration membranes have a smaller pore size than MF membranes and are commonly characterized by molecular weight cutoffs (MWCO). The main use of UF is for separation and purification of proteins in the biotech, food and pharmaceutical industries. In addition, UF is emerging as an excellent pretreatment technology for various reverse osmosis systems. RO and PV are dense membrane-based processes where sorption and diffusion characteristics are the dominating transport mechanisms. RO is a high pressure $(30-100$ bar) process, which has been conventionally used in the area of seawater desalination, but it has also gained attention for its environmental applications such as groundwater treatment and waste treatment (for pollution prevention) in the plating industry. RO is used for the removal of low to moderate concentrations of salts/organics from process streams. It cannot be used to treat aqueous feeds containing high concentrations of salts (>4 percent), solutes or solvents due to high feed osmotic pressure. The water recovery is highly dependent on the feed osmotic pressure, for example, with seawater recovery is typically limited to <50 percent. Recent studies are dealing with the use of crystallizer in the retentate stream and thus one could enhance water recovery.

The permeate in PV, unlike RO, is a vapor and theoretically, PV can be used to treat a variety of process streams with varying concentrations of salts, solutes and/or solvents. Also, PV membranes may be hydrophilic or organophilic in nature and hence depending on the need, water or organic solvents can be selectively transported across the membrane. The use of high feed pressures is not necessary in PV and in fact, most commercial installations are operated at feed side pressures between 1–1.5 bar and feed temperatures varying from 70–90°C. Flux values of PV are typically reported in terms of $kg/m^2/h$. Nanofiltration, also called "loose RO," is primarily used for the removal of high MW organics (>300) and salts (divalent/trivalent) from process streams. In terms of applications, NF bridges the gap between ultrafiltration (used to remove large (MW > 5,000) molecules) and RO (removal of small solutes and salts). Typical operating pressure for NF membranes is in the range of 4–17 bar. Nanofiltration membranes typically bear a weak negative charge and hence are able to reject multivalent anions along with organic materials of high molecular weight (300–1,000). Moreover, rejection of NaCl is low, reducing osmotic pressure limitations for separations aimed at recovering a moderate MW target organic without concentrating NaCl. The flow through MF and UF membranes is mainly convective whereas the flow through PV, RO and gas separation membranes is strictly diffusive. Both convective and diffusive flows contribute partially to the transport through NF membranes.

Membranes can find applications in many industrial areas. First, membranes can be used to produce a concentrated stream of the toxic pollutants. The organics can then be destroyed by thermal means, and the metals precipitated and sent to a hazardous waste landfill. Future technologies confronting these problems will attempt recovery/reuse of the metals. Several plating industries already practice the reuse of water and recycling of metals with RO. Second, membranes can be used for improving processing efficiency and economics of processes by making recycle/reuse possible. For instance, applied to in-process or in-plant recycling, PV can be used to remove organics preferentially through the membrane to provide a concentrated organic stream, which can be separated into the desired constituents by other technologies, such as distillation, for reuse.

### 9.1.1.2 Considerations for Pharmaceutical Processes

Pharmaceutical processes are typically low volume batch processes. The production of a synthetic organic medicinal chemical may involve chemical/biological modification of existing compounds (such as an antibiotic, or drug from plant/animal sources) or the complete synthesis of the drug molecules from the basic compounds. Pharmaceutical products are usually high-value products and hence optimum yield and recovery of these compounds is of utmost importance. The volume of the batches may also vary depending on the market demand of the product. Also, pharmaceutical streams are multi-component, complex streams mainly due to the fact that the synthesis of a product or an intermediate is usually comprised of multiple reaction and separation steps.

The amount of process waste generated per kg of product and the composition of the effluent stream will vary greatly depending on the number of reaction steps, yield of each step and the solvent used. For example, in the manufacture of aspirin, which requires only two steps with an overall yield of 80 percent, only 0.2 kg of organic residue is produced per kilogram of aspirin. On the other hand, in the production of vitamin A, which is a 13 step process with an overall yield of 15–20 percent, generates 7 kg of organic waste per kilogram of product.

Thus, pharmaceutical process streams may be aqueous or solvent-rich, containing a variety of solvents, inorganic salts (e.g., KCl, NaCl) and high MW compounds. The separation objective may also vary depending on the composition and the volume of the stream. Although a simple volume reduction of the stream sometimes suffices (mainly for waste minimization applications), in certain cases the recovery of a particular solvent or compound may be more desirable. Table 9.1 describes the different kinds of process streams and wastes typically generated in a pharmaceutical facility (USEPA, 1991). We shall concentrate mainly on the processing of liquid streams or effluents.

The conventional trend in the pharmaceutical industry has been towards incineration of concentrated mixed solvent streams and bio-treatment of dilute aqueous streams. The concentrated solvent streams are generally incinerated (onsite or offsite) because recovery of solvents using conventional separation systems is not economical. Moreover, the solvents, if recovered, have to be very pure in order to avoid potential contamination problems. This poses an additional strain on the economics of any recovery system. The bio-treatment of dilute streams may also be problematic due to high BOD/COD loading and potential toxicity of the waste stream. In recent times, the environmental agencies have become more stringent regarding the effluent discharge limits and incineration practices in a manufacturing facility. Another problem associated with offsite incineration is the possibility of hazardous waste spills and the associated liability. Hence there is an increased emphasis on

**TABLE 9.1. Pharmaceutical Process Streams/Effluents**

| Stream Description | Process Origin | Composition |
|---|---|---|
| Process liquors | Organic synthesis | Mixed solvents |
| Spent fermentation broth | Fermentation process | Contaminated water |
| Aqueous solutions | Solvent extraction process | Aqueous stream containing solvents |
| Volatile organic compounds | Chemical storage tanks/drums | Solvents |
| Spent solvents | Solvent extraction or wash practices | Contaminated solvents |

Adapted from USEPA (1991).

volume reduction of wastes and recovery of solvents. The list of solvents commonly used in pharmaceutical processing is a long one, which includes the following:

Acetone
Ethanol
Methanol
Isopropanol
Butanol
Methylene chloride
Dimethyl formamide (DMF)
Methyl isobutyl ketone (MIBK)
Tetrahydrofuran (THF)
Methyl acetate
Ethyl acetate
Butyl acetate
Pyridine

### 9.1.1.3 *Theory of Membrane Processes*

Of the existing technologies, reverse osmosis, nanofiltration, pervaporation and hybrid systems hold the maximum potential for organics and solvent waste remediation applications in the pharmaceutical industries and waste treatment applications. We briefly review transport models and solute separation mechanisms for these processes.

*Reverse Osmosis* The solution-diffusion model (SD) has been widely used for solvent and solute transport for RO membranes. For water transport, the driving force can be expressed as follows:

$$J_w = A(\Delta P - \Delta \pi) \qquad (9\text{-}1)$$

where $A$ is the water permeability coefficient, $\Delta P$ is the transmembrane pressure difference, and $\Delta \pi$ is the osmotic pressure difference. For dense RO membranes, $A$ is directly proportional to water diffusivity and concentration in the membrane and inversely proportional to membrane thickness. For dilute solutions, the osmotic pressures could be easily calculated using Van't Hoff's equation (e.g., with dilute solutions, $\pi$ is directly proportional to molar solute concentrations).

The solute flux for species $i$, $J_i$ is primarily due to the concentration difference and is defined as:

$$J_i = \frac{D_{im}K_{im}}{\delta}(c_{i,wall} - c_i) \qquad (9\text{-}2)$$

where $D_{im}$ is the solute diffusivity in the membrane, $K_{im}$ is the solute partition coefficient, $\delta$ is the membrane thickness, and $C_{i,wall}$ is solute species concentration on the feed side of the membrane surface, and $C_i$ is solute bulk concentration. In the absence of concentration polarization, the wall concentration in Eq. (9-1) is equal to the bulk concentration. The coefficient of the concentration driving force in the above equation is often collectively represented as a parameter, $B$, which is called the solute permeability coefficient.

The principal advantage of the SD model is that only two parameters are needed to characterize the membrane system. These parameters can be easily determined with pure water and wastewater experiments under minimal concentration polarization and fouling conditions. With defect-free RO membranes involving salt separations, the salt flux is independent of $\Delta P$, whereas water flux increases linearly with net applied pressure, as can be seen from Eqs. (9-1 and 9-2). With some organic systems, since organic-membrane interaction may occur, a modified SD model that includes an organic sorption term (Klassen, 2003) or a surface force-pore flow model (Mehdizadeh and Dickson, 1991) needs to be used as transport models. For example, with aromatic polyamide membranes, even a dilute solution of non-ionized 2,4,6-trichlorophenol showed as much as 50 percent flux decline (not due to concentration polarization or fouling). On the other hand, when the system operated at a pH where the organic was ionized, no flux drop was observed indicating insignificant organic-membrane interaction.

*Pervaporation*   Pervaporation involves the use of dense polymeric membranes to separate one or more components from a liquid feed mixture. There is a selective permeation of one or more components through a dense membrane followed by vaporization of the same species. The permeation and the evaporation steps thus control the degree of separation that can be achieved in the process. The selective permeation of one of the components through the membrane depends primarily on whether the membrane is hydrophilic or hydrophobic. The permeating species are evaporated by maintaining a vacuum (10–40 torr) on the permeate side of the membrane. A sweep gas may also be used on the permeate side for the same purpose. One of the greatest advantages of pervaporation is that it can be used in the separation of azeotropic mixtures.

The basic driving force for this process is the difference in the activities of the component on the feed side and the permeate side. A lower partial pressure on the permeate side and a higher feed temperature will lead to a greater difference in the activities of the component and hence a greater flux. The permeability of a component depends on the diffusivity and solubility of the component in the membrane. Diffusivity and solubility in turn, are functions of the concentration levels employed, swelling characteristics of the membrane and interaction between the permeating components.

Wijmans and Baker (1993) visualized the pervaporation process as two separate steps, a vapor liquid equilibrium step followed by a vapor permeation step. In the first part, the liquid feed is assumed to be in equilibrium with a hypothetical feed vapor which is in contact with the membrane. The hypothetical feed vapor is then assumed to sorb and permeate through the membrane. Based on this model, a very simple form of the flux equation can be derived as shown in Eq. (9-3), assuming no coupling effects and a constant diffusion coefficient.

$$J_i = \frac{P_i}{\delta}\left(x_i \gamma_i P_i^{vap} - P_{i,p}\right) \tag{9-3}$$

where $J_i$ is the flux of component $i$ through the membrane, $P_i$ is the permeability of the membrane, $\delta$ is the thickness of the membrane, $x_i$ is the feed mole fraction of component $i$, $\gamma_i$ is the activity of component $i$ in the feed, $P_i^{vap}$ is the pure component vapor pressure, and $P_{i,p}$ is the permeate partial pressure of component $i$.

Researchers have proposed a variety of different models (Mulder and Smolders, 1984; Dutta et al., 1996; Shieh and Huang, 1998; Molina et al., 1997) to predict the flux and separation behavior for pervaporation systems. However the model just described has the inherent advantage of simplicity. The concentration driving force is expressed in terms of the difference in the partial pressure of the component on the feed and the permeate side. The constant $P_i$ is known as the species permeability of the membrane and is a constant, which includes the diffusivity and the solubility of the species in the membrane. The constant $(P_i/\delta)$ is also called the normalized component flux because it is the ratio of the component flux to the vapor pressure driving force.

Two parameters are generally used to characterize a pervaporation process, namely, the flux of the permeating components and the separation factor of the membrane. The separation factor of the membrane can be expressed in terms of a separation factor $(\alpha_{ij})$,

$$\alpha_{ij} = \frac{y_i/y_j}{x_i/x_j} \tag{9-4}$$

where $x_i$ and $y_i$ are the concentration of species $i$ in the feed and permeate, respectively.

*Nanofiltration*   Nanofiltration membranes are more "open" than PV, RO and gas separation membranes. The pores of the NF membrane may be a few angstroms in size and are able to exclude molecules having molecular weights exceeding 300 (Williams et al., 1999; Rautenbach and Mellis, 1995). The water permeation mechanism follows the pore-flow model and hence the flux is highly dependent on the feed pressure. The water flux $J_w$ through the membrane is given by Eq. (9-1) in this case also. When the feed concentration of the rejected solute is higher, the osmotic pressure difference $(\Delta\pi)$ is higher and hence the water flux through the membrane will be lower. It should also be noted, that NF membranes are generally negatively charged and can be used to selectively remove low concentration organics while permeating salts, such as NaCl. Thus the osmotic pressure effect even from 5–10 percent NaCl solutions is minimal. The solute flux (Bhattacharyya and Williams, 1992) in NF membranes is as follows:

$$J_{sol} = J_w C_{sol(m)} + Z_{sol} C_{sol(m)} \frac{FE}{RT} - D_{sol(m)} \frac{d(\gamma_{sol(m)} C_{sol(m)})}{dz} \tag{9-5}$$

where $Z_{sol}$ is the charge on the solute, $C_{sol(m)}$ is the solute concentration in the membrane, $D_{sol(m)}$ is the diffusion coefficient of the solute in the membrane, and $E$ and $F$ are the Donnan potential and Faraday constants, respectively. The first term in the above Eq. (9-5) represents the solute flux through the membrane due to convection, the second term includes the contribution of the Donnan potential to the solute flux and the last term incorporates the diffusive solute flux.

It is evident from Eq. (9-5) that NF membranes are also able to reject salts and organic ions at low concentrations in the feed. The Donnan exclusion effect is responsible for this phenomenon. Most commercial NF membranes have negatively charged groups (carboxylic and sulfonic) present in the pores. The Donnan potential field is stronger in the presence of co-ions (negatively charged ion of the salt) with higher charge and hence the rejection of the salt is higher. At high salt concentrations in the feed, the Donnan potential field is no longer strong enough to resist the movement of the salt ions across the membrane and the rejection of the membrane decreases. It is also evident that the membrane rejection would be a strong function of the pH in the case of charged species. The rejection is expected to be lower below the pKa of the solute and higher above the pKa of the solute. In the case of charged solutes with multiple ionization states (such as oxalic acid), the rejection shows markedly different values at different pKa's of the system. The typical rejection characteristics of commercial NF membranes are shown in Figure 9.1.

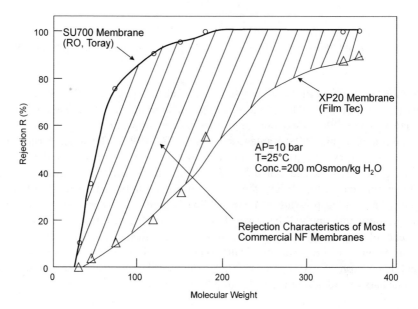

**Figure 9.1.** Rejection of non-ionized organic solutes with different molecular weights for commercial nanofiltration membranes (Rautenbach and Mellis, 1995).

### 9.1.1.4 Development of Membrane/Hybrid Processes for Treating a Stream

The process flowsheet can vary to a great extent depending on the separation goals. The determination of separation goals and the associated treatment strategy is usually a recursive process. The objective involves minimization of the capital/operating costs of the treatment strategy and meeting the discharge/recovery specifications at the same time. The optimal treatment strategy may be comprised of combinations of various unit operations and this is usually called a hybrid process. This section presents three mini-case studies in which membranes could be used effectively in a stand-alone or hybrid mode to tackle the separation aims.

*Mini-Case Study 9-1: Disposing of a Waste Stream High in Total Dissolved Solids and Solvent Concentrations*  Consider a dilute aqueous waste stream containing low concentrations (2–3 wt%) of solvents, high MW organic compounds and inorganic salts. The stream needs to be disposed but it cannot be directly discharged to the river/ocean due to its high TDS and solvent concentrations. Also, it may not be possible to treat this stream in a biological facility due to its high BOD/COD content. Reverse osmosis can prove to be an effective technique for the BOD/COD reduction of this stream.

Most commercial RO membranes, which have high ability to reject organic solutes in water, can be used for this application. The permeate stream from the RO unit would be a high quality water stream which could be further subjected to biotreatment or may be directly discharged in certain cases. The retentate stream, which would be enriched in the solvent/salts can either be incinerated completely or distilled to remove the solvents. The process schematic of the above treatment strategy is shown in Figure 9.2. Low pressure RO membranes/modules (Gerard et al., 1998; Okazaki et al., 1991) are available that exhibit a high flux at low pressures. These membranes can also be used for the treatment of the above stream due to lower membrane area requirement and higher recovery levels.

*Mini-Case Study 9-2: Recovering a Valuable Pharmaceutical Compound*
An aqueous process stream contains solvents in moderate concentrations (20–50 wt%), a high MW pharmaceutical compound in small quantities (1–2 wt%) and low concentrations (2–5 wt%) of a monovalent salt like NaCl. It is desired to recover as much of the high MW compound as possible because of its higher value and dispose the remaining stream. It is assumed that the compound has a limited solubility in the solvent mixture. To recover the pharmaceutical intermediately, one would need to concentrate it in the solvent mixture before crystallizing it out. One possible way of achieving this would be to distill the process stream. If the high MW compound is heat sensitive, however, the expensive vacuum distillation process would have to be used. In addition, the distillation bottoms would be an aqueous stream concentrated in both the salt and the high MW organic. Thus in future cycles of concentration and crystallization of the high MW compound the salt could act as a contaminant.

Nanofiltration membranes have a low rejection for monovalent salts and zero rejection of low MW organic solvents. Thus if the process stream is pumped at a

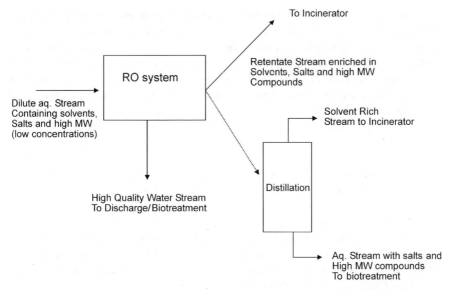

**Figure 9.2.** Process schematic involving RO for treatment of a waste system.

high pressure through a NF membrane unit, the membranes would permeate the solvents, partially rejecting NaCl and completely rejecting the high MW organics. As a result, the retentate stream would comprise primarily solvents concentrated in the high MW compound. The permeate stream would contain solvents and the inorganic salt. The retentate stream can be further sent to a crystallizer/evaporator for recovery of the valuable drug intermediate/product. The permeate stream may be further distilled to separate the solvents and water. The concentrated solvent stream can be incinerated while the aqueous stream containing salts may be biotreated. The flowsheet for this process is shown in Figure 9.3.

Solvent resistant NF membranes such as the MPF series marketing by Koch Membrane Systems, can be used for the transport of solvents. This is an important point because most of the commercially available NF membranes are designed to handle aqueous streams, but solvent-resistant RO/NF membranes are still in their infancy.

In the presence of solvents, NF membranes may show flux and rejection characteristics markedly different from those seen in the presence of water. Figure 9.4, comparing pure solvent fluxes through some commercial NF membranes clearly shows that the solvent flux through each membrane is strongly dependent on the solvent and the polymer interactions. Figure 9.4 also shows that Desal HL, a commercial hydrophilic solvent-resistant membrane exhibits a high methanol flux and a low hexane flux. This is an expected trend because methanol is a highly polar, water-like solvent whereas hexane is a non-polar solvent. The interactions of methanol with the membrane material are stronger than that of hexane and hence the high methanol flux. (Bhattacharyya et al., 1998a; Hestekin et al., 2001; Williams et al., 1999).

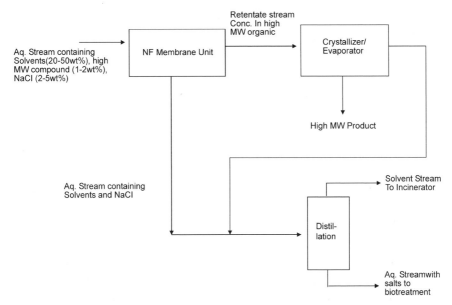

**Figure 9.3.** Flowsheet for recovery of valuable drug intermediate from aqueous waste stream.


*Mini-Case Study 9-3: Recovering a Solvent in Pure Form*   Consider a process stream primarily consisting of ethyl acetate, tetrahydrofuran or dimethyl formamide, or other solvent, containing small quantities of impurities (e.g., water, methanol and ethanol). It is desired to recover the solvent in a pure form for reuse. It often happens in the processing of such streams that the presence of water leads to azeotrope formation. In such cases, water must be removed from the stream before the solvents can be further separated by distillation. Pervaporation is an attractive technology for separation of azeotropic mixtures. Pervaporation membranes are available in

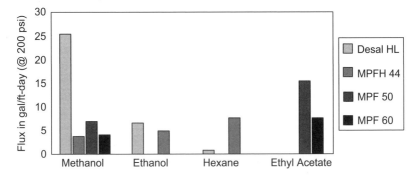

**Figure 9.4.** Organic solvent fluxes of four commercial various solvent-resistant NF membranes.

two forms: hydrophilic (based on polyvinyl acetates, PVA) and organophilic (based on polydimethylsiloxane, PDMS). As the names suggest, the hydrophilic membranes preferentially sorb and permeate aqueous or waterlike species whereas the organophilic membranes have an affinity for the organic/nonpolar compounds. The hydrophilic pervaporation membranes can be used for removal of small quantities of water from solvent streams.

The authors have studied the dehydration of a complex solvent stream (containing 74.6 wt% ethyl acetate, 14.5 wt% methyl acetate, 6.3 wt% ethanol, 0.5 wt% methanol, 0.5 wt% isopropyl alcohol, and 3.6 wt% water) by means of hydrophitic zeolite membranes based on sodium acetate (NaA), as indicated in Figure 9.5, which shows the variation of the total flux of the system with the water concentration of the complex solvent stream at 60°C. The water flux dropped sharply below 3 wt% water in the feed. It was possible to dehydrate the stream up to 0.4 wt% water. The quality of water obtained in the permeate varied from about 99.9 wt% water at 3.6 wt% feed concentration to 88.9 wt% water at 0.4 wt% water in the feed. Thus very high water selectivities could be obtained using the zeolite NaA membrane at a significantly high flux. The solvent stream, depleted of water, can be further distilled in a distillation column to recover the ethyl acetate. Although zeolite membranes have a higher flux than polymeric membranes under similar conditions, they are currently more expensive than the inorganic products. For corrosive solvents like DMF and THF, one would have to use inorganic membranes.

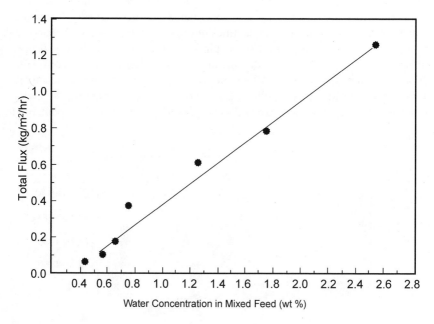

**Figure 9.5.** Variation of total flux through zeolite NaA membrane with the feed water concentration in the complex solvent stream at 60°C.

### 9.1.1.5 Selected Membrane Applications in the Pharmaceutical Industry

*Extractive Membrane Bioreactor*    The technology for the extractive membrane bioreactor (EMB) was developed by Livingston and co-workers (Livingston, 1994; Livingston et al., 1998; Ferreira et al., 2003) at Imperial College, in London. The process is mainly used to treat the organic contaminants, mainly toxic organic contaminants, low in water solubility, that are present in pharmaceutical industry wastewaters. The process combines extraction and biodegradation into one step. Figure 9.6 shows the principle of operation of an extractive membrane bioreactor. The unit employs dense organophilic membranes to separate the waste stream and the biomedium containing the bioorganisms. The membrane basically serves two purposes: it separates the waste stream from the biological medium, and it rejects salts, inorganic catalysts, and so on present in the waste stream while selectively sorbing the organic contaminants at the same time. Most of the biodegradation of the organic contaminant takes place in the biofilm that grows on the surface of the membrane. The removal efficiency of an EMB depends on the mass transfer of the organics through the membrane (governed by sorption and diffusion properties) and the degradation properties of the microbial culture.

The extractive membrane bioreactor technology shown in Figure 9.7 was implemented by SmithKline Beecham Pharmaceuticals on laboratory scale, to

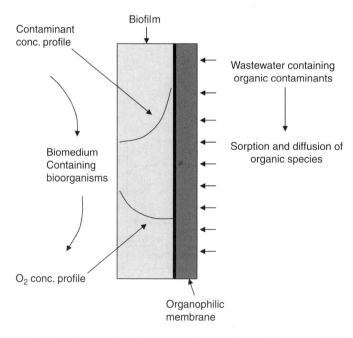

**Figure 9.6.** Principle of operation of an extractive membrane bioreactor. (Adapted from Livingston et al., 1998 and Ferreira et al., 2003.)

remove dichloromethane (DCM) from a wastewater stream (Freitas dos Santos et al., 1999). The lab scale trials showed that DCM, a toxic volatile organic compound, could be removed from the waste stream with a 90–100 percent removal efficiency over an extended period of 400 h. A comparison between the EMB and alternative treatment technologies such as carbon adsorption, steam stripping and pervaporation also showed that the EMB technology was better in terms of both capital and operating costs.

*Volume Minimization of Pharmaceutical Wastes by Pervaporation*   As mentioned earlier, it is common practice to incinerate pharmaceutical wastes containing solvents. When the major component of the waste stream is water, a major portion of the incineration cost is used in turning the water into steam. Shah et al. (1999) have suggested the use of hydrophilic pervaporation membranes for removal of water from such streams, thus minimizing the volume of the stream to be incinerated. The retentate emerging from the pervaporation system would be enriched in organics and solvents and hence would have a higher fuel value, which in turn would mean lower cost when the retentate is incinerated. The permeate from the pervaporation unit would be primarily a water stream, low in concentrations of solvents and organics. This stream can be further sent to a biotreatment facility before discharge.

The composition of an actual aqueous pharmaceutical waste containing moderate quantities of solvents and small quantities of high MW organics and inorganic salts is shown in Table 9.2. Shah and coworkers (2000) tested this stream against hydrophilic polymeric PVA membranes. A total flux of 0.9 kg/m$^2$/h was obtained for the waste at 60°C and the TOC of the permeate was eight times lower than that of the feed. A preliminary cost estimate showed that the total disposal cost of the waste

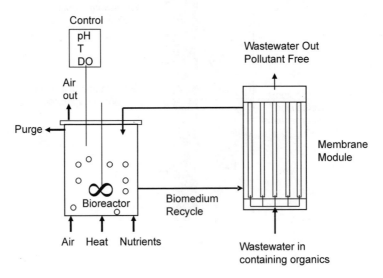

**Figure 9.7.** Experimental layout of an extractive membrane bioreactor (Freitas dos Santos and Lo Biundo, 1999).

TABLE 9.2. **Pharmaceutical Aqueous Waste Stream Composition**

| Compound | Approximate Composition |
|---|---|
| Water | 64 wt% |
| Ethanol | 24 wt% |
| Methanol | 2 wt% |
| Sodium chloride | 4 wt% |
| Other organic impurities (toluene, HCl, product) | 6 wt% |
| Total organic carbons | $\sim$145,000 mg/L |
| Chemical oxygen demand | $\sim$700,000 mg/L |

Adapted from Shah et al. (1999).

stream would be reduced by about 50 percent if the volume minimization before incineration, is carried out. Thus pervaporation seems to be an attractive technology for treatment of aqueous wastes.

*Membrane Technology for Pharmaceutical Product Recovery* Solvent extraction is commonly used in the pharmaceutical industry for purification and recovery of intermediates/products. Membrane based contactors can be used for the same purpose mainly because they provide a very high surface area to volume ratio and eliminates solvent contamination of aqueous streams. Other advantages (Klassen et al., 2003; Prasad and Sirkar, 1989) provided by membrane contactors are as follows: no flooding limitations, easy handling of particulates and systems susceptible to emulsification, and no need for density differences between phases.

Basu and Sirkar (1992) have used Celgard microporous hydrophobic hollow fiber membrane contractors to study the extraction of the pharmaceutical drug intermediate mevinolinic acid (MK-819). MK-819 is made by a fermentation process and hence there are various impurities present in the fermentation broth. MK-819 along with a few impurities is initially directly extracted into isopropyl acetate (IPAc). Further purification of MK-819 is done by first extracting MK-819 from IPAc solution into water at a pH exceeding 5.8 and then back extracting the drug intermediate into IPAc at low pH. A schematic of this process is shown in Figure 9.8. Prasad and Sirkar (1989) have suggested the use of membrane contactors for this separation as opposed to the conventional mixer/settler arrangement. Experiments showed that greater than 90 percent product recovery could be attained in both the steps using hollow fiber modules. The overall recovery of the product can be further optimized by appropriately arranging the modules in series or parallel fashion.

### 9.1.1.6 *Other RO/NF and Functionalized Membrane Applications for Heavy Metal Recovery and Water Recycling* Process streams containing heavy metals can come from plating operations, nonferrous metal production and

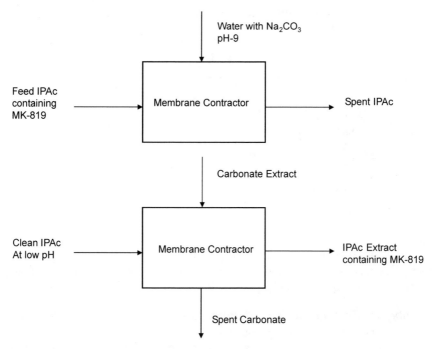

**Figure 9.8.** Process schematic for product recovery using membrane contractors. (Adapted from Basu and Sirkar, 1992.)

the manufacture of electronic equipment. For metal recovery and recycle of water (involving zero discharge requirements) ion exchange and membrane processes play a dominant role in various industries (Bhattacharyya and Hestekin, 1999). Pressure-driven membrane systems can be all the types described thus far: MF (after precipitation), RO for producing high quality permeate water and concentrated metals for recycle, NF for selective heavy metals separation from high NaCl or for plating wastes containing metal chelates, surfactant/polymer chelate enhanced UF, and functionalized ion exchange membranes for metal capture. Figure 9.9 is a schematic diagram for various membrane processes.

RO is the most well-established membrane technology for metal recovery and water recovery. For example, RO can provide greater than 99 percent rejections of various heavy metals ($Cd^{2+}$, $Ni^{2+}$, $Cu^{2+}$, etc.) with cellulose acetate and aromatic polyamide membranes. In some cases two-stage RO process has been used for obtaining high quality permeate water and recover metal in concentrated form. This includes a full scale plant for nickel acetate recovery and water recycle (Bhattacharyya and Hestekin, 1999). Bhattacharyya et al. (1998a) have provided a review of RO and applications. Wong et al. (2002) have successfully developed a hybrid process (pilot plant) for recycling spent final rinsing water from an electroless plating operation (in Singapore). The process consisted of a combination of MF, carbon adsorption, NF, and ion exchange. The plan operated at 90 percent water recovery and

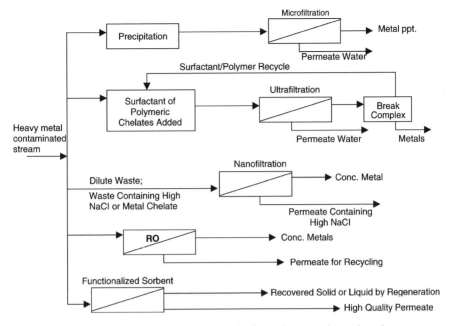

**Figure 9.9.** Metal recovery and water recycle for various membrane-based systems.

produced very high quality water ($<5$ $\mu$S/cm) for direct recycle to the plant. The design data for a 25 m$^3$/h plant showed a payback period of about 15 months.

Although RO/NF is well established for the removal of metal ions, appropriate functionalization (with polymeric chelate/ion exchange groups) of MF membranes also provides a technique for capturing metals at very high capacity with very low pressure ($>3$ bar) operation. Poly-aminoacid functionalized membrane sorbents (Bhattacharyya et al., 1998b; Ritchie and Bhattacharyya, 2000) offer a wide array of advantages for recovering heavy metal ions over conventional ion exchange and other high pressure membrane processes. The use of polymeric ligands with multiple binding sites leads to high metal sorption capacity with macroporous membranes. The interactions between these groups and the heavy metal ions are very rapid, since sorption proceeds under convective flow and transport resistance is minimized. Figure 9.10 is a schematic representation of a MF membrane pore containing covalently attached polymeric ligands with COOH groups. A polyglutamic acid is often used as a membrane sorbents as are polyarginine (amine group for anion capture) and polycysteine (thiol group for selective Hg capture). For heavy metal capture with applications in plating, polyligands with COOH groups will be most attractive. We have used inexpensive, commercially available cellulosic and silica-based membrane materials that can easily be functionalized with polymeric metal capturing ligands. In our studies involving various metals, such as $Cu^{2+}$, $Cd^{2+}$, and $Pb^{2+}$, metal sorption capacities have ranged between 0.5–1.9 g metal/g membrane.

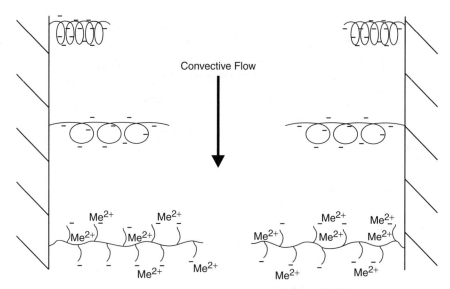

**Figure 9.10.** Schematic of metal sorption mechanisms for polyamino acid funtionalized MF membrane sorbents.

Functionalized MF membranes can also be used in hybrid operation with RO to further concentrate the retentate stream for possible direct recycle of recovered metals to manufacturing operations. Figure 9.11 shows such a scheme for recovering high quality permeate water and metal recycle. Researchers in Japan have also reported the incorporation of iminodiacetate and diethylamine functionality in membrane supports for metal capture (Sunaga et al., 1999; Li et al., 1994). Other methods include the use of immobilized multifunctional dyes in membranes for heavy metal capture (Denizli et al., 1998).

### 9.1.1.7 *Design and Economic Considerations*    The use of membrane processes such as, RO and NF for waste purification and material reuse has been demonstrated for applications ranging from desalting to high value material recovery. Overall design considerations include membrane selection, module type selection, and pretreatment requirements to minimize fouling. Brackish water type RO membranes are commonly used for feedstreams of low to moderate osmotic pressures (salt $<5000$ mg/L) and when high separation of low MW organics and salt is a key requirement. On the other hand, if one is interested in purification of water from low ppm to ppb range ultra low pressure RO (Ozaki and Li, 2002) membranes would be a suitable choice. Cross-linked aromatic polyamide membranes are the predominant RO membranes for aqueous systems, whereas nonpolar solvent transport (e.g., hexane) through RO/NF type membranes will require cross-linked hydrophobic (e.g., siloxane, polyimide) materials (Bhanushali et al., 2002; Gould et al., 2001; Machado et al., 1999).

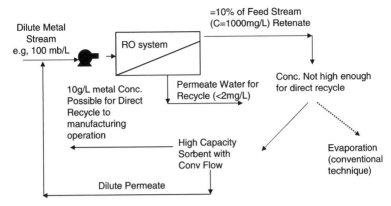

**Figure 9.11.** Hybrid RO/functionalized membrane system for water recovery and metal recycle.

Figure 9.12 shows (Bhanushali et al., 2002) solvent permeability (normalized with respect to ethanol) data as reported for Koch MPF 50 (siloxane-based) and our work with a recently developed silicone-based membrane D (Osmonics Membrane D), from General Electric. For example, membranes of these types would be very useful in the food industry for separating edible oils from hexanes, leading to considerable energy savings and direct recycling of solvent. As expected, with hydrophobic membranes the water flux is very much less than the pentane flux. With traditional RO/NF polyamide membranes, the flux trend will be exactly opposite. Hydrophilic pervaporation membranes (such as PVA, zeolite NaA) on the other hand are excellent for polar solvent recycle from highly concentrated solution (Shah et al., 2000). In addition, the commercial availability of various acid tolerant membranes, thin-film composite membranes, and design of modules for highly corrosive systems has provided many applications (besides sea water and brackish water treatment) in industries. For example, a low pressure RO membrane (Osmonics Membrane AK) is highly acid resistant and is stable to 20 percent sulfuric acid at 50°C.

The design of membrane module systems requires selection of the membrane material, module geometry, product flow rate and concentration, solvent (water) recovery, operating pressure, pretreatment requirements, and minimum tolerable flux. The effects of the operating variables must be established by laboratory or pilot experiments. Pretreatment may involve removal of particulates by MF and UF systems prior to RO/NF. Fouling is not a problem for PV operation, but the pressure drop in the permeate side must be low. Most commercial modules (8 in by 40 in) are spiral wound, with typical membrane area of about 315 ft$^2$. Users can shop for modules and build their own system, or they can buy complete custom-designed systems containing membrane modules (membrane elements in pressure vessels), all piping, pumps, basic control system, and so on.

Membrane systems operate around the world today with daily capacities ranging from few thousands liters to more than 500 million liters. The integration of RO/NF

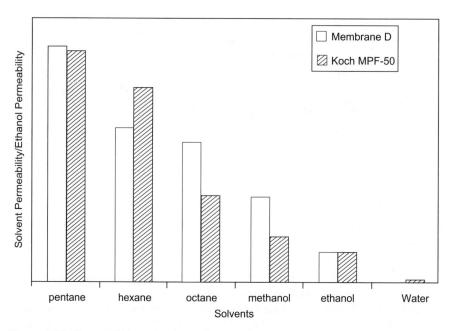

**Figure 9.12.** Permeabilities of polar and non-polar solvents through hydrophobic Koch MPF-50 membrane and hydrophobic silicone-based (Osmonics) membrane D.

type membrane processes and other retentate concentration methods can provide significant minimization of waste by simultaneous material recycling and water reuse. The journal *Environmental Progress* (2001) devoted an entire issue to pilot and full-scale applications involving both aqueous and non-aqueous systems. For membrane design considerations and cost of various operations (RO, UF, NF, MF, PV, etc.), *Comparative Cost of UF vs. Conventional Pretreatment for SWRO Systems* (Glueckstern and Priel, 2003), *New Insights into Membrane Science and Technology: Polymeric, Inorganic and Biofunctional Membranes* (Bhattacharyya and Butterfield, 2003), and *The Membrane Handbook* (Ho and Sirkar, 1992) contain extensive information.

Reverse osmosis is a fairly mature technology in the area of seawater and brackish water. In these applications, extensive water pretreatment (coagulation-flocculation, settling, microfiltration are standards) is included to provide reliable long term operation with membrane life (in terms of salt rejection and flux) ranging from 5–8 years. The current trend for RO pretreatment is the use of UF (Glueckstern and Priel 2003). Nanofiltration applications are quite extensive for partial softening of water, organic removal from large water systems, dye separations from textile industries, etc. For all membrane systems involving industrial wastes fouling minimization and the pretreatment needs must be established with lab scale membranes and modules.

For pressure-driven membrane processes such as RO/NF capital costs include pretreatment equipment, membrane modules + pumps, and utilities, whereas for

PV operations permeate side condensation unit (such as, a chiller) and vacuum pump is also required. RO/NF equipment costs fall in the range of $400 to $800/m$^3$-day capacity. For conventional RO/NF systems, typical module cost is about $800 (350 ft$^2$ membrane area). With industrial process streams involving very low pH operation the capital cost may be 20–30 percent higher because of enhanced module and pump contact material requirement. The dollar cost of PV modules (per unit area) is considerably higher than for RO. Operating costs include membrane replacement (thus membrane life), energy, chemicals, labor, capital investment cost. These results in overall production costs ranging from $0.50–$2.00/m$^3$. For membrane-based metal removal processes the reported operating costs for several applications (Bhattacharyya and Hestekin, 1999) range between $0.25–$0.88/m$^3$. For applications other than conventional desalination, economic calculations must also include post processing of retentate, and the benefits of permeate reuse and water recovery. In textile and paper industries, membranes can play an important role in water reuse. For example, Ciardelli et al. (2000) reported the economic feasibility of treating (by UF/RO) dyehouse effluents for reuse and their results indicated operating (with 3-year membrane life) and investments costs of $1.10/m$^3$ for a 1000 m$^3$/day plant.

In the area of solvent resistant membranes, ExxonMobil and W. R. Grace developed a full-scale membrane process (with reduced cooling water reuse, VOC emissions from dewaxing operation) for obtaining higher yields of lube oil at lower energy consumption (Gould et al., 2001). The polyimide –based membrane modules were operated at $-10°C$ in the presence of corrosive solvent blends, such as MIBK and toluene. The plant, which had a capital cost of $5.5 million and a throughput of 36,000 barrels/day, reportedly realized a payback period for the membrane system of less than a year. Solvent-resistant membranes are also very valuable for pharmaceutical product recovery and for the recycle of homogeneous, organometallic catalysts based on rhodium or palladium (Scarpello et al., 2002). Thus membranes and membrane processes offer an excellent avenue in downstream processing and material recycling with their versatility and low energy requirements.

## ACKNOWLEDGMENT

The authors recognize the partial support of Glaxo SmithKline and NIST-ATP for some of the work presented here. The authors like to thank Drs. Ritchie, Hestekin for their research contribution with the functionalized membrane work, and Dr. Bhanushali for the NF solvent resistant membrane results. We would also like to extend our sincere appreciation to Dr. Steve Kloos, GE-Osmonics Corporation.

## REFERENCES

Basu, R., and Sirkar, K. K. (1992). Pharmaceutical product recovery using a hollow fiber contained liquid membrane: a case study. *J. Membrane Science*, **75**, 131–149.

Bhanushali, D., Bhattacharyya, D., and Kloos, S. D. (2002). Solute transport in solvent-resistant nanofiltration membranes for non-aqueous systems: experimental results and the role of solute-solvent coupling. *J. Membrane Science*, **208**(1), 343–359.

Bhattacharyya, D., and Butterfield, A. (Editors) (2003). *New Insights into Membrane Science and Technology: Polymeric, Inorganic and Biofunctional Membranes*. Elsevier Science, New York.

Bhattacharyya, D., and Hestekin, J. (1999). Metal ion recovery from aqueous waste streams by membrane technologies, In Emerging Separation and Separative Reaction Technologies for Process Waste Reduction: Adsorption and Membrane Systems (edited by Radecki, Crittenden, Shonnard, and Bulloch), pp. 199–214, CWRT, AIChE, New York.

Bhattacharyya, D., and Williams M. E. (1992). Theory of Reverse Osmosis, In Membrane Handbook, Ho, W. S. and Sirkar, K. K., (eds), Van Nostrand Reinhold, NY, 275.

Bhattacharyya, D., Hestekin, J. A., Brushaber, P., Cullen, L., Bachas, L. G., and Sikdar, S. K. (1998a). Novel poly-glutamic acid functionalized microfiltration membranes for sorption of heavy metals at high capacity. *J. Membrane Science*, **141**(1), 121–135.

Bhattacharyya, D., Mangum, W. C., and Williams, M. E. (1998b). Reverse Osmosis, In Encyclopedia of Environmental Analysis and Remediation, Wiley, NY, pp. 4149–4166.

Ciardelli, G., Corsi, L., and Marcucci, M. (2000). Membrane separation for wastewater reuse in textile industry. *Resources, Conservation, and Recycling*, **31**, 189–197.

Denizli, A., Kesenci, K., Arica, M., Salih, B., Hasirci, V., and Piskin, E. (1998). Novel dye-attached macroporous films for cadmium, zinc and lead sorption: alkali blue 6B-attached macroporous poly(2-hydroxyethyl methacrylate). *Talanta*, **46**(4), 551–558.

Dutta, B. K., Ji,W., and Sikdar, S. K. (1996). Pervaporation: principles and applications. *Separation and Purification Methods*, **25**(2), 131.

*Environmental Progress*, (2001). Special issue on Membranes and Environmental Applications (Guest Editor: D. Bhattacharyya) Vol. 20, no. 1.

Ferreira, F. C., Livingston, A. G., Han, S., Boam, A., and Zhang, S. (2003). Membrane aromatic recovery system: a new process for recovering phenols and aromatic amines from aqueous streams, In *New Insights into Membrane Science and Technology: Polymeric, Inorganic and Biofunctional Membranes*, Elsevier Science, New York, pp. 165–181.

Freitas dos Santos, L. M., and Lo Biundo, G. (1999). Treatment of pharmaceutical industry process wastewater using the extractive membrane bioreactor. *Environmental Progress*, **18**(1), 34–39.

Gerard, R., Hachisuka, H., and Hirose, M. (1998). New membrane developments expanding the horizon for the application of reverse osmosis technology. *Desalination*, **119** (1–3), 47–55.

Glueckstern, P., and Priel, M. (2003). Comparative cost of UF vs conventional pretreatment for SWRO systems. *Desalination and Water Reuse*, **13**, 34–39.

Gould, R. M., White, S. L., and Wildemuth, C. R. (2001). Membrane separation in solvent lube dewaxing. *Environmental Progress*, **20**, 12–16.

Hestekin, J., Sikdar, S., Kim, B. M., and Bhattacharyya, D. (1998). Membranes for Treatment of Hazardous Wastes, In *Encyclopedia of Environmental Analysis and Remediation*. Wiley, New York, pp. 2684–2708.

Hestekin, J. A., Smothers, C. N., and Bhattacharyya, D. (2001). Nanofiltration of Charged Organic Molecules in Aqueous and Non-Aqueous Solvents: Separation Results and Mechanisms. In *Membrane Technology in the Chemical Industry* (Nunes and Peinemann, Editors), pp. 173–190, Wiley-VCH.

Ho, W. S., and Sirkar, K. K. (eds) (1992). *Membrane Handbook.* Chapman Hall, New York.

Klassen, R., Feron, P. H., van der Vaart, R., and Jansen, A. E. (2003). Industrial Applications and Opportunities for Membrane Contactors, In New Insights into Membrane Science and Technology: Polymeric, Inorganic and Biofunctional Membranes (Bhattacharyya and Butterfield, Editors), pp. 125–145, Elsevier Science, New York.

Koros, W. J., and Mahajan, R. (2000). Pushing the limits on possibilities for large scale gas separation: which strategies? *Journal of Membrane Science*, **175**, 181–196.

Li, G., Konishi, S., Saito, K., and Sugo, T. (1994). High collection rate of Pd in hydrochloric acid medium using chelating microporous membrane. *Journal of Membrane Science*, **95**, 63.

Livingston, A. G. (1994). A new process technology for detoxifying chemical industry wastewaters. *J. Chem. Tech. Biotech.*, **60**, 117–124.

Livingston, A., Arcangeli J-P., Boam, A. T., Zhang, S. F., Marangon, M., and Freitas dos Santos, L. M. (1998). Extractive membrane bioreactor for detoxification of chemical industry wastes: process development. *J. Membrane Sci.*, **151**, 29–44.

Machado, D. R., Hasson, D., and Semiat, R. (1999). Effect of solvent properties on permeate flow through nanofiltration membranes. Part 1: Investigation of parameters affecting solvent flux. *J. Membrane Science*, **163**, 93–102.

Mehdizadeh, H., and Dickson, J. M. (1991). Evaluation of surface force-pore flow and modified surface force- pore flow models for reverse osmosis transport. *Chemical Engineering Communications*, **103**, 65–82.

Molina, C., et al. (1997). Model for pervaporation: application to ethanolic solutions to aroma. *J. Membrane Sci.*, **132**, 119–129.

Mulder, M. H., and Smolders, C. A. (1984). On the mechanism of ethanol/water mixtures by pervaporation I. Calculations of concentration profiles. *J. Membrane Sci.*, **17**, 289.

Okazaki, M., Ikeda, T., Nakagawa, Y., and Bairinji, R. (1991). Recent progress of reverse osmosis membrane modules for ultrapure water production. Conference Proceedings-Annual Semiconductor Pure Water Conference Feb. 26–28, Published by Balasz Analytical Laboratory, 87–95.

Ozaki, H., and Li, H. (2002). Rejection of organics compounds and by ultra-low pressure reverse osmosis membranes. *Water Research*, **36**, 123–130.

Prasad, R., and Sirkar, K. K. (1989). Hollow fiber solvent extraction of pharmaceutical products: a case study. *J. Membrane Science*, **47**, 235–259.

Rautenbach, R., and Mellis, R. (1995). Hybrid processes involving membranes for the treatment of highly organic/inorganic contaminated waste water. *Desalination*, **101**, 105.

Ritchie, S. M., and Bhattacharyya, D. (2000). Polymeric ligand-based functionalized materials and membranes for ion exchange, In *Ion Exchange and Solvent Extraction Series* (ed. A. K. SenGupta), Vol. 14, pp. 81–118, Marcel Dekker.

Ritchie, S. M., Kissick, K. E., Bachas, L. G., Sikdar, S. K., and Bhattacharyya, D. (2001). Polycysteine and other polyamino acid functionalized microfiltration membranes for heavy metal capture. *Env. Sci. and Technology*, **35**, 3252–3258.

Scarpello, J. T., Nair, D, Freitas dos Santos, L., White, L. S., and Livingston, A. G. (2002). The separation of homogeneous organometalic catalysts using solvent-resistant nanofiltration membranes. *J. Membrane Science*, **203**, 71–85.

Shah, D., Ghorpade, A., Hannah, R., Kissick, K., and Bhattacharyya, D. (2000). Pervaporation of alcohol-water and dimethylformamide-water mixtures using hydrophilic zeolite NaA membranes: mechanisms and experimental results. *J. Membrane Science*, **179**, 185–205.

Shah, D., Bhattacharyya, D., Mangum, W., and Ghorpade, A. (1999). Pervaporation of pharmaceutical waste streams and synthetic mixtures using water selective membranes. *Environmental Progress*, **18**(1), 21–29.

Shieh, J. J., and Huang, R. Y. (1998). A pseudophase-change solution-diffusion model for pervaporation. 1. single component permeation. *Sep. Sci. Technol.*, **33**(6), 767.

Sunaga, K., Kim, M., Saito, K., Sugita, K., and Sugo, T. (1999). Characteristics of porous anion-exchange membranes prepared by cografting of glycidyl methacrylate with divinylbenzene. *Chemistry of Materials*, **11**, 1986.

USEPA (1991). *Guides to Pollution Prevention: the Pharmaceutical Industry*, Risk Reduction Engineering Laboratory and Center for Environmental and Research Information. Office of Research and Development (Cincinnati, Ohio), US EPA Report No.1.8:P76/11, pp. 9.

Wijmans, J. G., and Baker, R. W. (1993). A simple predictive treatment of the permeation process in pervaporation. *J. Membrane Sci.*, **79**, 101.

Williams, M. E., Hestekin, J. A., Smothers, C. N., Bhattacharyya, D. (1999). Separation of organic pollutants by reverse osmosis and nanofiltration membranes: mathematical models and experimental verification. *Industrial and Engineering Chemistry Research*, **38**(10), 3683–3695.

Wong, F. S., Qin, J. J., Wai, M. N., Lim, A. L., and Adiga, M. (2002). A pilot study on a membrane process for the treatment and recycling of spent final rinse water from electroless plating. *Separation and Purification Technology*, **29**, 41–51.

Zydney, A. L. (1998). Protein separations using membrane filtration: new opportunities for whey fractionation. *International Dairy Journal*, **8**(3), 243–250.

### 9.1.2 Zero Discharge Technologies Based on Supercritical $CO_2$ and $H_2O$

RAM B. GUPTA

Dept. of Chemical Engineering, Auburn University, Auburn, AL 36849-5127, USA

A fluid is in the supercritical state when it is above its critical temperature and pressure; in a phase diagram, the supercritical region is unique and distinct from the liquid or vapor phase (Figure 9.13). When a fluid changes from liquid to vapor phase, it undergoes a phase transition. But, when fluid is taken from liquid to gas through the supercritical region, no phase transition occurs.

Supercritical fluids have unique properties that can be beneficial in industrial processes. They have gas-like viscosities and diffusivities and liquid-like densities. Solubility in a supercritical fluid is a direct function of the fluid density, which is

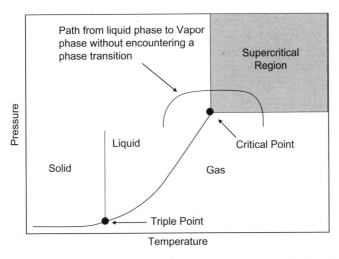

**Figure 9.13.** Pressure–temperature diagram showing supercritical region.

highly variable near the critical point and can be easily manipulated by a small change in the pressure or temperature. Table 9-3 shows critical points of various fluids.

Both carbon dioxide and water, the most environmentally benign fluids listed in Table 9.3, are abundant in nature and highly biocompatible. A technology can move toward zero discharge easily by substituting these two fluids of the organic solvents conventionally used in chemical processing. Obviously, when processing makes use of existing $CO_2$, no new $CO_2$ is added to the environment; hence there is no added contribution to the greenhouse effect.

**TABLE 9.3. Critical Points of Various Fluids**

| Fluid | Critical Temperature (°C) | Critical Pressure (bar) |
|---|---|---|
| Carbon dioxide | 31.1 | 73.8 |
| Ethane | 32.2 | 48.4 |
| Ethylene | 9.3 | 50.4 |
| propane | 96.7 | 42.5 |
| Propylene | 91.9 | 46.2 |
| Cyclohexane | 280.3 | 40.7 |
| Isopropanol | 235.2 | 47.6 |
| Benzene | 289.0 | 48.9 |
| Toluene | 318.6 | 41.1 |
| p-Xylene | 343.1 | 35.2 |
| Chlorotrifluoromethane | 28.9 | 39.2 |
| Trichlorofluoromethane | 198.1 | 44.1 |
| Ammonia | 132.5 | 112.8 |
| Water | 374.2 | 220.5 |

Unsurprisingly, interest in supercritical fluids is strong, and the number of scientific publications on the subject has increased exponentially in recent years. Cumulative publications listed in *Chemical Abstracts* for the period 1980–2000 exceeded 9,000 for $CO_2$ and 3,000 for $H_2O$.

### 9.1.2.1 *Properties of Supercritical Water*

Around the critical point, the density of water changes rapidly with both temperature and pressure and it is intermediate between the densities of liquid and gaseous water. This results in unique solvation properties for supercritical water. At the same time, supercritical water possesses high diffusivity and low viscosity (Table 9.4).

The dielectric constant of water is much lower than that of ambient water. Thus, water that has reached its supercritical state is no longer like a polar solvent but behaves like a nonpolar organic solvent. For example, at 25°C, benzene is sparingly soluble in water (0.07 wt%), but when the temperature and pressure are raised to 300°C and 250 bar, respectively, solubility can go as high as 35 wt%. Supercritical water also shows complete miscibility with gases such as nitrogen, oxygen, air, hydrogen, carbon dioxide and methane. In direct contrast, inorganic salts have very low solubility in supercritical water. For example, NaCl solubility is about 37 wt% at 300°C and about 120 ppm at 550°C and 250 bar. $CaCl_2$ solubility drops from 70 wt% at subcritical conditions to as low as 3 ppm at 500°C and 250 bar. Insolubility of inorganic salts is consistent with the low dielectric and ionic dissociation constants of supercritical water. In addition, near the critical point, solubilities are highly pressure dependent and strongly correlated with the density of supercritical fluid. A very useful characteristic of supercritical water is the continuous tunability of its properties from gas-like to liquid-like conditions, which in turn means that the nature of a solvent can be varied.

### 9.1.2.2 *Technologies for Zero Discharge Based on Supercritical Water*

The unique properties of supercritical water allow it to be substituted for organic solvents in chemical syntheses or in the complete oxidation of organic

**TABLE 9.4. Properties of Supercritical Water as Compared to Liquid and Gas Waters (Broll et al., 1999)**

| | Liquid Water | Supercritical Water | Gas Water |
|---|---|---|---|
| Temperature, °C | 25 | 400 | 400 |
| Pressure, MPa | 0.1 | 25 | 0.1 |
| Density, g cm$^{-3}$ | 0.997 | 0.17 | 0.0003 |
| Dielectric constant | 78.5 | 5.9 | 1 |
| Dissociation constant | $10^{-14}$ | $10^{-5.9}$ | — |
| Heat capacity, J g$^{-1}$K$^{-1}$ | 4.22 | 13 | 2.1 |
| Viscosity, cP | 0.89 | 0.03 | 0.02 |
| Thermal conductivity, W m$^{-1}$ K$^{-1}$ | 0.608 | 0.160 | 0.055 |

wastes. Thus it is an excellent medium for achieving zero discharge conditions in a plant.

*Supercritical Water Oxidation (SCWO)* Supercritical water oxidation is an emerging technology to achieve zero discharge goals of the industries. In this process (SCWO) (Figure 9.14), waste effluent is pumped into the reactor along with an oxidant such as air, oxygen or hydrogen peroxide. The inlet streams are pre-heated with heat from the exit stream for a higher thermal efficiency. For waste that contains more than 5 percent organics, external heating or fuel is seldom needed, except during start-up. In most cases, high destruction efficiencies are obtained within a minute of the reaction time, as shown in Table 9.5.

Most operational SCWOs in the United States are at military installations. Since 1994, however, Huntsman Corporation has been running an SCWO plant in Austin, Texas, designed by Eco Waste Technologies and supported by comprehensive work of Gloyna and Li (1993). General specifications of the process are (Schmieder and Abeln, 1999):

| | |
|---|---|
| Reactor length | 200 m |
| Temperature | 540–600°C |
| Pressure | 250–280 bar |
| Capacity | 1100 kg/h |
| Feed content | Mainly alcohols and amines, 50 g/L total organic carbon (TOC) |
| Oxidation efficiency | 99.99 |

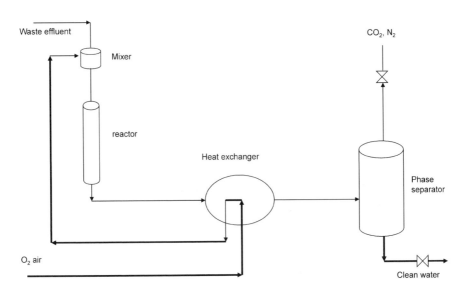

**Figure 9.14.** Schematic of supercritical water oxidation process (Schmieder and Abeln, 1999.)

**TABLE 9.5. Reported Destruction Efficiency for Wastes in Supercritical Water (Schmieder and Abeln, 1999)**

| Compound | Destruction Efficiency (%) | Waste | Destruction Efficiency (%) |
|---|---|---|---|
| Acetic acid | 99.9 | Activated sludge | 50–99.8 |
| Ammonia | 20–99.99 | Brewery effluents | > 99.99 |
| Aniline | 10 | Electronic scrub | >99.9 |
| Cyanide | 86 | Hanford waste | >99.9 |
| 3-Chlorobiphenyl | 99.99 | Sewage waste | 85–99.4 |
| 2-Chlorophenol | 99 | Municipal sludge | >99.99 |
| 4-Chlorophenol | 99.99 | Navy hazard. waste | 95–99.92 |
| 2,4-Dichlorobenzene | 99.7 | Paper mill effluent | 81–99.98 |
| Dichloromethane | 97–100 | Percolate | 96–99.8 |
| 2,4-Dinitrotoluene | 83–99 | Pharmaceutical | 99.997 |
| Dioxines | 99.99999 | Chemical | 85–99.99 |
| Ethanol | 99–99.997 | Polymer | 99.95 |
| Formic acid | 93 | Rocky flats | >99.9 |
| Hexachlorocyclohexane | 99 | | |
| Hydrogen | 23–99.9 | | |
| Phenol | 85–99.993 | | |
| Polychlorinated biphenyl | 99–99.999 | | |
| Pentachlorophenol | 99.99 | | |
| Pyridine | 4–99.99 | | |
| Polyvinyl chloride | 100 | | |

| | |
|---|---|
| Liquid effluent | 10 mg/L amine |
| | 3 mg/L TOC |
| | 6.4 mg/L nitrate |
| | 2.7 mg/L ammonia |
| Gas effluent | 0.6 ppm oxides of nitrogen |
| | 60 ppm carbon monoxide |
| | 200 ppm methane |
| | 0.12 ppm sulfur dioxide |
| Cost | $250/ton of raw waste, or $1,600/ton carbon |
| Limitations | Plant not suitable for wastes high in chlorine or salts |

The plant is being used for the oxidation of waste that does not contain salts or chlorinated compounds, which give rise to severe corrosion.

*Corrosion Challenge* Heteroatoms such as chlorine, fluorine, sulfur, and phosphorus can be highly corrosive when allowed to contact reactor walls, especially those made of 316 stainless steel, and much research has been devoted to using high-nickel alloys such as Inconel for lining the reactor walls. In addition to standing up to the oxidative environment, the reactor lining material must be resistant to temperatures up to approximately 500°C and pressures as high as 50 MPa.

Unfortunately, none of the metallic materials commercially available at present fulfills all these requirements (Kritzer et al., 1998).

Engineering solutions of the corrosion problem include carrying out the reaction in a sacrificial inner tubing fitted inside the pressure-bearing container and using water to create a transpiring flow pattern between the corrosive mixture and the reactor wall. Muthukumaran and Gupta (2000) have proposed a design in which a small amount of sodium carbonate dissolve in cold water is continuously injected in the reactor (Figure 9.15). Sodium carbonate, being highly insoluble in supercritical water, immediately precipitates out as microparticles with a surface area over 100-fold higher than that of the reactor wall. Most of the corrosive species adhere to the particle surface, and the reactor walls are saved from exposure to the corrosive chemicals. Heteroatoms are then carried out of the reactor on the particle surfaces. Upon cooling after exit, the particles dissolve back in water, without causing abrasion in the back-pressure regulator. In addition to corrosion protection, this method enhanced the rate of oxidation, because the newly generated surface of sodium carbonate is highly catalytic.

*Other Applications of Supercritical Water*  In addition to its use in waste oxidation, supercritical water is proposed for inclusion in schemes for the recovery of precious metals from waste (Collard et al., 2001), etching of silicon nitride and silica films (Morita, 2001), synthesis of organic chemicals (Savage, 1999; Broll et al., 1999), transformation of biomass (e.g., cellulose, chitin, starches, lipids, proteins, lignin) into hydrocarbon mixtures (Catallo and Junk, 2001), and upgrading of heavy hydrocarbons (Paspek, 1984).

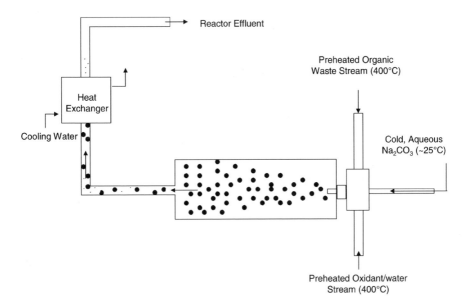

**Figure 9.15.** Sodium carbonate microparticle assisted SCWO for corrosion resistance (Muthukumaran and Gupta, 2000).

### 9.1.2.3 Technologies for Zero Discharge Based on Supercritical $CO_2$

Because of its mild critical temperature (31.1°C) and pressure (73.8 bar), supercritical $CO_2$ can be used with heat sensitive compounds. So far, most industrial applications of supercritical $CO_2$ have been in extraction and cleaning. Applications in chemical synthesis and microparticle formation are emerging. An excellent example of the movement toward zero discharge is the use of supercritical $CO_2$ in the textile industry.

*Discharge Problems With Textile Dyeing*   The textile industry discharges into the environment immense amounts of wastewater from yarn preparation, dyeing, and finishing operations. This wastewater includes wash from preparation and continuous dyeing, alkaline waste from preparation, and batch dye waste containing large amounts of residual dye, salt, acid, or alkali (USEPA, 1997). Of the 700,000 tons of dyes produced annually worldwide, about 10–15 percent is disposed of in effluent from dyeing operations (Snowden-Swan, 1995). The average wastewater generation from a dyeing facility is estimated at between 1 and 2 million gallons per day. Dyeing and rinsing processes for disperse dyeing generate about 12–17 gallons of wastewater per pound of product, and for reactive and direct dyeing generate about 15–20 gallons per pound of product (Snowden-Swan, 1995).

*Dyeing Based on Supercritical $CO_2$*   An alternative to conventional dyeing is the supercritical-fluid dyeing (SFD) process. Supercritical carbon dioxide shows adequate solubility for organic dyes, along with non-toxicity, low viscosity and high diffusivity. In addition, solubility is easily controlled by changing pressure or temperature (Figure 9.16). Polar dyes can be dissolved into supercritical $CO_2$ with the help of cosolvent such as ethanol or acetone. Upon release of pressure, the dye is completely precipitated, hence it is not released into the environment, but is easily recovered for reuse. In addition, the SFD gives better dyeing results due to high diffusivities of the dye in the fluid and in the textile made of both natural or synthetic polymers. Supercritical $CO_2$ swells polymer (Shim and Johnston, 1989; Kazarian et al., 1997) and has no surface tension which makes dye penetration easier than it is in the conventional dyeing process. SFD is also energy efficient since it does not involve the washing/drying steps that are normally required in conventional aqueous- or solvent-based dyeing. SFD has been investigated for dyeing polyester (Chang et al., 1996; Saus et al., 1993) and natural fibers (Gerbert et al., 1994; Ozcan et al., 1998). The solubilities of various dyes have been experimentally determined under supercritical conditions, since the solubility plays a key role in the process (Clifford et al., 1996).

*Cleaning Based on Supercritical $CO_2$*   Commercial operations such as machine parts cleaning and dry-cleaning can give rise to discharges of such liquid contaminants as petroleum, perchloroethylene, chloroflurocarbon or aqueous solvents, or toxic vapors, posing significant health risks due to ground water contamination. A zero discharge operation can be achieved by utilizing supercritical $CO_2$ as the cleaning solvent (Cooney, 1997; Weber et al., 1995). Commercial processes for degreas-

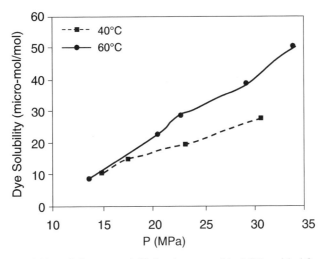

**Figure 9.16.** Solubility of disperse red-60 dye in supercritical $CO_2$ with 4.3 mol% ethanol cosolvent (Muthukumaran et al., 1999).

ing and drycleaning are now available that use a specially designed detergent (Desimone et al., 1999) to enhance the removal of cutting oil, drawing oil from the metal parts, and dirt from clothes. The low surface tension of $CO_2$ makes possible deep cleaning not always achieved with water or organic solvents. After the cleaning process, a decrease in the $CO_2$ pressure results in the separation of the contaminants from $CO_2$. Both $CO_2$ and detergent can be reused.

Similarly, supercritical $CO_2$ degreasing can be used in the leather industry (Marsal et al., 2000). Recently, supercritical $CO_2$ has been promoted in the semiconductor industry (Mullee, 2001) for physical vapor deposition, chemical vapor deposition, dry etching and photolithography, as well as for the cleaning of tool components. There are, however, some limitations on cleaning of inorganic contaminants (Laube, 2001).

*Using $CO_2$ to reduce Volatile Organic Compounds in Paint Industry* The coatings industry has a severe problem of solvent emission and must employ methods such as adsorption and incineration, which increase costs, even as the residual emissions produce smog and other health hazards (Ahuja et al., 2001). Supercritical carbon dioxide can be used to reduce the amount of organic solvent used in paint. For example, in the Unicarb process, a significant amount of organic solvent is replaced by supercritical carbon dioxide (Hoy, 1991). An additional benefit is that the coating is more uniform. The Unicarb process can result in a net reduction in volatile organic compounds without increasing other waste stream disposal products, raising costs, or negatively affecting product quality. Equipment costs and other factors that affect process return on investment can vary, but a 5-year payback period was estimated in a case study of applying nitrocellulose lacquer finish on a chair finishing line (Heater et al., 1994).

*Polymerization in Supercritical CO₂*   For several polymerization reactions, it is now possible to reduce both gaseous and liquid discharges by using $CO_2$ as a solvent. An excellent example is the synthesis of polycarbonate polymers, which are used in compact disks, digital video disks, military shields and bullet proof windows (Wells and DeSimone, 2001). The traditional synthesis, in which bisphenol A is reacted with phosgene in a two-phase methylene chloride/ alkaline water system calls for large amounts of environmentally harmful solvents. The alternate method of melt-phase transesterfication between bisphenol and diaryl carbonate is environmentally friendly but yields a poor grade product, unsuitable for data storage applications. Now, using supercritical $CO_2$, the polymerization can be carried out in solid phase at even low temperatures (Beckman and Porter, 1987).

## REFERENCES

Ahuja, A., Kumar, P., and Chandra, S. (2001). Environmental pollution and its abatement by paint industry. *Paintindia*, **51**(4), 55–66.

Beckman, E., and Porter, R. S. (1987). Crystallization of bisphenol: a polycarbonate induced by supercritical carbon dioxide. *J. Polym. Sci.*, Part B, **25**, 1511–1517.

Broll, D., Kaul, C., Kramer, A., Krammer, P., Pichter, T., Jung, M., Vogel, H., and Zehner, P. (1999). Chemistry in supercritical water. *Angew. Chem. Int. Ed.*, **38**, 2998–3014.

Catallo, W. J., and Junk, T. (2001). Transforming biomass into hydrocarbon mixtures in near-critical or supercritical water. U.S. Patent #6,180,845, January 30, 8 pp.

Chang, K. H., Bae, H. K., and Shim, J. J. (1996). Dyeing of PET textile fibers and films in supercritical carbon dioxide. *Korean Journal of Chemical Engineering*, **13** (3), 310–316.

Clifford, A. A., and Bartle, K. D. (1996). Supercritical fluid dyeing. *Textile and Technology International*, 113–117.

Collard, S., Gidner A., Harrison B., and Stenmark, L. (2001). Precious metal recovery from organics-precious metal composition with supercritical water reactant. Canada Patent #CA 2,407,605, Nov. 8.

Cooney, C. M. (1997). Supercritical CO₂-based cleaning system among green chemistry award winners. *Environ. Sci. Technol.*, **31**(7), 314A–315A.

Desimone, J. M., Romack, T. J., Betts, D. E., and McClain, J. B. (1999). Cleaning process using carbon dioxide as a solvent and employing molecularly engineered surfactants. U.S. Patent #5,866,005, 12 pp., Feb. 2, Cont.-in-part of U.S. 5,783,082.

Gerbert, B., Saus W., Knittel, D., Buschmann, H. J., and Schollmeyer, E. (1994). Dyeing natural fibers with disperse eyes in supercritical carbon dioxide. *Textile Research Journal*, **64**(7), 371–374.

Gloyna, E. F., and Li, L. (1993). Supercritical water oxidation: an engineering update. *Waste Management*, **13**, 379–394.

Heater, K. J., Parsons, A. B., and Olfenbuttel R. F. (1994). Evaluation of supercritical carbon dioxide technology to reduce solvent in spray coating applications. Report (EPA/600/R-94/043; Order No. PB94–160629), 59 pp.

Hoy, K. (1991). Unicarb system for spray coatings: a contribution to pollution prevention. *Coating & The Environment*, **181**(4288), 438.

Kazarian, S. G., Brantley, N. H., West, B. L., Vincent, M. F., and Eckert, C. A. (1997). In situ spectroscopy of polymers subjected to supercritical $CO_2$: plasticization and dye impregnation. *Applied Spectroscopy*, **51**(4), 491–494.

Kritzer, P., Boukis, N., and Dinjus, E. (1998). The corrosion of a alloy 625 in high-temperature, high-pressure aqueous solutions of phosphoric acid and oxygen. *Mater. Corros.* **49**(11), 831–839.

Laube, D. (2001). Critical cleaning trends limitations of $CO_2$ cleaning for semiconductor process tools. *A2C2 Magazine*, **4**(2), 9–12.

Marsal, A. P. Celma, J., Cot, J., and Cequier, M. (2000). Supercritical $CO_2$ extraction as a clean degreasing process in the leather industry. *Journal of Supercritical Fluids*, **16**(3), 217–223.

Morita, K. (2001). Method for etching silicon nitride and silica films by subcritical or supercritical water. U.S. Pat. Appl. Publ., 26 pp.

Mullee, W. H. (2001). Efficient Removal of Resist or Residue from Semiconductors using Supercritical Carbon Dioxide US Patent #6306564, 12 pp.

Muthukumaran, P., Gupta, R. B., Sung, H.-D, and Shim, J.-J. (1999). Dye solubility in super-critical carbon dioxide, effect of hydrogen bonding with cosolvents. *Korean J. Chem. Eng.*, **16**, 111–117.

Muthukumaran, P., and Gupta, R. B. (2000). Sodium-carbonate-assisted supercritical water oxidation of chlorinated waste. *Ind. Eng. Chem. Res.*, **39**(12), 4555–4563.

Ozcan, A. S., Clifford, A. A., Bartle, K. D., and Lewis, D. M. (1998). Dyeing of cotton fibres with disperse dyes in supercritical carbon dioxide. *Dyes and Pigments*, **36**(2), 103–110.

Paspek, S. C., Jr. (1984). Upgrading heavy hydrocarbons with supercritical water and light olefins. U.S. Patent #4,483,761, Nov. 20.

Saus, W., Knittel, D., and Schollmeyer, E. (1993). Dyeing of textiles in supercritical carbon dioxide. *Textile Research Journal*, **63**(3), 135–142.

Savage, P. E. (1999). Organic chemical reactions in supercritical water. *Chem. Rev.*, **99**, 603–621.

Schmieder, H., and Abeln, J. (1999). Supercritical water oxidation: state of the art. *Chem. Eng. Technol.*, **22**(11), 903–908.

Shim, J.-J., and Johnston, K. P. (1989). Adjustable solute distribution between polymers and supercritical fluids. *AIChE J.*, **35**(7), 1097–1106.

Snowden-Swan, L. J. (1995). Pollution Prevention in the Textile Industries, In *Industrial Pollution Prevention Handbook*, Freeman, H. M. (ed.), McGraw-Hill, Inc., New York.

USEPA, 1997: http://es.epa.gov/oeca/sector/sectornote/pdf/textilsn.pdf.

Weber, D. C., McGovern, W. E., and Moses, J. M. (1995). Precision surface cleaning with supercritical carbon dioxide: issues, experience, and prospects. *Met. Finish.*, **93**(3), 22–26.

Wells, S. L., and DeSimone, J. (2001). $CO_2$ technology platform: an important tool for environmental problem solving. *Angewandte Chemie, International Edition*, **40**(3), 518–527.

### 9.1.3   Ultraviolet Disinfection

TAPAS K. DAS
Washington Department of Ecology, Olympia, Washington 98504, USA

In rural areas of India, Bangladesh, Brazil, Peru, China, and other developing nations, waterborne diseases such as typhoid, cholera, hepatitis, and gastroenteritis, infect and kill many infants and children each day (WHO, 1996). Wastewaters or water can contain an incredibly large variety of microorganisms. Some are harmless; many, however, are disease causing. These microorganisms must be destroyed before wastewaters can be safely discharged into a receiving body of water or reclaimed or reused. With increasing emphasis on promoting a sustainable ecological future and concern over the presence of toxic chemicals in the water, the modern disinfection processes are increasingly leaning toward technologies that destroy pathogens while balancing the effects of introducing the disinfected wastewater into populations of aquatic biota or drinking water supplies.

*9.1.3.1   The Move Toward Ultraviolet Disinfection in North America*   In most areas of the United States and Canada, wastewater is disinfected by means of irradiation with ultraviolet (UV) light instead of by such older methods as chlorination and chlorination/dechlorination.

A traditional method of purifying drinking water, the addition of chlorine, long ago became impractical for large municipalities. Not only did fire codes begin to limit the amount of liquid chlorine that could be stored, but US regulations in place since 1985 restrict the amount of chlorine that may be discharged into receiving waters and established limits on the total residual chlorine (TRC) permissible in wastewater effluents. Compliance with these regulations could be achieved by incorporating a dechlorination step into the disinfection process, and indeed, this was the route usually chosen (USEPA, 1985).

Chlorination followed by dechlorination has ceased to be the treatment of choice however. Not only does the process destroy the aquatic biota in receiving waters and produce compounds that may be carcinogens, but the sulfur dioxide that removes chlorine from the effluent stream is itself an environmental pollutant.

Thus across the United States, environmental protection agencies and corporations began to look for alternative ways to disinfect wastewater. These efforts produced a literature indicating that ultraviolet (UV) disinfection systems were both effective and economical (Das, 2004; Das and Ekstrom, 1999; Loge et al., 1996a,b; LOTT, 1994; Scheible et al., 1986; Scheible, 1987; USEPA, 1992; Washington Department of Ecology, 1998; White, 1999). We shall discuss some of these alternative technologies, including a development from the early 1980s: parallel flow, open channel modular UV systems, usable both in the retrofit market and for new wastewater treatment plants.

*Mini-Case Study 9-4: The LOTT System*   After considering the available alternatives, the city of Olympia, Washington, decided to install and operate the first comprehensive UV disinfection system in the wastewater treatment plant

(WWWTP) on the west coast. Details of the system for the cities and counties of Lacey, Olympia, Tumwater, and Thurston (LOTT) will be presented later (see Figure 9.18). During the pilot study and through completion of the project, the state Department of Ecology worked with the counties, providing assistance and support (LOTT, 1994).

The LOTT system treats wastewater coming primarily from over 100,000 residences but also from a brewery and some light industries. The secondary treatment process is a biological nutrients removal system able to remove from the water over 90 percent of the biodegradable organic material (BOD) and total suspended solid (TSS), as well as nutrients including phosphorus, before discharging an average of 22.0 million gallons per day (mgd) into Puget Sound's Budd Inlet.

LOTT also treats storm water, and during the winter months the WWTP receives high storm water flow, sometimes totaling 55 mgd. Due to higher flow through the WWTP between November and February, total retention time in the clarifiers goes down and consequently TSS and turbidity level go up slightly. The modular components of the system can be adjusted to provide adequate year-round disinfection. Effluent grab samples are taken daily and analyzed for fecal coliform counts to determine the compliance with the permit from the EPA's National Pollutant Discharge Elimination System (NPDES). Twenty-four hour composite samples are taken daily and analyzed for TSS to determine the compliance with the permit. Some results of these studies are presented later (see Section 9.1.3.7).

### 9.1.3.2 UV Light and Its Mechanism of Germicidal Action

*9.1.3.2 UV Light and Its Mechanism of Germicidal Action* The power of sunlight to destroy microbial life has long been known and appreciated. Effective disinfection in air, on surfaces and in water has been accomplished by exposure to the direct rays of the sun. Sunlight is an important factor in the self-purification of water in streams and in impounding reservoirs. The effect of sunlight at destroying bacteria, particularly intestinal bacteria has been reported upon many times. The ordinary rays of sunlight play little part in this bactericidal action. The results are caused by ultraviolet rays. Sources of high intensity ultraviolet light have been developed which can be used to disinfect water, wastewater, air, etc.

The term "ultraviolet light" is applied to electromagnetic radiation emitted from the region of the spectrum lying beyond the visible light and before x-rays. The upper wavelength limit is 400 nanometres (1 nm $= 10^{-9}$ meter) and the lower wavelength limit is 100 nm, below which radiation ionizes virtually all molecules. The narrow band of UV light lying between wavelengths of 200 and 300 nm has often been called the germicidal region because UV light in this region is lethal to microorganisms including bacteria, protozoa, viruses, molds, yeasts, fungi, nematode eggs and algae. Figure 9.17 shows that the most destructive wavelength is 260 nm which is very close to the wavelength of 254 nm produced by germicidal low-pressure UV lamps. Figure 9.17 also shows the similarity between UV light's ability to kill the fecal coliform bacterium *Escherichia coli* and the ability of its genetic material (i.e., nucleic acid) to absorb UV light. UV light causes molecular rearrangements in the genetic material of microorganisms and this prevents them from reproducing.

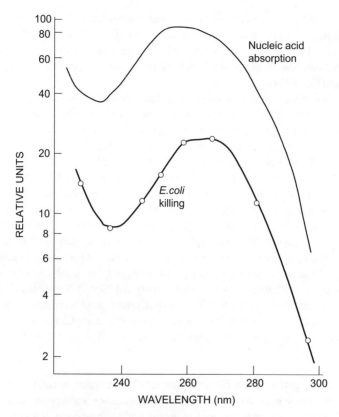

**Figure 9.17.** Comparison of the action spectrum for inactivation of *E. coli* to the absorption spectrum of nucleic acids (Harm, 1980).

If a microorganism cannot reproduce then it is considered to be dead (Das, 2001, 2004; USEPA, 1999).

Thus far we have spoken only of killing cells, using unqualified words like "germicidal," "lethal," and "dead." As intuition suggests, however, not all pathogens in a UV-treated effluent stream are killed, even in the most efficient WWTP.

We shall discuss germicidal efficiency later (see Section 9.1.3.6). For the moment, we merely point out that photochemical damage caused by UV may be repaired by some organisms. Studies show that the amount of cell damage and subsequent repair is directly related to the UV dose. The amount of repair will also depend on the dose (intensity) of photoreactivating light. For low UV doses the resulting minimal damage can be more readily repaired than for high doses where the number of damaged sites is greater (Lindenauer and Darby, 1994).

### 9.1.3.3 UV Lamps
Germicidal lamps operate electrically on the same principle as fluorescent lamps. UV light is emitted as a result of an electron flow through the ionized vapor between the electrodes of the lamps. The glass of the germicidal lamp

is made of quartz which transmits UV light and the glass of a fluorescent lamp is made of soft glass which absorbs all of the UV light at a wavelength of 254 nm. The bulb of the fluorescent lamp is coated with a phosphor compound which converts UV to visible light. A germicidal lamp produces about 86 percent of its total radiant intensity at a wavelength of 254 nm and about 1 percent at other germicidal wavelengths. Germicidal lamps with high quality quartz also produce UV light at a wavelength of 185 nm. This wavelength produces ozone which is corrosive to the UV equipment and the ends of the lamps. The UV lamps in UV equipment should not produce ozone. The medium pressure mercury lamp spectrum produces most of its light in the visible range. The medium pressure mercury lamp operates at very high temperatures (600–800°C) and the lifetime is about one-third of that of a low pressure mercury lamp (Trojan Technologies Inc., 2000).

*Slimline Instant Start Lamp*   This is the lamp of choice for 100 percent of the large UV wastewater systems. It is an instant start lamp that produces 26.7 watts of UV-C from 100 watts of power at 0.18 watts of UV-C per centimeter of arc length using a conventional core-coil ballast. Its optimum operating surface temperature is 40°C.

It is called instant start because a high voltage is applied to the cathodes to instantly strike an arc. According to the manufacturer, these lamps have a rated average useful life of 9,000 hours at which time the UV-C output has dropped by 30 percent. A pilot study with a full scale UV system at a wastewater treatment plant confirmed that the UV lamps drop in output by almost 50 percent but the lamps are reliable up to 40 months using a conventional magnetic ballast (Trojan Technologies Inc., 2000; LOTT, 1994).

*Rapid Start Lamps*   These are the lamps of choice in Europe. These lamps are almost identical to the instant start lamp except that a small voltage (approximately 4 volts) is continuously supplied by the ballast to the lamp cathodes to help strike the arc when the open circuit voltage is applied. This reduces the voltage to strike the arc thereby decreasing damage to the cathodes. This also reduces ballast size and losses and thus improves the efficiency of the system. Rapid start lamps may be dimmed as low as 0.2 percent. This would eliminate the need for duplicate banks of UV lamps. These lamps have the same UV-C output as instant start lamps.

*Low Pressure Flat Lamps*   This is a low pressure mercury lamp which uses a flat quartz envelope and a much higher current. This same principle is used in the General Electric "Power Groove" lamps. Decreasing the distance the ions and electrons must travel to reach the wall increases the UV output.

This flat lamp is reported to produce three times the UV-C output per centimeter of arc length at a twenty percent power saving but this has not been confirmed by an independent organization. This lamp is made excessively for a UV manufacturer in Germany.

*Ballasts and Power Supplies to UV Lamps*   The principal function of a ballast is to limit current to a lamp. A ballast also supplies sufficient voltage to start and operate the lamp. In the case of rapid start circuits, a ballast supplies voltage to heat the lamp cathodes continuously. A UV lamp is an arc discharge device. The more current in the arc, the lower the resistance becomes. Without a ballast to limit current, the lamp would draw so much current that it would destroy itself.

The most practical solution to limiting current is an inductive ballast. The simplest inductive ballast is a coil inserted into the circuit to limit current. This works satisfactorily for low wattage lamps. For most lamps the line voltage must be increased to develop sufficient starting voltage. Rapid start circuits require low voltage to heat the electrodes continuously to reduce the starting voltage. The pilot study report provides descriptions of different types of ballast systems, including details of construction, operation, and efficiency (LOTT, 1994).

### 9.1.3.4  Open Channel Modular UV Systems
Figure 9.18 is a schematic diagram of the modular disinfection system at the LOTT treatment plant introduced in Mini-Case Study 9-4.

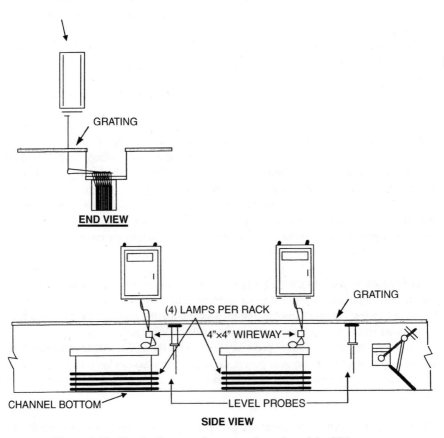

**Figure 9.18.** Simple diagram of an open channel modular UV system.

The design features racks of UV lights placed in an open channel so that the water flows parallel to the radiation source. Each rack is independent of every other rack and has its own group of ballasts. Every group of ballasts has an individual ground fault interrupter circuit. The level of the effluent over the lamps is controlled by a flow sensitive device.

The benefits of open channel modular UV systems are as follows:

- Because each major component is modular there is no need to shut down the entire UV system to replace or clean any part. This eliminates the need for a backup system.
- The flow of water by the UV lamps is by gravity thereby eliminating pumps.
- The effluent flows parallel to the UV lamps so that debris can only catch on the lamp holders and not on the UV lamps.
- The system can be sited outdoors or indoors.
- The UV system can be installed in an existing channel or contact chamber.
- Increases in system size can be accommodated by simply making the original channel long enough to contain more than one bank of UV racks.

*UV System for Wastewater at LOTT*   The subsections that follow describe major components of the most popular UV system for treating wastewater.

*UV Channel*   The channel is built to accommodate one or more banks of UV racks in series along with a water level control device. The optimum distance between UV banks is four feet.

If large variations in flow are anticipated, it is better to have more than one channel. Having multiple channels saves on electrical costs and increases lamp life because the channels can be turned on and off above or below predetermined flow rates. This is important because above a certain range the depth of water over the lamps will be too great and below the proper flow range UV lamps will be exposed to the air.

*UV Lamp Racks*   The lamp racks that hold the UV lamps parallel to the flow of wastewater also protect people from the UV light. The lamp racks at the LOTT facility are made of stainless steel. Anodized aluminum is sometimes used to hold electronic ballasts, but this metal is not resistant to acids used in cleaning.

The racks are sturdy enough to permit wastewater treatment plant personnel to walk on them for maintenance purposes. Each UV rack has its own power/communication cable, with a connector either on the rack itself or at the control panel/power distribution center.

*Level Controller*   The level controller serves to maintain a constant water depth of 1.9–2.54 centimeters (0.75–1.0 inch) over the top of the highest protective quartz sleeve at all the anticipated flow rates. If the wastewater were to exceed this depth, the UV intensity would be too low to destroy all the pathogens.

The two primary types of level control devices are the sharp-crested weir and flap gate. A flow proportional valve or sluice gate and a combination of a weir and flap gate could also be used. Normally weirs are used for small UV systems of less than twenty UV lamps and flap gates are used for all the larger UV systems.

*Flap Gate*   A flap gate operates through the use of gravity: as the water flowing through the channel hits the face of the gate, the gate opens, allowing water to pass. Weights are placed on the gate to limit the opening of the gate for a given flow. As flow increases, the force on the gate is greater and the gate opens further. A properly designed gate will maintain water levels within the specified limits over a wide range of flow rates. The disadvantage of a flap gate is that it leaks at or near zero flow thereby allowing UV lamps to come into contact with the air. When this happens, contaminants bake onto the protective quartz sleeve, forming a coating that prevents UV disinfection when the normal flow returns.

*Weir*   A weir will guarantee a maximum water level at peak flows due to the predictability of water crest elevation over a weir at a given flow. A weir can also be designed to keep the lamps submerged at zero flow. The main disadvantage of a weir is the considerable space required and the tendency for solids to accumulate at the bottom upstream side of the weir. A valve can be installed to flush the solids from the weir. Configuring the weir in a serpentine fashion will save on space, but even so the space between the edges of the weir must be large enough to prevent flooding.

*Power Distribution and Control Center (PDC Center)*   For every bank of UV lamp racks, there must be a power distribution and control (PDC) center to house the components that interface with any remote process control equipment or other banks of UV lamp racks.

If a chlorination building already exists beside a channel or chlorine contact chamber, that building can be used as the PDC center.

### 9.1.3.5   *Parameters Affecting the UV Disinfection of Wastewater*   The efficiency of a UV disinfection system strongly depends on effluent quality. The higher the level of contaminants, the more drastic is the intensity of the irradiation in wastewater. Following are the major parameters which must be considered in designing a UV disinfection system for wastewater:

UV transmittance (T) or absorbance

Total suspended solids (TSS)

Particle size distribution (PSD)

Flow rate

Iron

Hardness

Wastewater source

Equipment maintenance and worker safety

The customer or the consultant must provide this information to the UV manufacturer because each UV system is designed on an individual basis.

*UV Transmission or Absorbance*   UV lights' ability to penetrate wastewater is measured in a spectrophotometer at the same wavelength (254 nm) that is produced by germicidal lamps. This measurement is called the *Percent Transmission* or *Absorbance* and it is a function of all the factors which absorb or reflect UV light. As the percent transmission gets lower (higher absorbance) the ability of the UV light to penetrate the wastewater and reach the target organisms decreases.

The UV transmission of wastewater must be measured because it cannot be estimated simply by looking at a sample of wastewater with the naked eye. The system designer must either obtain samples of wastewater during the worst conditions or carefully attempt to calculate the expected UV transmission by testing wastewaters from plants that have a similar influent and treatment process. The designer must also strictly define the disinfection limits because this determines the magnitude of the UV dose.

The range of effective transmittances ($T$) will vary depending on the secondary treatment systems. In general, suspended growth-treatment processes produce effluent with $T$ varying from 60–65 percent. Fixed film processes range from 50–55 percent $T$ and lagoons 35–40 percent $T$. Industries that influence UV transmittance include textile, printing, pulp and paper, food processing, meat and poultry processing, photo developing, and chemical manufacturing.

Figure 9.19 illustrates the effect of a UV absorbing soluble compound on the disinfection ability of a parallel flow UV system with two banks of UV lamps in series. As the UV transmission decreases the number of fecal coliform counts increases. Therefore the applied dose of UV light required is dependent upon the disinfection standard and the UV transmission. Figure 9.19 also shows the results of doubling the UV dose as the wastewater passes from one bank of UV lights through a second identical bank of UV lamps. By doubling the UV dose, an UV transmission of 7.5 percent as compared to 24 percent can be treated to reach a fecal coliform limit of 200 per 100 milliliters. Therefore the UV system must be designed for the minimum UV transmission.

*Suspended Solids*   Suspended solids in biologically treated effluents are typically composed of bacteria-laden particles of varying number and size. Some of the suspended solids in wastewater will absorb or reflect the UV light before it can penetrate the solids to kill any occluded microorganisms. With longer contact times and higher intensities, UV light can penetrate suspended solids but its germicidal ability is limited.

Obtaining the proper information about the level of suspended solids is very important for the sizing of the UV system. If a wastewater treatment plant producing high levels of suspended solids is already in operation a pilot study will show how often the quartz sleeves must be cleaned to eliminate fouling by the suspended solids. Pilot testing will also determine whether the fecal coliform limit can be attained.

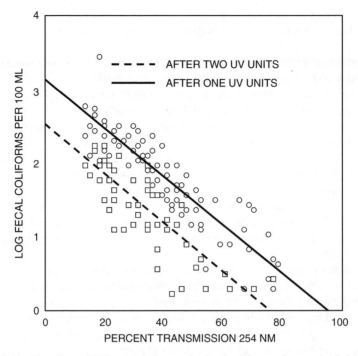

**Figure 9.19.** The effect of UV transmission on the fecal coliforms after one bank (○) and two banks of (□) of UV lights.

Filtration can reduce TSS levels, thus lowering the level of UV irradiation necessary to achieve a given disinfection target. If wastewaters were devoid of suspended solids, UV disinfection could be used almost universally.

*Particle Size Distribution* Particle size distribution (PSD) measurements of wastewater effluent are used as an indicator of filter and clarifier performance. Typically, particle sizes are related to the type of wastewater process and level of treatment, which in turn results in a decrease in both the number and mean size of particles. Table 9.6 illustrates the effect of large particle size on UV demand.

*Flow Rate* The United States Environmental Protection Agency provides an in-depth analysis of the effect of hydraulics on the UV disinfection of wastewater (USEPA, 1992).

The degree of inactivation by ultraviolet radiation is directly related to the UV dose applied to the water or wastewater. Dose is described as the product of the rate at which the energy is emitted (intensity) and the time the organism is exposed to the energy.

$$D = It \qquad (9\text{-}6)$$

**TABLE 9.6.  An Increase in Particle Size Directly Affects the UV Demand**

| Particle Size ($\mu$m) | UV Demand |
| --- | --- |
| <10 | Easily penetrated, low UV demand can be penetrated, UV demand increased |
| >40 | Will not be completely penetrated, high UV demand |

where

$D$ = dose, in microwatt-seconds per square centimeter ($\mu$W·s/cm$^2$)

$I$ = intensity or irradiation, in microwatts per square centimeter ($\mu$W/cm$^2$)

$t$ = time, in seconds

As the flow rate increases, the number or size of the UV lamps must be proportionately increased to maintain the required level of disinfection. Therefore the UV system must be designed for the maximum flow rate at the end of lamp life. To ensure that every microorganism is exposed to the specified average dose of UV light, the UV unit must be designed to provide as much sideways motion as possible with very little forward mixing. This is especially important when the water has a low UV transmission or high suspended solids. The open channel UV system where the wastewater flows parallel to the submerged lamps has a very good hydraulic profile (LOTT, 1994).

Since the height of the wastewater above the top row of UV lamps is rigidly controlled by a flap gate or weir at all flow rates, the system must be designed for the maximum flow rate. This is especially important if the wastewater treatment plant receives runoff water after storms.

The UV system design must also accommodate the minimum flow rate. Many smaller wastewater treatment plants approach zero flow at night. During this period of time the wastewater has a greater chance to warm up around the quartz sleeves and produce deposits on the sleeves. If the quartz sleeves are exposed to the air, not only will any compounds left on the sleeves bake onto the warm lamps, but water splashing onto the sleeves also will result in UV absorbing deposits. When the flow returns to normal, a layer of water will pass through the UV unit without being properly disinfected. The designer must select the flow device very carefully to compensate for this situation. A flap gate has a normal flow range of 1 : 5, and as mentioned earlier all these gates leak at low flow rates. It is possible to reach 1 : 10 but it is better to use two or more channels. A weir, which keeps the lamps fully submerged at zero flow may be a much better solution (LOTT, 1994).

*Iron*  Iron affects UV disinfection by absorbing UV light. It does this in three ways. If the concentration of dissolved iron is high enough in the wastewater the UV light will be adsorbed before it can kill any microorganisms. Regardless of concentration, however, some iron will precipitate out on the quartz sleeves and absorb the UV light before it enters the wastewater. In addition, iron that is absorbed onto suspended solids, clumps of bacteria and other organic compounds prevents UV light from piercing the suspended solids and killing the entrapped microbes. The UV industry has adopted a level of 0.3 ppm as the maximum allowable level of iron but there is no data to substantiate this limit. The level of iron should be measured in the wastewater and if it approaches 0.3 ppm a pilot study should be instituted to determine whether the disinfection level can be attained and what the cleaning frequency should be. An in-place cleaning system can be incorporated in the UV design. If possible a wastewater treatment plant should be designed with a non-iron method of precipitating phosphate.

*Hardness*  Calcium and magnesium salts, which are generally present in water as bicarbonates or sulfates, cause water hardness, which in turn produces the formation of mineral deposits. For example, when water containing calcium and bicarbonate ions is heated, insoluble calcium carbonate is formed:

$$Ca^{2+} + 2HCO_3 \rightarrow CaCO_3 \text{ (precipitate)} + CO_2 + H_2O$$

This product precipitates and coats on any warm or cold surfaces. The optimum temperature of the low pressure mercury lamp is 40°C or 104°F. At the surface of the protective quartz sleeve there will be a molecular layer of warm water where calcium and magnesium salts will be precipitated, preventing UV light from entering the wastewater.

Unfortunately no rule exists for determining when hardness will become a problem. Table 9.7 shows the classification of water hardness. Waters containing around 300 mg/L of $CaCO_3$ deposits may require pilot testing of a UV system. This is especially important if very low flow or no flow situations are anticipated because the water will warm up around the quartz sleeves, and excessive coating will result.

*Wastewater Source*  It should be determined whether the wastewater treatment plant receives periodic influxes of industrial wastewater which may contain UV absorbing organic compounds, iron or hardness which may affect UV performance. These industries may be required to pre-treat their wastewater.

For example, a textile mill may be periodically discharging low concentrations of dye into the municipal wastewater system. By the time this dye reaches the treatment plant it may be too diluted to detect without using a spectrophotometer. Yet even

TABLE 9.7.  **Classification of Water Hardness**

| Hardness Range (mg/L as $CaCO_3$) | Hardness Description |
|---|---|
| 0–75 | Soft |
| 75–150 | Moderately hard |
| 150–300 | Hard |
| >300 | Very hard |

low levels of dye can readily absorb ultraviolet light thereby preventing UV disinfection.

*Equipment Maintenance, Lamp Life, and Workers' Safety* Equipment maintenance factors affecting UV intensity include lamp age and sleeve fouling. The intensity of the radiation supplied by a lamp gradually decreases with use of the device, and this is factored into the design. The recommended low pressure lamp replacement time is approximately 5,000 hours, but some plants have disinfected successfully using lamps up to 8,000 hours. Medium pressure lamp replacement time is approximately 5,000 hours. The lamp life depends upon the number of ON and OFF cycles used for flow pacing during disinfection. Uniform intensity in a system can be managed with a staged lamp replacement schedule.

Because accumulations of inorganic and organic solids on the quartz sleeve decrease the intensity of UV light that enters the surrounding water, conventional low pressure technology systems include a fouling factor in the design. These systems require cleaning and maintenance by plant operators on a regular basis. The Trojan System UV4000 has an automatic wiping system in place that combines chemical and mechanical cleaning and does not require operator maintenance time (Trojan Technologies Inc., 2000).

UV is generated on-site and does not raise significant safety concerns in surrounding communities. Worker safety requires protection from exposure (primarily of the eyes and skin) from UV light, and electrical hazards, and safe handling and disposal of the expended lamps, quarts, ballasts, and cleaning chemicals.

### 9.1.3.6 Germicidal Efficiency

Investigations have shown that microorganisms vary widely in their sensitivity to ultraviolet energy. Kawabata and Harada (1959) reported the following contact times required to achieve a 99.9 percent kill (3-log reduction) at a fixed UV intensity for the following organisms:

| | |
|---|---|
| *E coli* | 60 s |
| *Shigella* | 47 s |
| *Salmonella typhosa* | 49 s |
| *Streptococcus faecalis* | 165 s |
| *Bacillus subtilis* | 240 s |
| *Bacillus subtilis* spores | 369 s |

Ultraviolet radiation has also been shown to be effective in the inactivation of viruses. Huff (1965) reported satisfactory results which included studies of several strains of polio virus, Echo 7, Coxsackie 9 viruses. The intensities varied from 7,000–11,000 $\mu$W.s/cm$^2$.

Current designs are for high intensity and lower exposure times—6–10 s. There is no doubt that the germicidal efficiency of UV is predictable for a given species of organism on the basis of the UV intensity-exposure time product (Eq. 9-1). In practice it will be necessary to prove and evaluate a given installation to confirm the design parameters (White, 1999).

The relationship between UV dose and bactericidal kill is characterized by a mathematical model that assumes second-order kinetics when the coliform concentrations are in the range where disinfection usually takes place, as follows:

$$\frac{dN}{dt} = kN^2I \qquad (9\text{-}7)$$

Integrated, this becomes

$$\frac{1}{N} - \frac{1}{N_o} = kIt \qquad (9\text{-}8)$$

where

$N$ = coliform counts, most probable number (MPN)/100 mL, at time $t$
$N_o$ = influent coliform counts, most probable number (MPN/100 mL)
$k$ = rate constant (counts/s)
$I$ = the average ultraviolet intensity (or irradiation) in the exposure chamber
$t$ = exposure time (s)

The influent coliform concentration is usually so much greater than the final concentration that the term $l/N_o$ becomes negligible and Eq. (9-8) can be simplified to:

$$\frac{1}{N} = kIt \qquad (9\text{-}9)$$

Equation (9-9) can be used to calculate coliform count at the end of UV exposure at a varying UV intensity and time of exposure for a known rate constant. On the basis of 350 samplings conducted throughout a one year pilot program Scheible and Bassell (1981) have been able to show quite favorable correlation between UV dose and coliform kill by the following empirical equation:

$$\text{Effluent fecal coliform} = (1.26 \times 10^{13})(\text{UV dose})^{-2.27} \qquad (9\text{-}10)$$

Loge et al. (1996a) used a probabilistic design approach in their pilot studies and developed an empirical formula with four coefficients that can be used to calculate the fecal coliform density after exposure to UV light. The resulting correlation can be used to predict reasonably well the number of lamps necessary to meet the permit requirements for WWTPs.

### 9.1.3.7 Disinfection Standards

The level of disinfection required by the USEPA National Pollutant Discharge Elimination System (NPDES) permit is commonly less than 200 fecal coliform counts/100 mL as a 30-day geometric mean. In general, a UV dose of 20–30 mW.s/cm$^2$ is required to achieve this level of disinfection in secondary-treated waste water with a 65 percent transmittance and TSS <20 ppm. The UV dose requirement to meet specific limits depends on the nature of the particle with respect to numbers, size, and composition. Therefore, UV dose requirements will vary.

A more stringent limit of $<2.2$ total coliform/100 mL is required for water re-use in California and Hawaii. In such cases, filtered effluents with TSS 2 ppm or less and 65 percent transmittance, may require UV dose as high as 120 mW.s/cm$^2$ to achieve this level of disinfection. The concentrations of solids, bacteria in the particles, and the PSD, are the main limiting factors when attempting to meet stringent disinfection limits.

It appears that the UV dose required to meet the traditional coliform limits will achieve better virus inactivation results than the comparable chlorine dose. Figure 9.20 compares the relative doses of UV and chlorine required to inactivate selected organisms compared to fecal coliform indicator (Trojan Technologies Inc., 2000).

Figure 9.21 presents the monthly average values of fecal coliform after the UV exposure and TSS in effluent for 1998 and 1999 obtained from LOTT WWTP (Mhatre, 2000). During the winter and rainy seasons (typically November through February in the Pacific Northwest), fecal coliform counts and TSS concentrations were found to be marginally higher than the other months of dry and warm seasons. At higher flow rates, efficiency of TSS removal in the secondary treatment system is lower, due to shorter retention times, which contribute to higher TSS and fecal coliform count. However, during dry and warmer seasons, the secondary treatment system removes a higher percentage of TSS, and as a result, fecal coliform counts and TSS levels are much lower.

Figure 9.22 presents values of effluent turbidity and fecal coliform counts for the LOTT facility. Higher effluent turbidity may indicate higher TSS level. Composed

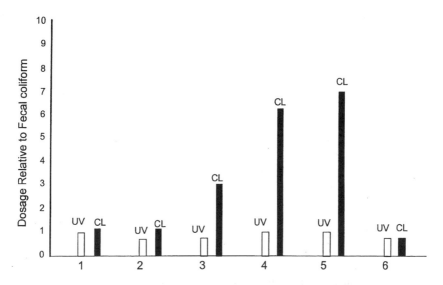

**Figure 9.20.** Comparison of the relative effectiveness of chlorine vs UV on bacteria and viruses: (1) *Escherichia coli*; (2) *Salmonella typhosa*; (3) *Staphylococcus aureus*; (4) *Polio virus type* 1; (5) *Coxsackie AZ virus*; (6) *Adenovirus type* (Trojan Technologies Inc., 2000).

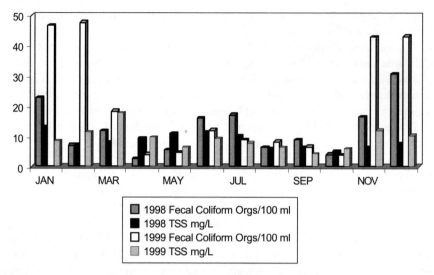

**Figure 9.21.** Results of TSS fecal coliform counts at the LOTT WWTP: monthly average 1998–1999 (Mhatre, 2000).

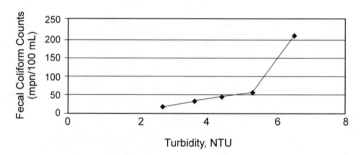

**Figure 9.22.** Turbidity in nephelometric turbidity units vs fecal coliform counts (Mhatre, 2000).

of bacteria-laden particles of varying number and sizes, suspended solids reduce the UV intensity in an effluent by absorbing and scattering UV light, with lower disinfection and higher coliform counts as results.

## ACKNOWLEDGMENTS

The author gratefully acknowledges the receipt of valuable information and support from Asha Mhatre of the city of Olympia's wastewater treatment plant and Larry P. Ekstrom (formerly at the city of Olympia's wastewater treatment plant).

# REFERENCES

Das, T. K. (2001). Ultraviolet disinfection application to a wastewater treatment plant. *Clean Products and Processes*, **3**(2), 69–80.

Das, T. K. (2004). *Disinfection*. In the Kirk-Othmer *Encyclopedia of Chemical Technology*, 5th Ed., Vol. 8, pp. 605–672, Hoboken, NJ: Wiley.

Das, T. K., and. Ekstrom, L. P (1999). UV application to a major wastewater treatment plant on the west coast. Paper No. 100d, American Institute of Chemical Engineers Spring Annual Meeting, Houston.

Harm, W. (1980). *Biological Effects of Ultraviolet Radiation*. IUPAB Biophysics Series, Cambridge University Press, pp. 29.

Huff, C. B. (1965). Study of ultraviolet disinfection of water and factors in treatment efficiency. *Public Health Report*, **80**(8), 695–705.

Kawabata, T., and Harada, T. (1959). The disinfection of water by the germicidal lamp. *J. Illuminating Eng. Soc.*, **36**, 89.

LOTT, Lacey, Olympia, Tumwater and Thurston (1994). Pilot study report on UV disinfection of wastewater. The city of Olympia, Washington.

Lindenauer, K. G., and Darby, J. L. (1994). Ultraviolet disinfection of wastewater: effects of doses on subsequent photoreactivation. *Wat. Res.*, **28**(4), 805.

Loge, F. J., Darby, J. D, and. Tchobanoglous, G. (1996a). "UV disinfection of wastewater: probabilistic approach to design. *J. Env. Eng.*, 1078.

Loge, F. J., Emerick, R. W., Heath, M., Jacangelo, J., Tchobanoglous, G., and Darby, J. L. (1996b). Ultraviolet disinfection of secondary wastewater effluents: prediction of performance and design. *Water Environmental Research*, **68**(5), 900.

Mhatre, A. (2000). Personal communication. City of Olympia, LOTT wastewater treatment plant, Olympia, Washington.

Scheible, O. K. (1987). Development of a rationally based design protocol for the ultraviolet light disinfection process. *J. Water Pollut. Control Fed.*, **59**(1), 25–31.

Scheible, O. K., and. Bassell, C. D (1981). Ultraviolet disinfection of secondary wastewater treatment plant. EPA Municipal Env. Res. Lab Report, EPA-600/S2–81-152, Cincinnati, Ohio.

Scheible, O. K., Casey, M. C., and Forndran, A. (1986). Ultraviolet disinfection of wastewaters from secondary effluent and combined sewer overflows. EPA-600/S2–86/005, USEPA, Cincinnati, Ohio.

Trojan Technologies, Inc. (2000). Overview of UV disinfection. 3020 Gore Road, London, Ontario, Canada N5V 4 TS.

US Environmental Protection Agency (1985). Ambient aquatic life water quality criteria for chlorine. EPA 440/5–84-030, Environmental Research Laboratories, Duluth, MN, Gulf Breeze, FL, Narragansett, RI.

US Environmental Protection Agency (1992). Second draft users manual for UVDIS, Version 3.1 UV disinfection process design manual, EPAG0703, Contract No. 68-C8–0023, USEPA, Cincinnati, OH.

US Environmental Protection Agency (1999). Wastewater technology fact sheet, ultraviolet disinfection.

Washington Department of Ecology (Ecology) (1998). Ecology's criteria for sewage works design. Chapter T5.

White G. C. (1999). *The Handbook of Chlorination and Alternative Disinfectants*, 4th Ed., Wiley, New York, pp. 1206–1236.

WHO (1996). World water project. World Health Organization.


## 9.2   SOME EMERGING AND INNOVATIVE PROCESSES

The sections that follow provide an introduction to major elements of industrial and municipal wastewaters reclamation, recycling, and reuse, groundwater remediation and demonstrated technologies. We discuss brine concentrators for recycling industrial wastewaters (Section 9.2.1), the membrane-electrode process (Section 9.2.2), advances in pollution prevention in the metal finishing industry (Section 9.2.3), profitable pollution prevention in electroplating: an in-process focused approach, (Section 9.2.4), and the use of advanced oxidation technology for pollution prevention at foundries (Section 9.2.5).


### 9.2.1   Brine Concentrators for Recycling Wastewater

JON DALAN

Chemical Engineering Consulting Services, 1711 North 163rd, Shoreline, Washington 98133, USA

Brine Concentrators are vapor compression evaporator systems that produce distilled water and a very small salt concentrate stream. These are ideal for water recycling because the concentrate stream is so low that wastewater can be treated economically with a very high recovery and with no liquid discharge.

*Mini-Case Study 9-5: Saving the Colorado River*   The market for brine concentrators initially arose because of federal clean water regulations. The Colorado River, a major source of drinking and irrigation water for the southwestern United States, had been growing increasingly saline as a result of human activities. It was to control such damage to the environment here and elsewhere that the US EPA, in the 1970s, promulgated regulations curtailing discharges to the Colorado and forbidding construction of new plants that could not achieve zero discharge of water into the river.

To comply with the regulations, both new and existing power plants that were using river water had to recycle their water wastes. Although vapor compression was not new to industry as an energy source, the technological fit with power plants was a natural because it allowed these facilities to use electricity they were generating as the source of the mechanical energy needed in recycling water.

Federal law was not the only factor prompting industry to turn to the use of brine concentrators for recycling wastewater. All over the country, there were local siting regulations, as well. Thus with the advent of the private power industry in the early

1990s, entrepreneurs turned to zero discharge water systems, which allowed them to use sites with a limited water supply, far away from discharge points. Similarly, the move to clean-burning natural gas diminished the importance of locating power plants near sources of fuel or water.

As of 2004, there were approximately 60 brine concentrators in the United States. They are sold as package plants, designed and constructed with energy conservation principles. All the original units were at coal fired power plants, but as metal smelters, manufacturers of chemicals and semiconductors, and other enterprises began to recognize the need to eliminate water discharges, the use of brine concentration has spread.

The sections that follow discuss brine concentrators in detail, beginning with process basics.

### 9.2.1.1  Evaporator Basics

In a continuous evaporator process the ratio of the feed to the waste (concentrate) flow rate is called the concentration factor (CF). Since by mass balance:

$$F = D + W$$

where   $F$ = Feed flow rate, lb/h
   $D$ = Distillate flow rate, lb/h
   $W$ = Waste flow rate, lb/h

It follows that for a concentration factor of $F/D$, the ratio of distillate flow to feed flow would be the reciprocal of $F/D$:

$$\frac{D}{F} = \frac{CF - 1}{CF}$$

This second quantity, $D/F$, is called the recovery ratio and has a maximum (although impossible) value of 1. This is the proportion of the treated water that is recovered as distilled water. For nonvaporizing species, the CF times the concentration of the species in the feed water becomes the concentration in the waste stream. The waste stream concentration becomes the determining factor in setting the CF, hence the evaporation rate for a given feed rate. The following tabulation is helpful in providing a feel for these numbers:

| CF  | Recovery |
|-----|----------|
| 2   | 0.5      |
| 5   | 0.8      |
| 10  | 0.9      |
| 100 | 0.99     |

In calculations of feed and concentrate streams, mass flow rates are frequently replaced by the volumetric flow rates. This equivalence is justified because most feed waters and distilled water are very close to each other in density, and the concentrate stream is usually quite small compared to the feed and distillate streams.

The brine concentrator is a single-step evaporator, and if steam is used, it is a single-effect (stage) evaporator. Multiple effects should be used if steam is the heat source used. In multiple effect (stages) evaporation technique, the steam evaporated out of one of the effects (stages) is used as the heating steam in the next effect (stage). The steam economy, defined as the ratio of the motive steam to the total energy required for evaporation, is approximately inversely proportional to the number of effects. For example, a 10-effect evaporator has a steam economy of 0.10.

Specific energy consumption is expressed as kilowatt-hours per thousand gallons of distillate. For high CF systems, where the feed and distillate are approximately equal, specific energy consumption is essentially equal to kW-h/1000 gallons of feed.

For brine concentrators, however, common specific energy consumption is 100 kW-h/1000 gallons of distillate. Upon doing the necessary conversions, we see that this is equivalent to a 12 effect evaporator. Thus the mechanical vapor compression (MVC) technique is considerably more energy efficient than using steam. Vapor compression, however, relies on a consistently adequate supply of electricity.

*Controlling the CF*   The CF in any given brine concentrator is limited by the concentrations obtained in the concentrate, which in turn are reduced by the presence of scaling compounds. However, if the concentrate produces scaling compounds, the evaporator requires constant cleaning, an extremely unreliable operating mode. The prevalent scaling compounds in brine concentrators are calcium sulfate, calcium carbonate, magnesium sulfate and silica, all which appear in brine concentrator applications because BC feed waters are typically concentrates of groundwater or river water.

Calcium Carbonate is eliminated as a precipitating compound by acidifying the feed, thus converting the carbonates to carbon dioxide. As shown in Figure 9.23, a pH below 6.5 will achieve the conversion. If the mass flow rate of suspended calcium sulfate going out the waste steam exceeds the mass flow rate precipitated out of the feed water, then the suspended calcium sulfate must be recycled. The recycling is done with a hydrocyclone on the waste flow, with the concentrated suspended solids (bottoms of the cone) stream being rerouted back to the evaporator and the dilute suspension (top flow of cone) sent to the waste stream (Hodel, 1993).

Calcium sulfate and silica saturations are such that high CF operation is possible only on the dilute feeds. The saturation concentration of calcium sulfate is about 3000 mg/L in brines. The saturation concentration of silica is 150 mg/L. Thus, for a typical water supply with a silica concentration of 5 mg/L feeding a cooling tower, with a tower CF of 7, the allowable CF in an evaporator treating cooling tower blow down is only about 2. With scale inhibitors, these solubility limits can probably be exceed by a factor of 2, providing the scale inhibitor is stable at 212°F, the boiling point of water (Dalan, 2000).

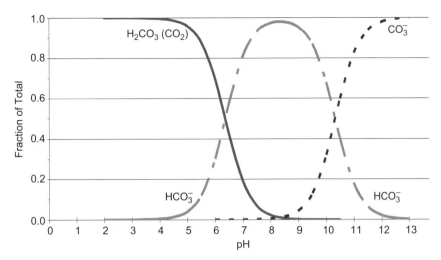

**Figure 9.23.** Alkalinity constituents with pH.

The seeded slurry method of scale control offers a way of overcoming the limitations with these two main scaling compounds. This technology, which has been around since the 1970s, uses a suspension of calcium sulfate to seed the circulating brine, which enables calcium sulfate, silica and other sparingly soluble salts to precipitate on the slurry particles instead of the heat transfer tubes (Dalan, 2000). The solubility limits of calcium sulfate (conservatively stated as 3000 mg/L in brines) and silica (conservatively stated as 150 mg/L) can be exceeded by a factor of 10–100.

If the feed to the evaporator is saturated at these quantities in a facility that does not have the seeded slurry system, the evaporator will immediately precipitate salts on the heat transfer tubes. In such cases, a crystallizer, which is a forced circulation evaporator, must be used instead of a regular brine concentrator type evaporator (Dalan, 2000).

*Falling Film Evaporation*   Brine concentrators use the heat transfer mechanism known as falling film evaporation. The water flow is vertical by gravity down a tube, and a film of water forms on the tubes' inner diameter. The vapor that is evaporated migrates to the center of the tube and travels downward with the water. To maintain the film over the whole length of the tube, only a small portion of the water is evaporated per length of tube and the rest is recirculated. The amount of evaporation per one tube travel, called the extraction per pass, is typically 3–5 percent, indicating that for even thin film and long tube travels, a film is maintained for the whole length of the tube. The water evaporated is replaced by fresh feed to the recirculation loop.

Distribution devices that produce falling films in all the tubes include patented tube inserts or distribution plates. Distribution plates are perforated plates located above the top tube sheet with strategically sized and located holes to produce a film in the top of each tubes (Hodel, 1993).

With a falling film, the required temperature difference between the heating vapor and brine is low, typically $5-10°F$, thus making this mechanism compatible with economical (low pressure ratio) compressors.

Overall heat transfer coefficients (based on the inside tube area) for the falling film mechanisms range from $300-600$ Btu-h$^{-1}$·ft$^{-2}$·°F$^{-1}$. When the seeded slurry mode is used, no fouling factor is necessary.

*Typical Feed Waters*   Two typical feed waters for a brine concentrator from a power plant are cooling tower blowdowns and regenerants from an ion exchange type demineralizer. Cooling tower blowdown compositions can be calculated from the makeup water chemistry by multiplying the makeup water concentration by the cooling tower CF. Some zero discharge facilities discharge their plant wastes, including the regenerant wastes, into their cooling tower basins (Dalan, 2000).

Boiler blowdown from a steam cycle is basically steam condensate, a dilute water stream. This flow typically is part of the brine concentrator feed stream.

Plant wash downs are also typically included. This water is the service water contaminated with local minerals and oil that are part of a typical plant. For oil fired plants, the main contaminant is oil, for coal fired plants, it is coal and ash dust and for gas fired plants, it is typically pretty close to the composition of the local service water, but still with suspended solids.

In coal fired plants, sulfur dioxide is removed from the flue gas by scrubbers. Scrubbers typically use solutions that convert the gases containing oxides of sulfur to ionic sulfate. Lime solutions therefore produce a scrubber blow down saturated in calcium sulfate, which cannot be used as brine concentrator feed unless treated with the seeded slurry method of evaporation.

In newer installations, using natural gas as the power plant fuel, a frequent brine concentrator feed is reverse osmosis (RO) concentrate. RO is frequently used as a pretreatment of cooling tower blow down, or as a first step in a demineralizer train (typically using service water as feed and ending with an EDI (electrodeionization) demineralizing step.

### 9.2.1.2   The Brine Concentrator Package

*Process Description*   Figure 9.24 shows a process flow diagram for a brine concentrator. Filtered feed (particles $<25$ μm in diameter) is introduced to feed tank, to which sulfuric acid and scale inhibitor are added; the acid serves to convert the carbonates and bicarbonates in the feed to dissolved carbon dioxide. After a short residence time, the feed is pumped from the tank through a control valve and into a plate-and-frame heat exchanger, which heats the feed while cooling the distillate from the evaporator. From the heat exchanger, the feed travels to the top of the deaerator, where carbon dioxide is steam-stripped (with steam from the evaporated water) and vented to the atmosphere. All other noncondensible gases in the feed

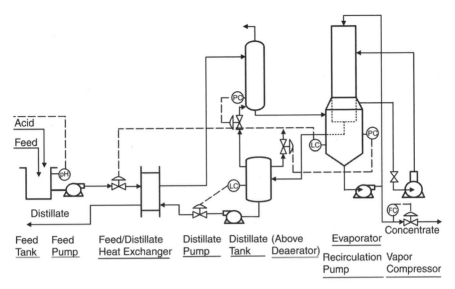

**Figure 9.24.** Process flow schematic for brine concentrator.

stream, the most significant of which is dissolved oxygen, are also steam stripped in the deaerator.

In the deaerator, the feed remaining after the stripping of the gases flows by gravity to the evaporator recirculation sump, the closed tank that receives the liquid from the falling film down the tubes and is the holding volume for the recirculation pump.

Further process details are described in the subsections that follow.

*Distillate Production*   From the liquid surface in the sump and from inside the tubes, evaporated steam flows up through the annulus formed by the tube bundle diameter and the sump diameter. Mist eliminators, located in this annular space, deflect and condense any liquid particles in the vapor stream and return them to the sump. After exiting the top of the mist eliminator, the vapor flows to the inlet of the compressor. In the compressor, the pressure is raised typically 3–5 psi, thus raising the saturated steam temperature 10–15°F. This discharge vapor is routed to the shell side of the tubes. As the vapor boils the water on the inside of the tubes, it condenses, and drains by gravity to the distillate tank. The distillate is pumped out, regulated by a level control loop, through the hot side of the feed/distillate heat exchanger and to its eventual destination, typically a storage tank.

*Concentrate Production*   The concentrate is produced by bleeding off the brine solution from the recirculation loop. The flow is regulated with a flow control loop, utilizing a flow meter downstream of a control valve. The concentrate is routed to a storage tank. Since the recirculation flow rate is much higher than the feed, distillate, or concentrate flows, the concentration of salts is the same in the whole recirculation loop, and the recirculating brine volume acts like a continuously stirred tank reactor.

*Control Loops*   There are seven analog control loops in this system: the feed tank pH, the evaporator sump level, the deaerator inlet steam pressure, the sump (tube side) vapor pressure, the outside steam pressure, the distillate tank level and the concentrate flow rate. The capacity of the system is remotely set by an operator, who adjusts the position of the inlet guide vanes of the compressor. This setting sets the steam flow to the shell side of the heater, thus the distillate flow. Since the feed flow is regulated by the level in the sump, a higher distillate flow induces a higher feed flow; a lower distillate flow induces a lower feed flow.

The feed pH is controlled by acid metering pumps which respond to a pH signal from the well mixed feed tank. The feed flow control valve responds to the sump level and distillate flow. The setting of a vent valve from the distillate tank controls the deaerator inlet pressure. The sump pressure and/or outside steam pressure is controlled by venting the distillate tank. Venting the distillate tank is equivalent to venting the shell side of the heater. A control valve downstream of the distillate pump controls the distillate tank level. The concentrate flow is regulated with a flow control loop. In order to keep the CF constant, the flow is set at a certain ratio with respect to the feed flow.

These systems typically are automated to the point where unattended operation is possible. Safety shut downs exist to shut the system down in case of process upsets. However, these systems do not have an automatic start feature.

### 9.2.1.3  *Concentrate Disposal Processes*   The following concentrate disposal processes and equipment have been used in brine concentrator systems:

- Mixing and wetting the concentrate stream with ash (used in coal fired power plants)
- Evaporation ponds (used in areas with high net evaporation)
- Spray dryers
- Drum dryers
- Forced circulation crystallizers
- Belt Filters

*Mixing the Concentrate Stream*   When the concentrate stream is small, it can feasibly be mixed with ash from a coal-wood burning boiler because the solid component is usually hot or can be subjected to natural evaporation. This disposal option is clearly the least expensive, but can only be used in the right circumstances. Because of inherent leachability of the salt components, mixing a concentrated salt with a solid waste is a heavily regulated endeavor.

*Evaporation Ponds*   Evaporation ponds are constructed much like stormwater detention ponds, with gently sloped walls, an impermeable liner, a semirigid liner foundation, monitoring wells and an inflow structure. Nowadays it is necessary to equip the monitoring wells with lifelong data collection devices.

Pond depth, which seldom exceeds 10 ft (for a 30-year life of a salt stream containing 30 percent by weight total solids) is determined by storm water flow events and, more importantly, the depth required to collect the salt from the concentrate for the life of the plant. These ponds necessarily cover a large area; for example, 92 acres is required for a site with an anticipated net evaporation of 10 inch per year and a concentrate stream of 2 gpm.

Evaporation ponds were used in the early days of brine concentrators because of their low operating costs and relatively inexpensive construction expense, but they are seldom used any more. Nevertheless, their low operating costs (the evaporation is done by solar energy) and relatively low capital costs make them attractive in remote arid locations, where rainfall is light and land is cheap.

*Spray Dryers*   Spray dryers are cyclone-like devices into which concentrate is atomized at the top by a centrifugal disk or pneumatic atomizer. Hot gas is introduced at a sufficient rate and temperature to evaporate the free water before any moist solids can reach the dryer vessel wall. The dry solids that form are carried out with the hot gas and water vapor out the bottom of the drying chamber and are delivered to a cyclone collector to remove to remove larger size particles. Entrained dust that is not removed in the cyclone is conveyed into a bag house and the dust-free off gas and water vapor from the concentrate stream is released to the atmosphere. Solids collected in the cyclone and baghouse are pneumatically conveyed to a storage hopper. From the storage hopper, the solids can be trucked to disposal (Dalan and Rosain, 1992). A schematic of a spray dryer system is shown in Figure 9.25.

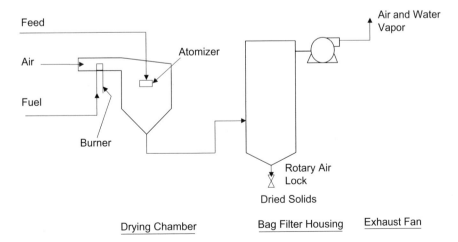

**Figure 9.25.** Schematic of spray dryer.

Spray drying itself is conceptually simple. However, combined with a hot air or combustion system, the feed system and the solids capture and pneumatic handling systems, the system becomes fairly complicated. The most common operating problem is plugging or erosion of the spray nozzles with the concentrate stream. In addition, not all brine concentrator concentrates can be spray dried. Pilot plant testing should precede the selection of spray drying equipment for an application.

The controls of the spray dryer consist of regulators that maintain inlet hot gas temperature at a constant gas flow rate and keep the inlet liquid flow rate and temperature within their allowable ranges. A cascade arrangement, in which the gas temperature depends on the inlet slurry flow and temperature, is sometimes used (Shaw, 1994).

The specific energy draw of a spray dryer is typically 2,000 Btu per pound of water evaporated. Spray dryers are not used in brine concentrator systems that have a concentrate flow rate above 10 gpm, both for capital cost reasons and because of the high energy consumption.

*Drum Dryers*  Drum dryers, with their simplicity and low capital cost, are gaining in popularity as a final concentrate disposal technique for brine concentrators, especially for low (<5 gpm) concentrate flow rates.

A drum dryer (Figure 9.26) consists of two rotating hollow drums, which rotate into each such that the top gap, or nips, holds the slurry to be dried. The feed is distributed to the length of the nip. The inside of the drums is heated with steam. As the slurry is evaporated in the nip space, a fan exhausts the water from the

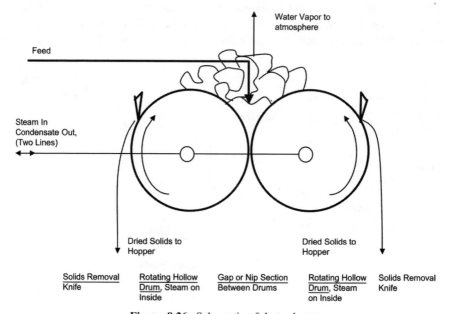

**Figure 9.26.** Schematic of drum dryer.

slurry. The dried solids stick to the drum and at approximately 90° from the nip, knives scrape off the solids, either into a conveyor or a hopper below the twin drums.

The steam pressure and flow rate, the nip spacing, the slurry level in the gap, the slurry feed flow rate, the drum speed, and the force on the knives are all independently controllable. All these controls are manipulated to produce a dry solid for a given feed flow and feed slurry composition.

Drum dryers typically are operated intermittently. Thus they are set up for one set of feed slurry conditions at a time and preparing the equipment for each new set of conditions requires much operator attention. However, the dryers are relatively inexpensive and energy economical, using as little as 1,300 Btu of steam per lb of water evaporated.

*Forced Circulation Crystallizers*  By far the most common concentrate disposal process makes use of forced circulation crystallizers, like the one schematically represented in Figure 9.27. The circulating brine is forced through the heat transfer tubes, into a vapor disengagement vessel, and returned to the suction of the recirculation pump. The slurry, typically 20 percent by weight suspended solids, rejuvenates its solid salt concentration as the feed enters and the evaporation occurs.

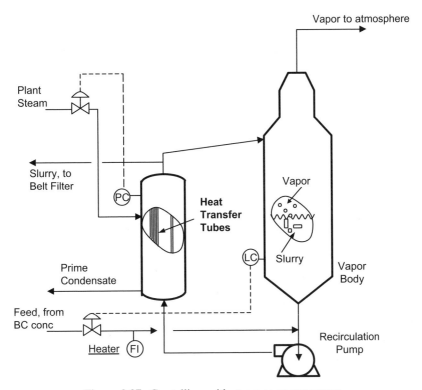

**Figure 9.27.** Crystallizer with steam as energy source.

The blowdown stream is sent to a solids separator, which can be a filter press, centrifuge or belt filter, to produce the nearly dry salt (crystallization).

The energy input to the system is regulated by the pressure on the shell side of the heater. Level in the vapor body is regulated by the feed flow. The suspended solids in the slurry are regulated by the frequency of the intermittent blow down.

For smaller units, outside steam is used, with the steam condensate returned to its boiler and the vapor from the slurry vented to the atmosphere.

In larger units, typically above 20 gpm of evaporation, a vapor compression cycle is used, as shown in Figure 9.28. In this cycle vapor body steam is compressed in a positive displacement rotary lobe blower, and compressed steam serves as the heating medium for the heater. The controls of this unit are much like the brine concentrator, except there is no deaerator or distillate heat exchanger. The evaporation capacity is regulated by the amount of bypass steam in the compression cycle, the vapor body level on feed forward control, and the tube side pressure, which is controlled by adjusting the distillate tank vent setting. The distillate is typically sent to the evaporator distillate tank, where it is mixed with the evaporator distillate and cooled by the incoming evaporator feed.

*Belt Filter*   Figure 9.29 shows a schematic of a belt filter, which is typically used as the solids dryer with forced circulation crystallizers.

The filtration is a batch process. A charge of slurry is gravity fed to a chamber which consists of the filter media sitting on a porous support plate as the floor and a closed box held down by pneumatic pressure as the wall and roof. When the chamber is full of slurry, the feed is shut off and air pressurizes the chamber. The air forces the water from the slurry into a receiving pan, and the water drains back to a holding tank. The solids remain on the filter medium, typically a tight non-woven sheet. The holding tank contents are pumped back to the crystallizer

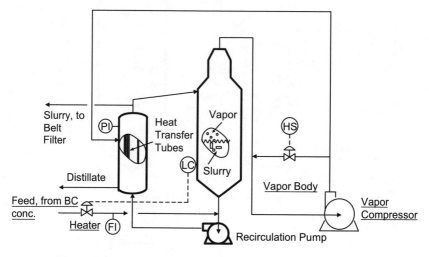

**Figure 9.28.** Crystallizer with vapor compressor as energy source.

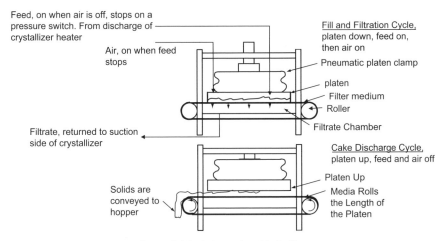

**Figure 9.29.** Schematic of belt filter.

recirculation loop. The when the cake is dry, as indicated by a decrease in air pressure to the chamber, the air is turned off. Then the walls and roof (called the platen) is raised and the filter medium is rolled off and the solids fall into a hopper. The batch cycle then starts again with the closing of the platen and the introduction of more feed.

The batch process capacity is governed by an interval timer, which sets how often a batch cycle is started, and by a duration timer, which sets how long the feed valve is open. The cycle duration is governed by how long it takes for the motive air pressure to go down. There is a trade off between the interval and duration timer settings, as far as producing the desired dryness cake at the right capacity. Cake dryness is typically specified at 80 percent solids. The parameters are set at the start up of the system and usually are not changed extensively after that except for the interval timer.

### 9.2.1.4 *Parametric Considerations*

*Parametric Energy Consumptions* Energy typically constitutes about 60 percent of the operating costs of brine concentration systems. In life-cycle cost analysis, the cost of energy typically exceeds the periodic capital charge attributed to the turnkey price of the equipment. Table 9.8 shows some typical energy con-

**TABLE 9.8. Typical Specific Energy Consumption Values for Brine Concentrator Systems**

| Process | Specific Energy Consumption |
| --- | --- |
| Vapor compression evaporator | 100 kW-h/1,000 gallons of distillate |
| Crystallizer, using vapor compression | 200 kW-h/1,000 gallons of crystallizer distillate |
| Crystallizer, using steam | 1.1 lbs of steam/lb of steam evaporated |
| Drum dryer | 1.5 lbs of steam/lb of steam evaporated |
| Spray dryer | 2,000 Btu/lb. of steam evaporated |

sumptions for the unit processes of the brine concentrators. However, to make the unit costs translate into a more complete cost picture, a selection chart like Figure 9.30 must be consulted. With a higher CF than 60, it does not make much of a difference which concentrating technology is used. Our selection chart shows that below a CF of 20, a crystallizer should be a mechanical vapor compression unit.

*Parametric Capital Costs*  Figures 9.31 and 9.32 show parametric capital costs for some the technologies discussed in Section 9.2.1.2. The costs are parametric; that means that they can vary ±30 percent from the chart figures.

As shown in Figure 9.31, crystallizers cost more than MVC evaporators when the flow rate is under 50 gpm. Rarely is a stand-alone crystallizer with a capacity above 50 gpm installed in a zero liquid discharge system. Figure 9.32 shows that between 5 and 10 gpm, spray dryers are less expensive than crystallizers. Frequently, a steam driven crystallizer is used between 5 and 20 gpm, and a vapor compression crystallizer above 20 gpm of crystallizer evaporation.

### 9.2.1.5 Zero Discharge Systems Using Brine Concentrators  For simple systems, such as the production of demineralized water for a steam cycle in a plant that needs to practice zero discharge, reverse osmosis (RO) is used as a first step in the demineralizing train. Reverse osmosis is frequently accomplished in two passes, with the permeate (clean stream) reprocessed in second RO system, thus producing water very low in total dissolved solids (TDS) as feed to demineralizing equipment. Demineralization is done by ion exchange, or a newer process called electrodeionization.

For zero discharge to be possible, the concentrate from the first pass of the RO must be treated with a brine concentrator. Figure 9.33 shows a block

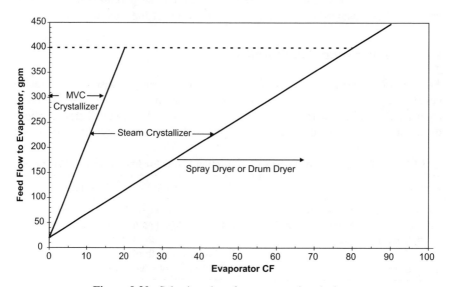

**Figure 9.30.** Selection chart for concentration devices.

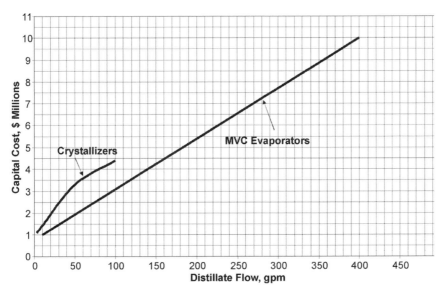

**Figure 9.31.** Parametric capital costs evaporators and crystallizers source (Dalan and Rosain, 1992; Haussman and Rosain, 1996), updated to 2002.

diagram of a generic demineralizer converted to perform as a zero discharge system.

For plants with cooling towers, almost exclusively power plants, cooling tower blow down is sometimes treated with RO, assuming the cooling tower CF is not

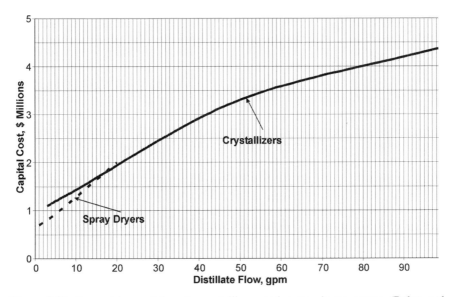

**Figure 9.32.** Parametric capital costs crystallizers and spray dryers source (Dalan and Rosain, 1992; Haussman and Rosain, 1996), updated to 2002.

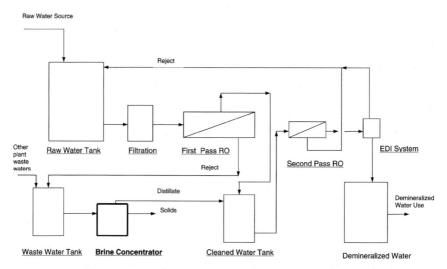

**Figure 9.33.** Schematic of brine concentrator as part of demineralized water production.

so high that the circulating cooling tower is saturated in either silica, calcium carbonate or calcium sulfate. The RO product receives a second pass and is fed to a demineralizer, while the first pass RO concentrate is fed to a brine concentrator system, as shown in Figure 9.34.

### 9.2.1.6 Alternative Zero Discharge Methods

We mention briefly staged cooling and high efficiency reverse osmosis (HERO) systems, two technologies that may, as time passes, supplant brine concentrators as zero liquid discharge systems. If, however, the waste stream flows are high enough, concentrating devices, including brine concentrators will still be needed in true zero liquid discharge systems.

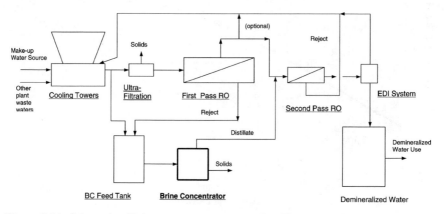

**Figure 9.34.** Schematic of brine concentrator as part of cooling tower blow down treatment.

*Staged Cooling*  In power plants with water cooled condensers, the effective cooling tower CF can be driven quite high if the scale formation chemicals in the cooling tower water are removed by softening (removing calcium and magnesium) a slip stream of the cooling. A variation of this idea, called staged cooling, has been patented by Eau Tech Partners, and is in operation in at least three plants. Figure 9.35 is a system schematic (Sanderson and Lancaster, 1988, 1989)

The system divides the plant cooling load into two condenser sets and two cooling towers, with the blow down from the first cooling tower being sent to a chemical softener and then as the feed to the second cooling tower. The second cooling tower then has a slip stream softener, thus enabling a high cooling tower CF on the second tower. A final concentrating device, perhaps a pond, must be used. The final solids concentrate of the staged cooling system is reported as 100,000 mg/L TDS, about 1/3 of what is possible in an evaporator system (Sanderson and Lancaster, 1988, 1989).

*High Efficiency RO Systems (HERO) Process*  A recently patented system overcomes the silica limit in reverse osmosis separation by softening the feed and operating the reverse osmosis system at high pH, where silica becomes soluble. Softening the feed removes calcium and magnesium, so that calcium carbonate, calcium sulfate or magnesium sulfate don't precipitate in the concentrate

A schematic of the proprietary operating system is shown in Figure 9.36.

The RO system CF is limited by the osmotic pressure exerted by RO concentrate. With the current state of RO membranes and pumps, a final TDS of 100,000 mg/L is possible. A final concentrating device is still needed. Figure 9.36 shows the use of an evaporation pond.

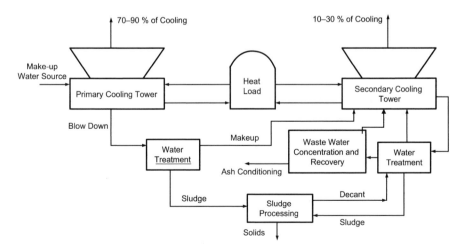

**Figure 9.35.** Schematic of stage cooling system (Dalan and Rosain, 1992; Sanderson and Lancaster, 1988, 1989).

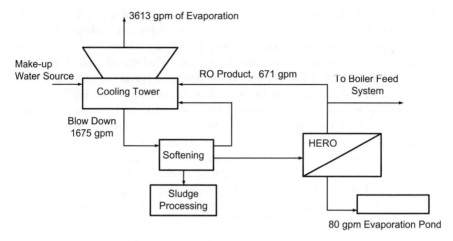

**Figure 9.36.** Schematic of HERO process (Duke Energy Inc.)

### 9.2.1.7 Economics of Brine Concentrator Systems   Based on the parametric cost information on brine concentrator systems given earlier (Section 9.2.1.4), we will now overview some economic aspects of these systems.

Brine concentrators, and by extension, integrated zero liquid discharge systems, only make sense in grass roots facilities only when:

- There is a shortage of surface or well water
- The facility needs water to operate
- There is no water discharge option, such as a source of potable water or a river
- A discharge permit is unobtainable

Zero liquid discharge takes away the siting constraint; that is, it is no longer necessary to locate the plant near a large usable water source or suitable discharge point. The economic advantage of such an effect is hard to generalize, being very specific to the actual circumstance.

The net present value of a zero discharge facility is negative. Dalan and Rosain (1992) found that a 265 gpm brine concentrator had operating costs of $8.94/1,000 gallons of water treated (total of $1,252, 600/year), while the avoided cost of extra demineralizer regeneration chemicals was $216,000/year. The avoided cost amounts to 1.54/1,000 gallons. Since any dollar amount here is an operating expense, the cash flow is negative.

A typical specific operating expense is in the $5–$7/1,000 gallons treated range. In this estimate is the price of electricity at a retail price of 0.05/kW-h. In grass roots power plant planning, the electricity is many times considered a parasitic load on the power plant (electricity needed to produce the power). In this accounting method, the cost of electricity is zero. The elimination of electricity as an operating cost brings the total treatment cost down to the $2–3/1,000 gallons range.

The typical energy load for a brine concentrator is 100 kW-h/1000 gallons of water produced.

The determination of water economics is very geography specific. For example, in western Washington, water and sewer bills typically are in the $1-2/1,000 gallon range, where as in eastern Washington, water costs more, if indeed it is economically available.

For non-power plant applications, local high prices for electricity can be overcome by seeking out alternate energy sources. Compressors (the majority energy user) in Brine Concentrator plants have been installed that are steam driven or natural gas (via a natural gas engine) driven. Table 9.9 presents breakdown of typical operating costs (Dalan and Rosain, 1992; Haussman and Rosain, 1996).

For existing facilities, installing a zero discharge plant makes sense when:

- The discharge permit conditions change, with the result that the existing facility can no longer be operated economically
- The cost of water rights plus treatment fees exceed the operating costs of a zero discharge facility

This last point is illustrated by Mini-Case Study 9-6.

*Mini-Case Study 9-6: Calculating Payback for a Zero Discharge System*  A power plant has a 300 gpm blow down stream from the plant. Preliminary estimates show that a zero discharge plant would cost $4.55/1,000 gallons to operate. The local utility raises the water and disposal fees to $3.00/1,000 gallons, at a total cost to the plant of $6.00/1,000 gallons.

The capital cost of a 300 gpm, from Figure 9.31, is approximately $7.5 million. The annual saving is ($1.45 × 300 × 1440 × 365)/1000, or $228,636/year. This represents a project with a simple ROI of 3 percent and a simple payback of 32 years.

**TABLE 9.9. Operating Cost Breakdown for a 265 gpm System Resulting in a Cost of $9.62/1,000 Gallons of Feed**

| Item | Consumption | Unit cost | Annual Cost |
|---|---|---|---|
| Operating labor | 1 mhr/h | $50/mhr | $438,000 |
| Maintenance (labor) | .2 mhr/h | $50/mhr | $87,600 |
| Maintenance (materials, including spare parts) | | | $80,000 |
| Electricity | 1,617 kW-h | $0.05/kW-h | $707,000 |
| Chemicals | | | |
| Sulfuric acid | 293 lb/day | $0.06/lb | $6,416 |
| Polymers | 40 lb/day | $1.50/lb | $21,900 |
| Total chemicals | | | $28,316 |
| Total annual operating cost | | | $1,340,916 |
| $/1,000 gallons of feed | | | $9.62 |

*Source:* Dalan and Rosain (1992), updated by Dalan to 2002.

## REFERENCES

Dalan, J. (2000). 9 things to know about zero liquid discharge. *Chem. Eng. Progress*, **98**(11), 71–76,

Dalan J., and Rosain, R. (1992). Zero discharge wastewater treatment facility for a 900 MW GCC power plant. Report No. EPRI TR-100375, Electric Power Research Institute, Palo Alto, CA.

Duke Energy Inc., Duke energy selects HERO membrane process for ZLD application. Project profile series #002, Aquatech, Canonsburg, PA, available on www.aquatech.com.

Haussman, C., and Rosain, R. (1996). Power plant wastewater treatment technology review report. Report No. EPRI TR-10781, Electric Power Research Institute, Palo Alto, CA.

Hodel, A. (1993). Evaporators spawn zero discharge. *Chem. Processing*, **56**(9), 26–30.

RCC, Resources Conservation Company, a division of Ionics, Inc, Watertown, MA (2003) *Crystallization*, available at www.ionicsrcc.com

Sanderson, W. G., and Lancaster, R. L. (1988). The staged high recycle cooling process— acceptance test operations at the Wheelabrator/Shasta Power Plant. ASME/IEEE Power Generation Conference, 88 JPGC/PWR-52, Philadelphia, PA.

Sanderson, W. G., and Lancaster, R. L. (1989). The water concentrating cooling tower—an application of staged cooling as a final concentration step in a zero discharge system. 50th Annual Meeting International Water Conference, IWC-89–12, Pittsburgh, PA.

Shaw, F. W. (1994). Fresh Options in Drying. *Chemical Eng*, July.

## 9.2.2  Membrane-Electrode Process

VIKRAM GOPAL

7585 Roxbury Drive, Ypsilanti, MI 48197, USA

The membrane-electrode (M-E) process is a novel "green technology" for selective heavy metal removal and recovery from aqueous industrial effluents. Selective cation recovery from dilute, aqueous, multi-cation waste streams addresses a solution method to a class of problems with worldwide implications. The absence of large scale systems to handle such waste streams and severe limitations of conventional technologies have been the motivating forces behind the development of the membrane-electrode (M-E) process (Gopal, 1998).

Current treatment methods in heavy metal recovery include pH adjustment, chemical oxidation, chemical reduction, electrochemical reduction, ion-exchange, adsorption, and flocculation. Individually or a combination of one or more of these well established methodologies allow for more than 99 percent removal of metals. However, the form(s) in which the metals are removed makes it virtually impossible to recycle or reuse them (Eilbeck and Mattock, 1987; Patterson, 1985).

The M-E process is a hybrid process combining "selective cation removal," a property of electrochemical reduction and cation exchange in dilute solutions, a behavior of ion-exchange materials. The M-E process has been effectively demonstrated for selective recovery of $Pb^{2+}$ and $Cu^{2+}$ ions from $Pb^{2+}/Fe^{2+}$, and

$Cu^{2+}/Ni^{2+}$ binary systems, respectively, and for $Cu^{2+}$ ion recovery from $Cu^{2+}/Ni^{2+}/Zn^{2+}$ ternary system.

### *9.2.2.1 Process Description*

A typical M-E process consists of a cation exchange membrane cathode, and a graphite plate anode (Figure 9.37). A dc voltage supply varying between 0 and 10 V is used to apply a potential difference (pd) between the electrodes. To minimize solution resistance, the electrodes are placed only 10 mm apart in the solution (Gopal et al., 1998).

The potential difference applied across the electrodes can be optimized to maximize selectivity in cation exchange. Typically, it is found that the cations can be recovered sequentially in order of their increasing electrochemical reduction potentials. In developing the M-E cell, the author began with a simulated waste stream having a batch volume of 50 mL; the setup was gradually expanded to use larger volumes (up to 400 mL) and semicontinuous processing. Temperature, pH, and conductivity were measured before and after ion-exchange. Care was taken to ensure that no electrical connections on the cathode were exposed to solution, and only the ion-exchange surface came into contact with solution. Samples for cation analysis were taken at periodic intervals of time from locations between the electrodes and from the bulk solution. The solution was continuously agitated in order to minimize concentration polarization.

In the M-E process, there is an enhancement in the migration of cations in the bulk solution, while at the membrane surface, a controlled ion-exchange takes place (Gopal et al., 1997; Gopal et al., 1998). The energy supplied at the cathode surface allows a preferential cation exchange of the more noble cation onto the membrane surface. At the end of each experimental run, the depleted membranes

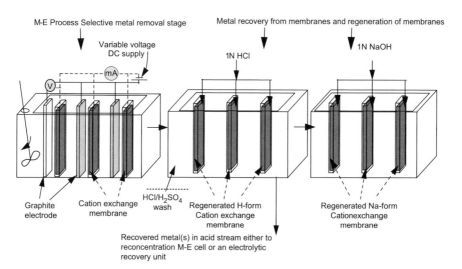

**Figure 9.37.** M-E process schematic–removal, and recovery of heavy metals and regeneration of membranes.

with the heavy metal cations are moved into a cation recovery system (1N HCl acid wash), where the cations from the membranes are exchanged for $H^+$ ions from the acid. If Na-form membranes are required, the H-form membranes are converted to Na-form by ion-exchange with NaOH or NaCl solution. As a rule, the regenerated membranes are reused in the M-E cell; in the case of Na-form polystyrene sulfonic acid membranes, however, it is necessary to convert $H^+$ ions into $Na^+$ by reacting the membranes with NaCl or NaOH.

***9.2.2.2 Membrane Description*** The membranes, which comprise polystyrene sulfonic acid (Na-form) cast on a conductive base, were prepared and characterized in the laboratory using conventional techniques (Gopal et al., 1998). A balance between crosslinking density and ion-exchange capacity was maintained in order to achieve stable membranes with maximum ion-exchange capacity.

The following are the design parameters considered while preparing the membranes:

- Crosslinking density of the membrane
- Ion-exchange capacity
- Film thickness
- Specific resistance
- Type of membrane base

*Definitions and Units of M-E Process Parameters*

RATE OF ION-EXCHANGE ($R_{IX}$)   It is defined as the amount of cations exchanged per square millimeter of membrane surface per minute. In Eqs. (9-11) through (9-15), membrane surface area is always expressed in square millimeters, and the amount of cations is always given in milligrams. The target cation is one that must be preferentially removed either due to environmental reasons or because of economics. All other cations in solution are referred to as "secondary" cations.

$$\text{Expressed as: } \left( \frac{\text{amount of cation exchanged}}{\text{membrane} \cdot \text{min}} \right) = \text{surface area/min} \quad (9\text{-}11)$$

where the amount of cations exchanged is determined by analyzing test solution at $t = 0$ minute and then at regular intervals of time. All metal analysis was performed on a flame atomic absorption spectrophotometer (FLAA).

CATION RECOVERY FLUX/CATION EXCHANGE FLUX ($C_{Xf}$)   The amount of cations exchanged per square millimeter of membrane surface.

$$\text{Expressed as: } \left( \frac{\text{amount of cation exchanged}}{\text{membrane surface area}} \right) \quad (9\text{-}12)$$

SELECTIVITY OF TARGET ION IN BINARY SOLUTIONS ($I_{sel}$)  The ratio of the fraction of target cation on the membrane surface to the fraction of the target cation in solution.

$$\text{Expressed as: } \frac{\left(\dfrac{\text{target cation on the membrane}}{\text{secondary cation on the membrane}}\right)}{\left(\dfrac{\text{target cation in solution}}{\text{secondary cation in solution}}\right)} \qquad (9\text{-}13)$$

where "target cation on the membrane" is determined after regeneration with HCl and FLAA analysis of HCl solution for the target cation.

LOADING OF THE MEMBRANE ($M_{load}$)  The ratio of ion-exchange sites used to the total available ion-exchange sites on the membrane, expressed as a percentage:  - Defined as:

$$\left(\frac{\text{amount of cations on the membrane}}{\text{maximum cation exchange capacity of the membrane}}\right)100\% \qquad (9\text{-}14)$$

where cation exchange capacity is in milligrams.

RECOVERY RATIO OF TARGET CATION ($I_{rec}$)  The ratio of the amount of target cation exchanged to the amount of the secondary cation of interest exchanged onto the membrane surface. The result is expressed as a real number.

$$\text{Expressed as:}\left(\frac{\text{target cation exchanged on the membrane}}{\text{secondary cation exchanged onto the membrane}}\right) \qquad (9\text{-}15)$$

### 9.2.2.3 Application of Membrane-Electrode Process to Binary Cation Systems

*System 1 : 100 ppm $Cu^{2+}$/ 100 ppm $Ni^{2+}$ Solution*  Figure 9.38 shows that increasing the potential difference on the $Cu^{2+}$ ion selectivity in a 100 ppm

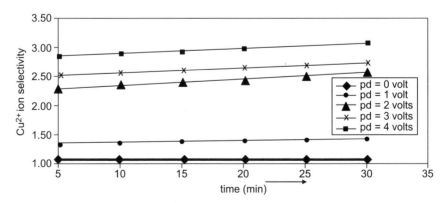

**Figure 9.38.** Effect of pd on $Cu^{2+}$ ion selectivity profile in a 100 Cu/100 Ni solution.

Cu/100 ppm Ni solution mixture increases the $Cu^{2+}$ ion selectivity (Gopal and April, 1998a). For instance, after 30 minutes of ion-exchange, and at a potential difference of 0 volt (conventional ion-exchange), the $Cu^{2+}$ ion selectivity is 1.08, while it is 1.48 at 1 volt, 2.50 at 2 volt, 2.52 at 3 volt, and 3.18 at 4 volt. Error bars indicate 95 percent confidence interval for all data points.

*System 2: 100 ppm $Pb^{2+}$/100 ppm $Fe^{2+}$ Solution*   Figure 9.39 shows that increasing the potential difference from 0 volt (conventional ion-exchange) to 2 volt causes $Pb^{2+}$ ion selectivity to increase from 1.2 to 1.6.

### 9.2.2.4   Mechanism of the M-E Process

*Existence of Reverse Potential (RP) Phenomenon*   Increasing the potential difference in the M-E process results in a lower but more selective rate of ion-exchange. The inverse relationship (potential difference vs rate of ion-exchange), called the RP phenomenon, provides a means of controlling the rate of ion-exchange in the M-E process that is not available in conventional ion-exchange processes (Gopal and April, 2000). Table 9.10 illustrates the effect of the RP phenomenon on selective $Cu^{2+}$ ion recovery from a Cu/Ni binary system (Gopal and April, 1998a; Gopal and April, 2000). After 30 minutes of ion-exchange, 65% $Cu^{2+}$ ions and 64 percent $Ni^{2+}$ ions were recovered at a potential difference of 0 volt, while 48 percent $Cu^{2+}$ ions and 40.5 percent $Ni^{2+}$ ions were recovered at a potential difference of 1 volt, and 34 percent $Cu^{2+}$ and 17 percent $Ni^{2+}$ at a potential difference of 2 volt. The corresponding values of $Cu^{2+}$ recovery ratios are 1.03, 1.21, and 2, respectively, which indicate a more selective $Cu^{2+}$ ion-exchange with increasing potential difference.

*Factors Affecting RP Phenomenon*   A 2 level, four factor experiment was designed to determine significant factors affecting the RP phenomenon. The

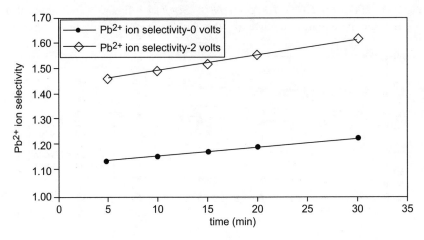

**Figure 9.39.** Effect of potential difference on $Pb^{2+}$ ion selectivity.

**TABLE 9.10. The Reverse Potential (RP) Phenomenon in (100 ppm Cu/100 ppm Ni) Binary Systems**

| Time (min) | 0 Volt % $Cu^{2+}$ Ion Recovery | 0 Volt % $Ni^{2+}$ Recovery | 1 Volt % $Cu^{2+}$ Ion Recovery | 1 Volt % $Ni^{2+}$ Recovery | 2 Volt % $Cu^{2+}$ Ion Recovery | 2 Volt % $Ni^{2+}$ Recovery |
|---|---|---|---|---|---|---|
| 0 | 0.00 | 0.00 | 0.00 | 0.00 | 0.00 | 0.00 |
| 5 | 15.61 | 15.27 | 9.87 | 8.29 | 6.22 | 3.05 |
| 10 | 28.83 | 28.92 | 19.29 | 15.89 | 12.41 | 6.01 |
| 15 | 40.05 | 38.63 | 26.55 | 22.86 | 17.43 | 8.88 |
| 20 | 51.19 | 49.55 | 35.03 | 29.25 | 23.93 | 11.66 |
| 30 | 65.39 | 64.35 | 47.91 | 40.49 | 33.61 | 16.97 |

factors varied were potential difference, initial cation ($Cu^{2+}$) concentration, thickness of coating, and type of membrane (PSM-10 or CPSM-10). Table 9.11 lists the coordinates defining the experimental space and the corresponding cation recovery flux after 30 minutes of ion-exchange. This experimental design also includes some additional points for validation purposes.

**TABLE 9.11. Effect of M-E Process Variables on Cation Recovery Flux**

| X1 pd (Volt) | X2 Initial Cation Conc. (ppm) | X3 Coating Thickness ($\mu$m) | X4 Type of Membrane | Y Cation Recovery Flux at 30 minutes (mg/mm$^2$ Membrane) |
|---|---|---|---|---|
| 0.00 | 30 | 25 | 1 | $5.03 \times 10^{-4}$ |
| 0.00 | 30 | 25 | 2 | $5.00 \times 10^{-4}$ |
| 0.00 | 30 | 50 | 1 | $5.10 \times 10^{-3}$ |
| 0.00 | 30 | 50 | 2 | $5.05 \times 10^{-3}$ |
| 0.00 | 100 | 25 | 1 | $2.72 \times 10^{-3}$ |
| 0.00 | 100 | 25 | 2 | $2.66 \times 10^{-3}$ |
| 0.00 | 100 | 50 | 1 | $2.75 \times 10^{-3}$ |
| 0.00 | 100 | 50 | 2 | $2.71 \times 10^{-3}$ |
| 1.00 | 30 | 25 | 1 | $3.97 \times 10^{-4}$ |
| 1.00 | 30 | 25 | 2 | $3.90 \times 10^{-4}$ |
| 1.00 | 30 | 50 | 1 | $4.10 \times 10^{-4}$ |
| 1.00 | 30 | 50 | 2 | $4.06 \times 10^{-4}$ |
| 1.00 | 100 | 25 | 1 | $1.89 \times 10^{-3}$ |
| 1.00 | 100 | 25 | 2 | $1.85 \times 10^{-3}$ |
| 1.00 | 100 | 50 | 1 | $1.95 \times 10^{-3}$ |
| 1.00 | 100 | 50 | 2 | $1.90 \times 10^{-3}$ |
| 2.00 | 100 | 50 | 2 | $1.37 \times 10^{-3}$ |
| 2.00 | 100 | 50 | 1 | $1.40 \times 10^{-3}$ |

[1]Coding for the "Type of membrane": 1 - PSM-10, 2 - CPSM-10. Further both the membranes were based on conducting rubber and had the same ion-exchange capacities (IEC).

Table 9.12 shows the results obtained by regressing the data in Table 9.13. It is determined that the $F_{calculated} = 143.574$, which is greater than $F_{tabulated} = 5.21$, indicating that the variations in the cation recovery flux are not by chance. Also, the significant factors affecting the cation recovery flux are potential difference and cation concentration.

### 9.2.2.5   System Independence of the RP Phenomenon

The technical feasibility of the M-E process in different cation systems was validated by showing that the RP phenomenon is independent of cation type (Gopal, 1998). In the experiment for which the results were shown in Table 9.13 numerical factors varied include conductivity of solution (a direct function of cation concentration) and potential difference, while the categorical factor varied was "cation solution type." The response studied was "cation recovery flux."

An increase in the potential difference decreased the cation recovery flux for all solutions. For example, in a solution with an initial conductivity 285 $\mu$S/cm, an increase in the potential difference by 2 volt decreased the cation recovery flux by 48 percent. This behavior is independent of initial cation concentration in solution as well as type of cation(s) in solution indicating that the RP phenomenon exists in all solutions.

Conductivity of solution increases with ion-exchange in all cases because the H-form of CPSM-10 membrane was used in this study. These membranes exchanged $H^+$ ions for the cations in solution and since $H^+$ ions have a much higher conductivity than heavy metal cations, the conductivity increased. It is interesting to observe the change in conductivity of the solution at different potential differences. At high potential difference, that is, when the cation recovery flux is the least, the differential solution conductivity is the smallest, and vice-versa. This is due to the "RP phenomenon."

**TABLE 9.12. ANOVA of the Data Listed in Table 9.11**

|  | Df | SS | MS | F (Calculated) |
|---|---|---|---|---|
| Regression | 4 | 5.717E-06 | 1.4293E-06 | 143.574 |
| Residual | 13 | 1.294E-07 | 9.955E-09 | F (tabulated at 1 percent level of significance) $= 5.21 < 143.57$, variations in cation recovery flux are significant |
| Total | 17 | 5.847E-06 |  |  |
|  | Coefficients | Standard error | $t$ stat (Calculated) | $t$ Stat (tabulated) at 1 percent level of significance $= 3.012$ |
| Intercept | 0.00054 | 0.00011 | 4.82 | A significant factor |
| X1 (pd) | $-0.00048$ | 3.699E-05 | $-12.91$ | A significant factor |
| X2 (conc.) | 1.558E-05 | 6.946E-07 | 22.42 | A significant factor |
| X3 (coating) | 4.849E-07 | 1.945E-06 | 0.2493 | Not a significant factor |
| X4 (membrane type) | $-1.993E-05$ | 4.703E-05 | $-0.424$ | Not a significant factor |

**TABLE 9.13. Factors Affecting RP Phenomenon**

| Type of Solution | pd (Volt) | Solution Conductivity Prior to Ion-Exchange ($\mu S/cm$) | Solution Conductivity Post Ion-Exchange ($\mu S/cm$) | Difference in Conductivity ($\mu S/cm$) | Total Cation Recovery Flux at 30 min. ($mg/mm^2$ Membrane) |
|---|---|---|---|---|---|
| 5 | 0 | 534 | 1345 | 811 | 4.56E-03 |
| 5 | 1 | 534 | 1241 | 707 | 3.78E-03 |
| *5* | *2* | *534* | *1405* | *871* | *6.98E-03* |
| 5 | 2 | 534 | 980 | 446 | 2.65E-03 |
| 4 | 0 | 98.5 | 299.5 | 201 | 7.22E-04 |
| 4 | 1 | 98.5 | 211.5 | 113 | 6.27E-04 |
| 4 | 2 | 98.5 | 233.5 | 135 | 3.01E-04 |
| 3 | 0 | 104 | 463 | 359 | 1.03E-03 |
| 3 | 1 | 104 | 226 | 122 | 4.87E-04 |
| 1 | 0 | 285 | 740 | 455 | 2.00E-03 |
| 1 | 1 | 285 | 689 | 404 | 1.93E-03 |
| 1 | 2 | 285 | 538 | 253 | 1.04E-03 |

Code for the "type of solution": 1–100 ppm Cu, 3–30 ppm Cu, 4–30 ppm Ni, 5–100 ppm Cu/100 ppm Ni.

In Table 9.13, the data set in the third row, displayed in bold italics, indicates a conductivity of solution after ion-exchange that is much higher than expected. Indeed, what we know of the RP phenomenon suggests that a higher potential difference should result in a lower cation recovery flux, while maintaining concentration constant, thereby resulting in a lower solution conductivity. On investigation, it was found that the surface of the membrane had pinhole defects, which had resulted in a loss in control of the rate of ion-exchange. This condition was rectified, but the discrepant data set is left as part of the tabulated information to illustrate the sensitivity of the membrane in the M-E process.

Table 9.14 tabulates the analysis of variance (ANOVA) of the data presented in Table 9.13. F-test values at the 1 percent significance level gave $F_{calculated=66.76}$ > ($F_{tabulated=9.55}$), indicating that the variations in cation recovery flux are not chance incidents. Also, both potential difference and cation concentration effectively represent the variation in data at the 10 percent significance level (that is, at 90 percent confidence level). This indicates that the recovery flux is dependent upon potential difference, and cation concentrations, and is independent of the type of cation in solution.

### 9.2.2.6  Kinetics of Ion-Exchange in the M-E Process

In the M-E process, the cation concentration in solution decreases exponentially (Gopal, 1998). The depletion in cation concentration is represented by Eq. (9-16).

$$C(t) = C_0 exp(-kt)_{potentialdifference} \tag{9-16}$$

**TABLE 9.14. Anova of the Data Presented in Table 9.13**

| | df | SS | MS | $F_{calculated}$ | $F_{tabulated}$ |
|---|---|---|---|---|---|
| Regression | 2 | 4.24E-05 | 2.1198E-05 | 66.762 | 9.55 |
| Residual | 8 | 2.54E-06 | 3.1751E-07 | Since $F_{calculated} > F_{tabulated}$, hence these factors are significant in controlling the loading of the membrane | |
| Total | 10 | 4.494E-05 | | | |
| | Coefficients | Standard error | $t_{calculated}$ | $t_{tabulated}$ with 95 percent probability = 2.306 | $t_{tabulated}$ with 90 percent probability = 1.860 |
| Intercept | −0.00126 | 0.00037 | −3.365 | Significant | significant |
| X1 (pd) | 0.00049 | 0.00021 | 2.263 | Significant | significant |
| X2 (solution conductivity) | 7.3292E-06 | 6.393E-07 | 11.463 | Significant | significant |

where $C(t)$ is the cation concentration in solution at any time $t$, $C_0$ is the initial cation concentration in solution, and $k$ is the ion-exchange rate constant. Equation (9-16) describes concentration profile at a constant potential difference. Upon comparing the rates of ion-exchange, $dC(t)/dt$, at different potential difference values, it is found that the rate of ion-exchange decreases with increasing potential difference. This results in lower cation recovery flux at higher potential difference. For instance, at a potential difference = 0 volt, and after five minutes of ion-exchange, $dC/dt = (-)$ 0.51 ppm/min, while at a potential difference = 1 volt, the $dC/dt = (-)$ 0.38 ppm/min. The minus signs indicate a decreasing cation concentration in solution with time. These observations likewise define the existence of the RP phenomenon.

The model represented by Eq. (9-16) fits the data within a 95 percent confidence interval; Table 9.15 shows how close the calculated values of the $Cu^{2+}$ ion-exchange are to the experimental ones.

**TABLE 9.15. Reverse Potential Phenomenon in a 30 ppm $Cu^{2+}$ Ion Solution**

| Time (Min) | pd = 0 Volt %Calc. $Cu^{2+}$ Ion Exchange | pd = 0 Volt %Expt. $Cu^{2+}$ Ion-Exchange (95% CI) | pd = 1 Volt %Calc. $Cu^{2+}$ Ion Exchange | pd = 1 Volt %Expt. $Cu^{2+}$ Ion-Exchange (95% CI) |
|---|---|---|---|---|
| 5 | 8.84 | 8.48 ± 0.42 | 6.57 | 6.44 ± 0.32 |
| 10 | 16.89 | 17.23 ± 0.86 | 12.72 | 12.21 ± 61 |
| 15 | 24.23 | 23.75 ± 1.19 | 18.45 | 19.01 ± 0.95 |
| 20 | 30.93 | 29.38 ± 1.47 | 23.81 | 22.86 ± 1.14 |
| 30 | 42.59 | 4 1.74 ± 2.09 | 33.50 | 32.83 ± 1.64 |

### 9.2.2.7  *Practical Limits of the M-E Process*

MAXIMUM POTENTIAL DIFFERENCE   Maximum selectivity in the M-E process is a function of potential difference and dissociation potential of solution (potential difference at which gas evolution starts taking place on membrane surface) (Gopal and April, 1998b).   The electrode reactions taking place in the M-E process for a $Cu^{2+}/Ni^{2+}$ binary system are as follows:

*Oxidation Reactions:*

$$2OH^- \Rightarrow H_2O + \frac{1}{2}O_2 + 2e^- \quad \text{(Graphite anode)} \qquad (9\text{-}17)$$

$$2RSO_3H \Rightarrow 2RSO_3^- + 2H^+ \quad \text{(Cation exchange membrane)} \qquad (9\text{-}18)$$

where R denotes polystyrene (polymer) "backbone" of the membrane

*Reduction Reactions:*

$$2H^+ + 2e^- \Rightarrow H_2 \quad \text{(Cation exchange membrane)} \qquad (9\text{-}19)$$

$$Cu^{2+} + 2RSO_3^- \Rightarrow 2RSO_3Cu \quad \text{(Cation exchange membrane)} \qquad (9\text{-}20)$$

$$Ni^{2+} + 2RSO_3^- \Rightarrow 2RSO_3Ni \quad \text{(Cation exchange membrane)} \qquad (9\text{-}21)$$

Typically, the circuit currents observed at operating potential difference are in the range of $10$–$100$ μA, and thus the effect of anodic oxidation on the graphite electrode [Eq. (9-21)], and hydrogen liberation on the cation exchange membrane [Eq. (9-18)] are neglected. Physical evidence of gas evolution has been observed at voltages above 4 volts, corresponding to currents greater than 7 milliamperes (Gopal and April, 1998b). However, this limitation can be overcome by cascading M-E cells such that target cation concentration increases geometrically in each subsequent cell.

*Effect of Cation Loading on Selectivity in the M-E Process*   All the results presented thus far have been for systems where the total area of ion-exchange provided is much higher than that needed, hence the observation that the rate of ion-exchange (and subsequently, the cation recovery flux) is independent of the cation loading on the membrane. This situation is valid until the loading reaches approximately 30 percent of the total available ion-exchange capacity (IEC). Thereafter, the rate constants for the different cations vary widely. To avoid this problem, membranes must be regenerated and cations recovered prior to 30 percent IEC. The problem also can be overcome by using the moving membrane-electrode process, illustrated in the next section.

### 9.2.2.8  *Potential Applications of the M-E Process*

*Recycling of Process Streams*   Typically, electroplating industries subject a workpiece to a series of rinse washes. During rinsing small amounts of plated metals and impurities (usually introduced into the system by the anode) (Varghese, 1993) are carried over in the rinse water. Often these rinse washes, containing dilute concentrations of plated heavy and transition metals from different stages, are mixed and piped into a treatment pond. The final outcome being precipitation of metals as sludges and discharge of treated water to publicly owned treatment works (POTW).

The M-E process can be effectively used for selective recovery of these metals from rinse water (Figure 9.40). For instance, a typical rinse wash for Cu plating contains Cu ($\sim$100 ppm) along with impurities of lead ($\sim$1 ppm) and bismuth ($\sim$1 ppm) (Varghese, 1993). This stream can be passed through a series of three "M-E cells" (M1, M2, and M3) which can virtually remove all the copper. When the loading of the membrane in cell M1 reaches a preset value, rinse water is diverted to cells M4, M5, and M6 through valve V3, and regeneration cycle (of cells M1, M2, and M3) is started by opening valve V1, and simultaneously closing valve V2. This operation is carried out repeatedly. During regeneration, M-E cells are washed either with hydrochloric acid or with sulfuric acid which leach the metal into solution, while the membranes are regenerated with H$^+$ ions. Membranes may require an additional regeneration with NaCl or NaOH if Na-form membranes are used. If concentration and flow rates of acid used for regeneration are properly adjusted, the acid leaving cell M3 or M6 can be introduced into the plating bath. The clean water can be reused in rinse washes, thus, permitting closed-loop operation (Gopal, 1998).

*Resource Conservation and Reconcentration in the Mining and Hydro-Metallurgical Industries* In nature, most heavy metals and transition metals occur as complex inorganic mixtures of sulfur. When these mined ores are washed with water (in order to separate particulate and dissolved matter), sulfur is converted into sulfuric acid. This causes some of the heavy metal to dissolve in wash water (Patterson, 1985; Eisenmann, 1981; Caroll, 1993). Normally this wash water is taken to a settling pond for pH treatment. This results in the precipi-

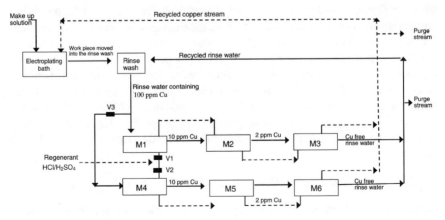

**Figure 9.40.** Application of the M-E canister prototype in recycling electroplating bath.

tation of metals as hydroxides that are subsequently dredged and landfilled. The treated water is released to surface water sources.

The M-E process can be effectively used for selective metal recovery from settling ponds. Figure 9.41 is a schematic representation of selective metal recovery from wash water from a mining operation (Gopal, 1998). Typically, water in the settling pond consists of a mixture of metals (Table 9.10) (typical coal mine drainage includes $Fe^{2+}$, $Cd^{2+}$, $Mn^{2+}$, $Mn^{4+}$, $Al^{3+}$, $As^{2+}$, $Zn^{2+}$). By maintaining a set of moving membrane-electrode (MM-E) cells (M-E-1, M-E-2, etc.) at different pd, metals can be recovered sequentially in the order of increasing electrochemical potentials. These recovered metals can then be reconcentrated (in cells RC-1, RC-2, etc.) following a scheme similar to that explained earlier. Following this, they can be subsequently recovered, either reduced electrolytically, or chemically converted into useful by-products.

Figure 9.42 is a schematic for the proposed large-scale operation of the MM-E prototype (Gopal, 1998). Such a system could be run either in batch or continuous mode. Similar to a typical M-E cell, it consists of a cation exchange membrane as the cathode, and a graphite anode. Ion specific electrodes, used for on-line cation analysis, continuously feed the data to a dedicated computer, which in turn controls a stepper motor. The stepper motor operates a series of flat belt pulleys that determine the linear speed of the membrane. Provision is made to ensure that the cation loading is always less than 30 percent. The depleted membrane is taken to an acid wash where the metals are leached, and the membrane regenerated to the Na-form in an NaOH wash. The metals in the acid wash are reconcentrated, either by passing through subsequent MM-E cells, or by electrodialysis. The concentrated metals are then electrolytically recovered.

### 9.2.2.9  Summary of Significant Findings

- M-E process has been validated in complex (binary and ternary) cation systems. The existence of the reverse potential phenomenon has been established in all the systems tested.

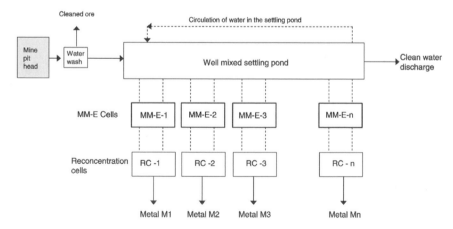

**Figure 9.41.** Application of the M-E process in the mining industry.

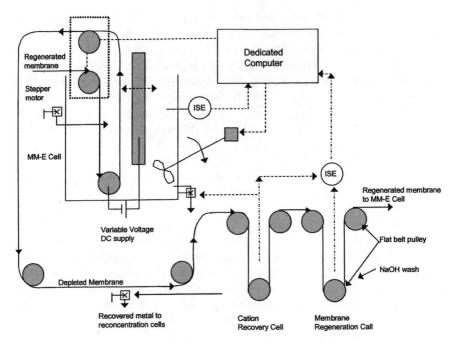

**Figure 9.42.** Application of the moving membrane-electrode (MM-E) process.

- The results from single cation systems can accurately predict the effect of potential difference on cation selectivity (RP phenomenon) in dilute aqueous binary and ternary systems. Consequentially, the rate constants in single cation solutions are found to closely approximate (within a 95 percent CI) binary and ternary solutions. This very important finding can be used to determine the ion-exchange rate constants of different cation systems, and then simulate the behavior of the different solution mixtures in the M-E process.
- Potential difference and cation concentration control the rate of ion-exchange.

## ACKNOWLEDGMENTS

The author would like to thank the Gulf Coast Hazardous Substance Research Center (GCHSRC), Beaumont, TX for funding this project. Also, the author thanks Dr. Gary C. April, The University of Alabama, Dr. S. S. Kulkarni, National Chemical Laboratory, Pune, India, and Mr. S. Gopala Krishnan, CEO, Pune Instrumentation, Pune, India for their technical insight during the course of this project.

## REFERENCES

Caroll, R. (1993). *Alabama Coal Data for 1992.* Geological Survey of Alabama, Inf. Ser. 58N, AL.

Eisenmann, J. L. (1981). Nickel recovery from electroplating rinsewaters by electrodialysis, EPA-600/S2–81-130, Industrial Environmental Research Laboratory, Cincinnati, OH 45268, EPA Project Summary.

Eilbeck, W. J., and. Mattock, G. (1987). *Chemical Processes in Waste Water Treatment.* 1st Ed., Ellis Horwood Limited, Chichester, UK.

Gopal, V. (1998). Development of the membrane-electrode (M-E) process for selective removal, recovery, and reuse of heavy metals from multi-component dilute aqueous solutions. Ph.D. dissertation, The University of Alabama.

Gopal, V., April, G. C., and Schrodt, V. N. (1998). Selective removal of copper from multi cation dilute aqueous solutions using the membrane-electrode process. *Journal of Separation Science and Technology*, **33**(5).

Gopal, V., April, G. C., and. Schrodt, V. N. (1997). Selective lead ion recovery from multiple cation waste streams using the membrane-electrode process. Euromembrane 1997, Enschede, The Netherlands.

Gopal, V, and April, G. C. (1998a). Heavy metal decontamination from dilute aqueous solutions using the membrane-electrode (M-E) process. *Chemical Technology of South Africa* (invited article, accepted Nov. 01, 1998, published December).

Gopal, V., and. April, G. C. (2000). Reverse potential phenomena in the membrane-electrode (M-E) process. *Chem. Eng. Comm.*, **180**, 127–143.

Gopal, V., and. April, G. C. (1998b). Selective lead ion recovery from multiple cation waste streams using the membrane-electrode process, *Journal of Separation and Purification Technology*, **14**, 85–93.

Patterson, J. W. (1985). *Industrial Wastewater Treatment Technology.* Butterworth Publishers, 2nd Ed., MA.

Varghese, C. D. (1993). *Electroplating and other Surface Treatments*, Tata McGraw-Hill Publishing Company Limited, New Delhi.

### 9.2.3 Advances in Pollution Prevention in the Metal Finishing Industry

JOHN O. BURCKLE (GRANTEE, SENIOR ENVIRONMENTAL EMPLOYMENT PROGRAM)
T. DAVID FERGUSON (SUSTAINABLE TECHNOLOGY DIVISION)
National Risk Management Research Laboratory, Office of Research and Development, U. S. Environmental Protection Agency, Cincinnati, Ohio, USA

Surface finishing of metal products is a major manufacturing operation performed in thousands of production shops to provide weather-resistant, wear-resistant, and aesthetically pleasing finishes for thousands of manufactured products. Surface-finishing technology involves direct atom-to-atom bonding between a base material (such as steel, aluminum, brass, or plastics) and a metal or organic surface top coating that provides the desired material performance and appearance properties. Surface pretreatment is crucial for proper performance and durability of the produced part. Cleaning and oxide removal is critical, so pretreatment processes usually involve many tanks with various purposes. Multi-step surface preparation processes are generally employed to remove oils, soiling and dirty materials, old coatings, corrosion products, residual cutting fluids, brazing residuals, pickling acid residuals, cleaner residuals, etc. The surface preparation process removes contaminants, preserves the cleaned surface, and/or modifies the surface for the next coating. After the finishing process, the parts require rinsing/cleaning to remove residual plating solution. These baths, both plating and cleaning, ultimately are exhausted because of depletion of strength or buildup of impurities and, as a result, become major waste streams (USEPA, 2001).

In the past, pollution control in the metal finishing industry has been achieved primarily through end-of-pipe treatment. Facilities added waste treatment systems as a final process step to meet regulatory discharge limits to municipal sewers or as direct dischargers. This approach results in the production of residual sludges contaminated with heavy metals that require appropriate pretreatment for environmentally safe disposal (the so-called "Hammer Provisions" of the Resource Conservation and Recovery Act that apply to hazardous waste disposal in the United States). In the mid-1980s, various facilities began integrating "pollution prevention" into their operations to reduce the amount of wastewater generated and treated. However, the speed of adoption has been relatively slow, most noticeably in smaller facilities (USEPA, 1982, 1985).

The most logical and efficient approach to effectively achieve environmental protection is by preventing pollution. Prevention is achieved using techniques that reduce, eliminate, or recycle/reuse waste materials so that the generation of waste is eliminated, avoiding the need for waste disposal, and wastes that are generated are handled responsibly, either through recovery and recycling to displace other material needs and/or to conserve energy, or minimized and made safe for disposal into the environment. The metal finishing industry began to apply pollution prevention principles to reduce process waste emissions in the mid-1980s, but there were no public statistics generated of the impact on pollution avoided or economic benefit accrued. In discussing the progress of domestic metal finishing industry in transitioning to operations based upon pollution prevention, we begin by describing the National Metal Finishing Strategic Goals Program (SGP) initiated by the US EPA in 1998 in cooperation with industry and public interest stakeholders (USEPA, 1998, 2001).

### 9.2.3.1 National Metal Finishing Strategic Goals Program
This was the first nationwide program initiated by the US EPA to build voluntary participation in a compliance program by providing incentives in return for a real commitment to reducing pollution to levels below those required for compliance. The SGP was created to address environmental problems and institutional barriers to achieving compliance in a number of industries. It was designed to provide a working relation that facilitates and encourages companies to go beyond environmental compliance to pollution prevention based operations. SGP member companies were offered incentives, resources, and a means for removing regulatory and policy barriers as they work to achieve specific environmental goals.

The SGP program brought together stakeholders to identify important issues, conduct demonstration projects, and develop consensus policy recommendations

that offered opportunities for reducing pollution. The stakeholders included indus-
try, labor, environmental groups, state and local government, other federal agencies.
The metal finishing industry was represented by the National Association of Metal
Finishers (NAMF), American Electroplaters and Surface Finishers Society (AESF),
Metal Finishing Suppliers Association (MFSA), and Surface Finishing Industry
Council (SFIC). Through the SGP (USEPA, 2001), the participants worked together
to improve environmental performance and the bottom line.

Participation required top management commitment to the implementation and
maintenance of an approved Environmental Management System. The EMS has
caused management to take careful note of the true costs of pollution and institute
policies and procedures to eliminate this waste at its source (USEPA, 2003a). The
results of these efforts have provided the first large-scale quantification of the
power of pollution prevention for achieving significant reductions in pollution and
the resulting economic benefits in the metal finishing industry.

The SGP program was designed to help a metal finishing company achieve its
goals — both environmental and economic. A wide variety of state and local
SGP resources (USEPA, 2001) were available to provide a company with the
tools need to get the job done. These include:

- Free, non-regulatory environmental audits
- Funding for environmental technologies
- On-site technical assistance in (evaluation and planning) achieving compliance
  and improving pollution prevention, safety and health measures
- Free assistance from interns to help fill out the SGP data worksheets
- Free workshops on energy, water and waste reduction
- Regulatory flexibility
- Environmental Management Systems (EMS) training
- Public recognition

All parties benefited from the use of these tools. A few of the advantages listed in
the project report are listed below.

- Companies received the incentives and resources needed to take the risk of
  going beyond compliance requirements.
- As the participating companies employ less polluting technologies, waste is
  reduced, less pollution is discharged to the environment, and the plant saves
  money, becoming more cost efficient and competitive.
- As the results demonstrated in the participating plants are employed by other
  plants to reduce pollution and improve competitiveness, the practices of pollution prevention are spread throughout the industry.
- The industry as a whole benefits from the positive action of SGP member
  companies.

- As the metal-finishing industry becomes increasingly more self-regulated, government regulators — from the EPA to the local publicly owned treatment works — save time and money.

The SGP has seven environmental performance goals (USEPA, 2001) that form the core of the program (see Table 9.16). Through the end of 2000, many participants had already made significant progress toward meeting the goals, which translated into real environmental gains:

- 380 million gallons of water conservation
- 120 million pounds of hazardous waste not sent to landfills
- 665,000 pounds of organic chemicals not released to the environment.

A comparison of the reductions achieved through 2000 and 2003 show continuing and significant improvements in pollution prevention. This continued progress of the participants is a positive indicator of the benefits of a program driven by the employment of the environmental management system. The SGP participants were provided free, non-regulatory environmental audits and access to technical support in achieving compliance and improving pollution prevention. Data like these readily demonstrate the power of the EMS management tool to reduce pollution. But to see how this progress was achieved, we must examine the role of pollution prevention (P2) technologies in enabling the reductions.

### 9.2.3.2 The Role of Pollution Prevention Technologies

*Moving Towards the Zero Discharge Goal* A highly desirable "state" for environmental protection would be zero discharge of pollutants to the air, water,

**TABLE 9.16. Progress Towards SGP Goals**

| SGP Goal | Average Achievement by SGP Metal Finishers (over all projects) | |
| --- | --- | --- |
| | 2000 | 2003 |
| 1. 50 percent reduction in water usage | 41 percent reduction | 56 percent reduction |
| 2. 25 percent reduction in energy use | 14 percent reduction | 41 percent reduction |
| 3. 90 percent reduction in organic toxic release inventory (TRI) releases | 77 percent reduction | 84 percent reduction |
| 4. 50 percent reduction in metals released to water and air | 58 percent reduction | 65 percent reduction |
| 5. 50 percent reduction in land disposal of hazardous sludge | 36 percent reduction | 48 percent reduction |
| 6. 98 percent metals utilization | 17 percent utilization factor | 64 percent utilization factor |
| 7. Reduction in human exposure to toxic materials in the facility and surrounding community | 51 percent of activities accomplished | 85 percent of activities accomplished |

and land. Today, this is a goal not yet realized. However, *approaching* zero discharge has been found to be realistic, as demonstrated by the continual improvement in environmental performance achieved in the SGP for the metal finishing sector. Approaching zero discharge can be defined as reducing wastes emitted from a process by a significant amount, with significant reduction ranging from a low defined by the regulatory standard and the high defined by the technology employed.

There are two key elements to achieving movement towards the state of zero discharge — implementation of an environmental management system and deployment of certain technologies based upon pollution prevention. The framework of an environmental management system is the *management tool* that provides a state of "continuous assessment and compliance of plant operations," while P2 technologies provide the *technological tools* needed to achieve significant improvements in performance and reductions in generation of waste.

*Planning and Implementation*    As part of the planning phase, a compliance assessment of a participating plant is necessary to establish a baseline against which progress and savings in incremental costs can be measured. The goal of the plant assessment is to identify the root causes of the most significant problems, to identify areas in which P2 options could save the most money, and to assign priorities to address the most significant problems. The plant assessment must include an inventory of all chemicals, wastes, bath chemistries; the overall plant layout; and a site inspection. Once the problems are identified and prioritized, various solutions can be proposed and the effects of their deployment evaluated. A comprehensive plan addresses housekeeping and maintenance issues in order to sustain any P2 efforts and also to establish good standard practice.

This planning approach is most efficiently implemented through an EMS. In addition, the imperatives of "total quality management" apply. Management must buy into the process and be willing to provide the necessary resources to achieve success. Experience has shown that often the best solutions come from those working most closely to the problem. It is important that employees be included in the improvement program and kept well informed so that they will become stakeholders in the process.

Implementation through an EMS has several advantages. First it is more effective because it provides a tool for involving management through its provision for continuous improvement in both environmental performance and worker health and safety. Second, it provides a mechanism to integrate process and product quality issues that influence reduction of waste and the improvement of productivity and profitability. The review of these considerations should be incorporated into the process and reviewed over time as a pathway for identifying future opportunities and establishing priorities. This process leads to the identification of the power of pollution prevention technologies to achieve these savings offered by waste reduction. The Agency provides an EMS template tool (USEPA, 2004) to assist those who are interested.

Many of the same elements that must be defined for the compliance assessment are also needed to establish an EMS. Certification under the ISO 14,001 — an EMS that is standardized worldwide — is being increasingly required of those companies engaged in the manufacture of products for export, including components in the supply chain of such products. Also the EPA offers Compliance Incentives — "policies and programs that eliminate, reduce or waive penalties under certain conditions for business, industry, and government facilities which voluntarily discover, promptly disclose, and expeditiously correct environmental problems" and is including a requirement for the implementation of environmental management systems in settlements. Information about these incentives, programs, and the environmental benefits achieved by such programs may be found on the Internet (http://www.epa.gov/compliance/incentives/index.html).

*Practicing Pollution Prevention* Technologies based on the precept of Pollution Prevention serve to eliminate the generation of wastes or to reduce the disposal of wastes through recycle/reuse. Pollution in the metal finishing industry is basically the discharge of some unwanted form of material or other resource (energy, labor, time, etc.). Loss of these resources equates to the loss of profit and economic productivity. It stands to reason that the more of a resource used above the minimum required by the process, the more this incremental use (read: "waste") adds to the "unnecessary" component of the total costs of the operation. In the case of water use, for example, this unnecessary cost is not just the cost of the excess water wasted. The cost of this waste is magnified by the costs for its overall management for treatment and disposal, including the capital and operating costs for moving the excess water through the process, its cleaning, and the disposal of the treated water and any residual wastes. The treatment and disposal costs are usually more costly than the initial raw material. Often these costs (process operations and compliance costs) are not tracked by a company because the accounting process used is inadequate to perform such tracking.

Companies participating in the SGP have found significant cost savings by implementing P2 practices (USEPA, 2001). Significant cost savings result in improved economic efficiency and an improved bottom line on the balance sheet, making the operation more competitive. Other advantages include the protection of employees, reduction of liabilities, and, at the same time, enhancement of the company's business image.

Success in achieving the goals of the Strategic Goals Program for the metal finishing sector has been attributed largely to the employment of pollution prevention techniques. A principal purpose of the program was to demonstrate the attractive environmental and economic advantages offered by P2 based approaches to waste reduction. The outcome anticipated was to "stimulate a keen awareness and appreciation" of these extremely valuable tools so that they would be adopted into everyday practice on a sector-wide basis. In addition, the adoption of an environmental management system and integration into company operations was ensures continued improvements (both in environmental performance and cost savings) within the entire fabric of the participating companies.

There are many approaches to reduce pollution in various unit processes that were investigated and documented by EPA sponsored research in the 1980s, including housekeeping and maintenance methods and management of process chemicals and rinse water. These methods are described in the EPA Capsule Report entitled *Approaching Zero Discharge in Surface Finishing* (USEPA, 2000) and the training course document entitled *P2 Concepts and Practices for Metal Plating and Finishing* (AEFS, American Electroplaters and Surface Finishers Society). These learning aids address a number of process design areas such as those listed in Table 9.17.

Transitions of manufacturing operations to P2 based systems have been demonstrated using a range of technology options from relatively simple improvements in existing process technology to more sophisticated approaches based upon significant process changes. There are numerous examples of the power of pollution prevention technologies, ranging from relatively simple improvements of water management to more sophisticated recovery processes, as documented in the results of the Strategic Goals Program and other programs as noted in Section 9.2.3.3.

**TABLE 9.17.  Pollution Prevention Technologies for Surface Finishers**

| To Improve in These Operations... | Employ these Technologies and Practices |
|---|---|
| Extend bath life<br>• reduce chemical consumption<br>• reduce bath dumping<br>• maximize on-line time<br>• maintain product quality | • improve bath chemical solution management<br>• reduce contamination by improved housekeeping<br>• add new continuous bath purification technologies |
| Reduce water consumption<br>• improve water utilization to decrease water demand and sewer charges<br>• increase in-plant recycling<br>• decrease water discharges and water intake | • employ a rinse method that uses less water<br>• reduce drag-in and drag-out<br>• employ new process analysis and planning methods for water conservation<br>• use advanced technologies for removal of chemicals to permit rinse water recycling |
| Minimize waste<br>• decrease usage of all chemicals and water<br>• decrease waste generation improve recovery and recycling in-plant | • improve process management and oversight<br>• implement an environmental management system<br>• improve rinsing operations as described above<br>• increase in-plant recycling of water |
| Reduce the use of hazardous chemicals<br>• reduce environmental risks and costs<br>• improve worker safety<br>• maintain or improve productivity and quality<br>• acquire new or maintain existing markets where clients are "greening the supply chain" | • move to less toxic chemistries<br>• use new finishing processes and coatings<br>• reduce the number of stages in a line |

### 9.2.3.3 The MERIT Partnership for Metal Finishers (USEPA, 2003b)
The Metal Finishing Association of Southern California, EPA Region 9, and the California Manufacturing Technology Center, a NIST Manufacturing Extension Center, established the Merit Partnership P2 Project for Metal Finishers. This project involved implementing P2 techniques and technologies at metal finishing facilities in southern California and documenting and sharing results. Technical support for this project was provided by Tetra Tech EM Inc. The project was funded by the Environmental Technology Initiative and EPA Region 9, and was implemented, in part, through the CMTC by the National Institute of Standards and Technology.

These studies addressed the following P2-based process technologies:

- Reverse Osmosis Applications for Metal Finishing Operations
- Innovative Cooling Systems for Hard Chrome Electroplating
- Modifying Tank Layouts to Improve Process Efficiency
- Reducing Rinse Water Use with Conductivity Control Systems
- Reducing Dragout with Spray Rinse Systems
- Metal Recovery and Wastewater Reduction Using Electrowinning
- Finding an Alternative to Solvent Degreasing
- Extending Metal Finishing Bath Life
- Extending Electroless Nickel Bath Life Using Electrodialysis

*Mini-Case Study 9-7 (USEPA, 1998): A Water Management P2 Example*   A California company implemented a number of relatively simple and low cost P2 operations: flow restrictors in rinse tanks to manage rinse water, drag-out reduction using drip boards between process tanks and hang bars above plating tanks. Between 1997 and 2000, the plant achieved a 50 percent reduction in water consumption, a 25 percent reduction in energy consumption, and 98 percent metals utilization. Since October 2001, the plant has increased savings in energy and water consumption by improving reductions to 62.5 percent and 70 percent less process water from the original baseline, respectively.

For other case studies, go to access on-line at:
http://es.epa.gov/cooperative/topics/metcasestudies.html.
http://www.epa.gov/ORD/NRMRL/std/mtb/metal_finishing.htm#nmmsfp

### 9.2.3.4 Case Study: An Emerging Pollution Prevention Technology
The USEPA recently completed a study of the proprietary Picklex process. Picklex® is a "non-polluting" pretreatment/conversion coat which replaces chromate conversion coating and zinc and iron phosphatizing in powder coating, paint and other organic finish applications" (Ferguson et al., 2001; Ferguson and Monzyk, 2003).

*Description*  Metal pretreatment is crucial for cleaning and oxide removal to obtain proper performance and durability of the produced part, and usually involve many tanks with various purposes. For example, aluminum anodizing may require eight pretreatment tanks; in conventional chromate conversion coating, there are twelve stages, while conventional zinc phosphatizing requires six. All of these baths contain heavy metals and must be dumped periodically.

The volume of hazardous/toxic waste streams produced from metal surface finishing operations is significant (USEPA–TRI, 1995a). The elimination of any of the surface processing steps is beneficial to the manufacturing process because it reduces processing costs, waste production, and energy consumption. Strong acids used in the pre-treatment of metals pose a great health and safety risk to workers. With this in mind, a no-waste or waste reducing surface-finishing agent designed to lower processing costs for metal finishing operations would be of great benefit.

The Picklex process (a proprietary process), is one such alternative to conventional metal surface pretreatment that offers significant reductions in waste generation. The process was developed by International Chemical Products, Inc. with assistance from the USEPA. The Picklex process works in a completely different way than conventional processes. It incorporates the corrosion products from the metal surface into the protective coating that is applied to that same metal surface. This new P2 approach "thinks outside the box" by solving the common environmental problem of metal buildup in the processing tanks while forming a protective surface coating on the work piece. A two-phase program was undertaken to evaluate the ability of Picklex to perform technically and economically as a pollution prevention based replacement for conventional metal pretreatment or pretreatment/conversion coat in finishing operations. The goal was to demonstrate its potential to eliminate or significantly reduce the amount of hazardous and toxic chemicals consumed and the amount of pollution generated while maintaining equal or better product performance properties.

A broad laboratory evaluation of Picklex was undertaken in which full multi-step, bench-scale, batch operation tests using side-by-side test lines of seven conventional processes and of Picklex were performed. The results of the laboratory tests were quite favorable, and field tests (Ferguson and Monzyk, 2003) were conducted to evaluate in-plant performance in side-by-side tests with conventional preparation technologies. Only the most promising applications — the replacing of chromate conversion coating and zinc phosphatizing — were taken to field evaluation. In these focused field tests, Picklex was used to prepare aluminum and steel substrates for powder coating only. Although the testing scope was narrowed, it was more detailed. A total of 41 different combinations of substrate, degreaser, pretreatment, conversion coat, and powder coat were tested in the field evaluation. Only uncontaminated panels and components, without corrosion products, were used. The results for all test cases demonstrated that the quality of the final product was equal in all respects to conventional practice. Product quality was targeted as the single most important acceptance criterion within all the comprehensive results, because if the final product quality was not acceptable, the pollution reductions would not matter. The field test results replicated those obtained earlier, indicating that the lab testing may be used to accurately predict results achievable in practice.

*Environmental Benefit*   Two commercial processes, chromate conversion coating and zinc phosphatizing, were employed in the field tests as the experimental control baseline Figures 9.43 and 9.44, respectively. Metal buildup in these conventional processes ruined the baths, and the baths had to be dumped. In the metal finishing sector, the discarding of such metal-containing waste streams is a major contributor to pollution. The Picklex chemical used in the process tank is used up on the product; however, with filtering, the bath does not become contaminated. As a result, it is not necessary to dump the bath to control impurity build-up, and the chemical make-up is simply added to the bath as needed. This process was demonstrated to have a strong potential as a leading technology for pollution prevention in surface preparation operations.

As can be seen in the comparison of the conventional practices to this emerging technology, the Picklex process eliminates up to eleven steps in metal finishing processes. The reductions of waste at the source were very significant. These reductions were accomplished by: the elimination of the hexavalent chromium tank and subsequent rinsing operations; the elimination of the acid baths used for etching and cleaning along with subsequent rinsing operations; the elimination of the zinc phosphatizing baths and subsequent rinsing operations; and the elimination of the deoxidizing bath and subsequent rinses. All of the steps eliminated are steps that emit pollution. In addition, the Picklex process significantly reduces the pollution from the remaining steps. It does this while providing the same high quality finish and wear resistant capabilities as the conventional processes.

Normally the Picklex bath does not need to be dumped. However, as part of the study, the bath was evaluated for waste treatment issues in the event of a spill or accidental contamination. Fresh, spent, and impurity-spiked spent Picklex samples were treated using the conventional pH 9 precipitation industrial waste water treatment method to produce samples for a waste disposal assessment for Picklex. The waste solids were assessed for toxic leachability. The treated water supernatants for discharge were also examined and found as most likely dischargeable. Waste treatment may not be frequent or necessary in many processes. Actual discharge

**TABLE 9.18. Potential Cost Savings (Ferguson and Wilmoth, 2000; Ferguson et al., 2001)**

| | Savings of Picklex over Conventional Process | |
| --- | --- | --- |
| Cost Type | Chromate Conversion Coat on Aluminum | Zinc Phosphatizing on Steel |
| Capital cost savings | $254,000 | $230,000 |
| Annual operating cost savings | $46,000 | $32,000 |

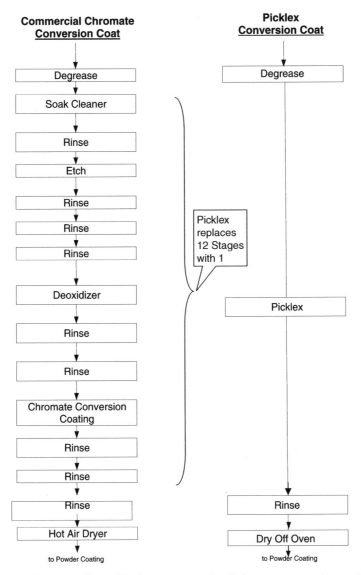

**Figure 9.43.** Comparison of Picklex to conventional chromate conversion coating.

limits from industrial waste treatment plant operations are site-specific and are determined on a case-by-case basis with local, state (and federal only where necessary) regulatory agencies. Hence, no exact classification of these potential waste solutions is possible until a specific location is known. All leachates passed with respect to the applicable federal regulations (USEPA, 1995b). Therefore, Picklex does not appear to present unusual waste treatment issues.

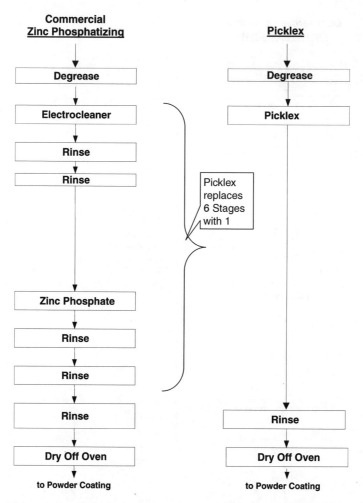

**Figure 9.44.** Comparison of Picklex to conventional zinc phosphatizing.

*Economic Benefit*   A cost comparison was made for two most promising processes — replacing chromate conversion coating and zinc phosphatizing. These were chosen based on the potential cost savings that industry could achieve. There would appear to be a significant economic incentive to migrate existing practice to this new technology, based on the significant capital and annual operating costs that is projected for the Picklex technology, summarized in Table 9.18. In fact, a new shop adopting the process in place of chromate conversion coat would save capital costs of $254,000, with the reduction of the process from 12 to one tank (Figure 9.43). For zinc phosphatizing on steel, the savings is estimated to be $230,000.

*Technology Transferability (Ferguson and Monzyk, 2003)* The Picklex technology is very robust and flexible. It is applicable from the very large to very small operations. It can be used in a new installation, in a retrofit for process modernization, or as a drop in replacement for any existing facility. Picklex can be applied by dipping the part in tanks or by spraying or brushing. Cost savings are very significant, and are a function of the size of the facility and the number of processing lines that are required. Transferability is enhanced by the prospective significance of the overall facility cost savings and productivity improvements offered.

## REFERENCES

Ferguson, D., and Wilmoth, R. (2000). Cost effective control of hexavalent chromium air emissions from functional chromium electroplating. *Environmental Health & Safety Solutions*, **1**(3), 10–19.

Ferguson, D., Hindin, B., Chen, A., and Monzyk, B. (2001). Picklex as a non-polluting metal surface finishing pretreatment and pretreatment/conversion coating. *Clean Products and Processes*, **2**, 220–227 (http://link.springer-ny.com/link/service/journals/10098/bibs/1002004/10020220.htm).

Ferguson, D., and Monzyk, B. (2003). Non-polluting metal surface finishing pretreatment and pretreatment/conversion coating. *Plating & Surface Finishing*, April, 66–75.

USEPA (1982). *Control and Treatment Technologies for the Metal Finishing Industry: In-Plant Changes.* US EPA, Office of Research and Development, EPA 625/8–82-008.

USEPA (1985). *Environmental Regulations and Technology—The Electroplating Industry.* US EPA, Office of Research and Development, EPA 625/10–85-001.

USEPA (1995a). *Toxic Release Inventory (TRI) Release Data for Fabricated Metals Facilities SIC 340.* Available on the Internet at: (http://www.epa.gov/triexplorer).

USEPA (1995b). 40CFR261 Identification and Listing of Hazardous Wastes. *Code of Federal Regulations*, US GPO. Internet (http://www.gpoaccess.gov/cfr/index.html).

USEPA (1998). *Metal Finishing/Electroplating Industry: Case Studies* Available on the Internet at: http://es.epa.gov/cooperative/topics/metcasestudies.html

USEPA (2000). *Approaching Zero Discharge in Surface Finishing.* US EPA, Office of Research and Development, EPA/625/R-99/008.

USEPA (2001). *Living the Vision - Accomplishments of the National Metal Finishing Strategic Goals Program.* EPA 240-R-00–007, January. http://www.strategicgoals.org/

USEPA (2003a). *Environmental Management Systems: Systematically Improving your Performance. MetFin_BizCase_15* on the internet at: http://www.epa.gov/ispd/metalfinishing/metfin_pdf/metfin_bizcase.pdf

USEPA (2003b). Merit Program Reports internet at: http://www.sectorstar.org/sector/MetalFinishing/program.cfm?ProgramID = 130

USEPA (2004). A Guide to Developing an Environmental Management System for Metal Finishing Facilities. http://www.epa.gov/sectors/metalfinishing/metfin_ems.html

## 9.2.4  Profitable Pollution Prevention in Electroplating: An In-Process Focused Approach

HELEN H. LOU

Department of Chemical Engineering, Lamar University, Beaumont, TX 77710, USA

YINLUN HUANG

Department of Chemical Engineering and Materials Science, Wayne State University, Detroit, MI 48202, USA

Industrial pollution prevention (P2) is one of the key steps in environmental protection. Over the past three decades, the electroplating industry has been implementing numerous basic P2 techniques that have greatly reduced the quantity and toxicity of end-of-plant waste. In recent years, various new P2 technologies have been developed for technology change, material substitution, in-plant recovery/reuse and treatment. This section reviews the development of what are called profitable P2 (or simply P3) technologies delineates strategies for their implementation.

### 9.2.4.1  P2 and the Metal Finishing Industry

The US Environmental Protection Agency defines "pollution prevention" as the maximum feasible reduction of all wastes (wastewater, solid waste, and air emissions) generated at production sites (USEPA, 1999). For the past two decades, the federal agency has been working closely with the electroplating industry for a cleaner environment. The National Metal Finishing Strategic Goal Program described in Section 9.2.3.1 is an industry-specific partnership for environmental protection whose goals include 50 percent reduction in water use, 90 percent reduction in organic toxic release inventory, 50 percent reduction in hazardous sludge generation and disposal, and 25 percent reduction in energy use (USEPA, 1999). Today, a large number of P2 technologies are available for the electroplating industry. Solvent substitution and technology change contribute significantly to the improvement of environmental quality (Noyes, 1993; Gallerani, 1996; USEPA, 1999). Chemical recovery, solution maintenance, on-site waste treatment and reuse are among others practiced in industries (Barnett and Harten, 2003; Cushnie, 1995). Many P2 technologies have not fully permeated the industry however, mainly because of concerns about costs, potential adverse impacts on productivity and product quality, and difficulties in implementation (Load et al., 1996). This is not hard to understand. As explained in Section 9.2.3, the electroplating industry generates huge amounts of waste in the forms of wastewater, spent solution, air emissions, and sludge. These wastes, which contain over 100 hazardous or toxic chemicals, metals, and other regulated contaminants, cost the electroplating industry hundreds of millions of dollars per year for waste treatment and disposal (Cushnie, 1995).

In principle, P2 should not prevent plants that install the new technology from realizing economic benefits. In deed, in practice, a considerable portion of plant waste is preventable, simply by dismantling the mechanism of generating it. This measure obviously eliminates the consumption of chemicals, water, and energy that had produced the targeted waste. Thus, not only can waste be minimized, but design and operating costs can be reduced as well. Such P2 efforts can lead to profit for plants. Lou and Huang (2000) have introduced a concept that extends conventional P2 by adding a new dimension, economics. Profitable P2 (P3) theory and technologies are developed based on fundamental principles in process systems engineering.

### 9.2.4.2 From P2 To P3

In cleaning tanks, which comprise the preprocessing stage of electroplating, most of the dirt (e.g., oil, soil, oxides) is removed from the parts to be electroplated by solvents and electrical energy; sludge accumulates at the bottom of the tanks. However, barrels of parts that have been cleaned contain a certain amount of cleaning solution, with loose dirt floating on the surfaces, which is carried to the succeeding rinsing system. A large amount of rinse water continuously flows through rinse tanks to wash away the loose dirt residue and the drag-in from the surface of parts. The drag-out from any tank equals the drag-in to the succeeding tank, and is a major source of pollutants in wastewater leaving rinsing systems. Similarly, parts from plating tanks also contribute a large amount of pollutants to the final rinsing operations. This means that spent batch solutions contain high levels of metals and compounds that are difficult to separate from the liquid waste stream.

Most P2 technologies available for the electroplating industry minimize not the waste from production lines, but the effluent waste streams of the entire facility. The use of such passive P2 technologies will not interfere with normal production. However, those technologies do nothing to reduce the waste being generated by the processes themselves. Thus, they neither help reduce operating costs nor lower the health risk to employees working in the lines.

Waste in electroplating lines can be divided into two classes: unavoidable and avoidable. To see that this is true, we review the process briefly. Although the amount of dirt and other undesirable substances from parts to be electroplated calls for the use of chemicals and water, and this kind of waste is unavoidable, improper cleaning, rinsing, and plating will result in avoidable waste, through the consumption of unnecessarily large amounts of chemicals and water. Indeed, the chemical loss in an electroplating operation can be as high as 60 percent of the total consumption. Almost all lost chemicals will enter the rinse systems. Then, in a chain reaction, more rinse water will be consumed to remove excess chemicals from part surfaces, whereupon more wastewater will be generated, and increased amounts of wastewater will flow into a wastewater treatment facility. The treatment facility, in turn, will have to be expanded, and then more chemicals will be needed. The waste related to chemical loss should be classified as avoidable waste and minimized.

The minimization of avoidable waste implies not only the true source reduction but also profit increment. Thus, any P2 technology focusing on minimization of avoidable waste should ultimately yield profits as well. This class of P2 technologies is what we are calling profitable P2 (P3) technologies. The use of the basic, low-cost P2 technologies that served in the early days of environmental regulation can reach only a limited waste reduction standard. The P2 technologies focusing on technology change, alternative materials, etc., are effective but costly. The P3 technologies, however, not only can achieve environmental targets, but also can lead to profits through improved production efficiency.

### 9.2.4.3  P3 Fundamentals and Strategies
As stated earlier, P3 focuses on the process in which waste is generated. The merit of P3 is the simultaneous realization of waste reduction (environmental impact) and improvement of production (economic incentive). This can be expressed as (Lou and Huang, 2000):

$$P3 = \text{Waste} \downarrow + \text{Production} \uparrow \tag{9-22}$$

The waste reduction in Eq. (9-17) results in the following series:

$$\text{Dirt removed} \downarrow + \text{Chemicals} \downarrow + \text{Water} \downarrow + \text{Energy} \downarrow = \text{Waste} \downarrow \tag{9-23}$$

And the rise in production represents the sum of the following events:

$$\text{Production} \uparrow = \text{Product quality} \uparrow + \text{Production rate} \uparrow + \text{Operating cost}$$
$$\downarrow + \text{Capital cost} \downarrow \tag{9-24}$$

Clearly, the reduction of chemicals, water, and energy consumed in a production line as represented by Eq. (9-23) will result in the direct reduction of operating costs and the indirect reduction of capital costs as suggested by Eq. (9-24). The key to waste reduction is control of production quality. Ensuring the environmental and economic benefit of P3 calls for a deep understanding of process fundamental and waste generation mechanisms. Process static and, particularly, dynamic behavior are the focus of system characterization. This will become the basis for developing P3 strategies for general process systems. To be effective, P3 technologies incorporate strategies of at least four types: for reducing waste generated in each processing unit, for reducing waste transferring among units, for reducing chemicals, water, and energy, and for ensuring cleaning, rinsing and plating quality. These strategies are closely related, and their functionalities associated with a general plating process are plotted in Figure 9.45. The central point is to uncover the cause-effect relationships between process operation and waste minimization. This can be accomplished through complete plant modeling, as we shall explain in Section 9.2.4.4. Although a detailed understanding of the equations presented there requires familiarity with the principles of mass and energy balance, thermodynamics, and kinetics, the text descriptions of the models will be clear to readers without this background.

### 9.2.4.4  Integrated Modeling and Optimization
Dynamic plant models permit the characterization, at any time, of such aspects of process behavior, parts processing status, solvent solution cleaning capability, rinsing water contamination level, and plating status. Hitherto, a variety of unit-based process models have been developed (Gong et al., 1997a,b; Luo et al., 1998a,b; Lou and Huang, 2000, 2001; Zhou et al., 2001), which include those for the operations of soak cleaning,

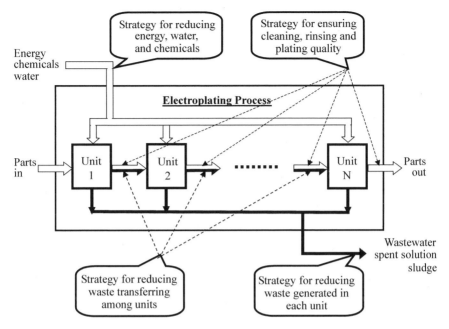

**Figure 9.45.** P3 strategies.

electrocleaning, pickling, single and countercurrent rinsing, basic plating, and sludge accumulation. The subsections that follow review the models used in P3 for the electroplating industry.

*Unit-Based Operation Modeling*  We present four models. The first three are general. Once they are applied to a process, specific cleaning, plating, and rinsing configurations should be differentiated. Thus, the model parameters should be adjusted accordingly.

CLEANING MODEL    In a cleaning tank, the dirt on part surface is removed by the application of a certain type of energy, such as mechanical, chemical, thermal, electrical, and/or radiation energy. A dirt removal model has the following form.

$$A_p \frac{dW_{Pc}}{dt} = -r_{pc} \tag{9-25}$$

$$r_{Pc} = \gamma_c C_a W_{pc} \tag{9-26}$$

$$\gamma_c = \gamma_0 \left[ 1 - e^{-\alpha(t-t_0)} \right] \tag{9-27}$$

where $A_p$ is the total surface area of the parts in a barrel; $C_a$ the chemical concentration in the cleaning tank; $r_{pc}$ the dirt removal rate in the tank; $W_{pc}$ the amount of dirt on the parts in the tank; $\gamma_c$ the loose dirt on parts in the tank.

The amount of chemical in the tank changes along the cleaning operations. In addition, the chemical is carried over through drag-out to succeeding tanks. A chemical concentration model can be established as follows:

$$V_c \frac{dC_a}{dt} = -\frac{r_{Pc}}{\eta} - C_a D_o + W_c \tag{9-28}$$

where $V_c$ is the capacity of the cleaning tank; $W_c$ the flow rate of the chemical added to the cleaning tank; $D_o$ the drag-out flow rate; $\eta$ the chemical capacity coefficient for dirt removal.

RINSING MODEL    After cleaning, the loose dirt remaining on the parts and mixed in the drag-out should be washed out in rinse operations. In each rinse tank, there are two operational modes: the rinse mode in which the parts are submerged in the tank, and the idle mode in which the parts are withdrawn while rinse water still continuously flows. The efficiency of the dirt removal is largely dependent on the difference between the cleanness of the rinse water, and the dirtiness of the parts. On the other hand, the configuration of a rinsing process and the water flow rates are directly related to the wastewater minimization and parts rinsing quality. To derive an optimal configuration and water flow rates, we need to know the cleanness of barrels of parts during rinsing, and the rinse water cleanness. This requires the following models for parts and water of each rinsing tank.

$$A_p \frac{dW_{pr}}{dt} = -r_{pr} \tag{9-29}$$

$$r_{pr} = k_r \gamma_r(t_e)\big(\theta\big(W_{pr} - W_{pc}(t_e)\big) - x_r\big) \tag{9-30}$$

$$V_r \frac{dx_r}{dt} = r_{pr} + F_r(x_r(t_{in}) - x_r) + D_i x_i - D_o x_r \tag{9-31}$$

where $W_{pr}$ is the dirt on parts during rinsing; $W_{pc}(t_e)$ the dirt on parts when leaving the cleaning tank; $r_{pr}$ the dirt removal rate; $x_r(t)$ the pollutant composition in the rinse tank; $\gamma_r(t_e)$ the loose dirt that leaves the cleaning tank; $\theta$ and $k_r$ are model parameters; $F_r$ the flow rate of the rinse water in the tank; $V_r$ the rinsing tank capacity; $x_r(t_{in})$ the pollutant concentration in influent rinse at time $t$; $D_i$ the drag-in flow rate; $D_o$ the drag-out flow rate; $x_i$ the dirt concentration in drag-in.

PLATING MODEL    A plating tank usually consists of a number of slots, each of which can accommodate a barrel of parts for plating. A number of plating operations can occur simultaneously in these slots. A plating deposition model can be developed based on basic electrochemical principles (Lou and Huang, 2000). The allowable amount of metal deposition can be found from knowledge of Faraday's law of electrolysis.

SLUDGE ACCUMULATION MODEL    Sludge can be dry and wet. By "dry sludge" we mean the net quantity of waste by weight. Wet sludge, however, is always measured by its volume, which varies with the type of sludge and treatment methods. In this work, only dry sludge is quantified. The base sludge (S) can be found in cleaning and rinsing tanks. According to the sludge sources, the base sludge in cleaning tanks includes the dirt (oil, soil, grease, solid particles, etc.) removed from the surface of parts ($S_d$) and that of the chemicals used to remove the dirt ($S_c$). In rinse tanks, the sludge due to natural contaminants in make-up water or rinse water ($S_w$) and that due to drag-out from cleaning tanks ($S_g$) should be considered. Thus, the total amount of base sludge is the sum of all of them, i.e.,

$$S = S_d + S_c + S_g + S_w \tag{9-32}$$

Each type of sludge can be quantified based on its sources and by utilizing the information from the cleaning and rinsing models (Luo et al., 1998a,b).

*Plant-Wide Integrated Modeling*  Cleaning, rinsing and plating operations are interrelated in a pre-specified operational sequence. Thus, a plant-wide integrated model should be generated, which consists of all the unit-based cleaning, rinsing, and plating models and a sludge model. These models utilize the process information according to the parts processing sequence. An integrated plant-wide model should be capable of depicting the parts operation status in any tank at any cycle time, providing the information of the chemical concentration of any cleaning tank at any cycle time, showing the pollutant concentration of rinse water in any rinse step (in both the rinse and idle modes) at any cycle time, characterizing the plating operation in the plating tank at any cycle time, and calculating the sludge generation and accumulation in any cleaning and plating tank at any cycle time. Lou and Huang (2000) demonstrated initial successful applications of the integrated modeling.

*Model-Based Optimization*  Plant-wide integrated modeling makes P3 possible. The success of P3 then depends on achieving the most efficient operation in terms of waste minimization, quality assurance, and cost reduction: that is, optimization. With optimization, the environmental and economic objectives can be consistent, since to minimize the quantity and toxicity of end-of-process waste, the consumption of chemicals, water, and energy must be minimized. However, minimization must not neglect the following process operational constraints.

QUALITY CONSTRAINTS  These include the restriction of dirt residue on the parts after each cleaning and rinsing step. The thickness of the metal coating on the surface of parts must be within a specific range, etc. These result in a number of inequality constraints.

PROCESS SPECIFICATION CONSTRAINTS  The upper and lower limits for solvent concentrations and those for water flowrates, and the processing time for each operation step should all be expressed as inequalities.

MODEL EQUALITY CONSTRAINTS  The prediction of process behavior must be based on the models described above. These models, therefore, become the basis of the optimization.

ENVIRONMENTAL CONSTRAINTS    Constraints on the concentrations of specific pollutants in wastewater must be specified. Note that there will be no need to have a constraint on the quantity of wastewater, since it need be minimized.

If the environmental constraints are too strict, there may be no feasible solution. In this case, installing the minimum required waste treatment facilities in a plant will be all that can be done to optimize that particular enterprise.

### 9.2.4.5   *P3 Applications*    This section briefly introduces four successful P3 technologies, each of which can give drastic reduction of waste, great improvement of production, and significant reduction of operating cost, with no or negligible capital cost. These technologies are:

- Operation technology for optimal cleaning and rinse time determination (Zhou and Huang, 2002).
- Design methodology for developing a steady state optimal water allocation network (WAN) (Yang et al., 1999, 2000).
- Design methodology for developing a dynamic switchable water allocation network (SWAN) (Zhou et al., 2001).
- Operating technology for minimum sludge generation (Luo et al., 1998a,b).

*Operation Technology for Optimal Cleaning and Rinse Time Determination*    The dynamic variation of the dirt to be removed from parts is largely determined by the chemical concentration setting in the tank and the cleaning time. The amount of dirt remaining on parts when they are dragged out of the cleaning tank is an indicator of cleaning quality. If other process variables remain constant, the chemical cost can be formulated as a function of a single variable — the cleaning time. Since a longer cleaning time permits a lower chemical concentration, the cost of chemicals can be reduced by increasing the cleaning time of parts in the tank, assuming the production schedule can accommodate the lengthening of this phase by a few seconds.

Similarly, the dynamic pollutant concentration in the rinse tank is determined mainly by rinse water flow rate and rinse time. Rinsing quality is defined as the concentration of the solution remaining on the parts and in their barrels when each barrel leaves the rinse tank. If the pollutant concentration in the drag-out is the same as that in the rinse tank, increasing the rinse time of a barrel can lower the rinse water flow rate, thus reducing the cost of rinse water and wastewater treatment alike.

In the usual cleaning and rinsing process, a cleaning step is followed by a rinsing step. Obviously, cleaning quality will influence rinse dynamics and rinse quality. But the capability of the rinsing system will also influence the cleaning quality setting. Therefore, a two-layer optimization utilizing an integrated cleaning and rinsing model is carried out to determine the optimal cleaning and rinse times as well as chemical concentration in each cleaning tank and water flow rate of each rinse tank. The objective is to minimize the consumption of chemical and fresh water, while ensuring both cleaning and rinsing quality.

This technique has been applied to a cleaning and rinsing system that consists of three cleaning steps followed by three single rinsing step each. The cleaning and rinsing durations have been optimized and the total operating cost can be reduced by 6.9 percent. The comparison of the original and optimized system is listed in Table 9.19.

*Design of a Steady-State Optimal Water Allocation Network*   In the traditional rinse system design, fresh water is frequently sent to each rinse tank. If a used water stream is to be used again, the amount to be reused is determined largely on the basis of experience. Yang et al. (1999, 2000) introduced a design method in which an optimal steady-state water allocation network (WAN) is installed for any plating lines. By systematically evaluating the feasibility of every used water stream for potential reuse, the optimal distribution of fresh and used water can be determined for each rinse step.

The design of a WAN is a superstructure-based optimization technique. Ideally, a superstructure is a process structure that consists of all possible connections among all process units. This implies that all feasible WANs are included in such a structure. Figure 9.46 depicts a superstructure of the rinse system that consists of N rinse steps, each of which requires M rinse tanks. In this superstructure, fresh water can be sent to every rinse tank $R_{i,j}$, and the effluent water stream from $R_{i,j}$ can be reused for all other rinse tanks or sent to an in-plant wastewater treatment facility. Thus, this superstructure contains all possible design solutions of water allocation networks (WANs). A nonlinear optimization technique (Floudas, 1995) is then used to remove all undesirable connections among the rinse tanks based on the design objective as well as process and engineering constraints.

The objective of the WAN design is to minimize the total annualized cost for the network. This cost covers freshwater consumption and expense for installing pipes for water distribution. The equality and inequality constraints include the basic mass

**TABLE 9.19.  Comparison of the Original and Optimal Operation by Cleaning and Rinsing Time Determination**

| Operations and Cost Type | Original System | Optimized System |
|---|---|---|
| Cleaning 1 | 4.5 min | 4.35 min |
| Cleaning 2 | 4.5 min | 4.35 min |
| Cleaning 3 | 4.5 min | 5.22 min |
| Total chemical cost | $89,916 | $82,975 |
| Rinse 1 | 1 min | 0.72 min |
| Rinse 2 | 1 min | 0.72 min |
| Rinse 3 | 1 min | 1.14 min |
| Total rinsing cost | $20,724 | $19,956 |
| Total operating cost | $110,640 | $102,931 |

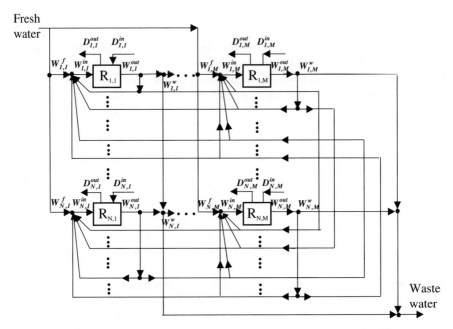

**Figure 9.46.** Superstructure for the rinse system of an electroplating line.

balances for stream mixing and splitting necessary for water redistribution, the component mass balances, and other process and environmental constraints. The optimization is solved using so-called network superstructure concept in order to guarantee the global optimality (Edgar and Himmelblau, 1988).

The WAN design technology has been used to successfully design a number of rinsing systems for different plating lines. Figure 9.47a illustrates the original plating line that contains three rinsing subsystems, each of which has two rinse tanks in series with countercurrent rinse water flow. The total freshwater flow rate is approximately 16 gpm. By using the design methodology, an optimal solution is identified as shown in Figure 9.47b. With the installation of the WAN, the fresh water consumption is reduced to 9 gpm, which is about 44 percent of water or wastewater reduction, while the rinsing quality is also guaranteed.

*Design of a Dynamically Switchable Water Allocation Network*   The WAN design methodology developed by Yang et al. (1999, 2000) does not require the knowledge of rinse dynamics. Any rinse tank runs in two modes: rinse mode and idle mode. A derived WAN can be operated by meeting the steady-state rinse performance of the whole rinse system. In on-site experiments that further investigated rinse dynamics, the authors found that new opportunities for end-of-process wastewater reduction were available.

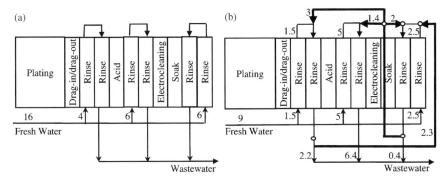

**Figure 9.47.** Water use and reuse in a plating line: (a) original process flowsheet; (b) new process flowsheet with an embedded water allocation network (WAN).

In normal operation, the operation cycle of a plating line is fixed. The operation cycle applies to every rinse tank. It is counted from the beginning of one rinse mode to the beginning of the next rinse mode. The time distribution of the two modes in a rinse cycle, which is determined by the hoist schedule for a given production rate, is fixed also.

In an idle mode, the quality of the rinse water in the tank reaches the standard quickly before the end of the mode. Figure 9.48 illustrates a practical example with a rinse cycle of 10 min: 2.5 min in operation (for rinsing) and 7.5 min in idle (for water replenishment). The requirement of pollutant weight concentration in the tank before next rinse is no higher than 2,750 mg/L. The figure shows that the concentration comes back to 2,750 mg/L after 2.7 min of water replenishment (between 2.5 min and 5.2 min). In normal operational practice, however, the rinse water continuously flows at the constant flow rate through the tank for the rest of

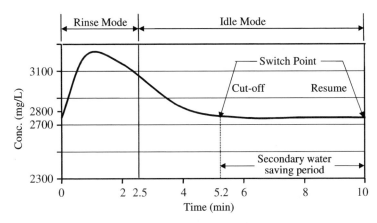

**Figure 9.48.** Rinse dynamics and identification of the secondary water-saving period for SWAN.

the cycle (i.e., the flow lasts for another 4.8 min) to ensure that the tank is ready for next quality rinse operation. Although such caution is understandable, continuing the flow after the contaminant concentration of the rinse water has come down to the standard generates excessive wastewater. A better strategy is to cut off the rinse water at 5.2 min, since the rinse tank is already able to accept the next barrel rinse. Note that once the rinse water is cut off, the water level in the tank is maintained since the out-going water is only due to overflow according to the rinse tank structure.

To realize maximum wastewater reduction, Zhou et al. (2001) proposed a switch-able water allocation network (SWAN). Such a network consists of a primary WAN and a secondary WAN. The two WANs can be operated smoothly in every operation cycle. The identification of the primary WAN is the same as that for designing a steady-state WAN. It is developed using a superstructure-based optimization tech-nique. The secondary WAN, as opposite to the primary WAN, is derived by the same optimization model with some additional constraints for disallowing (fresh and/or used) water flows into a certain number of rinse tanks for a certain time period. These constraints can be developed by analyzing the dynamic behavior of each rinse tank in the primary WAN using the system dynamic rinse models. The basic idea of cutting off water into a rinse tank and then resuming the water flow is elaborated in Figure 9.49. As shown, this second water-saving period starts when the pollutant concentration in the tank reaches, or at least is close enough to the stable point. The period ends when the next rinse job starts. Dynamic analysis should be performed on every rinse tank, which will generate $N \times M$ secondary water-saving periods to the maximum extent, assuming there are $N \times M$ tanks in the rinse system of the same plating line, where $N$ is the number of rinse steps,

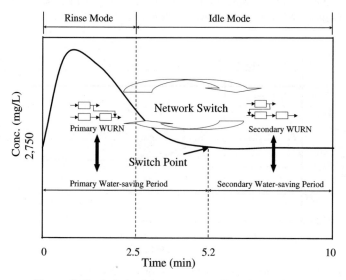

**Figure 9.49.** Switch of the two water allocation networks.

and $M$ is the number of rinse tanks in a rinse step. Then, the most robust switch strategy will be selected from these options. This period in which the rinse water to a number of rinse tanks is to be cut off is called the secondary system water-saving period.

In each operation cycle, the primary WAN will be operated in the primary system water-saving period, while the secondary WAN will be operated in the rest of the cycle (i.e., the secondary system water-saving period). At the time of switching networks, the contaminant concentrations of the relevant rinse tanks are stable. This ensures the transition stability of the entire rinse system. Figure 9.49 delineates the concept of switching of the two WANs.

The SWAN design technology has been successfully applied to several real-world problems. In one application, the original rinse system of an electroplating line consists of five rinse subsystems, as depicted in Figure 9.50a. Through the investigation of rinse dynamics, an optimal SWAN is constructed as shown in Figure 9.50b. The switch from primary WAN to the secondary WAN can be accomplished with only four valves. In each operation cycle, the primary WAN runs for the first 7.5 min, and the secondary WAN for the next 2.5 min. As summarized in Table 9.20, the original system costs \$56,829 annually, including the operating cost for wastewater treatment. By contrast, the SWAN costs only \$34,517 annually on the same basis, for a 39.3 percent reduction of the total annualized cost.

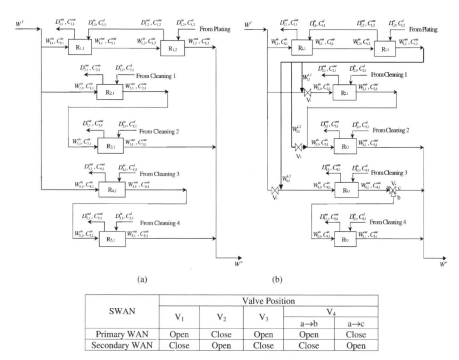

| SWAN | Valve Position | | | | |
|---|---|---|---|---|---|
| | $V_1$ | $V_2$ | $V_3$ | $V_4$ | |
| | | | | a→b | a→c |
| Primary WAN | Open | Close | Open | Open | Close |
| Secondary WAN | Close | Open | Close | Close | Open |

**Figure 9.50.** Original and optimized rinse system with SWAN (a) flowsheet of the original rinse system (b) flowsheet with the SWAN.

**TABLE 9.20. Cost Analysis of the Original and the Optimal Rinse Systems with SWAN**

| Cost | Original System ($/yr) | Switchable WAN (SWAN) ($/year) | | |
| | | Primary WAN (75 percent) | Secondary WAN (25 percent) | Subtotal |
| --- | --- | --- | --- | --- |
| Fresh water | 11,612 | 5,806 | 1,089 | 6,895 |
| Capital | – | 183 | 43 | 776* |
| Wastewater treatment | 45,217 | 22,607 | 4,239 | 26,846 |
| Total | 56,829 | | 34,517 | |
| Cost saving | – | | 22,312 | |
| Cost saving (%) | – | | 39.3% | |

*The capital cost for the SWAN includes that for not only re-piping, but also $550/year for the four valves for water flow change.

*Model-Based Sludge Reduction*   Eventually the mixture of dirt (removed by chemicals) and chemicals (removed by rinse water) becomes sludge. Generation of this so-called base sludge is unavoidable. Whereas most of the dirt/chemicals mixture is in the cleaning tanks, with the rest entering the remaining systems via drag-out from the cleaning tanks, the sludge is expected to be more localized. That is, the sludge generated in one tank should not be sent to the next tank. More specifically, it is highly desirable, whenever technically feasible and economically acceptable, to prevent the sludge generated in cleaning tanks from being carried into rinsing tanks.

Most plating plants generate unnecessarily high amounts of sludge from reasons such as improper use of chemicals, high rinse water flow rates, and excessive drag-out into rinsing tanks. The sludge generated from these sources, called avoidable sludge, should be quantitatively estimated and minimized. Mathematically, sludge reduction can be an optimization problem of minimizing $S$ as expressed in Eq. (9-32). This can be in turn interpreted as the minimization of three types of avoidable sludge ($S_c$, $S_g$, and $S_w$). Since the amount of dirt that must be removed from the parts cannot be minimized due to plating requirement, the term $S_d$ is excluded from optimization.

The sludge related to chemical solvents ($S_c$) can be minimized through reducing the chemical consumption by selecting cleaner with higher efficiency. The amount of chemicals required for dirt removal can be calculated by solving the dynamic cleaning model. The most undesirable sludge source is the drag-out from cleaning tanks to rinsing systems. Since the main function of rinsing is to carry away the cleaning solvents and the mixture of dirt and chemicals introduced through drag-out, if there is more drag-out, more rinse water is needed. Use of an increased amount of rinse water, in turn, makes it more difficult to treat wastewater and also cost more for sludge handling. Practically, drag-out related sludge ($S_g$) cannot be eliminated, but can be minimized through operational improvement, such as the use of dynamic models to adjust chemical concentration settings in the cleaning tanks. Drainage time and bath temperature are always factors in drag-out reduction.

In sludge reduction, the sludge due to natural contaminants in make-up water and fresh water for rinsing ($S_w$) is proportional to the volume of rinse water consumed. Therefore, the reduction in rinse water, with no diminution in rinsing quality, is highly desirable. Dynamic rinse models can be used to determine the optimal flow rate of rinse water for a given operation, as illustrated in Mini-Case Study 9-8.

*Mini-Case Study 9-8: Reducing Sludge in an Electroplating Facility*   Plant managers decide to reduce the amount of sludge generation through operational improvement: the reduction of drag-out and of chemical and rinse water consumption. For a total of 70 barrels of parts, the original settings of chemical concentrations in the presoak, soak, and electrocleaning tanks are all 8 percent. The water flowrate through the rinsing tank is set to 0.023 m³/min. This process is optimized based on the integrated dynamic models, which leads to the following chemical concentration settings in the presoak, soak, and electrocleaning tanks: 10 percent, 8 percent, and 6 percent, respectively. The drag-out rate is thus reduced from 0.012 g/cm² to 0.009 g/cm². This also allows the reduction of rinse water flow rate from 0.023 m³/min to 0.019 m³/min. With these changes, the total amount of sludge can be reduced to 66 kg, which means 15 percent of reduction. Figure 9.51 depicts the total sludge accumulation dynamics in the process before and after the optimization.

**9.2.4.6  *Concluding Remarks***   Many modern P2 technologies for electroplating require significant capital investments, which hindered wide application. The older P2 technologies result in passive reduction only (i.e., end-of-plant waste,

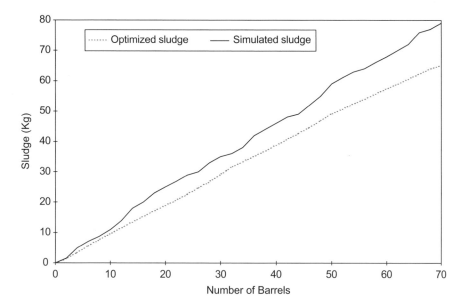

**Figure 9.51.** Comparison of the sludge accumulations before and after process optimization.

rather than end-of-line waste, is reduced). Profitable pollution prevention (P3) focuses on the process that generates waste, and minimizes the waste from that specific process. The successful P3 technologies described here give reason to believe that the development of P3 technologies offers another useful route to environmental protection.


## ACKNOWLEDGMENT

This work is in part supported by National Science Foundation through grants, DMI-0225844 and DMI-0225843.


## REFERENCES

Barnett, W., and Harten, P. (2003). Implementation of the U.S. EPA's metal finishing facility pollution prevention tool. *Proceedings of the 24<sup>th</sup> AESF/EPA Conference on Environmental Excellence*, Daytona Beach, FL, February 3–7.

Cushnie, G. C. (1995). *Pollution Prevention and Control Technology for Plating Operations.* NCMS, Ann Arbor, MI.

Edgar, T. F., and Himmelblau, D. M. (1988). *Optimization of Chemical Processes.* McGraw-Hill, New York, NY.

Floudas, C. A. (1995). Nonlinear and Mixed-Integer Optimization, Oxford University Press, UK.

Gallerani, P. (1996). AESF/EPA pollution prevention training course for metal finishing. *Plating & Surface Finishing*, **83**(1), 48.

Gong, J. P., Luo, K. Q., and Huang, Y. L. (1997a). Dynamic modeling and simulation for environmentally benign cleaning and rinsing. *J. of Plating and Surface Finishing*, **84**(11), 63–70.

Gong, J. P., Luo, K. Q., and Huang, Y. L. (1997b). Process optimization and pollution prevention via OP2EP-advisor, in *Proceedings of the SUN/FIN International Technical Conference*, pp. 485–490, Detroit, MI, June 9–13.

Load, J. R., Pouech, P., and Gallerani, P. (1996). Process analysis for optimization and pollution prevention. *Plating & Surface Finishing*, **83**(1), 28–35.

Lou, H. H., and Huang, Y. L. (2000). Profitable pollution prevention: concept, fundamentals, and development. *J. of Plating and Surface Finishing*, **87**(11), 59–66.

Lou, H. H., and Huang, Y. L. (2001). Qualitative and Quantitative Analysis for Maximum Reduction of Chemicals and Wastewater in Electroplating Operations. In *Green Engineering*, Anastas, P. T., L. G. Heine, and T. C. Williamson (ed.), Ch. 5, pp. 42–61, Oxford University Press, Cary, NC, 2001.

Luo, K. Q., Gong, J. P., and Huang, Y. L. (1998a). Modeling for sludge estimation and reduction. *J. of Plating and Surface Finishing*, **85**(10), 59–64.

Luo, K. Q., Yang, Y. H., Gong, J. P., and Huang, Y. L. (1998b). Model based sludge reduction in cleaning and rinsing operations, in *Proceedings of the 19th AESF/EPA Pollution Prevention & Control Conference*, pp. 328–333, Orlando, FL, January 26–28.

Noyes, R. (1993). *Pollution Prevention Technology Handbook.* Ch. 28, Noyes Data Corporation, Park Ridge, NJ.

USEPA (1999). *National Metal Finishing Strategic Goals Program: An Industry's Voluntary Commitment to a Cleaner Environment.* EPA 231-F-99–002, OPR/ISPD, EPA, Washington, DC.

Yang, Y. H., Lou, H. R., and Huang, Y. L. (1999). Optimal design of a water reuse system in an electroplating plant. *J. Plating and Surface Finishing,* **86**(4), 80–85.

Yang, Y. H., Lou, H. R., and Huang, Y. L. (2000). Synthesis of an optimal wastewater reuse network. *Int. J. of Waste Management,* **20**(4), 311–319.

Zhou, Q., Lou, H. H., and Huang, Y. L. (2001). Design of a switchable water allocation network based on process dynamics. *Ind. Eng. Chem. Res.,* **40**(22), 4866–4873.

Zhou, Q. and Huang, Y. L. (2002). Hierarchical optimization of cleaning and rinsing operations in barrel plating. *J. of Plating and Surface Finishing,* **89**(4), 68–71.

### 9.2.5 Advanced Oxidation Technology at Foundries

FRED S. CANNON, PH.D., P.E.
Associate Professor of Environmental Engineering, Department of Civil and Environmental Engineering, The Pennsylvania State University, University Park, Pennsylvania 16802, USA

JEFF GOUDZWAARD, P.E.
Manager, Plant and Environmental Engineering, Neenah Foundry Company, Neenah, Wisconsin 54597, USA

ROBERT W. PETERS, PH.D., P.E.
Professor of Environmental Engineering, Department of Civil and Environmental Engineering, University of Alabama at Birmingham, Birmingham, Alabama 35294–4440, USA

JAMES C. FURNESS, JR., PRESIDENT
Furness-Newburge, Inc., Versailles, Kentucky 40383, USA

ROBERT C. VOIGT, PH.D.
Professor of Manufacturing Engineering, Department of Industrial and Manufacturing Engineering, The Pennsylvania State University, University Park, Pennsylvania 16802, USA

CHARLES M. KURTTI AND JOHN H. ANDREWS
Neenah Foundry Foundry Co., Neenah, Wisconsin 54597, USA

The metal casting industry represents a major manufacturing segment in the United States, employing approximately 200,000 people nationwide. There are more than 3,000 foundries across the country; and foundries represent one of the nation's oldest manufacturing sectors. Roughly 10–16 million tons of metal castings are poured into molds annually, creating thousands of products that are vital to the US economy. About 60 percent of the total casting tonnage poured in the United States is produced in green sand molds. Green sand molds include silica sand, bentonite clay, finely crushed coal, water, cereal, and core binders. These ingredients are all essential to successful casting.

In conventional casting operations, the greensand is continually recycled; a portion of the green sand (perhaps 4–6 percent) is discarded through each cycle, however, to control the buildup of sand fines and dead clay. This discarded greensand amounts to about 3–5 million tons of waste material annually. The most costly materials wasted are the clay and coal; and the advanced oxidation (AO) process for treating mold cooling and stack emissions has decreased the make-up demands for these ingredients.

Most of the stack emissions from green sand foundries are generated from molding lines during pouring, cooling, and shakeout. When a green sand mold is exposed to the high temperatures of molten metal, this heat causes the coal and organic binders to pyrolyze and to release hazardous air pollutants (HAPs) and volatile organic compounds (VOCs). HAP and VOC emissions are strictly regulated by the US Environmental Protection Agency. Foundries face formidable challenges in lowering air emissions under the 1990 Clean Air Act Amendments (Trombly, 1995; Belyi, 1993), and in lowering energy requirements and materials use (Worrell et al., 2001). Several research teams have addressed some of these issues (Cole et al. 1996; Olenbush and Foti, 2000; Funken et al., 1999; Jans and Hoigne, 1998). The need exists in the foundry industry to develop pollution prevention strategies that economically comply with demanding air quality requirements and materials conservation constraints.

One promising approach for diminishing both air pollutants and materials use is to incorporate AO systems into green sand processing systems. Advanced oxidation systems have been installed in 15–20 full-scale foundry lines in the United States. since 1995. This process has been reported to decrease emissions by 20–75 percent as well as decrease indoor air emissions opacity by a visibly noticeable but non-quantified amount. The beneficial effects of AO processing on green sand system performance has also been reported to diminish clay and coal requirements by 20–35 percent; and this significantly reduces materials costs (Neill et al., 2001; Land et al., 2002; Cannon et al., 2000; Andrews et al., 2000; Voigt et al., 2000, 2002).

The AO system conditions the water, clay, and bituminous coal (seacoal) in a manner that effectively enhances the capture and/or decomposition of emissions. At the same time, the AO system, operating with a blackwater clarifier system, permits active clay to be separated from otherwise discarded sand system baghouse fines. The clay from these fines can then be recycled to the green sand system and fully activated during subsequent mulling cycles. In the foundry setting, AO is usefully viewed as a process modification that also influences the overall green sand system performance. The AO green sand system must be optimized not only to reduce emissions during pouring, cooling, and shakeout, but also to obtain improved green sand properties and casting performance, and to diminish materials use.

An alternative to reducing air pollution by means of the AO process is end-of-pipe incineration, which does effectively remove stack emissions. Given the great amount of fuel required to treat such large volumes of air, however, incineration would be excessively expensive in a foundry setting. In an era of increased environmental regulation and foreign competition, a cost-effective emissions treatment process is critical for the survival of US foundries.

### 9.2.5.1 Advanced Oxidation Processes for Green Sand Systems

Three basic configurations of this AO process has been adapted to processing foundry greensand: advanced oxidation-clear water (AO-CW), advanced oxidation-dry dust-to blackwater (AO-DBW) systems, and advanced oxidation-blackwater from wet collector (AO-BW-WC). For the AO-clearwater process, water (alone) is treated by sonication, hydrogen peroxide and ozone; and the resultant "AO water" is added to the sand coolers for cooling the hot sand; and into the mullers to provide make-up water to prescribed set points. The two AO-blackwater processes (AO-BW-WC and AO-DBW) are similar to this, but the AO process is used in conjunction with wet or dry particulate collection systems. This allows a foundry to return the useful portion of the collected fines to the sand system as a blackwater slurry. After AO treatment, the blackwater slurry goes to a clarifier tank where the silica fines and dead clay settle to the bottom and are removed as a sludge. The blackwater retains active clay and coal, which are returned to the sand system's coolers and mullers. A system in which the AO process returns dust from a baghouse is referred to as an AO-DBW system. When the sludge from a wet collector is returned, the system is referred to as an AO-BW system. The results from the foundry in Neenah, Wisconsin (see Section 9.2.5.5) are for an AO-Dry Dust-to Blackwater (AO-DBW) system.

Advanced oxidation (AO) systems represent a process modification to a conventional green sand foundry. The conventional foundry includes green sand mold formation, molten metal pouring, metal cooling, metal product separation, green sand shakeout, dust exhaust, dust collection, sand cooling, materials make-up, green sand mulling; and then another cycle of molding, pouring, cooling, shake-out, and mulling (see Figure 9.52). Conventionally, a vacuum collection system draws air emissions and fine particles from the pouring, metal cooling, and shakeout areas. These vent to either a wet collection scrubber or dry baghouse collection system that captures fine particulates. Neenah foundry uses a dry baghouse system. Particulates include clay, coal fines, and abraded silica sand fines.

In the Neenah AO-DBW system, the baghouse dust is blended with water to form a slurry. This slurry is conditioned with ozone (to near saturation), hydrogen peroxide (100–300 ppm), and sonication. The AO-dosed slurry then flows through a black water clarifier that separates the reusable active clay and coal that settles slowly, from the non-reusable silica sand fines and dead clay that settle more quickly to the bottom. The AO-treated effluent from the clarifier, which contain much of the active clay, is recycled back to the sand cooler and to the mullers. At Neenah, this effluent slurry has contained 10–17 percent solids by volume.

When a slurry of water plus baghouse dust passes through the AO system, several events occur. First, at ambient temperature, AO mineralizes a small fraction of organic compounds that are present in the water-dust mixture. However, the AO dose is not designed to have a high enough dose to mineralize more than a small fraction of the VOCs. Secondly, sonication and advanced oxidants assist in the

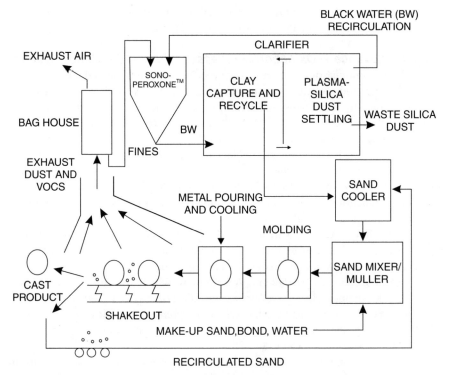

**Figure 9.52.** Foundry configuration with advanced oxidation-dust-to-blackwater system.

activation of clay particles, to prepare them to be cleaned of the VOCs that have coated them in a previous molding and heating cycle. Third, in the blackwater clarifier, sonication dislodges clay platelets from silica fines and from one another in a manner that allows the clay platelets to settle more slowly than the silica fines in the black water clarifier; and the reactivated clay can thus be drawn off for reuse via a side discharge port. Equally important, the AO processes initiate a host of chain reactions that create in the coal many radical scavengers that are only slowly reactive at ambient temperature, but become aggressively reactive at the high temperatures typical for green sand molds during exposure to molten metal. It appears that these radicals within the coal and VOC structure will cause the VOCs to polymerize with the coal surface in a manner that enables the coal to become an activated carbon adsorbent that contains increased porosity (refer to Wang et al., 2004a,b). The extent of this phenomenon is the subject of continued research at Penn State. The net effect, results herein and elsewhere show, is that AO causes VOC and HAP emissions to decrease (see also Glowacki et al., 2003; Cannon et al., 2000).

The bituminous coal (seacoal) in green sand is essential to the molding process, because of the changes it undergoes at high temperature: (1) the coal forms a malleable expanding coke that inhibits the penetration of molten iron into the mold; (2) it reacts with oxygen to form carbon monoxide and other species, reactions that preclude iron oxides from forming on the cast iron surface; and (3) the gases released from the heated coal create between the metal and mold a gaseous blanket that establishes a smooth surface finish. In AO systems, another change is suggested as well: (4) a portion of the coal may become an activated carbon, able to serve as a VOC and HAP adsorbent (refer to Wang et al., 2004a,b). Organic binders, such as phenolic urethanes, are also often included when core sands are required. The cores define the dimensions of interior voids within cast iron parts (such as the piston cavities in engine blocks).

Although the coal and core binders offer important properties and have dramatically improved the resultant casting surface finish, they can also represent a source of VOC emissions (LaFay et al., 1990; Dempsey et al., 1997; Kauffmann and Voigt, 1997; Jans and Hoigne, 1998). Volatile organic compounds are generated when molten metal pyrolyzes coal and core binders so as to release benzene/xylene/toluene type compounds, phenols, and other organics (see comprehensive list, Glowacki et al., 2003). The advanced oxidation process diminishes these VOC releases, as later sections demonstrate. Recent work (Wang et al., 2004a) indicates that AO can effectively influence these surface organics that accumulate on the surface of the clay platelets in a manner that permits more rapid and more complete clay activation, while also reducing emissions.

In conventional green sand systems, the volatiles and condensables that are released from heated coal and organic binders can also cause unfavorable effects on the green sand's strength properties by coating the clay in a manner that diminishes the ability of one clay platelet to bind to sand or to another clay platelet (refer to Sanders and Doclman, 1970; Smiernov et al., 1980; Geiseking, 1975; Jackson, 1979; Helmy et al., 1999; and AFS, 1997 for a fundamental discussion the strength of clays and clay activation in green sands). However, with AO green sand systems, this organic coating can be removed via advanced oxidants such as sonication, ozone ($O_3$) and hydrogen peroxide ($H_2O_2$). When these are mixed into the blackwater system, then into the muller and heated mold, they react at elevated temperatures with these organic deposits in a manner that apparently cause them to be released from the grain surfaces (refer to Brant and Cannon, 1996; Cannon and Brant, 1998; Zhang and Cannon, 1999, 2002). With AO, this "cleaning" of the contaminants allows for better clay and sand particle interaction and binding (Wang et al., 2004a,b). The cleaned clay surfaces are more hydrophilic (i.e., more water-loving), and thus they can engage more readily in clay-water-clay bonding that contributes to the strength of the green sand. The net effect observed in numerous foundries is that a given green compressive strength can be achieved with less clay (Land et al., 2002; Neill et al., 2001).

Bench-scale testing at Pennsylvania State University has also revealed that the AO-treated greensand adsorbed a considerably higher level of $m$-xylene than did non-AO-treated greensand. Indeed, when normalizing the mass of $m$-xylene adsorbed per pass of coal and/or clay, the AO-treated sample exhibited 1/10 to 1/20 the unit sorption capacity as would be offered by an equal mass of commercially manufactured activated carbon (Wang et al., 2004a,b). Also, the Penn State team observed that the pore volume distribution was higher for AO-treated green-sand than for non-AO greensand, and this trait also points to activated carbon formation from the coal component of green sand. A greensand that had undergone multiple cycles of re-use would be expected to act somewhat like an activated carbon because mold material that is roughly $\frac{1}{4}$ in. to 2 in. from the molten metal surface will experience similar temperatures (700–1,000°C) and a steam gasification environment similar to that used in the commercial manufacture of activated carbon (Cannon et al., 1993).

In yet further bench-scale tests, AO-treated greensand relinquished its mass less readily than non-AO treated green sand, when undergoing thermogravimetric analysis (TGA) experiments in the 430–500°C range. At this temperature, VOC's are most prevalently released (Wang et al., 2004a,b). Surface elemental analysis revealed that when a coal-clay mixture was heated to 400°C, carbon was volatilized from the coal and redeposited on the clay surface (Wang et al., 2004a,b). When this clay-coal mixture was washed in water treated by the proprietary Sonoperoxone AO process, then heated again to 105°C, most of this carbon coating was removed. However, when the clay-coal mixture was washed in non-AO water, then heated again to 105°C, far less of the carbon coating was removed. The coal-clay mixtures that had been AO-washed also exhibited an uptake for methylene blue dye and for water that was comparable to that for clay that been heated to 400°C without any coal present to mask its surface. These capacities were considerably higher than for the clay-coal mixtures that had been heated to 400°C, washed in non-AO water, then heated again to 105°C (Wang et al., 2004a). These bench scale tests indicated that the AO combination of sonication, hydrogen peroxide, and ozone will dislodge carbon residue off of clay platelets when heated to a mild temperature of 105°C. In full-scale operations, a considerable fraction of a green sand mold will experience temperatures of 105°C or higher during metal pouring.

### 9.2.5.2 Advanced Oxidation in a Foundry Setting

The AO treatment process at Neenah and other foundries uses hydrogen peroxide, ozone, and ultrasonics in the presence of metal catalysts to create highly reactive radical-containing species in the blackwater slurry. When heated in a green sand mold, these radicals and the radical scavengers that they create can react with organic compounds and organic solids changing the substances' physical and chemical properties to make them less likely to be released as emissions during mold pouring, cooling, and shakeout.

In general, advanced oxidation processes are characterized as ones that generate reactive radicals such as $H \cdot$, $\cdot OH$ and $HO_2 \cdot$, or highly oxidized species such as ozone. These species can degrade, break down, and/or mineralize organic matter at temperatures that are relatively low in comparison to combustion (refer to Kuo, 1986). Advanced oxidant systems now available can incorporate the following:

- Hydrogen peroxide
- Ozone

- Sonication
- Metal catalysis (iron, vanadium, and titanium dioxide)
- Underwater plasma
- Ultraviolet radiation
- Electron beam irradiation
- Pulsed electric discharge

The ability of these AO processes to oxidize, mineralize, or alter organic compounds has been investigated extensively; a number of the studies have also characterized advanced oxidation radical reaction mechanisms (Glaze and Kang, 1989; Glaze, 1995; Valentine and Wang, 1998; Strukul, 1992; Atkinson, 1985; Jans and Hoigne, 1998; Bolton et al., 1998; Choi and Hoffmann, 1997; Hua and Hoffmann, 1997; Mak et al., 1997; Joshi et al., 1995; Chen and Pignatello, 1997; Lin and Gurol, 1998; Gurol et al., 1997).

The AO system we shall describe, which has been developed for foundry applications, combines the first four of these AO generating methodologies into one unified process. Moreover, any one or a few of these advanced oxidant-generating processes can oxidize and/or mineralize organic compounds when the advanced oxidants are at high doses, i.e., at molar ratios of 2–30 moles of AO species applied/mole of organic contaminant mineralized (Takahashi, 1990). When these individual AO generating processes are combined together, they augment the influence of one another in a manner that produces an effect considerably greater than what would have been achieved if the individual components had been used individually and the results summed. For the AO process used in the foundry setting described here, it is neither intended nor anticipated that the doses of advanced oxidants would fully mineralize the VOC emissions that could otherwise be released. Instead, the effects of AO treatment are more subtle, as discussed shortly.

*Ozone and Ozone Plus Hydrogen Peroxide*    The use of ozone and ozone plus hydrogen peroxide has become well established as a water treatment process (Alaton and Balcioglu, 2002; Camel and Bermond, 1998; Chandrakanth and Amy, 1996; Chandrakanth et al., 1996; Chen et al., 2002; Dilmeghana and Zahir, 2001; Edwards and Benjamin, 1992; Esplugas et al., 2002; Jekel, 1998; Jyoti and Pandit, 2003; Neppolian et al., 2002; Perez et al., 2002; Prendiville, 1986; Qureshi et al., 2002; Siddiqui et al., 1997; Teo et al., 2003; Wang et al., 2002; Westerhoff et al., 1997). In water treatment, ozone and hydrogen peroxide serve to disinfect microorganisms, mineralize organic species (if the advanced oxidants are at high enough concentration [Takahashi, 1990]), and clean clay surfaces so that the clays can flocculate and settle more readily in flocculation tanks and sedimentation basins (refer to Georgeson and Karimi, 1988).

*Acoustic Cavitation*    An acoustic cavitation system combined with vapor stripping operations has been used to remove/destroy chlorinated solvents and benzene, toluene, ethylbenzene, and xylenes from solution (Peters et al., 2004,

1998; Mohammad and Peters, 2003a,b,c). As an example of the synergistic effects displayed between advanced oxidant processes, Dilmeghana and Zahir (2001) observed that the degradation rate of chlorobenzene when using UV-$H_2O_2$ and UV-$H_2O_2$-$O_3$ were an order of magnitude higher than that observed when using UV light alone, indicating the combined advanced oxidation processes operated in a synergistic manner.

The acoustic-energy technology involves the application of sound waves being transmitted through a liquid as a wave of alternating cycles. Sound is a mechanical wave that consists of a pressure disturbance transmitted through molecular collision. In fluids, sound (acoustic) waves are longitudinal (compressional), thus the particle motion is in the direction of wave propagation. Compression cycles exert a positive pressure on the liquid, pushing molecules together, while expansion cycles exert a negative pressure, pulling molecules away (rarefaction) from each other (see Figure 9.53). Microbubbles, the initial vapor phase in the liquid, develop when the pressure in a liquid drops below the vapor pressure. The cavitation bubbles that are formed can be as small as $10^{-10}$–$10^{-8}$ m in diameter. During the alternating cavitation cycles, the microbubbles grow to a critical size and implode (collapse to zero size), releasing a large amount of energy. Temperatures on the order of 5,000°K and pressures up to 500 to 1,000 atmospheres have been calculated/observed at the collapsing interface in microbubble implosions, while the bulk solution stays near ambient temperature and pressure. The collapsing bubble interface results in the formation of hydroxyl ($\cdot$ OH) and hydrogen (H $\cdot$ ) radicals. These radicals are extremely

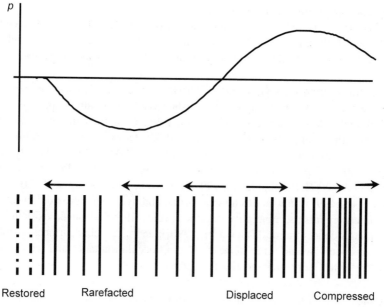

**Figure 9.53.** Schematic representation of acoustically based advanced oxidation with alternating expansion and compression.

effective for destroying and altering organic compounds. The intensity of cavity implosion (and hence the nature of the reactions involved) can be controlled by process parameters such as the sonic frequency, sonic intensity power per unit volume of liquid, static pressure, choice of ambient gas, and addition of advanced oxidants such as hydrogen peroxide ($H_2O_2$), ozone ($O_3$), and metal catalysts.

Following the military declassification of various sonication technologies and other materials and data, considerable scientific and innovative advances have been made in the field of sonication. Current applications of the technology incorporate these advances in the areas of much higher power densities (from watts to kilowatts), multiple frequency capabilities, miniaturization of acoustic cavitation equipment, and improved focusing and directionalizing of acoustic waves.

### 9.2.5.3 Possible Design Layouts for an Advanced Oxidation System

Sonoperoxone, an AO system manufactured by Furness-Newburge, has been implemented in various foundry layouts and applications; and a brief discussion of these follows.

#### AO Dry Dust Collector to Blackwater (AO-DBW) Clay Recovery System

This system (Figure 9.54) recycles the clay from the baghouse collector back into the sand system, as noted earlier. It is the system that Neenah Foundry Plant 2 has used. It has the following advantages: reducing new bond purchases, reducing solid waste disposal, increasing compressive strength and shear strength of green sand, and reducing smoke and odor dramatically within the foundry. The

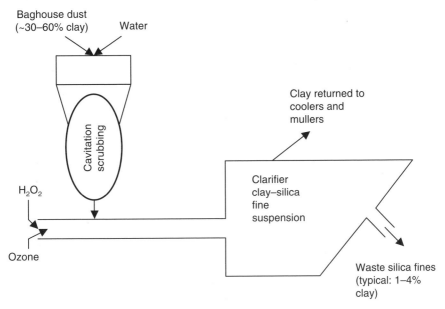

**Figure 9.54.** Schematic of AO dry dust collector to blackwater clarifier (AO-DBW) clay recovery system. [Adapted from Furness-Newburge Inc. 2003.]

AO-DBW system at Neenah Foundry Plant 2 will be examined in detail in the case study of Section 9.2.5.5.

### AO Blackwater from Wet Collector (AO-BW-WC) Clay Recovery System
This system (Figure 9.55), which enhances the recycle of wet collector via cavitational cleaning and separation of clay/silica fines, has the following advantages: reducing new coal and clay purchases, reducing solid waste disposal volume and cost, increasing compressive strength and shear strength of green sand, reducing smoke and odor dramatically, and reducing smoke stack opacity.

### AO Clear Water (AO-CW) Treatment System for Foundry Water
This system (Figure 9.56) diminishes the smoke and odor in a foundry by cleaning the sand system; it has been used by Grede-Reedsburg, in Reedsburg, Wisconsin, and by CERP-Technicon in Sacramento, California. This process has the following advantages: increasing compressive strength and shear strength of green sand, reducing bond usage, increasing production up-time, and reducing smoke and odor dramatically.

### AO Slurried Bond Feeding System
The AO Slurried Bond Feeding system (Figure 9.57) introduces a smooth, pre-worked bond (clay plus coal) into the system's mullers via controllable slurry. Advantages of this system include: providing a more consistent bond feed, reducing particulate losses from dry bond feed, increasing muller efficiency, increasing compressive strength and shear strength

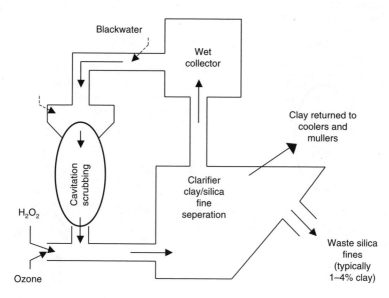

**Figure 9.55.** Schematic of AO blackwater from wet collector (AO-BW-WC) clay recovery system. [Adapted from Furness-Newburge Inc., 2003.]

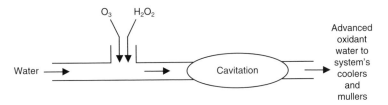

**Figure 9.56.** Schematic of AO clear water (AO-CW) treatment system for foundry make-up water. [Adapted from Furness-Newburge Inc., 2003.]

of green sand, reducing smoke and odor dramatically within the foundry, and providing easy maintenance with no large tanks to clean out.

*AO Core Room Scrubbing System*   This system (Figure 9.58) has the ability to remove core room odors without acid treatment. It offers the following advantages: improving odor and VOC removal (>95 percent), eliminating the need for sulfuric acid treatment, thereby resulting in no need to handle or transport acid, and creating advanced oxidant water for sand treatment.

### 9.2.5.4  *Performance at Some Foundries Using Full-Scale Advanced Oxidation*   The advanced oxidation process has been used in green sand foundries in a number of different system configurations. Following AO installation and sand system stabilization, some foundries have reported emissions reductions of 20–75 percent, with corresponding coal and clay consumption decreases of 20–35 percent when implementing a Sonoperoxone process by Furness-Newburge, Versailles, KY (Wang et al., 2004a,b; Land et al., 2004, 2002; Goudzwaard et al.,

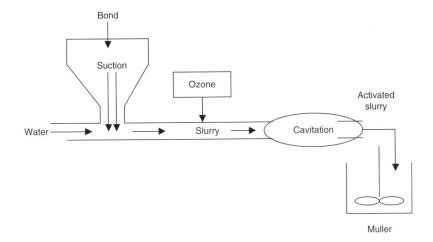

**Figure 9.57.** Schematic of AO slurried bond feeding system. [Adapted from Furness-Newburge Inc., 2003.]

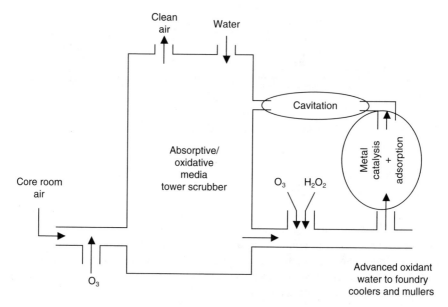

**Figure 9.58.** Schematic of AO core room scrubbing system. [Adapted from Furness-Newburge Inc. 2003].

2003; Glowacki et al., 2003; Neill et al., 2001; Cannon et al., 2000; Hairston, 1999). The increased sand strengths have resulted in several of these foundries also reducing their mold cracking, increasing their production up-time, and decreasing their scrap rates.

The Grede-Reedsburg Foundry, which has employed an AO-clear water system since 1995, has observed a reduction of approximately 55 percent in total VOCs during cooling and shakeout both for cored and non-core castings; a corresponding 45–50 percent drop in benzene emissions for non-cored jobs and a 35–40 percent drop in benzene emissions for cored castings was also observed (Cannon et al., 2000). These emission reductions and other results, summarized in Table 9.21, were realized via a gradual decline that progressed over the course of several years after the AO system had been placed in service. A portion of this decline can be attributed to on-going sand system modifications during the emissions testing time period. For example, lignite coal (20 percent) was blended in with the bituminous coal (80 percent) as part of the pre-mix addition midway during the trial period. Emissions had progressively declined before this lignite introduction, and they continued to progressively decline afterwards.

Wheland Foundry of Chattanooga, TN used the AO Sonoperoxone blackwater system at several of their facilities. After implementing the technology, there was a significant reduction in the requirements for clay and coal per ton of hot metal poured (Neill et al., 2001). The premix (clay and coal) requirement was reduced from 145 lbs/ton of iron poured prior to implementation of the AO technology to 115 lbs/ton Fe after incorporation of AO. There was also a significant improvement in the properties of the green sand used to make the molds, as manifest by the green

**TABLE 9.21. Summary of Emissions From Full-Scale Foundry, Grede-Reedsburg (Emissions Listed as Pounds of Emissions per Ton of Metal Poured)**

| | Grede Reedsburg, Reedsburg, WI (AO-CW) | | |
|---|---|---|---|
| | Baseline, Before AO | Optimized AO Sand System | Percent Change |
| Date of comparison | 1995 | 1999 | |
| Loss on ignition, 1,800°F, % | 3.0 | 3.4 | +13 |
| MB clay, % | 11.0 | 10.2 | −7 |
| Water-to-MB clay ratio, % | 31 | 33 | +6 |
| Total VOC, no core | 0.37 | 0.17 | −54 |
| Total VOC, with core* | 0.49 | 0.22 | −55 |
| Benzene, no core | 0.035 | 0.018 | −49 |
| Benzene, with core* | 0.081 | 0.05 | −38 |

*Core is phenolic urethane.

compressive strength, which climbed from 26 psi before AO to 31 psi after the AO system was stabilized. The loss on ignition content (i.e., organic content plus waters of hydration within clays) was 3.4 percent before AO and 3.0 percent after AO was stabilized. Silica sand additions dropped from 520 lb/ton Fe before AO to 300–400 lb/ton Fe after AO stabilized. Since implementation of this technology, cracked mold and swelling cast defects diminished markedly, resulting in a reduction of 16 percent of the castings that were previously rejected. There was also a significant reduction in the visible emissions and odors from the molding operation, which has been confirmed by limited stack testing. Together, these data indicate that with this AO system, Wheland Foundry was able to achieve better casting control while adding less coal and clay; and both these factors represented cost savings. Wheland Foundry estimated that the installation cost of the AO technology was recovered in 3–4 months.

At International Truck and Engine-Waukesha, WI, stack testing before and after AO-CW system installation has also been reported. This foundry observed that an AO-clearwater system resulted in a 74 percent drop in total VOCs (from 0.187–0.048 lb/ton of iron poured), a 65 percent drop in benzene (from 0.017–0.006 lb/ton Fe), and a 12 percent drop in carbon monoxide emissions (from 1.1–0.95 lb/ton Fe).

We turn now from a synopsis of results at two US foundries to a case study of the system with which the authors are most familiar.

### 9.2.5.5   Case Study: Methods and Procedures at Neenah Foundry

*Advanced Oxidation-Blackwater System Adaptation*   In January 2000, Neenah Foundry installed an Advanced Oxidation, Dry Dust-to-Blackwater (AO-DBW) system into their Plant 2 (Figure 9.59) The advanced oxidants included ozonation (to saturation), hydrogen peroxide (100–400 ppm), acoustical sonication (16–25 kHz), and hydrosonication. The electrical costs for operating the ozone and

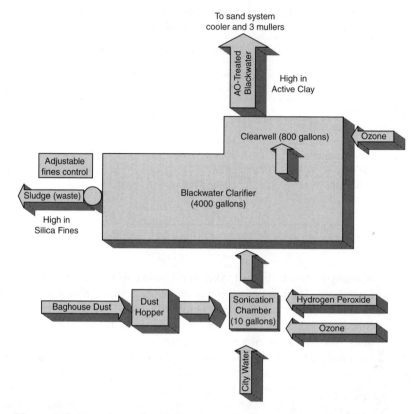

**Figure 9.59.** The advanced oxidation dust-to-blackwater system at Neenah Plant 2.

acoustical sonication were in the $3–20/day range. For 2 years before AO start-up, the operating and emissions performance of the green sand system was progressively optimized by, for example, reducing sand system coal to as low a level as possible without increasing scrap rates. The sand system was operated at the **non-AO-optimized** conditions during June–December 1999, and emissions were tested for both cored and non-cored castings; the results are discussed later and summarized in Tables 9.22 and 9.23.

At Neenah, the AO-DBW system was gradually brought on-line, starting in January of 2000, and it progressed to a fully AO-blackwater-stabilized process over the course of 6 months. It then maintained this **AO-stabilized** condition throughout the remainder of the study period. Changes in the green sand properties due to the AO-blackwater additions made it necessary to re-stabilize the sand system to ensure consistent performance. These **AO-stabilized** changes led to a significant overall reduction in sand system bond consumption (see Land et al., 2002).

A limited number of **baseline** emission tests were performed in 1995, prior to the initial non-AO sand optimization. During this testing, emissions samples were collected from mold cooling only. In comparison, for the *non-AO-optimized* (1999) and *AO-stabilized* (2000–2001) tests, emissions samples were collected from both mold cooling and shakeout. None of the emissions testing in our study included pouring. The *non-AO-optimized* and the *AO-stabilized* samples were monitored for total VOCs, benzene, formaldehyde, carbon monoxide (CO), carbon dioxide ($CO_2$), methane, ethane, and propane. Sand properties were also closely monitored and recorded during these time intervals (refer to Land et al., 2002).

*Initiating Advanced Oxidation at Neenah in Context of Wisconsin Air Regulations* Neenah Foundry began its evaluation of Hazardous Air Pollutant (HAP) formation mechanisms as a result of Wisconsin's Hazardous Air Pollution Rule, NR445 (1988). This rule affected facilities that were built or modified after 1988. Benzene is a "Group A" compound that required application of the Lowest Available Emission Rate (LAER) if greater than 300 pounds per year of benzene are emitted. Formaldehyde is a "Group B" compound that required application of Best Available Control Technology (BACT) if greater than 250 pounds per year are emitted. (The application of the LAER and BACT concepts was discussed in Chapter 5.)

Emissions in excess of these levels were anticipated at Neenah Foundry, as well as at other large Wisconsin Foundries, necessitating the installation of control technology. Incineration fuel costs for just one of Neenah's several production lines was estimated at 2.5 million dollars per year; and Neenah foundry personnel therefore chose to instead explore emissions reduction strategies that did not involve end-of-pipe treatment. Other Wisconsin Cast Metals Association foundries also collaborated with Neenah foundry in this endeavor.

In 2000, an advanced oxidation dry dust-to-blackwater (AO-DBW) system was installed and gradually phased into service at Neenah. The AO system operation began as a clear water system with only AO water additions to the sand system. After one month of clear water operation, gradual increases of baghouse dust additions and hydrogen peroxide dose increases were introduced over a six month period. The initial blackwater slurry concentration was 2 percent solids by volume. This was gradually increased to 5 percent, 8 percent, 10 percent and 12 percent solids by volume.

The AO-DBW system, illustrated in Figure 9.59, has four material inputs: water, baghouse dust, ozone, and hydrogen peroxide. There are two discharges from a blackwater clarifier: AO-enriched blackwater that is returned to the foundry sand system via a blackwater pump and sludge from a drag chain that is disposed of. Dry baghouse dust is introduced into the water just upstream of the ultrasonic transducers, creating a secondary benefit from ultrasonics activation. In the blackwater clarifier, ultrasonic energy facilitates the removal of adhered clay and coal from the other inert fines in the baghouse dust so that the clay can be reused. Finally, the water that has been enriched with clay and coal from the baghouse dust and exposed to additional advanced oxidants is pumped back into a sand cooler or into one of three 4,000-pound sand mullers.

*Emissions Analyses*   Throughout the long test period that compared non-AO-optimized and AO-stabilized conditions, emissions testing was performed during standard production runs of the same casting part numbers (when possible). The sand system emission analyses were performed by a certified outside laboratory.

Sand system emissions analyses were conducted in accordance with OSHA Method 7 and EPA Method 18. All emissions samples were collected using slip stream capture methods that were pumped from the mold cooling or shakeout exhaust ducts. Prior to commencing a given emissions test, the DISA molding line was allowed to reach near steady-state conditions for a given casting part (over several hours) before a cumulative one-hour emissions stack test was performed.

Total volatile organic carbon (VOC) analyses were conducted by extracting emissions species from the activated carbon tubes with carbon disulfide, and then injecting this extract into a gas chromatography-flame ionization detector (GC-FID), extended for 45 minutes, with chromatography column temperature ramping up to 200°C. Generally, no peaks were observed after 35 minutes GC time; and all peaks between 5 minutes and 25 minutes that exceeded 1,500 area counts were included in the total VOC number. This corresponded roughly to organic compounds that contain 4–16 carbon atoms (i.e., C4 to C16 range). These values were then compared to standard hexane peaks; and have thus been reported herein as "total VOCs as hexane (C4 to C16)." The benzene peak from the GC-FID response was also measured and compared to a benzene standard.

The formaldehyde analysis was performed separately by means of running air samples through a gas chromatograph in conjunction with a nitrogen-phosphorous selective detector (GC-NPD).

### Results

*Green Sand Properties*   As indicated in Table 9.22, the AO process impacted cooling and shakeout emissions, as well as green sand properties (the average values during the times of emissions monitoring). The tabulated emissions values are for mold cooling plus shakeout. Emissions levels are average values and interpolated average values at plant average core loadings. When comparing non-AO optimized to AO-stabilized conditions, one notes that the green compressive strength increased 6 percent with AO, and wet tensile strength increased 9 percent. Bond consumption dropped 33 percent, from 187 lb/ton of iron poured for the *non-AO optimized* condition, down to 125 lb/ton Fe for the *AO-stabilized* condition. Moreover, the ratio of green compressive strength (GCS) to methylene blue (MB) clay rose 26 percent, indicating that the AO system promotes clay activation during mulling cycles. The casting quality and casting defect rate was the same for the non-AO optimized and AO-stabilized conditions.

When comparing the AO-stabilized condition with the non-AO optimized condition, one observes that VOC emissions dropped 63 percent for no core castings and 47 percent at average core loading, while benzene, methane plus ethane, and propane also dropped. This large decrease in VOCs was greater than the 33

**TABLE 9.22. Operations Performance at Neenah Plant 2: (i) Baseline (Jan. 1995), (ii) Non-AO-Optimized (June–December, 1999), (iii) AO-Stabilized with Dry Dust-to-Blackwater (AO-DBW)**

| Parameter | Baseline, 1/95 | Non-AO-Optimized 6/99–12/99 | AO-DBW-Stabilized 7/2000–5/2001 | Percent Change: Non-AO Optimized vs. AO Stabilized |
|---|---|---|---|---|
| **Green sand properties:** | | | | |
| Green compressive strength (CS), (psi) | 25 | 33.3 | 34.6 | +4 |
| Bond addition: clay and coal, (lb/ton Fe) | 200 | 187 | 125 | −33 |
| Moisture, (%) | 3.0 | 3.04 | 2.68 | −12 |
| Methylene blue (MB), clay (%) | 9.8 | 10.1 | 8.3 | −20 |
| Moisture/MB clay ratio | 30.6 | 30.1 | 31.5 | +4.6 |
| Green CS/MB clay, (psi/%) | 2.55 | 3.30 | 4.17 | +26 |
| Green CS/clay added, (psi/lb/ton Fe) | 0.195 | 0.244 | 0.369 | +51 |
| Wet tensile strength, Newtons/cm$^2$ ($\times 10^{-3}$) | – | 377 | 408 | +8 |
| Loss on ignition (LOI), (%) | 5.1 | 5.44 | 3.65 | −33 |
| LOI/coal added, (%/lb/ton Fe) | 0.075 | 0.119 | 0.130 | +10 |
| Volatile content matter, (%) | | 2.3 | 2.0 | −13 |
| Core binder level in core sand, (%) | 1.75 | 1.0–1.1 | 1.0–1.1 | 0 |
| *Preblend*: % clay | 64 | 73 | 75 | +2.7 |
| % coal | 34.0 | 24.5 | 22.4 | −8.6 |
| % cereal | 2 | 2 | 2 | 0 |
| % soda ash | 0 | 0 | 0.7 | NA |
| **Emissions (mold cooling plus shakeout):** | | | | |
| *No core castings* | | | | |
| VOCs, (lb/ton metal) | SFT** | 0.60 | 0.22 | −63 |
| Benzene, (lb/ton metal) | SFT** | 0.055 | 0.030 | −45 |
| Carbon monoxide, (lb/ton metal) | SFT** | 2.05 | 2.38 | +14 |
| Methane + ethane + propane, (lb/ton metal) | SFT** | 0.61 | 0.39 | −36 |
| *Cored castings* | | | | |
| *Average core load** | | | | |
| VOCs, (lb/ton metal) | SFT** | 0.86 | 0.45 | −47 |
| Benzene, (lb/ton metal) | SFT** | 0.082 | 0.066 | −19 |
| Carbon monoxide, (lb/ton metal) | SFT** | 2.4 | 2.7 | +12 |
| Methane + ethane + propane, (lb/ton metal) | SFT** | 0.65 | 0.45 | −30 |
| *Very heavy core** | | | | |
| VOCs, (lb/ton metal) | SFT** | 1.9 | 1.4 | −26 |
| Benzene, (lb/ton metal) | SFT** | 0.20 | 0.22 | +10 |
| Carbon monoxide, (lb/ton metal) | SFT** | 3.9 | 4.3 | +10 |
| Methane + ethane + propane, (lb/ton metal) | SFT** | 0.85 | 0.72 | −15 |

*Average core load is 3.4 lb phenolic urethane binder per ton of metal poured (320 lb core sand/ton Fe for cores with 1.05 percent binder). Very heavy core load was 24.8 lb binder per ton of metal poured (2,360 lb core sand/ton Fe for cores with 1.05 percent binder). Interpolated values based on linear regression. Volatile organic compounds includes the sum of all gas chromatography peaks in the C4-C16 range.

**SFT = See figures and text: A limited number of non-AO-baseline (1995) emissions tests were conducted for mold cooling (only) VOCs and benzene; and these are compared to non-AO-optimized and AO-stabilized conditions in Figures 9.62 and 9.64.

percent decrease in loss on ignition material (LOI). This indicates that emissions were effectively diminished by direct AO effects for a given LOI, and AO-driven sand system changes that permit and require significant coal and clay reductions without changes in casting quality.

*Foundry Emissions*    Figures 9.60 through 9.66 plot emissions rates of several pollutants as a function of core loading for *baseline* (1995), *non-AO-optimized* (1999), and *AO-DBW-stabilized* (2000–2001) conditions [note that for conciseness, the figure labels omit the "DBW"]. All these emissions have been normalized to pounds emissions per ton of iron poured. For all three of these conditions, the emission rates for most pollutants increased with increasing core loading. The same phenolic urethane cold box core resin was used throughout these emission tests; and the core binder level of 1.0–1.1 percent was consistent throughout the 1999–2001 study period. Core loading strongly influenced stack emissions because of the significant creation of benzene and other VOCs from core binder decompositions during cooling and shakeout.

In each figure, a vertical line indicates the average core loading of 320 lb of core sand per ton of iron poured that has been used at Neenah Foundry Plant 2 during these years of operation. At a 1.05 percent core binder level, this amounts to 3.4 lb of resin per ton of metal poured. The emissions data included core loadings that extended up to 2,360 lb core sand per ton a metal poured (24.8 lb resin/ton metal). Emissions estimates for average core loading are shown in these graphs; the emissions reductions given in Table 9.22 were computed by comparing the linear regressions depicted in Figures 9.60–9.66, supplemented by data on other emissions (not included for reasons of space).

1. TOTAL VOC EMISSIONS    Total VOC emissions during cooling and shakeout for no-core castings decreased from 0.60 lb/ton Fe for *non-AO optimized* conditions, down to 0.22 lb/ton Fe for the *AO-DBW stabilized* condition — a 63 percent decrease (Figure 9.60 and Table 9.22). These differences between *non-AO optimized* and *AO-stabilized* are statistically significant at the 99 percent confidence level. At average core loading (320 lb core sand/ton of iron poured), total VOC emissions dropped from 0.86–0.45 lb/ton — a 47 percent decrease. The slopes of the regression lines for the *non-AO optimized* versus *AO-stabilized* conditions are nearly parallel. This indicates that AO processing diminishes VOC emissions by a nearly constant amount (about 0.4 lb/ton) regardless of core loading. This constant decrease in VOC emissions occurs even though the AO-containing green sand fraction of the mold decreased (slightly) as core loading increased.

Total VOC emissions from mold cooling alone were evaluated and are presented in Figure 9.61. For the *AO-stabilized* condition, mold cooling VOCs ranged from 0.09 lb/ton Fe for the non-cored tests to 0.47 lb/ton Fe for the most heavily cored castings, a drop of 48 percent vs the non-AO-optimized condition. By comparing these results with those of Figure 9.60, one observes that mold cooling emissions represented roughly 40 percent of the sum of mold cooling plus shakeout emissions for the no-core castings, and 25 percent of this sum for the very heavily cored castings.

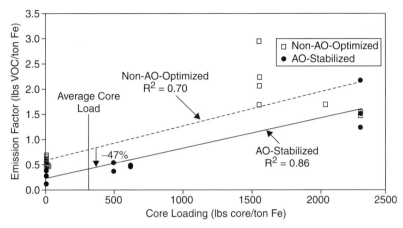

**Figure 9.60.** Total volatile organic compound (VOC) emissions during mold cooling plus shakeout: comparison of non-AO-optimized vs AO-stabilized sand systems at various core loadings.

In evaluating the effects of AO treatment, it is useful to compare the decreases in VOC emissions to the decreases in the green sand's Loss on Ignition (LOI) during the test period. Specifically, for the no-core castings, VOCs decreased 63 percent, when comparing the *non-AO optimized* to the *AO-stabilized* for mold cooling plus shakeout. The green sand's LOI was lowered 33 percent (from 5.44 percent for the *non-AO optimized* to 3.65 percent for the *AO-stabilized* condition).

The AO process creates operating conditions where this lower LOI can be employed while still maintaining casting quality. Indeed, the experience at Neenah Foundry and

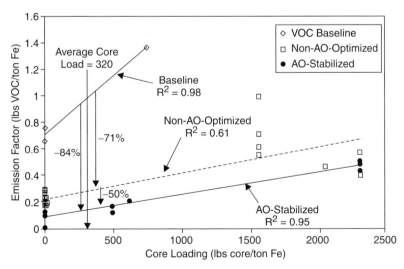

**Figure 9.61.** VOC emissions during mold cooling only: comparison of baseline (No AO), non-AO-optimized, and AO-stabilized sand systems at various core loadings.

other foundries is that the LOI, methylene blue clay, and bond addition rates must go down in order to achieve consistent sand system performance and maintain casting quality. Lower bond additions also result in lower sand system operating costs.

The AO-induced decrease in VOCs are most appropriately evaluated with respect to the sand system decreases in LOI. The data suggest that for no core castings, about half of the VOC decrease from AO treatment could be attributed to the drop in LOI that the AO process facilitated; and about half could be attributed to lower emissions that a given level of LOI would generate when AO is employed. This concept is developed further elsewhere (Glowacki et al., 2003).

The study team also conducted a limited number of mold cooling (only) emissions tests before optimizing the non-AO green sand system (i.e., in 1995), and these VOC results are also presented in Figure 9.61. At the average core loading, mold cooling VOCs for the 1995 *baseline* were 0.94 lb/ton metal. In contrast, the final *AO-stabilized* counterpart was 0.15 lb/ton; and this represented an 84 percent decline. This 84 percent drop reflects the impact that both the green sand optimization and subsequent advanced oxidation processing have had on emissions for Neenah Foundry, Plant 2.

2. BENZENE EMISSIONS    Non-core emissions of benzene from the mold cooling plus shakeout averaged 0.055 lb/ton metal for *non-AO optimized*, and 0.038 lb/ton metal for *AO-stabilized* (Figure 9.62). This change represents a 31 percent drop, which was statistically significant to the 99 percent confidence level. Other differences noted were not statistically significant.

In a comparison of mold cooling emissions only at average core loading (320 lb core sand/ton metal), *baseline* benzene emissions in 1995 (before sand optimization) were 0.048 lb/ton Fe (see Figure 9.63). This compared to 0.034 lb/ton Fe for the *non-AO optimized* case — a 29 percent drop. Moreover, mold

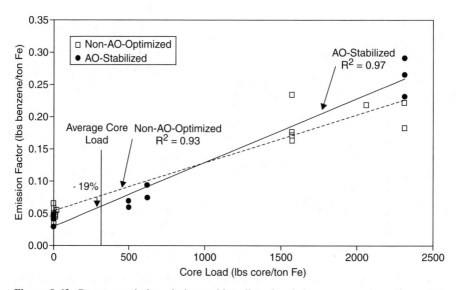

**Figure 9.62.** Benzene emissions during mold cooling plus shakeout: comparison of non-AO-optimized vs AO-stabilized sand systems at various core loadings.

cooling benzene emissions at average core loading for the *AO-stabilized* condition was 0.022 lb/ton metal. Thus, since 1995, benzene emissions from mold cooling have decreased a total of 54 percent at Neenah Plant 2 due to sand optimization and AO system changes combined.

It is clear that AO processing can have a direct effect on reducing emissions even for cored castings during mold cooling, since emissions from core binder decomposition must pass through the green sand mold before they are released to the atmosphere. However, core emissions during and after shakeout are released directly into the atmosphere. AO processing therefore could have little or no effect on such core-generated shakeout emissions; however, it produced some decrease in mold emissions.

3. FORMALDEHYDE   Formaldehyde emissions showed little change with AO treatment. For no-core castings, the *non-AO optimized* formaldehyde emissions averaged 0.005 lb/ton Fe, compared to an average of 0.004 lb/ton Fe for the *AO-stabilized* condition. However, the formaldehyde emission levels varied greatly for both conditions, and they were not statistically different from one another. At the highest core loadings (2,300 lb core sand/ton iron), formaldehyde emissions increased to 0.012–0.018 lb/ton metal, regardless of whether AO treatment was employed.

4. CARBON MONOXIDE AND CARBON DIOXIDE   Carbon monoxide emissions increased slightly when AO treatment was employed (Figure 9.64). For no-core castings, this amounted to a 14 percent increase in emissions from mold cooling and shakeout, although this increase was not statistically significant. As core loading increased, CO emissions increased; and the AO emissions paralleled those without AO. The carbon dioxide emissions levels were not influenced by AO processing. For no-core castings, $CO_2$ emissions amounted to 30 lb/ton Fe, and with the heaviest core loading, these rose to 42 lb/ton metal, regardless of AO.

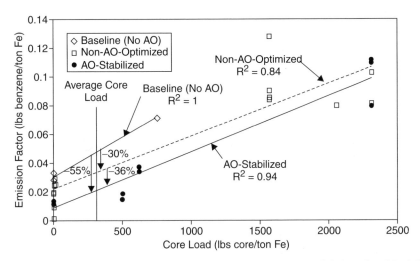

**Figure 9.63.** Benzene emissions during mold cooling: comparison of (i) baseline (No-AO), (ii) non-AO-optimized, and (iii) AO-stabilized sand systems at various core loadings.

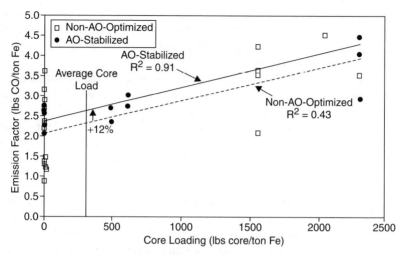

**Figure 9.64.** CO emissions during mold cooling plus shakeout: comparison of non-AO-optimized vs AO-stabilized sand systems at various core loadings.

5. METHANE, ETHANE, AND PROPANE    Analyses for these gases did not commence at Neenah until December 1999. The data collected show that methane plus ethane emissions levels were considerably lower for the *AO-stabilized* conditions than they were for *non-AO optimized* conditions (Figure 9.65). For no-core castings, the non-AO emissions of these two species was 0.61 lb/ton Fe, compared to 0.39 lb/ton Fe when AO was employed (cooling plus shakeout). This represented a 36 percent decrease, which was statistically significant to the 99 percent confidence interval. At average core loading, the *AO-stabilized* emissions were 43 percent below the *non-AO optimized* values. As core loading increased, emissions increased at the same rate for AO and non-AO emissions trials. As discussed previously for other emissions species, this is in spite of the fact that the AO-containing portion of the mold (i.e., the green sand portion) decreased slightly when the core loading increased. This suggests that some core emissions are effectively mitigated upon passing through the green sand mold. Nevertheless, the main effect of AO is on emissions from the green sand portion of the mold where the AO can be introduced.

In a somewhat similar manner, propane emissions without core loading were greater for *non-AO optimized* conditions (0.90 lb/ton Fe) than for *AO-stabilized* conditions (0.62 lb/ton Fe) (Figure 9.66), and these differences were statistically significant to the 99 percent confidence interval. However, with increasing core loading, the differences between non-AO and with-AO conditions became less pronounced.

6. EMISSIONS SUMMARY    The emissions of a number of species were lowered significantly when AO was employed. These reductions were most prominent for no-core castings. As core loading increased, the non-AO and with-AO regression values either paralleled one another (as for total VOCs and methane + ethane), or the percent emissions reductions via AO decreased somewhat (as for benzene,

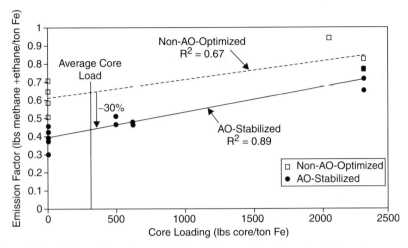

**Figure 9.65.** Methane plus ethane emissions during mold cooling plus shakeout: comparison of non-AO-optimized vs AO-stabilized sand systems at various core loadings.

formaldehyde, and propane). This means that the advanced oxidation reactions primarily impacted the green sand and least impacted the core sands. This perspective matches the operational manner in which AO is applied: the advanced oxidants from the blackwater become mingled with the green sand in the sand cooler and the muller; and they can therefore impact the green sand composition. In contrast, the advanced oxidants, which are water borne, cannot become intimately mingled with the core sand because the water that carries the advanced oxidants would eliminate the binding nature of the phenolic urethane. Also, this smaller margin of AO

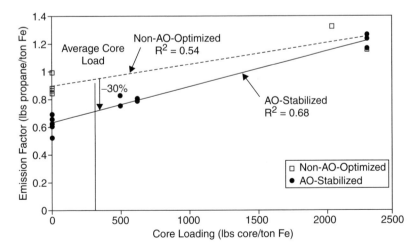

**Figure 9.66.** Propane emissions during mold cooling plus shakeout: comparison of non-AO-optimized vs AO-stabilized sand systems at various core loadings.

improvement when processing heavily cored castings is due in part to the (slightly) smaller fraction of AO-enhanced green sand within the mold at the higher core loadings. Even with cores present, the surrounding AO green sand can be expected to reduce emissions during cooling. However, during shakeout, the mold shape is broken up and internal core residuals, now exposed to the air, can thus produce significant emissions during shakeout. In conventional shake-out operations, the AO-treated green sand has limited contact with these emissions, whereas such contact could facilitate the adsorption of these emissions. The study team is actively exploring approaches for achieving yet further adsorption during shakeout.

MASS BALANCE ANALYSIS; EMISSIONS VS CORE BINDER  The authors conducted a mass balance on the monitored emissions and compared these to the phenolic urethane binder loading. After computing the difference between the non-cored and heavily cored (24.8 lb core resin/ton Fe) conditions for the AO-DBW case, they approximated the proportion of this difference that could be attributed to carbon. This calculation was computed for the case where advanced oxidation (AO-DBW) was employed. The net relative comparison would have been similar if it had been computed for *non-AO optimized* conditions.

The calculations indicate that roughly 30 percent of the 15 lb of carbon that was added into the mold within the phenolic urethane binder, became manifest as emissions (4.4 lb C/ton Fe). Of these binder-attributed emissions that were monitored, 43 percent (as carbon) was carbon dioxide, and the balance was roughly evenly split amongst VOCs, carbon monoxide, and the alkanes (methane + ethane + propane).

The 70 percent portion of the phenolic urethane binder that did not become manifest as emissions could have instead remained as a solid residue, become released as condensibles or tars, or become released as either low molecular weight or high molecular weight emissions that were not included in the above monitored parameters. The tests that were conducted as a part of this study were not designed to distinguish between these possibilities.

*Clay and Coal Efficiency; and Cost Considerations Pertaining to Advanced Oxidation*  In addition to diminishing emissions, the AO-DBW system has reduced the amount of pre-blended bond that the Neenah Foundry 2 has needed to use, where the pre-blend has primarily included western bentonite clay and bituminous coal, with a small amount of cereal and soda ash. As summarized in Table 9-22 and Figure 9.67, the *baseline* bond consumption in 1995–1998 was 200 lb preblend/ton Fe. This dropped to 187 lb/ton Fe during the *non-AO-optimized* period of 1998–1999; and this then dropped yet again to an average 125 lb/ton Fe during the *AO-stabilized* period of 2000–2002 (i.e., a 33 percent drop when comparing *non-AO-optimized* to *AO-stabilized* conditions.

These declines in clay and coal usage occurred while the green compressive strength increased slightly. The efficiency of the clay usage can be normalized via two approaches: The first approach is to normalize the ratio of green compressive strength (psi) to the methylene blue clay (percent) present in the green sand mold. The non-AO-optimized green sand provided 3.3 GCS psi/MB%, while the

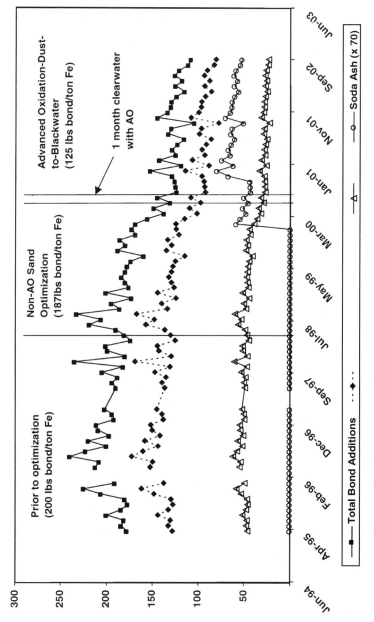

**Figure 9.67.** Clay and coal additions, Neenah Foundry Plant 2, April 1995–September 2002.

515

AO-stabilized green sand provided 4.17 psi/percent; and the AO system thus achieved a 26 percent increase in this normalized strength parameter (Table 9.22). The second approach is to normalize the green compressive strength (psi) to the weight of new clay additions (lb clay/ton of iron poured). The non-AO optimized green sand provided 0.244 psi/lb clay/ton Fe, whereas the AO-stabilized provided 0.369 psi/lb clay/ton Fe. This amounted to a 51 percent increase in this strength parameter that was achieved with AO.

Parenthetically, this 51 percent increase is of the same order of magnitude as the increase in water sorption capacity that AO has achieved in batch experiments. Wang et al. (2004a) observed that when a coal-clay mixture was heated to 400°C, then treated with non-AO water, then heated to 105°C, the product coal/clay mixture could adsorb 4.3 percent water moisture when it was subjected to air that contained 50 percent relative humidity. When this experiment was repeated with an AO water treatment, the coal/clay mixture could adsorb 6.5 percent water moisture; and this AO-related sorption value was 50 percent higher than for its non-AO counterpart. Greater water sorption is linked to greater green compressive strength, since the strength depends in part on the hydrogen bonding facilitated by moisture.

The efficiency of the coal usage can be normalized by considering the loss on ignition (percent) per weight of coal added (lb coal/ton Fe). The loss on ignition is linked to the mold's organic content. As shown in Table 9.22 for Neenah operations, this normalized ratio rose from 0.119 percent/lb/ton Fe for *non-AO-optimized*, to 0.130 percent/lb/ton Fe for the *AO-stabilized* conditions; and the AO thus incurred a 10 percent increase in this normalized parameter. This normalized LOI parameter, coupled with the emissions data above, indicates that through AO conditioning, the coal and other organic content within the green sand mold becomes less volatile and more suitable for achieving quality castings.

To the foundry industry, the above-documented declines in coal and clay consumption have represented lower materials purchasing costs and also lower waste materials handling costs. When these two cost savings are compared to the cost of installing and operating the Advanced Oxidation-Dust-to-Blackwater Clarifier system in the Neenah Plant 2 foundry, one finds that the AO-DBW system paid for itself within 15–20 months. This cost appraisal excludes the yet further costs that are avoided when AO treatment allows a foundry to avoid installing an end-of-pipe incineration system that might otherwise need to be included for emissions control, as discussed above.

### 9.2.5.6 Emissions: Comparison of Technikon-CERP and Penn State Results on Emissions

In parallel research at Technikon-CERP in collaboration with Penn State, Glowacki et al. (2003) appraised emissions changes that could be achieved with an advanced oxidation-clearwater (AO-CW) system. Table 9.23 compares the full-production tests at Technikon-CERP with the Neenah Foundry tests detailed in Section 9.2.5.5. Intriguingly, the two data sets are quite consistent with one another, both in the quantitative levels of these emissions, and in the relative reductions in emissions that were achieved with AO. For example, for the non-cored condition, both at Neenah and Technikon-CERP, non-AO VOCs were 0.52–0.6 lb/ton Fe,

**TABLE 9.23. Operations Performance at Neenah Plant 2 (AO-Dry Dust to Black Water) Compared to Full-Production Testing at Technikon-CERP (AO-Clear Water) Glowacki et al., 2003**

| Parameter | Non-AO | With AO | Percent Change |
|---|---|---|---|
| **Neenah no-core pours** | | | |
| Loss on ignition, (percent) | 5.44 | 3.65 | −33 |
| VOCs (lb/ton metal) | 0.60 | 0.22 | −63 |
| Benzene (lb/ton metal) | 0.055 | 0.030 | −45 |
| **Technikon no-core pours** | | | |
| Loss on ignition, (percent) | 5.00 | 3.60 | −28 |
| VOCs (lb/ton metal) | 0.519 | 0.187 | −64 |
| Benzene (lb/ton metal) | 0.159 | 0.069 | −57 |
| **Neenah average core** | | | |
| Phenolic urethane binder (lb/ton Fe) | 3.4 | 3.4 | 0 |
| Loss on ignition (percent) | 5.44 | 3.65 | −33 |
| VOCs (lb/ton metal) | 0.86 | 0.45 | −47 |
| Benzene (lb/ton metal) | 0.082 | 0.066 | −19 |
| **Technikon heavy core** | | | |
| Phenolic urethane binder (lb/ton Fe) | 12.7 | 8.7 | −31 |
| Loss on ignition, (percent) | 4.62 | 3.07 | −34 |
| VOCs (lb/ton metal) | 1.14 | 0.61 | −46 |
| Benzene (lb/ton metal) | 0.315 | 0.292 | −7 |
| **Neenah very heavy core** | | | |
| Phenolic urethane binder (lb/ton Fe) | 24.8 | 24.8 | 0 |
| Loss on ignition (percent) | 5.44 | 3.65 | −33 |
| VOCs (lb/ton metal) | 1.9 | 1.4 | −26 |
| Benzene (lb/ton metal) | 0.20 | 0.22 | +10 |

*VOCs in Neenah tests were total VOCs (C4 to C16) as Hexane, with peak areas greater than 1,500 area counts. VOCs in Technikon-CERP tests were sums of specific analytes. Technikon emissions included those from pouring, cooling, shakeout, and green sand conveyance. Neenah emissions included mold cooling and shakeout. All the with-AO tests presented in this table employed moisture/MB clay ratios of 30–33 percent.

while with-AO VOCs were 0.19–0.22 lb/ton Fe; and AO incurred a 63–64 percent drop in VOC emissions. This similarity may be partly fortuitous, since these two systems had some differences in how they operated (one was AO-CW, and the other was AO-DBW), and in how emissions were monitored (see table footnotes). Nonetheless, the consistency supports the perception that the AO-related trends characterized by these two studies represent reproducible phenomena.

## ACKNOWLEDGMENTS

This study involved numerous foundry personnel of the Neenah Foundry Company, and financial participation by Neenah. Funding has also come from the US Department of Energy, the National Science Foundation, and the US Environmental

Protection Agency. Emissions analyses were conducted and/or supervised by Steve M. Strebel of the Wisconsin Occupational Health Laboratory. Portions of the paper authored by Goudzwaard et al. (2003) were reprinted with permission from the American Foundry Society.

## REFERENCES

AFS Foundry Sand Additives Committee (4-H), (1997). Dust collector material: untapped green sand additives. *Modern Casting*, 44–46.

Alaton, I. A., and Balcioglu, I. A. (2002). The effect of pre-ozonation on the $H_2O_2$/UV-C treatment of raw and biologically pre-treated textile industry wastewater. *Water Science Technology*, **45**(12), 297–304.

Andrews, J., Bigge, R., Cannon, F. S., Crandell, G. R., Furness, Jr., J. C., Redmann, M., and Voigt, R. C. (2000). Advanced oxidants offer opportunities to improve mold properties/ emissions. *Modern Casting*, **90**(9), 40–43, (September).

Atkinson, R. (1985). Radical reaction kinetics. *Chemical Reviews*, **85**, 69–201.

Belyi, O. A. (1993). Reduction of dust- and gas emissions in foundries. *Liteinoe Proizvodstvo*, **5**, 32–33 (May).

Bolton, J. R., Valladares, J. E., Zanin, J. P., Cooper, W. J., Nickelwen, M.G., Kajdi, D. C., Waite, T. D., and Kurucz, C. N. (1998). Figures-of-merit for advanced oxidation technologies: A comparison of homogeneous $UV/H_2O_2$, heterogeneous $UV/TiO_2$ and electron beam processes. *Journal of Advanced Oxidation Technology*, **3**(2), 174–181.

Brant, F. R., and Cannon, F. S. (1996). Aqueous-based cleaning with hydrogen peroxide. *Environmental Health and Science*, **A31**(9), 2409–2434.

Camel, V., and Bermond, A. (1998). The use of ozone and associated oxidation processes in drinking water treatment. *Water Research*, **32**, 3208–3222.

Cannon, F. S., Snoeyink, V. L., Lee, R. G., Dagois, G., and DeWolfe, J. (1993). The effect of calcium in field-spent granular activated carbons on pore development during regeneration. *Journal of the American Water Works Association*, **85**(3), 76–89.

Cannon, F. S., and Brant, F. R. (1998). Aqueous-based cleaning of residues from surfaces. U.S. Patent No. 5,725.678 (March 10).

Cannon, F. S., Furness, Jr., J. C., and Voigt, R. C. (2000). Economical use of advanced oxidation systems for green sand emission reductions. 12th AFS International Environmental, Health & Safety Conference, Lake Buena Vista, FL, pp. 317–332 (October 9–11).

Chandrakanth, M. S., and Amy, G. L. (1996). Effects of ozone on the colloidal stability and aggregation of particles coated with natural organic matter. *Environmental Science and Technology*, **30**, 431–443.

Chandrakanth, M. S., Honeyman, B. D, and Amy, G. L. (1996). Modeling the interactions between ozone, natural organic matter, and particles in water treatment. *Colloids and Surfaces A: Physichemical and Engineering Aspects*, **107**, 321–342.

Chen, R., and Pignatello, J. (1997). Role of quinone intermediates as electron shuttles in Fenton and photoassisted Fenton oxidations of aromatic compounds. *Journal of Environmental Science and Technology*, **31**(8), 2399–2406.

Chen, Y. H., Chang, C.Y., Huang, S. F., Chiu, C. Y., Ji, D., Shang, N. C., Yu, Y. H., Chiang, P. C., Ku, Y., and Chen, J. N. (2002). Decomposition of 2-naphthalenesulfonate in aqueous solution by ozonation with uv radiation. *Water Research*, **36**(16), 4144–4154.

Choi, W., and Hoffmann, M. R. (1997). Novel photocatalytic mechanisms for $CHCl_3$, $CHBr_3$, and $CCl_3CO_2^-$ degradation and the fate of photogenerated trihalomethyl radicals on $TiO_2$. *Environmental Science and Technology*, **31**, 89–95.

Cole, G., Schuetzle, D., and Rogers, J. (1996). CERP program represents tomorrow's foundry. *Modern Casting*, **86**, 39–41, (July).

Dempsey, T., LaFay, V. S., Neltner, S. L., and Taulbee, D. N. (1997). Understanding the properties of carbonaceous additives and their potential to emit benzene. *AFS Transactions*, 105.

Dilmeghana, M., and Zahir, K. O. (2001). Kinetics and mechanism of chlorobenzene degradation in aqueous samples using advanced oxidation processes. *Journal of Environmental Quality*, **30**(6), 2062–2070.

Edwards, M., and Benjamin, M. M. (1992). Effect of preozonation on coagulant-NOM interactions. *Journal of the American Water Works Association*, **84**(8), 63–72.

Esplugas, S., Gimernez, J., Contreras, S., Pascual, E., and Rodriguez, M. (2002). Comparison of different advanced oxidation processes for phenol degradation. *Water Research*, **36**(4), 1034–1042.

Furness-Newburge, Inc. (2003). Schematics of Sonoperoxone systems.

Funken, K.-H., Pohlmann, B., Lüpfert, E., and Dominik, R. (1999). Application of concentrated solar radiation to high temperature detoxification and recycling processes of hazardous wastes. *Solar Energy*, **65**(1), 25–31.

Geiseking, J. E. (1975). *Soil Components, Volume II, Inorganic Components*. Springer-Verlag.

Georgeson, D. L., and Karimi, A. A. (1988). Water quality improvements with the use of ozone at the Los Angeles water treatment plant. *Ozone Science and Engineering*, **10**, 255–276.

Glaze, W. H. (1995). A kinetic model for the oxidation of 1,2-dibromo-3-chloropropane in water by the combination of hydrogen peroxide and uv radiation. *Industrial and Engineering Chemistry Research*, **34**, 2314–2323.

Glaze, W. H., and Kang, J. W. (1989). Advanced oxidation processes. Description of a kinetic model for the oxidation of hazardous materials in aqueous media with ozone and hydrogen peroxide in a semi batch reactor. *Industrial and Engineering Chemistry Research*, **28**, 1573–1580.

Glowacki, C. R., Crandell, G. R., Cannon, F. S., Clobes, J. K., Voigt, R.C., Furness, J. C., McComb, B. A., and Knight, S. M. (2003). Emissions studies at a test foundry using an advanced oxidation-clear water system. *American Foundry Society Transactions*, No. 03–152 (20 pp.).

Goudzwaard, J. E., Kurtti, C. M., Andrews, J. H., Cannon, F. S., Voigt, R. C., Firebaugh, J. E., Furness, J. C., and Sipple, D. L. (2003). Foundry emissions effects with an advanced oxidation blackwater system. *American Foundry Society Transactions*, No. 03–079 (20 pp.). Copyright 2003 American Foundry Society (www.afsinc.org). Reprinted with permission.

Gurol, M. D., Lin, S. S., and Bhat, N. (1997). Granular iron oxide as a catalyst in chemical oxidation of organic contaminants, pp. 9–21 in *Emerging Technologies in Waste Management, American Chemical Society*, W. Tedder, and F. Pohland (eds), Plenum Press, NY.

Hairston, D. (1999). Ultrasound makes waves in the CPI: Sound energy can knock debris off a catalyst or shove dye into a fabric. *Chem. Eng.*, **106**(8), 26–27.

Helmy, A. K., Ferreiro, E. A., and de Bussetti, S. G. (1999). Surface area evaluation of montmorillonite. *Journal of Colloids and Interface Science*, **210**, 167–171.

Hua, I., and Hoffmann, M. R. (1997). Optimization of ultrasonic irradiation as an advanced oxidation technology. *Environmental Science and Technology*, **31**, 2237–2243.

Jackson, M. L. (1979). Soil Chemical Analysis-Advanced Course, Publ. by the author, Univ. of Wisconsin, Madison, WI 53706.

Jans, U., and Hoigne, J. (1998). Activated carbon and carbon black catalyzed transformation of aqueous ozone into OH-radicals, *Ozone Science & Engineering*, **20**(1), 67–89.

Jekel, M.R. (1998). Effects and mechanisms involved in preoxidation and particle separation processes. *Water Science & Technology*, **37**(10), 1–7.

Joshi A. A., Locke, B. R., Arce, P., and Finney, W. C. (1995). Formation of hydroxyl radicals, hydrogen peroxide and aqueous electrons by pulsed streamer corona discharge in aqueous solution. *Journal of Hazardous Materials*, **44**, 3–30.

Jyoti, K. K., and Pandit, A. B. (2003). Hybrid cavitation methods for water disinfection. *Biochemical Engineering Journal*, **14**(1), 9–17.

Kauffmann, P., and Voigt, R. C. (1997). Empirical study of impact of casting process changes on VOC and benzene emission levels and factors. *American Foundry Society Transactions*, **105**, 297–303.

Kuo, K. K. (1986). *Principles of Combustion*. Wiley, New York, NY.

LaFay, V. S., Neltner, S. L., Taulbee, D. N., and Wellbrock, R. (1990). Carbonaceous additives and emission of benzene during the metalcasting process. *American Foundry Society Transactions*, **98**, 293–299.

Land J. D., Voigt, R. C., Cannon, F. S., Furness, J. C., Goudzwaard, J., and Luebben, H. (2002). Performance and control of a green sand system during the installation and operation of an advanced oxidation system. *Am. Foundry Soc. Transactions*, **110**, 705–715.

Land, J. D., Cannon, F. S., Voigt, R. C., and Goudzwaard, J. (2004). Perspectives on foundry air emissions: A statistical analysis approach. Accepted for publication in *Am. Foundry Soc. Transactions*.

Lin, S. S., and Gurol, M. M. (1998). Catalytic decomposition of hydrogen peroxide on iron oxide: Kinetics, mechanisms and implications. *Environmental Science and Technology*, **32**(10), 1417–1423.

Mak, P., Zele, S. R., Cooper, W. J., Kurucz, C. N., Waite, T. D., and Nickelsen, M. G. (1997). Kinetic modeling of carbon tetrachloride, chloroform, and methylene chloride removal from aqueous solution using the electron beam process. *Water Research*, **31**(2), 219–228.

Mohammad, J., and Peters, R. W. (2003a). Combined sonication + vapor stripping for treatment of petroleum hydrocarbon-contaminated groundwater. Paper presented at the 2003 American Institute of Chemical Engineers (AIChE) Meeting, San Francisco, CA (16–21 November).

Mohammad, J., and Peters, R. W. (2003b). In-situ remediation of petroleum-contaminated groundwater using air sparging and sonication. Paper presented at the 2003 National Ground Water Association Remediation Conference on Site Closure and the Cost of Cleanup, New Orleans, LA (13–14 November).

Mohammad, J., and Peters, R. W. (2003c). Performance comparison of integrated AOP systems for removal of BTEX from water. Paper presented at the Mississippi/Alabama Section of the American Water Works Association Meeting, Biloxi, MS (9–11 October).

Neill, D. A., Cannon, F. S., Voigt, R. C., Furness, J. C., and Bigge, R. (2001). Effects of advanced oxidants on green sand system performance in a black water system. *American Foundry Society Transactions.*, **109**, 937–955.

Neppolian, B., Jung, H. Choi, H., Lee, J. H., and Kang, J. W. (2002). Sonolytic degradation of methyl *tert*-butyl ether: the role of coupled Fenton process and persulphate ion. *Water Research*, **36**(19), 4699–4708.

Olenbush, E., and Foti, R. (2000). Metal casting industry leads the way in environmental initiatives. *Engineered Casting Solutions (USA).* **2**(2), 39–41 (Spring).

Perez, M., Torrades, F., Domenech, X., and Peral, J. (2002). Fenton and photo-fenton oxidation of textile efffuents, *Water Research.* **36**(11), 2703–2710.

Peters, R. W., Manning, J. L., Ayyildiz, O., and Wilkey, M. L. (2004). Use of sonication for in-well softening of semivolatile organic compounds. Invited paper presented at the 225th American Chemical Society (ACS) National Meeting, New Orleans, LA (23–27 March, 2003); Paper accepted for publication in an *ACS Symposium Series Volume.*

Peters, R. W., Wilkey, M., Ayyildiz, O., Quinn, M., Pierce, L., Hoffmann, M., and Gorelick, S. (1998). Use of sonication for in-well softening of semivolatile organic compounds. Poster paper presented at the DOE Environmental Management Science Program Workshop, Chicago, IL (27–30 July).

Prendiville, P. W. (1986). Ozonation at the 900 cfs Los Angeles water purification plant. *Ozone Science & Engineering*, **8**, 77–93.

Qureshi, T. I., Kim, H. T., and Kim, Y. J. (2002). UV-catalytic treatment of municipal solid-waste landfill leachate with hydrogen peroxide and ozone oxidation. *Chinese Journal of Chemical Engineering*, **10**(4), 444–449.

Sanders, C. A., and Doclman, R. L. (1970). Clay technology, durability of bonding clays parts VI-X. *American Foundry Society Transactions*, **78**, 233–251.

Siddiqui, M. S., Amy, G. L., and Murphy, B. D. (1997). Ozone enhanced removal of natural organic matter from drinking water sources. *Water Research*, **31**, 3098–3106.

Smiernov, G. A., Doheny, E. L., and Kay, J. G. (1980). Bonding mechanisms in sand aggregates. *American Foundry Society Transactions*, **88**, 659–682.

Strukul, G. (1992). *Catalytic Oxidations with Hydrogen Peroxide as Oxidant.* Kluwer Academic Publisher, Boston, MA.

Takahashi, N. (1990). Ozonation of several organic compounds having low molecular weight under ultraviolet irradiation. *Ozone Science and Engineering*, **12**, 1–18.

Teo, K. C., Yang, C., Xie, R. J., Goh, N. K., and Chis, L. S. (2003). Destruction of model organic pollutants in water using ozone, uv, and their combination (Part I). *Water Science Technology*, **47**(1), 191–196.

Trombly, J. (1995). Recasting a dirty industry. *Environmental Science and Technology*, **29**(2), 76A–78A (February).

Valentine, R. L., and Wang, H. C. A. (1998). Iron oxide surface catalyzed oxidation of quinoline by hydrogen peroxide. *Journal of Environmental Engineering*, **124**(1), 31–38 (January).

Voigt, R. C, Cannon, F. S., et al. (2002). Final Report: Non-Incineration Treatment to Reduce Benzene and VOC Emissions from Green Sand Systems, Department of Energy Report No. DE-FC0799ID13719.

Voigt, R. C., Cannon, F. S., and Furness, J. C. (2000). Advanced oxidation for reducing air emissions in foundries. American Foundry Society Conference on Environmental Health and Safety, Orlando, FL (October).

Wang, S. P., Shiraishi, F., and Nakano, K. (2002). Synergistic effect of photocatalysis and ozonation of formic acid in an aqueous solution. *Chemical Engineering Journal*, **87**(2), 261–271.

Wang, Y. Cannon, F. S., Komarneni, S., and Voigt, R. C. (2004a). Mechanisms of advanced oxidation processing on reducing bentonite consumption in foundries. Submitted to *Colloids and Surfaces.*

Wang, Y., Cannon, F. S., Neill, D., Crawford, K., Voigt, R. C., Furness, J. C., and Glowacki, C. R. (2004b). Effects of advanced oxidation treatment on green sand properties and emissions. Accepted for publication in *American Foundry Society Transactions.*

Westerhoff, P., Song, R., Amy, G., and Minear, R. (1997). Applications of ozone decomposition models. *Ozone Science & Engineering*, **19**, 55–73.

Wisconsin Department of Natural Resources, Air Pollution Rule NR445, (1988). http://folio.legis.state.wi.us/cgi-bin/om_isapi.dll?clientID = 76435&infobase = code.nfo& jump = ch.%20NR%20445

Worrell, E., Price, L., and Martin, N. (2001). Energy efficiency and carbon dioxide emissions reduction opportunities in the U.S. iron and steel sector. *Energy*, **26**(2001), 513–536.

Zhang, X. Y., and Cannon, F. S. (2002). Hydrogen peroxide cleaning of asphalt from surfaces: The accelerating rate calorimetry (ARCTM) study: The effect of silica gel and solution pH on hydrogen peroxide decomposition. *Journal of Advanced Oxidation Technology*, **5**(2), 198–210.

Zhang, X. Y., and Cannon, F. S. (1999). Hydrogen peroxide cleaning of asphalt from surfaces: effect of temperature. *Journal of Advanced Oxidation. Technologies*, **4**(4), 434–446.

## 9.3   GROUNDWATER QUALITY

Water entering the soil either by rainfall or irrigation may be taken up by plants, evaporated into the atmosphere, or held within soil pores. The excess percolates downward and eventually becomes groundwater. This percolating water, called recharge, passes downward through the root zone until it reaches the water table. Below the water table is the saturated zone, where the groundwater collects.

The geologic formation through which groundwater moves is called an aquifer. Aquifers can vary in size, from smaller ones supplying water to several wells, to larger ones, capable of supplying water to thousands of families.

### 9.3.1   How Aquifers Become Contaminated

Groundwater is polluted in numerous ways, including, mainly as the direct or indirect result of human activities. Some of the principal routes of pollution are as follows:

- Fertilizer and manure surplus, leading to nitrate leaching
- Inappropriate use of pesticides

- Mining activities and deposition of hazardous wastes
- Atmospheric deposition of pollutants

Soil has the ability to filter some of these contaminants. However, the filtration process and the movement of chemicals that ultimately pollute the groundwater itself are also affected by the texture of the soil, the organic matter it contains, and its pH. We describe these properties briefly.

Soil organic matter, which includes completely and partially decayed remains of plants and animals, influences how much water the soil can hold before movement occurs. A higher organic matter content will increase the water holding capacity of the soil. Molecules of some chemicals may also adhere to the organic matter particles.

Soil pH, the relative acidity/alkalinity (sometimes called sour/sweet) can have an effect on the movement of chemicals, especially fertilizers through the soil, by influencing their availability to plants. At very low pH (acidic), some major nutrients, such as nitrogen, phosphorous, and potassium can become unavailable to plants and may even be transported down through the soil and reach the water table, a process called leaching.

Pollutants that are carried downward by water percolating through the soil pass first to the water table, then into the saturated zone, and finally to the groundwater, where they form a region of contaminated water called a plume. Because groundwater moves only a few feet a month, or even more slowly, a plume may not appear in well water or another water supply some distance away for years after the introduction of the contaminant.

As the experience of the past decade has shown, fertilizers, pesticides, and other chemical and metallic pollutants that reach an aquifer may make it unusable as a drinking water source. Therefore, prevention of contamination of groundwater is of primary importance.

Even groundwater that is unaffected by human activities contains some impurities. The types and concentrations of natural impurities depend on the nature of the geological material through which the groundwater moves and the quality of the recharge water. Ground water moving through sedimentary rocks and soils may pick up a wide range of compounds such as magnesium, calcium, and chlorides. Some aquifers have high natural concentration of dissolved constituents such as arsenic, boron, and selenium.

In the past, the slow movement of groundwater and the relative purity of rainwater have resulted, over time, in the natural dilution of many such pollutants. Today, with the heavy demands made on aquifers worldwide by growing populations and expanding industrial enterprises, substances that formerly posed little threat to human are being recorded at danger levels in groundwater. The effect of these natural sources of contamination on ground water quality depends on the type of contaminant and its concentration. Many contaminants occur naturally include aluminum, arsenic, barium, chloride, chromium, coliform bacteria, copper, manganese, mercury, nitrate, selenium, silver, sodium, sulfate, total dissolved solids, zinc. Section 9.3.2 discusses newly developed technologies for removing dissolved arsenic from groundwater.

## 9.3.2   Investigating New Arsenic Removal Technologies

JOHN E. GREENLEAF AND ARUP K. SENGUPTA
Department of Civil and Environmental Engineering, Lehigh University, Bethlehem, PA 18015, USA

The presence of dissolved arsenic in contaminated groundwater has emerged as a major concern on a global scale (AWWA, 2001; Bagla and Kaiser, 1996). In the United States alone, complying with an Environmental Protection Agency standard that becomes effective on 23 January, 2006 would require corrective action for over 4,000 water supply systems serving an approximate population of 20 million. A vast majority of these systems use groundwaters.

A world away, over seventy million people in Bangladesh and other regions of the Indian subcontinent are routinely exposed to arsenic poisoning through drinking groundwater (Bagla and Kaiser, 1996; Nickson et al., 1998; Bearak, 1998). According to current estimates, the adverse health effects caused by arsenic poisoning in this geographic area are far more catastrophic than any other natural calamity in recent times. Although the genesis of arsenic contamination in groundwater for this area is yet to be fully understood, natural geochemical weathering of subsurface soil, and not industrial pollution, is the sole contributor of dissolved arsenic in groundwater. In principle, the arsenic leaching mechanism in contaminated groundwaters is the same regardless of geographic location. However, the difference in magnitude and severity of arsenic contamination may vary with subsurface soil composition, groundwater withdrawal rate, application of fertilizer and other related human activities.

### 9.3.2.1   A Proposed Method for Removing Arsenic from Contaminated Water   The presence of dissolved arsenic in groundwater is due primarily to geochemical soil leaching. In all such contaminated groundwaters, dissolved arsenic exists as inorganic compounds in the oxidation states of +III and +V; we shall refer to these as As(III), or arsenites, and As(V), or arsenates. Environmental separation of dissolved arsenic from groundwaters essentially involves removals of these inorganic arsenic species from contaminated water bodies. Because organoarsenical compounds are less toxic than their inorganic counterparts and result primarily from industrial discharges, we focus on the chemistry of inorganic arsenic compounds and engineered processes for their removal from the aqueous phase. There remain significant differences between As(III) and As(V) compounds with respect to ligand characteristics, dissociation behavior, sorption affinity, and toxicity. Consequently, their mobilities in subsurface environment and removability from groundwater by physical-chemical treatments such as, coagulation, ion exchange, sorption and membrane processes are influenced by the oxidation states, as reported extensively in the open literature (USEPA, 1998; Pierce and Moore, 1982; Ghosh and Yuan, 1987; Huang and Fu, 1984; Clifford, 1999; Hering et al., 1997; Lackovic et al., 2000; Masscheleyn et al., 1991; Waypa et al., 1997; Jain et al., 1999).

The broad objective of this section is to present a new hybrid polymeric/inorganic sorbent that: is compatible with fixed-bed column processes with excellent mechanical strength and attrition resistance properties; is selective toward both As(III) and As(V) species; does not warrant any pre- or post-treatment such as pH adjustment or oxidation; does not alter the water quality besides arsenic removal; does not produce fines during lengthy column runs; and lastly, is amendable to efficient regeneration and reuse.

### 9.3.2.2 Arsenic: Its Chemistry and Natural Occurrence

The natural occurrence of arsenic and its fate and mobility in the aqueous and soil environment are seemingly different from the engineered processes often employed for efficient removal of arsenic from contaminated water. However, these two diverse phenomena are quite similar in much underlying chemistry. For this reason, we will first discuss the chemistry of arsenic in relation to its presence and mobility in the natural environment. Redox conditions, pH and the chemistry of accompanying materials are three primary variables in this regard. Pertinent redox reaction and acid dissociation constants are as follows (Drever, 1988; Ferguson and Gavis, 1972; Morel and Hering, 1993).

1. $H_3AsO_4$: $pK_{a,1} = 2.2$; $pK_{a,2} = 7.0$; $pK_{a,3} = 11.5$
2. $HAsO_2$ or $H_3AsO_3$: $pK_{a,1} = 9.2$ (others not known)
3. $H_3AsO_4 + 2H^+ + 2e^- \rightarrow H_3AsO_3 + H_2O$; $E_h = +0.56$ V

Figure 9.68 shows the distribution of As(V) and As(III) species as a function of pH. Note that at near-neutral pH, monovalent $H_2AsO_4^-$ and divalent $HAsO_4^{2-}$ are the predominant As(V) species. An electrically neutral $HAsO_2$ (or hydrated $H_3AsO_3$) is the major As(III) species under identical conditions. Note also the similarity to Figure 9.23, given earlier for carbonate species.

Figure 9.69 represents the predominance or pe-pH diagram, from which we can find the concentrations for $e^-$ and $H^+$ for various As(III) and As(V) species. For meaningful interpretation of arsenic mobility or immobility in the environment, some commonly encountered solid phase arsenic compounds are superimposed. Under a highly oxidizing environment and moderate alkaline condition, ferric arsenate precipitate, $FeAsO_{4(S)}$, is the predominant solid phase. On the contrary, under a highly reducing environment, realgar (AsS), orpiment ($As_2S_3$) and arseno-pyrites (FeAsS) are the most thermodynamically favorable solid phases.

Around neutral pH under atmospheric conditions (pe values greater than 13.0), As(V) is by far the most predominant species. Oxyanions of As(V) are thus the most commonly encountered arsenic species in surface waters (rivers, lakes, etc.). For groundwaters, prevailing redox and pH conditions as shown by the shaded rectangular box in Figure 9.69 favor the presence of As(III) species along with As(V). Recently gathered data from multiple groundwater wells in Bangladesh and in the city of Hanford, California show evidence of As(III) constituting well

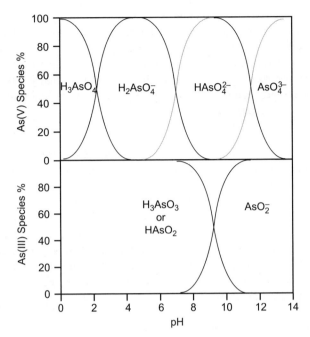

**Figure 9.68.** Distribution of As(V) and As(III) oxyacids and their conjugate anions as a function of pH.

over fifty percent of total dissolved arsenic in many wells (Hering and Chiu, 2000; Safiullah et al., 1998).

### 9.3.2.3 Mobility and Immobility in the Natural Environment

Under highly reducing conditions within the subsurface soil, arsenic exists primarily as $As_2S_{3(S)}$, $AsS_{(S)}$ and $FeAsS_{(S)}$. If these solid phases are exposed to a relatively oxidizing environment, sulfides will tend to be oxidized into more soluble sulfate species causing geochemical leaching or oxidative dissolution of arsenic in groundwater. In Bangladesh and in the eastern part of India, groundwater has been indiscriminately withdrawn during the last twenty years for agricultural irrigation. As a result, the recharging of aquifers occurred essentially through the percolation of surface water containing dissolved oxygen. It has been postulated but not proved that the continuous exposure of arsenic-containing solid phases to the oxidizing conditions generated by aerated water has afforded a major pathway for elevation of arsenic concentration in groundwater through soil leaching (Das, 1995).

Any solute, dissolved gas or solid phase, which tends to alter the redox condition of the aqueous phase, strongly influences the speciation of dissolved arsenic, as illustrated in Figure 9.70. Note that in the presence of oxygen or $MnO_{2(S)}$ i.e., when $O_2/H_2O$ or $MnO_2/Mn^{2+}$ is the prevailing redox pair, As(V) is the

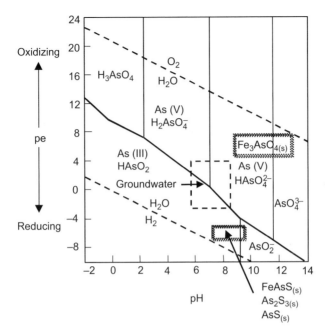

**Figure 9.69.** Predominance or pe-pH diagram of various As(V) and As(III) species including precipitates.

predominant species (Drever, 1988). At near-neutral pH, hydrated Fe(III) oxides (HFO) favor oxidation of As(III) to As(V) thermodynamically. However, oxidation of As(III) by iron solids has not been observed as yet (Oscarson et al., 1981; Pierce and Moore, 1982).

In the presence of dissolved oxygen and/or manganese dioxide, relatively soluble Fe(II) is easily oxidized to insoluble Fe(III) oxides or HFO. Both As(V) oxyanions and As(III) oxyacid are fairly strong ligands that are selectively sorbed onto HFO particles through formation of inner sphere complexes, that is, covalent linkages between the adsorbed ion and the sorption site with no water of hydration between the adsorbed ion and the surface functional group (Manning and Goldberg, 1997; Manning et al., 1998). The fate and transport of arsenic in subsurface environments are, therefore, controlled not only by pH and redox conditions but also by sorption behaviors of As(V) oxyanions and neutral As(III) molecules. Figure 9.71 illustrates the fate of inorganic arsenic species under varying redox conditions and in the presence of iron.

In many industrial waste sites, the fate and transport of arsenic and other trace metals in subsurface system are closely linked to biogeochemical reactions. These reactions occur as a result of organic carbon being degraded by different microorganisms using a series of terminal electron acceptors (Smith and Jaffe, 1998; Masscheleyn et al., 1991). Arsenic contaminated sites may change over the short term (i.e., decrease of organics due to enhanced bioremediation) or longer term

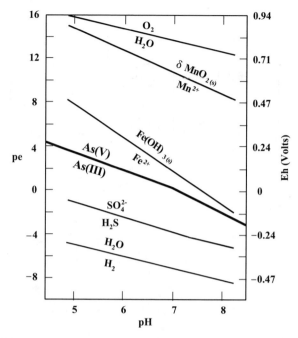

**Figure 9.70.** Hierarchy and stability of As(III)/As(V) system in comparison with other redox pairs as a function of pH.

(decrease of organic contaminant load due to natural attenuation). Geochemical conditions will continue to change and consequently, arsenic compounds can be mobilized/immobilized via processes such as oxidation/reduction, sorption/desorption, precipitation/dissolution, and/or the formation of complex coordinate compounds. Figure 9.72 illustrates how arsenic speciation and partitioning in groundwater systems change with the changes in electron acceptors responsible for biological reactions. In principle, the integrity of hazardous sites containing biodegradable organic wastes along with inorganic arsenic contaminants is amenable to more reducing environments and thus different from natural subsurface environments free of organic matters.

The mobility of arsenic in the subsurface environment may also be influenced directly by microbially mediated biological reactions. Although the exact role of microbially mediated arsenic release from soil is deemed relatively insignificant, the reduction of As(V) to As(III) is one mechanism by which arsenic may be mobilized. To this effect, the strain MIT-13, a respiratory As(V)-reducing bacterium has been isolated from the Aberjona Watershed (Ahmann et al., 1994). Since sorption affinity of arsenites onto amorphous iron-oxide materials is less than arsenates, arsentate-reducing bacteria may play a significant role in mobilizing arsenic through the direct reduction of arsenate.

As other plausible microbially mediated mechanism is through reductive dissolution of Fe(III) to Fe(II) without any change in the oxidation state of As(V).

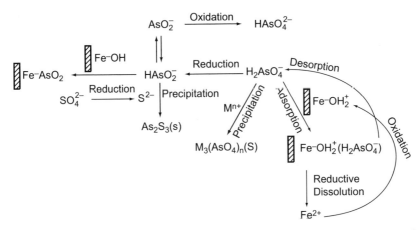

**Figure 9.71.** Schematic illustration of the fates of inorganic arsenic in a subsurface environment under varying redox conditions.

Cummings et al. (1999) found that the BrY strain of a dissimilatory iron-reducing bacterium, *Shewanella alga*, promotes arsenic mobilization from a crystalline ferric arsentate as well as from sorption sites within sediments. Spectroscopic analyses confirmed that Fe(III) was reduced to Fe(II) but the oxidation state of As(V) remained unchanged in both liquid and solid phases.

### 9.3.2.4 *Principles of Arsenic Removal Technologies* Removal of dissolved arsenic from natural groundwater and surface water essentially constitutes removals of As(V) oxyanions and neutral As(III) oxyacid. Some of the principal existing and emerging arsenic removal technologies are as follows:

- Activated alumina sorption
- Polymeric anion exchange
- Sorption on iron oxide coated sand (IOCS) particles
- Enhanced coagulation with alum or ferric chloride dosage
- Ferric chloride coagulation followed by microfiltration
- Pressurized granulated iron particles
- Polymeric ligand exchange
- Iron-doped alginate
- Sand with zero-valent iron
- Reverse osmosis or nano-filtration

The technical details of the foregoing processes and their relative advantages and disadvantages are recorded in the open literature (Horng and Clifford, 1997; Clifford, 1999; Edwards, 1994; Ghosh and Yuan, 1987; Cheng et al., 1994; Driehaus et al., 1995, 1998; McNeil and Edwards, 1995; Sorg and Logsdon, 1978; Hering

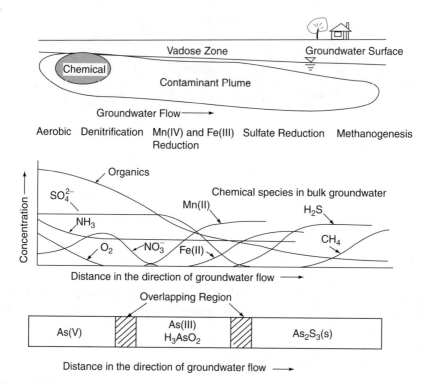

**Figure 9.72.** An illustration of the change in arsenic speciation with the change in biogeochemistry of the subsurface environment. (Adapted from Bouwer and Zehnder, 1993, *Trends in Biotechnology.*)

et al., 1997; Min and Hering, 1998; Ramana and SenGupta, 1992; Vagliagandi et al., 1995; Shen, 1973; Lackovic et al., 2000; Bajpai and Chaudhury, 1999; Torrens, 1999; Kartinen and Martin, 1995; Frank and Clifford, 1986; Clifford et al., 1986; Waypa et al., 1997). Since the arsenic to be removed from contaminated groundwater often is present in trace concentrations ($<500\ \mu g/L$), precipitation is rarely a major removal mechanism. While the equipment configuration and operational protocol for the arsenic removal processes just listed are often quite different, the underlying chemistry essentially rests on the following two types of interactions. First, As(V) oxyanions (e.g., $H_2AsO_4^-$ and $HAsO_4^{2-}$) possess negative charges and can, therefore, undergo coulombic or ion-exchange (IX) type interactions. In this instance, conventional anion exchange processes are seemingly quite suitable for removals of As(V) oxyanions. In contrast, As(III) exists as a non-ionized oxyacid around neutral pH and is not amenable to removal by ion exchange processes. Second, As(V) and As(III) species are fairly strong ligands or Lewis bases, i.e., they are capable of donating lone pairs of electrons. Thus, they can participate in Lewis acid-base (LAB) type interactions and often exhibit high sorption affinity toward solid surfaces with Lewis acid characteristic. In the development and consideration of new arsenic removal technologies, these interactions play an important role.

### 9.3.2.5 Development of a Polymeric/Inorganic Hybrid Sorbent for Selective Arsenic Removal

It is universally recognized that a fixed-bed sorption process is operationally simple, requires virtually no start-up time and is forgiving toward fluctuations in feed compositions. However, in order for the fixed-bed process to be viable and economically competitive, the sorbent must exhibit high selectivity toward the target contaminant, be amenable to efficient regeneration and durable. Amorphous and crystalline Hydrated Fe Oxide (HFO) precipitates show strong affinity toward both As(III) and As(V) oxyanions through ligand exchange in the coordination spheres of structural Fe atoms. Extended x-ray absorption fine structure (EXAFS) spectroscopy studies have confirmed that As(III) and As(V) species are selectively bound to the oxide surface through formation of inner sphere complexes (Manning and Goldberg, 1997; Manning et al., 1998). Note that strongly basic polymeric anion exchangers or activated alumina are not suitable here because they lack the ability to remove As(III) species at near-neutral pH. Understandably, hydrated iron oxides including ferrihydrites, hematites and goethites possess the requisite attributes to be effective arsenic-selective sorbents. However, the current synthetic process, although straightforward, produces only very fine (submicrometer-sized) particles, which are unusable in fixed beds as explained shortly.

To address the foregoing problems, we have perfected a simple chemical-thermal technique to produce a hybrid (polymeric/inorganic) sorbent which is selective towards arsenic(III) and (V) compounds and at the same time compatible with fixed-bed column operation (SenGupta et al., 2000). The sorbent is referred to as Hybrid Ion Exchanger (HIX).

*Underlying Concept and Premise* Precipitated hydrated Fe(III) oxide or HFO submicrometer particles are not suitable for column usage in plug-flow configuration because their poor mechanical strength results in column blockage after periods of excessive pressure drop. To circumvent this problem, we prepared a hybrid sorbent which is essentially a macroporous cation exchanger within which fine colloid-like HFO particles have been uniformly and irreversibly dispersed. The new sorbent, the HIX, combines excellent mechanical strength and hydraulic characteristics of spherical polymer beads with the high arsenic sorption properties of HFO microparticles over a wide range of pH. A commercially available macroporous cation exchanger with polystyrene matrix and sulfonic acid functional group was used as the parent cation exchanger. The exchanger beads varied in diameter from 400–800 μm.

The hybrid ion exchanger or HIX is described in detail shortly, in the subsection entitled Experimental Procedures. The procedure consists of three steps, illustrated in Figure 9.73, and summarized as follows:

**Step 1:** Loading of Fe(III) onto the sulfonic acid sites of the cation exchanger by passing 4 percent $FeCl_3$ solution at an approximate pH of 2.0;

**Step 2:** Desorption of Fe(III) and simultaneous precipitation of Fe(III) hydroxides within the gel and pore phase of the exchanger through passage of a solution containing both NaCl and NaOH, each at 5 percent W/V concentration.

**Step 3:** Rinsing and washing with a 50/50 ethanol–water solution followed by a mild thermal treatment (50–60°C) for 60 min.

At the conclusion of Step 3, submicron HFO particles form agglomerates and are irreversibly encapsulated within the spherical exchanger beads. Turbulence and mechanical stirring did not result in any loss of HFO particles (approximately 12 percent Fe by weight) within the polymeric material. Figure 9.74 diagrams the interactions resulting in the selective sorption of $H_2AsO_4^-$ onto dispersed HFO particles. We note parenthetically that investigators in Germany (Driehaus et al., 1998) have developed and tested a pressurized granular ferric hydroxide material designed for high As(V) removal capacity. These materials, however, must be disposed of after one run without regeneration and reuse, whereas our hybrid exchangers are

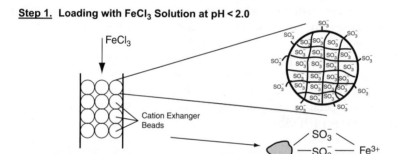

**Step 1.** Loading with $FeCl_3$ Solution at pH < 2.0

**Step 2.** Desorption and simultaneous precipitation in the gel phase and pores

**Step 3.** Alcohol wash and mild thermal treatment

**Figure 9.73.** Step-wise procedure for the preparation of arsenic selective hybrid ion exchanger (HIX).

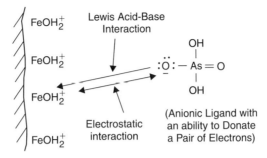

**Figure 9.74.** Electrostatic and Lewis acid-base interactions for selective sorption of $H_2AsO_4^-$ onto HFO particles (for $H_2AsO_3$, only Lewis acid-base interaction is present).

recyclable (see later subsection entitled Regeneration, Rinsing, and Reuse). Additional details pertaining to the synthesis of HIX particles are available elsewhere (SenGupta et al., 2000; DeMarco, 1998).

### *Experimental Procedures*

*Column Run, Regeneration and Rinsing*  Fixed-bed column runs were carried out using glass columns (11 mm in diameter), constant-flow stainless steel pumps and a commercial fraction collector. The ratio of column diameter to exchanger bead diameter was approximately 20 : 1; earlier work on chromate and phosphate removals with similar setups showed no premature leakage due to wall effects under identical conditions (SenGupta and Lim, 1988; Zhao and SenGupta, 1998). The superficial liquid velocity (SLV) and the empty bed contact time (EBCT) were recorded for each column run. Deionized water and analytical grade reagents were used for preparing feed solutions. For As(III) column runs, nitrogen gas was continuously sparged into the feed solution to guarantee that no As(III) was oxidized to As(V). Independent analyses of feed samples at various stages of the column run confirmed that no oxidation had occurred. In a blank test of As(III) feed solution kept open to laboratory atmosphere, there was only partial (20 percent) oxidation to As(V) after two weeks.

Exhausted HIX beads were regenerated after lengthy column runs using 10% w/v sodium hydroxide solution. Empty bed contact time (EBCT) during regeneration was greater (at least 6.0 min) than that used during the sorption cycle. Following regeneration, the bed was rinsed for about ten to fifteen bed volumes with water sparged by carbon dioxide.

*Kinetic Test*  In the kinetic test for the sorption of As(V), HIX beads were loaded into the cell of a stirrer, whose speed could be adjusted and read directly from a digital output.

The centrifugal action during rotation of the stirrer assembly maintained a vigorous hydrodynamic environment for the HIX beads in which diffusional resistances in the liquid film around the beads was essentially absent. At different time intervals, small volumes (<5 mL) of samples were collected from the solution and analyzed. Arsenic concentration in the HIX phase was calculated from the mass balance. The solution pH was 7.2–7.5 during the course of the kinetic test, and sulfate and bicarbonate were present as competing ions.

*Chemical Analyses* Total dissolved arsenic analyses of samples were carried out using an atomic absorption spectrophotometer (AAS) with graphite furnace accessories and an electrodeless discharge lamp (EDL). Filtration of samples through 0.4 μm filters and subsequent analyses confirmed that arsenic during fixed-bed column run was present only in the dissolved state. The stock solutions for calibration were prepared using analytical grade arsenite ($NaAsO_2$) and arsenate ($KH_2AsO_4$). For quality control, arsenic concentrations of stock solutions were intermittently checked against an arsenic standard solution. Chloride and sulfate were analyzed chromatographically, and the size of the HIX beads was determined by means of a particle counter incorporating a binocular optical system. Bicarbonate or inorganic carbon was determined as well.

*Fixed-Bed Column Runs: As(V) in the Feed* Figure 9.75 shows complete effluent histories of these target species during a fixed-bed column run using HIX. Influent pH was slightly above neutral and no pH adjustment was made. The following observations are noteworthy: First, the nonarsenic anions, namely, sulfate and

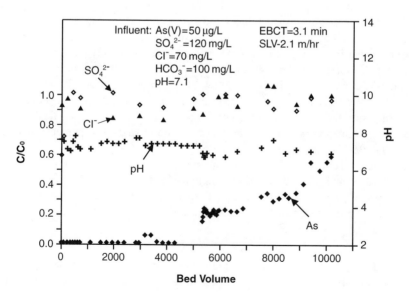

**Figure 9.75.** A complete normalized effluent history of sulfate, chloride, arsenic and pH during a fixed-bed column run with HIX. EBCT = empty bed contact time, SLV = superficial liquid velocity.

chloride, broke through almost immediately after the start of the column run, while 10 µg/L As breakthrough was observed after about 4,000 bed volumes. Mass-balance calculations confirmed that in the absence of kinetic limitations, HIX may treat well over what the data of Figure 9.75 imply: 8,000 bed volumes of contaminated water prior to arsenic breakthrough. Second, the effluent pH of 7.1 during the entire column run was essentially equal to the influent pH.

Thus, besides the absence of dissolved arsenic, the influent composition and pH remained essentially unchanged following the passage of contaminated water through HIX column. Figure 9.76 compares effluent histories of As(V) during separate column runs with two different sorbents, namely, a strong-base polymeric anion exchanger and the HIX. While complete arsenic breakthrough for the commercial material (IRA-900, from Rohm and Haas started at less than 300 bed volumes, HIX was removing arsenic well after 5,000 bed volumes. Also, arsenic breakthrough from IRA-900 underwent chromatographic elution, i.e., the concentration of arsenic at the exit of the column became significantly greater than its influent concentration. This phenomenon is attributed to anion exchanger's greater sulfate selectivity over As(V) oxyanions and similar observations were also made by other investigators (Das, 1995).

*Fixed-Bed Column Runs: As(III) in the Feed*    Figure 9.77 provides As(III) effluent histories of two separate column runs using the same feed composition with an As(III) concentration of 100 µg/L. For one run, the anion exchanger IRA-900 was used in the fixed-bed column while HIX was the sorbent during the second run. At near-neutral pH, As(III) is nonionized (i.e., $HAsO_2$ or $H_3AsO_3$) and, therefore,

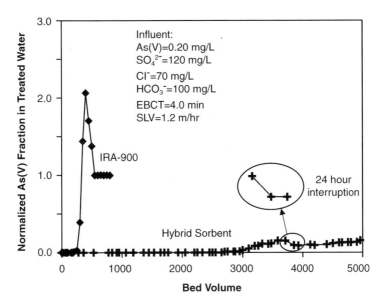

**Figure 9.76.** Comparison of As(V) effluent histories between a strong base anion exchanger (IRA-900) and HIX under otherwise identical conditions.

IRA-900 was unable to remove As(III) as shown in Figure 9.77. In comparison, As(III) was removed over a long period of time by HIX. Total dissolved arsenic breakthrough after 2000 bed volumes was less than ten (10) percent of its influent concentration. Nitrogen was continuously sparged in the influent storage tank to eliminate any possible As(III) oxidation to As(V). Intermittent analyses of the influent according to the protocol prescribed by Ficklin (Ficklin, 1983) and Clifford (Clifford et al., 1983) confirmed that As(V) was altogether absent in the feed. For HIX run with As(III), treated water composition with respect to pH and other electrolytes remained essentially the same as the influent. Earlier studies carried out in the continuous stirred tank reactor (CSTR) mode indicated that hydrated Fe(III) oxides are unable to oxidize As(III) to As(V). Such a reaction is thermodynamically favourable, however, and during the lengthy runs with a fixed bed HIX column, which approximates a plug flow reactor configuration, As(III) was indeed oxidized to As(V) at the exit of the column. Detailed experimental results and scientific analysis of the observations in this regard are reported later. During all the fixed-bed column runs, the pressure drop across the HIX columns was essentially identical to that in anion exchange columns. No generation of fines or a sudden decrease in flow rate was observed for month-long column runs with HIX.

*Sorption Kinetics: Mechanism and Effective Intraparticle Diffusivity*   HIX column runs were subjected to interruption tests during the time periods circled in

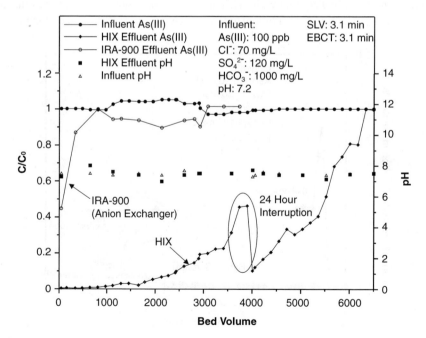

**Figure 9.77.** Comparison of As(III) effluent histories between a commercial anion exchanger (IRA-900) and HIX under otherwise identical conditions.

Figures 9.76 and 9.77. As the dissolved arsenic gradually started exiting from the column, the influent flow was deliberately discontinued for 24 h. When the flow was resumed, the effluent arsenic concentration dropped significantly as can be seen from the two figures. Following the passage of several hundred bed volumes of influent solution after the restart, arsenic at the exit of the column reached the concentration recorded prior to interruption. For an intraparticle diffusion controlled sorption process, the concentration gradient within the sorbent particle serves as the driving force and governs the overall rate. With the progress of the column run, this concentration gradient attenuates. The interruption allows the sorbed arsenic to evenly spread out within the spherical bead. As a result, the concentration gradient, and thus the uptake rate immediately after the column restart is greater than the uptake rate prior to the interruption. That is, a faster uptake immediately after the interruption and consequent drop in aqueous-phase exit concentration of the solute supports the contention that intraparticle diffusion is the primary rate-limiting step. Figure 9.78 depicts the changes in solid phase arsenic concentration gradients within a single spherical HIX bead during various stages of the interruption test. Several previous studies also confirmed that kinetics in selective sorption processes is governed by intraparticle diffusion (Li and SenGupta, 2000a, Crank, 1975).

A single spherical HIX bead may be viewed as an ensemble of tiny HFO agglomerates with an interconnected network of pores. Thus, to determine the intraparticle pore diffusivity for As(V), a finite volume batch kinetic test was carried out using the set up described earlier in the subsection entitled Kinetic Test. The stirrer assembly was rotated at 1,500 revolutions per minute. It was confirmed independently that the liquid-phase film diffusion was not the rate-limiting step under this experimental condition.

Figure 9.79A shows the decrease in the aqueous phase As(V) concentration with time, and Figure 9.79B plots fractional uptake for the batch kinetic test. Dissolved arsenic concentration after seventy two hours was taken as the equilibrium value. For finite solution volume system and appropriate initial and boundary conditions, fractional uptake rates are available in the open literature (Crank, 1975).

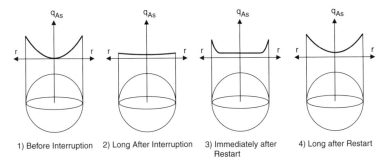

**Figure 9.78.** Concentration gradient of arsenic within a single HIX bead during the various stages of interruption test.

The solid lines in Figure 9.79 represent the model predictions of the kinetic test results and the best-fit effective intraparticle diffusivity, $D_{\text{eff}}$, was computed to be $5.5 \times 10^{-11}$ cm$^2$/s. The magnitude is comparable to other highly preferred solutes for selective sorbents studied earlier (Li and SenGupta, 2000a,b) and is representative of high sorption affinity. This effective intraparticle diffusivity value can be used in modeling arsenic effluent history during fixed bed column runs (Zhao, 1997; Suzuki, 1990).

*Regeneration, Rinsing and Reuse*   In order for the sorption process to be viable, the hybrid exchanger (HIX) has to be amenable to efficient regeneration and reuse. To this end, a portion of the exhausted HIX after the lengthy run in Figure 9.75 was regenerated using 10 percent NaOH. A concentration profile of arsenic during the desorption process and other salient hydrodynamic parameters are provided in Figure 9.80. Note that in less than eight bed volumes, almost the entire amount of arsenic(V) was completely desorbed from the bed. Mass balance calculations confirmed that over 90 percent arsenic was recovered during regeneration. Similar high efficiency of regeneration was also obtained for HIX column used for As(III) removal. The observation that arsenic desorption was completed in less than ten bed volumes demonstrates that sorption sites of HFO microparticles are easily accessible through the network of pores, i.e., pore blockage and consequent increase in the tortuosity of dissolved solutes did not result from the dispersion of submicron HFO particles within the porous beads. At high alkaline pH, HFO sorption sites are all deprotonated and negatively charged; so are all arsenite and arsenate species. Thus, the Donnan co-ion exclusion effect is very predominant, resulting in efficient desorption.

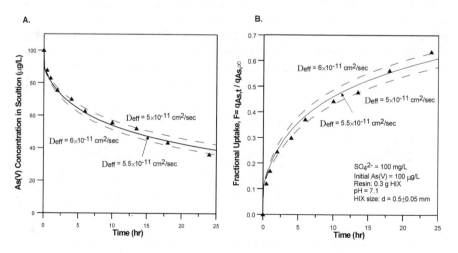

**Figure 9.79.** Results of batch kinetic test data and computed (solid lines) best-fit effective intraparticle diffusivity $D_{\text{eff}}$: (A) plot of aqueous-phase As(V) concentration vs time; and (B) plot of fractional uptake vs time.

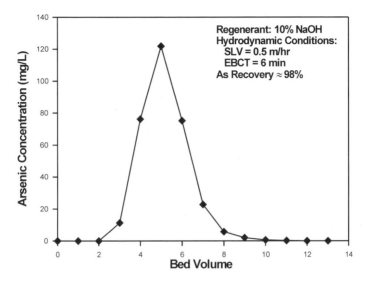

**Figure 9.80.** Dissolved arsenic concentration profile during regeneration with caustic soda (NaOH).

Following regeneration, the bed was reconditioned using water sparged with carbon dioxide to bring the pH from alkaline to near-neutral condition. Figure 9.81 plots pH versus bed volume for three successive postregeneration rinsing processes.

In less than eight bed volumes, pH was lowered to neutral value. Absolutely no dissolution of HFO particles, i.e., the presence of dissolved Fe was observed during the regeneration and rinsing processes. Following rinsing, HIX bed is essentially ready for the next sorption cycle and no further pH adjustment is necessary.

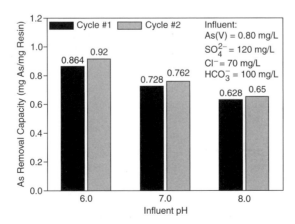

**Figure 9.81.** As(V) and As(III) removal capacities of HIX at different pH values.

Regeneration with caustic soda and rinsing with $CO_2$-sparged water essentially involve deprotonation and protonation of the surface sorption sites of HFO particles as shown in Figure 9.82. The modified rinsing process avoided use of strong hydrochloric acid or sulfuric acid altogether. During the three-year period of laboratory investigation, the same HIX particles were used repeatedly following sorption, desorption and rinsing; no noticeable deterioration was observed. In order to further validate that HIX retains its arsenic removal capacity over long-term usage, separate mini-column tests at two different pHs were carried out for two successive cycles using the same HIX particles. The results of the study (Figure 9.83) reveal that there was only a marginal change in the As(V) removal capacity between these two successive cycles.

Figure 9.84 illustrates the cyclicity of arsenic removal with HIX, highlighting three major steps: sorption or removal of As; desorption or regeneration of As; and rinsing or conditioning. While the sorption step may last for thousands of bed volumes, the regeneration and rinsing together require approximately twenty bed volumes.

### 9.3.2.6 Conclusions

A host of inorganic and polymeric sorbent materials have been tested in both laboratory and field-scale systems for ability to remove trace arsenic solutes from contaminated waters. In almost every such application,

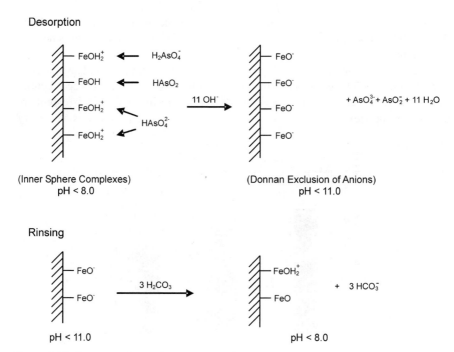

**Figure 9.82.** Deprotonation and protonation of HFO particles at their surface sorption sites during regeneration.

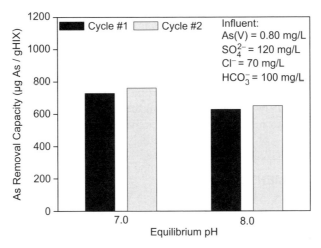

**Figure 9.83.** Arsenic removal capacities of HIX during two successive cycles at pH 7 and 8.

the goal was to increase the number of bed volumes of treated water prior to arsenic breakthrough. This treatment methodology often warranted adjustment of pH and/ or oxidation of contaminated waters prior to sorption. In certain cases, post pH- and alkalinity-adjustments were necessary and the composition of the treated water with respect to other solutes varied significantly. For granular inorganic medium, lengthy column runs produced fines due to attrition leading to increased pressure drops in columns. Partial dissolution during chemical regeneration was also recorded.

An extensive laboratory investigation was carried out to evaluate the performance of a new reusable polymeric/inorganic sorbent material for selective arsenic

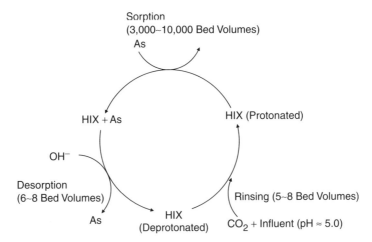

**Figure 9.84.** Schematic depiction of the three major steps of the cyclic process with HIX: sorption, desorption, and rinsing.

removal. The hydrated Fe(III) oxide microparticles offered high sorption affinity toward dissolved arsenic species, and the spherical polymer beads provided excellent hydraulic characteristics in fixed beds. Mechanical agitation and repeated column usage did not result in any loss of HFO particles from the beads.

In addition to the advantages described in the preceding subsections, the, polymeric cation exchanger is one of the most inexpensive and durable of sorbents. This material is widely available throughout the world, and its use as the primary substrate is likely to make HIX more attractive economically than other arsenic removal sorbents.

## ACKNOWLEDGMENTS

The authors are thankful to Matthew DeMarco for his work regarding the development of HIX material. We are also thankful to the United States Environmental Protection Agency (USEPA) and the Pennsylvania Infrastructure Technology Alliance (PITA) for their partial financial support.

## REFERENCES

Ahmann, D., Roberts, A. L., Krumholz, L. R., and Morel, F. M. M. (1994). Microbe grows by reducing arsenic. *Nature*, **371**, 750.

American WaterWorks Association (AWWA) (2001). *Mainstream*, **45**(2), February.

Bagla, P., and Kaiser, J. (1996). India's spreading health crisis draws global arsenic experts. *Science*, **274**, 174–175.

Bajpai, S., and Chaudhury, M., (1999). Removal of arsenic from manganese dioxide coated sand. *Jour. Env. Engr., ASCE.* **125**(8), 782–784.

Bearak, D. (1998). New Bangladesh disaster: wells that pump poison. *The New York Times*, Nov. 10.

Bouwer, E. J., and Zehnder, A. J. B. (1993). Bioremediation of organic compounds-putting microbial metabolism to work. *Trends in Biotechnology*, **11**(8), 360–367.

Cheng, R. C., Liang, S., Wang, H. C., and Beuhler, M. D. (1994).Enhanced coagulation for arsenic removal. *JAWWA*, **86**(9), 79–90.

Clifford, D. (1999). *Water Quality and Treatment*, R. D. Letterman (ed.), 5th Ed., McGraw-Hill Inc., New York, NY, Chapter 9.

Clifford, D., Ceber, L., and Chow, S. (1983). As(III)/As(V) separation by chloride-form ion-exchange resins. Proceedings AWWA WQTC, Norfolk, VA.

Clifford, D., Subramonian, S., and Sorg, T. (1986). Removing dissolved inorganic contaminants from water. *Environ. Sci. Technol.* **20**, 1072–80.

Crank, J. (1975). *The Mathematics of Diffusion*, 2nd Ed., Oxford Science Publications: Oxford, UK.

Cummings, D. E., Caccavo, F., Fendorf, S., and Rosenzweig, R. F. (1999) Arsenic mobilization by the dissimilatory Fe(III)-reducing bacterium *Shewanella alga BrY. Envir. Sci. Technol.*, **33**, 723–729.

Das, D. (1995). Ph.D. dissertation. Jadavpur University, India.

DeMarco, M. J. (1998). A new hybrid polymeric sorbent for arsenic removal. MS thesis, Lehigh University, Bethlehem, PA.

Drever, J. I. (1988). *The Geochemistry of Natural Water.* Prentice-Hall, Englewood Cliffs, NJ.

Driehaus, W., Jekel, M., and Hildebrandt, U. (1998). Granular ferric hydroxide — a new adsorbent for the removal of arsenic from natural water. *J. Water SRT Aqua*, **47**(1), 30–35.

Driehaus, W., Seith, R., and Jekel, M. (1995). Oxidation of arsenic(III) with manganese oxides in water treatment. *Wat. Res.* **29**(1), 297–305.

Edwards, M., (1994). Chemistry of arsenic removal during coagulation and Fe-Mn oxidation. *JAWWA* **76**, 64–78.

Ferguson, J. F., and Gavis, J. (1972). A review of arsenic cycle in natural waters. *Wat. Res.* **6**, 1259–1274.

Ficklin, W. H. (1983). Separation of arsenic(III) and arsenic(V) in ground waters by ion-exchange. *Talanta* **30**(5), 371–373.

Frank, P., and Clifford, D. (1986). As(III) oxidation and removal from drinking water. *EPA Project Summary, Report No. EPA/600/S2-86/021.* Water Eng. Res. Lab., Envir. Protection Agency, Office of Research and Development, Cincinnati, OH.

Ghosh, M. M., and Yuan, J. R. (1987). Adsorption of inorganic arsenic and organicoarsenicals on hydrous oxides. *Envir. Progress* **3**(3), 150–157.

Hering, J. G., Chen, P.-Y., Wilkie, J. A., and Elimelech, M. (1997). Arsenic removal from drinking water during coagulation. *Jour. Env. Engr.* **126**(5), 471.

Hering, J. G., and Chiu, V. Q. (2000). Arsenic occurrence and speciation in municipal groundwater-based supply system. *Jour. Envir. Engr.*, ASCE, **126**(5), 471–474.

Horng, L. L., and Clifford, D. (1997). The behavior of polyprotic anions in ion-exchange resins. *Reactive and Functional Polymers*, **35**, 41–54.

Huang, C. P., and Fu, P. L. (1984). *J. Wat. Poll. Control. Fed.*, **56**(3), 233–242.

Jain, A., Raven, K. P., and Loeppert, R. H. (1999). Arsenite and arsenate adsorption on ferrihydrite: surface charge reduction and net $OH^-$ release stoichiometry. *Envir. Sci. Technol.* **33**(8), 1179–1184.

Kartinen, E. O., and C. J. Martin (1995). An overview of arsenic removal processes. *Desalination*, **103**, 79–88.

Lackovic, J., Nikolaidis, A. and Dobbs, G. M. (2000). Inorganic arsenic removal by zero-valent iron. *Environmental Engineering and Science*, **17**(1), 29–39.

Li, P., and SenGupta, A. K. (2000a). Intraparticle diffusion during selective ion exchange with a macroporous exchanger. *React. Func. Polym.*, **44**, 273–287.

Li, P., and SenGupta, A. K. (2000b). Intraparticle diffusion during selective sorption of trace contaminants: the effect of gel versus macroporous morphology. *Envir. Sci. Technol.*, **34**, 5193–5200.

Manning, B. A., Fendorf, S. E., and Goldberg, S. (1998). Surface structures and stability of As(III) on goethite: spectroscopic evidence for inner-sphere complexes. *Envir. Sci. Technol.* **32**(16), 2383–2388.

Manning, B. A., and Goldberg, S. (1997). Adsorption and stability of arsenic(III) at the clay mineral-water interface. *Environ. Sci. Technol.* **31**, 2005–2011.

Masscheleyn, P. H., Delaune, R. D., and Patrick, W. H. (1991). Effect of redox potential and pH on arsenic speciation and solubility in a contaminated soil. *Environ. Sci. Technol.* **25**, 1414–1419.

McNeil, L. S., and Edwards, M. (1995). Soluble arsenic removal at water treatment plants. *JAWWA*, **87**(4), 105–114.

Min, J. M., and Hering, J. G. (1998). Arsenate sorption by Fe(III)-doped alginate gels. *Wat. Res.*, **32**(5), 1544–1552.

Morel, F. M. M., and Hering, J. G. (1993). *Principles and Applications of Aquatic Chemistry.* Wiley-Interscience, New York.

Nickson, R. et al. (1998). Arsenic poisoning of Bangladesh groundwater. *Nature*, **395**, 338.

Oscarson, D. W. et al. (1981). Oxidative power of Mn(IV) and Fe(III) oxides with respect to As(III) in terrestrial and aquatic environments. *Nature*, **291**, 50-51.

Pierce, M. L., and Moore, C. B. (1982). Adsorption of As(III) and As(V) on amorphous iron hydroxide. *Wat. Res.*, **6**, 1247.

Ramana, A., and SenGupta, A. K. (1992). Removing selenium (IV) and arsenic(V) oxyanions with tailored chelating polymers. *Jour. Env. Eng. Div., ASCE*, **118**(5), 755-775.

Safiullah, S. et al. (1998). Proc. Int. Conf. On arsenic in Bangladesh (Dhaka). 8–12 February.

SenGupta, A. K., DeMarco, M., and Greenleaf, J. (2000). A new polymeric/inorganic hybrid sorbent for selective arsenic removal. pp 142-149. Proceedings of IEX 2000: Ion Exchange at the Millennium (J. A. Greig, ed); Churchill College, Cambridge University, England; 16–21 July, 2000. Imperial College Press, London.

SenGupta, A. K., and Lim, L. (1988). Modeling chromate ion-exchange processes. *AIChEJ*, **34**, 2019.

Shen, Y. S. (1973). Study of arsenic removal from drinking water. *J. Amer. Wat. Works Assoc.*, **65**(8), 543.

Smith, S. L., and Jaffe, P. R. (1998). Modeling the transport and reaction of trace metals in water saturated soils and sediments. *Wat. Res.*, **34**, 3135–3147.

Sorg, T. J., and Logsdon, G. S. (1978). Treatment technology to meet the interim primary drinking water regulations for inorganics: Part 2. *JAWWA*, **70**(7), 379.

Suzuki, M. (1990). *Adsorption Engineering.* Elsevier Science, Amsterdam, NL.

Torrens, K. D. (1999). Evaluating arsenic removal technologies. *Pollution Engineering*, July, pp. 25–28.

USEPA (United States Environmental Protection Agency) (1998). *Research Plan for Arsenic in Drinking Water.* EPA/600/R-98/042, Office of Research and Development, Cincinnati, OH.

Vagliagandi, F. G. A., et al. (1995). Adsorption of arsenic by ion exchange, activated alumina and iron-oxide coated sand (IOCS). Water Quality Technology Conference, *AWWA*, 12–16 Nov., New Orleans, LA.

Waypa, J. J., Elimelech, M., and Hering, J. G. (1997). Arsenic removal by RO and NF membranes. *JAWWA*, **89**(10), 102–114.

Zhao, D., and SenGupta, A. K. (1998). Ultimate removal of phosphate using a new class of anion exchanger. *Wat. Res.*, **32**(5), 1613–1625.

Zhao, D. (1997). Ph.D. dissertation, Department of Civil and Environmental Engineering, Lehigh University, Bethlehem, PA, USA.

# CHAPTER 10

# TECHNOLOGIES FOR POLLUTION PREVENTION: SOLID WASTE

R. W. DYSON
Ex-Senior Lecturer, School of Polymer Technology, Metropolitan University London, London N7 8DB, UK (b.dyson@unl.ac.uk)

I. M. MUJTABA
Reader in Computational Engineering, School of Engineering Design & Technology, University of Bradford, Bradford, BD7 1DP (I.M.Mujtaba@bradford.ac.uk)

## 10.1 SOLID WASTE TREATMENT: SOME PERSPECTIVES ON RECYCLING

A. HAQUE
Ex-Postdoctoral Research Associate Waste Minimisation and Pollution Prevention Group, Centre for Process Systems Engineering, Imperial College London, London SW7 2AZ, UK (a.haque@ic.ac.uk)

N. N. HAQUE
Department of Geography, King's College London, Strand, London WC2R 2LS, UK (anis.nasreen@talk21.com)

J. N. B. BELL
Professor of Environmental Pollution and Director of MSc Studies, Centre for Environmental Technology, T. H. Huxley School of Environment, Earth Science and Engineering, Imperial College London, London SW7 2BP, UK (n.bell@imperial.ac.uk)

In 2000, Americans generated 232 million tons of municipal solid waste (MSW), an increase of 13 percent over 1990 and 53 percent over 1980 (USEPA, 2002). Thus management of MSW continues to be an important challenge facing the United States and other highly industrialized nations in the twenty-first century. Solid waste management is critical in the developing world, as well, where age-old

*Toward Zero Discharge: Innovative Methodology and Technologies for Process Pollution Prevention,* edited by Tapas K. Das.
ISBN 0-471-46967-X   Copyright © 2005 John Wiley & Sons, Inc.

traditions of wasting nothing are often followed without regard to the significance and implications of the zero discharge concept.

## 10.1.1   Why Recycle?

Each ton of solid waste diverted from disposal, whether reused, recycled, converted in a waste-to-energy program, or composted, is one less ton of solid waste requiring disposal. To see the value of reusing, recycling and composting solid waste, one need simply consider the amount of disposal space required to accept that material. By implementing environmentally benign waste management strategies (as well as resource-management strategies), a population can reduce its dependence on incinerators and landfills. And when recycled materials are substituted for virgin plastics, metal ores, minerals, glass, and trees, there is less pressure to expand the chemical, mining, and forestry industries. Supplying industry with recycled materials is preferable to extracting virgin resources from mines and forests not only because it conserves scarce natural resources but because it reduces dangerous air and water pollutants, such as greenhouse-gas emissions, and saves energy.

Saving energy is an important environmental benefit of recycling because generating energy usually requires fossil-fuel consumption and results in emissions that pollute the air and water. The energy required to manufacture paper, plastics, glass and metal from recycled materials is generally less than the energy required to produce them from virgin materials. Additionally, the collection, processing and transportation of recycled materials typically uses less energy than the extraction, refinement, transportation, and processing steps to which virgin materials must be subjected before industry can use them.

As is well known, a great amount of energy used in industrial processes and in transportation comes from the burning of fossil fuels. Recycling helps stem the dangers of global climate change by reducing the amount of energy used by industry, thus reducing greenhouse-gas emissions, as well.

## 10.1.2   What is Recycling?

Recycling is the industrial process in which mechanical, chemical, and/or biological means are used to reprocess or convert materials in discarded products into recyclates, feedstock, or energy that can be used again. Recycling is more than a waste-management strategy; it is also an important strategy for reducing the environmental effects of industrial production.

The aim of recycling is to close the loop of a material's flow through the stages of manufacturing, marketing, consumption and disposal. In closed-loop recycling, an assigned set of operations is repeated in each of successive cycles, and the operations vary according to the type of waste material being recycled. Thus recycling is different from reuse, which is the further use, without any change in properties, of discarded products that have been recovered at the end of their service life.

There are three major ways of recycling waste materials:

1. *Mechanical recycling*, which can be divided into *primary recycling*, the in-house processing of clean, single-graded production offcuts into products

equivalent in quality to the original items, and *secondary (post-consumer) recycling*, the reprocessing of discarded wastes from various sources into recycled products of inferior quality (for plastics and paper) or equivalent quality (for aluminum, glass, and metal) to the original items.

2. *Tertiary or chemical or feedstock recycling processes*, such as composting biodegradable wastes, transforming mixed plastics into monomer or low molecular weight oligomers in liquid or gaseous form, and application of mixed plastic flakes as a reduction agent for steelmaking.

3. *Quaternary recycling processes*, such as waste incineration with energy recovery, in which part of the energy content of combustible municipal or agricultural wastes is recouped and used, for example, to raise steam (Van Beukering, 1999; Frisch et al., 1999; Brophy et al., 1997; Mader, 1997; Walker, 1995; see also Section 10.3).

Since recycling is an industrial process, its level of success depends highly on economic, technical, social, and political factors, which in turn vary greatly country to country (Bell, 1998; Haque, 1998). However, mechanical recycling of wastes is long-established and tends to be the most popular practice in both the developed and developing parts of the world (Gandy, 1994; Haque, 1998).

### 10.1.2.1 A Brief Overview of Recycling in the United States and Britain

The history of primary recycling of discarded products, particularly of plastic scrap, goes back nearly as far as the industrial manufacture of products from virgin materials (Saba and Pearson, 1995). In fact, waste management systems in New York City and London routinely collected for secondary recycling reusable items (mainly paper, cloth, glass bottles and metals) from street cleaning enterprises, commercial premises, garbage dumps, and refuse transfer stations. The first refuse sorting plant in the United States opened in New York in 1898, and in the early 1920s London boasted a model sorting plant equipped with a long traveling belt from which recyclable materials could be removed (Gandy, 1994).

In the years that followed World War II, planners who thought beyond the economic value of waste materials realized that recycling could reduce the volume of the waste streams and that separation of incombustible glass and metals could increase the operational efficiency of incineration plants.

In the United States today most locations recycle about 20–30 percent of their trash, approaching, meeting, or exceeding the national recycling target set decades ago. In the United Kingdom wastes generated by commerce and industry are more readily recycled than domestic waste, a discrepancy reflecting the higher level of purity of the wastes of the former sector. The country recycles 25 percent of industrial wastes and 5 percent of domestic wastes (Haque, 1998). The national government has set a revised target of 45 percent recovery of municipal wastes, including 30 percent recycling and composting, by 2010 (DETR, 1999).

Schemes for waste collection and separation in both countries included providing households with outdoor receptacles for recycling, use of strategically located bottle and can banks, and sorting in a central material reclamation facility. Cans, paper products, glass, lead-acid batteries, cars, and textile goods have well established recycling routes. In contrast, discarded plastic products and packaging—an increasing component of the waste stream—present many problems in recycling. Technical difficulties and high costs associated with identifying, separating, and recycling plastics of different types are compounded by the lack of markets for recyclate.

***10.1.2.2 Recycling Today***  Today in the United States and the United Kingdom, pressures for recycling arise largely from environmental concerns and from industrial policies minimizing consumption of energy and materials, aimed at lowering the raw materials and production costs. Further support for recycling activities is generated by concerns over the environmental pollution stemming from landfill and incineration, the conventional disposal treatments for solid waste.

Therefore, in many developed countries, the recycling of post-consumer waste is being actively encouraged by new legislation, as well as by economic incentives in place since the 1970s. A German law requiring retailers to take back their packaging has resulted in the availability of enormous quantities of material for recycling. Indeed, the success of this initiative created surpluses that depressed prices of recovered materials in Europe, thus hindering recycling efforts elsewhere. The German experience was perhaps anticipated by a former US Environmental Protection Agency administrator, who called recycling "a good idea that has gone too far" (quoted in Bickerstaffe, 1997).

In contrast to developed countries, the generation of waste per capita is substantially less than in less industrialized nations, which nevertheless post a recycling percentage for certain materials of close to 100 percent. Recycling was introduced to developing countries in the 1920s but did not take hold on a large scale until at least the late 1950s (Nahar, 1990; Ratra, 1994). From the start, these waste recycling systems have been driven by economic necessity, since the poor earn a living by scavenging discarded glass, plastics, paper, metals, and other materials which they then sell via an informal system for recycling in facilities that vary in their sophistication.

***10.1.2.3 Recycling as a Route to Sustainable Productivity and Growth***  Despite the massive recycling programs and pollution-prevention efforts now under way, the amount of waste to be disposed of, as noted at the outset, is on the rise. We are using natural resources much faster than nature can replenish them. After we have used resources extracted from the earth, we discard them, often in more harmful forms, into the same environment we depend on to provide air to breathe, water to drink, and soil to grow food. Concerns about this practice are voiced repeatedly and are well known.

Another concern that is related to waste, although it has broader ramifications as well, is community safety. The public is becoming more aware of how many tons of hazardous and non-hazardous materials alike are hauled over road, rail, and sea. Given the risk of accidents, especially when hazardous wastes are involved, the volume of transported substances continues to grow, concerns mount.

Plans for sustainable production and growth focus on leading the way to a future in which few materials and toxic substances will be required to support the needs of society, a future in which materials are reused or recycled rather than being incinerated, buried, or dumped. This shift will take place as we redesign processes, consumer and corporate behaviors, reuse more materials, and improve technologies. Industry, business, and government at all levels will need to capitalize on successful programs and invest in infrastructure and activities to meet these goals.

Interestingly, the economics of recycling is inversely related to the level of economic development of the recyclers, and in more general terms, the propensity to recycle is also inversely related to their socio-economic status. In developed nations, facilities for reprocessing plastic wastes operate under known conditions and yield products of known quality. In developing countries, however, plastic wastes generally are recycled several times under undocumented conditions and exposed to the cumulative effects of extraneous contaminants. The deleterious effects on material quality of such uncontrolled and repeated processing are not hard to imagine.

Thus it is suggested that limits to the maximum number of times that plastics may feasibly be recycled vary with the socio-economic conditions in different countries. Section 10.2 considers the plastics recycling systems in a developing country, Bangladesh. This extended case study is followed in Section 10.3, by a review of the process of catalytic steam gasification of poultry litter, a means of obtaining energy from waste.

## REFERENCES

Bell, J. N. B. (1998). Recycling, *Microsoft® Encarta® 98 Encyclopedia.* © 1993–1997 Microsoft Corporation.

Bickerstaffe, J. (1997). The effect of environmental consideration and legislative developments on packaging, In *Chemical Aspects of Plastics Recycling*, Hoyle, W. and Karsa, D. R. (eds), The Royal Society of Chemistry, Cambridge, pp. 3–12.

Brophy, J. H., Hardman, S., and Wilson, D. C. (1997). Polymer cracking for feedstock recycling of mixed plastics waste, In *Chemical Aspects of Plastics Recycling*, Hoyle, W. and Karsa, D. R. (eds), The Royal Society of Chemistry, Cambridge, pp. 214–219.

DETR, Department of the Environment, Transport and the Regions (1999). *A Way with Waste: A Draft Waste Strategy for England and Wales*, Part One, DETR, pp. 24–25.

Frisch, K. C., Klempner, D., and Prentice, G. (1999). *Recycling of Polyurethanes: Advances in Plastics Recycling*, Volume 1, Technomic Publishing Company Inc., Lancaster, Pennsylvania, pp. ix–xi.

Gandy, M. (1994). *Recycling and the Politics of Urban Waste*. Earthscan Publication Limited, London, pp. 38–76.

Haque, A. (1998). Limits of plastics recycling in the developed and developing economy. Ph. D. thesis, T. H. Huxley School of Environment, Earth Science and Engineering, Imperial College of Science, Technology and Medicine, University of London.

Mader, F. (1997). Plastics waste management in Europe—the need for an integrated approach. In *Chemical Aspects of Plastics Recycling*, Hoyle, W. and Karsa, D. R. (eds), The Royal Society of Chemistry, Cambridge, pp. 13–27.

Nahar, N. (1990). *Solid Waste Management in Dhaka City*, M.Sc. thesis, Department of Geography, University of Dhaka, Dhaka, Bangladesh, pp. 35–38.

Ratra, O. P. (1994). Plastic waste and the environment, In *Conference Proceedings on Plastics and the Environment*, Organised by Federation of Indian Chambers of Commerce and Industry and Indian Plastics Institute, New Delhi, 7–8 September, pp. 18–28.

Saba, R. G., and Pearson, W. E. (1995). Curbside recycling infrastructure: a pragmatic approach. In *Plastics, Rubber, and Paper Recycling: A Pragmatic Approach*, Rader, C. P., Baldwin, S. D., Cornell, D. D., Sadler, G. D. and Stockel, R. F. (eds), American Chemical Society, Washington, DC, pp. 2–10.

USEPA-U.S. Environmental Protection Agency, (2002). *Municipal Solid Waste in the United States: Facts and Figures*, EPA/530/R-002/001, Office of Solid Waste and Emergency Response, Washington, DC.

Van Beukering, P. (1999). Plastics recycling in China: an international life cycle approach, CREED final report, Vrije Universiteit Amsterdam and International Institute for Environment and Development (IIED), Amsterdam, pp. 9–22.

Walker, P. M. B. (1995). *Dictionary of Science and Technology*. Larousse Plc., Edinburgh, pp. 859.

## 10.2 PLASTIC RECYCLING IN A DEVELOPING COUNTRY: A PARADOXICAL SUCCESS STORY

A. HAQUE
Ex-Postdoctoral Research Associate Waste Minimisation and Pollution Prevention Group, Centre for Process Systems Engineering, Imperial College London, London SW7 2AZ, UK (a.haque@ic.ac.uk)

N. N. HAQUE
Department of Geography, King's College London, Strand, London WC2R 2LS, UK (anis.nasreen@talk21.com)

J. N. B. BELL
Professor of Environmental Pollution and Director of MSc Studies, Centre for Environmental Technology, T. H. Huxley School of Environment, Earth Science and Engineering, Imperial College London, London SW7 2BP, UK (n.bell@imperial.ac.uk)

R. W. DYSON
Ex-Senior Lecturer, School of Polymer Technology, Metropolitan University London, London N7 8DB, UK (b.dyson@unl.ac.uk)

I. M. MUJTABA
Reader in Computational Engineering, School of Engineering Design & Technology, University of Bradford, Bradford, BD7 1DP (I.M.Mujtaba@bradford.ac.uk)

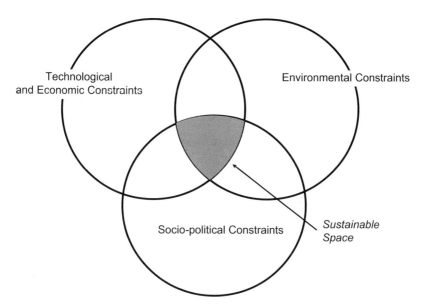

**Figure 10.1.** Conflicting objectives of sustainable development.

Sustainable development, in particular sustainable industrialization, is achievable through careful juxtaposition of three sets of conflicting constraints: techno-economic, environmental and socio political (Figure 10.1) (Koenig and Cantlon, 1998; Clift, 1998). As a result of such balancing, and good practice, plastic waste management systems in Bangladesh have been operating with tremendous success at a recycling rate of 95 percent, since before the beginning of the 1990s (Haque, 1998). This rate, which is very close to "zero discharge," is achieved by means of informally developed industrial ecological areas of waste recycling units in the major cities of this country without any understanding of their implications for the environment or for the economy.

Zero discharge is an essential element in the progression toward sustainable development for the manufacturing, processing, and reprocessing/recycling industries. It is enhanced by the development, within the discipline of industrial ecology (IE), of innovative ideas and concepts including closed loop recycling and byproduct synergy (Haque, 1998; Allen, 1999a). While the first concept advocates a systematic approach to manufacturing, use, reuse, recycling and disposal, byproduct synergy results in the organization of new interactions, to promote uses of wastes as raw material, among the complementary industrial units regardless of corporate structure and geographic location, or niche in an industrial sector (Haque, 1998; Pizzocaro, 1998).

One common but crucial problem for any economy is to define the *ultimate limit* for closed-loop mechanical recycling of waste plastics, in particular, how many times these plastics can be recycled before the recycled materials are useless, having undergone so much quality degradation that no market value remains. The maximum number of recyclings will vary depending on the relevant constraints of technology, the market place, social and environmental ethos, and legislative factors in the context of an overall economy.

Some thousand plastic products manufacturing and recycling units, covering almost all states of the closed loop recycling chain and located in a square-kilometer area of Dhaka, Bangladesh, have formed a "plastic village" by integrating different stages in the waste material chain and also by bringing other supplementary industrial units onto the same platform. The result, which this section explores in detail, has been a "plastics ecology" that yields both technical and economic benefits.

In this section, we investigate byproduct synergy potential and other industrial ecological features of the closed-loop plastic recycling systems in a developing country, Bangladesh, following a life cycle analysis (LCA) approach, introduced in Chapter 3 and graphically represented in Figure 10.2. The "clean technology" cited in Figure 10.2 is not a set of hard technologies but rather represents a new generation of thinking (Clift, 1997) that leads to the more advanced concepts in the diagram. We take the discipline of industrial ecology (IE) a step beyond most present usage by incorporating some key technical, geographical, and environmental issues of the recycling systems concerned.

### 10.2.1 Scope of the System in Bangladesh

The visual appearance of recycled products in Bangladesh indicates that deterioration in the quality of plastic resulting from the recycling process itself, may be

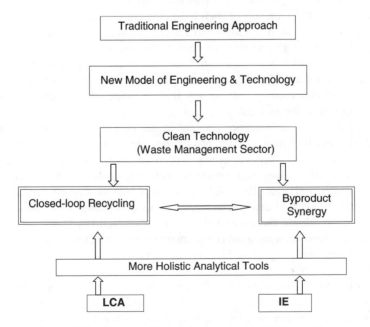

**Figure 10.2.** Conceptual development of sustainable recycling.

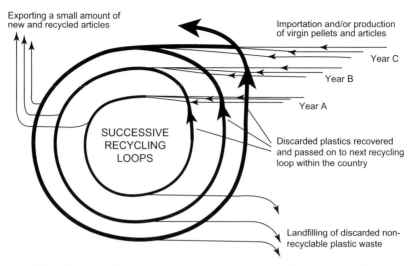

**Figure 10.3.** Schematic diagram of closed-loop plastics recycling system in Bangladesh.

exacerbated if the plastic, even after cleaning and washing, remains contaminated. This is a problem that is very often noted in the recycled products of developing countries (Haque, 1998; Ganguli, 1995). Under the current effectively closed-loop recycling system in Bangladesh (Figure 10.3), more material comes into the system to be recycled in the usual sequence of manufacturing-marketing-consumption-discarding-recovery-recycling. Hence, the size of the *recycling loop* is continually expanding to bring additional materials into the system. Each generation of recycled plastics, however, shows the cumulative effects of repeated processings, eventually deteriorating until they are not suitable for further recycling.

Consequently, at some point in the foreseeable future in cities like Dhaka, the existing system of crude landfills may be one of a few available options for the disposal of nonrecyclable plastics. This possibility reflects the conditions embodied in the following hypothesis.

- Depending on the rate and nature of quality deterioration over the course of successive recyclings, the average duration between successive recyclings, and the levels of contamination in the feedstocks, recycling of postconsumer plastic wastes may, in the foreseeable future, become increasingly worthless for production of consumer products.
- This serious deterioration in quality of recycled products may have economic significance for Dhaka's closed-loop plastic recycling systems, which recycles a higher percentage of contaminated plastic waste.

## 10.2.2 Aims and Objectives

The aim of the study, we shall describe next, was to determine the future scenarios of plastic recycling in terms of "byproduct synergy" potentials of plastic waste and the rate of deterioration in their quality over the successive recycling cycles, taking into account the appropriate socio-economic, technical, and legislative factors as these vary with the levels of economic development.

There were four related objectives:

- To represent how plastic waste material flows within closed-loop manufacturing and recycling systems in developing nations, these being integrated to form a particular Industrial Ecology.
- To determine the effects of successive *secondary* and *primary* recyclings on the properties of virgin and general scrap plastics (post-consumer high density polyethylene—HDPE) collected from the Dhaka waste streams.
- To develop mathematical models for the future scenarios of plastic recycling in terms of these effects of recyclings, taking into account all possible limitations of recycling in the case study country.
- To highlight the implications of such novel examples of plastic recycling under 'industrial ecology' for use as alternative solution tools to overcome the existing barriers of plastics recycling in developed countries.

### 10.2.3 Industrial Ecology and Developing Countries

Although the history of industrial ecological thinking in the developed world goes back at least to the 1950s (Nicholas, et al., 1999; Erkman, 1997), it re-emerged in the early 1990s as an established field of research (JCP, 1997; JIE, 1999; ISI, 1999) with a range of "interesting visions and thought provoking concepts" (Allen, 1999b). Following the positive perspectives of Allen (1999b), Nicholas et al. (1999), Pizzocaro (1998), Ehrenfeld (1997) and many others, IE can be viewed not just as a metaphor or niche of intricately/tightly networked industrial processes, but is indeed "a new set of systems" or new generation of "design tools," which will help to build up such process networks integrated both internally and externally at three levels:

- Within an industrial plant (a meso-scale problem)
- Within an industrial sector of different complementary industrial plants (a macro-scale problem)
- Across different potentially interdependent industrial sectors (a mega-scale problem)

In any of these scales, the fundamental engineering and environmental significance and benefits of IE are associated with its core issue, knowledge about byproduct synergy and how a corporate synergy system can make use of wastes as raw material and thus promote cleaner production at all scales of manufacturing and recycling (Chiu et al., 1999; Allen, 1999a; Frosch et al., 1997).

While in developed countries IE is still in its infancy or perhaps has advanced to the case study level, Bangladesh, India and Taiwan (Haque, 1998; Chiu et al., 1999;

Luken and Freij, 1995) provide examples of IE areas. Indeed, the developing world has been enjoying the benefits of industrial ecology currently practices for more than four decades (Haque, 1998). In the latter part of the world, this system has been developed albeit on an unplanned informal basis, following a natural deterministic process driven by economic necessity and conforming to the existing sociopolitical structure, without again recognizing its positive environmental implications.

## 10.2.4  Description of the Case Study

Details of the fieldwork undertaken in Dhaka, the molding methods used and the laboratory experiment procedures followed are described fully in Haque (1998). In brief, in a five-month fieldwork program (November 1995–March 1996) researchers in Dhaka searched the relevant literature, made direct observations of the plastics recycling activities and units involved within the municipal area of the city, collected field data and fresh information on technical and socio-economic issues, and molded sample containers, in the form of 2-liter jerry-cans, using both virgin and scrap HDPE.

The fieldwork started with a feasibility study on the Dhaka plastics markets for both virgin and recycled products, followed by an investigation aimed at (1) elucidating all possible flow routes of the post-consumer plastic waste streams; (2) understanding the fate of different batches of scrap HDPE products; and (3) identifying the locations of different operations in the recycling chain and the precise nature of each operation, the type of person involved, the items being recycled, the types of plastic involved, and successive reprocessing steps in each recycling cycle.

To develop bases for distinguishing different batches of HDPE present in the Dhaka waste streams, we conducted informal interviews with 270 people. Among the questions asked were the following:

- Are you able to differentiate between different batches of HDPE general scraps and their number of processings?
- How and on what basis do you identify processing numbers?
- Which general scrap [from samples presented] has been processed and how many times?
- What color does a particular batch of scraps usually have after [a given] number of processings?
- Why do plastic processors use color pigments with their feedstocks?
- Is there any tradition or common pattern that recyclers have followed in using color pigments over the years?
- What are your opinions about quality degradation of recycled plastics? What factors affect their quality?

Based on the answers to these questions, a batch of virgin HDPE and eight different representative batches of scrap HDPE recycled different numbers of times were collected from the Dhaka plastic markets. Control portions were set aside from each batch. Sample molds were produced by a single-screw extrusion-blow-molding machine and subjected to successive reprocessings according to local technique. During every reprocessing cycle, sample jerry-cans were collected. Test specimens were prepared from the sample materials as well as from the control portions, and were sent to England for testing their rheological and mechanical properties at the School of Polymer Technology, University of North London. These samples were examined in relation to the *number of times [they had been] processed* using the following test criteria:

Melt flow rate (MFR), in grams per 10 min

Stress at yield (maximum load required to initiate deformation or elongation),

Stress at break (maximum load required to break the specimens at the maximum elongation point),

Strain at break (deformed length over original length at the break point), and

Impact resistance, which was assessed in terms of the following impact parameters:

force to initiate fracture in kilonewtons,

stiffness (resistance to deformation) factor,

depth of deflection in millimeters,

energy required for deflection of fracture, in joules,

energy required for crack propagation, in joules, and

toughness (resistance to crack propagation) assessment factor (BSI, 1997a,b).

Finally, to determine future scenarios for plastics recycling and landfilling in Bangladesh, a number of mathematical models were developed at the School of Engineering Design & Technology, University of Bradford, UK (see Section 10.2.7).

### 10.2.4.1 Industrial Ecological Issues: Spatial Profile

By the beginning of the 1990s, almost all the plastic processing machines in Dhaka were power driven but operated manually (Haque, 1998). These molding machines could produce from reclaimed materials products of finished quality equal to the output of far more sophisticated machinery (Vogler, 1985; Haque, 1998).

Much credit for the high finishing quality goes to another vigorous and skilful manual industry in Dhaka that turns out dies with carefully smoothed and chromium plated surfaces for the plastics molders. These factories for dies, along with other complementary and supplementary industrial and commercial units are clustered in an *industrial ecological area* of plastic waste recycling in the Islambagh section of Dhaka, Bangladesh's capital.

While Dhaka is the collecting and accumulating center for recyclable scrap plastics from the whole country, *Islambagh*, located in the old part of the city on the northern bank of the Buriganga River, is the site of many plastic products manufacturing and recycling units including complementary and supplementary facilities. In fact, this small area serves as the center for plastic waste processing for the whole country. Of more than 600 factories making plastic products in Bangladesh, about 500 are located in and around Dhaka. Within the city, more than 60 percent of plastics recycling factories are concentrated in Islambagh and its periphery, thus forming an industrial ecological area. The majority of the inhabitants of Islambagh make their living from trading, processing and molding plastic waste, and thus this community can be designated as a "plastic community," like the "plastic village" around Hanoi, Vietnam (Simpson, 1994). This clustering maximizes the byproduct synergy potentials of the waste plastics, which can be used effectively as resource raw materials in Islambagh's uniquely organized cooperative arrangements between different groups of highly specialized and skilled workers–plastic waste merchants, sorters, cleaners, washers, dryers, shredders, agglomerators, pelletizers, molders, die makers, molding machines and spare parts manufacturing and repairing workers, electricians, and wholesalers and retailers of recycled products.

The flow of the plastics recycling system in Bangladesh is one-directional and toward Islambagh, after passing through a number of processing and reprocessing stages, recycled products are delivered to almost all corners of the country and abroad, by rickshaws, rickshaw vans, carts, boats, and sometimes trucks.

### 10.2.4.2 Industrial Ecological Issues: Socioeconomic Issues  Along with the gradual advancement in this informal sector of plastics recycling over the last few decades, this whole system, particularly the byproduct synergy potentials of plastic waste in Bangladesh is sustained by the local socioeconomic dynamics. Multiple economic and social forces have acted to concentrate all stages of plastics recycling in the Islambagh district of Dhaka, enabling recyclers to turn discarded plastics into a useful resource at minimal cost. Some of the most important forces are as follows (Haque and Bell, 1998):

1. The need to reduce as much as possible the costs of transporting materials from one stage to another. In Islambagh, materials can be hand-carried or loaded into carts or rickshaws.
2. Economic benefits to the recycling units from receiving support from next-door factories which conveniently produce and repair recycling equipment and spare parts and offer other services.
3. Availability of unloading and loading facilities for recycled products and recyclable wastes, respectively, at the quays of the nearby *Buriganga* River.
4. Availability, nearby, of wholesale markets for recyclate and recycled products.
5. Availability of skilled recycling workers from the experienced inhabitants of this area.

However, the national success in successive recyclings of plastics is also a result of a number of key push factors:

1. A large difference in the costs of imported virgin pellets and recycled pellets/ flakes (the price of recycled pellets is less than half of that of virgin material).
2. Economic necessity, which insures a plentiful supply of low-wage waste pickers and recycling workers.
3. Local availability of technology and skills needed to recycle and use plastic waste.
4. A widespread market for recycled plastic products in a country of nearly 150 million people, mostly of low-income groups.

Of course, the formation of the Islambagh industrial ecological area, itself a function of items 2 and 3 just listed, has played a large role in Bangladesh's success.

The recycling system is favored by two other factors as well: no restrictions on application of recyclable scrap plastics, and a lack of public awareness and regulatory measures on environmental standards and health and safety at work. Moreover, the large demand for relatively low cost recycled plastic products from the increasingly growing urban and rural population of the countries across Southeast Asia acts as a further stimulus for future sustainability of the recycling in Bangladesh.

Plastic waste collection, separation and recycling activities in Bangladesh, therefore, present a socio-economically and technologically very different picture from that in Western countries. Table 10.1 shows that countries with developing economies such as Bangladesh and India, have a low per capita plastic consumption and waste generation coupled with higher rates of reuse and recycling, while the reverse is true in the economically developed countries. In the UK and elsewhere in the western world, the per capita plastics consumption and waste generation are very high but there are remarkably low rates of reuse and recycling with most waste (5–6 percent in Western Europe in 1998) being incinerated or buried in landfill (PRF and CPRR, 1992; Rader and Stockel, 1995; DOE, 1995; APME, 1997; ENDS, 2000).

The content of plastic in municipal solid waste (MSW) in Bangladesh varies between 1 and 2 percent, compared to 5 to 6 percent in the cities of India, 11.2 percent for UK cities, and 8.0 percent for both Western Europe and the US cities. The presence of such a small fraction of plastics in MSW of developing countries clearly indicates that the role of recycling in these countries is extremely significant in achieving near zero waste levels. Nevertheless, in the subsections that follow, we assess and review the long-term future of recycling in developing countries.

### 10.2.5 Supply Chain Issues

Imported virgin pellets are the main original source of plastics in Bangladesh (Haque and Bell, 1998). The decade from 1985–1995 witnessed a steady increase in annual imports of all categories of virgin plastics. The amount of plastic imported in 1994–1995 was about 12 times greater than the imports in 1985–1986.

Along with the country's cultural and economic development as reflected by the steadily increasing trend of plastic imports and consumption over the last few

**TABLE 10.1. Mechanical Recycling of Post-Consumer Plastics by Country/Region (1 tone = 1.016 tonnes)**

| Country/Region | Total Post-Consumer Plastic Waste Generated Locally [Units: ×1000 tonnes/year] | Per Capita Generation of Post-Consumer Plastic Waste (kg/year)[a] | Proportion of Plastics in Municipal Solid Wastes | Amount Recycled Mechanically | References |
|---|---|---|---|---|---|
| Bangladesh | 12 | 0.01 | Dhaka: 1.8% | 95–98% | Haque and Bell (1998) |
| India | 300 | | Calcutta: 5.3% | 60–80% | Ratra (1994), Ganguli (1995) |
| Austria | 279 | | | 15.3% | APME (1997) |
| Switzerland | 342 | | | 11.9% | APME (1997) |
| Germany[3] | 3,181 | 39.11 | | 10.9% | APME (1997) |
| Netherlands | 847 | | | 7.9% | APME (1997) |
| UK | 2,193 | 37.62 | UK Cities: 11.2% | 4.6% | APME (1995) |
| France | 2,997 | 51.58 | | 4.4% | APME (1997) |
| Western Europe | 16,313 | 42.46 | 8% | 7.6% | APME (1997) |
| | (17.6 million tonnes in 1998[b]) | | | (11.2% in 1998[b]) | |
| USA | | | US Cities: 8% | | PRG and CPRR (1992) |

*Data sources:* [a] Haque (1998); [b]ENDS (2000).

decades, the rapid growth in population, which more than doubled between 1961 and 2004, is a major factor in the higher demand for commodity products such as household plastic products. Since the country does not produce virgin plastics, huge amounts of plastic pellets must be imported.

A variety of new products are produced from these virgin pellets and marketed for diverse applications, mostly within the country. After their useful service life (first-time use and multiple re-uses), scrap plastics are disposed of in the mixed MSW streams of the country. These plastic wastes arise from sources in Dhaka and also from other cities and rural areas. Since there is no large scale mechanised waste recovery facility in any of the cities of Bangladesh, hand picking by individual scavengers at the city garbage cans as well as the dumping sites is carried out on a massive scale. Despite the physical condition of recyclable plastics, which are mixed with other MSW including biohazardous materials from hospitals, the waste pickers, sorters and other recycling workers in this informal sector carry out their strenuous jobs very effectively and efficiently, under extremely difficult conditions (Haque et al., 1997).

Once the materials have been brought to Islambagh for recycling, dirty cross-contaminated and mixed post-consumer plastics are processed and converted into feedstocks for the production of recycled articles, thus completing a recycling cycle. Figure 10.4 is a flowchart showing the many stages in the process. It is reasonable to suggest that the greater the number of stages required for collection, transportation, sorting, cleaning and reprocessing the recyclables, the higher the cost of recycling overall. All these activities are performed over and over again during the course of each closed-loop cycle of successive recyclings as shown earlier (Figure 10.3).

However, the number of intermediate stages between the source and the molding factory varies considerably, depending on the amount and types of dirt present in the recovered materials as well as the degree of mixture of different grades of plastic in a given batch. Similarly, the number of units operating at different stages of recycling also varies in accordance with the volume of work at each particular stage.

In Bangladesh in 1994, 38 percent of the annual imports of virgin thermoplastics (31,900 tonnes) was recycled (Figure 10.5). Only about 2–5 percent, consisting of those fractions which were small or difficult to collect, were destined to end up in landfill sites and low lying ditches. The major portion of the rest (62 percent of the total imports) presumably currently remains in use, having a long life. The rate of discarding of plastics products (in terms of the percentage recycled) was high in the case of those produced from PP (58 percent of PP imports) and LDPE (44 percent of LDPE imports) due to their short service life compared to other plastic products. Lower discarding rates were experienced for PS (15 percent), followed by PVC (20 percent). This is because PS was generally used in producing certain long-life items required in this developing country rather than disposable items as in the developed countries.

The HDPE commodity products, frequently used by consumers in the country at the household level, have a service life averaging 3 years for first time use, multiple

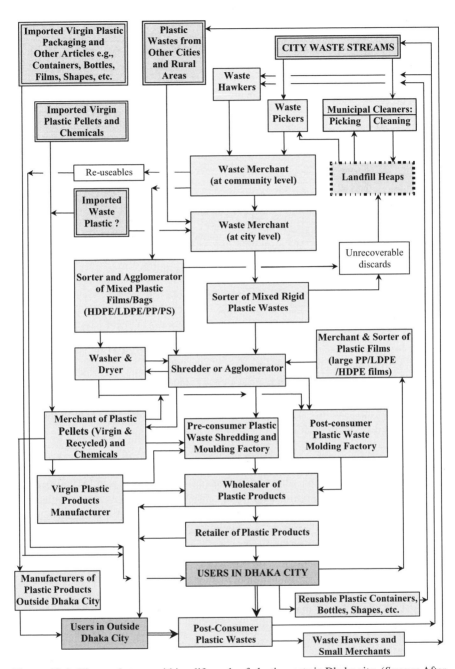

**Figure 10.4.**  Flow and stages within a life cycle of plastic waste in Dhaka city. (*Source*: After Hague and Bell, 1997.)

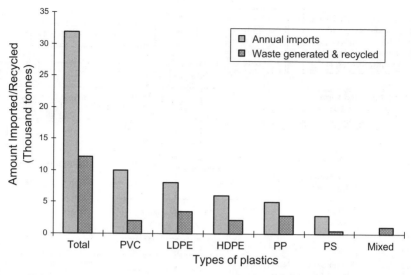

**Figure 10.5.** Comparison between the amount of plastics imported and recycled in Bangladesh in 1994.

reuses were estimated on the basis of the responses from 100 housewives interviewed informally in Dhaka and shown in Table 10.2.

The following facts and characteristics associated with life cycles of HDPEs as recycled in Dhaka (Figure 10.6) were used in characterization, identification and collection of nine batches of HDPE feedstocks to be molded into sample products as mentioned in Section 10.2.4.

### 10.2.5.1 The Recycling Cycle

Each recycling cycle starts with collection of discards, followed by processing stages (sorting, shredding, washing, and pelletizing/agglomerating) and ending with molding (manufacturing). Then recycled products are marketed, whereupon they again enter the cycle of use, discarding, and collection. This means that each step in the reprocessing of plastics involves at least two successive melting or processing steps (pelletizing/agglomerating and molding). However, the number of processing steps in a single recycling cycle increases as the materials attain higher numbers of recyclings. For example, if a batch of plastics has already been recycled five times (i.e., passed five recycling loops), it is more likely that the batch experienced 13 or 14 processing/melting steps (Figure 10.6).

### 10.2.5.2 The Color Change Sequence

When a new plastic product made of virgin pellets (denoted Pro-0) is discarded after its service life (both use and reuse), it is likely to have undergone at least six successive "closed loops" in the existing recycling system of postconsumer plastic waste in Dhaka. Over these successive recycling loops, discarded plastic products are usually found to have increasingly

**TABLE 10.2. Service Life of Different HDPE Plastic Articles in the Developing Economic Context of Bangladesh**

| Name of Articles[1] | Sizes or Quality[2] | Range of Service Life Lengths | Average Service Life Length | Overall Average |
|---|---|---|---|---|
| Carrier bags | All sizes | 1 h–1/2 y | 0.08 y | 2.9 y |
| Mugs | All sizes | 1–3 y | 1.5 y | |
| Buckets | All sizes | 1–5 y | 2.8 y | |
| Bowls | All sizes | 1/2–6 y | 2.6 y | |
| Jerry-cans (*Gallons*) | Small: 1001–5000 mL | 1/6–7 y | 1.7 y | |
| | Medium: 5001–10000 mL | 1–7 y | 3.7 y | |
| | Large: 10001–15000 mL | 3–8 y | 4.9 y | |
| | V. large: >15000 mL | 2–6 y | 3.7 y | |
| Containers (*Boiyum*) | V.V. small: <50 mL | 7 days–1 y | 0.3 y | |
| | V. small: 50–1000 mL | 1 month–5 y | 2.5 y | |
| | Small: 1001–5000 mL | 1–5 y | 3.5 y | |
| Water pots (*Badna*) | Good quality | 2–5 y | 3.2 y | |
| | Poor quality | 1–2 y | 1.4 y | |
| Bottles | V. small: 50–1000 mL | 1 day–1 y | 0.5 y | |
| | Small: 1001–5000 mL | 1/2–2 y | 1.3 y | |
| Baskets | All sizes | 6 months–2 y | 0.9 y | |
| Furniture | All sizes | 1–4 y | 2.0 y | |

[1]Typical local name of articles are in *italic* font.
[2]Scale for article's sizes: V.V. small = <50 mL; V. small = 50–1,000 mL; Medium = 5,001–10,000 mL; Large = 10,000–15,000 mL; Small = 1,001–5,000 mL; V. large = >15,000 mL.

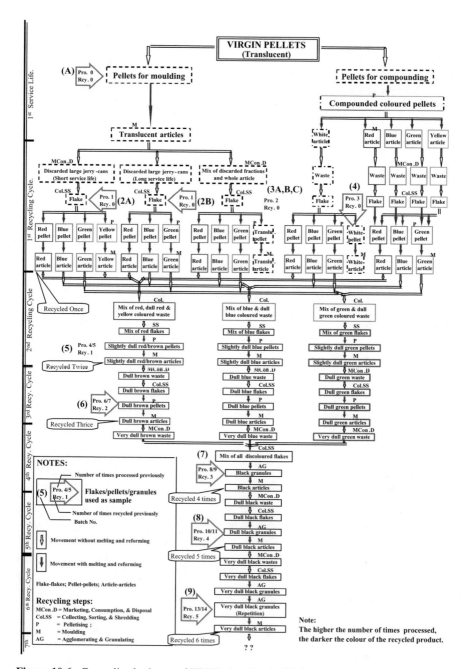

**Figure 10.6.** Generalized scheme of HDPE recycling in Dhaka city: trends in color changes as the plastic is recycled.

darker colors as the number of recyclings/processings rises. This darkening trend, which can be noted in Figure 10.6, is as follows: transparent/translucent/white ⇒ low grade translucent/white ⇒ off white/yellow/red/green/blue ⇒ brown/dark green/dark blue ⇒ dull brown/dull green/dull blue ⇒ black ⇒ dull black ⇒ very dull black.

The number of recycling loops and the recycled products' patterns of color changing were determined on the basis of interviews with the 270 recyclers mentioned earlier. These results were coupled with comprehensive understanding of how changes in external appearance and colour of recycled plastic products occurred over the course of successive recyclings. This understanding, based on earlier experience of the system by one of the authors (Haque, 1998) for more than a decade, was also confirmed by the information obtained from the recyclers in Dhaka.

With some exceptions at the lower end of the socioeconomic ladder, the plastics recyclers were aware of the deteriorations in color, external appearance, and overall quality of recycled plastic products that typically occur at each cycle of recyclings. They attributed these effects to repeated processings of plastics on manually operated recycling machines, which sometimes burn the items as a result of overheating and longer retention time; much exposure to sunlight during the items' service life; and the extraneous contaminants inevitably present in later generations of feedstocks. For an attractive external appearance of recycled products, plastics recyclers in the city use increasingly darker colors for their feedstocks that are processed a higher number of times

### 10.2.5.3 Impurities

It has been mentioned that the more often a piece of plastic is recycled, the greater the number of processings that will be needed in a particular recycling loop. This is because increasing amounts of impurities incorporated into feedstocks containing materials that have undergone several successive "production/consumption" cycles. These impurities are generally screened out by passing the feedstocks repeatedly through a processing extruder and then through a number of filters set in the extruder's barrel. Irrespective of the number of times processed, a batch of waste materials is washed manually only once after being cut into flakes.

### 10.2.6 Effects of Successive Reprocessing On Recyclate

Superficially the system functioning in the industrial ecology area of Islambagh, in Dhaka, Bangladesh, would seem to be a very environmentally friendly recycling arrangement, despite some serious doubts over worker health and future implications of the degraded quality of recycled products. Workers' health and safety issues have been addressed by Haque (1994). Our major theme here is to investigate the effects of successive "secondary recyclings" and "primary recyclings" (in-house reprocessings) on the properties of discarded HDPE recovered from the waste streams of Bangladesh.

A manually operated extrusion blow molding machine does not normally produce a parison of consistent properties, principally because of the inevitable variations in time between stopping and rotating the screw. The polymer melt is, as a consequence, of inconsistent temperature. However, such machines can produce a bottle or jerry-can of acceptable quality for local household consumption. Moreover, effective sorting and categorizing of different types of plastics in a lot is neither economically feasible nor technically possible in the context of Dhaka city. Therefore, the end products of different colored and mixed plastic waste (e.g., containers for tobacco preparation, betel leaf mixture, herbal medicine) are of very low quality, their deep colors, mainly black, reflecting the heterogeneity of the stock from which they were made.

With the understanding obtained from a study of the existing plastics recycling scheme (Figure 10.6), the nine batches of representative HDPE plastic wastes mentioned in Section 10.2.4; were collected from the Dhaka plastics market; the different batches of feedstocks had been processed between zero to thirteen/ fourteen times. Twelve categories of sample feedstocks were prepared in the form of pellets and granules, as shown in column 2 of Table 10.3. The samples were categorized according to the number of times processed previously and level of impurity contents estimated visually. All the sample feedstocks were then used for in-house successive reprocessings and thus to produce and collect sample products in the form of jerry-cans. To avoid cross-contamination, the molding machine was purged before starting each cycle of in-house successive reprossessing.

The virgin pellets were successively reprocessed up to seven times, following the same procedures of successive in-house molding and sample collections (molded into jerry cans, shredded into flakes, sampled, and remolded) in all cases. As indicated however, different batches of scraps were reprocessed successively for different numbers of times, depending on such local factors as availability of the molding machine, time, and resources. The same molding machine, molding die and shredding machines were used in all cases, but the nozzle temperatures of the molding machine ranged from 180°C for the virgin material to 130°C for the material already processed 13 times. That a lower temperature could be used for the material processed a higher number of times indicates a reduction in molecular weight through thermal degradation as a result of constant recycling. This is substantiated by the fact that the melt flow rate (MFR) values also increased.

Figure 10.7 shows the MFR of the scraps against number of times processed previously under the existing recycling system. There is a steady increase, allowing for experimental error, in the MFR of the general scraps as the number of processings increases in the existing recycling system. With the cumulative effect of this system, the MFR value of the very dull black scrap, (9)/14, which has probably passed through 14 melting processes and five cycles of recycling under the tropical conditions, has been increased by 29 percent of the MFR value measured for the once processed virgin HDPE, (1)/1 (data not shown). On the basis of the laboratory test results with low standard errors and the distribution of the error bars in the Figure 10.7, it is clear that this increase is statistically significant.

TABLE 10.3. Descriptions of the Sample HDPE Batches Collected from the Dhaka Plastic Waste Streams and Notations of the Sample Feedstocks and Test Specimens Punched from the Extrusion-Blow-Moulded Sample Products (in the Form of Jerry-Cans)

| Batch No. | Descriptions of the Sample HDPE Batches and Sample Feedstocks Prepared | NTPP$^{\otimes}$ | Notations* of Sample Feedstocks | Notations* of the Test Specimens Punched from the Moulded Products |
|---|---|---|---|---|
| (1) | Virgin translucent pellets | 0 | (1)/0 | (1)/1 |
| (2) | Discarded translucent jerry-cans made of virgin pellets | 1 | (2)/1 | (2)/1 |
| | (A) Translucent rod pellets, from the jerry-cans (i.e., No. 2) of short service life | 1 | (2A)/1 | (2A)/2 |
| | (B) Translucent rod pellets from the jerry-cans (i.e., No. 2) of long service life | 1 | (2B)/1 | (2B)/2 |
| (3) | (A) Translucent rod pellets (from discarded articles/fractions of both short & long service life) | 2 | (3A)/2 | (3A)/3 |
| | (B) Same batch of (3A) but washed twice and color flakes removed before pelletising | 2 | (3B)/2 | (3B)/3 |
| | (C) Same batch of (3A) but washed twice and color flakes removed before pelletising | 2 | (3C)/2 | (3C)/3 |
| (4) | Milky colored rod pellets | 3 | (4)/3 | (4)/4 |
| (5) | Brown colored granules | 4 or 5 | (5)/4 | (5)/5 |
| (6) | Dull brown colored granules | 6 or 7 | (6)/6 | (6)/7 |
| (7) | Black colored granules | 8 or 9 | (7)/8 | (7)/9 |
| (8) | Dull-black colored granules | 10 or 11 | (8)/10 | (8)/11 |
| (9) | Very dull-black colored granules | 13 or 14 | (9)/13 | (9)/14 |

Note: $^{\otimes}$NTPP–Number of times processed previously (i.e., number of passes through melting process).
*Format of the notations: (3A)/5 = (Batch number, with sub-category number if there is any)/Number of times processed.

567

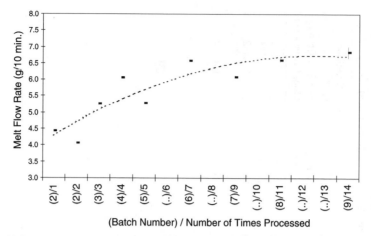

**Figure 10.7.** Melt flow rates of the HDPE **general scraps** against their number of times processed by the existing secondary recycling system in Dhaka, Bangaladesh ($\pm$ SE, standard error).

Two HDPE batches out of the 12 categories of sample feedstocks tested for their MFR also experienced an increase in value, following the effects of in-house successive mouldings (primary reprocessing). For example, MFR values of milky coloured pellets, (4)/3, and dull brown granules, (6)/6 have increased by 14 percent and 19 percent within a low number of in-house reprocessings, respectively (Figure 10.8).

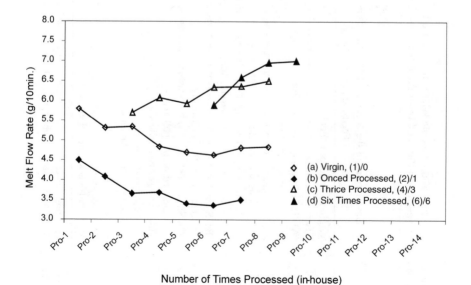

**Figure 10.8.** A comparison among the MFR vs successive number of processing of (a) translucent virgin pellets, (1)/0; (b) translucent short life flakes, (2A)/1; (b) (c) milky coloured pellets, (4)/3; and (d) dull-brown granules, (6)/6.

However, although in the typical case the MFR of a polymeric material increases with higher processing number, some contradictory results were obtained for two materials—the virgin HDPE, (1)/0, and translucent flakes, (2A)/1. For these materials, MFR *decreased* with increasing number of in-house successive extrusion-blow-moldings (Figure 10.8). The apparent differences in behavior probably reflect experimental errors of measurement, meaning that the materials are similar and show little change in MFR with recycling. Indeed, the individual materials themselves are very similar. Except for these four batches [(1)/0, (2A)/1, (4)/3 and (6)/6], the MFR values of all other batches of scraps remained the same, irrespective of the effect of successive extrusion-blow-moldings.

A comparison of a batch of single grade translucent short-life once-processed flakes, (2A)/1, and a batch of long life flakes, (2B)/1, shows MFR values of both materials to be more or less the same, but the flakes behave differently with respect to the number of processings. Figure 10.9 compares the MFR values of the single washed translucent twice processed pellets (3A)/2 with the batches' two derivatives, which were cleaned by double washing, (3B)/2, and winnowing, (3C)/2. However, there is no significant trend in any set of points, nor is there any significant difference between any of these three materials. In other words, all three materials exhibit essentially the same MFR behavior over the range of measurements.

Overall, melt flow test results of HDPE scraps indicate a small increase in MFR with increasing number of times processed previously under the existing recycling system. In the case of in-house successive extrusion-blow-moldings of scraps performed in Dhaka, a similar increase in MFR was experienced by only two

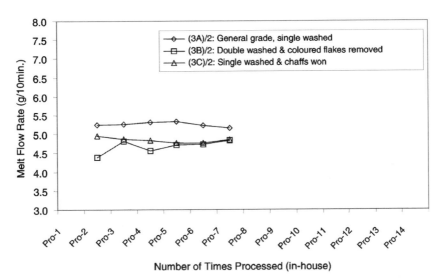

**Figure 10.9.** A comparison among the MFR vs successive number of processing of (a) translucent virgin pellets, (1)/0; (b) translucent short life flakes, (2A)/1; (b) (c) milky coloured pellets, (4)/3; and (d) dull-brown granules, (6)/6.

batches. However, there was a small but statistically significant increase in their MFR from $4\,g/10\,min$ to $7\,g/10\,min$ after 13/14 processings. A contradictory result, i.e., decrease in MFR, which is observed in the case of in-house successive reprocessings of the virgin pellets [(1)/0] and once processed translucent flakes [(2A)/1], needs further investigation.

Details of the tensile strength and impact resistance testing procedures and test results are available in Haque (1998) and will be published elsewhere. In brief, the tensile strength test results of the scrap HDPE indicate no significant deterioration in stress at yield, stress at break, and strain at break up to four to five the number of times processed. There is experimental evidence of a significant change in the properties of moldings from HDPE material processed more than about five or six times. Up to this number, tensile tests show typical HDPE behaviour; viz. a sharp increase in force to a peak at relatively low strain (a few percent) that by a drop in force and followed by an increasing force as the material is pulled to break. This is typical of a material which shows a yield point followed by cold drawing to break.

HDPE materials processed more than five to six times show quite different behavior. The most noticeable difference is that there is no yield point. The material yields and flows immediately that a force is applied. Similar behavior is shown by conventional LDPE materials. The most likely explanation is that the HDPE materials have been severely degraded and the molecular weight is now so low that there is little resistance to deformation under very low load. This explanation is supported by impact resistant test results which suggest that material becomes less resistant to crack initiation in bending. Moreover, compared with LDPE, which is a branched-chain polymer, HDPE is very linear and relies on its long chains (high molecular weight) and their ability to crystallize to up to 80–85 percent for its strength properties. Reducing chain length will therefore reduce strength properties and make the material more ductile and more like LDPE. This phenomenon is one of the key motivating issues for determining future scenarios of the closed-loop recycling systems in Bangladesh.

Sustaining the HDPE recycling system in Bangladesh since the 1960s has been the scrap material's toughness. There is some improvement in strain at break properties, and the decrease in stress at break, though significant, is not catastrophic.

In addition, irrespective of the conditions of feedstock and quality of recycled products, plastics recyclers in Bangladesh keep using all variety of scrap plastics for a range of applications by making process adjustments on the basis of the available feedstocks. For instance, 11 times processed dull black HDPE granules are generally used for producing jerry-cans by extrusion-blow-moulding and give jerry-cans processed 12 times. If the jerry-cans fail to hold air blown at the molding stage, more material is allowed to extrude, producing jerry-cans with thicker walls. These jerry-cans still possess some value, not entirely due to their polymeric performance, but to their relatively strong structural capacity to act as containers. If the molds are unable to hold air blown even after extrusion of additional material, this lot will be removed and transported to other recycling units to produce lower grade products by injection molding, the feedstocks for which are not required to be as strong. In the next cycle of recycling, this lot of 12 times processed injection molded products may be or may not be recycled any more times, depending on overall physical condition.

However, as this is the case for all recycled plastics in Bangladesh, an important question needs to be asked: What would be the ultimate destination for the recycled plastics whose properties are so degraded that the material is unsuitable for any more recycling? To elucidate this little studied problem in plastics recycling, mathematical models were developed to determine future scenarios of plastics recycling and landfilling in the country.

### 10.2.7 Future Scenarios: Mathematical Modeling

In a country like Bangladesh, which lacks a comprehensive industrial database, it is difficult to ascertain the amounts of plastics to be imported, recovered/recycled, and landfilled per year and the composition and quality of different types of recycled plastics in terms of number of times processed, and predicting the data for imports, recovery/recycling, and landfilling for the future is equally problematic. Mathematical modeling could be a tool for solving these problems as well as for helping the future decision makers in the plastics recycling arena, because it permits consideration of a very large number of variables and factors. Modeling also provides the opportunity to perform many simulations over a range of future scenarios (in the order of years) in a relatively short computation time (in the order of seconds). However, modeling of these scenarios requires extensive, nontrivial, multiple iterated calculations to be considered quantitatively.

In this study, the modeling task started with identifying and defining the existing *processes* of plastics recycling and landfill in the country. Other fundamentals of modeling i.e. *parameters* and *assumptions* were resolved. Determination of the modeling parameters was followed by setting a number of assumptions needed in building the models. Some efforts were made to establish the basis of the assumptions made in the modeling.

***10.2.7.1 Fundamentals of Modeling*** The following fundamental aspects are inherent in modeling:

> *Processes* that give structure of the model, (in this study about one industrial sector of Bangladesh, "processes" refers to the dynamic aspects of each sequence in a closed-loop system of plastics)
>
> *Parameters* defining the rates or characteristics of these processes
>
> *Assumptions* needed to define the processes and the parameters

*Processes*
- Importation is the only primary source of virgin plastics in the country. What are the statistics for virgin plastics importation? Because of a lack of available information on the imports statistics since mid 1960s, annual

imports data of virgin polyethylene (HDPE and LDPE) are available only for the years of 1990–1991 to 1994–1995. No separate data were available. Importation of polyethylene in Bangladesh is increasing linearly over the years according to the following the equation:

$$y = 2.08x + 6.23$$

where, $x$ is the number of years passed from the base year and $y$ is imports for the corresponding year, $x$.

- Of the plastic products marketed in any year, about 95 percent are recovered and recycled within periods ranging from 2–6 years, depending on their service life span. The remaining 5 percent as the maximum of non-recyclable plastics are normally landfilled within the same length of time.

- The service life of a range of HDPE articles varied from a minimum of 1 hour (for carrier bags) to a maximum of 8 years (for rigid containers); the overall average service life was 2.9 years.

- A large variety of HDPE scraps are present in the system, with different processing numbers ranging from never processed round virgin pellets (Pro-0) to 13/14 times processed very dull-black scraps or granules (Pro-13/14). As noted earlier, over the course of five to six closed-loop recycling cycles operating in Bangladesh, HDPE scraps are generally passed through 13 or 14 melting stages (for processing: pelletizing, agglomcrating-granulating or molding). Depending on the average service life span of HDPE articles, or in other words, length of "recovery time step"' (see later subsections), discarded HDPE material needs a total of around 15/18 years to become 5/6 times recycled (or 13/14 times processed) material (Figure 10.5).

- As environmental awareness and economic development increase in Bangladesh, legislative measures may be enacted to limit the number of times waste plastics can be recycled. This would greatly increase the burden on landfilling facilities, potentially causing serious pollution of land and groundwater.

- In view of the foregoing circumstances, the results of these modeling exercises could be useful in planning plastic waste management for the future in developing countries like Bangladesh.

*Parameters*    The parameters used in building the scenario models are: trends in PE imports, recovery percentages, recovery time steps, landfilling percentages, landfilling time steps, year of imports, year of recovery and year of landfilling, and calculation year. These parameters are defined and elaborated with examples in the following subsections (see also Haque et al., 2000; Haque, 1998).

*Assumptions*    The following assumptions were made:

- Although plastic importation into the country began in the mid-1960s, the lack of data for the early years made it necessary to select 1990–1991 as the base year for the models: Year 1 ($Y_1$). It is assumed that there were no other forms/types of plastic existing in the market except the virgin PE imported in that year and that PE imports into Bangladesh will keep increasing linearly over the years of the next two decades following the equation given earlier ($y = 2.08x + 6.23$)

- It is assumed that all plastic materials or products made of either virgin pellets or recyclate stay in Bangladesh. None of these materials were exported in any form during the period investigated.

- It is assumed that no time gap exists between the periods of importation, manufacturing, and marketing. Similarly, we assume no time gap between the recycling stages of recovery, processing, and marketing. As soon as a lot of the materials has been imported or recovered, processing into products and marketing begins.

- It is assumed that the processes of recovery and landfilling are carried out successively, and the percentages of recovery and landfilling in each time step are fixed in time.

Since the future always involves some degree of uncertainty, allowances were made to accommodate the minimum and maximum ranges of the uncertainty in the amount of recovery and landfilling PE wastes. These were done by repeating the analysis with the optimistic and pessimistic values for the parameters subject to uncertainty. Therefore, to address the two extremes of service life span (1 hour to 8 years) of PE products, *a faster scenario* and *a slower scenario* were derived for each of the recovery and landfilling activities. The assumptions for these alternatives (see Figure 10.10) were as follows:

- *Faster recovery scenario (1 year time step)*: In the rapid discarding and recovery pattern, it is presumed that at any stage of the closed-loop recycling system, 95 percent recovery for any lot of products requires a total of 2 years since the lot has been marketed. It is also postulated that a 1 year period is required for the products to be used up and discarded only. Out of this total 95 percent:
  - First 40 percent is discarded and recovered in the *first 1 year time step* (e.g., R40 in Year 2) since these materials were marketed (e.g. in Year 1);
  - Another 55 percent is discarded and recovered in the *second 1 year time step* (e.g. R55 in Year 3) since the lot of materials were marketed.
- *Faster landfilling scenario (1 year time step)*: In the faster scenario, the remaining 5 percent of the lot just described is landfilled within the same 2 years of the 95 percent recovery:
  - First 2 percent is landfilled in the *first 1 year time step* (e.g. L2 in Year 2) since the materials were marketed (e.g., in Year 1).
  - Another 3 percent is landfilled in the *second 1 year time step* (e.g. L3 in Year 3) since the lot of materials were marketed.

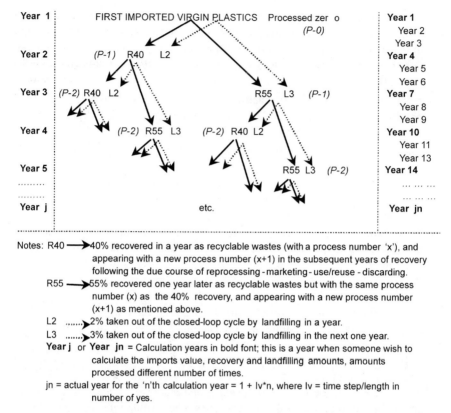

Notes: R40 ——➤40% recovered in a year as recyclable wastes (with a process number 'x'), and
       appearing with a new process number (x+1) in the subsequent years of recovery
       following the due course of reprocessing - marketing - use/reuse - discarding.

   R55 ——➤55% recovered one year later as recyclable wastes but with the same process
       number (x) as the 40% recovery, and appearing with a new process number
       (x+1) as mentioned above.

   L2  .......➤2% taken out of the closed-loop cycle by landfilling in a year.

   L3  .......➤3% taken out of the closed-loop cycle by landfilling in the next one year.

   **Year j** or **Year jn** = Calculation years in bold font; this is a year when someone wish to
       calculate the imports value, recovery and landfilling amounts, amounts
       processed different number of times.

   jn = actual year for the 'n'th calculation year = 1 + lv*n, where lv = time step/length in
       number of yes.

**Figure 10.10.** Faster and slower scenarios of the successive recovery and landfilling of waste plastic in Bangladesh. *Source*: After Haque et al., 2000 and Haque, 1998.

- *Slower recovery scenario (3 years time step)*: In the slower discarding and recovery pattern, it is assumed that completion of the 95 percent recovery of any lot of products requires a total of 6 years considering a 3 year time step instead of a 1-year time step as for the faster scenario:
  - First 40 percent of any lot of products is discarded and recovered in the *first 3 year time step* (e.g., R40 in Year 4) since these materials were marketed (e.g., in Year 1).
  - Another 55 percent is discarded and recovered in the *second 3 year time step* (e.g., R55 in Year 7) since the materials were marketed.
- Slower landfilling scenario (3 years time step): It is assumed that along with the 95 percent recovery, the remaining 5 percent of the same lot is landfilled within a total of 6 years, following a 3 years time step of landfilling:
  - First 2 percent is landfilled in the *first 3 year time step* (e.g., L2 in Year 4) since these materials were marketed (e.g., in Year 1).

- Another 55 percent is landfilled in *second 3 year time step* (e.g., L3 in Year 7) since the materials were marketed.

### 10.2.7.2 *Developing and Programming the Mathematical Models* Based on the understanding of the fundamentals of the faster and slower recovery and landfilling scenarios, mathematical models sufficiently detailed to encompass the enormous target system were developed by analyzing and synthesizing the functional aspects of the plastics recycling and landfilling systems currently operating in Bangladesh. These models, which are not presented here, can be found in Haque et al. (2000) and Haque (1998).

Along with descriptions of a series of model building concepts and mathematical equations, Haque et al. (2000) and Haque (1998) explained the methodology of computer programming to meet the modeling objectives—solving the problems, providing input data, and then generating and storing outputs/results in a readily available form for further use.

Briefly, all the mathematical models were programmed separately and then combined in the FORTRAN computer language, to produce a single mathematical modeling program. The program, which has a run time in the order of seconds requires a minimum of manual intervention in changing the input parameters for studying the alternative scenarios for plastic waste recycling and landfilling and for identifying the trade-off between recycling and landfilling. This trade-off is determined according to a hypothetical limit on "number of times processed" that may be legislated at some time in the relatively near future. There are only four input variables to be specified, depending on the desire of the future researcher or following the suggestions for further research: amount of first virgin plastic imports; number of years (i.e., next 10, 15 or 20 years) for which data are required; rates of recovery or landfilling (e.g., 45 and 50 percent instead of 40 and 55 percent, respectively); and length of time steps of recovery and landfilling which drives the system either as a slower or faster process (e.g., 2 year and 4 year time steps instead of 1 year and 3 year, respectively).

The program is particularly useful because it offers potential links to future dynamic planning and modeling efforts in that every input and output datum of any kind is stored in matrix form in multiple dimensions. For instance, a single recovery output datum can tell us the following about a material: its year of recovery, how many times it has been processed, its path of recovery (either through 40 or 55 percent; whether it has been through the faster or slower recycling system). All this information is crucially important not only for future planning and policy formulation purposes but to get a clear understanding of the materials' present quality by looking at its origin or path of its previous life-cycle(s).

Alternative attempts have also been made to program the mathematical models mentioned earlier using Excel Spreadsheets. Obviously, this spreadsheet model (Haque, 1998) is simple and relatively more understandable to nonspecialists. The spreadsheet model has its limits, but it provides the same set of results given by the FORTRAN model as discussed in the following subsections.

### 10.2.7.3   Results from Mathematical Modeling

Table 10.4 summarizes the basic findings obtained from the mathematical modeling of both the faster and slower scenarios of polyethylene (PE) imports, recovery and landfilling in Bangladesh for the years 1990–1991 to 2004–2005. Importation of PE, starting with an amount of 6.23 million kilograms (Mkg) in 1990–1991, rises to a seven fold higher amount of 35.35 Mkg in the year 2004-05. This rise in imports is the same for both the faster and slower models, but substantial differences have occurred in the cases of the amounts recovered and entered into the system, and hence, the total amount (imports and recovery) present in the recycling system.

For example, the total amount of PE recovered in the year 2004–2005 is 140.89 Mkg for the faster recovery model, in comparison to 42.42 Mkg for the slower recovery model. As a result, in that year, the total amount of materials present in the system differs by about 99 Mkg between the two models. Other major differences are observed in the cases of the total number of new batches recovered and entered into the system and the total number of recovered batches present in the system. Although both the models start with no recovered batch but only one batch of virgin materials in the base year of 1990–1991 (as assumed earlier), these numbers grow to large numbers (233 and 608, respectively) in the year 2002–2003 under the faster recovery, whereas they are much lower (only 5 and 11, respectively) for the slower model (Table 10.4). Table 10.4 also shows that the increase in the number of batches and amount of materials present in the system for each scenario follow a nonlinear pattern.

As noted earlier, the mathematical models were also able to distinguish the amount of material according to the material's number of times reprocessed. It becomes apparent that new materials are recovered and recycled each year and they appear into the closed-loop recycling system with a new processing number. For example, processed zero (Pro-0) materials (6.23 Mkg) in 1990–1991 becomes processed nine (Pro-9) materials (24.95 Mkg) in 1999–2000. A much longer period is needed by the slower recovery model to reach this stage, which will be in the year 2017–2018. In other words, deterioration in the quality of the recycled plastic materials due to the effects of successive reprocessings and extraneous particles (added naturally from the system) will be delayed substantially if materials are recovered and recycled following the slower recovery model developed on the basis of the longer life span of the plastic products.

Taking into consideration the effects of the faster and slower recovery patterns, the amounts of materials landfilled also vary between the faster and slower landfill models. As in the slower recovery model, discarding used up products and hence landfilling the unrecoverable scrap portions take place after longer time steps due to their prolonged life span in the slower landfill model. Therefore, within the same length of time, the amount of discard landfilled is lower than the amount landfilled under the faster model. For example, a total of 7.4 Mkg and 2.2 Mkg of materials will be landfilled in the year of 2004–2005, following the faster and slower landfill models, respectively.

Haque et al. (2000) and Haque (1998) have detailed and presented the variations in the findings obtained from the mathematical modeling (shown in Table 10.4) as a

TABLE 10.4. Summary for Faster and Slower Scenarios of Polyethylene (PE: LDPE and HDPE) Imports, Recovery, Total Materials Present in the Recycling System and Amount Landfilled in Bangladesh, 1990–1991 to 2004–2005. (In Mkg = million kilogram)

| | | Faster Scenarios | | | | | | Slower Scenarios | | | | |
|---|---|---|---|---|---|---|---|---|---|---|---|---|
| | | Recovery (R) | | | | | | Recovery (R) | | | | |
| Year | Imports (I) | Total Amount of Materials Recovered | Total Number of New Batches Recovered and Entered into the System | Total Number of Recovered Batches Present in the System | Total amount of materials (I + R) present in the system | Landfilling | Imports (I) | Total Amount of Materials Recovered | Total Number of New Batches Recovered and Entered into the System | Total Number of Recovered Batches Present in the System | Total Amount of Materials (I + R) Present in the System | Landfilling |
| 1990–91 | 6.23 | 0.00 | 0 | 0 | 6.23 | 0.000 | 6.23 | 0.00 | 0 | 0 | 6.23 | 0.000 |
| 1991–92 | 8.31 | 2.49 | 1 | 1 | 10.80 | 0.125 | 8.31 | 0.00 | 0 | 0 | 8.31 | 0.000 |
| 1992–93 | 10.39 | 7.74 | 2 | 3 | 18.14 | 0.403 | 10.39 | 0.00 | 0 | 0 | 10.39 | 0.000 |
| 1993–94 | 12.47 | 13.20 | 3 | 6 | 25.67 | 0.687 | 12.47 | 2.49 | 1 | 1 | 14.95 | 0.125 |
| 1994–95 | 14.55 | 20.24 | 5 | 11 | 34.79 | 1.057 | 14.55 | 3.32 | 0 | 1 | 17.87 | 0.166 |
| 1995–96 | 16.63 | 28.03 | 8 | 19 | 44.66 | 1.466 | 16.63 | 4.16 | 0 | 1 | 20.79 | 0.208 |
| 1996–97 | 18.73 | 37.00 | 13 | 32 | 55.71 | 1.937 | 18.73 | 9.41 | 2 | 3 | 28.12 | 0.486 |
| 1997–98 | 20.79 | 46.85 | 21 | 53 | 67.68 | 2.454 | 20.79 | 11.72 | 0 | 3 | 32.51 | 0.607 |
| 1998–99 | 22.87 | 57.70 | 34 | 87 | 80.57 | 3.024 | 22.87 | 14.03 | 0 | 3 | 36.90 | 0.727 |
| 1999–00 | 24.95 | 69.43 | 55 | 142 | 94.38 | 3.641 | 24.95 | 19.48 | 3 | 6 | 44.43 | 1.011 |
| 2000–01 | 27.03 | 82.07 | 89 | 231 | 109.09 | 4.305 | 27.03 | 22.83 | 0 | 6 | 49.87 | 1.186 |
| 2001–02 | 29.11 | 95.54 | 144 | 375 | 124.66 | 5.013 | 29.11 | 26.19 | 0 | 6 | 55.30 | 1.362 |
| **2002–03** | **31.19** | **109.86** | **233** | **608** | **141.05** | **5.766** | **31.19** | **33.24** | **5** | **11** | **64.43** | **1.732** |
| 2003–04 | 33.27 | 124.98 | 377 | 377 | 158.25 | 6.561 | 33.27 | 37.83 | 0 | 11 | 71.10 | 1.973 |
| 2004–05 | 35.35 | 140.89 | 610 | 610 | 176.23 | 7.397 | 35.35 | 42.42 | 0 | 11 | 77.77 | 2.213 |

577

result of addressing the hypothetical situation of a law banning the recycling of materials already processed more than, say, six times by incorporating a "quality factor" for recycled materials. It is postulated that if any lot of materials is taken out of the system and rendered useless by legislated restrictions, that amount would be added to the existing amount of materials generally landfilled each year. It was estimated that the amount of 4.31 Mkg landfill projected for the year 2000–2001 could be increased by 45 percent in the event of a prohibition of recycling recovered materials that had already been processed more than six times.

In addition, if importation of virgin plastics into Bangladesh should stop in the future, the following would happen:

- Both the virgin and recycled materials then present would be intensively recycled.
- Existing stocks of plastics would be diminished by the heavy recycling and ultimately would be removed from the recycling loop, with the seriously degraded plastics consigned to landfill.
- The time would come, sooner in the case of the faster landfill scenario, when there would be no plastics for recycling.

These modeling exercises were run in late 1990s. Before the models could be applied to a current situation, the degree of uncertainty associated with the original set of parameters and assumptions would have to be evaluated.

### 10.2.8 Conclusions

The plastic waste recycling systems in Bangladesh, though primitive by Western standards, have many merits. Participation in the recycling systems is high in comparison to developed countries due to specific socio-economic compulsions and cultural factors. The average consumer in this developing country, particularly the housewife, reflexively supports an eco-friendly product management system by retaining empty containers for storing food or non-food items for very long periods. The country has a highly efficient plastic waste collection and recycling system and many difficulties of plastics recycling are minimized by the existence of the unique industrial ecology area centered in the Islambagh district of Dhaka.

The recycling system is also supported by its economic viability, being propelled by economic forces and sustained by very low cost manual labor in the collection and sorting phases. Post-consumer waste is recycled on a large scale in the country for the following reasons: The greater expense of importing or producing virgin plastics, the economic need of those who perform the more disagreeable tasks (waste pickers and recycling workers), a large market/demand for cheap recyclable waste materials and recycled products, and lack of public awareness of and regulatory measures addressing environmental standards and workers health and safety. Overall, all these factors act as the "push forces" to keep recycling plastic waste in a country that has a recycling rate close to 100 percent. Certainly, this means that if no constraints are introduced by the legislature, more materials will enter into the system, the size of the *recycling loop* will expand increasingly to accumulate additional materials into the system, and these will require a larger work-force and more processing units for their recycling.

The poorly maintained waste disposal system found in the cities of Bangladesh, however, cast doubt on the sustainability of recycling in the future (Haque et al., 1997; Rashid et al. 1995). There is also an almost total lack of legislative concern about the environmental and hygiene issues related to waste disposal and recycling. Since there are few systematic programs of house-to-house collection and little sense of civic consciousness in this matter, discarded plastics lie in individual and communal heaps along the roadside where they become contaminated by other refuse. Even hospital plastic waste is disposed of with other hospital and municipal wastes and subsequently collected for recycling (Haque, 1994; Kumar, 1996). Although the country has achieved nearly a zero waste level for plastic wastes, with minimal pollution from transportation by vehicles powered by fossil fuels, there is nevertheless potential cost to human health.

However, depending on the level of contaminants and the rate and nature of the deleterious effects of successive recyclings on the properties of plastics, the quality of some materials recycled several times deteriorate to a stage that they become unsuitable for further recycling as was postulated at the beginning of this study. Consequently, at some point in the foreseeable future in developing country cities, such as Dhaka, the existing crude landfilling may be one of the few available options for the disposal of those non-recyclable useless plastics. This requires carefully assessed plans and strategies for plastics recycling, incorporating local circumstances and factors. Otherwise, the propensity for cross-contamination and its effects on the quality of recycled products will be increased to a greater extent from the current levels, and thus will continue to cause substantial problems in their processability down the chain of successive recyclings. The mathematical models developed and presented in this study could be a foundation building block to initiate the process of making sustainable progress in this informal sector of recycling.

## ACKNOWLEDGMENTS

The authors are grateful to: (a) the Ford Foundation for the financial support of A. Haque in the form of a PhD bursary, (b) Professor R. M. Ahsan, Department of Geography and Environment, University of Dhaka, Bangladesh, (c) Professor M. R. Ashmore, Department of Environmental Science, University of Bradford, UK, (d) the School of Polymer Technology, University of North London, UK, (e) the School of Engineering, University of Bradford, UK, and (f) all of the personnel who supported this study by involving themselves in the fieldwork in Dhaka in 1995–1996.

## REFERENCES

Allen, D. T. (1999a). Case study in industrial ecology: application of material flow analysis, a paper presented to *1999 Seminar Series: New Systems, New Tools* at the Centre for Process Systems Engineering, Imperial College of Science, Technology and Medicine, University of London, 24 June.

Allen, D. T. (1999b). Industrial ecology, Invited guest editorial. *Environmental Progress*, **18**(2), S3.

APME (1997). Association of Plastics Manufacturers in Europe, *Information System on Plastic Waste Management in Western Europe: European Overview, 1995 Data*, APME Technical and Environmental Centre and SOFRES Conseil, France.

BSI (1997a). The British Standard Institute, British standard methods of testing plastics, Part 3: Mechanical properties: Methods 320A to 320F: Tensile strength, elongation and elastic modulus: 1976, in *CD-Rom: British Standards*, British Standard Service, BSI, UK.

BSI (1997b). The British Standard Institute, British standard methods of testing plastics, Part 7: Rheological Properties: Method 720A: Determination of melt flow rate of thermoplastics: 1979, in *CD-Rom: British Standards*, British Standard Service, BSI, UK.

Chiu, S.-H., Huang, J. H, Lin, C., Tang, Y., Chen, W., and Su, S. (1999). Applications of a corporate synergy system to promote cleaner production in small and medium enterprises. *Journal of Cleaner Production*, **7**(5), 351–358.

Clift, R. (1997). Clean technology—the idea and the practice. *Journal of Chemical Technology and Biotechnology*, **68**(4), 347–350.

Clift, R. (1998). Engineering for the environment: the new model engineer and her role. *Process Safety and Environmental Protection*, **76**(B2), 151–160.

DOE (1995). The Department of Environment and the Welsh Office, *Making Waste Work: A Strategy for Sustainable Waste Management in England and Wales*, HMSO, London.

Ehrenfeld, J. R. (1997). Industrial ecology: a framework for product and process design. *Journal of Cleaner Production*, **5**(1–2), 87–95.

ENDS (2000). The ENDS Reports, rise in energy recovery of plastic waste in Europe. *The ENDS Reports*, Issue No. 303, 1 April.

Erkman, S. (1997). Industrial ecology: a historical view. *Journal of Cleaner Production*, **5**(1–2), 1–10.

Frosch, R. A., Clark, W. C., Crawford, J., Sagar, A., Tschang, F. T, and Webber, A. (1997). The industrial ecology of metals: a reconnaissance, *Philosophical Transaction of the Royal Society of London Series—A Mathematical Physical and Engineering Sciences*, **355**(1728), 1335–1347.

Ganguli, R. K. (1995). *Plastic Waste Recycling and Associated Health Impacts*. A dissertation submitted for the post-graduate diploma in public systems management, Indian Institute of Social Welfare and Business Management, Management House, College Square West, Kolkata (formerly Calcutta) 700 073, India.

Haque, A. (1994). *Plastic Waste Recycling and Associated Health Hazards in Dhaka City*. M.Sc. thesis, Imperial College Centre for Environmental Technology, Imperial College of Science, Technology and Medicine, University of London, London.

Haque, A. (1998). *Limits to Plastics Recycling in the Developing and Developed Economies*, Ph.D. thesis, Centre for Environmental Technology, T.H. Huxley School of Environment,

Earth Science and Engineering, Imperial College of Science, Technology and Medicine, University of London.

Haque, A., and Bell, J. N. B. (1998). Plastics recycling in a developing country: a case study in Dhaka city, Bangladesh. *Regional Development Studies (RDS)*, **4**, 181–192.

Haque, A, Bell, J. N. B., Islam, N., and Ahsan, R. M. (1997). Towards a sustainable plastic recycling in developing countries: a study of Dhaka, Kolkata (formerly Calcutta) and Delhi. *Proceedings of the sixth IRNES conference on Technology, the Environment and Us*, The Sixth IRNES Conference Committee, 22–23 September.

Haque, A., Mujtaba, I. M., and Bell, J. N. B. (2000). A simple model for complex waste recycling scenarios in developing economies. *Waste Management*, **20**: 625–631.

ISI (1999). Institute of Scientific Information Inc., BIDS Key Words Search on '*Industrial Ecology*' (http://www.bids.ac.uk).

JCP (1997) Journal of Cleaner Production. *Journal of Cleaner Production Special Issue: Industrial Ecology*, **5**(1–2), 1–132.

JIE (1999). *Journal of Industrial Ecology*, Co-edited by. Allen, D. T., the Reese Professor of Chemical Engineering at the University of Texas at Austin, Published by MIT Press, USA [In *Environmental Progress*, **18**(2), S3].

Koenig, H. E., and Cantlon, J. E. (1998). Quantitative industrial ecology, *IEEE Transactions on System, Man and Cybernetics Part C–Applications and Review*, **28**(1), 16–28.

Kumar, S. (1996). Deadly trade in Delhi's hospitals, *New Scientist* (UK), 11 May.

Luken, R. A., and Freij, A. C. (1995). Cleaner industrial production in developing countries: market opportunities for developed countries. *Journal of Cleaner Production*, **3**(1–2), 71–78.

Nicholas, M., Azapagic, A., and Clift, R. (1999). Industrial ecology in Europe. *Journal of Industrial Ecology*, **2**(4), 3–5.

Pizzocaro, S. (1998). Steps to industrial ecology: reflections on theoretical aspects. *International Journal of Sustainable Development and World Ecology*, **5**(4), 229–237.

PRF and CPRR-Plastics Recycling Foundation and Centre for Plastics Recycling Research (1992). *Plastics Recycling: From Vision to Reality*. New Jersey Commission on Science and Technology and Council for Solid Waste Solution, USA, p. 1.

Rader, C. P., and Stockel, R. F. (1995). Polymer recycling: an overview. In Rader, C. P., Baldwin, S. D., Cornell, D. D., Sadler, G. D., and Stockel, R. F. (eds), *Plastics, Rubber, and Paper Recycling: A Pragmatic Approach*. American Chemical Society, Washington, DC, pp. 2–10.

Rashid, H. U., Sadeque, S. Z., and Haider, I. (1995). Community involvement in solid waste management (Cover Story). *The Daily Bangladesh Observer: Observer Magazine*, 1 December.

Ratra, O. P. (1994). Plastic waste and the environment. *Conference Proceedings on Plastics and the Environment*, Organised by Federation of Indian Chambers of Commerce and Industry and Indian Plastics Institute, New Delhi, 7–8 September.

Simpson, I. (1994). Survival in scrap metal suburb: Hanoi's complex recycling network. *The Financial Times*, Business on the Environment, 18 May.

Vogler, J. (1985). Standing up for technology in Bangladesh. *Materials Reclamation Weekly*, **146**(18), 22–25.

## 10.3 FROM WASTE TO ENERGY: CATALYTIC STEAM GASIFICATION OF POULTRY LITTER

ATUL C. SHETH
Professor, Chemical Engineering, The University of Tennessee Space Institute, Tullahoma, TN 37388, USA

---

Approximately 7.6 billion broiler chickens were produced in the United States in 1996. This number has risen steadily, and experts predict that the industry will continue to expand, with more than 75 percent of the production coming from the southeastern United States. The amount of litter produced is substantial, and much of it ends up polluting the rivers and surface waters.

This section describes catalytic steam gasification a potentially useful option for the management of broiler litter. The process converts the feedstock to a fuel gas, which can be used as an energy source or as a chemical feedstock.

An experimental bench scale study carried out at the University of Tennessee Space Institute (UTSI), has demonstrated the technical and economic viability of this concept. Although a pilot-plant scale and continuous-mode studies remain to be done, in view of the sky-rocketing cost of fuel and natural gas, such sources of alternate energy, fuel, and chemical feedstocks merit close examination. They offer promising options to help alleviate the shortage of affordable energy and at the same time also help to reduce water pollution problems related to the disposal of solid waste.

### 10.3.1 Poultry Litter and Its Uses

The waste that must be removed from a broiler-growing house is a combination of manure and litter material such as wood shavings, sawdust, rice hulls or peanut hulls. From one average sized broiler house about 144 tons of manure would be removed each year. In 1992, the Southeastern region alone produced over 5 million tons of broiler litter.

Ideally, all broiler litter would be recycled for use as fertilizer or converted, with minimum processing and transportation, to animal feeds and all within EPA guidelines. However, despite the industry's best efforts, more poultry litter is being generated than can be utilized. Excess then becomes waste, which is particularly problematic in regions where production is highly concentrated. Solutions are expected primarily from source reduction technologies and from recycling to provide new products with value added features. Under the latter option, new uses for the poultry waste have been found as enhanced fertilizers, horticultural and mushroom growing medium, and feed products for farm and domestic animals. However, these relatively high value uses are slow to develop and the new firms specializing in such products will need substantial financial support.

Currently, manure from broilers (as well as egg-laying hens, broiler breeders, and turkeys) is spread as a fertilizer on pasture, hay, small grains and corn producing fields. A ton of broiler litter contains about 60 lb of nitrogen, 60 lb of phosphate and 40 lb potash. Recommended fertilization rates for most crops are three to four tons per acre per year. This means that about 36 acres of land must be available to handle the manure from one broiler house. Relatively few broiler growers have nearby land resources large enough to utilize the manure at recommended fertilization rates. If poultry manure is used in excess, however, it can pollute the surface and/or ground water. An outbreak of *Pfiesteria piscicida* in the Chesapeake Bay in the late 1990s poisoned a lot of fish and was linked to improper broiler litter disposal techniques. Other environmental accidents in Arkansas, Oklahoma, and Alabama have also been linked to poor disposal techniques (Jones, 1998).

Poultry manure/litter is a valuable nutrient for grain and fiber crops, forage crops, fruits, and vegetables. Without proper protective handling, however, manure, litter, dead birds, and/or wastewater can result in water contamination due to the premature release of nitrogen and phosphorus into the environment (Poultry Water Quality Consortium, 1994). Nitrogen is an essential plant nutrient but, in excess, it can be harmful. High concentrations of nitrate (dissolved nitrogen) in drinking water can affect human health, especially in infants and children. Ammonia in small quantities is toxic to fish and aquatic organisms.

### 10.3.2 Environmental Effects of Excess Poultry Litter

Nitrogen and phosphorus concentrations may reach unhealthy levels in waterbodies contaminated by poultry litter. Symptoms of this condition include: takeover by algae and rooted aquatic plants, prematurely ageing and choking of waterbody and creating unpleasant odors, offensive taste, and discoloration. All of these can make the water unfit for consumption or recreational and aesthetic use. Further, these eutrophic conditions can kill fish, clog water treatment plant filters, and lead to the growth of blue-green algae, a species that can be fatal to livestock (Poultry Water Quality Consortium, 1994).

Because nitrate-nitrogen is highly mobile, it can leach into groundwater and flow with stormwater runoff into surface waters. Use of excessive amounts of poultry manure and litter as fertilizer, is likely to cause high nitrogen and phosphorus concentrations in nearby waters. Soil erosion also increases the amount of phosphorus in surface waters. Excessive phosphorus in soil ($>800$ mg/L), may become soluble and move into groundwater (Poultry Water Quality Consortium, 1994).

Another source of environmental contamination consists of the calcium and sodium salts added to poultry feeds to help the birds maintain the correct chemical balance. Excess salts pass through the animals and are eliminated in manure. Sometimes, when the waste accumulates, the salts leach into groundwater and enter surface through unprotected runoff. There they alter the water's taste or harm freshwater plants and animals (Poultry Water Quality Consortium, 1994).

When suspended matter from poultry wastes reach surface water, where it looks most unattractive and water quality invariably suffers. In addition, the suspended material reduces the penetration of sunlight and therefore slows the production of oxygen. The result is an oxygen demand that reduces the levels of dissolved oxygen in the water. It also clogs fish gills, makes it difficult for sight-feeding fish to find food, and interferes with breeding cycles by settling over fish spawning areas (Poultry Water Quality Consortium, 1994).

In a natural environment, organic matter, such as poultry waste, is converted to simple compounds by naturally occurring micro-organisms. These simple compounds may be other forms of organic matter or they may be nonorganic compounds or gases, such as nitrates, orthophosphates, ammonia, and hydrogen sulfide. A biological reaction occurs when manure or other organic matter is added to water and aerobic organisms (oxygen requiring organisms) begin the decaying process. The bacteria consume free oxygen and produce carbon dioxide gas. Under anaerobic conditions (without oxygen), methane, amines, and sulfides are produced (Poultry Water Quality Consortium, 1994).

In addition to generating breakdown products that favour the growth of bacteria, animal waste itself is a potential source of some 150 disease-causing organisms or pathogens. These organisms include bacteria, viruses, fungi, protozoa, and parasites. Examples of undesirable microorganisms include Salmonella, Listeria, coliform bacteria, New Castle (virus), ringworm, coccida, and Ascaris (Poultry Water Quality Consortium, 1994). When these pathogens contaminate water or wastes, they can infect humans and other animals that drink the water, allow the wastes to contact the skin, or eat fish or other aquatic animals from the polluted waterbody. Most pathogens die relatively quickly, when their environment changes drastically. However, under favorable conditions, they can live long enough to cause problems. They may persist longer in groundwater than in surface water (Poultry Water Quality Consortium, 1994).

### 10.3.3 Biogas and Direct Combustion

Sustainable development of the poultry industry calls for the implementation of alternative technologies for disposing of poultry litter. In the past some farmers have attempted to produce a combustible gas from the poultry manure. Under anaerobic conditions, the manure yields a gas containing primarily methane ($CH_4$) and carbon dioxide ($CO_2$) plus small amounts of water vapor and other gases such as hydrogen sulfide ($H_2S$), and ammonia. This gas, called biogas, was used to heat poultry houses or other non-residential farm buildings. Attempts also were made to use biogas in conjunction with other fuels to generate electricity. At present, generating biogas from the anaerobic decomposition of the poultry waste is considered impractical for use on the farm.

#### 10.3.3.1 Disadvantages of Biogas and Direct Combustion
There are several factors working against the biogas option including the following (Payne and Donald, 1990).

- Biogas is not easily compressed and cannot be used to operate tractors or other moving equipment.
- Nearly all nutrients which enter the digester will also leave the digester. That is, the solid nutrients, and any water, added to create the necessary slurry, must still be disposed of on land.
- Biogas generation requires a very high level of management.
- Biogas is most effective when the non-essential gases are removed through scrubbing.
- Separating the methane from the other gases requires a great deal of skill.
- Biogas can be corrosive to metal parts.
- Biogas is not a practical, money-saving option unless the cost of conventional fuel increases dramatically.

In some European countries, poultry waste is directly burned in a combustor/ boiler and either the hot exhaust gas or steam is used for heating poultry houses. This approach produces a low quality energy that is not efficient and can be used only for heating non-residential farm buildings. Moreover, fouling of the internal parts of the boiler/combustor and slagging conditions in the grate area due to presence of the alkaline ash constituents make operation very difficult. In any event, with most of the nutrients gone, the ash/residue has very limited industrial uses.

### 10.3.3.2 Eliminating the Disadvantages of Biogas

Some of the disadvantages of the biogas option are eliminated by carrying out a catalytic steam gasification of the poultry waste. Originally, this process was developed by the Exxon Corporation for the conversion of coal to clean fuel. However, it can be applied to any lignocellulosic material such as poultry litter, biomass, agricultural waste etc. Catalytic steam gasification utilizes an alkali metal salt as a catalyst and steam as a source of oxygen to convert a carbonaceous feedstock to a gas mixture composed primarily of carbon monoxide, carbon dioxide, hydrogen, and methane. At low pressures and high temperatures, the process produces synthesis gas, a gas rich in carbon monoxide and hydrogen. The gas produced has a low to medium heat content and may be utilized as an energy source or chemical feedstock. The solid residue can be used as a fertilizer or in cement/concrete manufacturing.

Broiler litter is an attractive candidate for catalytic steam gasification for several reasons. The char produced from biomass is typically more reactive than the one produced from coal. The large amount of potassium and other minerals in the litter may also exhibit an inherent catalytic activity. Moreover, the very low sulfur content of broiler litter limits the production of hydrogen sulfide, an undesirable constituent of gaseous fuels. Thus, catalytic steam gasification of poultry litter has the potential of generating a valuable product and at the same time has the capability of protecting the environment by utilizing the litter in an environmentally benign manner.

### 10.3.4 Coal Gasification Process Concept

Since both coal and poultry litter (which is mostly biomass) are of carbonaceous composition, many of the processes utilizing coal can be modified to accommodate a biomass feedstock. Coal gasification converts solid coal to a mixture of combustible gases through controlled partial oxidation. The development of catalytic steam gasification offered a route to produce gases of higher heating value at significantly lower temperatures. A gasification agent such as steam, carbon dioxide, or air provides the necessary oxygen. In 1924, while testing the catalytic effects of various alkali salts, Taylor and Neville found that $K_2CO_3$ and $Na_2CO_3$ were particularly active (Hirsch et al., 1982). In the 1970s Exxon developed a continuous catalytic coal gasification process that relies on steam and recycled synthesis gas for gasification (Anderson and Smith, 1982). The overall reaction catalyzed by either KOH or $K_2CO_3$ yields the product, methane, and a by-product, carbon dioxide.

The application of the catalytic steam gasification concept to biomass is a relatively new area of research. Working with hybrid poplar, W. B. Hauserman demonstrated gasification rates comparable to those of lignites (Hauserman, 1994). The poplar was catalyzed with wood ash at a loading of 10 wt percent. Timpe and Hauserman concluded that both hybrid poplar and *Typha augustifolia* possess reactivities similar to low-rank coals in catalyzed and uncatalyzed conditions alike (Timpe and Hauserman, 1992). Addition of a potassium-rich catalyst to their experiments on a laboratory scale thermogravimeteric analyzer increased the reactivities of hybrid poplar and *Typha augustifolia* by factors as large as 28.

To realize the poultry litter steam gasification (Figure 10.11), the poultry waste or any other animal waste (source of carbon, hydrogen, and potassium) is mixed with other biomass waste (if needed as additional source of carbon and hydrogen), and suitable source of additional potassium (if required).

The resulting mixture can be gasified in "as-is" slurry form at 1300–1500°F and at 50–150 psi pressure in a gasifier. The geometry of the gasifier will depend upon the nature of the feedstock. For aggregates, a fluidized-bed gasifier may be suitable. However, for the "as-is" slurry form the fixed-bed downdraft type of geometry may be better. The steam can be produced externally and supplied to the gasifier (if dry

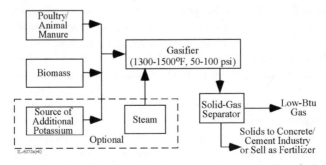

**Figure 10.11.** Schematic of steam gasification process for poultry litter/animal manure.

feed is used) or in situ from the wet/slurried feedstock. Depending upon the pressure used, the resulting gas will be rich in methane or in CO and $H_2$ and, after separating from the solid/char residues, can be used as a fuel for heating purposes or to produce electricity. The volume of the solid/char residue derived from the gasification will be significantly smaller than that of the starting waste, and therefore, can be used in cement or concrete manufacturing industries or can be sold as fertilizer. It is believed that the potassium present in poultry and certain animal wastes such as from swine, cows, horses, and sheep (see Table 10.5) can provide the necessary catalyst. However, if necessary, additional supplemental potassium can be obtained from potassium feldspar, ash from wood-burning stoves, cement-kiln dust, spent sorbents from acid gas scrubbing operations, or similar sources.

The conversion of coal or biomass to a gaseous product is achieved via a two-step process. Primary gasification or devolatilization involves the thermal decomposition of the feedstock to char and volatiles.

$$\text{Biomass} \rightarrow \text{Char} + \text{Volatiles} \qquad (10\text{-}1)$$

In general, devolatilization and pyrolysis reactions are endothermic. Volatiles typically generated by primary gasification include tars, phenols, methane, oils, naphtha, $H_2S$, and some CO and $H_2$ (Turner, 2001). The char produced by the primary gasification is the reactant for the secondary gasification reactions.

**TABLE 10.5. Specifications of Poultry and Other Animal Wastes[a] (Approx. lb. of Nitrogen (N), Phosphate ($P_2O_5$) and Potash ($K_2O$) per Ton of Animal Waste)**

| Animal | Percent Moisture | Nitrogen lb/ton | Phosphate lb/ton | Potash lb/ton |
|---|---|---|---|---|
| Poultry caged layers | 10 | 76 | 92 | 34 |
| | 35 | 47 | 60 | 24 |
| Fresh manure | 75 | 27 | 28 | 14 |
| | 90 | 11 | 11 | 5 |
| Broiler–litter | 25–30 | 34 | 37 | 30 |
| Swine-fresh manure | 75 | 10 | 7 | 13 |
| lagoon | 99 | 1–1.2 | 0.25–0.5 | 1–1.2 |
| Dairy-fresh manure | 84 | 12 | 5 | 12 |
| above ground storage | 88 | 11.5 | 2 | 8.2 |
| lagoon | 99 | 1 | 0.5 | 1 |
| Beef-fresh manure | 80 | 14 | 9 | 11 |
| Horses-fresh manure | 60 | 12 | 5 | 9 |
| Sheep-fresh manure | 65 | 21 | 7 | 19 |

[a]*Source*: US Department of Agriculture, Agriculture Stabilization and Conservation Service, Winchester, TN 37398-1740.

If we represent the poultry litter as $C_uH_vN_wO_xS_y$, the overall gasification reaction in the presence of excess steam and catalyst can be written as:

$$C_uH_vN_wO_xS_y + \text{Ash} + mH_2O \xrightarrow{K-catalyst}$$
$$CO + CO_2 + CH_4 + H_2 + H_2S + NH_3 + H_2O + \text{Ash} \qquad (10\text{-}2)$$

Jones (1998) carried out thermodynamic equilibrium analysis of reactions between coal and steam, and various product gases as functions of gasification temperature, pressure and hydrogen-to-oxygen (H/O) ratio. His results are shown in Figures 10.12–10.14. His results suggested that to produce CO- and $H_2$-rich synthesis gas, gasification operations should be carried out at low pressures ($\sim$100 psi), moderate temperatures ($\sim$1400–1500°F) and low H/O ratio ($\sim$1–2).

### 10.3.5   Testing Gasification Processes

#### 10.3.5.1 Experiments and Results of Bench Scale Tests of the UTSI Process
Jones (1998) and Turner (2001) have done bench-scale evaluations of the production of synthesis gas from steam gasification of poultry litter. The high pressure, high temperature fixed-bed gasification system used in their work is shown schematically in Figure 10.15. To avoid tar formation during the pyrolysis step and subsequent clogging of the gas discharge lines, they used a separate

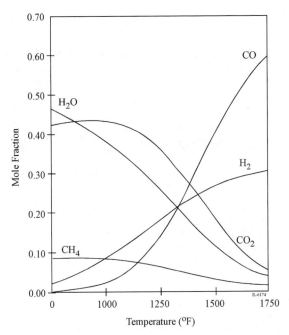

**Figure 10.12.** Effect of temperature on equilibirum compositio-150 psia, H/O = 1.00.

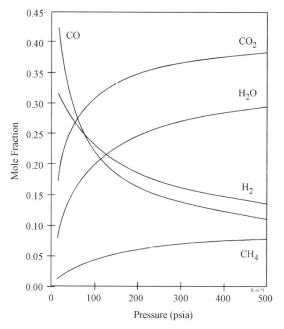

**Figure 10.13.** Effect of pressure on equilibrium composition-1300°F, H/O = 1.00.

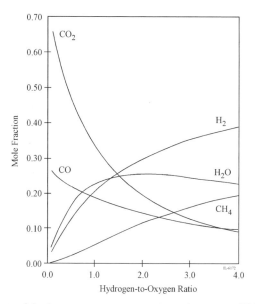

**Figure 10.14.** Effect of hydrogen-to-oxygen atomic ratio on equilibirum composition-1300°F, 150 psia.

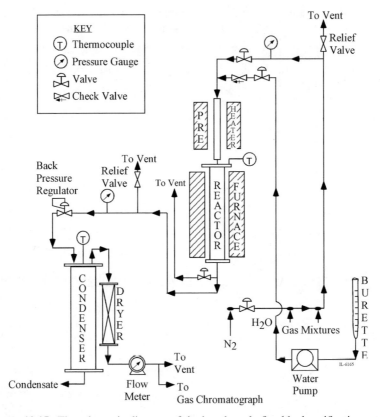

**Figure 10.15.** The schematic diagram of the bench-scale fixed bed gasification setup.

muffle furnace setup for pyrolysis of the litter. Table 10.6 summarizes the moisture, volatiles, fixed carbon, and ash contents of the fresh ("as-received") litter, dry litter, and char.

Major results from their study are summarized as follows:

- Higher gasification temperature provides higher conversion, and >95 percent conversion of carbon can be achieved in about 1 hour at approximately 1350°F–1400°F (732°C–760°C).
- Higher catalyst loading provides higher carbon conversion in the same amount of time. However, increasing the catalyst loading beyond 10–15 wt percent does not seem to enhance the gasification performance significantly.
- Acceptable carbon conversion can be achieved even at 50 psig (0.44 MPa) pressure.
- Fuel gas fraction, $\phi$, that is, the fraction of $CH_4$, $H_2$ and CO in the gasifier efflu-ent (on a dry basis), is consistently higher at 100–300 psig (0.88–2.64 MPa) pressure in comparison to that at 50 psig (0.44 MPa) pressure. The average

**TABLE 10.6. Broiler Litter Composition (in wt%)**

| Fraction | Fresh Litter | Dry Litter | Pyrolyzed Char |
|---|---|---|---|
| Moisture | 28.5 | – | – |
| Volatiles | 39.5 | 55.2 | – |
| Fixed carbon | 17.5 | 24.5 | 54.7 |
| Ash (ignited basis) | 14.5 | 20.3 | 45.3 |
| Insolubles | 2.46 | 3.45 | 7.70 |
| Potassium | 1.86 | 2.61 | 5.82 |
| Calcium | 0.63 | 0.88 | 1.96 |
| Sodium | 0.53 | 0.74 | 1.65 |
| Magnesium | 0.21 | 0.30 | 0.67 |
| Sulfur | 0.10 | 0.14 | 0.31 |
| Nickel | 102 ppm | 143 ppm | 319 ppm |
| Iron | 93 ppm | 131 ppm | 292 ppm |

value of $\phi$ in 100–300 psig (0.88–2.64 MPa) range varied from 0.61–0.64, while at 50 psig (0.44 MPa) it was around 0.53. Thus, there is only a slight advantage in operating the gasifier at a high pressure.

- At 100 psig (0.88 MPa) pressure and higher (15 wt percent) catalyst loading, the fuel gas fraction, $\phi$ is significantly higher than at other loadings. The average value of $\phi$ changes from 0.13 with no catalyst, to ~0.5 at 5 wt percent loading and reaches a value of 0.62 at 15 wt percent loading. Thus, besides improving the gasification rate and carbon conversion, the addition of catalyst up to 15 wt percent loading also increases the fraction of the reducing gases in the fuel gas.

### 10.3.5.2 Emissions Estimates for a Large-Scale Facility Using a Similar Process

Since large-scale testing of the UTSI concept has yet to be performed, measured plant emission data are not available. However, the Electric Power Research Institute has estimated emissions for a 100 MW integrated wood gasification, combined cycle power plant that feeds about 2,670 tons/day of wet wood feedstock (Appel Consultants Inc., 1993). This estimate is given in Table 10.7 and is believed to be similar to poultry litter fed plant. Nevertheless, because of the use of alkali metal and alkaline earth metal salts as gasification catalysts to supplement the ones inherent in the poultry litter itself, the $SO_2$ and $Cl_2$ (also includes HCl) emissions may be significantly lower. Total solid waste (including dry ash and scrubber waste treatment sludge containing 95 percent water) would be about 14 wt% of the input feed stock and will contain mostly wood ash, $K_2O$ and MgO and phosphates. The unburned carbon content of the solid waste will be very low. As a result, this waste would be very ideal for use as a fertilizer (to provide K and P but not much $N_2$ species) or in a cement/concrete manufacturing as a pozzolanic material. The direct contact waste water (from steam condensate and scrubber) containing ammonia and phenol type compounds can be mixed with the feed stock to slurry it and recycled back to the gasifier.

**TABLE 10.7.   100 MW Wood Gasification/Combined Cycle Power Plant Emissions(a)**

|  | Tons/day | lb/MBtu |
|---|---|---|
| Air pollutants |  |  |
| $SO_2$ | 1.17 | 0.079 |
| $NO_x$ | 0.55 | 0.037 |
| CO | 0.55 | 0.037 |
| $CO_2$ | 3,071 | 207 |
| Particulates | 0.017 | 0.0011 |
| Solid waste |  |  |
| Ash | 115 | 7.75 |
| Waste treatment sludge | 258 | 17.4 |

Taken from EPRI report (Appel Consultants, Inc., 1993).

### 10.3.6   Process Economics

**10.3.6.1 Preliminary Assessments**   To assess the economic feasibility of the catalytic steam gasification option for poultry litter. Jones and Sheth (Jones, 1998; Jones and Sheth, 1999) developed a preliminary design for a transportable gasification system to be constructed on a flatbed trailer. The proposed system had a capacity to process 1.5 tons of litter per hour in two parallel fixed bed, downdraft gasifiers, each 1 m in diameter and 2.3 high. The residence time of solids in each reactor was tentatively set at 180 minutes, during which drying, pyrolysis and steam gasification of the litter would be carried out in the same vessel at 700°C (1,300°F) and 345 kPa (50 psia) pressure. Schematic of this transportable gasification system is shown in Figure 10.16. The fuel gas produced from this scheme was estimated to have a heating value of 10,400 kJ/m$^3$ (279 Btu/scf) based on the equilibrium gas composition at the reactor operating conditions. This does not include the heating value of pyrolysis products. The estimated hourly gas production

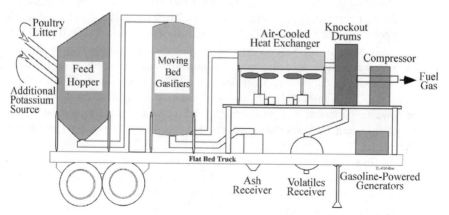

**Figure 10.16.** Schematic of a transportable poultry litter gasification unit.

was about 630 m³ (22,200 scf). This also does not include pyrolysis product gas contribution. A breakdown of the capital investment for this scheme is given in Table 10.8. The authors of the report (Jones, 1998; Jones and Sheth, 1999) estimated the value of the product gas at $17.06 per hour based on the price of $2.75 per million Btu for natural gas at that time in 1998. They estimated the operating costs for this transportable system to be about $2.33 per hour. Based on this, they estimated a payback period of 2.27 years (27 months) to recover the capital investment.

### 10.3.6.2 Assessments of a Stationary System

Later, Turner (2001) developed similar process economics for a considerably more complex but stationary system. He assumed a stationary plant operating for 330 days of the year and processing 100 tons per day of poultry litter. The gasifier was operated at 1,350°F and 100 psig pressure with 10 wt percent catalyst loading. He assumed that inherent level of total catalytic material in the "as-received" litter is about 3 wt percent. As a result, additional 7 wt percent catalyst was assumed to be physically mixed with the litter in the form of fertilizer grade langbeinite ($K_2Mg_2(SO_4)_3$). Litter was assumed to be acquired at $10/ton ($11/metric ton) from the poultry farmers. The cost of transporting it to the gasification facility was assumed to be about $1.00/ton ($1.1/metric ton). Labor costs for associated clean out, loading and unloading along with the operation of the well instrumented and automated system were lumped together and estimated as if there were nine plant workers, one manager/supervisor and one office employee per day. Breakdowns of the total capital investments and operating costs are given in Table 10.9. A plant life of 10 years was assumed. Utilities, including electricity, process water, and cooling water, were costed on the basis of a rough estimate of the facility's needs, in accordance with the standard guidelines used in chemical processing plant design (Peters and Timmerhaus, 1991).

To determine the sensitivity of estimated costs to the price of natural gas and to plant capacity, standard scale-up techniques were used to estimate similar costs for 500 tons/day (454.5 metric tons/day) and 1,000 tons/day (909 metric tons/day)

**TABLE 10.8. Total Capital Investment for a Transportable Broiler Litter Gasification unit**

| Item | Cost (In Dollars) |
| --- | --- |
| Delivered equipment | 64,290 |
| Purchased equipment installation | 16,700 |
| Instrumentation and controls | 5,570 |
| Piping | 13,280 |
| Electrical | 4,280 |
| Total initial capital investment | 104,100 |
| Contingency | 26,000 |
| Total initial investment | 130,100 |

**TABLE 10.9. Cost Breakdown for a Stationary Broiler Litter Steam Gasification Unit**

| Item | Cost (In Dollars) |
| --- | --- |
| **Total capital investment** | |
| Delivered equipment | 989,000 |
| Equipment installation | 514,280 |
| Instrumentation and controls | 118,680 |
| Piping | 178,020 |
| Electrical | 128,570 |
| Building, land, yard improvements | 34,000 |
| Engineering and supervision | 356,000 |
| Construction cost | 360,000 |
| Contractor's fee | 63,500 |
| Process/Project Contingency | 324,500 |
| **Total** | **3,066,550** |
| **Total annual operating costs** | |
| Labor | 634,400 |
| Materials | 452,000 |
| Transportation | 33,000 |
| Utilities | 98,000 |
| **Total** | **1,217,400** |

plants and for gas prices of $14.52 and $9.85 per million Btu ($13.83/GJ to $9.38/GJ) prevailing during the 2000–2001 time period (Peters and Timmerhaus, 1991). Also, gasifier char/residue and other solids collected from the process would amount to about 10 wt percent of the feed and were assumed to be sold at $29 per ton ($31.9/metric ton) as a fertilizer or to cement/concrete manufacturing plant (as a source of pozzolanic material).

A consultant's report for the lower Delmarva Peninsula, which forms the eastern boundary of Chesapeake Bay, puts this value as high as $50–$85 per ton ($55–$93.5 per metric ton) of such solids as fertilizer (Antares Group et al., 1999). Estimates of annual cost, and payback period to recover total capital investment for these different situations are given in Table 10.10.

Clearly, payback time for a bigger plant and the higher of two energy prices is attractive. Payback times could be shortened even more if the federal government were to offer energy incentives for energy production from renewable sources or fuels, or to assess dumping fees from farmers who failed to reprocess their litter.

To determine the effect of such options, additional estimates were made. Assuming poultry litter would be free of charge (instead of $10/ton or $11/metric ton charged in the base case here), but still incurring the transportation cost (at $1/ton or $1.1/metric ton), the payback time at high energy price for a 100 tons/day (90.9 metric tons/day) plant would drop from 6.4 years to about 3.8 years. Similarly, if the fertilizer value of the residue is assumed to be $75/ton or $82.5/metric ton based on the Delmarva Peninsula study (Antares Group et al., 1999), and still

**TABLE 10.10.  Comparison of Annual Cost and Payback Period for Stationary Poultry Litter Steam Gasification Plant at Different Feed Rate and Energy Price**

| Litter Feed Rate (tons/day) | Annual Cost ($/year) | At High Energy Price[a] | | At Low Energy Price[b] | |
|---|---|---|---|---|---|
| | | Gross Revenue ($) | Payback Period (years) | Gross Revenue ($) | Payback Period (years) |
| 100 | 1,984,037 | 2,460,900 | 6.4 | 1,700,000 | —[c] |
| 500 | 8,100,593 | 12,304,500 | 1.9 | 8,500,970 | 20.1 |
| 1,000 | 15,226,037 | 24,609,000 | 1.3 | 17,000,000 | 6.9 |

[a]High energy price corresponds to $14.52 per million Btu ($13.83/GJ).
[b]Low energy price corresponds to $9.85 per million Btu ($9.38/GJ).
[c]Payback period will be negative as there is no net profit under these conditions.

assuming $10/ton or $11/metric ton for litter and $1/ton or $1.1/metric ton for transportation, the payback time for a 100-tons/day (90.9 metric tons/day) plant at high-energy price would drop to about 4.9 years from 6.4 years. The results from this study indicate that the process economics developed by Jones and Sheth for a transportable system will be also attractive if the price of energy increases possibly causing the payback period for a transportable system to drop to a lower value than that estimated earlier. Thus, steam gasification of poultry litter is believed to offer a technically and economically viable option under certain circumstances.

## 10.3.7  Toward Zero Discharge in the Poultry Industry

The conversion of poultry litter to fuel gas is both technically feasible and economically viable. Moreover, the solid residue left after catalytic steam gasification can be used in cement/concrete manufacturing or as fertilizer supplying potassium, magnesium, and phosphorus. Process wastes would not be problem because the contact water and steam condensates are recycled to the gasifier, and fuel gas is very low in $NH_3$ and $H_2S$ type pollutants. Thus the proposed system offers an environmentally acceptable option with very little or even zero discharge to ground and surface waters.

## REFERENCES

Anderson, G. H., and Smith, G. P. (1982). The Exxon process for catalytic coal gasification. Proceedings of the American Gas Association (AGA) Transmission Conference.

Antares Group Inc., T. R. Miles Technical Consulting Inc., and Foster Wheeler Development Corporation (1999). Economic and technical feasibility of energy production from poultry litter and nutrient filter biomass on the lower Delmarva Peninsula. Final Report-Northeast Regional Biomass Program, Washington, DC, CONEG Policy Research Center, Inc.

Appel Consultants Inc. (1993). Strategic analysis of biomass and waste fuels for electric power generation. Report No. TR-102773, prepared for Electric Power Research Institute, Palo Alto, California 94304.

Hauserman, W. B. (1994). High-yield hydrogen production by catalytic gasification of coal or biomass. *International Journal of Hydrogen Energy*, **19**, 413–419.

Hirsch, R. L., et al. (1982). Catalytic coal gasification: an emerging technology. *Science*, **8**, 121–127.

Jones, J. A., and Sheth, A. C. (1999). From waste to energy—catalytic steam gasification of broiler litter. presented and published in the proceedings of the Renewable and Advanced Energy Systems for the 21st Century, an International Conference, sponsored by ASME/ KSME/JSME/SAREK/JSES/KSES, Maui, Hawaii.

Jones, J. A. (1998). From waste to energy-catalytic steam gasification of broiler litter. M.S. thesis, Chemical Engineering, University of Tennessee, Knoxville.

Payne, W. E., and Donald, O. J. (1990). Poultry, waste management and environmental protection manual. Circular ANR-580, The Alabama Co-operative Extension Service, Auburn University, Alabama 36849–5612.

Peters, M. S., and Timmerhaus, K. D. (1991), *Plant Design and Economics for Chemical Engineers*, 4th Ed., McGraw-Hill, Inc., New York, NY.

Poultry Water Quality Consortium (1994). *Poultry Water Quality Handbook.* Chattanooga, Tennessee 37402–280.

Timpe, R. C., and Hauserman, W. B. (1992). The catalytic gasification of hybrid poplar and common cattail plant chars. Proceedings of the Energy from Biomass and Wastes XVI, Institute of Gas Technology (IGT), Chicago, 903–919.

Turner, A. (2001). Determining the optimum conditions for catalytic steam gasification of broiler litter. M. S. thesis, Chemical Engineering, University of Tennessee, Knoxville, TN.

# CHAPTER 11

# MINIMIZATION OF ENVIRONMENTAL DISCHARGE THROUGH PROCESS INTEGRATION

GAUTHAM PARTHASARATHY

Research Specialist, Performance Products Technology, Solutia Inc., 730 Worcester Street, Springfield, MA 01151, USA. GGPART@solutia.com, Tel: 413-730-2255; Fax: 413-730-3610

RUSSELL F. DUNN

Consulting Chemical Engineer, Polymer and Chemical Technologies, LLC, 1431 Glenmore Drive, Cantonment, FL 32533, USA. rdunn@PolymerChemTech.com, Tel: 850-937-8311; Fax: 850-937-8309

In response to the staggering environmental and energy problems associated with manufacturing facilities, the chemical process industry has dedicated much attention and many resources to mitigating the detrimental impact on the environment, conserving resources, and reducing the intensity of energy usage. These efforts have gradually shifted from a unit-based approach to a systems-level paradigm. The past decade has seen significant industrial and academic efforts devoted to the development of holistic process design methodologies that target energy conservation and waste reduction from a systems perspective. Implicit in the holistic approach, however, is the need to realize that changes in a unit or a stream often propagate throughout the process and can have significant effects on the operability and profitability of the process. Furthermore, the various process objectives (e.g., technical, economic, environmental, and safety) must be integrated and reconciled. These challenges call for the development and application of a systematic approach that transcends the specific circumstances of a process and views the environmental, energy, and resource-conservation problems from a holistic perspective. This approach is called *process integration*. It emphasizes the unity of the process and it is broadly divisible into the categories of *mass integration* and *energy integration*.

*Toward Zero Discharge: Innovative Methodology and Technologies for Process Pollution Prevention,* edited by Tapas K. Das.
ISBN 0-471-46967-X   Copyright © 2005 John Wiley & Sons, Inc.

The following sections provide overviews of these topics and their implications in environmental process design and minimization of discharge of pollutants to the environment. While energy integration is briefly introduced, the concept of mass integration is explored in detail.

## 11.1   ENERGY INTEGRATION AND HEAT EXCHANGE NETWORKS

Energy integration deals with all forms of energy such as heating, cooling, power generation/consumption, pressurization/depressurization, and fuel. Much of the effort in this area has been directed toward increasing heat recovery and reducing expensive utility costs in chemical processes. Industrial heat exchange networks "HENs" are of particular importance because of their role in recovering process heat. A HEN is a network consisting of one or more heat exchangers that collectively satisfy the energy conservation and recovery task. Therefore, in most chemical process industries it is necessary to synthesize cost-effective HENs that can transfer heat among the hot and cold streams. During the design stage, temperature specifications for the hot and cold streams must be met and a decision must be made regarding the use of a process stream or an external utility (e.g., steam, cooling water, etc.) to accomplish the required heat duty. Even in relatively simple situations, the problem of pairing and sequencing of exchanger streams becomes a large one calling for application of systematic techniques. Figure 11.1 represents the HEN synthesis task schematically.

For a given system, the synthesis of HENs entails answering several questions including:

- Which heating/cooling utilities should be employed?
- What is the optimal heat load to be removed/added by each utility?
- How should the hot and cold streams be matched (i.e., stream pairings)?
- What is the optimal system configuration (e.g., how should the heat exchangers be arranged? Is there any stream splitting and mixing? etc.)?

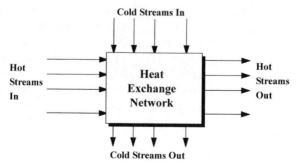

**Figure 11.1.**   Heat exchange network (HEN) synthesis.

Linnhoff and Flower (1978) introduced the HEN problem. Linnhoff and Hindmarsh (1983) summarized the pinch concept as applied to HEN designs, defined the problem, presented solution strategies and analyzed test cases to demonstrate the value of the approach. Subsequently, numerous methods have been developed for the synthesis of HENs. The first step in the heat pinch analysis is the construction of "hot" and "cold" composite curves. A "hot" stream is any stream in the process of interest that needs to be cooled down. A "cold" stream is any stream in the process of interest that needs to be heated up. The composite curves are obtained by the vector addition of individual hot and cold streams in the process. The minimum heating and cooling duties as well as the pinch temperature can be determined from the composite curve diagram. The pinch temperature (or pinch point) corresponds to the most thermodynamically constrained point in the whole system. In order to minimize utility usage, no heat should be passed through the pinch point. Mini-case study 11-1 illustrates the concept.

*Mini-Case Study 11-1: Heat Exchange Network Targeting of Minimum Utilities*  In the chemical process described by Figure 11.2, feed A is pre-heated from 25–125°C and fed to the reactor. Once-through non-contact cooling water is used in the condenser for the reactor overhead vapor stream (i.e., product B) and the cooler for the hot bottoms product stream (i.e., product D). The overhead stream is cooled from 125–37°C while the hot bottoms stream is cooled from 188–30°C. There is a solid–liquid separator following the reactor. Effluent from

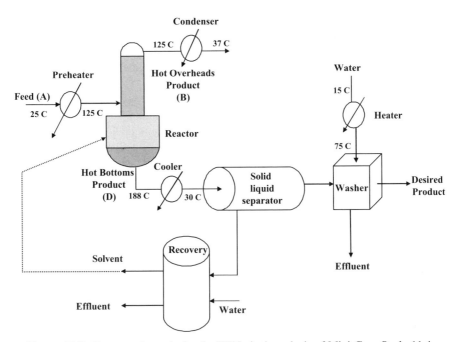

**Figure 11.2.** Process schematic for the HEN pinch analysis of Mini-Case Study 11-1.

the separator is sent to the recovery area for solvent recovery. There is a washer that requires fresh water at a higher temperature as washing is conducted at higher temperatures.

Water is heated from 15–75°C. All heating is provided by steam at 140°C and cooling is provided by cooling water entering at 15°C and exiting around 25°C. The first step in the analysis is to identify the hot and cold streams, whose properties are given in Table 11.1. The initial duties for heating (which is provided by steam) and cooling (which is provided by cooling water) are 2,020 and 2,144 KW, respectively. The pinch diagram for this example is given in Figure 11.3. Figure 11.4 represents a grand composite curve which is another representation of the same information and this diagram is particularly useful for selecting the appropriate utility types needed for the heating and cooling tasks. It may be noted that the minimum heating and cooling duties are 0 and 124 KW, respectively. This is a significant reduction from the initial heating and cooling requirements and demonstrates the potential offered by heat pinch analysis in optimizing the energy flows in a given process.

## 11.2  MASS INTEGRATION

Mass integration is a holistic approach to the generation, separation, and routing of species and streams throughout the process. It allows users to identify global insights, synthesize strategies, and address the root causes of the environmental and mass-processing problems at the heart of the process. Mass integration is a systematic methodology that provides a fundamental understanding of the global flow of mass within the process and employs this understanding in identifying performance targets and optimizing the allocation, separation, and generation of streams and species (El-Halwagi, 1997). In the environmental context, the development of the methodologies for waste reduction (mass integration) has been driven by the promulgation of more stringent environmental regulations coupled with the desire to improve industrial competitiveness. In the past, for waste reduction tasks, the industrial goal was to identify a recovery system that would effectively allow the recycling of certain wastes. This goal was generally accomplished by postulating a variety of process system configurations and operating conditions and then individually screening these alternatives to evaluate their overall economic impact to the

**TABLE 11.1. Stream Data for Hot and Cold Streams for Mini-Case Study 11-1**

| Stream | Description | T Supply (K) | T Target (K) | Enthalpy (KW) | mcp (kW/K) |
|--------|-------------|--------------|--------------|---------------|------------|
| 1 | Hot 1 | 398 | 310 | −880 | 10 |
| 2 | Hot 2 | 461 | 303 | −1,264 | 8 |
| 3 | Cold 1 | 298 | 398 | 1,300 | 13 |
| 4 | Cold 2 | 288 | 348 | 720 | 12 |

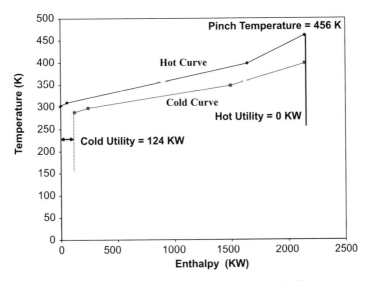

**Figure 11.3.** Pinch diagram for Mini-Case Study 11-1.

company (operating cost, capital investment, etc.). Recent efforts have enabled development of systematic design methodologies that are able to identify a system that accomplishes waste reduction cost-effectively. The primary focus of these efforts has been toward the development of systematic design methodologies for end-of-the-pipe separation and recycle systems (Section 11.2.1) and in-plant separation and allocation systems (Section 11.2.2). Another related area is the

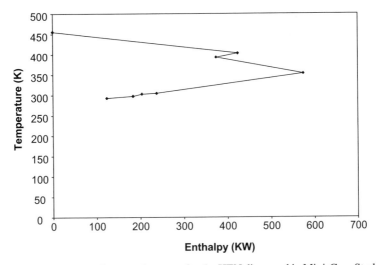

**Figure 11.4.** Process grand composite curve for the HEN discussed in Mini-Case Study 11-1.

development of water optimization networks (Section 11.3) that handle wastewater flows and minimize fresh water demands in a process.

The first papers on mass exchange networks dealt mainly with the design and optimization of end-of-pipe solutions for single and multicomponent ideal systems (El-Halwagi and Manousiouthakis, 1989a). El-Halwagi and co-workers introduced several related problems and provided solution methodologies. In 1996, the concept of waste interception networks (or WIN) was created (El-Halwagi et al., 1996). The WIN synthesis methodology is based on tracking process streams containing an undesirable species and identifying the optimal location(s) to intercept one or more streams with mass-exchangers to achieve a specified waste reduction task. The WIN design technique features the use of direct-contact *mass separating agents* to intercept the undesirable species. WINs were intended to be applied in-process, and they expanded the scope of the mass integration design methodology from separation to allocation possibilities (such as segregation, recycle and mixing).

In 1994, Dunn and co-workers introduced the synthesis of heat-induced separation networks and these methodologies combined aspects of mass integration and heat integration in the design task for end-of-the pipe systems (Dunn and El-Halwagi, 1994a,b; El-Halwagi et al., 1995). In 1997, heat-induced waste minimization networks "HIWAMINs" was introduced (Dunn and Srinivas, 1997). The HIWAMIN methodology is also based on tracking process streams containing an undesirable species and identifying the optimal location(s) to intercept one or more streams with heat-induced separators and heat exchangers to achieve a specified waste reduction and heat integration task. However, the HIWAMIN design technique features the use of indirect-contact, *energy separating agents* to intercept the undesirable species. Furthermore, the HIWAMIN design approach simultaneously addresses the waste minimization and process heat integration (heat exchange network) design tasks. HIWAMINs were meant to be applied in-process and they expanded the scope of the mass and energy integration design methodology. This concept was further expanded to include pressurization/depressurization effects (Dunn and El-Halwagi, 1994b; Dunn et al., 1999). Most of the systems handled were ideal in nature. Parthasarathy and El-Halwagi introduced the problem of mass integration and heat integration as applied to multi-component non-ideal systems. Condensation (Parthasarathy and El-Halwagi, 2000), evaporation and crystallization (Parthasarathy et al., 2001a,b) were studied as part of this effort.

## 11.2.1 End-of-pipe Separation Network Synthesis

### 11.2.1.1 *Mass Exchange Networks*   Separation systems (a combination of unit operations) designed to allow the recycle/reuse of waste streams, or certain constituents in waste streams, are generally envisioned as an end-of-the-pipe process system. The initial thrust to identify the most cost-effective waste separation system from a large group of process options (multiple technologies and/or separating agents for the separation task) resulted in the notion of synthesizing a mass-exchange network "MEN" (El-Halwagi and Manousiouthakis, 1989a;

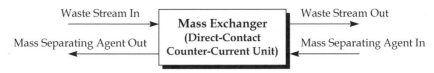

**Figure 11.5.** Schematic of a single mass exchanger for environmental process design.

1990a). A MEN is a network consisting of one or more mass exchangers that collectively satisfy the waste recovery task.

A single mass exchanger, schematically illustrated in Figure 11.5, is a direct-contact, countercurrent unit such as a stripper or ion exchange device that employs a mass-separating agent "MSA" to affect the transfer of the pollutant from the waste stream to the MSA. The design task of synthesizing a MEN is to systematically identify a cost-effective network of mass exchangers for the selective transfer of a certain undesirable species from a set of "rich" (waste) streams to a set of "lean" (MSA) streams. This undesirable species generally represents a pollutant if it is discharged to the environment, but may be a valuable raw material if it can be recovered for reuse within the plant. Figure 11.6 is a general representation of the MEN synthesis task for end-of-the-pipe waste minimization process design and is analogous to the HEN representation in Figure 11.1. It is noted before proceeding that an MSA is any stream that can accept pollutants from a stream. There are *process MSAs* which exist in a process flow sheet that could be used for a mass exchange via direct contact with little to no cost to the company and *external MSAs* such as liquid absorbents, adsorbents and extractants (such as solvents, activated carbon, liquid extractants, etc) that must be purchased.

For a given system, the synthesis of MENs entails answering several questions including:

- Which mass separating agents should be employed?
- What is the optimal mass load to be removed/added by each MSA?

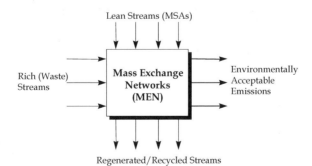

**Figure 11.6.** Mass exchange network (MEN) synthesis (El-Halwagi and Manousiouthakis, 1989a).

- How should the waste and MSA streams be matched (i.e., stream pairings)?
- What is the optimal system configuration (e.g., how should the mass exchangers be arranged? is there any stream splitting and mixing? etc.)?

Mini-case study 11-2 addresses some of these questions. More detailed responses can be found in the literature, which includes additional important categories of the MEN synthesis task:

- MENs for multiple component systems (El-Halwagi and Manousiouthakis, 1989b; Gupta and Manousiouthakis, 1994)
- MENs with regeneration systems (El-Halwagi and Manousiouthakis, 1990b; Garrison et al., 1995)
- MENs with chemical reactions (El-Halwagi and Srinivas, 1992; Srinivas and El-Halwagi, 1994a; Dunn and El-Halwagi, 1993; Warren et al., 1995)
- MENs with heat integration (Srinivas and El-Halwagi, 1994b)
- MENs via a structure-based approach (Papalexandari et al., 1994; Papalexandari and Pistikopoulos, 1994)
- MENs for wastewater reduction (Wang and Smith, 1994; Dunn and El-Halwagi, 1996)
- MENs with flexibility (Zhu and El-Halwagi, 1995)
- MENs for fixed load removal (Kiperstock and Sharratt, 1995)
- MENs with controllability (Huang and Edgar, 1995; Huang and Fan, 1995)

*Mini-Case Study 11-2: Benzene Recovery from Wastewater Emissions in a Chemical Manufacturing Plant* An ethylene/ethyl benzene manufacturing plant has two wastewater emission streams (Dunn and El-Halwagi, 1996). The first wastewater stream, designated $R_1$, is the quench water recycle stream for the cooling tower. The second wastewater stream, designated $R_2$, is the wastewater from the ethyl benzene portion of the plant. Benzene must be removed from stream $R_1$ prior to its recycle back to the cooling tower. Benzene must be removed from stream $R_2$ prior to sending this stream to bio-treatment; hexane ($S_1$) and heptane ($S_2$) are available as process MSAs for the benzene separation

**TABLE 11.2. Data for the Wastewater Streams for Mini-Case Study 11-2**

| Stream | Description | Flow rate (kg/s) | Supply Composition (ppm) | Target Composition (ppm) | Stream Disposition |
|--------|-------------|------------------|--------------------------|--------------------------|--------------------|
| $R_1$ | Wastewater from settling | 80 | 800 | 150 | Recycled to cooling tower |
| $R_2$ | Wastewater from ethyl benzene separation | 140 | 1,500 | 300 | Bio-treatment |

**TABLE 11.3. Data for the Process MSA's Available for the Benzene Separation Task of Mini-Case Study 11-3**

| Stream | Description | Flow Rate (kg/s) | Supply Composition (ppm) | Maximum Target Composition (ppm) | Minimum Mass Transfer Driving Force (ppm) |
|--------|-------------|------------------|--------------------------|----------------------------------|-------------------------------------------|
| $S_1$ | Hexane | 0.5 | 5 | — | 25,000 |
| $S_2$ | Heptane | 0.6 | 10 | 170,000 | 15,000 |

task (Tables 11.2 and 11.3). One external MSA, activated carbon, is also available for benzene separation from the wastewater streams.

The first step toward identifying the cost-effective benzene separation system is to generate the mass-exchange pinch diagram. First, the two waste streams are individually plotted and then combined to form a composite stream as shown by Figure 11.7. Next, the process MSAs are plotted in a similar manner using the equilibrium data and mass transfer composition driving forces provided to relate benzene composition in the process MSAs to benzene composition in the waste streams, also shown on Figure 11.7. Finally, the wastewater composite stream and MSA composite stream are combined in the same figure. The MSA composite stream is moved

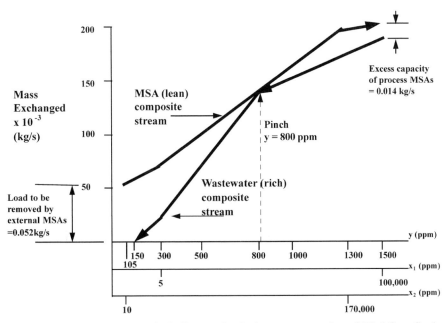

**Figure 11.7.** Mass exchange pinch diagram for the benzene separation of Mini-Case Study 11-2. Equilibrium data for the transfer of benzene from wastewater ($y$) to hexane ($x_1$) and heptane ($x_2$) are as follows: $y = 0.012x_1$ and $y = 0.007x_2$.

vertically such that the stream touches but does not cross the wastewater composite stream. This allows the designer to determine the amount of integrated mass exchange (removed by the process MSAs) and the amount to be removed by the external MSA (activated carbon), again as indicated by Figure 11.7. From Figure 11.7, the cost-effective benzene separation system is determined and a schematic of the separation network is shown as Figure 11.8. It is worth pointing out that this network features the minimum operating cost. Systematic techniques (e.g., mass-load paths, El-Halwagi and Manousiouthakis, 1989a,b) can be used to reduce the number of the units and the fixed cost at the expense of increasing operating cost.

### 11.2.1.2 Heat-Induced and Energy-Induced Separation Networks

Synthesis techniques have also been developed for other separations systems that are traditionally used for waste minimization via end-of-the-pipe applications. For instance, there is a wide class of separation systems that employ energy separating agents "ESAs" (hot and cold process streams and/or utilities such as steam and cooling water) to separate species from a waste stream via a phase change and are analogous to the process streams in MSAs. These include indirect contact unit operations, which employ an energy-separating agent (ESA) for the separation of certain species via phase change, such as condensers, evaporators, dryers and crystallizers collectively grouped as heat-induced separators. A general schematic of a single heat-induced separator is included as Figure 11.9. An ESA is any hot or cold stream that can be used in an indirect-contact unit operation to heat or cool another stream and/or bring about changes in the pressure imposed on a stream. There are *process ESAs* or streams existing within a process flow sheet that could

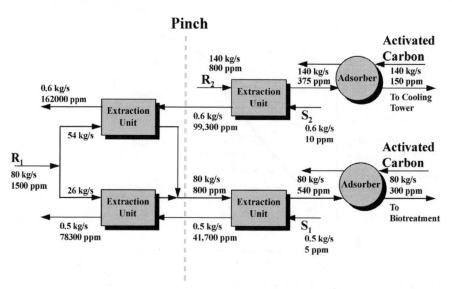

**Figure 11.8.** Separation network for the benzene separation process discussed in Mini-Case Study 11-2.

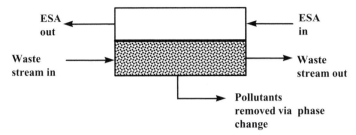

**Figure 11.9.** Schematic of a single heat-induced separator for environmental process design. (Reprinted with permission from Advances in Environmental Research, Vol. 2, Dunn et al, "A Spreadsheet-Based Approach to Identify Cost-Effective Heat-Induced and Energy-Induced Separation Networks for Condensation-Hybrid Processes," pp. 270–281, Copyright 1998, with permission from Elsevier)

be used for a heating or cooling task with no energy cost to the company and *external ESAs* such as hot and cold utilities (steam, hot oil, cooling water, refrigerants, etc.) and pressure (accomplished via compressors, turbines, vacuum pumps, etc.) that have a cost associated with their production and use. The notion of synthesizing heat-induced separation networks (HISENs) for recycle/reuse waste minimization process design encompasses the task of identifying a cost-effective system of heat-induced separators and heat exchangers that can achieve a specified waste reduction task (single component or multiple component waste streams) by heating/cooling the streams to produce a phase separation (Dunn and El-Halwagi, 1994a,b; Dye et al., 1995; El-Halwagi et al., 1995).

The original HISEN synthesis methodology has been generalized to include pressurization/depressurization equipment in conjunction with heat-induced separation specifically to address gaseous emissions containing volatile organic compounds "VOCs" and the resulting recovery system was referred to as an energy-induced separation network "EISEN" (Dunn et al., 1995). One of the simplest techniques for VOC recovery is to use heat-induced separation networks to affect condensation via cooling (El-Halwagi et al., 1995; Richburg and El-Halwagi, 1995; Dunn and El-Halwagi, 1994a,b). However, VOC condensation is a function of both temperature and pressure. Hence, a more general condensation design task should capitalize on the synergism between cooling and pressurization/depressurization. This synergism was addressed by the design methodology of Dunn and co-workers whose objective was to create a cost-effective network of heat-induced separators, heat exchangers and pressurization/depressurization devices, which can separate one or more species from a set of waste gas streams via phase change.

For a given system, the synthesis of HISENs/EISENs entails answering several questions including:

- Which energy separating agents should be employed?
- Should stream pressurization or depressurization be employed and, if so, to what level?
- What is the optimal mass and heat load to be removed/added by each ESA?

- How should the waste and ESA streams be matched (i.e., stream pairings)?
- What is the optimal system configuration (e.g., how should the heat-induced separators, heat exchangers and compressors/turbines be arranged? is there any stream splitting and mixing? etc.)?

Over the past few years, several important categories of the HISEN and EISEN synthesis task have been identified and addressed as summarized below:

- HISENs for single component VOC condensation systems (Dunn and El-Halwagi, 1994a)
- A shortcut graphical approach for HISENs for single component VOC condensation systems (Richburg and El-Halwagi, 1995)
- HISENs for multiple component VOC condensation systems (Dunn and El-Halwagi, 1994b)
- HISENs for fixed load removal (El-Halwagi et al., 1995)
- Hybrid HISEN and membrane systems (Crabtree et al., 1995; 1998)
- A spreadsheet-based approach for identifying cost-effective HISENs and EISENs for condensation-hybrid processes (Dunn and Dobson, 1998; Dobson, 1998)
- HISENs for crystallization systems (Multicomponent, non-ideal) (Parthasarathy et al., 2001a)
- HISENs for infinite component VOC condensation systems using clusters (Shelley and El-Halwagi, 2000)

## 11.2.2   In-Process Separation Network Synthesis

*11.2.2.1 WIN, HIWAMIN, and EIWAMIN*   Thus far, the design and optimization of end-of-pipe separation systems have been considered. There have also been several systematic methodologies developed for the design of cost-effective waste reduction systems based on in-plant modifications, such as the synthesis of waste interception and allocation networks "WINs" (El-Halwagi et al., 1996; El-Halwagi, 1997), the synthesis of heat-induced waste minimization networks "HIWAMINs" (Dunn and Srinivas, 1997), and the synthesis of energy-induced waste minimization networks "EIWAMINs" (Dunn et al., 1999). The WIN synthesis methodology, schematically represented in Figure 11.10, is based on tracking streams containing an undesirable species within a process, and identifying the optimal location(s) to intercept one or more in-plant streams with mass-exchangers to achieve a specified waste reduction task. The WIN design technique features the use of direct-contact *mass separating agents* to intercept the undesirable species.

A particularly useful framework for representing mass integration strategies is to represent the flow sheet from the species perspective as shown in Figure 11.11 (Garrison et al., 1995; El-Halwagi et al., 1996; El-Halwagi and Spriggs, 1996; and Hamad et al., 1996). This representation focuses on the role of species, sources and sinks in the process. Sources are process streams that carry the targeted

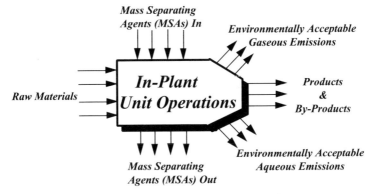

**Figure 11.10.** WIN synthesis representation.

species. Sinks are process units that are capable of processing the sources carrying the targeted species. Sinks include reactors, separators, heaters, coolers, pumps, compressors, pollution-control facilities, discharge media, and the like. Streams leaving the sinks, in turn, become sources. Therefore, sinks also are generators of the targeted species. Each sink/generator can be manipulated via design or operating changes to alter the flow rates and compositions that it can accept and that it discharges. It may be necessary to modify compositions of sources to prepare them for the sinks. This is done in a network of separation units referred to as the waste interception network (WIN). Mass integration, therefore, involves a combination of stream segregation, mixing, interception, recycle from sources to sinks (with or without interception), and sink/generator manipulation. *Segregation* is simply avoiding stream mixing. If streams with different compositions are segregated, it may later be possible to recycle them directly to sinks without further costly processing.

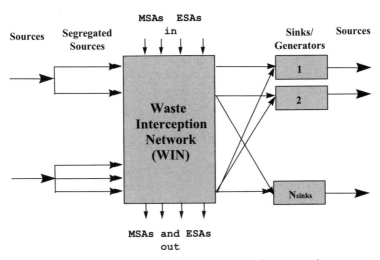

**Figure 11.11.** Mass integration from a species perspective.

*Mixing* can be used to reach appropriate flow rates and compositions. *Interception* is the use of separation technologies to adjust the compositions of the species-laden streams to make them acceptable for the sinks. As has been mentioned earlier, these separations may be effected by the use of MSAs, ESAs or membranes. Identifying the right combination for a WIN can be a large and complex problem because numerous streams typically must be processed, many separation technologies may be applicable, and initially it is not known how much of a species must be removed to make that stream suitable for a sink. Therefore, a systematic technique is needed to screen the candidate separating agents and separation technologies to find the optimal WIN. *Recycle* refers to the routing of a source to a sink. Each sink has a number of constraints on the flow rates and compositions of feeds that it can process. If a source satisfies these constraints, it may be recycled directly to the sink. If the source violates these constraints, segregation, mixing, or interception may be used to prepare the stream for recycle. Recycling of streams also results in the build up of species in the process. This is modeled by means of a graphical tool called a path diagram (El-Halwagi et al., 1996). This diagram relates streams containing the pollutant to the performance of the unit operations in which they are processed by plotting the composition of the pollutant versus its mass load (i.e., the product of composition and flow rate). *Sink/generator manipulation* involves design or operating changes that alter the flow rates and compositions of sources entering or leaving the sinks. These measures can include temperature or pressure changes, unit replacement, catalyst alteration, feedstock, solvent and product substitution (Joback and Stephanopoulos, 1989; Constantinou et al., 1995; Vaidyanathan and El-Halwagi, 1996; Hamad and El-Halwagi, 1998), reaction changes (Rotstein et al., 1982; Crabtree and El-Halwagi, 1995; Hildebrandt and Biegler, 1995).

The HIWAMIN methodology is also based on tracking streams containing an undesirable species within a process and identifying the optimal location(s) to intercept one or more streams with heat-induced separators and heat exchangers to achieve a specified waste reduction and heat integration task. However, unlike the WIN methodology, the HIWAMIN design technique features the use of indirect-contact, energy separating agents to intercept the undesirable species. Furthermore, the HIWAMIN design approach simultaneously addresses the waste minimization and process heat integration (heat exchange network) design tasks. The EIWAMIN design technique extends the HIWAMIN approach to include the use of pressurization and/or depressurization devices in addition to heat-induced separators and heat exchangers to further improve separation efficiencies and the cost-effectiveness of the final design.

The operating cost of an EIWAMIN includes heating utilities, cooling utilities, and the electricity required to support the plant heating and cooling requirements and additional pressurization/depressurization devices. The minimum annual operating cost is determined using a mixed-integer nonlinear programming (MINLP). The second stage of the design procedure is to identify the minimum number of units (heat-induced separators, heat exchangers, compressors and turbines) that satisfy the minimum operating cost solution determined previously.

The minimum number of heat-induced separators plus heat exchangers can be identified using the HISEN/HIWAMIN formulation for minimizing the number of units (El-Halwagi et al., 1995; Dunn and Srinivas, 1997), with modifications to include pressurization and/or depressurization units.

### 11.2.2.2 Case Study: HIWAMIN and EIWAMIN Synthesis in a Polymer Manufacturing Facility

The polymeric resin production plant represented in Figure 11.12 uses a reaction/separation process (Dunn et al., 1999). Monomers and solvent are fed to a polymerization reactor where the polymer, water and a by-product (BP) are formed and are then present, with the remaining solvent, in the reactor exit stream. This stream is then processed via a series of separation processes to allow the separation and purification of the various constituents in this stream.

It is desired to design a waste minimization network that will result in an overall solvent discharge in the Stripper wastewater stream that is 99 percent by weight less than the current discharge levels. Additionally, to meet more stringent wastewater discharge regulations, it is necessary to reduce the overall wastewater hydraulic load by at least 70 percent. The solution should also achieve maximum heat integration within the process. Total heating and cooling requirements for the initial

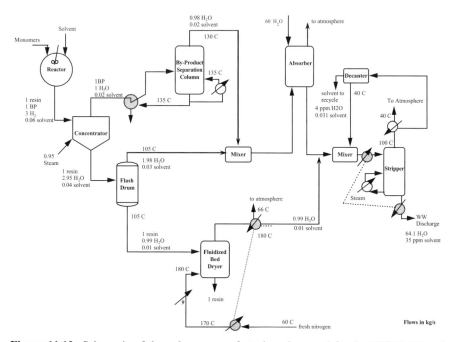

**Figure 11.12.** Schematic of the polymer manufacturing plant used for the HIWAMIN and EIWAMIN case study. (Reprinted with permission from Clean Products and Processes, Vol. 1, Dunn et al, "Synthesis of Energy-Induced Waste Minimization Networks (EIWAMINs) for Simultaneous Waste Reduction and Heat Integration," Figure 12, pp. 91–106, Copyright 2001, with permission from Springer-Verlag.)

process are 32,063 kW and 27,744 kW, respectively. Application of the HIWAMIN methodology for the polymer plant results in the process design solution presented in Figure 11.13. The solution includes the following features:

1. The overhead product stream of the BP Separation Column is totally condensed and the heat is used to provide the heat duty to the bottoms of the stripper (simultaneous separation and heat integration).
2. The overhead product stream of the Flash Drum is partially condensed (37 percent).
3. The flow rate of absorbent (water) in the absorber is reduced by 78 percent as a result of the above condensation practices.
4. The flow rate of wastewater from the stripper is reduced by 72 percent with a composition of 1 ppm by weight.

The new heating and cooling requirements for the plant are 2,350 kW and 2,719 kW, respectively. This solution represents a 92 percent reduction in the total plant utility costs, in addition to reducing the plant wastewater emissions to meet discharge regulations. The solution of the case study using the EIWAMIN

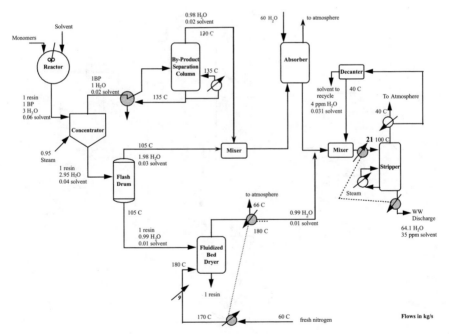

**Figure 11.13.** Optimal HIWAMIN solution for the polymer plant case study. (Reprinted with permission from Clean Products and Processes, Vol. 1, Dunn et al, "Synthesis of Energy-Induced Waste Minimization Networks (EIWAMINs) for Simultaneous Waste Reduction and Heat Integration," Figure 14, pp. 91–106, Copyright 2001, with permission from Springer-Verlag.)

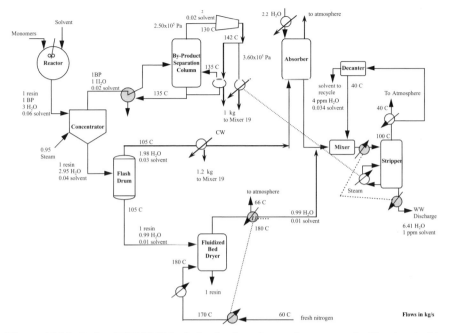

**Figure 11.14.** Optimal EIWAMIN solution for the polymer plant case study. (Reprinted with permission from Clean Products and Processes, Vol. 1, Dunn et al, "Synthesis of Energy-Induced Waste Minimization Networks (EIWAMINs) for Simultaneous Waste Reduction and Heat Integration," Figure 16, pp. 91–106, Copyright 2001, with permission from Springer-Verlag.)

design technique is presented in Figure 11.14. The solution includes the following features:

1. The overhead product stream of the BP Separation Column is compressed, elevating the stream temperature from 130–142°C.

2. The overhead product stream of the BP Separation Column is totally condensed and the heat duty from this condensation is used to provide reboiler duty for the BP Separation Column and the Stripper (simultaneous separation and energy integration).

3. The overhead product stream of the Flash Drum is partially condensed (60 percent).

4. The absorbent (water) in the Absorber is cooled down to 15°C using chilled water to enhance the thermodynamic effectiveness of the absorption process.

5. The flow rate of absorbent (water) in the absorber is reduced by 96.3 percent as a result of the previously mentioned condensation and cooling practices.

6. The flow rate of wastewater from the stripper is reduced by 90 percent with a composition of 1 ppm by weight.

The new heating and cooling requirements are 1,000 kW and 3,038 kW, respectively. The additional electricity requirement for the compressor is 0.03 kW. Hence, the total utility cost of the EIWAMIN solution is $184,000/year. This represents a total utility cost reduction of 39 percent over the HIWAMIN solution and 95 percent reduction in utility cost over the original plant operating conditions. Details of the HIWAMIN design methodology can be found in literature (Dunn and Srinivas, 1997).

### 11.2.3   Multicomponent Nonideal Mass Integration

Although significant advances have been accomplished in developing systematic techniques for mass integration, most of these techniques have been developed to address only single component ideal systems. Very few attempts have targeted mass integration for *multi-component non-ideal* systems. In these few attempts, it was generally assumed that each component behaves independently of the other species. In many industrial applications, there is a strong interaction among the species. This non-ideal interaction can govern the process performance and must be taken into account. In response to this need, Parthasarathy and El-Halwagi, 2000 and Parthasarathy et al., 2001a,b presented condensation of multiple VOCs and evaporation crystallization of ternary wastewater (i.e., water and two salts) systems. Both these systems consist of multiple species in non-ideal systems. In fact, the presence of non-idealities drives the final design of the network. The solution strategies include several techniques such as active constraints, graphical tools and bounding.

***11.2.3.1 VOC Condensation and Allocation***   Earlier efforts at designing VOC condensation networks were limited by a focus on condensing VOCs from the gas perspective only. However multiple VOCs whose interaction is non-ideal call for a more complex treatment. Therefore, for the system schematically represented in Figure 11.15, separation tasks were defined, and outlet gas compositions chosen, independent of potential usage of the liquid VOC condensate. The proposed framework offers the following merits:

- It provides a global tracking of the different species (e.g., VOCs) throughout the process and describes the interaction among the multiple components.
- It integrates the usage of condensed VOCs with the definition of the condensation tasks.
- It reconciles the purchase of fresh VOCs with the condensation of optimal quantities from gaseous sources.

The problem statement is described subsequently. The process under consideration has a set of gaseous sources, each containing a carrier gas and a set of targeted VOCs. The total flow rate of each gaseous source is known along with its supply composition and target composition. These target compositions are based on process considerations that can be technical (e.g., to meet composition requirements

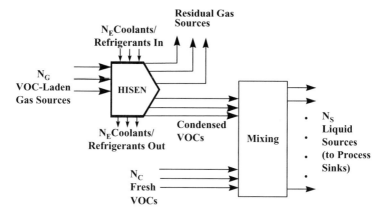

**Figure 11.15.** Multicomponent nonideal VOC condensation and allocation problem. (Reprinted from Chemical Engineering Science, 55, Parthasarathy, G., El-Halwagi, M., Optimum Mass Integration Strategies for Condensation and Allocation of Multicomponent VOCs, 881–895 © 2000 with permission from Elsevier)

of units accepting residual gas streams), environmental (e.g., to satisfy emission standards) and safety (e.g., to stay away from flammability limits). These target compositions represent upper bounds on the outlet compositions of residual gaseous streams. The separation of VOCs from the gas sources is induced via condensation using a set of candidate coolants and refrigerants. The cooling temperature of each ESA and the cost of the refrigerant are known. The condensed VOCs can be utilized in a set of process sinks (units). For each sink, the required inlet flow rate and composition are given. Also available for service is a set of fresh liquid VOCs that can be purchased at a known unit cost. The condensed and purchased VOCs are to be mixed and allocated to meet the flow rate and composition requirements of each sink. The objective of the design task is to identify optimum strategies for condensation and allocation of the condensed VOCs for reuse in process sinks as well as strategies for mixing condensed and purchased VOCs so as to minimize the total cost of VOC purchase and condensation.

*Mini-Case Study 11-3: Design of a System for the Separation and Allocation of Multicomponent Nonideal VOCs in an Adhesive Tape Manufacturing Process*  The manufacture of adhesive and magnetic tapes involve the use of significant amounts of VOCs (Dunn and El-Halwagi, 1994b). The adhesive tape manufacturing process represented in Figure 11.16 uses a mixture consisting mainly of methyl iso-butyl ketone (MIBK) with some methanol (MeOH). A slurry is formed by mixing the solvent with various coating ingredients. The slurry, suspended with resin binders and other additives, is deposited on a base film. Nitrogen gas is used to induce evaporation of the mixed solvent from the film during the coating and drying operations. The dry film is finally sent to finishing.

It is desired to integrate the process and synthesize a minimum cost solvent recovery system based on condensation that will optimize fresh solvent usage and

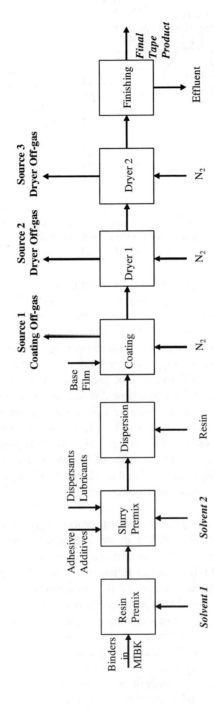

**Figure 11.16.** Adhesive tape manufacturing process used in Mini-Case Study 11-3. (Reprinted from Chemical Engineering Science, 55, Parthasarathy, G., El-Halwagi, M., Optimum Mass Integration Strategies for Condensation and Allocation of Multicomponent VOCs, 881–895 © 2000 with permission from Elsevier)

simultaneously meet all other constraints (environmental, process, safety etc.). There are two liquid sinks in the process; the resin premix stage and the slurry premix stage which require mixed liquid solvent at different compositions and flow rates. There are three potential VOC-laden gaseous sources. These are the streams leaving the coating operation and the two dryers. The final tape product and the finishing effluent also contain the solvent but this cannot be recovered and recycled. There is a trace impurity present in the gaseous streams that has a freezing point of $-14°C$. The source flow rates and compositions and process constraints on different sources and sinks are given. Two external refrigerants are considered; liquid nitrogen and cooling water. The optimal network will meet environmental constraints and optimize the fresh-solvent requirement simultaneously. Following several iterations of the global optimization algorithm (Parthasarathy and El-Halwagi, 2000), the solution represented by Figure 11.17 was obtained. The final optimal HISEN for the gaseous sources has a minimum total annualized cost of \$358,715/year. Since the process initially consumed 248 kg/h fresh MIBK and 98 kg/h fresh MeOH at a cost of \$803,300/year, the solution demonstrates substantial savings. After the analysis, the fresh solvent requirement was reduced to 2 kg/h fresh MIBK and 46 kg/h fresh MeOH.

### 11.2.3.2. Design and Optimization of a Crystallization Evaporation Network for a Ternary Wastewater System
Crystallization has been extensively studied as a unit operation and finds broad application in various industries (Kirk-Othmer, 1993) including separation (e.g., separation of p-xylene from o-xylene and m-xylene), concentration (e.g., concentration of fruit juice), solidification (e.g., modification of the appearance of sugar), purification (e.g., separation of an essential amino acid, l-isoleucine from a fermentation broth) and analysis (e.g., determination of the molecular structure). Crystallization operations play a vital role in product recovery and are capable of producing very high purity products from impure solutions. Another attractive feature of crystallization is that it generally requires less energy for separation as compared to distillation or other commonly used methods of purification. Another important application of crystallization is in its value in lowering pollution (Sharratt, 1996). For instance, crystallization can be used to separate salts from aqueous effluent streams so that these streams may be recycled and re-used in the process.

Let us consider the ternary system shown in Figure 11.18; it consists of water and two salts which may be pollutants. Since most chemical plant operation personnel would prefer not to implement multiple interceptions of multiple streams, interception (via evaporation and crystallization) is assumed to be implemented one node at a time (El-Halwagi, 1997). Simultaneous interception of multiple nodes may be considered, however, by allowing intercepted compositions of multiple nodes to be treated as optimization variables in the same formulation (El-Halwagi et al., 1996; El-Halwagi, 1997). Each stream upon interception results in three product streams namely, a pure water stream, a salt stream, and a mother liquor stream. The objective of the design task is to minimize the total cost of fresh water and evaporation and crystallization. This entails identifying optimum strategies for

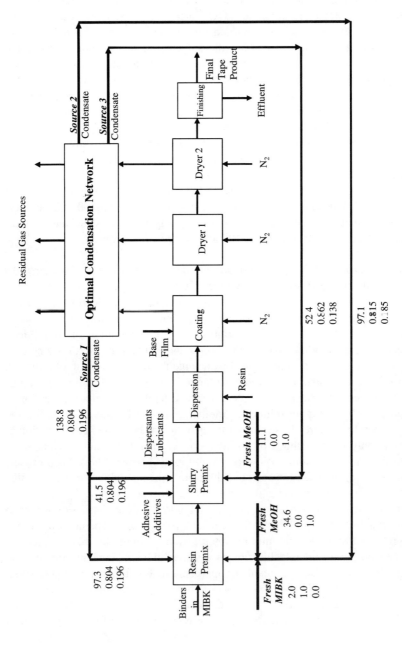

**Figure 11.17.** Optimal solution to Mini-Case Study 11-3. (The numbers for each stream are ordered as flows in kg/h, then mass fraction of MIBK, and finally mass fraction of MeOH.) (Reprinted from Chemical Engineering Science, 55, Parthasarathy, G., El-Halwagi, M., Optimum Mass Integration Strategies for Condensation and Allocation of Multicomponent VOCs, 881–895 © 2000 with permission from Elsevier)

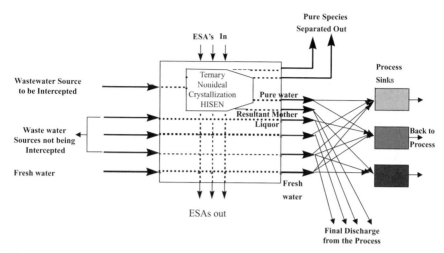

**Figure 11.18.** Problem representation for evaporation crystallization network design and optimization.   (Reprinted with permission from Industrial & Engineering Chemistry Research, Parthasarathy, G., Dunn, R. F., El-Halwagi, M., "Development of Heat-Integrated Evaporation and Crystallization Networks for Ternary Wastewater Systems – Part II Interception Task Identification for the Separation and Allocation Network", 40(13), 2842–2856 © 2001, American Chemical Society)

evaporation and crystallization of the wastewater sources to meet pre-specified separation tasks and developing strategies for mixing sources from evaporation and crystallization with external fresh water to meet allocation tasks.

For a process with several wastewater sources containing two salts and several sinks (with known inlet flow rate and composition constraints for both salts) that can accept the potential sources that may be recycled and re-used, it is desired to determine minimum cost strategies for segregation, mixing, recycle and interception (via evaporation and crystallization) one node at a time, that would reduce the discharge of water and the two salts from the process to a pre-specified level. It is expected that implementing the mass integration strategies of segregation, mixing, recycle and interception (via evaporation and crystallization) will lower the final discharge of all species of interest. This reduction in mass load of water and salts is based on some combination of technical, environmental and operational process considerations. Additionally, there are allocation requirements to be satisfied by the sources present. The wastewater sources are processed through a set of process equipment (or sinks) that is a subset of the entire flow sheet.

Process sinks that require fresh water and which are considered as part of the overall allocation problem form another critical subset. The sources out of the evaporation and crystallization network and external fresh water are to be mixed and allocated to meet the flow rate and composition requirements of each sink. The separation of the species of interest from the wastewater streams is accomplished via evaporation and crystallization using a set of energy separating agents consisting of candidate coolants and heating media, respectively. A case study is presented subsequently in Section 11.2.3.3 to demonstrate this approach.

### 11.2.3.3 Case Study: Ternary Wastewater System In an Ammonium Nitrate Manufacturing Process

In the process (depicted in Figure 11.19) for making ammonium nitrate, a widely used fertilizer, nitric acid is neutralized with anhydrous gaseous ammonia in an exothermic reaction (Chadwick, 1980). Some nitric acid laden gas streams are scrubbed in scrubber 2 with fresh water. The acid-laden wastewater stream from scrubber 2 is sent to neutralization. Following reaction, the ammonium nitrate solution contains between 78 percent and 84 percent of the salt. The solution is concentrated in a concentration step. Following addition of appropriate additives, the ammonium nitrate melt is sent to the prilling tower where it is sprayed against a hot air stream resulting in the formation of prills. The exiting gas streams, containing water, ammonium nitrate and sodium nitrate, are sent to scrubbing before being discharged to the atmosphere. The scrubbing is carried out in two scrubbers operating in series. The first scrubber (scrubber 3)

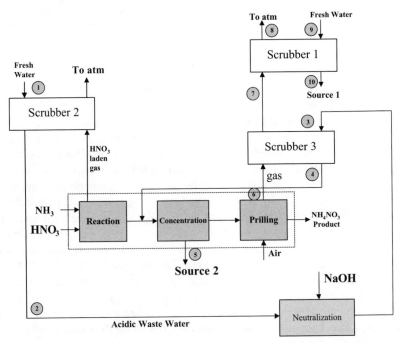

**Figure 11.19.** Ammonium nitrate manufacturing process in Section 11.2.3.3. (Reprinted with permission from Industrial & Engineering Chemistry Research, Parthasarathy, G., Dunn, R. F., El-Halwagi, M., "Development of Heat-Integrated Evaporation and Crystallization Networks for Ternary Wastewater Systems – Part II Interception Task Identification for the Separation and Allocation Network", 40(13), 2842–2856 © 2001, American Chemical Society)

uses the water stream from neutralization of the acid-laden wastewater stream from scrubber 2 while fresh water is used in the other scrubber (scrubber 1). The scrubber 3 effluent is concentrated to recover any ammonium nitrate present in the stream.

The species of interest are water, ammonium nitrate (salt A) and sodium nitrate (salt B). There are two main wastewater streams being discharged from the process namely, the effluent from scrubber 1 and the condensed water stream from the concentration step. There are four gaseous sources, three in-plant liquid sources and two terminal liquid sources in the flow sheet in Figure 11.19. Since stream 2 contains nitric acid, this in-plant liquid stream cannot be considered for potential interception. The two sinks in the process requiring fresh water have known bounds on acceptable flow rates (kg/h) and compositions of water streams going into each of them. It is desired to lower the discharge as follows: water 75 percent; salt A 65 percent and salt B 50 percent. To achieve these discharge reductions, we invoke the mass integration strategies of segregation, mixing, recycle and interception (via evaporation and crystallization) at minimum cost. The non linear optimization problem in this case study was solved using the LINGO™ software. The first program identified reductions achievable using various allocation strategies (i.e., segregation, mixing and recycling) without any interception. The results were reductions in discharge as follows: water 60 percent; salt A 50 percent and salt B 7 percent.

Because the desired targeted reductions were not achieved, interception of two in-plant sources (streams 3 and 4) and two terminal sources (streams 5 and 10) was considered one stream at a time. Iterations of the solution algorithm demonstrated that two of the sources (3 and 10) gave infeasible solutions, i.e., these sources could not be intercepted in a way that would achieve the overall required reductions in discharge of all species of interest. However, both the remaining two sources; one in-plant (stream 4) and one terminal (stream 5); were feasible. The former stream was selected for interception as interception of the latter stream required removal of all ammonium nitrate present in the stream resulting in an ammonium nitrate composition of near zero in the exit stream. Besides, the initial concentration of ammonium nitrate present in the terminal stream was so low that the solution would be economically unfavorable versus the solution for the in-plant stream. Thus, the benefits of in-plant interception, as compared to traditional end-of-pipe treatment, are apparent. Generally, although the loads being separated may be the same, in-plant interception offers the significant benefit of higher stream concentrations of the species of interest thus leading to a cheaper interception.

Thus, the optimal solution lies in using evaporation and crystallization to intercept one of the in-plant streams. The solution strategy for separation and allocation which satisfied overall required reductions in discharge targets is given in Figure 11.20. It may be noted that terminal stream 5 is discharged with zero composition of ammonium nitrate in the solution. The in-plant interception resulted in reduction of undesirable discharge (and loss) of product to the environment. Another useful consequence is the reduction in fresh water demand. The initial fresh water requirement of 136 kg/h was reduced to 59 kg/h signifying a reduction

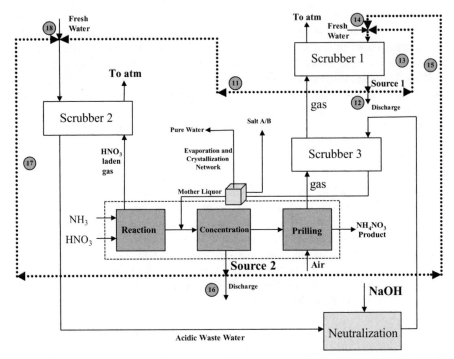

**Figure 11.20.** Solution to the ammonium nitrate case study of Section 11.2.3.3. (Reprinted with permission from Industrial & Engineering Chemistry Research, Parthasarathy, G., Dunn, R. F., El-Halwagi, M., "Development of Heat-Integrated Evaporation and Crystallization Networks for Ternary Wastewater Systems – Part II Interception Task Identification for the Separation and Allocation Network", 40(13), 2842–2856 © 2001, American Chemical Society)

of about 57 percent. Also, interception of the in plant stream will result in a pure water stream. This pure water stream can be used to replace fresh water and reduce the demand of fresh water further.

## 11.3  WATER OPTIMIZATION AND INTEGRATION

Among the various species in a chemical process, water is critical and is widely used across several industries. Water is a relatively inexpensive and useful solvent which is easy to handle and presents minimal toxicity and safety concerns. In a typical chemical process, water may play several roles including:

- It may act as a solvent for reaction (such as in single or multiple phase reactions).
- It may be a solvent for separations (such as in decanters and liquid-liquid extraction systems).

- It may be used for heat removal operations (such as a non-contact stream in condensers).
- It may be used in post-production operations (such as in washing operations wherein, the final product may be washed to remove undesirable solvents, by-products and unreacted raw materials).
- It may be used in environmental systems (such as in scrubbers that handle gaseous emissions from vents and process units).
- It may be used in safety systems (such as to provide seals, heels and in pump seal systems).
- It may be used to generate utilities in the process (such as steam generation).

Wastewater reduction and water conservation are becoming increasingly more important issues in process industries. More stringent environmental regulations, concerns over long-term health effects on humans and nature, and the future availability of "clean" water resources are just a few of the factors that are driving efforts toward improvements in water conservation and wastewater reduction in manufacturing processes. As these issues continue to receive intense government scrutiny and to raise concern among community-organized environmental groups, industries that fail to address the problems to the satisfaction of such outside stakeholders may find present operations threatened and the sustainability of future operations in doubt.

These critical concerns have refocused efforts over the past decade toward identifying cost-effective wastewater reduction and water conservation process designs, involving direct recycle and reuse of water that can be implemented within a variety of process industries. Several recent research efforts have yielded process design methodologies and tools that are generic and can be utilized by petroleum processors, pulp and paper manufacturers, plastics producers, food producers, makers of pharmaceuticals and specialty chemical, industrial laundries and fiber dyers, and many others. The design methodologies range from graphical approaches, including "water pinch" analysis and the source-sink graphical methodology to approaches based on mathematical optimization.

Wang and Smith who introduced the concept of water pinch analysis (1994) later incorporated flow rate constraints into the analysis (1995). Also of note are techniques to minimize water demand and wastewater generation (Dhole et al., 1996) and a new procedure for the design on water networks for single contaminant systems based on use of a load table (Olesen and Polley, 1997). Kuo and Smith (1997) addressed the problem of distributed effluent treatment systems for both single and multiple contaminant cases. Alva-Argaez et al., (1999) described a multi-component transshipment model for wastewater minimization problems.

Polley and Polley (2000), who provide generic rules to design better water networks, describe several simple mathematical equations which can be easily incorporated into a spreadsheet to converge onto the minimum water consumption target. Hallale (2002) introduced a new graphical targeting method for water minimization using water composite curves and the concept of water surplus. The source sink methodology discussed earlier in this chapter has also applied to the reduction of water demand and wastewater generation (El-Halwagi, 1997; Parthasarathy and Krishnagopalan, 2001; Dunn et al., 2001). Several papers have also considered application of various design methodologies towards optimization of water and wastewater streams in industries including pulp and paper (Hamad et al., 1998; Parthasarathy and Krishnagopalan, 1999, 2001), textiles (Ujang et al., 2002), chlor-alkali processes (Gianadda et al., 2002). For a given system, the synthesis of wastewater reduction and water conservation networks entails answering several questions including:

- Which wastewater streams utilities should be recycled or reused?
- What is the optimal load of each wastewater stream to be recycled or reused?
- What is the optimal allocation of wastewater streams that are to be routed to process water users?
- What is the optimal system configuration (e.g., how should the water allocation system be arranged? is there any stream splitting and mixing? etc.)?

Non-linear optimization programs may be developed to identify the recycle network that will minimize the wastewater discharged (or maximize the amount of recycled wastewater). These programs are based on general water allocation principles and use the transshipment model theory to allow the "shipment" of wastewater (referred to as sources or warehouses) to process water users (referred to as sinks, demands or customers). The optimization program is specifically designed to identify the optimum routing (allocation) of wastewater sources to water sinks. The case studies presented in Sections 11.3.1 and 11.3.2 illustrate the proposed methodology, and various process design scenarios are evaluated to highlight the general applicability of the proposed optimization program.

### 11.3.1   Case Study: Minimization of Wastewater Discharge Using Land Treatment Technology

A concern of increasing importance in water-consuming industries is the ultimate destination of wastewater streams. In the past, wastewater was often discharged to natural bodies of waters (rivers, bays, etc.), relying on dilution and transport to reduce the detrimental effect of these wastes on the environment. However, new environmental rules and legislation, and increased concern from communities and environmental groups constrain the possibilities for wastewater disposal resulting in a range of technologies and methods to minimize the impact of wastewater on the environment. One such method is the beneficial reuse of wastewater in land application technology. Detailed information on the design of land treatment systems for wastewater discharges is available in literature (Overcash and Pal, 1972).

   The land application technology of interest for this case study is slow rate irrigation or "spray fields." The water streams distributed in this way must obviously

**TABLE 11.4. Summary of Case Study Wastewater Streams (Sources) and Their Constituents**

| Source Stream Designations | Flow liter/s | TKN ppm | SO4-S ppm | Cl − ppm | Cu ppm | Na + ppm | Ca2 + ppm | Mg2 + ppm |
|---|---|---|---|---|---|---|---|---|
| CTB1 Stream | 17.350 | 1.26 | 116.01 | 72.45 | 0.14 | 91.33 | 7.96 | 6.46 |
| CTB4 Stream | 0.435 | 1.11 | 57.08 | 134.38 | 0.14 | 94.33 | 7.58 | 5.72 |
| CTB6 Stream | 2.524 | 0.72 | 96.67 | 61.2 | 0.14 | 121 | 10.84 | 7.12 |
| RX Stream | 18.927 | 0.82 | 10.69 | 2000 | 0 | 18.75 | 209 | nd |
| YP Stream | 28.391 | 220 | 35.87 | 32.18 | 0.16 | 49.74 | 3.48 | 2.48 |
| CS Stream | 37.854 | 28 | 178.39 | 315.99 | 4.90 | 0.16 | 0.82 | nd |
| NZ Stream | 14.511 | 0.99 | 3199.4 | 169 | 0 | 1643 | 23.67 | 17 |

TKN = total Kjeldahl nitrogen.
(Reprinted with permission from Clean Products and Processes, Vol. 3, Dunn et al., "Process Integration Design Methods for Water Conservation and Wastewater Reduction in Industry, Part 2: Design for Multiple Contaminants," Table 1, pp. 319–329, Copyright 2001, with permission from Springer-Verlag.)

observe a range of limitations with respect to the flow and content of certain substances. The determination of the wastewater constituents of relevance and their threshold values is based on the specific characteristics of the vegetation and soil of the land, such that the irrigated land and groundwater is preserved. The land available at the site for this case study can be classified as type 1 or type 2; according to the burden of flow and constituent loading from wastewater tolerated by the soil present. In addition, two process sinks (streams leading to unit operations that have some demand for water) are available for water recycling. Several wastewater streams (source streams) are generated at an industrial chemical plant. This scenario is based on existing manufacturing processes and is sufficient to illustrate the usefulness of the proposed design methodology for designing water recycle networks for streams containing multiple contaminants. The flow rates and compositions of the wastewater streams generated at this site are summarized in the Table 11.4. Streams flowing to land of types 1 and 2 represent sinks for our case study, and the constraints or limitations of the flow rates and constituents applied to the land are based on principles reported in literature (Overcash and Pal, 1972). These constraints are summarized below.

*Type 1 Land Constraints:*

- 283,290 m$^2$ of type 1 land is available (NP)
- Total flow rate of wastewater (liter/s) to the land $\leq 1.091 \times 10^{-4}/m^2$
- Total nitrogen (TKN) applied to the land per year $\leq 0.03062$ kg/m$^2$
- Total copper applied to the land per year $\leq 0.00334$ kg/m$^2$
- Total Cl$^-$ (ppm) applied to the land $\leq (m^2/(109.05*$flow rate applied in liter/s)) $+ 250$
- Total SO$_4$-S (ppm) applied to the land $\leq (m^2/(109.05*$flow rate applied in liter/s)) $+ 250$

- Sodium adsorption ratio (SAR) applied to the land $\leq 6.0$

*Type 2 Land Constraints:*

- 404,700 $m^2$ of type 2 land is available (P)
- Total flow rate of wastewater (liter/s) to the land $\leq 3.520 \times 10^{-5}/m^2$
- Total nitrogen (TKN) applied to the land per year $\leq 0.03062$ kg/$m^2$
- Total copper applied to the land per year $\leq 0.00334$ kg/$m^2$
- Total $Cl^-$ (ppm) applied to the land $\leq (m^2/(109.05^*$flow rate applied in liter/s)) + 250
- Total $SO_4$-S (ppm) applied to the land $\leq (m^2/(109.05^*$flow rate applied in liter/s)) + 250
- Sodium adsorption ratio (SAR) applied to the land $\leq 6.0$

There are two process water users in the plant. These unit operations (e.g., reactors, mixers, decanters, cooling towers, boilers, etc.) are coded to protect the proprietary nature of these operations. They have known flow and composition constraints on their inlet streams. A schematic representation of the source and sink streams for the case study is included in Figure 11.21. Mathematical optimization is used to identify the optimal allocation (water recycle network) of source streams and/or a portion of source streams to sink streams. The optimization program will specifically identify:

- The maximum flow rate of wastewater (source streams) to be land applied on types 1 and 2 land (sinks)
- The amount (acres) of land types 1 and 2 to be used
- The flow of each individual wastewater stream to land application for type 1 and type 2 land
- The flow of each individual wastewater stream to process units 1 and 2
- The land limiting constituent (LLC) for type 1 and type 2 land
- The limiting contaminant for process units 1 and 2
- The amount of freshwater reduction via wastewater recycling

Prior to evaluating recycle and reuse options the design team should identify opportunities for source reduction of water flows and constituents of the streams provided in Table 11.4. Any reduction in flow rates and/or compositions of any of the streams may have an impact on the ultimate water allocation design identified via the mathematical optimization program. This is a critical step since source reduction opportunities are largely dependent on the specific unit operations existing in any water-consuming plant. Moreover, process changes resulting in the source reduction of wastes are often much more cost-effective than recycle, reuse and interception designs for these wastes. In the scenarios that follow; we assume that the data in

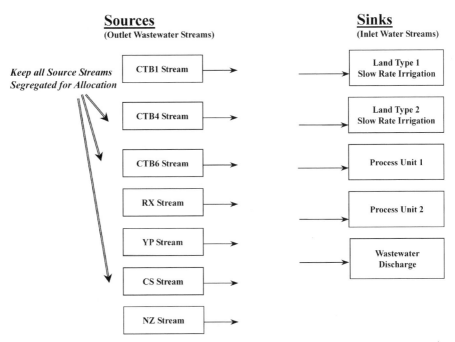

**Figure 11.21.** Source-sink stream representation for the case study of Section 11.3.1. (Reprinted with permission from Clean Products and Processes, Vol. 3, Dunn et al, "Process Integration Design Methods for Water Conservation and Wastewater Reduction in Industry, Part 2: Design for Multiple Contaminants," Figure 1, pp. 319–329, Copyright 2001, with permission from Springer-Verlag)

Table 11.4 represents stream conditions after all source reduction modifications to the plant processes.

The first scenario studied involves the identification of the optimal water routing and allocation from the sources to the sinks when no more than 404,700 total m$^2$ (100 acres) of combined land types 1 and 2 can be used for slow rate irrigation. The distribution of the flows for the optimal solution is depicted in Figure 11-22. The limiting constituent(s) for the combined flow to any sink are listed on the diagram below the arrow of the combined flow to any sink. For example, the limiting constituents to slow rate irrigation on land type 1 are flow, TKN, Cl and SAR. All of these parameters for the mixed stream are the maximum that can be tolerated by that sink.

The second scenario studied also involves the identification of the optimal water routing and allocation from the sources to the sinks when no more than 404,700 total m$^2$ of combined land types 1 and 2 can be used for slow rate irrigation. In addition, the RX stream must be totally recycled to meet environmental discharge restrictions. When the optimization program was run with no relaxation of constraints, recycling the entire RX stream to the sinks provided was shown to be infeasible because of the

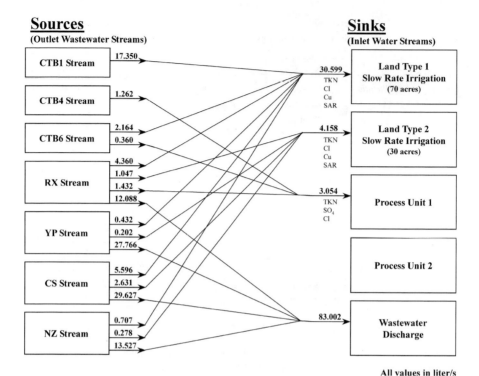

**Figure 11.22.** Optimal allocation of the flows, Scenario 1. (Reprinted with permission from Clean Products and Processes, Vol. 3, Dunn et al, "Process Integration Design Methods for Water Conservation and Wastewater Reduction in Industry, Part 2: Design for Multiple Contaminants," Figure 3, pp. 319–329, Copyright 2001, with permission from Springer-Verlag)

excessive chloride content in this stream. Therefore, a constraint was added to the optimization program to identify the minimum amount of chloride to be removed from the RX stream to allow the total stream flow to be recycled to the sinks. The distribution of the flows for the optimal solution, including the insertion of an interception technology(s) is depicted in Figure 11.23, including the listing of limiting constituent(s) for the combined flow to any sink.

### 11.3.2   Case Study: Optimization of Water Flows in a Kraft Pulp and Paper Process

The process for large-scale manufacture of paper has undergone very few changes over the last several decades. The pulp and paper industry today is highly capital intensive and there is an ongoing drive towards process integration thereby increasing the efficiency of utilization of available process resources. This case study deals with the kraft process (Figure 11.24), though the general principles described can be

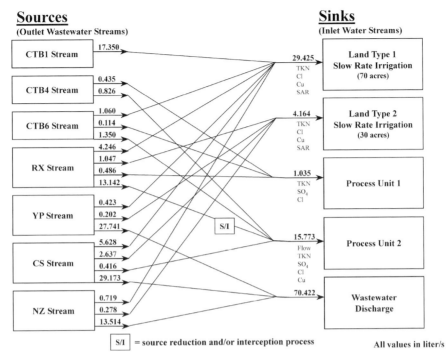

**Figure 11.23.** Optimal allocation of the flows, Scenario 2. (Reprinted with permission from Clean Products and Processes, Vol. 3, Dunn et al, "Process Integration Design Methods for Water Conservation and Wastewater Reduction in Industry, Part 2: Design for Multiple Contaminants," Figure 4, pp. 319–329, Copyright 2001, with permission from Springer-Verlag)

extended to other pulping processes. In the kraft process, the wood chips are reacted with the white liquor (alkaline solution) in a digester. Following reaction (or cooking), pulp is separated from the spent liquor (referred to as black liquor) in the brown stock washers (washers/screens). It is then directed to the bleaching operation for treatment with $ClO_2$ in an acidic environment followed by alkali extraction and caustic washing. The weak black liquor (WBL) is concentrated in a six-effect evaporator resulting in strong black liquor (SBL), which is burned in the recovery furnace to yield inorganic smelt. The smelt forms green liquor when dissolved, which goes to the green liquor clarifier and then is subjected to causticizing which includes a slaker, causticizer, white liquor clarifier, lime mud washer filter and limekiln. CaO gets converted to $Ca(OH)_2$ in the slaker. Next, in the causticizer, the green liquor is converted to the white liquor which goes to the white liquor clarifier, followed by recycle back to digester to complete the recovery cycle. The lime mud ($CaCO_3$) formed is sent to a lime mud washer followed by a lime mud filter. The washed and filtered lime mud is calcined in a lime kiln to form CaO and $CO_2$. The CaO thus formed is recycled back to the slaker thus completing the

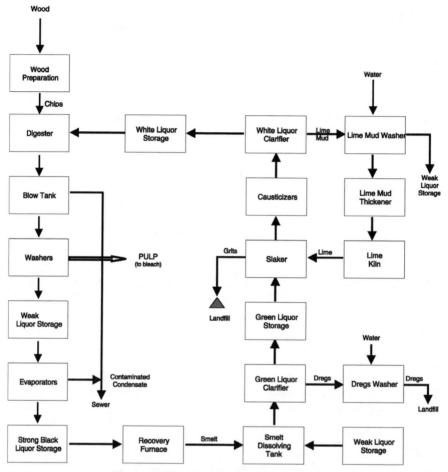

**Figure 11.24.** Schematic of the kraft process.

lime cycle. Some waste solids, called dregs and grits, are produced and must be discarded. The bleach plant is considered as a black box with the various $ClO_2$ stages and the alkali extraction stages lumped together respectively. The alkaline and acid bleach effluents are treated before discharge.

The pulp and paper industry has traditionally been a major consumer of fresh water (up to $70 \, m^3/t$) (Galloway et al., 1994). With increasing water scarcity, mills are being forced to consider means of reducing water usage. This has led to the drive to increase water recycling, which results in a simultaneous build up of non process elements or NPEs (Keitaanniemi and Virkkola, 1978), which play no part in the pulping process. Some sources for NPEs are wood chips (Cl, Si, Mg, Fe), makeup lime (Si, Mg, Al, Fe), mill water (K, Mg), and spent acid (Cl). Depending on their relative solubilities, the NPEs will accumulate in the liquor cycle, lime cycle or the bleach plant. Increased concentrations of NPEs in different process streams will lead to different problems including increased corrosion (K, Cl, Mg), plugging (K, Cl), scale formation (Al, Si, Ca, Ba), deposit formation (Al, Si, Ca, Ba), accumulation of inerts in lime cycle (P, Mg, Al, Si), operational problems in the bleach plant (Mn, Fe, Cu) and adverse environmental impacts (N, P, Cd, Pb). Some of the equipment affected include the recovery boiler (corrosion and plugging), digesters (corrosion and scaling), brown stock fiber line (plugging), evaporator (scales and deposits formation) and digester liquor heater (scales and deposits formation). All these factors lead to a decrease in overall process efficiency and can potentially cause significant operational and maintenance problems. Clearly, optimal allocation and prevention of build-up of NPEs in the different process streams will have beneficial effects. Among all NPEs, chlorides are present in relatively higher concentrations, and their buildup causes a range of operational difficulties including corrosion and plugging. We now consider a kraft process with known overall material and chloride balances and known process and environmental constraints. Chloride is the NPE of interest. It is desired to synthesize a series of water re-use strategies that can facilitate water loop closure by reallocating the various aqueous streams in the process and simultaneously distribute chloride optimally such that all process and environmental constraints are fully satisfied.

### 11.3.2.1 The Problem
This kraft process with bleaching uses a pulp production basis of 1,000 tons per day or tpd (1 tpd = 909 kg/day). The acid and alkaline bleach effluents are being treated and the chloride in these streams is separated via ion exchange using porous strongly basic anionic resin. The objective is to reduce the fresh water requirement in the given process by means of recycling and reusing process effluent streams systematically. The sinks (where fresh water is being used) and the effluent streams (potential sources) are identified in Figure 11.25. For every source and sink under consideration, there are bounds for total input flow rate and input compositions of chloride, which have to be satisfied, before a source can be sent to the sink. All the values listed in Figure 11.25 were established based on representative operating data (Sittig, 1977; Jones, 1973). Note that the fresh water demand of the sinks can be reduced by up to 78 percent via recycle of all available sources. This makes a mass integration analysis (to reduce fresh water demand) attractive.

### 11.3.2.2 Solution Without Interception Included
The problem can be solved via mathematical optimization or by using a graphical solution strategy. The graphical technique is easy to apply and provides insight into the process. The graphical technique relies on the lever arm rule that is derived from material balance. When two sources are mixed to give a resultant source, the following relations can be derived from material balance (Figure 11-26):

$$F_A + F_B = F_{Total} \qquad (11\text{-}1)$$

$$F_A \cdot x_A + F_B \cdot x_B = F_{Total} \cdot x_{Total} \qquad (11\text{-}2)$$

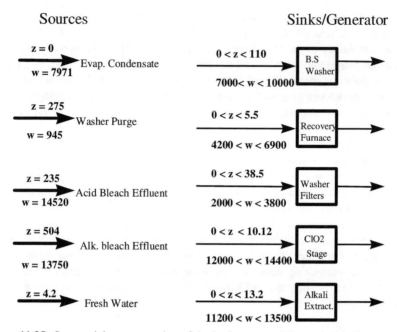

**Figure 11.25.** Source sink representation of the kraft process. (The total sink flow rates (w) are in units of tons/day or tpd (1 tpd = 909 kg/d) while the sink chloride compositions (z) are in units of *ppm*.) (Reprinted from Advances in Environmental Research, 5, Parthasarathy, G., Krishnagopalan, G., Systematic Reallocation of Aqueous Resources using Mass Integration in a typical Pulp and Paper Process, 61–79 © 2001 with permission from Elsevier)

Equations 11-1 and 11-2 yields,

$$\frac{F_A}{F_B} = \frac{x_{Total} - x_B}{x_A - x_{Total}} \tag{11-3}$$

This can be given as

$$\frac{F_A}{F_B} = \frac{Arm_A}{Arm_B} \tag{11-4}$$

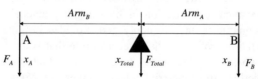

**Figure 11.26.** Representation of the lever arm balance. (Reprinted from Advances in Environmental Research, 5, Parthasarathy, G., Krishnagopalan, G., Systematic Reallocation of Aqueous Resources using Mass Integration in a typical Pulp and Paper Process, 61©79 © 2001 with permission from Elsevier)

Therefore, if $F_A$ is the flow rate of fresh water, minimization of this flow rate would be at minimum lever arm ($Arm_A$).

The source sink diagram for this process is given in Figure 11.27. Since all the available sources are already segregated, we can begin with mixing and direct recycle opportunities. Note that to minimize the fresh water used for a sink, the source with the smallest possible lever arm to that sink must be considered first for recycle. This comes from the lever arm rule according to which the amount of fresh water used will depend on the lever arm from the source to the sink. In Figure 11.27, the acid bleach effluent has the least lever arm with respect to the sink called B.S Washer and so it must be considered first for recycle. Also, as the evaporator condensate is pure, it can be used instead of fresh water. If $F_S$ and $F_C$ are the source flow rate and evaporator condensate flow rate respectively, then according to material balance:

$$F_S \cdot (x_S - x_{Sink}) = F_C \cdot (x_{Sink} - x_C) \tag{11-5}$$

$$F_S + F_C = F_{Sink} \tag{11-6}$$

where, $F_{Sink}$ is the total flow rate going to the sink under consideration. This is taken as the lower limit of allowable flow rate from Figure 11.25 (i.e., $F_{Sink} = 7,000$). Using source composition, $x_S = 235$ ppm, sink composition, $x_{Sink} = 110$ ppm, and

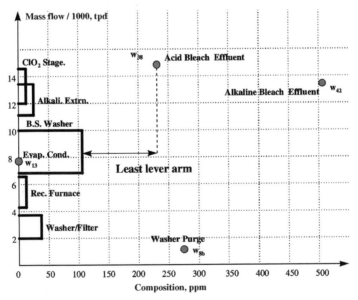

**Figure 11.27.** Initial source sink diagram. (Note that stream numbers start with the letter "w" and corresponding number from Figure 11.29.) (Reprinted from Advances in Environmental Research, 5, Parthasarathy, G., Krishnagopalan, G., Systematic Reallocation of Aqueous Resources using Mass Integration in a typical Pulp and Paper Process, 61–79 © 2001 with permission from Elsevier)

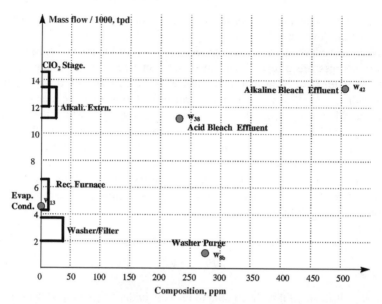

**Figure 11.28.** Source sink diagram after mixing. (Note that stream numbers start with the letter "w" and corresponding number from Figure 11.29.) (Reprinted from Advances in Environmental Research, 5, Parthasarathy, G., Krishnagopalan, G., Systematic Reallocation of Aqueous Resources using Mass Integration in a typical Pulp and Paper Process, 61–79 © 2001 with permission from Elsevier)

evaporator condensate composition, $x_C = 0$ ppm, $F_S$ and $F_C$ are calculated to be 3,276.6 and 3,723.5 tpd, respectively. This would totally satisfy the fresh water requirement of the B.S Washer and replace 7,000 tpd of fresh water being used currently. The source sink diagram can now be redrawn without the B.S Washer as a sink in Figure 11.28. Note that the available flow rates of the evaporator condensate source and the acid bleach effluent source are reduced while their respective compositions stay the same.

### 11.3.2.3 Solution With Interception Included
Recycle and mixing, as described earlier, are limited to avoid build-up of chloride. This may result in the violation of some composition constraint in the process. Interception can be used to lower the mass load of chloride in a source. A number of external separation techniques that have been specifically developed for the kraft process and are thermodynamically feasible for the stream being intercepted are listed subsequently. Additionally, we list the corresponding stream number from Figure 11-29 where each technique can be applied.

1. A method for limiting the concentration of sodium chloride in pulping liquor via the leaching of smelt (Reeve and Rapson, 1970) (stream 19 in Figure 11.29)

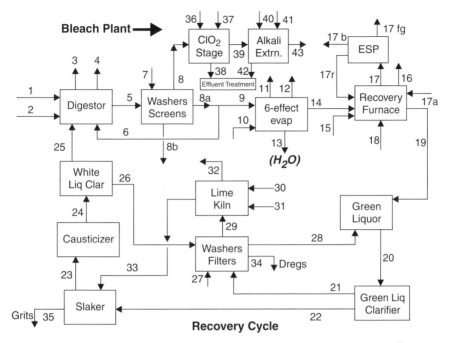

**Figure 11.29.** Stream numbers for a typical kraft pulp and paper process corresponding to the case study of Section 11-3-2. (Reprinted from Advances in Environmental Research, 5, Parthasarathy, G., Krishnagopalan, G., Systematic Reallocation of Aqueous Resources using Mass Integration in a typical Pulp and Paper Process, 61–79 © 2001 with permission from Elsevier)

2. A continuous process for selectively dissolving NaCl from precipitator dust (Moy et al., 1974) (stream 17r in Figure 11.29)

3. The salt recovery process, which provides a multi-step purification procedure producing high purity NaCl which may be used to generate bleaching chemicals (Reeve et al., 1974)

4. Chloride removal during cooling crystallization of sodium carbonate decahydrate in green liquor followed by white liquor evaporation (pilot plant study: Sainiemi et al., 1979) (stream 20 in Figure 11.29)

5. Salt removal by white liquor evaporation with a wide range of salt removal capacities from 50–250 pounds per ton of pulp (Pryke et al., 1979) (stream 25 in Figure 11.29)

6. Recovery of chloride containing bleach filtrates in a conventional Kraft recovery system by (technical and economic assessment: Maples et al., 1994) (stream 17r in Figure 11.29)

7. Selectively removing chloride and potassium from electrostatic precipitator dust (Shenassa et al., 1995) (stream 17r in Figure 11.29)

8. Bleach plant filtrate recycle with chloride removal (Full scale commercial process: Fleck et al., 1996) (stream 17r in Figure 11.29)

9. Some commonly used arrangements of treatment technologies to produce water of desired, industry standard quality are described (DeSilva, 1996) (streams 38 and 42 in Figure 11.29)

10. Conversion of chloride in flue gas dust to gaseous hydrogen chloride, which can be used to expel chloride from the liquor system (Warnqvist and Bernhard, 1975) (stream 17r in Figure 11.29)

A number of solution runs were carried out. The non-linear optimization is solved on GINO (Liebman et al., 1988). It is observed that the fresh water demand is lowered by 11,300 tpd through an initial optimization via the strategies of segregation, mixing and recycle. This is followed by an optimization that includes interception. One stream interception is considered at a time. Feasible solutions were obtained for three potential techniques are namely interception of stream 19 in Figure 11.29 (via leaching of smelt), interception of stream 25 in Figure 11.29 (via white liquor evaporation), and interception of stream 20 in Figure 11.29 (via green liquor crystallization). The solution strategy for interception of stream 20 is given in Figure 11.30. Inclusion of interception into the analysis results in lowering the fresh water demand by an additional 15,800 tpd. Assuming that the cost of interception of a particular stream in the process is related to the mass load of chloride

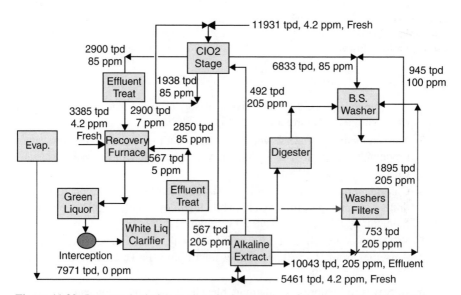

**Figure 11.30.** Integrated solution strategy for interception of the green liquor stream (stream 20 in Figure 11.29). (Reprinted from Advances in Environmental Research, 5, Parthasarathy, G., Krishnagopalan, G., Systematic Reallocation of Aqueous Resources using Mass Integration in a typical Pulp and Paper Process, 61–79 © 2001 with permission from Elsevier)

**TABLE 11.5. Cost Estimates for Interception of Respective Streams in the Kraft Process (All costs in 1999 Dollars, Operating Cost Estimated from Steam Usage at $0.002/lb.)**

| Stream Being Intercepted | Capital Cost ($) | Operating Cost ($/year) |
|---|---|---|
| Smelt (19 in Figure 11.29) | 7,200,000 | 2,050,000 |
| White liquor (25 in Figure 11.29) | 3,600,000 | 1,300,000 |
| Green liquor (20 in Figure 11.29) | 4,600,000 | 880,000 |

(Reprinted from Advances in Environmental Research, 5, Parthasarathy, G., Krishnagopalan, G., Systematic Reallocation of Aqueous Resources using Mass Integration in a typical Pulp and Paper Process, 61–79, Copyright 2001 with permission from Elsevier.)

separated out, the estimated cost of interception of streams 19, 20 and 25 are given in Table 11.5.

Since the three different techniques led to similar reductions in the fresh water requirement and mass loads of chloride to be separated, the technique to be used may be selected on the basis of relative costs and other factors such as ease of application to existing process. Corresponding to the lowering of fresh water usage, a net decrease in effluent discharge as well as load of chloride to be separated by effluent treatment is observed. For the three solution strategies, it may be noted that the entire acid bleach effluent is recycled and the only effluent exiting the process is the alkaline bleach effluent. The integrated solutions obtained in this work have some of the following advantages:

1. Various streams from the recovery cycle are used in the bleach plant. Also, several bleach plant streams are utilized in the recovery cycle.
2. The acid bleach effluent is used in the lime cycle (Washers filters sink) and the recovery cycle as described in previous works (Gleadow and Hastings, 1991; Teder et al., 1989; Hastings et al., 1994). In the proposed systems, however, the solutions have a complete recycle and reuse of the entire acid bleach effluent with no discharge.
3. The solutions include optimal flow rates going from each effluent source to respective sinks besides the chloride concentrations in these streams.
4. The total chloride mass load going to effluent treatment is lowered.
5. In the integrated solution for interception of stream 20 (i.e., green liquor crystallization), the alkaline bleach effluent is used to replace fresh water to the acid bleach stage. This is contrary to conventional practice that recommends keeping the acid and alkali bleach stages independent of each other. The use of the mass integration framework allows one to consider all possible recycle opportunities. It must be noted that the alkaline bleach effluent stream must be suitable treated (e.g., pH adjustment) before recycle to the acid bleach stage.

## 11.4 INDUSTRIAL APPLICATIONS OF HEAT INTEGRATION AND MASS INTEGRATION

Solutia (formerly a part of Monsanto Chemical Company) and General Electric Plastics, among other companies, have been particularly active in applying the tools described in this chapter to conserve energy and address waste reduction issues at various sites. Sections 11.4.1 and 11.4.2 offer a snapshot summary of the positive results achieved through these design activities.

### 11.4.1 Solutia

This analysis was conducted at Florida, Texas, Alabama, and Massachusetts sites in 1996, 1997, 1998, and 2003/2004 respectively and required 2–6 months to complete. Site wide analysis was conducted with optimization of water and energy flows. A total of around 250 designs were proposed for all sites with around 50 projects (having payback periods of 1.5 years or less) being implemented. These designs included energy savings, detailed economic analysis and waste water discharge reductions. Some of the highlights from each site analysis are described subsequently. Wastewater discharge was lowered at the Florida and Texas sites by ~1600 gallons per minute. At the Florida site, two selected designs that yielded energy savings would offset future utility expansions (1.6 years payback excluding the savings from avoiding the purchase of additional boilers/cooling towers). At one of the Alabama sites, the selected designs for short term implementation supply energy needs for two future plant expansions with no capital investment needed for new utility systems (boilers and/or cooling towers). At the Texas and Florida sites, projects for reduction of deep well injection wastewater offset wastewater generation that will result from a future plant expansion. Additionally, at the Florida site, specific process streams were identified to be eliminated from being discharged to satisfy modified NPDES permit requirements. The Massachusetts site was awarded a plant wide assessment from the Industrial Technologies Program of the US Department of Energy. This assessment was conducted for ~6 months and resulted in the development of various projects with total utility savings in excess of $3 million per year.

### 11.4.2 General Electric Plastics

This analysis was conducted at an Alabama site in 1995, 1998 and 1999. A site wide analysis was conducted with optimization of water, energy and caustic flows. Total time for the analysis varied from 4 months to a year. Additionally, software tools for enabling stable operations via tracking of material flows in the process were developed and applied. Around 40 different designs with a range of payback periods (some less than an year) were proposed. The analysis resulted in a 10 percent Decrease in Site Utility Costs and increased capacity by 4 percent (worth >$1 million/yearr additional revenue).

## 11.5 FURTHER READING

Several review papers and textbooks deal with the area of process integration. Heat integration and heat exchange networks have been extensively studied and documented. Gundersen and Naess (1988) who review developments in HEN research from an industrial point of view, refer to three main directions being followed: use of thermodynamic concepts, mathematical programming and the application of knowledge based systems. The relative merits of the thermodynamic approach versus the mathematical programming methodology are presented along with a summary of historical developments in both schools of thought. The authors list several available computer software tools to accomplish HEN design and optimization. Furman and Sahinidis (2002) have compiled HEN problems and solution strategies and individually discuss 461 related works. These authors describe several chronological milestones in the field and offer a critical assessment of the current status of research in this area and present suggestions and directions for future work. In another comprehensive look at different methods for HENs, Shenoy (1995) presents several test problems along with corresponding mathematical programming code. This is a useful introductory text for the beginner and expert practitioner.

The field of mass exchange networks and mass integration is relatively more recent. El-Halwagi (1997) is a comprehensive textbook focused on summarizing various design problems and solution methodologies. It is a useful introductory text for the beginner and expert practitioner. El-Halwagi and Spriggs (1998), a helpful synopsis of the developments in the mass integration arena, presents applicability beyond design by briefly describing process operation and simplification. Successful applications in the petrochemical, specialty chemicals and chemicals and polymers industries are included. Dunn and Bush (2001) present an overview of the CLEANER (or Combining Lower Emissions And Networked Energy Recovery) approach, a brief summary of several systems analysis tools and a summary of several process integration technology design tools. This paper summarizes all tools available to the process integration expert and presents a step-by-step approach to implementing a process integration analysis at a typical industrial process.

A review by Dunn and El-Halwagi (2003) includes tables summarizing various design methodologies and detailed descriptions of the techniques, along with example technologies targeted by each of them. The authors present tips for completing a process integration analysis successfully and summarize ten industrial applications of process integration tools and techniques.

## 11.6 FINAL THOUGHTS

The chemical process industry is facing increased pressure to develop processes and products that are energy efficient, environment-friendly and less expensive. Besides, the globalization of the world economy has led to increased competition. Another trend is an increased need to minimize environmental discharges resulting in "greener" processes. These drivers have resulted in significant focus being placed on the concepts of "life cycle analysis," "sustainability" and "industrial ecology." The key to tackling all these challenges lies in process integration, which involves leveraging all process resources in an optimal fashion to reduce overall cost and increase productivity while simultaneously minimizing energy use and lowering adverse environmental impact. However, previous attempts at process integration were arbitrary or based on experience. There is a need for accurate and systematic ways to reduce costs and improve overall process and utility performance in the chemical process industries. In response to this demand, the area of process systems engineering (which involves the application of computer based optimization and design techniques) has blossomed with several applications. Along with the creation and application of optimization algorithms, several unique problem statements have been formulated and solved resulting in significant cost savings and efficiency improvements. The authors have summarized developments in this area over the previous two decades. Integration problems and solution strategies for both mass and energy flows are described. Various problem definitions and solution algorithms (such as MEN, HEN, HISEN, WIN, HIWAMIN, and EIWAMIN etc.) are presented. Example case studies are provided to give the reader a flavor of the different types of problems and applications. A section is dedicated to describing several industrial applications of process integration technologies. It is expected that the next decade will witness significant growth in this area. It is hoped that this chapter will provide the beginner as well as experienced practitioner with enough information to appreciate the value of the process integration approach and encourage increased and widespread usage of these techniques to improve process and utilities efficiencies, reduce environmental discharges, reduce costs and optimize their respective processes.

## REFERENCES

Alva-Argaez, A., Vallianatos A., and Kokossis, A. C. (1999). Multi-contaminant transshipment model for mass exchange networks and wastewater minimization problems. *Computers and Chemical Engineering*, **23**, 1439–1453.

Chadwick, J. L. (1980). *Fertilizers: Nitrogen (Interim report)*. Process Economics Program Report No. 127A1, SRI International, Menlo Park, CA.

Constantinou, L., Jacksland, C., Bagherpour, K., Gani, R., and Bogle, L. (1995). Application of group contribution approach to tackle environmentally-related problems. *AIChE Symposium Series*, **90**(303), 105–116.

Crabtree, E. W., Dunn, R. F., and El-Halwagi, M. M. (1995). Design and analysis of membrane-hybrid systems for solvent recovery. *AIChE Annual Meeting*, Miami, FL.

Crabtree, E. W., Dunn, R. F., and El-Halwagi, M. M. (1998). Synthesis of hybrid gas permeation membrane/condensation systems for pollution prevention. *Journal of the Air and Waste Management Association* **48**, 616–626.

Crabtree, E. W., and El-Halwagi, M. M. (1995). Synthesis of environmentally acceptable reactions. *AIChE Symposium Series*, **90**(303), 117–127.

DeSilva, F. (1996). Tips for process water purification. *Chemical Engineering*, 72–82 (Aug.).

Dhole, V. R., Ramchandani, N., Tainsh, R. A., and Wasilewski, M. (1996). Make your process water pay for itself. *Chemical Engineering*, **103**(1), 100.

Dobson, A. M. (1998). *Shortcut Design Methodologies for Identifying Cost-Effective Heat-Induced and Energy-Induced Separation Networks for Condensation-Hybrid Processes*. Masters thesis, Auburn University, AL.

Dunn, R. F., and El-Halwagi, M. M. (1993). Optimal recycle/reuse policies for minimizing wastes of pulp and paper plants. *Journal of Environmental Science and Health*. **A28**, 217–234.

Dunn, R. F., and El-Halwagi, M. M. (1994a). Optimal design of multi-component VOC-condensation systems. *Journal of Hazardous Materials*. **38**, 187.

Dunn, R. F., and El-Halwagi, M. M. (1994b). Selection of optimal VOC-condensation systems. *Waste Managemen*, **14**, 103.

Dunn, R. F., and El-Halwagi, M. M. (1996). *Design of Cost-Effective VOC Recovery Systems*, Published by TVA Department of Economic Development and EPA Center for Waste Reduction. http://www.owr.ehnr.state.nc.us/ref/00034.html

Dunn, R. F., and Srinivas, B. K. (1997). Synthesis of heat-induced waste minimization networks (HIWAMINs). *Advances in Environmental Research*, **1**, 275.

Dunn, R. F., and Dobson, A. M. (1998). A spreadsheet-based approach to identify cost-effective heat-induced and energy-induced separation networks for condensation-hybrid processes. *Advances in Environmental Research*, **2**(3), 269–290.

Dunn, R. F., and Bush, G. E. (2001). Using process integration technology for cleaner production. *Journal of Cleaner Production*, **9**(1), 1–23.

Dunn, R.F., and Wenzel, H. (2001). Process integration design methods for water conservation and wastewater reduction in industry, Part 1: Design for single contaminant. *Clean Products and Processes*, **3**, 307–318.

Dunn, R. F., and El-Halwagi, M. M. (2003). Process integration technology review: background and applications in the chemical process industry. *Journal of Chemical Technology and Biotechnology*, **78**(9), 1011–1021.

Dunn, R. F., Zhu, M., Srinivas, B. K., and El-Halwagi, M. M. (1995). Optimal design of energy induced separation networks for VOC recovery. *AIChE Symposium Series*, **90**(303), 74–85.

Dunn, R. F., Hamad, A. A., and Dobson, A. M. (1999). Synthesis of energy-induced waste minimization networks (EIWAMINs) for simultaneous waste reduction and heat integration. *Clean Products and Processes* **1**, 91.

Dunn, R.F., Wenzel, H., and Overcash, M. (2001). Process integration design methods for water conservation and wastewater reduction in industry, Part 2: Design for multiple contaminants. *Clean Products and Processes* **3**(3), 319–329.

Dye, S. R., Berry, D. A., and Ng, K. M. (1995). Synthesis of crystallization-based separation schemes. *AIChE Symposium Series* **91**, 238–241.

El-Halwagi, M. M. (1997). *Pollution Prevention through Process Integration: Systematic Design Tools*. San Diego, CA: Academic Press.

El-Halwagi, M. M., and Manousiouthakis, V. (1989a). Synthesis of mass exchange networks. *AIChE Journal* **35**(8), 1233–1244.

El-Halwagi, M. M., and Manousiouthakis, V. (1989b). Design and analysis of multicomponent mass exchange networks. *AIChE Annual Meeting*, San Francisco, CA.

El-Halwagi, M. M., and V. Manousiouthakis (1990a). Automatic synthesis of mass exchange networks with single-component targets. *Chemical Engineering Science*, **45**(9), 2813–2831.

El-Halwagi, M. M., and Manousiouthakis, V. (1990b). Simultaneous synthesis of mass exchange and regeneration networks. *AIChE Journal*, **36**(8), 1209–1219.

El-Halwagi, M. M., and Srinivas, B. K. (1992). Synthesis of reactive mass exchange networks. *Chemical Engineering Science*, **47**, 2113–2119.

El-Halwagi, M. M., and Spriggs, H. D. (1996). An integrated approach to cost and energy efficient pollution prevention. *Proceedings of the Fifth World Congress of Chemical Engineering* 344, San Diego, CA.

El-Halwagi, M. M., and Spriggs, H. D. (1998). Employ mass integration to achieve truly integrated process design. *Chemical Engineering Progress*, 22–44 (Aug.).

El-Halwagi, M. M., Srinivas, B. K., and Dunn, R. F. (1995). Synthesis of heat-induced separation networks. *Chemical Engineering Science*, **50**, 81.

El-Halwagi, M. M., Hamad, A. A., and Garrison, G. W. (1996). Synthesis of waste interception and allocation networks. *AIChE Journal*, **42**(11), 3087–3101.

Fleck, J. A., Earl, P. F., and Fagan, M. J. (1996). Kraft mill bleach filtrate recycle and the commercial demonstration of chloride and potassium removal. International Environmental Conference and Exhibits, *TAPPI Proceedings*, 655–665.

Furman, K. C., and Sahinidis, N. V. (2002). A critical review and annotated bibliography for heat exchanger network synthesis in the 20th century. *Industrial and Engineering Chemistry Research*, **41**, 2335–2370.

Galloway, L., Gleadow, P., Hastings, C., and Lownertz, P. (1994). *Closed-cycle technologies for bleached kraft pulp mills*, National Pulp Mills Research Program, Technical Report No. 7, CSIRO, Canberra.

Garrison, G. W., Hamad, A. A., and El-Halwagi, M. M. (1995). Synthesis of waste interception networks. Paper 77f, *AIChE Annual Meeting*. Miami, FL.

Gianadda, P., Brouckaert, C. J., Sayer, R., and Buckley, C. A. (2002). The application of pinch analysis to water, reagent and effluent management in a chlor-alkali facility. *Water Science and Technology*, **46**(9), 21–28.

Gleadow P., and Hastings, C. (1991). Steps towards kraft mill closed cycle. *Pacific Paper Expo Preprints*, Vancouver BC, Canada, 130.

Gundersen, T., and Naess, L. (1988). The synthesis of cost optimal heat exchanger networks. *Computers and Chemical Engineering*, **12**(6), 503–530.

Gupta, A., and Manousiouthakis, V. (1994). Waste reduction through multicomponent mass exchange network synthesis. *Computers and Chemical Engineering*, **18**, S585–S590.

Hallale, N. (2002) A new graphical targeting method for wastewater minimization. *Advances in Environmental Research*, **6**, 377–390.

Hamad, A. A., Crabtree, E. W., Garrison, G. W., and El-Halwagi, M. M. (1996). Optimal design of hybrid separation systems for waste reduction. *Fifth World Congress of Chemical Engineers*. Paper 40c, San Diego, CA.

Hamad, A. A., and El-Halwagi, M. M. (1998). Simultaneous synthesis of Mass separating agents and interception networks. *Transactions of the Institution of Chemical Engineers*, **76**, 376.

Hamad, A. A., Varma, V., Krishnagopalan, G., and El-Halwagi, M. M. (1998). Mass integration analysis: a technique to reduce methanol and effluent discharge in pulp mills, *TAPPI*, **81**(10).

Hastings, C. R., Gleadow, P. L., Galloway, L. R., and Thorpe, A. (1994). Towards closed cycle kraft: case studies of closed cycle ECF and TCF bleached eucalypt kraft pulp mills. *Appita*, 431–437.

Hildebrandt, D., and Biegler, L. T. (1995). Synthesis of chemical reactor networks. In *AIChE Symposium Series*; Biegler, L., Doherty, M. (eds), American Institute of Chemical Engineers, NY, Vol. 91, 52.

Huang, Y. L., and Fan, L. T. (1995). Intelligent Process Design and Control for In-Plant Waste Minimization. In *Waste Minimization through Process Design*, Rossiter, A. P. (ed.), 165–180. New York, NY: McGraw-Hill.

Huang, Y. L., and Edgar, T. F. (1995). Knowledge Based design Approach for the Simultaneous Minimization of Waste Generation and Energy Consumption in a Petroleum Industry, In *Waste Minimization through Process Design*, Rossiter, A. P. (ed.), 181–196. New York, NY: McGraw-Hill.

Joback, K. G., and Stephanopoulos, G. (1989). Designing molecules possessing desired physical property values, *Proceedings, 3rd Conference on Foundations of Computer Aided Process Design* 363–387. Elsevier, NY, New York.

Jones, H. R. (1973). *Pollution Control and Chemical Recovery in the Pulp and Paper Industry*. Pollution Technology Review No. 3, Noyes Data Corporation.

Keitaanniemi, O., and Virkkola, N. E. (1978). Amounts and behavior of certain chemical elements in Kraft pulp manufacture: results of mill scale study. *Paperi Puu*, **60**(9), 507.

Kiperstock, A., and Sharratt, P. N. (1995). On the optimization of mass exchange networks for removal pollutants. *Transactions of the Institution of Chemical Engineers*, November, 73, Part B: 271–277.

Kirk-Othmer (1993). *Encyclopedia of Chemical Technology*; 7, 4th Ed., 683–691, Wiley, NY.

Kuo, W. J., and Smith, R. (1997). Effluent treatment system design. *Chemical Engineering Science*, **52**(23), 4273–4290.

Liebman, J., Lasdon, L. Schrage, L., and Waren, A. (1988). *Modeling and Optimization with GINO*. San Francisco, CA: The Scientific Press.

Linnhoff, B., and Flower, J. R. (1978). Synthesis of heat exchanger networks, *AIChE Journal*, **24**(4), 633–642.

Linnhoff, B., and Hindmarsh, E. (1983). The pinch design method for heat exchanger networks. *Chemical Engineering Science*, **38**(5), 745–763.

Maples, G. E., Ambady, R., Caron, J. R., Statton, S. C., and Canovas, R. E. V. (1994). BFR$^{TM}$: A new process towards bleach plant closure. *TAPPI*, **77**(11), 71–80.

Moy, W. A., Joyce, P., and Styan, G. E. (1974). Removal of sodium chloride from kraft recovery systems, *Pulp & Paper Magazine Canada*, **75**(4), 88–90.

Olesen, S. G., and Polley, G. T. (1997). A simple methodology for the design of water networks handling single contaminants. *Transactions of the Institution of Chemical Engineers*, 75, Part A, May: 420–426.

Overcash, M., and Pal, D. (1972). *Design of Land Treatment Systems for Industrial Wastes-Theory and Practice*, 684. Ann Arbor Science Publishers.

Papalexandri, K. P., Floudas, C., and Pistikopoulos, E. N. (1994). Mass exchange networks for waste minimization: a simultaneous approach. *Transactions of the Institution of Chemical Engineers*, **72**, 279–294.

Papalexandri, K. P., and Pistikopoulos, E. N. (1994). A multiperiod MINLP model for the synthesis of heat and mass exchange networks. *Computers and Chemical Engineering*, **18**(12), 1125–1139.

Parthasarathy, G., Dunn, R. F., and El-Halwagi, M. M. (2001a). Development of heat-integrated evaporation and crystallization networks for ternary wastewater systems — Part I: design of the separation system. *Industrial & Engineering Chemistry Research*, **40**(13), 2827–2841.

Parthasarathy, G., Dunn, R. F., and El-Halwagi, M. M. (2001b). Development of heat-integrated evaporation and crystallization networks for ternary wastewater systems – Part II interception task identification for the separation and allocation network. *Industrial & Engineering Chemistry Research*, **40**(13), 2842–2856.

Parthasarathy, G., and Krishnagopalan, G. (1999). Effluent reduction and control of non-process elements towards a cleaner kraft pulp and paper process. *Clean Products and Processes*, **1**(4), 264–277.

Parthasarathy, G., and Krishnagopalan, G. (2001). Systematic reallocation of aqueous resources using mass integration in a typical pulp and paper process. *Advances in Environmental Research*, **5**, 61–79.

Parthasarathy, G., and El-Halwagi, M. M. (2000). Optimum mass integration strategies for condensation and allocation of multicomponent VOCs. *Chemical Engineering Science*, **55**, 881.

Polley, G. T., and Polley, H. L. (2000). Design better water networks. *Chemical Engineering Progress*, **2**, 47–52.

Pryke, D. C., Lukes, J. A., and Reeve, D. W. (1979). Salt recovery by white liquor evaporation, *International Pulp Bleaching Conference*, 205–207.

Reeve, D. W., Lukes, J. A., French, K. A., and Rapson, W. H. (1974). The effluent free bleached kraft pulp mill Part IV. alt recovery process. *Pulp & Paper Magazine Canada*, **75**(8), 67–70.

Reeve, D. W., and Rapson, W. H. (1970). The recovery of sodium chloride from bleached kraft pulp mills. *Pulp & Paper Magazine Canada*, **71**(13), 48–54.

Richburg, A., and El-Halwagi, M. M. (1995). A Graphical Approach to the Optimal Design of Heat-Induced Separation Networks for VOC Recovery. *AIChE Symposium Series* Biegler, L., Doherty, M. (eds), Vol. 91, 256. American Institute of Chemical Engineers, New York, NY.

Rotstein, E., Resasco, D., and Stephanopoulos, G. (1982). Studies on the synthesis of chemical reaction paths—I. reaction characteristics in the (DG,T) space and a primitive synthesis procedure. *Chemical Engineering Science*, **37**(9), 1337–1352.

Sainiemi, J, Sannholm, K., and Ahlstrom, R. (1979). Salt Removal Method: Pilot Plant Experiences. *International Pulp Bleaching Conference*, 201–204.

Sharratt, P. N. (1996). Crystallization-meeting the environmental challenge. *Transactions of the Institution of Chemical Engineers*, **74**, 732.

Shelley, M. D., and El-Halwagi, M. M. (2000). Component-less design of recovery and allocation systems: a functionality-based clustering approach. *Computers and Chemical Engineering*, **24**, 2081–2091.

Shenassa, R., Reeve, D. W., Dick, P. D., and Costa, M. L. (1995). Chloride and potassium control in closed kraft mill liquor cycles. *International Chemical Recovery Conf.*, B177–B185.

Shenoy, U. V. (1995). *Heat Exchanger Network Synthesis: Process Optimization by Energy and Resource Analysis*. Gulf Publishing Company, Houston, TX.

Sittig, M. (1977). *Pulp and Paper Manufacture, Energy Conservation and Pollution Prevention*. Pollution Technology Review No. 36, Noyes Data Corporation.

Srinivas, B. K., and El-Halwagi, M. M. (1994a). Synthesis of reactive mass exchange networks with general nonlinear equilibrium functions. *AIChE Journal*, **40**, 463–472.

Srinivas, B. K., and El-Halwagi, M. M. (1994b). Synthesis of combined heat and reactive mass exchange networks. *Chemical Engineering Science*, **45**, 2059–2074.

Teder, A., Andersson, U., Littecke, K., and Ulmgren, P. (1989). The recycling of acidic bleach plant effluents and their effect on white liquor preparation. *TAPPI Proceedings*, Pulping Conference, 751–759.

Ujang, Z., Wong, C. L., and Manan, Z. A. (2002). Industrial wastewater minimization using water pinch analysis: a case study on an old textile plant. *Water Science and Technology*, **46**(11–12), 77–84.

Vaidyanathan, R., and El-Halwagi, M. M. (1996). Computer-aided synthesis of polymers and blends with target properties. *Industrial & Engineering Chemistry Research*, **35**, 627.

Wang, Y. P., and Smith, R. (1994). Wastewater minimization. *Chemical Engineering Science*, **49**(7), 981–1006.

Wang, Y. P., and Smith, R. (1995). Wastewater minimization with flow rate constraints, *Transactions of the Institution of Chemical Engineers*, **73**(A), 889–904.

Warnqvist, B., and Bernhard, R. (1975). Removal of chloride from recovery systems by reactions with sulfur dioxide. *Svensk Papperstidning* (11), 400.

Warren, A., Srinivas, B. K., and El-Halwagi, M. M. (1995). Optimal design of waste reduction systems for coal-liquefaction plants. *Journal of Environmental Engineering*, **121**, 742–746.

Zhu, M., and El-Halwagi, M. M. (1995). Synthesis of flexible mass exchange networks. *Chemical Engineering Communications*, **183**, 193–211.

# CHAPTER 12

# PROCESS POLLUTION PREVENTION IN THE PULP AND PAPER INDUSTRY

TAPAS K. DAS

Washington Department of Ecology, Olympia, Washington 98504, USA

The pulp and paper industry plays a crucial role in sustainable development because its chief raw material, wood fiber, is renewable. In this chapter we look briefly at how the industry is managing this resource to provide the continuous supply required to meet society's current and future needs. Then we review progress in preventing pollution from the industry's principal gaseous, effluent, and solid by-products. The chapter concludes with a detailed look at resource recovery, an enterprise that often results in pollution prevention as well as other benefits.

Our discussion centers on chemical pulping processes, which supply more than two-thirds of world's wood pulp. The most widely used chemical pulping process in paper making is the kraft, or sulfite, process. Other chemical-pulping process (mainly using acid sulfite and soda) are sometimes combined with various chemical-recovery sub-processes. First used over a century ago, these processes now result in recovery of 90 percent of the inorganics, which are used in pulping process. Nearly 100 percent of the dissolved organics are converted to energy. The process was covered in Chapter 11, where a flow diagram was presented (Figure 11.24).

## 12.1 ENVIRONMENTAL MANAGEMENT IN THE PULP AND PAPER INDUSTRY

In the pulp industry, management of the amount and type of discharges to soil, water, and air begins in the field, when the fiber source is planted. Specifically,

*Toward Zero Discharge: Innovative Methodology and Technologies for Process Pollution Prevention*, edited by Tapas K. Das.
ISBN 0-471-46967-X   Copyright © 2005 John Wiley & Sons, Inc.

planners select high yield trees that can be harvested with minimum impact to soil resources and surrounding areas. Management of clear cutting and protection of critical habitat are considerations and techniques of silviculture. Biogenetic engineering researchers are working to develop fiber resources that grow quickly and produce high quality fibers. In addition, the rapid development of processes to use secondary fiber effectively has placed the industry at the forefront of global efforts to recycle waste paper, conserve forest resources, and protect watersheds.

Most industrialized nations regulate emissions to the environment, principally solids, toxic substances, and other chemicals measured as biological and chemical oxygen demand (BOD and COD). At the facility and in the woodyard, nonpoint sources of runoff are contained and treated. As the fiber source passes through consecutive stages of processing, each operation is designed to minimize waste, energy use, and chemical discharges by means of efficient use of water and raw materials, recycling, substituting chemicals such as oxygen for chlorine, scrubbing, and other process controls.

Discharges to receiving water bodies must meet water quality standards to protect drinking water and aquatic life. Effluents contain lignin derivatives from pulping and bleaching operations and carry nutrients, suspended solids, color bodies, and factors that affect pH, dissolved oxygen, foaming, and other chemicals bearing upon water quality in the receiving water bodies. After the toxicity of dioxin was discovered, the industry made significant efforts to reduce and in some cases to eliminate the use of elemental chlorine by substituting chlorine dioxide, oxygen, ozone, hydrogen peroxide, or alternative bleaching agents in various sequences. Methods in common practice include extended delignification in the digester and manipulating the extent of delignification at succeeding stages, enzyme treatment, better brownstock washing, and optimal use of countercurrent washing. These process changes have significantly decreased the discharge of organic compounds measured as BOD or COD.

## 12.2 POLLUTION PREVENTION IN THE PULP AND PAPER INDUSTRY

Data collected from a large number of mills across the United States show that although there has been a significant increase in pulp production during the last 20–25 years, discharges to the environment have gone down. Over this period emissions of total reduced sulfur (TRS) gases have dropped by 85 percent, $SO_2$ emissions have gone down by over 40 percent, effluent flow has decreased by over 30 percent, and BOD and total suspended solids (TSS) have decreased by 75 percent and 45 percent, respectively. Paper recovery has increased from 20–50 percent. Cluster Rule, which links federal air and water pollution control regulations, has been in effect since 2001 and is expected to further reduce air and water discharges of hazardous air pollutants and chlorinated organics to the environment (Cluster Rule 40 CFR Part 63; USEPA, 2001). The Cluster Rule resulted in revised effluent guidelines for bleached papergrade kraft and soda mills and papergrade sulfite mills and established hazardous air pollution standards for pulp and paper mills. New and emerging technologies such as enzyme bleaching and black liquor gasification/ gas turbine cogeneration, production of ethanol from mill wastewater treatment plant sludges, beneficial land application of boiler ash, and vitrification of hog fuel boiler ash are accelerating pollution prevention, meeting the long-term industry goal of operating minimum impact mills.

We shall explore a number of options in the pulping, bleaching, and chemical recovery processes, as well as solid waste handling, in terms of the best available/emerging technologies and their performances, quality products, waste minimization, recycling, and pollution prevention.

Advancements in technologies and process changes implemented at mills in response to regulatory pressures, energy conservation and desire for pollution prevention have contributed to significant changes in the operating practices at pulp and paper mills. This section reviews some of the advances made by the industry over the past two decades in the areas of air pollution, effluent discharges, and solid waste (Das, 1997, 1999a,b, 2001). The opportunities that exist to establish realistic pollution prevention plans at bleached kraft pulp and paper mills for achieving maximum increases in in-plant reuse and recycling of water, chemicals and fibers are noted, as well. We begin by outlining the types of pollution generated by the industry.

## 12.2.1  Air Pollution

The kraft industry produces many of the substances that are subject to air pollution regulations concerned with minimizing particulates, odor, ground-level ozone precursors, acid rain, and smog. Thus acetone, methanol, sulfuric acid, sulfur dioxide, hydrogen sulfide and other reduced sulfur compounds, and nitrogen dioxide emissions from kraft plants are monitored closely. Volatile organic compounds such as alcohols, terpenes, and phenols can be controlled by incineration. Sour condensates are collected, steam stripped, and reduced sulfur compounds destroyed by incineration in the lime kiln. Similarly, sulfur-bearing off-gases are collected and destroyed by incineration. Monitoring is often performed on-line using continuous monitoring devices for stack effluents as well as ambient air. Pollution control equipment includes catalytic and thermal systems to destroy pollutants, electrostatic precipitators to remove particulate matter, cyclonic separators, venturi scrubbers, and fabric filters. Pollutants in molecular form are removed from the process stream primarily by adsorption (i.e., dry scrubbing) and by packed bed and venturi-type (i.e., wet) scrubbers for reuse or for destruction.

In the United States, the Clean Air Act of 1990 required plants to reduce emissions of 189 toxic and carcinogenic substances such as chlorine, chloroform, and dioxin (see Section 12.2.2.1) by 90 percent over the 1990s. The US Environmental Protection Agency is working to develop standards based on maximum achievable control technologies and the industry has invested billions of dollars in capital investments to retrofit or rebuild plant equipment to meet these measures. Sections 12.2.1.1 through 12.2.1.5 report progress made in the United States on preventing environmental contamination from some of the major pollutants produced by the kraft industry.

### 12.2.1.1 Reduced Sulfur Gases

The objectionable odor historically associated with the kraft process is that of the reduced sulfur compounds, hydrogen sulfide, methyl mercaptan, dimethyl sulfide, and dimethyl disulfide. In the kraft process, these compounds are measured as total reduced sulfur. Over the past 25 years a number of changes, including installation of black liquor oxidizers, conversion of direct contact evaporator (DCE) furnaces to non-DCE furnaces, improved combustion control, improved lime mud washing, and installation of non-condensible gas collection and combustion systems, have been implemented to reduce kraft mill TRS emissions. Black liquor oxidation minimizes TRS emissions from recovery furnaces in which vent gases come into direct contact with black liquor. Non-DCE furnaces eliminate contact between flue gases and black liquor and generally have better combustion control which result in lower TRS emissions. Improved lime mud washing reduces the level of sulfide in the lime mud and results in lower lime kiln TRS emissions. Non-condensible gases from digesters and evaporators contain high levels of TRS. When these gases are collected and combusted, the TRS gases are oxidized to $SO_2$, resulting in odor reduction. The resulting $SO_2$ is generally captured in alkaline scrubbers, and the captured sulfur can be returned to the process. The emissions of reduced sulfur compounds, which are measured as TRS, declined by 85 percent between 1970 and 1995 (Pinkerton, 1999).

### 12.2.1.2 Sulfur Dioxide and Oxides of Nitrogen

The pulp and paper industry is one of the major energy-consuming manufacturing industries in the United States. Energy is needed at pulp and paper mills to generate steam, electricity, and process heat. The combustion of energy-yielding fossil fuels, wood residues, and spent pulping liquors in turn produces $SO_2$ and $NO_x$. A report by the National Council of the Paper Industry for Air and Stream Improvement (NCASI) showed that from 1980–1985, US pulp and paper mill $SO_2$ emissions declined from 0.9 million tons to 0.5 million tons, while $NO_x$ emissions increased from 275,000 tons to 316,000 tons per year (NCASI, 1997). During this period kraft pulp production increased from $38.6–54.3 \times 10^6$ tons/year, paper and paperboard production increased from $61.0–89.5 \times 10^6$ tons/year, and sulfite pulp production decreased from $1.8–1.3 \times 10^6$ tons/year. These results are shown in Figures 12.1 and 12.2. $SO_2$ emission reductions have been achieved through improved combustion efficiency and higher solids firing in kraft recovery furnaces, through increased use of natural gas and biomass as fuel, increased $SO_2$ control at sulfite mills, and reductions in sulfite pulp production.

### 12.2.1.3 Chloroform Emissions

Chloroform is produced in bleach plants in the chlorine, chlorine dioxide, extraction, and especially hypochlorite stages. Over the last few years the industry has greatly reduced the use of hypochlorite in pulp bleaching. This, combined with reductions in chlorine usage, has resulted in very significant reductions in pulp and paper mill chloroform emissions. From

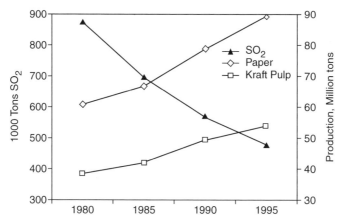

**Figure 12.1.** Industry $SO_2$ emissions and pulp production—1980–1995. (From NCASI, 1997.)

1987–1995 the reported emissions of chloroform decreased from 0.35 kg/ton to 0.15 kg/ton (Someshwar and Jain, 1999.)

### 12.2.1.4 Chlorine and Chlorine Dioxide

At pulp and paper mills, chlorine and chlorine dioxide emissions result primarily from one of these chemicals in bleach plants. Since 1990 there has been a significant reduction in the use of elemental chlorine which has been replaced with chlorine dioxide. Mills have also been using more hydrogen peroxide in the bleach plant. Ozone is sometimes used. As a result of these changes in the bleaching process and installation of more efficient $Cl_2/ClO_2$ scrubbers on bleach plant vents, reported point source emissions of $Cl_2$

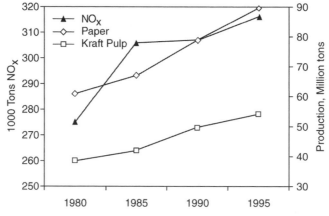

**Figure 12.2.** Industry $NO_x$ emissions and pulp production—1980–1995. (From NCASI, 1997.)

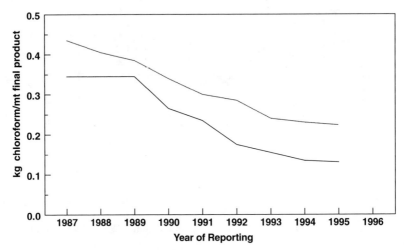

**Figure 12.3.** Reported pulp mill chloroform source emissions, 1987–1995: upper curves, mean releases; lower curves median.

and $ClO_2$ from pulp and paper mills have gone down significantly, as shown in Figures 12.3 and 12.4.

**12.2.1.5 Greenhouse Gases**   In 1999, NCASI assembled data on fossil fuel and purchased electricity used by the pulp and paper industry and used these data to estimate the emission of $CO_2$, a large component of greenhouse gases. The results plotted in Figure 12.5 show slight reductions in energy usage and significant

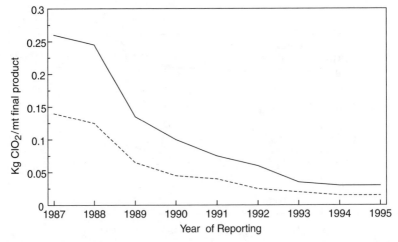

**Figure 12.4.** Reported pulp mill chlorine dioxide source emissions, 1987–1995: solid curve, mean releases; dashed curve, median.

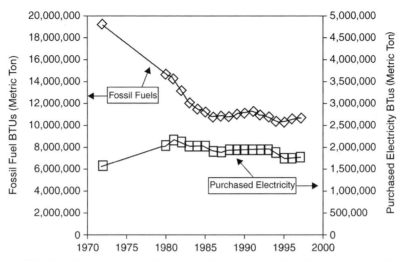

**Figure 12.5.** Fossil fuel and purchased electricity consumption for the pulp and paper industry in the United States: 1990–1998. (From NCASI, 1999a.)

diminution in $CO_2$ emissions per unit of production between 1970 and 1998 (NCASI, 1999a). These reductions were achieved through installation of more energy efficient equipment and better process controls.

## 12.2.2 Effluent Discharges

Before pulp mill effluents can be released to the environment they must be treated. Primary treatment involves the use of settling ponds or tanks in which suspended solids settle out of the liquid effluent. Solids can be composted and spread on land, converted to other useful products, or incinerated. Secondary effluent treatment includes oxidation and aeration in shallow basins having wide areas or in smaller areas using mechanical agitators and spargers to oxygenate fluids before release. Biological filter systems can be used to remove organic compounds and heavy metals, and often the process can be accelerated by adding nutrients and by using oxygen rather than air. In some areas natural wetland systems have been designed to achieve this. Other means of treatment include lime coagulation and the elevation of pH to precipitate organic color bodies as calcium lignates. Once precipitated, the sludge is dewatered and incinerated to destroy organics.

It has been reported by NCASI that there was a 30 percent reduction in effluent flow from mills between 1975 and 1988. During the same period final effluent 5-day BOD and TSS decreased by 75 and 45 percent (NCASI, 1991). Data for 1975, 1985, and 1988 are presented in Table 12.1. These reductions will continue as pulp and paper mills implement the so-called best management plans (BMPs) required by the Cluster Rule. BMPs will require better management of process losses and spills and are expected to reduce effluent discharges. For example, in roughly

**TABLE 12.1. Effluent Discharges from Pulp and Paper Mills (Das, 1997)**

|                        | 1975   | 1985   | 1995   |
|------------------------|--------|--------|--------|
| Effluent flow, gal/ton | 22,800 | 17,200 | 16,000 |
| BOD, lb/ton            | 18.0   | 4.8    | 4.4    |
| TSS, lb/ton            | 13.0   | 8.3    | 7.1    |

10 years, Simpson Tacoma kraft mill reduced its freshwater consumption by approximately 50 percent through various pollution prevention methods (Narum, 2001; USEPA, 1992a,b).

Simpson Tacoma now uses about 17 million gallons of freshwater per day (mgd), versus 32 mgd in 1989. This reduction in freshwater usage saves about $1.92 million per year ($350 per million gallons, city of Tacoma charges per average fixed and variable cost). The plant also saves money through reducing sodium hydroxide (NaOH) losses to sewers and to product fiber, by stopping overflowing weak washing dilute NaOH solution and through recycling processed water. Otherwise, NaOH would be needed in the process to make up for soda (sodium) lost. The savings is about 2.4 million per year (based on $335 per ton of NaOH and 20 ton of NaOH savings per day). Simpson also saves $0.18 million per year by reducing losses of black liquor sulfur to sewer and stack (Narum, 2001).

Other firms save water by installing savealls (devices which separate fiber from process water), heat exchangers, and other equipment which permit more reuse of process water. Internal water cleaning systems make it possible to substitute filtered white water for clean water. Separating process cooling and clean water is often necessary to achieve balance in operations and water use.

### 12.2.2.1 Dioxin and Furans

In early 1988, a study of effluents from bleached pulp mills showed significant levels of dioxins and furans. As a result of these findings, the industry implemented a series of process changes including (1) eliminating the use of certain defoamers which contained dioxin and furan precursors; (2) decreasing the use of chlorine as bleaching chemical; and (3) increasing the use of chlorine dioxide for pulp bleaching. Between 1988 and 1996, there was a very significant reduction in effluent 2,3,7,8-tetrachloro dibenzo-p-dioxin (TCDD) and 2,3,7,8-tetrachloro dibenzo-p-furan (TCDF) concentrations. For example, in 1988 40 percent of the mill effluent samples contained less than 10 parts per quadrillion (ppq) of 2,3,7,8-TCDD. In 1990, 70 percent of the samples contained less than or equal to 10 ppq of 2,3,7,8-TCDD. By 1996, all but two mill effluent samples showed 2,3,7,8-TCDD levels below 10 ppq. This record exemplifies how pollution prevention leading to zero discharge of dangerous environmental contaminants can be achieved at the source through process changes and substituting chemicals.

### 12.2.2.2 Adsorbable Organic Halides

Adsorbable organic halide (AOX) is produced in bleach plants primarily as a result of using elemental chlorine. It has been estimated that the loading of AOX in effluents from the first two bleaching stages, which are the major contributors to bleach plant AOX, are a function of the atomic chlorine charge and pulp kappa number (NCASI, 1993). Thus as chlorine use as bleaching chemical has been reduced, AOX discharge from pulp mill has gone down. Luthe (1998) reported that between 1988 and 1995, chlorine use at Canadian pulp mills decreased by 84 percent from 607,500 tons/year to 95,200 tons/year. During the same period the average AOX content of the effluents decreased by 81 percent from 4.8 kg/ton to 0.9 kg/ton. Another report showed that at mills using complete substitution, the average AOX levels were 0.16 kg/ton and 0.435 kg/ton, respectively, when oxygen delignification was and was not practiced (NCASI, 1994). These results suggest that lower effluent AOX levels would be expected from pulp and paper mills in the future as bleach plant modifications aimed at complete elimination of chlorine as a bleaching chemical are implemented.

### 12.2.3  Solid Wastes

The main potential solid wastes generated in wood pulping operations consist of wood and fibrous rejects, residue, and process losses; caustic waste solids, mainly calcium-based grits, dregs, and lime mud; and yard, wood yard, ashes, and general mill solid waste. The bark and wood rejects usually are processed and used in bark boilers on the site or are sold as hog fuel. Rejects from pulping operations are minimized by recooking knots and recovering good fiber and either adding the fiber to the hog fuel stream or selling it to pulping operations for making lesser grades. The balance of the lost fiber is collected in effluent primary clarifiers. It is then pressed and used as hog fuel. Effluent sludges from secondary treatment plants, if continuously recovered from the process, is dewatered with the primary sludge in a screw press and also is used as hog fuel. As an alternative, a recent study showed the possibility of converting mill wastewater treatment plant sludges into ethanol fuels.

Since the early 1970s, economic and regulatory factors, as well as the desire for pollution prevention, have encouraged mills to implement beneficial use of solid wastes (NCASI, 1999b). The NCASI survey shows that the US mills generally produce about 15 million tons of solid waste per year, and about 90 percent of these potential solid wastes are being recycled, reused, and utilized in waste-to-energy processes. The balance of remaining solid waste is mostly dirt and some general trash, which is disposed of in landfills.

*12.2.3.1 Wastewater Treatment Residuals*  Since most pulp and paper mills treat their own wastewater, they generate wastewater treatment system residuals. It was estimated that between 1988 and 1995 the quantity of residuals increased from 4.6 to 5.8 million dry tons (NCASI, 1999b). The increase was likely due to a 17 percent increase in production and increased use of recovered fiber in the industry. NCASI reported a significant decrease in landfilling as well as increases in such beneficial uses as land application, burning, and recycling between 1979 and 1995 (Figure 12.6).

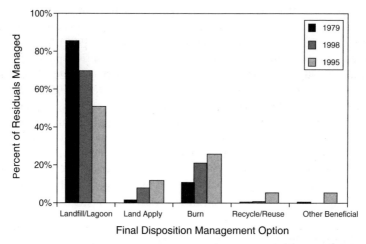

**Figure 12.6.** Industry management of residuals. (From NCASI, 1999b.)

***12.2.3.2 Ash*** The pulp and paper industry burns coal, wood/bark, and wastewater treatment residuals. It is estimated that between 1988 and 1995 the rate of ash generation in the pulp and paper industry remained unchanged at 2.86 million tons of ash annually (NCASI, 1999b). During this period, however, beneficial use of ashes (in soil conditioning, cement and concrete manufacturing) increased (NCASI, 1999b).

***12.2.3.3 Recyclable Paper and Fiber*** In the United States, recycling of paper has been practiced for many years. The practice, however, got greater emphasis in the early 1980s as landfill space got scarce and there was increased commitment to pollution prevention. In 1990 the US paper industry announced a national goal of increasing the paper recovery rate from 33–45 percent by 1997. This goal was later changed to 50 percent by 2000, with higher rates slated for coming years. Now recycled fibers comprise the fastest-growing sector in the industry, and technology developments in paper machines have allowed these fibers to be used in a wider range of product grades. Waste-to-energy technologies also are being developed for low-grade waste papers, and recycled papers not suitable for reuse in paper and board (NCASI, 2000). Table 12.2 presents data from the American Forest and Paper Association (AFPA) on the recovery rate of fiber from 1970–1997 (AFPA 1998). The recovery rate in Table 12.2 is calculated as ratio of recovered paper collected to new supply of paper and paperboard.

### 12.2.4 Emerging Technologies

Research is continuing on efforts to modify the process so that fewer pollutants are generated. Some companies are now at work developing new bleaching systems and enzymes that will optimize performances and lower loading requirements for ozone,

**TABLE 12.2. US Paper Recovery Rate, 1970–1997**

| (Percent) (AFPA, 1998) | |
| --- | --- |
| 1970 | 22.4 |
| 1975 | 24.2 |
| 1980 | 26.7 |
| 1985 | 26.8 |
| 1990 | 33.5 |
| 1995 | 43.8 |
| 1997 | 45.2 |

chlorine dioxide, and other chemicals in bleaching systems that are free of both elemental chlorine and chlorine compounds. Hardier and more efficient enzymes are also being developed to reduce chemical use for biopulping applications. One example is the proprietary substance Lignozym, which can mimic natural lignin degradation by white-rot fungi in a controlled reactor (Chem. Eng., 1997; Scott et al., 1998a,b).

Alternative uses of solid residues from pulp and paper mills are being explored. A study of the feasibility of converting mill wastewater treatment plant sludge into ethanol fuels assessed sludges from eleven pulp and paper mills in the Pacific Northwest to determine their chemical characteristics, ethanol assays, and economic potential (Kerstetter et al., 1997). A few mills processing old corrugated containers (OCC) are practicing fiber recovery from rejects for use in paperboard manufacturing.

Mills processing polyethylene-coated papers are examining the use of rejects to manufacture wood-plastic composite lumber and fuel pellets (NCASI, 1999b).

Unlike the areas of pulping and bleaching, which have posted significant process changes resulting greater energy efficiencies and environmental benefits, very little has changed in the field of black liquor gasification.

In kraft mills, black liquor is burned in Tomlinson furnaces to recover the cooking chemicals and produce steam for use in the process. The Tomlinson furnaces, which have been in use since the 1930s, have a low thermal efficiency, are susceptible to explosions, and are expensive. Thus since the 1970s there has been research to find an alternative. At present, a considerable amount of research is directed at developing a black liquor gasifier (Iisa et al., 1999a,b; Stone and Webster, 1996). A number of different designs are being considered and full-scale systems have been installed. It is possible that in the next 10–20 years, as the current generation of Tomlinson furnaces reaches the end of their useful life, new black liquor gasifiers/gas turbine cogeneration units may be installed to achieve higher energy efficiency, better safety, and lower emissions. Later, in our discussion of resource recovery, we shall highlight a low-temperature stream re-forming process that is emerging to provide an alternative to kraft recover (see Section 12.3.6).

Another innovative process, developed by Scott Lynn and other scientists at the University of California, featured the chemical conversion of hydrogen sulfide to elemental sulfur and syngas. This process is applicable to many commercially important gases, including kraft TRS emissions (mostly consists of $H_2S$), natural gas, most coal-derived gas, and refinery gases generated by hydrodesulfurization processes. The atmospheric emissions of sulfur-containing compounds are effectively zero for this recovery process (Towler and Lynn, 1995).

Hydrogen sulfide ($H_2S$) is a frequently found contaminant of industrial gas streams. Generally, concentrations of $H_2S$ must be reduced below 4 parts per million (ppm) before the gas stream may be used or discharged to the atmosphere. Conventional technology for $H_2S$ removal is expensive, consumes a great deal of energy, and emits sulfur-containing compounds to the atmosphere. The current method typically involves absorbing the hydrogen sulfide from the sour gas, then stripping the $H_2S$ from the absorbent and sending it to the Claus process. The Claus plant reaction is equilibrium-limited and does not go to completion even when two or three reaction stages are employed. Therefore, the gas exiting the Claus plant contains 2,000 to 6,000 ppm of unreacted $H_2S$ and $SO_2$ and must be treated in a tail gas unit before being released to the atmosphere. The tail gas unit is about as expensive as the Claus plant, uses a substantial amount of energy, and does not completely eliminate sulfur emissions.

The process of Towler and Lynn hinges on a newly discovered reaction chemistry for the partial oxidation of $H_2S$ to sulfur, in which $H_2S$ and carbon dioxide are reacted at high temperature to form water vapor, carbon monoxide, hydrogen and elemental sulfur. A recycle stream ensures that sulfur leaves the process only in the product stream of elemental sulfur. The final product gas contains only CO and $H_2$ and is valuable either as a fuel or as a chemical feedstock (Towler and Lynn, 1995).

## 12.3   RESOURCE RECOVERY AND REUSE

### 12.3.1   Value-Added Chemicals from Pulp Mill Waste Gases

Methanol, formed during the pulping of wood and contaminated with reduced sulfur compounds and terpenes, is the largest single source of VOC emissions from kraft pulp mills, accounting for 70–80 percent of total emissions. The Cluster Rule limits methanol emissions for all pulp mills in the U.S. Canada faces similar legislation.

Methanol emissions can be reduced by collecting condensate streams from the digesters, evaporators, and other sources in the mill. The collected condensate streams can then be steam-stripped to concentrate the methanol for incineration. A few mills are "hard-piped" to send this methanol-laden stream to bio-treatment plants, thus avoiding incineration.

Stripper overhead gas (SOG) contains roughly equal part of methanol and water (40–50 wt%) and roughly equal parts of TRS and terpenes (1–5 wt%).

In the United States, the SOG is likely to be burned in an incinerator, kiln, or boiler. Recently, some mills have found it advantageous to rectify the SOG to about 80 percent methanol and collect it as a liquid, which has a higher fuel value. With this concentrated (70–80 percent) methanol stream available, plants can cut down on the amount of natural gas that must be bought.

A process has been developed at Georgia-Pacific that converts the methanol and TRS (mostly mercaptans) in the rectified SOG into formaldehyde. Based on work by Professor Israel Wachs of the Chemical Engineering Department of Lehigh University in Bethlehem, PA, this patented catalytic process (Burgess et al., 2002; Wachs, 1999a,b) has achieved commercially viable yields of formaldehyde (70–80 percent) from a typical pulp mill SOG feedstock containing methanol, water, and TRS compounds.

The conversion of methanol and TRS to formaldehyde presents kraft mills with a more profitable alternative for SOG than incinerating the gas as a fuel. The formaldehyde produced can be used by resin manufacturers to produce thermosetting resins commonly used in plywood and other structural panels. A typical pulp mill of 2,000 ADTPD (air-dried tons per day) output may achieve a payout of 2–4 years, depending on the price of methanol and local economics.

The process also produces two levels of low-pressure stream, 60 and 70 psig, usable within a paper mill, by reducing most of the methanol to formaldehyde, rather than $CO_2$. The reduction in $CO_2$ emissions is about 80–85 percent of the amount otherwise generated by incineration or by the bio-treatment plant. For a 2,000 APTDP mill, this equates to 28 tons per year of $CO_2$ emissions that are avoided.

A typical itemization of income, costs, and earnings is given in Table 12.3. For a 2,000 ADTPD mill, a payout of three to four years is calculated based on formaldehyde prices of $0.06/lb (50 percent) basis. This assumes a customer-shipping radius of 500 miles from the mill producing the formaldehyde. A heat value credit to the mill is included (equivalent of natural gas fuel) for the methanol that would otherwise have been incinerated and used to generate steam via heat recovery exchangers.

**TABLE 12.3.  Annualized Income, Costs, and Earnings 2,000 ADTPD Mill, 14 lb Methanol/ton SOG (Burgess et al., 2002)**

| | |
|---|---|
| Income 15,000,000 lb/year formaldehyde (50 percent) at $0.06/lb, FOB mill | $900,000 |
| Operating cost | |
|    Direct labor | $110,000 |
|    Utilities | $25,000 |
|    Methanol in feed at $2.25/MMBTU fuel value | $260,000 |
|    Misc., catalyst, supplies, etc. | $70,000 |
|    Total operating cost | $465,000 |
| Gross margin | $435,000 |
| Depreciation | $210,000 |
| Earning before tax | $225,000 |
| Credit for steam generated (depending on the mill's steam balance) | $95,000 |
| Credit for terpene recovered and sold | $110,000 |
| Credit for $SO_2$ recycled | $45,000 |
| Total byproduct credits | $250,000 |
| Total earnings, including byproduct credits | $$475,000 |

## 12.3.2 Recovery and Control of Sulfur Emissions

Sulfur is often considered one of the four basic raw materials in the chemical industry. It can be recovered as a byproduct from sulfur removal and recovery processes (Kirk-Othmer, 1996). Historically, sulfur recovery processes focus on the removal and conversion of hydrogen sulfide ($H_2S$) and sulfur dioxide ($SO_2$) to elemental sulfur, as these species represent significant emission from pulping process. Various processes for the removal of $SO_2$ in the combustion gases are available. Direct catalytic oxidation of $SO_2$ to $SO_3$, and subsequent absorption of $SO_3$ in water to produce sulfuric acid, is an alternative method (Paik and Chung, 1996).

Ten to 20 percent of total mill TRS emissions are contributed by vent streams from brownstock washers, foam tanks, black-liquor filters, oxidation tanks, and storage tanks that are not typically collected in the noncondensible gas system. The emissions from these sources can be collected and combusted for energy recovery, reducing the atmospheric emissions of TRS and VOCs.

## 12.3.3 Delignification

Improved delignification processes allow modification of the subsequent bleach sequence by reducing the quantities of chemicals required to produce pulps of the commercially requisite brightness. Nevertheless, the degree to which lignin can be removed by extended cooking of the woodchips alone is limited. Overcooking results in a reduction of the strength qualities of the pulp. One route to delignification pulping uses anthraquinone as an additive. This approach allows cooking to lower kappa numbers without loss of pulp quality, but the additive is expensive and this drawback has restricted its use. Extended delignification without strength loss can be achieved by ensuring that the alkali concentration stays as constant as possible throughout the cook (Gullichsen, 1991). The use of methods such as modified concurrent–countercurrent cooking or rapid displacement heating can reduce the lignin content of softwood kraft pulp from 5 percent to less than 4 percent.

An earlier approach to delignification, the use of oxygen in combination with alkali was first practiced commercially in 1970. After steady growth during the 1980s and early 1990s it became one of the main delignification technologies. By 1994, according to industry figures (Reeve, 1996a,b) some 50 percent of world capacity for kraft pulp incorporated an oxygen delignification stage. A great advantage of this process, is that unlike chlorine based bleach effluents, the effluent can be recycled, concentrated and sent to the chemical recovery system. Most new mills and mill expansions, especially in Scandinavia, include it, while the use of oxygen to enhance the effectiveness of the first caustic extraction stage is universal.

Delignification is generally limited to between 40 and 50 percent since oxygen lacks the selectivity needed to take the process further. Accordingly, research has been directed at developing chemical pretreatments to improve oxygen selectivity and hence avoid pulp strength loss. An example is the PRENOX process (Gullichsen, 1991), although this has not evolved to the stage of commercial development and application.

## 12.3.4 Solvent Pulping

Solvent pulping, which has been receiving increased attention in recent years, offers a number of potential advantages over conventional pulping techniques. These include relatively low chemical and energy consumption coupled with low capital costs and low environmental impact. Aqueous organic solvents such as methanol and ethanol are used for delignification to produce a bleachable pulp which can be bleached with non-chlorine chemicals. Pilot scale tests have given high pulp yields with strength properties similar to sulphite and kraft pulps (Sierra-Alvarez and Tjeerdsma, 1995).

## 12.3.5 Biopulping: A Review of a Pilot-Plant Project

Biopulping is the treatment of wood chips and other lignocellulosic materials with lignin-degrading fungi prior to pulping. Ten years of industry-sponsored research has demonstrated the technical feasibility of the technology for mechanical pulping at a laboratory scale. Two 50-ton outdoor chip pile trials conducted at the US Department of Agriculture Forest Service, Forest Products Laboratory (FPL) in Madison, Wisconsin, established the engineering and economical feasibility of the technology. After refining the control and the fungus-treated chips through a thermomechanical pulp (TMP) mill, the resulting pulps were made into papers on the pilot-scale paper machine at FPL. In addition to the 30 percent savings in electrical energy consumption during refining, improvements in the strength of the resulting paper were attributed to the fungal pretreatment. Because of the stronger paper, at least 5 percent kraft pulp could be substituted in a blend of mechanical and kraft pulps. This recent work has clearly demonstrated that economic benefits can be achieved with biopulping technology through both the energy savings and substitution of the stronger biopulped TMP for more expensive kraft, while maintaining the paper quality.

### 12.3.5.1 Early Work at FPL   Although mechanical pulping, with its high yield, is viewed as a way to extend scarce resources, the process uses much electrical energy and yields paper of lower strength than that produced by chemical pulping. Bio-pulping, which uses natural wood decay organisms, has the potential to overcome these problems. Fungi alter the lignin in the wood cell walls, which has the effect of "softening" the chips. This substantially reduces the electrical energy needed for mechanical pulping and leads to improvements in the paper strength properties. The fungal pretreatment is a natural process; therefore, no adverse environmental consequences are foreseen (Kirk et al., 1993, 1994; Sykes, 1994).

Results of previous work and discussions with industry experts led to a proposal for a fungal treatment system that fits into existing mill operations with minimal disturbance. Figure 12.7 is a conceptual overview of the biotreatment process in relation to existing wood yard operations. Wood is harvested and transported to the mill site for debarking, chipping, and screening. Chips are decontaminated by

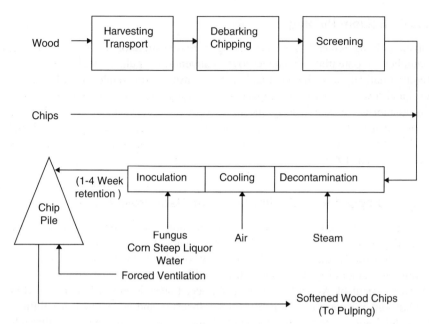

**Figure 12.7.** Overview of the biopulping process showing how the biotreatment process fits into an existing mill's wood-handing systems.

steaming, maintaining a high temperature for a sufficient time to decontaminate the wood chip surfaces, and then cooled so that the fungus can be applied. The chips are then placed in piles that can be ventilated to maintain the proper temperature, humidity, and moisture content for fungal growth and subsequent biopulping. The retention time in the pile is 1–4 weeks. Recent efforts have focused on bringing the successful laboratory-scale procedures up to the industrial level. Our laboratory process treats approximately 1.5 kg of chips (dry weight basis) at one time. In scale-up experiments that took biopulping from this lab scale to larger scales, it was demonstrated that chips can be decontaminated and inoculated in a continuous process rather than as a batch process and that the process scaled as expected from an engineering standpoint (Kirk et al., 1993, 1994).

### 12.3.5.2 Large Scale Implementation of Biopulping

On a large scale, decontamination and inoculation must be done on a continuous basis and not batch-wise as in a laboratory trial. To achieve this, the FPL investigators built a treatment system based on two screw conveyers that transport the chips and act as treatment chambers. Figure 12.8 is an overview of the continuous process equipment used in 5- and 50-ton trials. Steam is injected into the first screw conveyer, which heats and decontaminates the chip surface. A surge bin, located between the two conveyers, acts as a buffer. From the bottom of the surge bin, a second screw conveyer removes the chips, which are subsequently cooled with filtered air. In the second half of the second screw conveyer, the inoculum suspension containing

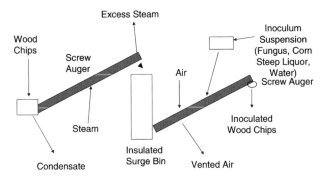

**Figure 12.8.** Continuous treatment system to decontaminate, cool, and inoculate wood chips. Wood chips are steamed in the first screw conveyor before being placed into a surge bin. The second screw conveyor then picks up the chips, cools them, and applies the inoculum. Pile Storate and Ventilation.

fungus, unsterilized corn steep liquor, and water is applied and mixed with the chips through the tumbling action. From the screw conveyer, the chips fall into the pile or reactor for a 2-week incubation. In the first scale-up trial, 5 tons of spruce wood chips were inoculated and incubated at a throughput of approximately 0.5 tons per hour.

In successful outdoor trials with the biopulping fungus *Ceriporiopsis subvermispora*, nearly 50 tons of spruce was treated at a throughput of about 2 tons per hour (dry weight basis) continuously for over 20 hours. During the 2 weeks, the chip pile was maintained within the temperature growth range for the fungus, despite the outdoor exposure to ambient conditions.

### 12.3.5.3 Decontamination
In pilot scale trials, the screw conveyer used for steam decontamination was essentially open to the atmosphere. This design was adequate for the 5-ton pilot scale, but additional steam was added to the surge bin to decrease the decontamination time for the chips. In the subsequent 50-ton outdoor trials, the larger screw conveyor eliminated the need for additional steaming in the surge bin. That is, the chips were suitably decontaminated after steaming in the screw conveyor. In fact, in the second 50-ton trial, the surge bin was used for additional cooling capacity by blowing air into the bottom of the surge bin. Because of steam temperatures that can be obtained in pressurized vessels, the surge bin will not be needed for decontamination. It must be kept in mind that good surface exposure of the chips to the steam is important for achieving decontamination in a short time, some surge capacity between operations may still be desirable to isolate the effects of short-term shutdowns and process variations in the sequence. Also, the flexibility if being able to use the surge bin for additional cooling or steaming can make the process more robust.

### 12.3.5.4 Economic and Environmental Benefits of the Process
In an economic analysis of biopulping, which yielded very favorable results, the

primary criterion for the effectiveness of biopulping was energy reduction at the refiner. For a 2-week process, the savings should be a minimum of 25 percent under the worst-case conditions of wood species and minimal process control, whereas under some circumstances up to nearly 40 percent can be achieved. Additionally, biopulping may allow mills that are throughput-limited as a result of refiner capacity to achieve total capacity increases.

With respect to product quality, the improved strength of the biomechanical pulps would allow the use of a lower percentage of kraft pulp in some blended pulps. Finally, only benign materials are used, and no additional waste streams are generated. Even greater benefits can be realized when the other benefits of biopulping— such as increased throughput and substitution for kraft—are considered. Throughput increases brought the simple payback period of the process to less than one year. Substituting this increased production for kraft pulp in blended products results in additional savings. From this analysis, biopulping can produce substantial economic savings for TMP producers.

We have presented results from a 200 ton/day TMP mill (Scott et al., 1998a). Under different scenarios and assumptions for utility costs, equipment needs, and operating costs, the net savings can range from $10 to more than $26 per ton of pulp produced, with an estimated capital investment of $2.5 million. Mills that are refiner-limited can experience throughput increases of over 30% from the reduction in refining energy by running the refiners to a constant total power load. Even a modest throughput increase of 10 percent, coupled with the energy savings of 30 percent, results in a payback in less than 1 year. This is equivalent to a savings of $34 per ton at a 15 percent rate of return on capital. Furthermore, many mills blend mechanical pulps and kraft pulps to achieve the optical and strength properties desired. Additional benefits of over $10 per ton can be realized when the anticipated stronger biopulped TMP is partially substituted for kraft at a 5 percent rate. A preliminary analysis of results from a 600 ton/day mill are also available (Das and Houtman, 2004; Scott et al., 1998b).

Although the environmental benefits of substituting biopulping for mechanical pulping have not been quantified, we can discuss the benefits qualitatively. Sykes (1994) determined that there were no environmental problems with the effluent from a biopulping process. In terms of greenhouse gases, there is a net benefit to the biopulping process. The process itself is oxidative, producing carbon dioxide as the fungus metabolizes the various wood components. However, based on the typical generation efficiency of electrical energy, the reduced energy requirements (approximately 30 percent), significantly lowers the overall generation of carbon dioxide when the entire system is analyzed. Finally, there are benefits to using biopulping fungus in the wood yard because biopulping fungus outcompetes the normal cohort of fungi and other organisms that will grow in an uncontrolled wood chip pile. Some of these other organisms, such as *Aspergillus*, can be detrimental to human health. The biopulping fungi occur naturally in the forests of the world and have no known health effects on humans.

A streamlined environmental life cycle assessments comparing chemical (kraft, sulfite), mechanical (or thermomechanical), and biopulping processes was reported by Das and Scott (2003). This LCA can assist those charged with evaluating the industry's current experience and practices in terms of environmental stewardship, regulatory and non-regulatory forces, life cycles of its processes and products, and future developments (Das and Houtman, 2004).

### 12.3.5.5 The Future of Biopulping

Despite much research effort to bring bio-pulping technology to commercialization, many questions remain unanswered. For example, the molecular mechanism of biopulping has not been elucidated. An understanding of this mechanism will facilitate the optimization of the process for both mechanical and chemical pulping. In addition, most of the work has focused on the use of the biotreatment for mechanical pulping, with some investigation of sulfite pulping; the use of biopulping as a pretreatment for the kraft process is still an open research issue. Finally, the use of this technology for other sub-strates—nonwoody plants such as kenaf, straw, and corn stalks—is an area offering opportunities for the future (Scott et al., 1998a,b).

### 12.3.6 New Process for Recovery of Chemicals and Energy from Black Liquor

An exciting emerging technology that uses low temperature steam re-forming instead of traditional combustion to recover chemicals and energy from black liquor boosts high thermal efficiency, low emissions (including particulate matters and odorous TRS gases), and low power consumption. This versatile process, a very good example of pollution prevention at the source through process changes, is now available for commercial scale processing of black liquors generate at the pulp and paper industry. The black liquor steam reforming process utilizes indirect heating of a steam fluidized bed of sodium carbonate solids to recover cooking chemical and energy (Figure 12.9). Black liquor is sprayed directly into the bed where the liquor droplets uniformly coat the bed solids, resulting in high rates of heating, pyrolysis and steam reforming. The process steam reforms black liquor in the absence of oxygen, producing a hydrogen-rich, medium-Btu gas with a heating value of 300–350 Btu/SCF. Bed temperatures are maintained at 1059 to 1150°F (565–620°C), thereby avoiding liquid smelt formation and the associated smelt-water explosion hazards. In this dry (no smelt) recovery process, the sodium sulfate in the liquor is reduced by reaction with the steam-reforming products, principally carbon monoxide and hydrogen. The reduced sulfur form is unstable in the reformer environment and decomposes rapidly to gaseous hydrogen sulfite ($H_2S$) and solid sodium carbonate ($Na_2CO_3$). The product gas is scrubbed with a solution derived from the sodium carbonate bed solids to produce green liquor for re-use in the pulping cycle.

Heat required to reach operating temperature and for the endothermic steam-reforming reactions is supplied by heat exchangers immersed in the fluidized bed. The heat exchangers consist of bundles of pulsed heater resonance tubes. A portion of the product gas is burned in the pulsed heaters to supply the necessary heat, thus making the steam reformer self-sufficient on its own fuel. Pulsations in

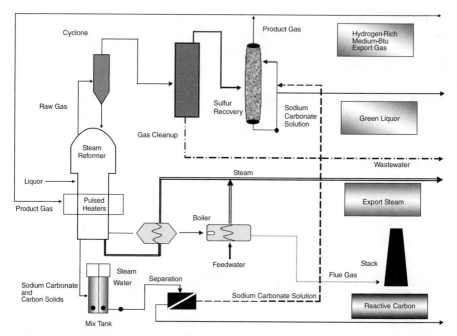

**Figure 12.9.** Low temperature steam reforming process to recover chemicals and energy—a schematic diagram of kraft recovery spent liquor steam reforming. (From Stone and Webster Engineering Corp., 1996.)

the resonance tubes produce a gas-side heat transfer coefficient which is up to five times greater than a conventional fire-tube heater. This efficient heat transfer reduces the size and cost of the heat exchangers and reformer vessel. The hot combustion gases leaving the pulsed heaters are sent to a waste heat boiler to generate steam and to preheat the pulsed heater combustion air. The product gas from the reformer is passed through a cyclone to remove particulate matter, then cooled in a waste heat steam generator, scrubbed and quenched.

By separating the sodium and sulfur elements, this process provides an imaginative and chemically attractive alternative to conventional kraft recovery. The process separates the individual unit operations of the conventional recovery boiler and allows independent control of each operation under its own optimized conditions, producing a well-controlled, more efficient modular process plant. It is applicable to both incremental and full-scale production facilities. The process also provides attractive energy alternatives, since the product gas is compatible with combined-cycle cogeneration operations (Stone and Webster Engineering Corp., 1996; Iisa et al., 1999a,b)

The US pulp and paper industry has made significant progress towards minimizing total pollutant discharges to the environment over the past two decades. Continued research and development of new technologies for the manufacture of pulp and paper will, no doubt, further enhance the recovery of materials from wood and continue to reduce the environmental impacts from these processes.

## REFERENCES

American Forest and Paper Association (AFPA) (1998). *1998 statistics — data through 1997 — paper, paperboard & wood pulp.* Washington, DC.

Burgess, T. L., Gibson, A. G., Furstein, S. J., and Wachs, I. E. (2002). Converting waste gases from pulp mills into value-added chemicals. *Environmental Progress*, **21**(3), 137–141.

*Chemical Engineering* (1997). Closing the pulp bleaching cycle. pp. 33–37.

Cluster Rule Regulation: 40CFR Part 63, Subpart S (foul condensates), Section 63, 446.

Das, T. K. (1997). A pollution prevention strategy in the kraft pulp and paper manufacturing process technologies. Paper 188c, American Institute of Chemical Engineers National Meeting, Los Angeles, CA.

Das, T. K. (1999a). Process pollution prevention advances in the pulp and paper industry. Paper 274b, American Institute of Chemical Engineers National Meeting, Dallas, TX.

Das, T. K. (1999b). Process technology advances in the pulp and paper industry. Paper 273e, American Institute of Chemical Engineers National Meeting, Dallas, TX.

Das, T. K. (2001). Pollution prevention advances in pulp and paper processing. *Environmental Progress*, **22**(2), 87–92.

Das, T. K., and Houtman, C. (2004). Evaluating chemical-, mechanical-, and bio-pulping processes and its sustainability characterization using life cycle assessment. *Environmental Progress*. Special Issue on Sustainability (in press).

Das, T. K., and Scott, G. M. (2003). Life cycle assessment comparisons of chemical-, mechanical-, and bio-pulping processes. American Institute of Chemical Engineers National Meeting, San Francisco, Paper #158d.

Gullichsen, J. (1991). Process internal measures to reduce pulp mill pollution load. *Water Science and Technology*, **24**(3–4), 45–53.

Iisa, K., Sinquefeld, S., and Sricharoenchaikul, V. (1999a). Pressurized gasification of black liquor at institute of paper science and technology. American Institute of Chemical Engineers National Meeting, Dallas, Texas.

Iisa, K., and Jing, Q. (1999b). Gasification of spent pulping liquor combined with mill wastes. American Institute of Chemical Engineers National Meeting, Dallas, Texas.

Kirk, R. E., and Othmer, D. F. (1996). *Encyclopedia of Chemical Technology*, 4th Ed. Wiley, New York, NY.

Kirk, T. K., Koning Jr., J. W., Burgess, R. R., Akhtar, M., et al. (1993). *Biopulping: A Glimpse of the Future?* Res. Rep. FPL-RP-523. US Department of Agriculture, Forest Service, Forest Products Laboratory, Madison, WI.

Kirk, T. K., Akhtar, M., and Blanchette, R. A. (1994). *TAPPI PRESS*, Biopulping: seven years of consortia research. 1994 *Tappi Biological Sciences Symposium*, 57–66, Atlanta, GA.

Kerstetter, J. D., Lynd, L., Lyford, K., and South, C. (1997). Assessment of potential for conversion of pulp and paper sludge to ethanol fuel in the Pacific Northwest. Washington State University, Cooperative Extension Energy Program, Olympia, WA.

Luthe, C.E. (1998). Progress in reducing dioxins and AOX: a Canadian perspective. *Chemosphere*, **36**(2), 225–229.

Narum, G. P. (2001). Simpson Tacoma Kraft Mill, Tacoma, WA, personal communication.

National Council of the Paper Industry for Air and Stream Improvement, Inc. (NCASI), (1991). *Progress in Reducing Water Use and Wastewater Loads in the U.S. Paper Industry*, Technical Bulletin No. 603. Research Triangle Park, NC: National Council of the Paper Industry for Air and Stream Improvement, Inc.

National Council of the Paper Industry for Air and Stream Improvement, Inc. (NCASI) (1993). *Regression Models to Estimate Adsorbable Organic Halide (AOX) Levels in Bleached Kraft Mill Wastewaters*, Technical Bulletin No. 654. Research Triangle Park, NC: National Council of the Paper Industry for Air and Stream Improvement, Inc.

National Council of the Paper Industry for Air and Stream Improvement, Inc. (NCASI). (1994). *Characterization of Adsorbable Organic Halide (AOX) Discharge Rates at Kraft Mills Employing Complete Chlorine Dioxide Substitution,* Technical Bulletin No. 667. Research Triangle Park, NC: National Council of the Paper Industry for Air and Stream Improvement, Inc.

National Council of the Paper Industry for Air and Stream Improvement, Inc. (NCASI). (1997). *Trends in Emissions of Sulfur Dioxide and Nitrogen Oxides from Pulp and Paper Mills, 1980–1995* Special Report No. 97–02. Research Triangle Park, NC: National Council of the Paper Industry for Air and Stream Improvement, Inc.

National Council of the Paper Industry for Air and Stream Improvement, Inc. (NCASI). (1999a). *Estimated Costs for the U.S. Forest Products Industry to Meet the Greenhouse Gas Reduction Target in the Kyoto Protocol*, Special Report No. 99–02. Research Triangle Park, NC: National Council of the Paper Industry for Air and Stream Improvement, Inc.

National Council of the Paper Industry for Air and Stream Improvement, Inc. (NCASI). (1999b). *Solid Waste Management Practices in the U.S. Paper Industry*, Technical Bulletin No. 793. Research Triangle Park, NC: National Council of the Paper Industry for Air and Stream Improvement, Inc.

National Council of the Paper Industry for Air and Stream Improvement, Inc. (NCASI). (2000). *Beneficial Use of Secondary Fiber Rejects*, Technical Bulletin No. 806. Research Triangle Park, NC, National Council of the Paper Industry for Air and Stream Improvement, Inc.

Paik S. C., and Chung, J. S. (1996). Selective Hydrogenation of Sulfur Dioxide with Hydrogen to Elemental Sulfur Over Transition Metal Sulfide Supported on $AL_2O_3$. *Applied Catalysis B: Environment*, **8**, 267–279.

Pinkerton, J. E. (1999). Trends in U.S. kraft mill TRS emissions. *Tappi Journal*, **82**(4):166–169.

Reeve, D. W. (1996a). Introduction to the principles and practice of pulp bleaching. Section 1, Chapter 1. In Dence, C.W. and Reeve, D.W. (eds), Pulp Bleaching: Principles and Practice. Tappi Press, Atlanta pp. 2–24.

Reeve, D. W. (1996b). Chlorine dioxide in bleaching stages. Section 4, Chapter 8. In Dence, C.W. and Reeve, D.W. (ed.), Pulp Bleaching: Principles and Practice. Tappi Press, Atlanta, pp. 379–394.

Scott, G. M., Akhtar, M., Lentz, M. J., and Swaney, R. E. (1998a). Engineering aspects of fungal pretreatment for wood chips. In *Environmentally friendly pulping and bleaching methods*. Wiley, New York, NY.

Scott, G. M, Akhtar, M., Lentz, M. J., Kirk, T. K., and Swaney, R. E. (1998b). New technology for papermaking commercialization biopulping. *Tappi Journal*.

Sierra-Alvarez, R., and Tjeerdsma, B. F. (1995). Organosolv pulping of wood from short rotation intensive culture plantations. *Wood and Fiber Science*, **27**(4), pp. 395–401.

Someshwar, A. V., and Jain, A. K. (1999). SARA 313 reporting: the impact of changing emission estimates. In *TAPPI International Environmental Conference Proceedings*. April 18–21, Nashville, TN.

Sykes, M. (1994). Environmental compatibility of effluents of aspen biomechanical pulps. *Tappi Journal*, **77**(1), 160–164.

Stone and Webster Engineering Corporation (1996). Pulse enhanced spent liquor chemical recovery — technology overview, Atlanta, Georgia.

Towler, G. P., and Lynn, S. (1995). U.S. Patent 5,397,556, Process for recovery of sulfur from acid gases.

US Environmental Protection Agency, (1992a). *Facility Pollution Prevention Guide*, EPA/600/R-92/088, Office of Research and Development, Washington, DC.

US Environmental Protection Agency, (1992b). *Model Pollution Prevention Plan for the Kraft Segment of the Pulp and Paper industry*, EPA/910/9-92-030, EPA Region 10, Seattle, WA.

US Environmental Protection Agency, (2001). *Pulp & Paper MACT II Standards*, CFR Subpart MM 66FR3180. Research Triangle Park, NC.

Wachs, I. E. (1999a). U.S. Patent 5,907,066, Treating methanol-containing waste gas streams.

Wachs, I. E. (1999b). U.S. Patent 5,969,191, Production of formaldehyde from methyl mercaptans.

# CHAPTER 13

# PROGRESS TOWARD ZERO DISCHARGE IN PULP AND PAPER PROCESS TECHNOLOGIES

TAPAS K. DAS

Washington Department of Ecology, Olympia, Washington 98504, USA

The US pulp and paper manufacturing industry is the country's fourth largest consumer of process water. The pulp and paper manufacturing industry has a long history of recycling and reuse. The kraft pulping process is unique in that most of the chemicals from spent liquor can be recovered for reuse in subsequent cooks, and therefore, this process can be a good example of a closed-loop and/or a zero discharge manufacturing system. Having exhausted many gains achievable through end-of-pipe pollution control, the pulp and paper industry is inventing and implementing process changes, process modifications, and retrofits, to improve wastewater quality, lessen air pollution, and achieve better solid waste management by means of recycling and reuse.

Technological advances aimed at reducing the formation of dioxins and furans during pulp bleaching have led to a series of process changes, including (1) eliminating the use of certain defoamers which contained dioxin and furan precursors, (2) decreasing the use of elemental chlorine as bleaching chemical; and (3) increasing the use of chlorine dioxide for pulp bleaching. As the industry has implemented these changes, dioxin and furan concentrations in bleaching effluent have dropped well below detection limits established by the USEPA. Indeed, dioxin has been zeroed out. Moreover, the process changes have made recovery of energy, process water and bleaching chemical a feasible approach to energy conservation, water conservation, water reuse, and pollution prevention.

*Toward Zero Discharge: Innovative Methodology and Technologies for Process Pollution Prevention,* edited by Tapas K. Das.
ISBN 0-471-46967-X   Copyright © 2005 John Wiley & Sons, Inc.

The recycling and recovery of all pulping and bleaching process wastewater is termed "closed cycle," and this chapter discusses the positive results obtained in the pulp and paper industry by doing just that.

The prevalent technologies used for zero-liquid-discharge systems are membrane processes, primarily reverse osmosis (RO), followed by evaporation, and then by crystalization. In the pulp and paper industry, to date, filtration followed by evaporation has been used (see Section 9.2.1). A zero-discharge system can produce from industrial wastewater a clean stream suitable for reuse in the plant and a concentrate stream that can be disposed of in an environmentally benign manner, or further reduced to a solid. For pulp and paper, the concentrate streams, if they are black liquor, are used as fuel in heat recovery boilers.

While there are several definitions of zero-discharge, in practice the term most commonly means that no water effluent stream will be discharged from the processing site. Zero-discharge systems have several advantages:

- Minimum consumption of fresh water
- Capability of recovering valuable resources
- Reduction in volume of sludge
- Better water quality
- Flexibility in facility site selection, since no receiving waterway is needed for wastewater treatment

Disadvantages include maintenance problems (e.g., scaling and corrosion), reduced plant reliability, and the presence of certain trace chemicals not found in wastes from the more traditional processes. The cost of installing a zero discharge system can probably only be justified for grass roots mills. However, for closed loop operations using existing equipment, cost is not prohibitive.

## 13.1 THREE CASE STUDIES

### 13.1.1 Louisiana-Pacific Corporation: Conversion to Totally Chlorine Free Processing

To take advantage of the benefits of a zero-discharge system, the Louisiana-Pacific Corporation's (L-P) Samoa pulp mill, located on the Northern California Coast, converted to totally chlorine free ("TCF") pulp processing. The mill, constructed in 1964, produces an average of 650 tons of bleached kraft pulp per day from waste wood chips generated by local sawmills. In January 1994, the mill became the only North American kraft mill to replace chlorine or chlorine containing compounds with hydrogen peroxide and oxygen in all its bleaching agents. The non-corrosive chemistry of these TCF bleaching chemicals and the similarity of pH and temperature conditions between bleaching stages makes it easier to recycle TCF bleaching wastewaters. Other key technologies enabling high recycle rates are extended digester cooking, improved brown-stock washing, closed screening, oxygen delignification, a high efficiency recovery boiler and advanced green liquor filtration. These pollution prevention technologies, coupled with innovative process changes, have enabled L-P to push the technical limits of CC operation, and dramatically reduced the environmental impact of the mill (Bicknell and Holdsworth, 1995; Jaegel and Spengel, 1996).

Data collected from a large number of mills across the United States and reported by the National Council of the Paper Industry for Air and Stream Improvement (NCASI, 1997) and Das (1999) show that as a result of these changes, between 1988 and 1996, effluent concentrations of 2,3,7,8-tetrachlorodibenzo-p-dioxin (TCDD) and 2,3,7,8-tetrachloro dibenzo-p-furan (TCDF) were very significantly reduced, with most of the data clustered in the region of concentration.

## 13.1.2 The World's First Zero Effluent Pulp Mill at Meadow Lake: The Closed-Loop Concept

The $250 million Millar Western Meadow Lake Mill is located on a 247-acre site about 200 miles northwest of Saskatoon, Saskatchewan. It uses mechanical action supplemented by mild chemicals to turn aspen wood chips into bleached chemi-thermomechanical pulp (BCTMP), about 240,000 metric tons per year. More efficient than the kraft process, this approach uses half the trees to make the same amount of pulp, producing almost one ton of pulp for each ton of wood on a water-free basis. The Millar Western BCTMP process also eliminates chlorine compounds and odorous sulfur-based impregnation chemicals. This environmentally-friendly mill uses hydrogen peroxide to increase the brightness of the pulp, making it suitable for printing and writing grades of paper as well as for tissue and paper towels.

The plant is the first pulp mill in the world to operate a successful zero liquid discharge system. Effluent from the thermomechanical pulping process is concentrated from 2 percent solids to 35 percent solids by three falling film vapor compression evaporators, followed by two steam-driven concentrators which further concentrate the effluent to about 70 percent solids. Of the 1760 gallons per minute of effluent sent to the system, 1720 is recovered as high purity water for reuse in the pulping process. Solids are burned in the boiler; the smelt is cast into ingots and stored on site for future chemical recovery.

In early 1990s, Millar Western Pulp (Meadow Lake) Ltd. announced plans to build a mill in northern Saskatchewan, the community was concerned about the pollution it would generate, especially effluent discharged to the Beaver River. Though a biological treatment system planned at the mill would have made the effluent cleaner than river water, Millar Western decided to go one step further and eliminate all effluent discharge from the pulp mill. The zero effluent system at Meadow Lake is the first of its kind in the world. The evaporator system, the key equipment in the water recovery process, was designed and supplied by Resources Conservation Company (RCC) (Fosberg, 1992). All effluent coming out of the mill is treated in the water recovery plant. As a result, the mill only needs about 300 gpm of makeup water to replace water lost to the atmosphere by evaporation. The same type of pulp mill without a water recovery plant would need about 2,500 gpm of raw water makeup. The effluent treatment system started up in January 1992, when the mill went on line.

Millar Western's Meadow Lake BCTMP mill is an example of successful closure of the water cycle in a mechanical pulp mill. An earlier attempt in Canada to close a kraft mill by recycling bleach plant effluents through the kraft chemical recovery process had failed on account of the buildup of corrosive materials (Smook, 1992). Thus, it is useful to study in detail the advanced system in place at Meadow Lake.

The effluent produced by the BCTMP process At Meadow Lake is discharged at a rate of almost 1,800 gpm. It has a temperature of 150°F, a pH of about 8 and contains about 20,000 ppm dissolved solids. Figure 13.1 shows a more detailed view of the water recovery portion of the system, consisting of five stages: clarification, evaporation, concentration, stripping, and incineration.

### 13.1.2.1 Clarification

The first unit operation to receive pulp mill wastewater is the floatation clarifiers. Since removal of fiber is very important to the performance of the evaporators, the mill decided to install two clarifiers instead of one. This allows for maximum removal efficiency and flexibility. Chemicals are added to aid in flocculation and floatation of the solids.

To ensure that upsets in the pulp mill do not directly affect the evaporators, an on-line meter measures suspended solids in the clarifier accepts stream. When the suspended solids exceed 900 ppm, the clarifier accepts are directed to the settling ponds. Clarifier accepts normally go directly to the evaporators in the winter to conserve heat. In the summer the accepts go preferentially to the settling pond to dump heat since the heat balance changes from season to season.

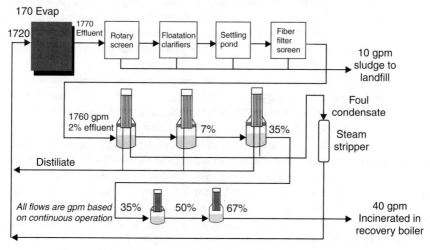

**Figure 13.1.** Effluent treatment system like that at Meadow Lake.

***13.1.2.2 Evaporation***    The heart of the zero effluent system is three vertical-tube, falling-film vapor compression evaporators which operate as explained earlier (Section 9.2.1.1). At 100 feet tall, and with thousands of square feet of heat transfer surface, this is the largest train of mechanical vapor recompression evaporators in the world. The evaporators concentrate effluent from 2 percent solids to 35 percent solids, by means of an energy-efficient mechanical vapor-compression process that recovers distilled water from the effluent. The evaporator consists principally of a heating element, vapor body, recirculation pump, and a vapor compressor.

The effluent is pumped from the vapor body sump to the top of the heating element (tube bundle). A distributor is installed in the top of each tube, causing the effluent to flow down the inside of each tube in a thin film. The distributor helps prevent fouling of the heat transfer tubes by keeping them evenly and constantly wet. It also allows the mill to operate at reduced capacity if desired, since the heating surfaces will remain wet regardless of the amount of effluent being processed. (The evaporators are also capable of handling 1.2 times more than design flow rates from the pulp mill which gives the mill a significant amount of catch-up ability.) When the effluent reaches the bottom of the tubes, the recirculation pump sends it back to the top for further evaporation.

As the effluent flows through the heated tubes, a small portion evaporates. The vapor flows down with the liquid. When it reaches the bottom of the tube bundle, the vapor flows out of the vapor body through a mist eliminator and then to the compressor. The compressed steam (at a few psi) is then ducted to the shell side of the tube bundle, where it condenses on the outside of the tubes. As it does so, it gives up heat to the tubes, resulting in further evaporation of the liquid inside. A large amount of heat transfer surface is provided, which minimizes the amount of energy consumed in the evaporation process. Operation of the vapor compression evaporator system requires only 65 kWh per 1,000 gallons of feed.

As the vapor loses heat to the tubes, it condenses into distilled water, which flows down the outside of the tubes. Because the water that first condenses out of the steam is cleaner than water condensing later, baffles are provided within the heating element to create two separate regions for condensing. Steam flows first through the clean condensate region where most condenses. The remaining vapor, which is rich in volatile organics such as methanol, condenses in the foul condensate region of the heating element.

A major portion (70 percent) of the clean condensate is sent directly to the pulp mill for use as hot wash water at the back end of the mill. The balance of the clean condensate goes to the distillate equalization pond where it is combined with makeup water from Meadow Lake and serves as the cold water supply to the mill. The foul condensate, which contains the volatile organic materials, is reused after stripping in a steam stripper. The steam stripper top product (which contains the concentrated organics) is incinerated.

***13.1.2.3 Concentration***    Like the three evaporators, the two concentrators are vertical-tube, falling-film design. Rather than using a vapor compressor to drive the system, the concentrator is operated with steam generated by the recovery boiler. The evaporation process in the concentrators is essentially the same as in the evaporators, but the effluent is concentrated further, to about 67 percent solids. The concentrated effluent is incinerated in the recovery boiler. The lead concentrator takes the liquor from 35–50 percent, while the lag concentrator goes from 50–67 percent solids.

### 13.1.2.4 Stripping
The foul condensate, only about 10 percent of the total condensate, is stripped of volatile organic compounds in a packed column stripper. The VOCs are selectively concentrated in the foul condensate because of the condensate segregation features built into the evaporator heating elements. Process steam from the concentrator is sent to a reboiler, which generates stripping steam from a portion of the stripped condensate. The stripped condensate is combined with the clean condensate and reused in the mill. The concentrated VOCs are incinerated in the recovery boiler as a concentrated vapor.

### 13.1.2.5 Incineration
At the recovery boiler, the organic components of the effluent are incinerated, a process that also generates steam to operate the concentrators. Inorganic chemicals in the effluent are recovered in the smelt from the boiler, which is cast into ingots and stored on site. The mill is considering recovering the sodium carbonate, which would then be converted to sodium hydroxide, a major chemical used in the BCTMP process.

## 13.1.3 Successful Implementation of a Zero Discharge Program

In July 1996, a paper company located on the west bank of the Mississippi River undertook a program to eliminate the discharge of industrial wastewater to the river. A wastewater recycling system consisting of pumps, surge tank, and filtration system reduced discharges by 99 percent. The successful pollution prevention includes the annual elimination of 562 million gallons of wastewater, 149,000 lb of total suspended solids, and 57,000 lb of biochemical oxygen demand. The plant primarily manufactures colored construction-grade paper from a mixture of secondary fiber, stone ground-wood pulp, and kraft pulp.

The company initiated a Zero Discharge Program, described in detail by Klinker (1996), having two goals:

- Eliminate the discharge of wastewater into the Mississippi River.
- Improve the efficiency of water use in manufacturing to reduce the mill's dependence on river water.

Recycling treated wastewater into the mill's water supply system would accomplish these goals.

Besides the regulatory motivation for reusing wastewater, the company had concerns about periodic interruptions in the flow of water to the mill. A local power company's hydroelectric plant located immediately upstream caused these interruptions. On occasion, the utility lowered the river level by halting water flow through a canal that also feeds water from the Mississippi to the paper mill. When this occurred, the mill had to stop its manufacturing process until there was sufficient volume of water to run the mill. By reusing wastewater, the company could reduce its dependency on river water and avoid this disruption to production.

The nucleus of the zero discharge program, a closed-loop wastewater recycling system in the mill, offers environmental benefits, but also generates difficulties because of the increase in the volume of recycled wastewater used in manufacturing and the expenses associated with addressing these problems.

**13.1.3.1 Closing the loop**  Figure 13.2 is a diagram of the company's closed-loop wastewater recycling system. Before the zero discharge program began, the Mississippi River supplied all the process and cooling water for the mill. Fresh water from the river entered the mill, passed over a fine mesh screen, and entered a 2,700-gal freshwater tank. The house pump directed it to process and cooling water demand points in the mill.

The resulting process wastewater underwent treatment in the company-owned, activated sludge, wastewater treatment plant. Discharge was through a process-wastewater outfall designated outfall 01. Cooling wastewater discharge was at outfall 02. Wastewater sludge underwent dewatering on a belt filter press followed by land application on company owned agricultural land.

Before the zero discharge program, the plant discharged an average of 607,000 gallons of process wastewater and 1.14 million gallons of cooling wastewater into

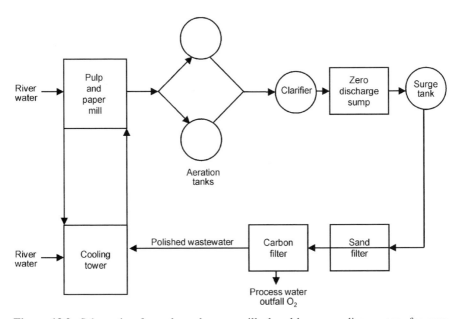

**Figure 13.2.** Schematic of a pulp and paper mill closed-loop recycling system for zero discharge system.

the river each day. By using the closed-loop system to pump increased amounts of treated wastewater into the freshwater tank, the mill was able to state that its process and cooling water consisted of nearly 100 percent recycled wastewater. The total volume of wastewater discharged to the river decreased by 82 percent (Klinker, 1996).

## 13.2 ADDITIONAL CONCERNS ADDRESSED BY ZERO DISCHARGE TECHNOLOGY

From environmental and safety points of view, as well as in terms of economics, data from zero discharge plants make it clear that there is no better option for pulp mills than an expedient move towards oxygen-based TCF, closed-loop pulp production. We turn now to a consideration of these two important areas.

### 13.2.1 Environment Discharges

The most studied form of pulp pollution is the effluent discharged by mills into the aquatic environment. As substantial reductions were made in the dioxin content of this discharge, it became apparent that the numerous other chemicals still in mill effluent were also capable of damaging aquatic life. It is being recognized that effluent from oxygen-bleaching mills is significantly less toxic that from mills that have removed only elemental chlorine from their process stream. It is therefore imperative for mills to move as quickly as possible to closed-loop production, which would reuse water and chemicals and eliminate effluent discharges.

The need to "close the loop" has been recognized by industry, although so far its efforts have focused almost exclusively on inventing a way to eliminate effluent and still use chlorine dioxide as a bleaching agent.

Far less studied are the characteristics of air emissions and sludge (solid waste). The fact is, we have little or no idea what is in either. The limited data available suggest that any improvement in air quality from oxygen-based, closed-loop mills is modest or nil. However, it is clear that the steps needed to move mills quickly towards zero effluent are the same steps required for improving overall air quality. With the removal of chlorinated compounds from the bleaching process, chlorinated air emissions cease to be a concern. It is also much easier to install important air stack control devices if there are no corrosive chlorine emissions.

Conversion to TCF closed-loop pulp production would eliminate creation of waste fiber, which could be safely burned for energy in the recovery boiler where mills recover many of their pulping chemicals.

### 13.2.2 Safety

Regrettably, the health and safety of workers and the communities near pulp mills are seldom factored into decisions about bleaching methods.

From a worker safety point of view, TCF bleaching is infinitely preferable. Chlorine dioxide is highly volatile, cannot legally be transported and must be manufactured on site. Gassing of workers during this and other stages of the elemental-chlorine-free (ECF) bleaching process is the most common form of injury. Even if industry succeeds in its efforts to devise a method of closing the loop with chlorine dioxide, the concerns related to health and safety of workers will remain.

Concern about workplace air quality in pulp mills led the US Environmental Protection Agency to require all kraft mills — *except TCF mills* — to monitor and collect bleaching vent gases, which are usually just vented to drift around at worker level.

This issue has been well researched and it is clear that, while oxygen bleaching is not without some risks, it is less hazardous to workers and communities than ECF processes.

## 13.3  OTHER EMISSION RECOVERY AND CONTROL PROCESSES

Over the past two decades, the pulp and paper industry has successfully reached pollution prevention goals through process modifications, retrofits, innovative green technologies, resource recovery and reuse, and converting waste streams to valuable chemical feedstocks and energy. These newer methods make better sense economically and ecologically over the old processes of incineration or biodegradation in wastewater ponds or landfills. The following section describes pulp and paper mills' best plans in moving toward zero discharge manufacturing, or minimum impact industry.

### 13.3.1  Recovery of Sulfur from Acid Gases

Scientists at the University of California have developed a new chemical process for converting hydrogen sulfide to elemental sulfur and syngas. This process is applicable to many commercially important sour gases, including pulping process digester gas, natural gas, most coal-derived gas, and refinery gases generated by hydrodesulfurization processes. The atmospheric emissions of sulfur-containing compounds are effectively zero for this recovery process.

Hydrogen sulfide ($H_2S$) is a frequently found contaminant of industrial gas streams. Generally, concentrations of $H_2S$ must be reduced below four parts per million (ppm) before the gas stream may be used or discharged to the atmosphere. Conventional technology for $H_2S$ removal is expensive, consumes a great deal of energy, and emits sulfur-containing compounds to the atmosphere. The current method typically involves absorbing the hydrogen sulfide from the sour gas, then stripping the $H_2S$ from the absorbent and sending it to the Claus process. The Claus plant reaction is equilibrium-limited and does not go to completion even when two or three reaction stages are employed. Therefore, the gas exiting a Claus plant contains 2,000 to 6,000 ppm of unreacted $H_2S$ and $SO_2$ and must be treated in a tail gas unit before being released to the atmosphere. The tail gas unit is about as expensive as the Claus plant, uses a substantial amount of energy, and does not completely eliminate sulfur emissions.

The process developed in California hinges upon a newly discovered reaction chemistry for the partial oxidation of $H_2S$ to sulfur. In summary, $H_2S$ and carbon dioxide are reacted at high temperature to form water vapor, carbon monoxide, hydrogen and elemental sulfur. A recycle stream ensures that sulfur leaves the process only in the product stream of elemental sulfur. The final product gas contains only CO and $H_2$ and is valuable either as a fuel or as a chemical feedstock (Towler and Lynn, 1993a,b).

Advantages offered by the process are as follows:

- It should be substantially less expensive to construct and operate than a Claus plant with a tail gas unit.
- The chemical (or fuel) value of the hydrogen from the $H_2S$ will be removed.
- Net emissions of sulfur-containing compounds to the atmosphere will be negligible.

### 13.3.2   Greenhouse Gases

The forest products industry has an important and complex role in the global carbon cycle. Since forests supply the primary raw material of the pulp and paper industry, sustainable management of these resources sequester massive amounts of carbon and ensure the availability of products that contribute to significant carbon pools during their use and after being discarded.

Efforts to expand the amount of forested land are increasing carbon storage in most of the developed world, and new plantations are being established in many areas of the developing world. Research is ongoing to identify forest management practices capable of optimizing carbon storage in existing forests while maintaining or enhancing forest productivity and protecting the environment. Recycling, too, is an important part of the carbon cycle, since it can help extend the time during which carbon is stored in products. It has been estimated that, on a global basis, the amount of carbon stored in forest products is increasing by 139 million metric tons of carbon per year (Winjum et al., 1998).

We now examine three important ways in which the forest products industry is working to reduce greenhouse gases: combined heat and power (CHP) systems, burning biomass fuels, and precipitated calcium carbonate.

#### 13.3.2.1 *CHP Systems*   The pulp and paper industry is one of the global leaders in the use of combined heat and power systems, also called cogeneration systems. CHP systems produce electrical power and thermal energy from the same fuel, yielding two or more times as much usable energy from the fuel than normal methods for generating power and steam. This reduces GHG emissions by reducing the demand for fossil fuels. The pulp and paper industries in some countries derive more than half of their energy from CHP systems.

### 13.3.2.2 Climate-Neutral Carbon Dioxide Emissions from Burning Biomass Fuels
The forest products industry also relies heavily on climate-neutral biomass fuels, which reduce atmospheric levels of greenhouse gases by displacing fossil fuels. In a number of countries, more than half of the industry's energy requirements are met using biomass fuels such as wood waste and black liquor. Forest products that cannot be economically recycled provide another source of biomass fuels.

Wood waste and black liquor, recovered from the industry's waste and process streams, are carbon neutral; that is, when burned they cause no net change in the carbon content of the biosphere over the harvest cycle and, therefore, do not contribute to the formation of greenhouse gases (GHGs). Energy-rich biomass — derived from wood chips, bark, sawdust, and pulping liquors recovered from the harvesting and manufacturing processes — is formed when atmospheric carbon dioxide is sequestered by trees during growth and transformed into organic carbon substances. When these biomass fuels are burned, the $CO_2$ emitted during the manufacturing and combustion processes is the atmospheric carbon dioxide that was sequestered during growth of the tree; hence, there is no net contribution to the atmospheric $CO_2$ level. This carbon cycle is a closed loop. New tree growth keeps sequestering atmospheric carbon dioxide and maintains the cycle.

Any increases or decreases in the amount of carbon sequestered by the forests are accounted for in the comprehensive forest accounting system. This is the approach generally prescribed for national inventories by the United Nations Framework Convention on Climate Change. Most international protocols, including that of the Intergovernmental Panel on Climate Change (IPCC), have adopted the convention set out by the United Nations. The IPCC has stated that emissions from biomass do not add to atmospheric concentrations of carbon dioxide (IPCC, 1997a,b,c).

The IPCC provides a list of biomass fuels (IPCC, 1997a,b,c). Unsurprisingly, it contains wood and wood waste and charcoal (biogas from wood waste and other biomass is not specifically listed by the international body, but it clearly falls within the general definition of biomass). Also listed are agricultural residues and wastes, municipal and industrial wastes containing organic material that is biological in origin which n turn allows the use of wastewater treatment sludges from pulp and paper mills (e.g., bagasse, bioalcohol, black liquor, landfill gas, sludge gas).

Forest products that cannot be economically recycled, though not listed by the IPCC, would seem to comprise another source of biomass fuels.

### 13.3.2.3 Use of Gases Rich in Carbon Dioxide to Manufacture Precipitated Calcium Carbonate
Calcium carbonate ($CaCO_3$) pigment, manufactured by grinding limestone or marble, or by chemical precipitation, is used as a coating and filler material in the production of some grades of paper and paperboard. At some mills, the correct characterization of lime kiln emissions is complicated by the now-common practice of using the lime kiln stack gas carbon dioxide, (or gas from another boiler), to manufacture precipitated calcium carbonate (PCC) at satellite plants. In this process, $CO_2$-rich lime kiln gas is reacted with purchased

calcium oxide (lime) to produce PCC. Although other mill stack gases are sometimes used, the gas from the lime kiln stack is favored, primarily because of its higher $CO_2$ content. Carbon released from $CaCO_3$ is climate-neutral biomass carbon that originates in wood and should be accounted for further $CO_2$ reduction.

While the amounts of $CO_2$ used in PCC manufacturing at satellite plants appear to be small compared to the industry's overall emissions, the quantities can be significant at individual mills because 50 percent or more of the $CO_2$ in the stack gas can be consumed in the PCC manufacturing process. Indeed, some kraft mills use biomass to supply almost all of their energy needs, with the only significant use of fossil fuel being the lime kiln. At such mills, the amounts of $CO_2$ captured in PCC manufacture can actually exceed the amounts of fossil $CO_2$ emitted by the mill.

Because the $CO_2$ exported to PCC plants is not emitted by the mill, nor by any source owned by the company, this $CO_2$ is reported as an export rather than an emission. Where $CO_2$ is exported from lime kilns, the gas consists of a combination of fossil fuel-derived $CO_2$ and climate-neutral biomass-derived $CO_2$. In the suggested reporting format for the calculation tools, these two types of $CO_2$ exports are reported separately.

With respect to PCC manufactured from mill GHG emissions, it is important to note that the calculation tools are intended to help characterize a facility's or company's greenhouse gas emissions and not the fate of these emissions. Nor are they intended to address the life cycle tradeoffs associated with use of mill emissions as a raw material in PCC manufacture. These are questions that require a much broader analysis than is possible within the scope of this inventory guidance.

### 13.3.3 Solid Waste Management and Conserving Energy

Reduction of greenhouse gases is by no means the forest product industry's only environmental concern. Some mills have developed system to mix dried primary and secondary waste-treatment sludge with wood waste and bark as fuel. In most instances, the mills have carefully analyzed the waste and the combustion products to isolate potential problems with contaminants and have taken the necessary steps to eliminate them. Sometime, opposition to incineration is stronger than opposition to landfills. Yet, scientifically acceptable incineration systems are necessary for environmentally sound waste management in all industrial sectors.

Pulp mill sludges are also used to amend soil, animal bedding, or to produce ethanol and other chemicals. Source reduction through such steps as improved retention of pulp fines can also reduce waste. These and other potential applications would negate altogether the need for new landfill capacity.

Energy conservation and the use of fossil-fuel alternatives play an indirect but significant role in environmental stewardship in the forest products industry. The standard practice of using bark and wood waste and black liquor as fuel eliminates more than 60 percent of the demand for fossil fuel in the US forest-products industry as a whole, including integrated pulp and paper mills (mills in which the papermaking operation is contiguous with the pulping operation) and nonintegrated mills (AFPA, 1994). Modern kraft pulp mill operations, with the exception of the lime kilns, can satisfy their total steam and electrical energy requirements using black liquor and wood waste and therefore do not require fossil fuels. Other key energy conservation measures commonly used today involve reduced water usage; energy recycling and reclaiming in digester areas; systems to improve management and reclamation of low-level heat, for example, from recovery systems; and improved insulation.

Concern about deforestation (and public misperception that the forest products industry is a contributor to it) and about the worldwide production of solid waste have put paper at the top of the list for recycling efforts in developed countries (AFPA, 1994). Of course, internal (intrafirm) recycling has long been practiced. Indeed, every paper mill has a repulper to handle its own waste. The key interest has been in the post-consumer recycling. The expansion of paper recycling has increased interest in deinking and debleaching technology, used to remove color from fiber, and in lowering the cost of doing so. Improvement will also be needed in the separation and disposal of deink sludge. Innovation in ink and copier-toner technology also hold promise for reducing the toxicity and amount of deink sludge.

## 13.4 CONCLUSIONS

The US pulp and paper industry has made significant progress towards reducing water consumption and increased water recycling and reuse through innovative technologies and process modification. Some mills have implemented processes that "close the loop" and proved to be successful zero discharge bleach plant systems. The effects of the EPA's Cluster Rule, which became applicable on April 15, 2001, have yet to be fully felt. It is reasonable to expect, however, that the new linkage of federal regulations aimed at reducing air and water pollution will result in an even higher level of processed water recycling, reuse within the mills, as well as greater pollution prevention and zero discharge in water, air and solid waste areas the next decade or two.

## REFERENCES

American Forest and Paper Association (AFPA) (1994). Pollution prevention report. Washington, DC.

Bicknell, B., and Holdsworth, T. (1995). Comparison of pollutant loadings from ECF, TCF and ozone/chlorine dioxide bleaching. International Non-Chlorine Bleaching Conference Proceedings.

Das, T. (1999). Process technology advances in the pulp and paper industry. American Institute of Chemical Engineers Annual National Meeting, Paper No. 273e, Dallas, TX.

Fosberg, T. (1992). Case study: water recovery systems at the world's first zero effluent pulp mill. International Water Conference—51st Annual Meeting, Pittsburgh, USA.

IPCC, Intergovernmental Panel on Climate Change (1997a). *Revised 1996 IPCC guidelines for national greenhouse gas inventories: Reporting instructions (Vol. 1).* IPCC National Green-house Gas Inventory Program, 1997a, http://www.ipcc-nggip.iges.or.jp/public/gl/invs4.htm

IPCC, Intergovernmental Panel on Climate Change (1997b). *Revised 1996 IPCC guidelines for national greenhouse gas inventories: Workbook (Vol. 2)*. IPCC National Greenhouse Gas Inventory Program, 1997b, http://www.ipcc-nggip.iges.or.jp/public/gl/invs5.htm

IPCC, Intergovernmental Panel on Climate Change. (1997c). *Revised 1996 IPCC guidelines for national greenhouse gas inventories: Reference manual (Vol. 3)*. IPCC National Greenhouse Gas Inventory Program, 1997c, http://www.ipcc-nggip.iges.or.jp/public/gl/invs6.htm

Jaegel, A., and Spengel, D. (1996). Multimedia environmental performance of TCF closed bleach plant kraft pulp production. International Non-Chlorine Bleaching Conference Proceedings.

Klinker, R. (1996). Successful implementation of a zero discharge program. *TAPPI Journal*, **79**(1), pp. 97–102.

NCASI, National Council of the Paper Industry for Air and Stream Improvement, Inc. (1997). Progress in Reducing the TCDD/TCDF Content of Effluents, Pulps, and Wastewater Treatment Sludges from the Manufacturing of Bleached Chemical Pulp: 1996 NCASI Dioxin Profile. Special Report No. 97-04. Research Triangle Park, NC, National Council of the Paper Industry for Air and Stream Improvement, Inc.

Resources Conservation Company (RCC), 3006 Northup Way, Bellevue, WA 98004. A Division of Ionics, Inc., Watertown, MA, available on the World Wide Web at http://www.ionicsrcc.com

Smook, G. A., (1992). *Handbook for Pulp and Paper Technologies*. 2nd Ed., Angus Wilde Publications, Bellingham, WA.

Towler, G. P., and Lynn, S. (1993a). Development of a zero-emissions sulfur-recovery process. 1. thermochemistry and reaction kinetics of mixtures of $H_2S$ and $CO_2$ at high temperature. *Ind. & Eng. Chem.* **32**(11), 2800.

Towler, G. P., and Lynn, S. (1993b). Development of a zero-emissions sulfur-recovery process. 2. sulfur recovery process based on the reactions of $H_2S$ and $CO_2$ at high temperature. *Ind & Eng. Chem.* **32**(11), 2812.

Winjum, J. K., Brown, S., and Schlamadinger, B. (1998). Forest harvests and wood products: sources and sinks of atmospheric carbon dioxide. *Forest Science*, **44**(2), 274.

*Epilogue — Final Thoughts*  The United Nations University's Zero Emissions Research Institute (ZERI) was introduced in Chapter 2. It is important to realize, however, that implementation of this methodology is not limited to academic studies. As we have shown throughout the book, government, foundations, industries, and individuals are all playing a role. We conclude by expanding a bit on this valuable program and by issuing a challenge for engineers everywhere.

- **International Implementation**
  Projects to implement Zero Emission (or Discharge) are already under way in over a dozen countries, funded from several United Nations organizations and Japan's Ministry of International Trade and Industry (MITI). Namibia has committed $1 million (in US dollars) for projects. China has introduced electron beam-ammonia (EB-A) technology to turn wastes from coal-fired electric plants into feedstock for fertilizer; its few initial projects indicate the potential

for wide implementation of the EB-A process in this energy-poor country.

In Japan MITI is developing [a1] technologies in which all of Southeast Asia can aim for a transition to Zero Emission. The goal is to develop facilities for the collection of a wide variety of municipal and industrial solid wastes: recovery technologies will be used to extract metals from liquid and solid wastes; gasification treatment will generate electricity and produce ammonia, methane, and other commodity gases (return on investment calculations for gasification are provided in the Appendix). Processors will play an intermediary role by collecting wastes of smaller producers and decomposing them to generate economically useful quantities.

- **Regulatory Agencies**

Government agencies will come under pressure to relinquish regulatory controls that hinder Zero Emission. This is already happening in Japan. In the United States, such policy changes are not expected soon. Modification of the Resource Conservation and Recovery Act, with its broadly inclusive definition of hazardous waste, would help greatly, however, as would revisions of the tax structure and government policies on virgin resource extraction. In addition, redrafting of regulations that are based on results instead of on the use of a specified technology accommodate innovation now stymied by out dated rules. Materials, energy, and industrial policies reflective of Zero Emissions will be the strongest catalyst for regulatory change.

- **The Sciences**

Existing energy efficiencies need to be taken off the shelf and employed. Further gains can be achieved through additional research and development. Chemistry and material sciences — particularly those employing thermodynamic technologies such as heat exchange — will be significant contributors to Zero Emission. Material sciences will contribute not only to dematerialization, but also re-materialization — a drive to materials that are Zero Emissions in their production and use and can either be easily refurbished or recycled, or easily converted to another material flow. A return to chemurgy, a reduced reliance on petrochemicals, and a better understanding of the reactions required for efficiency are all critical.

- **Individuals**

Zero Emission will not happen solely through the efforts of engineers. ZERI's Gunter Pauli clearly states that the drive is competitiveness, which comes from senior executives and managers. Their drive to cut costs and improve market share will set the first directives to engineering firms. Product designers will also need to be educated in design for the environment. Only then will engineers be able to employ concurrent engineering and industrial ecology to formulate industrial clusters.

- **The Challenges**

Engineers have always faced design constraints. Historically these constraints were the laws of physics, availability of materials, and energy. Modern engineers still face the limitation of the laws of physics but have been granted larger amounts of energy and a wide variety of materials. Modern society has added design parameters that include safety, durability, convenience, regulatory compliance, attractiveness, and price.

A true engineer does not view regulation as an obstruction, but rather as a design constraint like efficiency and durability. The goal has always been to develop the optimal design within given constraints, whether they are the laws of nature or society. Unfortunately, the actions of our modern society have placed undue burdens on nature. Nature's ability to absorb excessive amounts of pollutants and stressors while still providing critical services of acceptable water quality, clean air, food, and biodiversity is limited.

Industrial ecology is western society's response to meeting the challenge of sustainable development. To manufacturers falls the challenge of attaining Zero Emissions. They in turn pass this directive to their engineers. To engineers, the advance of technology has meant increasing degrees of freedom with regard to design. The collective body of knowledge and our harnessing of materials and energy has been the source of these freedoms. Safety was the first man-made design constraint that society imposed under the name of social good. Engineers responded to meet the challenge. Now society recognizes the need to impose a design parameter of Zero Emissions. Engineers will meet this challenge and accept it as they have the laws of physics — as a given.

# INDEX

Absorbance measurements, ultraviolet disinfection:
performance evaluation, 73
wastewater treatment, 423
Absorbent, 355–357
Absorption processes, membrane-based VOC removal and recovery, 355–359
Accident aftermath:
chlorine gas release, sulfur dioxide dechlorination, 65–67
environmental and health risk communication and, 151
Accidental Release Information Program (ARIP) database, 153
Accident prevention, inherently safer chemistry, 196–197
Acid deposition, health risk assessment, 108–109
Acid gases:
sulfur emissions, pulp and paper industry, 679–680
volatile organic compounds, electron-beam-generated plasma destruction, thermal oxidation and, 317–318
Acid Rain, Atmospheric Deposition and Precipitation Chemistry database, 153–154

Acoustic cavitation, foundry emissions, advanced oxidation systems, 497–499
Acute toxicity, ecological risk assessment, 124–125
protection levels, 126–127
two-dimensional chemical model, 140–144
Adhesive tape manufacturing, VOC condensation and allocation, multicomponent nonideal mass integration, 615–617
Adsorbable organic halides (AOX), pulp and paper industry, effluent discharges, 654–655
Adsorption mechanisms, carbon sequestration technology, flue gas capture, 371–372
Advanced oxidation (AO) processes, foundry emissions:
acoustic cavitation, 497–499
AO dry dust collector to blackwater clay recovery system, 499–500, 503–504
blackwater from wet collector clay recovery system, 500, 503–504
clear water treatment system, 500
core room scrubbing system, 501
full-scale operations, 501–503
green sand systems, 493–496
Neenah case study, 503–506
ozone and ozone plus hydrogen peroxide, 497
slurried bond feeding system, 500–501

*Toward Zero Discharge: Innovative Methodology and Technologies for Process Pollution Prevention*, edited by Tapas K. Das.
ISBN 0-471-46967-X   Copyright © 2005 John Wiley & Sons, Inc.

Advanced treatment and processing, zero water discharge systems, 227
AEROSOL code, dioxin/furan ash, sorbent and carbon injection removal, 293
Aesthetic value, risk management, 133
Agency for Toxic Substances and Disease Registry, 156
Air and Waste Management Association (AWMA) database, 154
AIRNow pollution database, 154
Air pollution prevention technologies:
  biodiesel emissions, 330–333
  carbon sequestration, 369–379
    biological carbon dioxide, algae fixation, 377–378
    chemical looping, 377
    direct carbon dioxide air capture, 378–379
    dry ice co-generation, 377
    flue gas capture, 371–372
    gas separation membranes, 373–374
    low-temperature distillation, 372–373
    potassium carbonate process, 375–376
    power cycles, 376–377
    small pore membranes, 374
    sodium carbonate process, 374–375
    zero emissions coal, 379
  current technologies, 264–265
  dioxin/furan removal, sorbent and carbon injection, 289–298
  electron-beam-based oxidation, VOC-contaminated aqueous phase waste streams, 316–329
  electron-beam-generated plasma, volatile organic carbon destruction, 265–278
  fuel cells, 334–351
    alkaline fuel cells, 340–341
    cost and volume reduction, 348
    CPO control, 348
    de-sulfurization limitations, 348–349
    de-sulfurizer components, 343
    direct methanol fuel cells, 341–342
    durability/life, 348
    efficiency and performance, 347
    fuel selection, 349–350
    future issues, 350–351
    molten carbonate fuel cells, 341
    part-load and transient carbon monoxide control, 347–348
    phosphoric acid fuel cells, 340
    polymer electrolyte membrane fuel cell, 337–339
    preferential oxidation, 346–347
    processing system, 342–343

  reforming and catalytic partial oxidation, 343–345
    research background, 334–337
    social and environmental benefits, 350
    solid oxide fuel cells, 339–340
    water gas shift, 345–346
  membrane-based VOC removal and recovery, 352–368
    absorption and stripping, 355–359
    commercial membrane modules, 362–364
    design equations, principles and strategies, 359–362
    gas station recovery, 365
    olefin polymerization vents, monomer recovery, 365–367
    tank farm gas recovery, 364–365
    vapor permeation, 352–355
  mercury oxidation in barrier discharge, 311–314
  nonthermal plasma, coal-fired utility boiler pollutant control, 300–311
  pulp and paper industry, 649–653
    chlorine/chlorine dioxide, 651–652
    chloroform emissions, 650–651
    greenhouse gases, 652–653
    sulfur dioxide/nitrogen oxides, 650
    sulfur gas reduction, 650
  research background, 263–264
  vapor biofiltration, 279–288
Air toxics, health risk assessment, 109–110
Algae-based carbon dioxide fixation, carbon sequestration technology, 377–378
Alkaline fuel cell (AFC), basic technology, 340–341
Ammonia:
  best available control technology, selective catalytic reduction, 174–175
  selective catalytic reduction and release of, nitrogen oxide emission control, 241
Ammonia slip/ammonium sulfur salts, nitrogen oxide emissions reduction, 244–245
Ammonium nitrate manufacturing, ternary wastewater systems, mass integration, discharge minimization, 620–622
Anaerobic digestion, biogas production, 186
Anomaly assessment, life cycle assessment interpretation, 55
AOS electron beam system, volatile organic compound destruction, 320–322
AO-stabilized conditions, foundry emissions:
  advanced oxidation systems, 504–506
  benzene emissions, 510–511
  cost and efficiency analysis, 514–516

green sand properties, 493–496, 506–508
methane, ethane, and propane, 512
total VOC emissions, 508–510
Applicable or relevant and appropriate
    requirements (ARARs), ecological risk
    assessment, 114
Aquatic life, sodium dioxide dechlorination,
    effects on, 67–69
Pacific Northwest case study, 70–72
Aqueous phase waste streams, volatile organic
    compounds, electron-beam-generated
    plasma destruction, 316–329
alternative oxidation solutions, 326–328
AOS system, 320–322
basic chemistry, 318–319
economic issues, 328–329
Florida International University/University of
    Miami system, 322–324
oxidation limitations, 317–318
UTSI pilot facility, 323–326
Aquifer contamination, pollution prevention
    technology, 522–523
Arsenates, groundwater contamination:
fixed-bed column runs, 534–535
removal technology, 524–525
Arsenic compounds, groundwater contamination,
    524–542
chemical analysis, 534
chemical properties and distribution, 525–526
column runs, regeneration and rinsing, 533
fixed-bed column runs, 534–536
kinetic testing, 533–534
mobility and immobility, natural environment,
    526–527
polymeric/inorganic hybrid sorbent, 531–533
regeneration, rinsing and reuse, 538–540
sorption kinetics, 536–538
Arsenites, groundwater contamination:
fixed-bed column runs, 535–536
removal technology, 524–525
Ash particle scavenging, dioxin/furans, sorbent
    and carbon injection removal, 293
Ash residuals, pulp and paper industry, 656
Atom economy, green chemistry, 194–195
Atomic absorption spectophotometer (AAS),
    arsenic removal, groundwater
    contamination, chemical analysis, 534
ATSDR Primer on Health Risk Communication,
    156
Auto-thermal reforming (ATR), fuel cell
    processing, 342–343
catalytic partial oxidation, 344–345
Auxiliary substances, green chemistry and, 195

Average exposure, health risk assessment,
    102–103
Averaging time (AT), health risk assessment,
    receptor dose studies, 98–101

Baghouse filler, volatile organic compounds,
    electron-beam-generated plasma
    destruction, 318
Ballast equipment, UV lamps, wastewater
    treatment, ultraviolet disinfection,
    419–420
Barriers to communication, risk assessment and
    determination of, 149–150
Baseline emissions testing, foundry emissions,
    advanced oxidation systems, 504–505
"Battery Limits" cost, pollution prevention
    economics, total capital investiment
    (TCI), 164–165
Belt filters, brine concentrators, wastewater
    recycling, concentrate disposal,
    442–443
Benchmark criteria, ecological risk assessment,
    118–119
Bench scale testing, poultry litter, catalytic steam
    gasification, 588, 590–591
Benthic macroinvertebrates, ecological risk
    assessment, 116–117, 119
Benzene emissions:
foundry emissions, advanced oxidation
    processes, 510–511
mass integration, discharge minimization, mass
    exchange networks, 604–606
Best available control technology (BACT):
air pollution prevention technologies, current
    techniques, 265
foundry emissions, advanced oxidation systems,
    505–506
pollution prevention economics, 170–179
carbon monoxide review, 176–178
catalytic oxidation, 177
turbines and duct burners, 178
nitrogen oxides review, 171–176
combustion turbines and duct
    burners, 171
lean-premix/dry-low nitrogen oxides, 175
$SCONO_x$ technology, 171–173
selective catalytic reduction, 174–175
steam/water injection, 175–176
XONON technology, 173–174
particulate matter ($PM_{10}$) control
    technologies, 178
volatile organic carbons, 179
zero discharge technology, 246–247

Best management plans (BMPs), pulp and paper industry, effluent discharges, 653–655

B1-H$_2$ scenario, hydrogen-based sustainable economy, 208–214

Binary cation systems, membrane-electrode process, wastewater recycling, 453–454

Bioassays, ecological risk assessment, toxicity testing, 124–129

Bio-based chemicals, sustainability, 214

Bio-based manufacturing, sustainable economics, 184–188

Bioconcentration factor (BCF), exposure assessment, health risk assessment, 93–95

Biodiesel:
emissions, air pollution prevention technologies, 330–333
availability and use, 332–333
cost analysis, 331–332
emissions data, 330–331
fuel economy, 333
handling protocols, 333
warranty, 333
from vegetable oils, 187–188

Bio-fill material, biofiltration of organic vapors, 284

Biofiltration, volatile organic vapors, 279–288
basic principles, 280–282
cost analysis, 288
diffusion limitation, wet biobarrier, 285–286
industrial applications, 286–287
kinetics and design, 285–288
microorganisms, 282
moisture and temperature effects, 283
oxygen and other nutrients, 282–283
packing materials for, 283–285
pressure drop, 283
process parameter variation, 286
reaction limitation, 286

Biogas production:
anaerobic digestion, 186
poultry litter, catalytic steam gasification, 584–585

Biological materials:
carbon sequestration technology, algae-based carbon dioxide fixation, 377–378

Biological oxygen demand (BOD):
pulp and paper industry, 648
zero water discharge system, DaimlerChrysler case study, 233–234

Biomass technology:
electricity generation:
combustion, 185
gasification, 185–186
ethanol production, 186–187
fuel sources, sustainable economics and development, 201–203
methanol from syngas, 187
poultry litter, catalytic steam gasification, 583–588
pulp and paper industry, climate-neutral carbon dioxide emissions, 681

Bio-oil, fast pyrolysis production, 187

Biopolymers, zero discharge industrial applications, 32

Biopulping, pulp and paper industry, 661–665
decontamination, 663
economic and environmental benefits, 663–665
Forest Products Laboratory research, 661–662
large-scale implementation, 662–663

Biorenewable energy sources:
life cycle assessment, 77–84
air emissions, 81
global warming potential, 80–81
resource consumption, 81–82
sensitivity analysis, 82
system energy balance, 77–80
sustainable economics and development, 201–206
biomass fuel, 201–203
hydropower, 205
solar thermal power, 203
tidal power, 206
wave energy, 205–206
wind power, 203–204

Bio Strata materials, biofiltration of organic vapors, 284

Black liquor, pulp and paper process:
chemical and energy recovery, 665–667
discharge minimization, water optimization and integration, 629–637
process pollution prevention technologies, 657–658

Blackwater, 500–504

Bleached chemi-thermomechanical pulp (BCTMP) process, pulp and paper industry, closed-loop effluent management, 673–676

Block Group Progration method, accidental chlorine gas release, population exposure data, 67

Brine concentrators, wastewater recycling, 432–449
capital costs, 444
concentrate disposal, 438–443
belt filter, 442–443
drum dryers, 440–441
evaporation ponds, 439
forced circulation crystallizers, 441–442
spray dryers, 439–440
stream mixing, 438
concentrate production, 437–438
control loops, 438
distillate production, 437
economics of, 448–449
energy consumption, 443–444
evaporator principles, 433–436
high efficiency RO process, 447–448
process engineering and design, 436–437
staged cooling, 447
zero discharge systems, 444–446
Brundtland Commission. *See* World Commission on Environment and Development
Business practices:
sustainable economics and development metrics, 206–207
zero discharge technology, pulp and paper industry, 685
Byproduct synergy potentials, plastics recycling, developing country case study, 553–554

Calcium carbonate precipitates, pulp and paper industry, carbon dioxide-rich gas manufacturing of, 681–682
Calcium concentrations, poultry litter, environmental impact of, 583–584
Capital costs:
brine concentrators, wastewater recycling, 444
poultry litter, catalytic steam gasification, 592–595
Captive power plant, zero water discharge system, 228–230
Carbon bed absorption, volatile organic compound removal, 275–278
Carbon dioxide ($CO_2$):
air pollution prevention:
carbon sequestration technology, 369–379
biological carbon dioxide, algae fixation, 377–378
chemical looping, 377
direct carbon dioxide air capture, 378–379
dry ice co-generation, 377
flue gas capture, 371–372
gas separation membranes, 373–374
low-temperature distillation, 372–373

potassium carbonate process, 375–376
power cycles, 376–377
small pore membranes, 374
sodium carbonate process, 374–375
zero emissions coal, 379
dielectric barrier discharges, 307
foundry emissions, advanced oxidation processes, 511
hydrogen-based sustainable economy, 211–214
pulp and paper industry:
biomass fuels, climate-neutral emissions from, 681
calcium carbonate precipitate manufacturing, 681–682
wastewater treatment, supercritical fluids, zero discharge techniques, 406–414
cleaning operations, 412–413
corrosion challenge, 410–411
dyeing, 412
metal recovery applications, 411
paint industry VOCs, 413
polymerization, 414
textile dyeing, 412
water oxidation, 409–410
water properties, 408
Carbon injection technology, dioxin/furan removal, 289–298
ash particle scavenging, 293
carbon sources, 291–292
control technologies, 293–297
deacon reaction, 290–291
de-novo synthesis, 290
dioxin/furan formation mechanisms, 290–291
dioxin/furan sources, 289–290
inhibition mechanisms, 292–293
mercury removal, 295–297
metals, 292
sulfuric acid treatment, 297
Carbon monoxide (CO):
best available control technology, combustion turbines and duct burners, 176–178
dielectric barrier discharges, nitrogen oxide conversions, 307–310
foundry emissions, advanced oxidation processes, 511
fuel cell technology, 337
polymer electrolyte membrane fuel cell, poisoning risk, 339
processing systems, 342–343
transient control problems, 347–348
health risk assessment, 105–106

Carbon sequestration technology, air pollution
  prevention, 369–379
  biological carbon dioxide, algae fixation,
    377–378
  chemical looping, 377
  direct carbon dioxide air capture, 378–379
  dry ice co-generation, 377
  flue gas capture, 371–372
  gas separation membranes, 373–374
  low-temperature distillation, 372–373
  potassium carbonate process, 375–376
  power cycles, 376–377
  small pore membranes, 374
  sodium carbonate process, 374–375
  zero emissions coal, 379
Carbon sources, dioxin/furan formation, 291–292
Carcinogenic risk, health risk assessment:
  common pollutants, 104
  exposure studies, 103
Carcinogen potency factor (CPF), health risk
    assessment:
  exposure studies, 95
  mechanistic models, 96–97
Cash flow analysis, pollution prevention
    economics, 168–169
Catalytic burner (CB), fuel cell processing, 347
Catalytic oxidation, best available control
    technology:
  carbon monoxide reduction, 177
  volatile organic carbons, 180
Catalytic partial oxidation (CPO), fuel cell
    processing, 342–343
  limitations of, 348
  reforming, 343–345
Catalytic reagents, green chemistry and, 196
Catalytic steam gasification, poultry litter,
    582–595
  applications, 582–583
  bench scale testing, UTSI process, 588–591
  biogas and direct combustion, 584–585
  coal gasification process concept, 586–588
  economics of, 592–595
  environmental impact, 583–584
  large-scale facility, emissions estimates,
    591–592
  zero discharge goals, 595
Catalytic steam reforming (CSR), fuel cell
    processing, 342–343
Cation loading, membrane-electrode process,
    wastewater recycling, 459
Cation recovery flux/cation exchange flux ($C_{Xf}$),
    membrane-electrode process, 452
Cavitation, 497–498

Center for Exposure Assessment Modeling
    (CEAM), 155
Characterization factors:
  ecological risk assessment, 129–130
  health risk assessment, 101–103
  life cycle impact assessment, 50–53
Charge transfer process, volatile organic
    compounds, electron-beam-generated
    plasma destruction, 267–269
Chemical Accident Histories and Investigations
    database, 153
Chemical adsorption, carbon sequestration
    technology, flue gas capture, 371–372
Chemical looping, carbon sequestration
    technology, 377
Chemical models, ecological risk assessment,
    137–145
  one-dimensional models, 137
  three-dimensional models, 144–145
  two-dimensional models, 137–144
Chemical oxidation, volatile organic compounds,
    electron-beam-generated plasma
    destruction, comparison of techniques,
    326–328
Chemical oxygen demand (COD):
  pulp and paper industry, 648
  volatile organic compounds, electron-beam
    destruction, 324–326
  zero water discharge system, DaimlerChrysler
    case study, 233–234
Chemical process industries, sustainable
    production and growth, 198–199
Chemical recycling:
  pulp and paper process, black liquor recovery
    process, 665–667
  solid waste pollution prevention, 547
Chemurgy, zero discharge industries, 32
China, industrial cluster development in, 21–22
Chlorination (disinfection process), life cycle
    assessment, 62–72
  human health and environmental impact,
    63–64
  limitations, 63
  sulfur dioxide dechlorination, 64–72
    accidental chlorine gas release, case study,
      65–67
    aquatic life, effects on, 67–69
    environmental impact, 65
    limitation, 65
    marine life sensitivity, chlorinated seawater,
      70–72
Chlorine dioxide, pulp and paper industry,
    651–652

Chlorine emissions:
  pulp and paper industry, 651–652
    zero discharge, totally chlorine free (TCF)
      processing, 672–673
Chloroform emissions, pulp and paper industry,
  650–651
Chronic toxicity, ecological risk assessment, 125
  protection levels, 127
  two-dimensional chemical model, 140–144
Clarification technology, pulp and paper industry,
  closed-loop effluent management,
  673–674
Claus process, pulp and paper industry, sulfur
  recovery, 679–680
Clay recovery systems, foundry emissions:
  advanced oxidation systems, efficiency and
    costs, 514–516
  AO blackwater from wet collector, 500
  AO dry dust collector to blackwater, 499–500
Clean energy technologies, pollution prevention
  economics, 181–182
Cleaning operations:
  process pollution prevention, electroplating in-
    process techniques, 479–480, 482–483
  supercritical carbon dioxide, zero-discharge
    technology, 412–413
Clear Skies Initiative (CSI), zero discharge
  technology, smokestack emissions, 248
"Closed cycle" technology, pulp and paper
  industry, zero discharge, research
  background, 672
Closed-loop systems:
  plastics recycling, developing country case
    study, 551–554
    supply chain issues, 558–565
    sustainability, 578–579
  pulp and paper industry:
    effluent discharge, 673–676
      clarification process, 674
      concentration, 675–676
      evaporation, 675
      incineration, 676
      stripping, 676
    environmental issues, 678
    implementation case study, 676–678
Cluster Rule, pulp and paper industry, 648–649
Coal emissions:
  carbon sequestration technology, zero emissions
    coal plants, 379
  foundry systems:
    advanced oxidation processes, 493–496
    advanced oxidation systems, efficiency and
      costs, 514–516

Coal-fired utility boilers, nonthermal plasma
    pollution control, 300–311
  carbon dioxide effects, 307
  carbon monoxide, methane and ethylene effects,
    307–310
  dielectric barrier discharge reactor,
    302–303
  electric frequency selection, 304
  gas analytical instrumentation, 303
  HV power measurement, 303
  mercury vapor generation, 303
  nitrogen oxide conversions, 304
  nitrous oxide by-products, 310
  oxygen effects, 304–306
  sulfur dioxide conversion geometry, 311
  sulfur dioxide oxidation, 310–311
  water vapor effects, 306–307
Coal gasification process, poultry litter,
    586–588
Color change sequence, plastics recycling,
    developing country case study,
    562–564
Column runs, arsenic removal, groundwater
    contamination:
  fixed-bed arsenite feed, 534–535
  regeneration and rinsing, 533–534
Combined heat and power (CHP) systems, pulp
    and paper industry emissions,
    greenhouse gas recovery, 680
Combining Lower Emissions And Networked
    Energy Recovery (CLEANER)
    technique, 639
Combustion control:
  best available control technology,
    carbon monoxide reduction,
    177–178
  biomass combustion, 185
Combustion turbines:
  best available control technology:
    carbon monoxide reduction, 176–178
    nitrogen oxide emissions, 171
  pollution prevention economics, best available
    control technology, nitrogen oxide
    reduction, 176
Commercialization efforts:
  biodiesel technologies, 332–333
  mercury oxidation, barrier discharge,
    312–313
  pulp and paper industry, biopulping, 665
Commercial membrane modules,
    membrane-based VOC removal
    and recovery, 362–364
Commodity values, risk management, 133

Communication issues, risk assessment, 146–153
  accident aftermath, 151
  case study, 151–153
  costs of noncommunication, 151–152
  public response determinants, 146–148
  sustainable strategies, 148–150
Communication program, guidelines for, 150
Completeness check, life cycle assessment,
    55–56
Concentrate disposal, brine concentrators,
    wastewater recycling, 438–443
  belt filter, 442–443
  drum dryers, 440–441
  evaporation ponds, 439
  forced circulation crystallizers, 441–442
  spray dryers, 439–440
  stream mixing, 438
Concentrate production, brine concentrators,
    wastewater recycling, 437–438
Concentration dimension, ecological risk
    assessment:
  one-dimensional chemical model, 137
  three-dimensional chemical model, 144–145
Concentration/duration, ecological risk
    assessment, two-dimensional chemical
    model, 141–144
Concentration factor (CF), brine concentrators,
    wastewater recycling, 433–436
Concentration/response, ecological risk
    assessment, two-dimensional chemical
    model, 137–138
Concentration technology, pulp and paper
    industry, closed-loop effluent
    management, 675–676
Confidence interval (CI) limits, chlorination
    (disinfection) process, aquatic life
    effects, 70–72
Consistency check, life cycle assessment,
    56–57
Contaminant intake, health risk assessment,
    receptor dose studies, 99–101
Contaminants of potential ecological concern
    (COPEC), ecological risk
    assessment, 114
Contingency costs, pollution prevention
    economics, total capital investment
    (TCI), 164–165
Continuous monitoring, health risk assessment,
    92–93
Continuous process systems, pulp and paper
    industry, biopulping, 662–663
Contribution analysis, life cycle assessment
    interpretation, 55

Control technologies:
  brine concentrators, wastewater recycling,
      control loops, 438
  dioxin/furan removal, sorbent and carbon
      injection technologies, 293–298
  zero discharge:
    nitrogen oxides, 240–244
      dry low combustor, 243–244
      SCONO$_x$ system, 242–243
      selective catalytic reduction, 240–242
      water-steam injection, 240
    pulp and paper industry, 679–683
      greenhouse gases, 680–682
      solid waste management and energy
          conservation, 682–683
      sulfur from acid gases, 679–680
Conversion technology:
  dielectric barrier discharges, sulfur dioxide
      geometry, 311
  zero discharge industry, 22
Copper catalysts, dioxin/furan formation, sorbent
    and carbon injection removal, 292
Copper cation selectivity, membrane-electrode
    process, wastewater recycling,
    453–454
Core binder loading, foundry emissions, advanced
    oxidation systems, 514
Core room scrubbing system, foundry emissions,
    advanced oxidation, 501
Corrosion problems:
  pulp and paper process, discharge minimization,
      water optimization and integration,
      630–637
  wastewater treatment, zero discharge
      technologies, supercritical fluids,
      410–411
Cost analysis. *See also* Total annual cost; Total
    captal investment
  biodiesel technologies, 331–332
  biofiltration of organic vapors, 288
  brine concentrators, wastewater recycling:
    capital costs, 444
    economics, 448–449
  electron-beam plasma destruction, volatile
      organic compounds, 328–329
  environmental and health risk communication,
      noncommunication costs, 151–152
  foundry systems, advanced oxidation systems,
      514–516
  fuel cell processing, 348
  metal finishing industry, water pollution
      prevention, 474–475
  nitrogen oxide emissions reduction, 244

pollution prevention economics, 162–163
  emissions reduction technology, 182
pulp and paper industry:
  discharge minimization, water optimization
    and integration, 636–637
  resource recovery and reuse, 659
ultraviolet disinfection performance evaluation,
  74–76
wastewater treatment, membrane-based
  technology, 400–403
zero discharge technology, Ebara process,
  smokestack emissions, 250–251
Cost-effectiveness calculation:
  emissions reduction, 182
  pollution prevention economics, 168–169
Cost savings opportunities, zero discharge
  industries, 29–30
"Cradle to cradle" paradigm, sustainability and
  zero discharge, 16
"Cradle to grave" paradigm:
  life cycle assessment, 35–36
  sustainability and zero discharge, 16
Criterion continuous concentration (CCC),
  ecological risk assessment, toxicity
  testing, 126–129
Criterion maximum concentration (CMC),
  ecological risk assessment, toxicity
  testing, 126–129
Critical initial dilution (CID), ecological risk
  assessment, acute toxicity levels, 127
Crystallization evaporation network, ternary
  wastewater systems, 617–620
Cultural and ethical value, risk management, 134
Cycle of recycling, plastics recycling, developing
  country case study, 562–564
Cycles of concentration (COC), zero water
  discharge system, captive power plant,
  229–230

DaimlerChrysler case study, zero water discharge
  system, 230, 233–234
Data accuracy requirements, life cycle
  assessment, 41
Data analysis, ecological risk assessment, toxicity
  testing, 125–126
Data collection plan, life cycle inventory, 43–45
Data organization and display:
  life cycle assessment, goal definition and
    scoping, 39–40
  life cycle impact assessment, 52–53
  life cycle inventory, 42–45
  zero water discharge systems, database
    guidelines, 226

Deacon reaction, dioxin/furan formation,
  290–291
Decomposer niche:
  limits of, 32
  zero discharge industry, investment
    recovery, 27
Decontamination, pulp and paper industry,
  biopulping, 663
Degradation design, green chemistry and, 196
Delignification, pulp and paper industry, 660
De manifestis risk, characterization of, 135–136
Dematerialization, zero discharge industry and,
  26–27
De minimus risk:
  characterization of, 135–136
  conceptual basis, 136
De-novo synthesis, dioxin/furan formation, 290
Derivatives, green chemistry and, 196
Design criteria:
  biofiltration systems, organic vapors kinetics,
    285–288
  membrane-based technologies:
    VOC removal and recovery, 359–362
    wastewater treatment, 400–403
  zero discharge technology, pulp and paper
    industry, 684–685
Designer wastes, zero discharge industry,
  22–23
Design for the environment (DFE) paradigm, zero
  discharge industries, 28
Destruction and removal efficiency (DRE), volatile
  organic compounds, electron-beam-
  generated plasma destruction:
  comparison with other technologies,
    276–278
De-sulfurizer, fuel cell processing system, 343
  limitations of, 348–349
Dielectric barrier discharges, 301
  carbon dioxide effects, 307
  carbon monoxide, methane and ethylene,
    307–310
  electric frequency selection, 304
  gas analysis instrumentation, 303
  HV power measurement, 303
  inlet gas compounds, nitrogen oxide
    conversions, 304
  mercury vapor generation, 303
  oxygen effects, 304–306
  pollution control, inlet/outlet concentrations,
    303–304
  reactor design, 302–303
  water vapor effects, 306–308
Dielectric constant, supercritical water, 408

Diffusion limitation, biofiltration systems, organic
vapors kinetics, 285–286
Dioxin/furan removal:
pulp and paper industry, effluent discharges, 654
sorbent and carbon injection techniques,
289–298
ash particle scavenging, 293
carbon sources, 291–292
control technologies, 293–297
deacon reaction, 290–291
de-novo synthesis, 290
dioxin/furan formation mechanisms,
290–291
dioxin/furan sources, 289–290
inhibition mechanisms, 292–293
mercury removal, 295–297
metals, 292
sulfuric acid treatment, 297
Direct capture techniques, carbon sequestration
technology, air pollution prevention,
378–379
Direct combustion, poultry litter, catalytic steam
gasification, 584–585
Direct contact evaporator (DCE), sulfur gas
reduction, pulp and paper industry, 650
Direct costs (DC), pollution prevention economics,
165–166
Direct methanol fuel cell (DMFC), basic
technology, 341–342
Dirty air, economic consequences of, 180–181
Disaster Communication (Indiana Law
University), 156
Discharge minimization:
information sources, 639
process integration, 6
heat exchange networks, 598–600
industrial applications, 638–640
mass integration, 600–622
end-of-pipe separation network synthesis,
602–608
in-process separation network synthesis,
608–614
multicomponent nonideal systems,
614–622
research background, 600–602
research background, 597–598
water optimization and integration, 622–637
land treatment technology, 624–628
pulp and paper process, 628–637
treatment technologies, 3
Disinfection standards:
ultraviolet disinfection performance
evaluation, 73

wastewater treatment, ultraviolet disinfection,
428–430
Disposal process, zero water discharge
systems, 227
Dissipative distribution of goods, zero discharge
industry, 24
Distillate production, brine concentrators,
wastewater recycling, 437
Documentation:
life cycle assessment, assumptions about, 41
life cycle impact assessment, 52–53
life cycle inventory results, 45–46
Dominance analysis, life cycle assessment
interpretation, 55
Donnan potential, membrane-based technology,
nanofiltration, 390
Dose-response relationship:
ecological risk assessment, two-dimensional
chemical model, 144
health risk assessment, receptor dose studies,
97–101
Drum dryers, brine concentrators, wastewater
recycling, concentrate disposal,
439–440
Dry acid deposition, health risk assessment,
108–109
Dry ice co-generation, carbon sequestration
technology, 377
Dry-low nitrogen oxide reduction:
best available control technology, 175
zero discharge technology, 243–244
Duct burners:
best available control technology:
carbon monoxide reduction, 176–178
nitrogen oxide emissions, 171
sulfur dioxide and $H_2SO_4$
emissions, 179
pollution prevention economics,
best available control technology,
nitrogen oxide
reduction, 176
Durability parameters, fuel cell processing, 348
Dyeing industries, supercritical carbon dioxide,
zero-discharge technology, 412

Earth ecosystem, sustainable economy,
183–188
Ebara Mebius system, smokestack emissions,
251–252
Ebara process, zero discharge technology,
smokestack emissions, 249–252
Eco-efficiency, sustainable development, 214–216
Eco-industrial parks:

development criteria, 255–256
infrastructure development, 29–30
sustainable development and, 214–216
zero discharge models, 252–259
  basic characteristics, 255
  development guidelines, 255–256
  India, mini-case studies, 256–259
  industrial ecology, 252–255
Ecological risk assessment (ERA):
communication issues, 146–153
  accident aftermath, 151
  case study, 151–153
  costs of noncommunication, 151–152
  public response determinants, 146–148
  sustainable strategies, 148–150
effects assessment, 118–120
  benthic community, 119
  fish community, 119
  soil invertebrates and plants, 119–120
  wildlife, 120
eutrophication, 123–124
exposure assessment, 114–118
  benthic macroinvertebrate community,
    115–117
  fish community, 116
  soil invertebrates, 117
  terrestrial plants, 117
  terrestrial wildlife, 117–118
global warming, 122–123
health risk assessment, 114–115
information sources, 122
internet information access, 153–157
management strategies:
  characterization issues, 135–136
  chemical models, 137–145
    one-dimensional models, 137
    three-dimensional models, 144–145
    two-dimensional models, 137–144
  de minimis risk concepts, 136
  ecological resources valuation, 132–134
    nonutilitarian values, 133–134
    utilitarian values, 133
  modeling techniques for, 134–135
  research background, 130–132
risk characterization, 129–130
risk management:
sampling and surveys, 120–121
speciation, 122
technical aspects, 110–114
  endpoint identification, 113
  measurement endpoint selection, 113–114
  site conceptual models, 111–113
toxicity testing, 124–129

application of results, 126–129
test results evaluation, 124–126
toxic units, 126
uncertainties, 130
visibility and regional haze, 123
Ecology. *See also* Industrial ecology
EcoManager tool, life cycle assessment and life
  cycle impact assessment, 58, 60
Economics of pollution prevention. *See also*
  Sustainable economics and
  development
metal finishing industry, water pollution
  prevention, 474–475
plastics recycling, developing country case
  study, 557–558, 578–579
poultry litter, catalytic steam gasification,
  592–595
pulp and paper industry, biopulping,
  663–665
research overview, 6
sustainability and, 192
volatile organic compounds, electron-beam-
  generated plasma destruction, 274
wastewater recycling, brine concentrators,
  448–449
wastewater treatment, membrane-based
  technologies, 400–403
EcoPro 1.5 software, life cycle assessment and life
  cycle impact assessment, 60
Ecosystem balance:
pollution prevention economics, 180
sustainable economy, 183–188
EECO-it 1.0 database, life cycle assessment and
  life cycle impact assessment, 58
E factor, green chemistry, 194–195
Effluent discharges:
membrane-based technology, pharmaceutical
  applications, 386–387
pulp and paper industry, 653–655
  adsorbable organic halides, 654–655
  dioxin/furans, 654
  zero discharge technology, closed-loop
    concept, 673–676
    clarification process, 674
    concentration, 675–676
    evaporation, 675
    incineration, 676
    stripping, 676
  zero water discharge, captive power plant:
    reuse options, 229–230
    treatment plant design, 230–232
Electric frequency, dielectric barrier discharges,
  selection criteria, 304

Electricity generation:
  biomass combustion, 185
  biomass fuel, sustainable economics and
     development, 201–203
  biomass gasification, 185–186
  hydrogen-based sustainable economy, 207–214
  life cycle assessment, biorenewables *vs.* fossil
     fuels, 77–84
  air emissions, 81
  global warming potential, 80–81
  resource consumption, 81–82
  sensitivity analysis, 82
  system energy balance, 77–80
Electrodeless discharge lamp (EDL), arsenic
     removal, groundwater contamination,
     chemical analysis, 534
Electrolytes, fuel cell technology, 335–337
Electron-beam-generated plasma, volatile organic
     compound destruction, 265–278
  aqueous phase waste streams, 316–329
  alternative oxidation solutions, 326–328
  AOS system, 320–322
  basic chemistry, 318–319
  economic issues, 328–329
  Florida International University/University
     of Miami system, 322–324
  oxidation limitations, 317–318
  UTSI pilot facility, 323–326
  comparison with other techniques, 274–278
  preliminary economics, 274
  process concept, 267–269
  research background, 266–267
  UTSI pilot facility, 269–274
Electron charge transfer, electron-beam chemistry,
     volatile organic compound destruction,
     319–320
Electroplating in-process techniques, profitable
     pollution prevention, 476–490
  integrated modeling and optimization,
     478–482
  metal finishing industry, 476–477
  P3 applications, 482–490
  process pollution prevention fundamentals and
     strategies, 478
  P2 to P3 transition, 477–478
Electrostatic precipitator (ESP), volatile organic
     compounds, electron-beam-generated
     plasma destruction, 318
Elemental-chlorine-free (ECF) process,
     pulp and paper industry,
     zero discharge, 679
Emergency Response Planning Guidelines
     Level 2 (ERPG-2) values,

     accidental chlorine gas release,
     toxic endpoint distance, 66
Emissions reduction:
  air emissions:
  electricity generation, biorenewables *vs.*
     fossil fuels, 81
  pollution prevention, economic criteria, 182
  biodiesel, 330–331
  cost-benefit analysis, 182
  poultry litter, catalytic steam gasification,
     testing of, 591–592
  zero discharge technology:
  gas turbine case study, 239–240
  pulp and paper industry, 679–683
  greenhouse gases, 680–682
  solid waste management and energy
     conservation, 682–683
  sulfur recovery, 679–680
Empty bed contact time (DBCT), arsenic removal,
     groundwater contamination, 533–534
End-of-pipe separation networks, mass integration,
     discharge minimization, 602–608
  heat-induced/energy-induced separation
     networks, 606–660
  mass exchange networks, 602–606
Endpoint identification, ecological risk
     assessment, 113–114
Energy conservation, pulp and paper industry, zero
     discharge technologies, 682–683
Energy consumption:
  biorenewables *vs.* fossil fuels, life cycle
     assessment, 77–80
  brine concentrators, wastewater recycling,
     443–444
Energy density, mercury oxidation, barrier
     discharge, 312
Energy efficiency, green chemistry and
     design for, 195
Energy-induced separation networks (EISENs),
     mass integration, discharge
     minimization, 606–608
Energy-induced waste minimization networks
     (EIWAMIN), mass integration,
     discharge minimization:
  in-process separation network synthesis,
     608–611
  polymer manufacturing, 611–614
Energy integration:
  defined, 6
  heat exchange networks, 598–600
  process integration and, 597–598
Energy recovery, pulp and paper process, black
     liquor recovery process, 665–667

Energy-separating agents (ESAs), mass
  integration, discharge minimization:
development of, 602
heat-induced and energy-induced separation
  networks, 606–608
VOC condensation and allocation,
  multicomponent nonideal systems,
  614–617
Energy sources:
fuel cells, 334–351
  alkaline fuel cells, 340–341
  cost and volume reduction, 348
  CPO control, 348
  de-sulfurization limitations, 348–349
  de-sulfurizer components, 343
  direct methanol fuel cells, 341–342
  durability/life, 348
  efficiency and performance, 347
  fuel selection, 349–350
  future issues, 350–351
  molten carbonate fuel cells, 341
  part-load and transient carbon monoxide
    control, 347–348
  phosphoric acid fuel cells, 340
  polymer electrolyte membrane fuel cell,
    337–339
  preferential oxidation, 346–347
  processing system, 342–343
  reforming and catalytic partial oxidation,
    343–345
  research background, 334–337
  social and environmental benefits, 350
  solid oxide fuel cells, 339–340
  water gas shift, 345–346
ultraviolet disinfection performance
  evaluation, environmental
  impact studies, 75–76
Energy value, recovery in waste of, 3
Engineered materials, sustainability, 214
Engineering. *See also* Green chemistry and
  engineering
Environmental impact:
chlorine (disinfection) process, 63–64
electroplating in-process techniques, plant-wide
  integrated modeling, 482
fuel cell technology, 350
metal finishing industry, water pollution
  prevention, 472–474
poultry litter, 583–584
process integration and minimization of:
  heat exchange networks, 598–600
  industrial applications, 638–640
  mass integration, 600–622

end-of-pipe separation network synthesis,
  602–608
in-process separation network synthesis,
  608–614
multicomponent nonideal systems,
  614–622
research background, 600–602
research background, 597–598
water optimization and integration,
  622–637
  land treatment technology, 624–628
  pulp and paper process, 628–637
pulp and paper process:
  biopulping, 663–665
  water optimization and integration, 628–637
  zero discharge technology, 678
sulfur dioxide dechlorination, 65
sustainability, 193
ultraviolet disinfection performance evaluation,
  75–76
Environmental management systems (EMS):
metal finishing industry, water pollution
  prevention, 467–469
pulp and paper industry, 647–648
Environmental Protection Agency (EPA):
Center for Exposure Assessment Modeling, 155
Chemical Accident Histories and
  Investigations, 153
Integrated Risk Information System, 153
life cycle inventory guidelines, 42–46
National Service Center for Environmental
  Publications (NSCEP), 156
pollution prevention definition, 2
Regulatory Air Models, 155
Risk Management Consequence Analysis, 156
Environmental vulnerability, risk assessment and
  determination of, 147–148
Enzymes, zero discharge industrial
  applications, 32
Equality constraints, electroplating in-process
  techniques, plant-wide integrated
  modeling, 481
Ethane, foundry emissions, advanced oxidation
  processes, 512
Ethanol, biomass production, 186–187
Ethylene, dielectric barrier discharges, nitrogen
  oxide conversions, 307–310
European Community (EC), pollution prevention
  definition, 2
Eutrophication:
ecological risk assessment, 123–124
poultry litter, environmental impact of,
  583–584

Evaporation ponds, brine concentrators, wastewater recycling, concentrate disposal, 439

Evaporation technology, pulp and paper industry, closed-loop effluent management, 675

Evaporator principles, brine concentrators, wastewater recycling, 433–436

Existence value, risk management, 133

Exposure assessment:
ecological risk assessment, 114–118
  benthic macroinvertebrate community, 115–117
  fish community, 116
  soil invertebrates, 117
  terrestrial plants, 117
  terrestrial wildlife, 117–118
health risk assessment, 93–101
  bioconcentration prediction, 94–95
  contaminant intake calculation, 99–100
  dermal soil contact, average daily intake, 100–101
  mechanistic models, 96–97
  partition coefficient and bioconcentration factor, 93–95
  point concentrations, 95–96
  receptor doses, 97–101
  slope factors, 95

Exposure point concentrations (EPCs):
ecological risk assessment, terrestrial wildlife, 117–118
health risk assessment, exposure studies, 95–96

External energy-separating agents, mass integration, discharge minimization, heat-induced and energy-induced separation networks, 607–608

External mass-separating agents, mass integration, discharge minimization, mass exchange networks, 603–606

Extractive membrane bioreactor (EMB), wastewater treatment, pharmaceutical industry, 395–396

Falling film evaporation, brine concentrators, wastewater recycling, 435–436

Faraday potential, membrane-based technology, nanofiltration, 390

Fast landfilling scenario, plastics recycling, developing country case study, mathematical modeling, 573–578

Fast pyrolysis, bio-oil production, 187

Fast recovery scenario, plastics recycling, developing country case study, mathematical modeling, 573–575

Feed pressure, membrane-based VOC removal and recovery, 354–355

Feedstocks:
green chemistry and renewable feedstocks, 196
plastics recycling, developing country case study, reprocessing effects, 566–571
recycling, solid waste pollution prevention, 547

Feed water characteristics, brine concentrators, wastewater recycling, 436

Fiber outside diameter (OD), membrane-based VOC removal and recovery, 357–359

Fick's law, membrane-based VOC removal and recovery, 352–353

Field sampling, ecological risk assessment, 120–121

Field surveys, ecological risk assessment, 121

Fish community, ecological risk assessment, 116, 119

Flap gate construction, ultraviolet disinfection, wastewater treatment, 422

Florida International University/University of Miami electron-beam system, volatile organic compound destruction, 322–323

Flow diagram, life cycle inventory, 43

Flow rate:
wastewater treatment, ultraviolet disinfection, 424–425
water optimization and integration, land treatment technology, 625–628

Flue gas capture, carbon dioxide emissions, carbon sequestration technology, 371–372

Flue gas desulfurization (FGD):
best available control technology, sulfur dioxide and $H_2SO_4$ emissions, 180
pollution prevention economics, total capital investiment (TCI), 165
volatile organic compounds, electron-beam-generated plasma destruction, 318

Forced circulation crystallizers, brine concentrators, wastewater recycling, concentrate disposal, 441–442

Formaldehyde emissions, foundry emissions, advanced oxidation processes, 511

Fossil fuels, life cycle assessment, 77–84
air emissions, 81
global warming potential, 80–81
resource consumption, 81–82
sensitivity analysis, 82
system energy balance, 77–80

Foundry oxidation technologies, wastewater recycling and pollution prevention, 491–517
acoustic cavitation, 497–499
advanced oxidation conditions, 496–499
benzene emissions, 510–511
carbon monoxide/carbon dioxide emissions, 511–512
clay and coal efficiency, 514–516
design layouts, 499–501
emissions characteristics, 508–514
formaldehyde emissions, 511
full-scale advanced oxidation, 501–503
green sand properties, 506–508
green sand systems, 493–496
mass balance analysis, emissions vs. core binder, 514
methane, ethane, and propane emissions, 512
Neenah foundry case study, 503–506
ozone and hydrogen peroxide, 497
research background, 491–493
Technikon-CERP and Penn State emissions comparisons, 516–517
volatile organic compound emissions, 508–510
Free radicals:
electron-beam chemistry, volatile organic compound destruction, 319–320
volatile organic compounds, electron-beam-generated plasma destruction, 267–269
Freshwater usage, pulp and paper industry, effluent discharges, 654
Fuel cells:
hydrogen-based sustainable economy, 211–214
pollution prevention with, 334–351
alkaline fuel cells, 340–341
cost and volume reduction, 348
CPO control, 348
de-sulfurization limitations, 348–349
de-sulfurizer components, 343
direct methanol fuel cells, 341–342
durability/life, 348
efficiency and performance, 347
fuel selection, 349–350
future issues, 350–351
molten carbonate fuel cells, 341
part-load and transient carbon monoxide control, 347–348
phosphoric acid fuel cells, 340
polymer electrolyte membrane fuel cell, 337–339
preferential oxidation, 346–347
processing system, 342–343

reforming and catalytic partial oxidation, 343–345
research background, 334–337
social and environmental benefits, 350
solid oxide fuel cells, 339–340
water gas shift, 345–346
Fuel economy, biodiesel, 333
Fuel processor efficiency, fuel cell technology, 347
Fuel sources, fuel cell technology, 349–350
Functionalized membranes, wastewater treatment, heavy metal recovery and water recycling, 399–400
Furan. See Dioxin/furan removal

GaBi, life cycle assessment and life cycle impact assessment, 60
Gas analysis, dielectric barrier discharges, instrumentation, 303
Gas conditioning towers, zero water discharge system, captive power plant, 228–230
Gasoline recovery, membrane-based VOC removal and recovery:
gas stations, 365
tank farms, 364–365
Gas separation membranes, carbon sequestration technology, 373–374
Gas turbines:
best available control technology, sulfur dioxide and $H_2SO_4$ emissions, 179
zero discharge technology, nitrogen oxide emissions, 238–246
General Electric plastics case study, heat integration and mass integration, 638
Generation of waste, minimization, 2
Geographic location, industrial cluster development, 21–22
Germicidal action, wastewater treatment, ultraviolet disinfection:
action mechanism, 417–418
efficiency of, 427–428
GINO programming, pulp and paper process, discharge minimization, water optimization and integration, 636–637
Global warming potential (GWP):
ecological risk assessment, 122–123
electricity generation, biorenewables vs. fossil fuels, life cycle assessment, 80–81
pollution prevention economics, 181

Goal definition and scoping, life cycle assessment, 38–42
assumptions documentation, 41
data accuracy requirements, 41
data organization and results display, 39–40
defined, 36
information needs assessment, 39
manufacturing stage, 41
project goals, 38–39
quality assurance procedures, 41–42
raw materials acquisition, 40
recycle/waste management, 41
reporting requirements, 42
use/reuse/maintenance phase, 41
Good combustion practices,
best available control technology:
carbon monoxide reduction, 177
volatile organic carbons, 180
Graphical solution strategy, pulp and paper
process, discharge minimization, water
optimization and integration, 631–634
Gravity analysis, life cycle assessment, 56
Great Britain, recycling history, 547–548
Green by design process, basic principles, 199
Green chemistry and engineering:
sustainable economics and development,
193–199
twelve principles of, 194–197
Greenhouse gases:
carbon sequestration technology, 370–371
ecological risk assessment, 122–123
electricity generation, biorenewables *vs.* fossil
fuels, global warming potential, 80–81
pulp and paper industry emissions, 652–653
recovery and control, 680–682
calcium carbonate precipitates, carbon-
dixode-rich gases, 681–682
climate-neutral carbon dioxide, biomass
fuels, 681
combined heat and power systems, 680
Green sand systems, foundry emissions, advanced
oxidation processes, 493–496, 506–508
GREET model, fuel cell technology, 350
Ground-level ozone, health risk assessment,
104–105
Groundwater contamination:
ecological risk assessment, site conceptual
models, 111–113
pollution prevention technology, 522–542
aquifer contamination, 522–523
arsenic removal technologies, 524–542
chemical analysis, 534

chemical properties and distribution,
525–526
column runs, regeneration and rinsing, 533
fixed-bed column runs, 534–536
kinetic testing, 533–534
mobility and immobility, natural
environment, 526–527
polymeric/inorganic hybrid sorbent,
531–533
regeneration, rinsing and reuse, 538–540
sorption kinetics, 536–538
poultry litter, environmental impact of,
583–584
Grouping techniques, life cycle impact assessment,
51–52

Handling protocols, biodiesel, 333
Hanover principles, sustainable development and,
217–218
Hardness properties, wastewater treatment,
ultraviolet disinfection, 426
Hazard exposure, health risk characterization, 102
Hazardous air pollutants (HAPs):
biofiltration of, 281–282
health risk assessment, 109–110
Hazards Analysis for Toxic Substances, Version 3
(HATS3), 155
Haze, ecological risk assessment, 123
Health risk assessment:
cancer risk assessment, 104
chlorine (disinfection) process, 63–64
common pollutants, 103–110
acid deposition, 108–109
air toxics, 109–110
carbon monoxide, 105–106
ground-level ozone, 104–105
lead and mercury, 106–107
particulate matter, 107–108
communication issues, 146–153
accident aftermath, 151
case study, 151–153
costs of noncommunication, 151–152
public response determinants, 146–148
sustainable strategies, 148–150
continuous monitoring example, 92–93
ecological risk assessment *vs.*, 114–115
exposure assessment, 93–101
bioconcentration prediction, 94–95
contaminant intake calculation, 99–100
dermal soil contact, average daily intake,
100–101
mechanistic models, 96–97

partition coefficient and bioconcentration
  factor, 93–95
point concentrations, 95–96
receptor doses, 97–101
slope factors, 95
internet information access, 153–157
management strategies:
  characterization issues, 135–136
  chemical models, 137–145
    one-dimensional models, 137
    three-dimensional models, 144–145
    two-dimensional models, 137–144
  de minimis risk concepts, 136
  ecological resources valuation, 132–134
    nonutilitarian values, 133–134
    utilitarian values, 133
  modeling techniques for, 134–135
  research background, 130–132
monitoring example, 90–92
plastics recycling, developing country case
  study:
  regulatory issues, 579
  reprocessing effects, 565–571
problem formulation, 90–93
regulatory statutes concerning, 91
risk characterization, 101–103
  average and minimum exposure, 102–103
  carcinogenic risk, 103
total weighted average concentration,
  105–106
toxicity assessment, 101
Heat exchange networks (HENs):
  energy integration and, 598–600
  information sources, 639
  minimum utilities targeting, 599–600
Heat-induced separation networks (HISENs), mass
  integration, discharge minimization,
  606–608
Heat-induced waste minimization networks
  (HIWAMINs), mass integration,
  discharge minimization:
  development of, 602
  in-process separation network synthesis,
    608–611
  polymer manufacturing, 611–614
Heat integration, mass integration and, 638
Heat recovery steam generator (HRSG), nitrogen
  oxide emission control, selective
  catalytic reduction, 240–242
Heavy metals, wastewater treatment, membrane-
  based applications, recovery and water
  recycling, 397–400

Henry's law, membrane-based VOC removal and
  recovery, 358–359
High-density polyethylene (HDPE), recycling
  process, developing country case study:
  color change sequence, 562–565
  high-density polyethylene characteristics,
    555–556
  high-density polyethylene service life, 564
  impurities problems, 565
  industrial ecology, 554–555
  mathematical modeling, 571–578
  recycling cycle, 562–564
  socioeconomic issues, 557–558
  spatial profile, 556–557
  successive processing, effects on recyclate,
    565–571
  supply chain issues, 558–565
High efficiency reverse osmosis (HERO),
  wastewater recycling, brine
  concentrators, 446–448
$H_2SO_4$ emissions, best available control
  technology, 179
Human health. See Health risk assessment
HV power measurement, dielectric barrier
  discharges, 303
Hybrid ion exchanger (HIX), arsenic removal,
  groundwater contamination, 531–533
  fixed-bed arsenate feeds, 534–535
  fixed-bed arsenite feeds, 535–536
  kinetic testing, 533–534
  regeneration, rinsing and reuse, 538–540
  sorption kinetics and intraparticle diffusivity,
    536–538
Hydrated Fe oxide (HFO) precipitates, arsenic
  removal, groundwater contamination,
  531–533
  regeneration, rinsing and reuse, 538–540
Hydrocarbons, health risk assessment,
  104–105
Hydrogen-based sustainable economy:
  fuel cell technology, 334–336
  guidelines for, 207–214
Hydrogen peroxide systems:
  foundry emissions, advanced oxidation
    systems, 497
  volatile organic compounds, electron-beam-
    generated plasma destruction, 317–320
Hydrogen sulfide, pulp and paper process:
  conversion technologies, 657–658
  sulfur recovery, 679–680
Hydrogen-to-oxygen (H/O) ratio, poultry litter,
  catalytic steam gasification, 588–589

Hydro-metallurgical industries, conservation and reconcentration, membrane-electrode process, 460–461
Hydropower, sustainable economics and development, 205
Hydroxyl radical, electron-beam chemistry, volatile organic compound destruction, 319–320

IDEMAT, life cycle assessment and life cycle impact assessment, 60
Immobility, arsenic compounds, groundwater contamination, 526–529
Impact assessment:
  chemical process industries, sustainable production and growth in, 198–199
  life cycle assessment:
    defined, 36
    principles of, 46–53
Impact categories, life cycle impact assessment, 46–49
Impact resistance testing, plastics recycling, developing country case study, reprocessing effects, 570–571
Impurities, plastics recycling, developing country case study, 565
Incineration:
  poultry litter, catalytic steam gasification, 587–588
  pulp and paper industry, closed-loop effluent management, 676
  volatile organic compound removal, 275–278
India eco-industrial parks, zero discharge technology case study, 256–259
Indirect costs (IC), pollution prevention economics, 165–166
Industrial applications:
  biofiltration systems, volatile organic vapors, 286–287
  heat integration and mass integration, 638
  poultry litter, 582–583
  zero discharge:
    conversion technology development, 22
    dematerialization, 26–27
    design wastes, 22–23
    emissions methodology, 20–23
    industrial cluster construction, 21–22
    inventory input-output, 20–221
    investment recovery, 27
    Kalundborg case study, 14–15
    limits, 31–32
    material-product recycling and reuse, 23–28

opportunity development, 29
paradigm change, 28–30
process engineers and engineering firms, 16–20
process innovations, 28
productivity of raw materials, beer to mushroom case study, 19–20
regulatory reform, 23
research background, 11–12
return on investment, provision for, 30
separation technology, 30–31
sustainability, ecology and emissions, 12–15
throughput analysis, 20
toxicity reduction of materials and chemicals, 27–28
transition strategies, 23–28
Industrial cluster development:
  conversion technology and, 22
  zero discharge and development of, 21–22
Industrial ecology (IE):
  plastics recycling, developing country case study, 554–555
  socioeconomic issues, 557–558
  spatial profile, 556–557
  sustainable development, 214–216
  zero discharge technology:
    models, 252–255
    sustainability, 12–15
Information assessment, life cycle assessment, goal definition and scoping, 39
Information sources:
  ecological risk assessment, 122
  process integration and discharge minimization technology, 639
  risk assessment, internet sources, 152–157
Inherently safer chemistry, accident prevention, 196–197
In-process separation network synthesis, mass integration, discharge minimization, 608–614
  polymer manufacturing, 611–614
  WIN/HIWAMIN, and EIWAMIN systems, 608–611
Integrated gasification combined cycle (IGCC):
  life cycle assessment:
    biorenewables vs. fossil fuels, 77–84
    global warming potential, 80–81
    system energy balance, 77–80
  sustainable economics and development, biomass fuel, 202–203

Integrated Risk Information System (IRIS), 153
Interception process:
  mass integration, discharge minimization,
    in-process separation network
    synthesis, 610–611
  pulp and paper process, discharge minimization,
    water optimization and integration,
    631–637
Intergovernmental Panel on Climate Change
  (IPCC), life cycle impact assessment,
  51
International community, zero discharge
  technologies, pulp and paper industry,
  683–684
Internet, risk assessment information sources on,
  152–157
Interpretation techniques, life cycle assessment:
  completeness check, 55–56
  consistency check, 56–57
  defined, 36
  impact assessment, 53–58
  results interpretation, 54–55
  sensitivity check, 56
Intraparticle diffusivity, arsenic removal,
  groundwater contamination, 536–538
Introduction of waste, minimization, 2
Inventory analysis:
  life cycle assessment, defined, 36
  life cycle inventory, 42–46
  zero discharge industries and, 20–21
Investment recovery:
  pollution prevention economics, 166–167
  zero discharge industry, 27
Ion-exchange kinetics, membrane-electrode
  process, wastewater recycling, 457–458
Ionization potential:
  electron-beam plasma, volatile organic
    compound destruction, 319–321
  volatile organic compounds,
    electron-beam-generated
    plasma destruction, 267–274
Iron concentrations:
  membrane-electrode process, wastewater
    recycling, cation selectivity, 454
  wastewater treatment, ultraviolet disinfection,
    425–426
ISO 14001, zero effluent systems, Formosa Plastic
  Manufacturing case study, 234–235

Kalundborg case study:
  industrial cluster development, 21–22
  industrial ecology and zero discharge industries,
    14–15

Kinetic testing, arsenic removal, groundwater
  contamination, 533–534
Kraft pulp and paper process:
  air pollution prevention, 649–653
    chlorine/chlorine dioxide, 651–652
    chloroform emissions, 650–651
    greenhouse gases, 652–653
    sulfur dioxide/nitrogen oxides, 650
    sulfur gas reduction, 650
  discharge minimization, water optimization and
    integration, 628–637

Laboratory bioassays, chlorine (disinfection)
  process, aquatic life effects, 69
Lactic acid, bioproduction, 188
Lamp components, ultraviolet disinfection,
  wastewater treatment, 418–421, 427
Land treatment technology, water optimization
  and integration, 624–628
Large-scale facilities:
  poultry litter, catalytic steam gasification,
    testing of, 591–592
  pulp and paper industry, biopulping, 662–663
Latency period, pollutant exposure, health risk
  assessment, 104
LCAD tool, life cycle assessment and life cycle
  impact assessment, 60
LCAiT tool, life cycle assessment and life cycle
  impact assessment, 60
Lead cations, membrane-electrode process,
  wastewater recycling, 454
Lead exposure, health risk assessment, 106–107
Lean-premix technology, best available control
  technology, nitrogen oxide
  reduction, 175
Least hazardous chemical syntheses, green
  chemistry and, 195
Level control devices, ultraviolet disinfection,
  wastewater treatment, 421–422
Lewis acids, arsenic compounds, groundwater
  contamination, 530–531
Life cycle assessment (LCA):
  basic components of, 36
  benefits of, 36–37
  chlorination (disinfection process), 62–72
    human health and environmental impact,
      63–64
    limitations, 63
    sulfur dioxide dechlorination, 64–72
      accidental chlorine gas release, case study,
        65–67
      aquatic life, effects on, 67–69
      environmental impact, 65

Life cycle assessment (LCA) (*Continued*)
  limitation, 65
    marine life sensitivity, chlorinated
      seawater, 70–72
  defined, 5, 35–36
  disinfection technologies, environmental
    performance evaluation, 62–76
  electricity generation, biorenewables *vs.* fossil
    fuels, 77–84
    air emissions, 81
    global warming potential, 80–81
    resource consumption, 81–82
    sensitivity analysis, 82
    system energy balance, 77–80
  flowchart for, 37–38
  goal definition and scoping, 38–42
    assumptions documentation, 41
    data accuracy requirements, 41
    data organization and results display, 39–40
    information needs assessment, 39
    manufacturing stage, 41
    project goals, 38–39
    quality assurance procedures, 41–42
    raw materials acquisition, 40
    recycle/waste management, 41
    reporting requirements, 42
    use/reuse/maintenance phase, 41
  interpretation, 53–58
  limitations of, 37
  plastics recycling, developing country case
    study, 552, 558–565
  software tools, 58–61
  sustainable product development, 200–201
  ultraviolet disinfection performance evaluation,
    72–76
    cost analysis, 74–75
    disinfection standards, 73
    environmental impact, 75–76
    government research on, 62–63
    operation, maintenance, and worker study,
      73–74
    transmission/absorbance evaluation, 73
Life cycle impact assessment (LCIA), 46–53
  software tools, 58–61
Life cycle inventory (LCI), 42–46
  data collection, 42–45
Life Cycle Product/Process Design (lCPPD),
    sustainable product development,
    200–201
Lignozym, pulp and paper industry, pollution
    prevention technologies, 657
Limits of zero discharge, industrial applications,
    31–32

Linearized multi-stage model, health risk
    assessment, exposure studies, 96–97
Liquid matrix streams, zero discharge industries,
    separation technology, 30–31
LOTT system case study, wastewater treatment,
    ultraviolet disinfection,
    416–417, 421
Louisiana-Pacific Corporation case study, pulp and
    paper industry, zero discharge, totally
    chlorine free (TCF) processing,
    672–673
Lower heating value of hydrogen, fuel cell
    technology, 347
Lowest available emission rates (LAER):
  foundry emissions, advanced oxidation systems,
    505–506
  pollution prevention economics, best available
    control technology, 171
  selective catalytic reduction, 174–175
  zero discharge technology, 246–247
Low pressure flat lamp, wastewater treatment,
    ultraviolet disinfection, 419
Low-temperature distillation, carbon sequestration
    technology, 372–373


Maintenance guidelines, ultraviolet disinfection:
  performance evaluation, 73–74
  wastewater treatment, 427
Manufacturing:
  biobased products, sustainable economics,
    184–188
  industrial ecology, 13–15
  life cycle assessment, goal definition and
    scoping, 41
  zero waste facilities, 11–12
Mass balance analysis, foundry emissions,
    advanced oxidation systems, 514
Mass exchange networks (MENs), mass
    integration, discharge minimization,
    end-of-pipe separation network
    synthesis, 602–606
Mass integration:
  defined, 6
  discharge minimization, 600–622
    end-of-pipe separation network synthesis,
      602–608
    heat-induced/energy-induced separation
      networks, 606–660
    mass exchange networks, 602–606
    in-process separation network synthesis,
      608–614
    polymer manufacturing, 611–614

WIN/HIWAMIN, and EIWAMIN
    systems, 608–611
multicomponent nonideal systems,
    614–622
    ammonium nitrate manufacturing, ternary
        wastewater system, 620–622
    crystallization evaporation network,
        ternary wastewater system, 617–620
    VOC condensation and allocation,
        614–617
    research background, 600–602
industrial applications, 638
process integration, 597–598
Mass-separating agents (MSAs), mass integration,
    discharge minimization:
benzene recovery, 604–606
development of, 602
in-process separation network synthesis,
    608–611
mass exchange networks, 603–606
Mass transfer coefficient, membrane-based VOC
    removal and recovery, 358–359
Material balance, pulp and paper process,
    discharge minimization, water
    optimization and integration, 631–634
Mathematical modeling:
health risk assessment, exposure point
    concentrations, 95–96
plastics recycling, developing country case
    study, 571–578
    development and programming issues, 575
    fundamentals of, 571–575
    results analysis, 576–578
water optimization and integration, land
    treatment technology, 626–628
Maximally exposed individual (MEI), health risk
    characterization, 102
Maximum achievable control technology
    (MACT):
ecological risk assessment, two-dimensional
    chemical model, 141
health risk assessment, air toxic analysis,
    109–110
Maximum potential difference, membrane-
    electrode process, wastewater
    recycling, 459
Measures of ecosystem and receptor
    characteristics, ecological risk
    assessment, 114
Measures of effect, ecological risk assessment,
    endpoint measurement, 113–114
Measures of exposure, ecological risk assessment,
    endpoint measurement, 113–114

Mechanical recycling:
plastics recycling, developing country case
    study, 558–559
solid waste pollution prevention, 546–547
Mechanical vapor compression (MVC), brine
    concentrators, wastewater recycling,
    434–436
Mechanistic models, health risk assessment,
    exposure studies, 96–97
Median lethal concentration ($LC_{50}$), chlorination
    (disinfection) process, aquatic life
    effects, 70–72
Melt flow rate (MFR), plastics recycling,
    developing country case study,
    reprocessing effects, 566–571
Membrane-electrode process, wastewater
    recycling, 450–462
binary cation systems, 453–454
binary solution target selectivity, 453
cation recovery flux/cation exchange flux, 452
ion-exchange kinetics, 457–458
limitations, 459
loading properties, 453
mining and hydrometallurgical industries
    conservation and reconcentration,
    460–461
process stream recycling, 459–460
rate of ion-exchange, 452
reverse potential phenomenon, 454–456
system independence, reverse potential
    phenomenon, 456–457
target cation recovery ratio, 453
Membrane technologies:
air pollution prevention:
    carbon sequestration technology:
        gas separation membranes, 373–374
        small pore diameters, 374
    volatile organic compound removal and
        recovery, 352–368
        absorption and stripping, 355–359
        commercial membrane modules,
            362–364
        design equations, principles and strategies,
            359–362
        gas station recovery, 365
        olefin polymerization vents, monomer
            recovery, 365–367
        tank farm gas recovery, 364–365
        vapor permeation, 352–355
wastewater recycling, membrane-electrode
    process, 450–462
binary cation systems, 453–454
binary solution target selectivity, 453

Membrane technologies (*Contd.*)
  cation recovery flux/cation exchange
    flux, 452
  ion-exchange kinetics, 457–458
  limitations, 459
  loading properties, 453
  mining and hydrometallurgical industries
    conservation and reconcentration,
    460–461
  process stream recycling, 459–460
  rate of ion-exchange, 452
  reverse potential phenomenon, 454–456
  system independence, reverse potential
    phenomenon, 456–457
  target cation recovery ratio, 453
  wastewater treatment, 384–403
    design and economics issues, 400–403
    heavy metal recovery and water recycling,
      397–400
    hybrid processes, 391–394
    membrane types and properties, 384–385
    nanofiltration, 389–390
    pervaporation, 388–389
    pharmaceutical processes, 385–387
      extractive membrane bioreactor,
      395–396
      product recovery, 397
      volume minimization, 396–397
    reverse osmosis, 387–388
  zero discharge industries, 31
Mercury emissions:
  exposure, health risk assessment, 106–107
  oxidation, barrier discharge, 311–314
  removal, carbon injection technology:
    dioxin and, 295–297
    sulfuric acid, 297
  vapor, dielectric barrier discharges, 303
MERIT partnership, metal finishing industry,
    water pollution prevention, 470
Metal finishing industry:
  pollution prevention in, 476–477
  water pollution prevention, 463–475
    economic benefits, 474
    environmental benefits, 472–473
    MERIT partnership for metal finishers,
    470
    Picklex case study, 470–472
    strategic goals program, 464–466
    technology transferability, 475
    zero discharge goal planning and
    implementation, 466–469
Metal ions, dioxin/furan formation, sorbent and
    carbon injection removal, 292

Metal recovery:
  heavy metals, wastewater treatment, membrane-
    based applications, 397–400
  supercritical water, zero-discharge
    technology, 411
Methane:
  dielectric barrier discharges, nitrogen oxide
    conversions, 307–310
  foundry emissions, advanced oxidation
    processes, 512
Methanol:
  biomass-derived syngas production, 187
  pulp and paper industry, resource recovery and
    reuse, 658–659
  VOC condensation and allocation,
    multicomponent nonideal mass
    integration, 615–617
Methyl iso-butyl ketone (MIBK), VOC
    condensation and allocation,
    multicomponent nonideal mass
    integration, 615–617
Microbial mediation, arsenic compound mobility,
    groundwater contamination, 528–529
Microchannel reaction, sustainable development,
    216
Microfiltration (MF), wastewater treatment,
    membrane-based technology, 384–385
Microorganisms:
  biofiltration of organic vapors, 282
  poultry litter, environmental impact of, 584
Minimum exposures, health risk assessment,
    102–103
Minimum utilities targeting, heat exchange
    networks, 599–600
Mining industry conservation and
    reconcentration, membrane-electrode
    process, 460–461
Mixed-integer nonlinear programming (MINLP),
    mass integration, discharge
    minimization, in-process separation
    network synthesis, 610–611
Mixing process, mass integration, discharge
    minimization, in-process separation
    network synthesis, 610–611
MK-819 compound, membrane-based technology,
    product recovery, 397
Mobility in natural environment, arsenic
    compounds, groundwater
    contamination, 526–529
Modeling techniques. *See also* Mathematical
    modeling
  chemical models, ecological risk assessment,
    137–145

Modeling techniques. (*Continued*)
  one-dimensional models, 137
    three-dimensional models, 144–145
    two-dimensional models, 137–144
  risk management, 134–135
Moisture effects, biofiltration of organic
    vapors, 283
Molecular weight cutoffs (MWCO), wastewater
    treatment, membrane-based technology,
    384–385
Molten carbonate fuel cell (MCFC), basic
    technology, 341
Monitoring procedures, health risk assessment,
    90–93
Monomer recovery, membrane-based VOC
    removal and recovery, olefin
    polymerization vents, 365–367
Multicomponent nonideal mass integration,
    discharge minimization, 614–622
  ammonium nitrate manufacturing, ternary
    wastewater system, 620–622
  crystallization evaporation network, ternary
    wastewater system, 617–620
  VOC condensation and allocation, 614–617
Multistage modeling, health risk assessment,
    exposure studies, 96–97
Municipal solid waste (MSW), plastics recycling,
    developing country case study, 558
  supply chain issues, 558–565

Namibia case study, zero discharge industries and
    productivity of, 19–20
Nanofiltration, wastewater treatment,
    membrane-based technology,
    384–385, 389–390
  design and economics issues, 400–403
  heavy metal recovery and water recycling,
    397–400
  valuable pharmaceutical compound retrieval,
    391–393
National Academies' Reports Online, 156
National ambient air quality standards (NAAQS),
    pollution prevention economics, best
    available control technology,
    170–179
National Ambient Water Quality Criteria,
    ecological risk assessment, 119
National Oceanic and Atmospheric Administration
    (NOAA), Chemical Reactivity
    worksheet, 154
National Service Center for Environmental
    Publications (NSCEP), 156

Natural gas combined cycle plant (NGCCP), life
    cycle assessment, global warming
    potential, 81
Neenah foundry case study, advanced oxidation
    technologies, 503–506
Net present value (NPV), pollution prevention
    economics, 166–167
Nickel cations, membrane-electrode process,
    wastewater recycling, 453–454
Nitrogen concentrations, poultry litter,
    environmental impact of, 583–584
Nitrogen oxides ($NO_x$):
  best available control technology, pollution
    prevention economics, 171–176
    combustion turbines and duct burners, 171
    lean-premix/dry-low nitrogen oxides, 175
    $SCONO_x$ technology, 171–173
    selective catalytic reduction, 174–175
    steam/water injection, 175–176
    XONON technology, 173–174
  dielectric barrier discharges:
    carbon monoxide, methane and ethylene
      effects, 307–310
    chemical compounds, conversion effects,
      304
    inlet-outlet concentrations, 303–304
    oxygen effects, 304–306
  health risk assessment:
    acid deposition, 108–109
    ground-level ozone, 104–105
  pulp and paper industry, 650
  zero discharge technology:
    ammonia slip/ammonium-sulfur salts,
      244–245
    cost analysis, 244
    dry low combustor, 243–244
    Ebara process, smokestack emissions,
      249–252
    gas turbine case study, 238–246
    regulatory background, 239, 246–247
    $SCONO_x$ system, 242–243
    selective catalytic reduction, 240–242
    spent catalyst, 245
    sulfur-bearing fuels, 245–246
    water-steam injection, 240
Nitrous oxide, dielectric barrier discharges, 310
No/lowest observed adverse effects level
    (NO/LOAEL), ecological risk
    assessment, 120
Non-AO optimized conditions, foundry emissions:
  advanced oxidation systems, 504–506
  benzene emissions, 510–511
  cost and efficiency analysis, 514–516

Non-AO optimized conditions, foundry emissions (*Continued*)
  green sand properties, 493–496, 506–508
  total VOC emissions, 508–510
Noncommunication costs, environmental and health risk communication and, 151–152
Non-linear optimization, pulp and paper process, discharge minimization, water optimization and integration, 636–637
Non process elements (NPEs), pulp and paper process, discharge minimization, water optimization and integration, 630–637
Nonthermal plasma. *See also* Dielectric barrier discharges
  coal-fired utility boiler pollutant control, 300–311
    carbon dioxide effects, 307
    carbon monoxide, methane and ethylene effects, 307–310
    dielectric barrier discharge reactor, 302–303
    electric frequency selection, 304
    gas analytical instrumentation, 303
    HV power measurement, 303
    mercury vapor generation, 303
    nitrogen oxide conversions, 304
    nitrous oxide by-products, 310
    oxygen effects, 304–306
    sulfur dioxide conversion geometry, 311
    sulfur dioxide oxidation, 310–311
    water vapor effects, 306–307
  mercury oxidation in barrier discharge, 311–314
Nonutilitarian values, risk management, 133–134
No observable effect concentration (NOEC), ecological risk assessment, toxic units measurement, 126
Normalization, life cycle impact assessment, 51

Olefin polymerization vents, membrane-based VOC removal and recovery, monomer recovery, 365–367
Oligotrophic conditions, ecological risk assessment, 123–124
Onboard refueling vapor recovery (ORVR) system, membrane-based VOC removal and recovery, 365
One-dimensional chemical model, ecological risk assessment, 137
Open channel modular UV systems, wastewater treatment, ultraviolet disinfection, 420–422

Operation guidelines, ultraviolet disinfection performance evaluation, 73–74
Opportunity development, zero discharge industries, 29–30
Oscillating water column (OWC), electricity generation, sustainable economics and development, 206
Outrage factors, risk assessment and determination of, 147–148
Oxidation:
  dielectric barrier discharges, sulfur dioxide, 310–311
  membrane-electrode process, wastewater recycling, 459
  volatile organic compounds, electron-beam-based treatment, 317–318
    comparison of techniques, 326–328
Oxygen concentrations:
  biofiltration of organic vapors, 282–283
  dielectric barrier discharges, 304–306
  mercury oxidation, barrier discharge, 311–312
  pulp and paper industry, delignification, 660
Ozone systems:
  foundry emissions, advanced oxidation systems, 497
  pulp and paper industry emissions, chlorine and chlorine dioxide, 651–652
  volatile organic compounds, electron-beam-generated plasma destruction, 317–320

Packing materials, biofiltration of organic vapors, 283–284
Palladium membranes, carbon sequestration technology, 374
Palm oil extraction case study, return on investment in, 30
Paper recycling, conversion technology, 22
Parametrization, plastics recycling, developing country case study, mathematical modeling, 572
Particle size distribution (PSD), wastewater treatment, ultraviolet disinfection, 424
Particulate matter (PM):
  best available control technology, pollution prevention economics, 178
  health risk assessment, 107–108
Partition coefficient, exposure assessment, health risk assessment, 93–95
Part-load issues, fuel cell technology, 347–348
Payback time calculations:
  brine concentrators, wastewater recycling, 449

poultry litter, catalytic steam gasification,
594–595
Penn State testing, foundry emissions, advanced
oxidation systems, 516–517
Percent transmission measurements, ultraviolet
disinfection:
performance evaluation, 73
wastewater treatment, 423
Permeability, 353
Permeate pressure, membrane-based VOC
removal and recovery, 354–355
Permeation flux equations:
membrane-based VOC removal and recovery,
353–354
wastewater treatment, pervaporation
models, 389
Pervaporation process:
wastewater treatment, membrane-based
technology, 384–385, 388–389
pharmaceutical waste volume minimization,
396–397
zero discharge industries, 31
Pharmaceutical waste, wastewater treatment,
membrane-based technology,
384–403
extractive membrane bioreactor, 395–396
product recovery, 397
valuable compound retrieval, 391–393
volume minimization, 396–397
Phosphoric acid fuel (PAFC), basic
technology, 340
Phosphorus concentrations, poultry litter,
environmental impact of, 583–584
Physical adsorption, carbon sequestration
technology, flue gas capture, 371–372
Picklex process case study, metal finishing
industry, water pollution prevention,
470–475
Plasma based destruction. *See* Electron-beam-
generated plasma
Plastics recycling, developing country case study:
Bangladesh infrastructure characteristics,
552–554
color change sequence, 562–565
high-density polyethylene characteristics,
555–556
high-density polyethylene service life, 564
impurities problems, 565
industrial ecology, 554–555
mathematical modeling, 571–578
recycling cycle, 562–564
socioeconomic issues, 557–558
spatial profile, 556–557

successive processing, effects on recyclate,
565–571
supply chain issues, 558–565
Plating model, process pollution prevention,
electroplating in-process techniques, 480
Pollutants, health risk assessment, 103–110
acid deposition, 108–109
air toxics, 109–110
carbon monoxide, 105–106
ground-level ozone, 104–105
lead and mercury, 106–107
particulate matter, 107–108
Pollution, waste as, 2
*Pollution Prevention Act of 1990,* 264
Pollution prevention (P2). *See also* specific types
of pollution, e.g. Air pollution
prevention technologies
defined, 2–4
economics of, 6
metal finishing industry, 476–477
real-time analysis to, 196
sustainable economics:
best available control technology, gas turbine
power plant application, 170–179
cash flow calculations, 168–169
clean smokestacks case study, 179–182
cost estimates, 162–163
earth-based sustainable economy, 183–188
research background, 161–162
responsible balance, 169–170
technology comparison criteria, 166–167
total annual cost, 165–166
total capital investment, 163–165
Polyactide (PLA) resin, lactic acid
production, 188
Polydimethylsiloxane (PDMS), membrane-based
VOC removal and recovery, 352–353
Polymer electrolyte membrane fuel cell (PEMFC):
basic technology of, 337–339
carbon monoxide poisoning, 339
freeze condition startup, 339
membrane dry-out, 349
starvation problems, 339
Polymeric/inorganic hybrid sorbent, arsenic
removal, groundwater contamination,
531–533
Polymerization, supercritical carbon dioxide,
wastewater treatment, 414
Polymer manufacturing, mass integration,
discharge minimization, HIWAMIN/
EIWAMIN synthesis, 611–614
Population exposure data, accidental chlorine gas
release, 66–67

Potassium carbonate process, carbon sequestration technology, 375–376

Potassium hydroxide (KOH), alkaline fuel cells, 340–341

Potential commodity values, risk management, 133

Poultry litter, catalytic steam gasification, 582–595
applications, 582–583
bench scale testing, UTSI process, 588–591
biogas and direct combustion, 584–585
coal gasification process concept, 586–588
economics of, 592–595
environmental impact, 583–584
large-scale facility, emissions estimates, 591–592
zero discharge goals, 595

Power cycles, carbon sequestration technology, 376–377

Power distribution and control center (PDC), ultraviolet disinfection, wastewater treatment, 422

Power supplies, wastewater treatment, ultraviolet disinfection, 419–420

Precipitated calcium carbonate (PCC), pulp and paper industry, carbon dioxide-rich gas manufacturing of, 681–682

Predict Mixed Chemicals' Reactions database, 154

Preferential oxidation (PrOx) reactors, fuel cell processing, 345–346

Pressure drop, biofiltration of organic vapors, 283

Pressure effects, poultry litter, catalytic steam gasification, 587–589

Pressure swing adsorption (PSA) process:
carbon sequestration technology, flue gas capture, 372
membrane-based VOC removal and recovery, 352
tank farm gas recovery, 364–365

Prevention of significant deterioration (PSD), pollution prevention economics, best available control technology, 170–179

Primary recycling:
plastics recycling, developing country case study, 554
reprocessing effects, 565–571
solid waste pollution prevention, 546–547

Probit regression line parallelism, chlorination (disinfection) process, aquatic life effects, 72

Process energy-separating agents, mass integration, discharge minimization, heat-induced and energy-induced separation networks, 606–608

Process engineers and engineering firms:
biofiltration systems, volatile organic vapors, 286–287
brine concentrators, 436–438
fuel cell technology, 342–343
efficiency and performance evaluation, 347
life cycle assessment, biorenewables vs. fossil fuels, system energy balance, 77–80
plastics recycling, developing country case study, mathematical modeling, 571–575
zero discharge industries and, 16–20
improvement strategies, 28
separation technology, 30–31
throughput analysis, 20

Process integration, discharge minimization, 6
heat exchange networks, 598–600
industrial applications, 638–640
information sources, 639
mass integration, 600–622
end-of-pipe separation network synthesis, 602–608
in-process separation network synthesis, 608–614
multicomponent nonideal systems, 614–622
research background, 600–602
research background, 597–598
water optimization and integration, 622–637
land treatment technology, 624–628
pulp and paper process, 628–637

Process intensification, sustainable development and, 216

Process mass-separating agents, mass integration, discharge minimization, mass exchange networks, 603–606

Process pollution prevention (P3):
electroplating in-process techniques:
fundamentals and strategies, 478
integrated modeling and optimization, 478–482
optimal cleaning and rinse time, operational technology, 482–483
plant-wide integrated modeling, 481
pollution prevention transition to, 477–478
unit-based operation modeling, 479–481
pulp and paper industry:
air pollution, 649–653
chlorine/chlorine dioxide, 651–652
chloroform emissions, 650–651
greenhouse gases, 652–653

sulfur dioxide/nitrogen oxides, 650
sulfur gas reduction, 650
effluent discharges, 653–655
adsorbable organic halides, 654–655
dioxin/furans, 654
emerging technologies, 656–658
environmental management history,
647–648
research background, 648–649
resource recovery and reuse, 658–667
biopulping, 661–665
black liquor chemical and energy recovery,
665–667
delignification, 660
solvent pulping, 661
sulfur emissions recovery and control, 660
value-added chemicals, pulp mill waste
gases, 658–659
solid wastes, 655–656
ash, 656
recyclable paper and fiber, 656
wastewater treatment residuals, 655
zero discharge paradigm, 1, 4
Process-specific constraints, electroplating
in-process techniques, plant-wide
integrated modeling, 481
Process stream recycling, membrane-electrode
process, 459–460
Product development, life cycle assessment and
sustainability, 200–201
Products of incomplete combustion (PICs),
volatile organic compounds, electron-
beam-generated plasma destruction,
thermal oxidation, 317–318
Product stages, life cycle assessment, goal
definition and scoping, 40–41
Programming protocols:
mass integration, discharge minimization,
in-process separation network synthesis,
610–611
mathematical modeling, plastics recycling,
developing country case study, 575
Propane, foundry emissions, advanced oxidation
processes, 512
Public health issues, pollution prevention
economics, 180
Public policy, sustainable economics and
development metrics, 206–207
Public response, risk assessment and
determination of, 146–148
Pulp and paper industry:
discharge minimization, water optimization and
integration, 628–637

process pollution prevention technologies:
air pollution, 649–653
chlorine/chlorine dioxide, 651–652
chloroform emissions, 650–651
greenhouse gases, 652–653
sulfur dioxide/nitrogen oxides, 650
sulfur gas reduction, 650
effluent discharges, 653–655
adsorbable organic halides, 654–655
dioxin/furans, 654
emerging technologies, 656–658
environmental management history,
647–648
research background, 648–649
resource recovery and reuse, 658–667
biopulping, 661–665
black liquor chemical and energy
recovery, 665–667
delignification, 660
solvent pulping, 661
sulfur emissions recovery and
control, 660
value-added chemicals, pulp mill waste
gases, 658–659
solid wastes, 655–656
ash, 656
recyclable paper and fiber, 656
wastewater treatment residuals, 655
zero discharge:
chlorine-free processing case study,
672–673
closed-loop effluent management,
673–676
clarification process, 674
concentration, 675–676
evaporation, 675
incineration, 676
stripping, 676
design constraints, 684–685
emission recovery and control,
679–683
greenhouse gases, 680–682
solid waste management and energy
conservation, 682–683
sulfur recovery, 679–680
environment discharges, 678
implementation case study, 676–678
individual entrepreneurs, 684
international implementation, 683–684
regulatory issues, 684
research background, 7, 671–672
safety issues, 678–679
science community involvement, 684

Quality assurance procedures:
electroplating in-process techniques, plant-wide integrated modeling, 481
life cycle assessment, 41–42
plastics recycling, developing country case study, mathematical modeling, 578
Quality of life criteria, pollution prevention economics, 181
Quaternary recycling, solid waste pollution prevention, 547
Quotient method, ecological risk assessment, one-dimensional chemical model, 137

Radiant energy systems, volatile organic compounds, electron-beam-generated plasma destruction, 317–318
comparison of techniques, 327–328
Rapid start lamps, wastewater treatment, ultraviolet disinfection, 419
Rate of ion-exchange ($R_{IX}$), membrane-electrode process, 452
Raw materials. *See also* Engineered materials
life cycle assessment, acquisition, 40
recycling and reuse of, 24–26
zero discharge industries:
inventory inputs and outputs, 20–21
productivity of, 19–20
toxicity reduction, 27–28
Raw water consumption, zero water discharge systems, design criteria, 227
RBLC database, best available control technology:
good combustion practices, 177
particulate control technologies, 178–179
Reaction limitation, biofiltration systems, organic vapors kinetics, 286
Real-time analysis, to pollution prevention, 196
Receptor doses, health risk assessment, exposure studies, 97–101
Recovery credits (RC), pollution prevention economics, 165–166
Recovery process:
pulp and paper industry:
black liquor recovery process, 665–667
zero discharge technologies, 679–683
greenhouse gases, 680–682
solid waste management and energy conservation, 682–683
sulfur from acid gases, 679–680
wastewater treatment, membrane-based technology, pharmaceutical compounds, 397
Recovery ratio, membrane-electrode process, target cations, 453

Recreation values, risk management, 133
Recyclable paper and fiber, pulp and paper industry, 656
Recyclates, plastics recycling, developing country case study, successive reprocessing, effects of, 565–571
Recycling:
life cycle assessment, goal definition and scoping, 41
mass integration, discharge minimization, in-process separation network synthesis, 610
plastics recycling, developing country case study:
Bangladesh system, 552–554
color change sequence, 562–565
high-density polyethylene characteristics, 555–556
high-density polyethylene service life, 564
impurities problems, 565
industrial ecology, 554–555
mathematical modeling, 571–578
recycling cycle, 562–564
socioeconomic issues, 557–558
spatial profile, 556–557
successive processing, effects on recyclate, 565–571
supply chain issues, 558–565
pollution prevention and, 3
solid waste pollution prevention, 545–549
benefits, 546
current trends in, 548
mechanical recycling, 546–547
sustainable productivity and growth, 548–549
tertiary, chemical, or feedstock recycling, 547
U. S. and British recycling, 547–548
zero discharge industry and, 24–26
Redox conditions, arsenic chemistry, groundwater contamination, 525–526
Reduction reactions, membrane-electrode process, wastewater recycling, 459
Reforming mechanisms, fuel cell processing, catalytic partial oxidation and, 343–345
Regeneration mechanisms, arsenic removal, groundwater contamination, 538–540
Regional haze, ecological risk assessment, 123
Regulatory Air Models, 155
Regulatory policies:
foundry emissions, advanced oxidation systems, 505–506
health risk assessment, 91
zero discharge technology, 23

future issues, 246–247
gas turbine case study, 239
pulp and paper industry, 684
Rematerialization, zero discharge industry, 26–27
Renewables:
feedstocks, green chemistry and, 196
ultraviolet disinfection performance evaluation, 75–76
REPAQ tool, life cycle assessment and life cycle impact assessment, 61
Reporting requirements, life cycle assessment, 42
results reporting, 58
Reprocessing effects, plastics recycling, developing country case study, 565–571
Residual systems, pulp and paper industry, wastewater treatment, 655
Resource conservation and reconcentration, membrane-electrode process, mining/ metallurgical industries, 460–461
Resource Conservation and Recovery Act of 1976 (RCRA), zero discharge industry and, 23
Resource consumption, electricity generation, biorenewables *vs.* fossil fuels, 81–82
Resource recovery and reuse, pulp and paper industry, 658–667
biopulping, 661–665
black liquor chemical and energy recovery, 665–667
delignification, 660
solvent pulping, 661
sulfur emissions recovery and control, 660
value-added chemicals, pulp mill waste gases, 658–659
Resource valuation, risk management and, 132–134
Results interpretation, life cycle assessment, 57–58
Return on investment (ROI), zero discharge industries, 30
Reuse of materials:
arsenic removal, groundwater contamination, 538–540
pulp and paper industry pollution, resource recovery and reuse, 658–659
zero discharge industry, 24–26
Revenue opportunities, zero discharge industries, 29–30
Reverse osmosis (RO):
wastewater recycling, brine concentrators:
feed water characteristics, 436
high-efficiency models, 446–448
zero discharge systems, 444–446

wastewater treatment, membrane-based technology, 384–385, 387–388
design and economics issues, 400–403
heavy metal recovery and water recycling, 397–400
zero water discharge system, DaimlerChrysler case study, 233–234
Reverse potential (RP) phenomenon, membrane-electrode process:
system independence, 456–457
wastewater recycling, 454–456
Reverse water gas shift (RWGS), fuel cell processing, 346
Rinsing models:
arsenic removal, groundwater contamination, 538–540
electroplating in-process techniques, 480, 482–483
Risk assessment:
chemical process industries, sustainable production and growth in, 198–199
communication issues, 146–153
accident aftermath, 151
case study, 151–153
costs of noncommunication, 151–152
public response determinants, 146–148
sustainable strategies, 148–150
ecological risk:
communication issues, 146–153
accident aftermath, 151
case study, 151–153
costs of noncommunication, 151–152
public response determinants, 146–148
sustainable strategies, 148–150
effects assessment, 118–120
benthic community, 119
fish community, 119
soil invertebrates and plants, 119–120
wildlife, 120
eutrophication, 123–124
exposure assessment, 114–118
benthic macroinvertebrate community, 115–117
fish community, 116
soil invertebrates, 117
terrestrial plants, 117
terrestrial wildlife, 117–118
global warming, 122–123
information sources, 122
internet information access, 153–157
management strategies:
characterization issues, 135–136
chemical models, 137–145

Risk assessment (*Contd.*)
    de minimis risk concepts, 136
    ecological resources valuation, 132–134
    modeling techniques for, 134–135
    research background, 130–132
    risk characterization, 129–130
    sampling and surveys, 120–121
    speciation, 122
    technical aspects, 110–114
        endpoint identification, 113
        measurement endpoint selection, 113–114
        site conceptual models, 111–113
    toxicity testing, 124–129
        application of results, 126–129
        test results evaluation, 124–126
        toxic units, 126
    uncertainties, 130
    visibility and regional haze, 123
  health risks:
    cancer risk assessment, 104
    chlorine (disinfection) process, 63–64
    common pollutants, 103–110
        acid deposition, 108–109
        air toxics, 109–110
        carbon monoxide, 105–106
        ground-level ozone, 104–105
        lead and mercury, 106–107
        particulate matter, 107–108
    communication issues, 146–153
        accident aftermath, 151
        case study, 151–153
        costs of noncommunication, 151–152
        public response determinants, 146–148
        sustainable strategies, 148–150
    continuous monitoring example, 92–93
    exposure assessment, 93–101
        bioconcentration prediction, 94–95
        contaminant intake calculation, 99–100
        dermal soil contact, average daily intake,
            100–101
        mechanistic models, 96–97
        partition coefficient and bioconcentration
            factor, 93–95
        point concentrations, 95–96
        receptor doses, 97–101
        slope factors, 95
    internet information access, 153–157
    management strategies:
        characterization issues, 135–136
        chemical models, 137–145
        de minimis risk concepts, 136
        ecological resources valuation, 132–134
        modeling techniques for, 134–135

        research background, 130–132
    monitoring example, 90–92
    problem formulation, 90–93
    regulatory statutes concerning, 91
    risk characterization, 101–103
        average and minimum exposure, 102–103
        carcinogenic risk, 103
        total weighted average concentration,
            105–106
        toxicity assessment, 101
    internet information access, 153–157
    research overview, 5–6
    sulfur dioxide dechlorination, accidental
        chlorine gas release, 65–67
Risk management:
  characterization issues, 135–136
  chemical models, 137–145
    one-dimensional models, 137
    three-dimensional models, 144–145
    two-dimensional models, 137–144
  chemical process industries, sustainable
    production and growth in, 198–199
  de minimis risk concepts, 136
  ecological resources valuation, 132–134
    nonutilitarian values, 133–134
    utilitarian values, 133
  modeling techniques for, 134–135
  research background, 130–132
Risk Management Consequence Analysis
    (RMP*Comp), 156
RMP*Comp software, 156
  accidental chlorine gas release, toxic endpoint
    distance, 66

Safe disposal, pollution prevention and, 3
Safety guidelines:
  plastics recycling, developing country case
    study:
      regulatory concerns, 579
      reprocessing effects, 565–671
  pulp and paper process, zero discharge
    technology, 678–679
  ultraviolet disinfection, wastewater
    treatment, 427
  ultraviolet disinfection performance evaluation,
    73–74
Sampling techniques, ecological risk assessment,
    120–121
Saveall devices, pulp and paper industry, effluent
    discharges, 654
Scavenging mechanisms, dioxin/furan ash,
    sorbent and carbon injection
    removal, 293

Scientific value:
  risk management, 133–134
  zero discharge technology, pulp and paper
    industry, 684
SCONO$_x$:
  best available control technology:
    gas turbines and duct burners, 176
    pollution prevention economics, 171–173
  nitrogen oxide emission control, zero discharge
    technology, 242–243
Secondary recycling:
  plastics recycling, developing country case
    study, 554
  reprocessing effects, 565–571
  solid waste pollution prevention, 546–547
Segregation and reuse:
  mass integration, discharge minimization,
    in-process separation network
    synthesis, 609–611
  pollution prevention, 2–3
Selective catalytic reduction (SCR):
  nitrogen oxide emission control, zero discharge
    technology, 240–242
  pollution prevention economics:
    best available control technology, 174–175
    total capital investiment (TCI), 165
Selectivity parameters:
  membrane-based VOC removal and recovery,
    360–362
  membrane-electrode process, binary solution
    target ions, 453
Sensitivity analysis:
  electricity generation, biorenewables *vs.* fossil
    fuels, 82
  health risk assessment, mechanistic models, 97
  life cycle assessment, 56
Sensitivity check, life cycle assessment, 56
Separation technology:
  mass integration, discharge minimization:
    end-of-pipe separation network synthesis,
      602–608
    heat-induced/energy-induced separation
      networks, 606–660
    mass exchange networks, 602–606
    in-process separation network synthesis,
      608–614
    polymer manufacturing, 611–614
    WIN/HIWAMIN, and EIWAMIN
      systems, 608–611
  zero discharge industries, 30–31
Services of nature values, risk management, 133
Sherwood plots, recycling and reuse of materials,
    25–26

Short-rotation woody crop (SRWC) practices,
    biomass fuel, sustainable economics and
    development, 202–203
Significant emission rate (SER), pollution
    prevention economics, best available
    control technology, 170–179
SimaPro 4, life cycle assessment and life cycle
    impact assessment, 61
Simultaneous saccharification and fermentation
    (SSF), ethanol from biomass, 187
Sink/generator manipulation, mass integration,
    discharge minimization, in-process
    separation network synthesis,
    610–611
Sinks:
  mass integration, discharge minimization:
    in-process separation network synthesis,
      609–611
    ternary wastewater systems, crystallization
      evaporation network, 619–620
    wastewater treatment, pulp and paper process,
      633–634
    water optimization and integration, land
      treatment technology, 625–628
    zero discharge industries, 15–16
Site conceptual models (SCMs), ecological risk
    assessment, 111–113
Slimline instant start lamp, wastewater treatment,
    ultraviolet disinfection, 419
Slope facators, health risk assessment, exposure
    studies, 95
Slow landfilling scenario, plastics recycling,
    developing country case study,
    mathematical modeling, 574–578
Slow recovery scenario, plastics recycling,
    developing country case study,
    mathematical modeling, 574–578
Sludge accumulation model, process pollution
    prevention, electroplating in-process
    techniques, 480–481
Sludge reduction, process pollution prevention,
    electroplating in-process techniques,
    488–489
Slurried bond feeding system, foundry emissions,
    advanced oxidation, 500–501
Small-pore membranes, carbon sequestration
    technology, 374
Smog components, health risk assessment,
    104–105
Smokestack emissions, zero discharge technology,
    247–252
  Clear Skies Initiative, 248
  Ebara process, 249–252

Social benefits:
 fuel cell technology, 350
 plastics recycling, developing country case
  study, 557–558
Sodium carbonate process, carbon sequestration
  technology, 374–375
Sodium concentrations, poultry litter,
  environmental impact of, 583–584
Software tools, life cycle assessment and life cycle
  impact assessment, 58–61
Soil contact, average daily intake, dermal contact,
  receptor dose studies, 100–101
Soil invertebrates, ecological risk assessment,
  117, 119–120
Soil pH, aquifer contamination, pollution
  prevention technology, 523
Solar-thermal power, sustainable economics and
  development, 203
Solid oxide fuel cell (SOFC), basic technology,
  339–340
Solid waste pollution prevention:
 plastics recycling, developing country case
  study, 550–579
  Bangladesh system, 552–554
  color change sequence, 562–565
  high-density polyethylene characteristics,
   555–556
  high-density polyethylene service life, 564
  impurities problems, 565
  industrial ecology and, 554–555
  mathematical modeling, 571–578
  recycling cycle, 562–564
  socioeconomic issues, 557–558
  spatial profile, 556–557
  successive processing, effects on recyclate,
   565–571
  supply chain issues, 558–565
 poultry litter, catalytic steam gasification,
  582–595
  applications, 582–583
  bench scale testing, UTSI process, 588–591
  biogas and direct combustion, 584–585
  coal gasification process concept, 586–588
  economics of, 592–595
  environmental impact, 583–584
  large-scale facility, emissions estimates,
   591–592
  zero discharge goals, 595
 pulp and paper industry:
  process pollution prevention technologies,
   655–656
   ash, 656
   recyclable paper and fiber, 656

    wastewater treatment residuals, 655
   recovery technologies, 657
   zero discharge technologies, 682–683
 recycling technology, 545–549
  benefits, 546
  current trends in, 548
  mechanical recycling, 546–547
  sustainable productivity and growth,
   548–549
  tertiary, chemical, or feedstock recycling, 547
  U. S. and British recycling, 547–548
Solute flux, membrane-based technology,
  nanofiltration, 390
Solutia case study, heat integration and mass
  integration, 638
Solution-diffusion (SD) model, membrane-
  based technology, reverse osmosis,
  387–388
Solvent extraction:
 volatile organic compound removal, 275–278
 wastewater treatment, membrane-based
  technology, 391
  pharmaceutical applications, 386–387
  pure form recovery, 393–394
Solvent pulping, pulp and paper industry, 661
Solvents, green chemistry and auxiliaries to, 195
Sonication, 493–497
Sonoperoxone process, foundry emissions,
  advanced oxidation system, 501–503
Sorbent injection technology:
 arsenic removal, groundwater contamination,
  531–533
 dioxin/furan removal, 289–298
  ash particle scavenging, 293
  carbon sources, 291–292
  control technologies, 293–297
  deacon reaction, 290–291
  de-novo synthesis, 290
  dioxin/furan formation mechanisms,
   290–291
  dioxin/furan sources, 289–290
  inhibition mechanisms, 292–293
  mercury removal, 295–297
  metals and, 292
  sulfuric acid treatment, 297
Sorption kinetics, arsenic removal, groundwater
  contamination:
 basic principles, 531–533
 intraparticle diffusivity, 536–538
Sources:
 wastewater treatment:
  pulp and paper process, 633–634
  ultraviolet disinfection, 426–427

water optimization and integration, land treatment technology, 625–628
zero discharge industries, 15–16
zero water discharge systems, 227
Spatial profiling, plastics recycling, developing country case study, 556–557
Speciation, ecological risk assessment, 122
Species permeability, wastewater treatment, pervaporation models, 389
Spent catalyst disposal:
nitrogen oxide emission control, selective catalytic reduction, 242
nitrogen oxide emissions reduction, risks and benefits, 245
Spray dryers, brine concentrators, wastewater recycling, concentrate disposal, 439–440
Spray fields, water optimization and integration, land treatment technology, 624–628
Staged cooling, wastewater recycling, brine concentrators, 447
Stationary system costs, poultry litter, catalytic steam gasification, 593–595
Steam/water injection, best available control technology, pollution prevention economics, 175–176
Strategic goals program (SGP), water pollution prevention, metal finishing industry, 464–466
Stream mixing:
brine concentrators, wastewater recycling, concentrate disposal, 438
mass integration, discharge minimization, in-process separation network synthesis, 610–611
Stripper overhead gas (SOG), pulp and paper industry, resource recovery and reuse, 658–659
Stripping process:
membrane-based VOC removal and recovery, 355–359
pulp and paper industry, closed-loop effluent management, 676
Succinic acid, bioproduction, 188
Sulfur-bearing fuels, nitrogen oxide emissions:
reduction, risks and benefits, 245–246
selective catalytic reduction and, 241–242
Sulfur/chlorine ratio, dioxin/furan formation, sorbent and carbon injection removal, 292–293
Sulfur dioxide emissions:
best available control technology, pollution prevention economics, 180

dechlorination, life cycle assessment, 64–72
accidental chlorine gas release, case study, 65–67
aquatic life, effects on, 67–69
environmental impact, 65
limitation, 65
marine life sensitivity, chlorinated seawater, 70–72
dielectric barrier discharges:
barrier discharge oxidation, 310–311
conversion geometry, 311
health risk assessment, acid deposition, 108–109
pulp and paper industry, 650
Sulfur gas emissions, pulp and paper industry:
recovery and control, 660
from acid gases, 679–680
reduction technologies, 650
Sulfuric acid, mercury adsorption, carbon injection technology, 297
Supercritical fluids, wastewater treatment, zero discharge techniques, 406–414
carbon dioxide dyeing, 412
carbon dioxide polymerization, 414
cleaning operations, 412–413
corrosion challenge, 410–411
metal recovery applications, 411
paint industry VOCs, 413
supercritical water oxidation, 409–410
supercritical water properties, 408
textile dyeing discharge, 412
Supercritical water oxidation (SCWO), wastewater treatment, zero discharge technologies, 409–410
Superfund sites, ecological risk assessment, 114
Supply chain issues, plastics recycling, developing country case study, 558–565
Surveys, ecological risk assessment, 120–121
Suspended solids. See also Total suspended solids (TSS) standards
poultry litter, environmental impact of, 583–584
wastewater treatment, ultraviolet disinfection, 423–424
Sustainable economics and development
bio-based chemical and engineered materials, 214
biorenewable energy sources, 201–206
biomass fuel, 201–203
hydropower, 205
solar thermal power, 203
tidal power, 206

Sustainable economics and development (*Contd.*)
  wave energy, 205–206
  wind power, 203–204
  business practices and public policy, 206–207
  defined, 183–184
  eco-efficiency and eco-industrial parks, 214–216
  environmental and health risk communication, 148–150
  green by design process, 199
  green chemistry and engineering, 193–199
  Hanover principles, 217–218
  hydrogen-based economy, 207–214
  Kalundborg case study, 14–15
  life cycle assessment, product development and, 200–201
  plastics recycling, developing country case study, 557–558, 578–579
  pollution prevention:
    best available control technology, gas turbine power plant application, 170–179
    cash flow calculations, 168–169
    clean smokestacks case study, 179–182
    cost estimates, 162–163
    defined, 6
    earth-based sustainable economy, 183–188
    research background, 161–162
    responsible balance, 169–170
    technology comparison criteria, 166–167
    total annual cost, 165–166
    total capital investment, 163–165
  process intensification and microchannel reaction, 216
  recycling, 548–549
  research background, 191–193
  zero discharge as critical component of, 15–16
  zero discharge industries, ecology and emissions, 12–15
Switchable water allocation network (SWAN), process pollution prevention, electroplating in-process techniques, 486–488
Syngas, biomass-derived methanol production, 187
System characteristics, zero discharge industries, 15–16
System energy balance, life cycle assessment, biorenewables *vs.* fossil fuels, 77–80

Tank farms, membrane-based VOC removal and recovery, 364–365
Target cation recovery ratio ($I_{rec}$), membrane-electrode process, 453

Target ion selectivity ($I_{sel}$), membrane-electrode process, binary solution target ions, 453
TEAM tool, life cycle assessment and life cycle impact assessment, 61
Technikon-CERP analysis, foundry emissions, advanced oxidation systems, 516–517
Technology comparisons, pollution prevention, economic criteria, 166–167
  clean energy technologies, 181–182
Technology development, zero discharge industry, 27–28
Temperature effects:
  biofiltration of organic vapors, 283
  poultry litter, catalytic steam gasification, 587–589
Temperature swing adsorption (TSA), carbon sequestration technology, flue gas capture, 372
Tensile strength testing, plastics recycling, developing country case study, reprocessing effects, 570–571
Ternary wastewater systems, mass integration, discharge minimization:
  ammonium nitrate manufacturing, 620–622
  crystallization evaporation network design and optimization, 617–620
Terrestrial plants, ecological risk assessment, 117, 119–120
Terrestrial wildlife, ecological risk assessment, 117–118, 120
Tertiary recycling, solid waste pollution prevention, 547
Textile dyes, supercritical carbon dioxide, zero-discharge technology, 412
Thermal incineration:
  poultry litter, catalytic steam gasification, 587–588
  volatile organic compound removal, 275–278
Thermal oxidation, volatile organic compounds, electron-beam-generated plasma destruction, 317–318
  comparison of techniques, 326–328
Thermomechanical pulp (TMP) mill, pulp and paper industry, biopulping, 661
Thiele modulus, biofiltration, organic vapors, 285
Three-dimensional chemical model, ecological risk assessment, 144–145
Threshold limit values (TLVs), health risk assessment, 90–93
Throughput analysis, zero discharge technology, 20

Tidal power, sustainable economics and development, 206

Time-concentration function, ecological risk assessment:
three-dimensional chemical model, 144–145
two-dimensional chemical model, 138–140

Time-response function, ecological risk assessment:
one-dimensional chemical model, 137
two-dimensional chemical model, 138

Tolerance distribution models, health risk assessment, 96–97

Tomlinson furnace, pulp and paper process, process pollution prevention technologies, 657

Total annual cost (TAC), pollution prevention economics, 165–166

Total capital investiment (TCI), pollution prevention economics, 163–165

Total dissolved solids (TDS):
wastewater recycling, brine concentrators, zero discharge systems, 444–446
wastewater treatment, membrane-based technology, 391
zero water discharge system, DaimlerChrysler case study, 233–234

Total hydrocarbon (THC) analyzer measurements, electron-beam-generated plasma destruction, volatile organic compounds, pilot test results, 272–274

Totally chlorine free (TCF) processing, pulp and paper industry, zero discharge, 672–673
environmental issues, 678
safety issues, 678–679

Total reduced sulfur (TRS) emissions, pulp and paper industry, 648–649
air pollution reduction, 650
recovery and control technologies, 660
resource recovery and reuse, 659

Total residual oxidant (TRO) component, chlorination (disinfection) process, aquatic life effects, 70–72

Total suspended solids (TSS) standards:
poultry litter, environmental impact of, 583–584
pulp and paper industry, 648–649
ultraviolet disinfection performance evaluation, 73
wastewater treatment, ultraviolet disinfection, 423–424

Total weighted average (TWA), health risk assessment:
carbon monoxide exposure, 105–106
monitoring case study, 90–93

Toxic endpoint distance, sulfur dioxide dechlorination, accidental chlorine gas release, 66

Toxicity assessment:
ecological risk assessment, 120
testing techniques, 124–129
health risk assessment, 101

Toxicity criteria, chlorine (disinfection) process, 63–64
aquatic life effects, 67–69

Toxicity reduction, zero discharge industry, chemicals and materials, 27–28

Toxic units (TUs), ecological risk assessment, 126

TRACI tool, life cycle assessment and life cycle impact assessment, 61

Transmission measurements:
ultraviolet disinfection, wastewater treatment, 423
ultraviolet disinfection performance evaluation, 73

Trichloroethylene (TCE):
biofiltration of, 279–280
electron-beam-generated plasma destruction:
comparison with other technologies, 275–278
pilot test results, 271–273
research background, 266–267
electron-beam plasma destruction, 316–329

Tsumeb case study:
industrial cluster development, 21–22
zero discharge industry, 19–20

Two-dimensional chemical model, ecological risk assessment, 137–144

Ultimate limit definition, plastics recycling, 551–552

Ultrafiltration (UF), wastewater treatment, membrane-based technology, 384–385

Ultralow sulfur diesel (ULSD), biodiesel blended with, 331

Ultraviolet disinfection:
performance evaluation, life cycle assessment, 72–76
cost analysis, 74–75
disinfection standards, 73
environmental impact, 75–76
government research on, 62–63
operation, maintenance, and worker study, 73–74
transmission/absorbance evaluation, 73
wastewater treatment, 416–430

Ultraviolet disinfection (*Contd.*)
   absorbance/transmission, 423
   disinfection standards, 428–430
   efficiency parameters, 422–427
   equipment maintenance, 427
   flow rate, 424–425
   germicidal efficiency, 427–428
   germicidal mechanisms, 417–418
   iron concentrations, 425–426
   lamp components, 418–420
   North American case studies, 416–447
   open channel modular systems, 420–422
   particle size distribution, 424
   slimline instant start lamp, 419
   suspended solids, 423–424
   wastewater source characteristics, 426–427
Ultraviolet oxidation systems, volatile organic
      compounds, electron-beam-generated
      plasma destruction, 317–318
Ultraviolet transmission/absorbance, ultraviolet
      disinfection, wastewater treatment, 423
Umberto 3.0, life cycle assessment and life cycle
      impact assessment, 61
Uncertainty analysis:
   ecological risk assessment, 130
   life cycle assessment, 56
United States, recycling history in, 547–548
University of Tennessee Space Institute System
      (UTSI):
   poultry litter, catalytic steam gasification,
      588, 590–591
   volatile organic compound destruction:
      aqueous phase waste streams, 323–326
      pilot facility, 269–274
Use/reuse/maintenance, life cycle assessment,
      goal definition and scoping, 41
Utilitarian values, risk management, 133

Valuable compounds, retrieval, wastewater
      treatment, membrane-based technology,
      391–393
Value-added chemicals, pulp and
      paper industry, resource
      recovery and reuse, 658–659
Van't Hoff equations, membrane-based
      technology, reverse osmosis, 387–388
Vegetable oils, biodiesel production, 187–188
Visibility:
   ecological risk assessment, 123
   pollution prevention economics, 180
Volatile organic compounds (VOCs):
   air pollution prevention technologies:
      biofiltration, 279–288

      basic principles, 280–282
      cost analysis, 288
      diffusion limitation, wet biobarrier,
         285–286
      industrial applications, 286–287
      kinetics and design, 285–288
      microorganisms, 282
      moisture and temperature effects, 283
      oxygen and other nutrients, 282–283
      packing materials for, 283–285
      pressure drop, 283
      process parameter variation, 286
      reaction limitation, 286
   current techniques, 265
   electron-beam-generated plasma destruction,
      265–278
      aqueous phase waste streams, 316–329
         alternative oxidation solutions, 326–328
         AOS system, 320–322
         basic chemistry, 318–319
         economic issues, 328–329
         Florida International University/
            University of Miami system, 322–324
         oxidation limitations, 317–318
         UTSI pilot facility, 323–326
      comparison with other techniques,
         274–278
      preliminary economics, 274
      process concept, 267–269
      research background, 266–267
      UTSI pilot facility, 269–274
   membrane-based removal and recovery,
      352–368
      absorption and stripping, 355–359
      commercial membrane modules, 362–364
      design equations, principles and strategies,
         359–362
      gas station recovery, 365
      olefin polymerization vents, monomer
         recovery, 365–367
      tank farm gas recovery, 364–365
      vapor permeation, 352–355
  foundry emissions:
   advanced oxidation processes, 493–496
   Neenah case study, 505–506
   total emissions results, 508–509
  health risk assessment, 104–105
  mass integration, discharge minimization:
   heat-induced and energy-induced separation
      networks, 607–608
   multicomponent nonideal systems, 614–617
  pollution prevention economics, best available
      control technology, 180

wastewater treatment, supercritical carbon
dioxide, paint industry discharge, 413
Voltage characteristics, fuel cell technology,
336–337
Volume reduction:
fuel cell processing, 348
wastewater treatment, membrane-based
technology, pharmaceutical compounds,
396–397

Warranty issues, biodiesel, 333
Waste:
energy value recovery, 3
green chemistry and prevention of, 194
pollution as, 2
Waste interception networks (WIN), mass
integration, discharge minimization:
development of, 602
in-process separation network synthesis,
608–611
Waste management, life cycle assessment, goal
definition and scoping, 41
Waste-steam injection, nitrogen oxide
emission control, zero discharge
technology, 240
Wastewater recycling:
membrane-based applications, heavy metal
recovery, 397–400
pollution prevention technologies, 432–517
brine concentrators, 432–449
capital costs, 444
concentrate disposal, 438–443
concentrate production, 437–438
control loops, 438
distillate production, 437
economics of, 448–449
energy consumption, 443–444
evaporator principles, 433–436
high efficiency RO process, 447–448
process engineering and design, 436–437
staged cooling, 447
zero discharge systems, 444–446
electroplating in-process techniques,
476–490
integrated modeling and optimization,
478–482
metal finishing industry, 476–477
P3 applications, 482–490
process pollution prevention fundamentals
and strategies, 478
P2 to P3 transition, 477–478
foundry oxidation technologies, 491–517
acoustic cavitation, 497–499

advanced oxidation conditions, 496–499
benzene emissions, 510–511
carbon monoxide/carbon dioxide
emissions, 511–512
clay and coal efficiency, 514–516
design layouts, 499–501
emissions characteristics, 508–514
formaldehyde emissions, 511
full-scale advanced oxidation, 501–503
green sand properties, 506–508
green sand systems, 493–496
mass balance analysis, emissions *vs.* core
binder, 514
methane, ethane, and propane emissions,
512
Neenah foundry case study, 503–506
ozone and hydrogen peroxide, 497
research background, 491–493
Technikon-CERP and Penn State
emissions comparisons, 516–517
volatile organic compound emissions,
508–510
membrane-electrode process, 450–462
binary cation systems, 453–454
binary solution target selectivity, 453
cation recovery flux/cation exchange
flux, 452
ion-exchange kinetics, 457–458
limitations, 459
loading properties, 453
mining and hydrometallurgical industries
conservation and reconcentration,
460–461
process stream recycling, 459–460
rate of ion-exchange, 452
reverse potential phenomenon, 454–456
system independence, reverse potential
phenomenon, 456–457
target cation recovery ratio, 453
pulp and paper industry, zero discharge
technology, 676–678
Wastewater treatment. *See also* Aqueous phase
waste streams
discharge minimization:
mass integration, 600–622
end-of-pipe separation network synthesis,
602–608
heat-induced/energy-induced separation
networks, 606–660
mass exchange networks, 602–606
in-process separation network synthesis,
608–614
polymer manufacturing, 611–614

Wastewater treatment (*Contd.*)
  WIN/HIWAMIN, and EIWAMIN
      systems, 608–611
  mass exchange networks, 603–606
  benzene recovery, 604–606
  multicomponent nonideal systems,
      614–622
  ammonium nitrate manufacturing, ternary
      wastewater system, 620–622
  crystallization evaporation network,
      ternary wastewater system, 617–620
  VOC condensation and allocation,
      614–617
  process integration and optimization,
      622–637
  land treatment technology, 624–628
  pulp and paper process, 628–637
membrane-based applications, 384–403
  design and economics issues, 400–403
  heavy metal recovery and water recycling,
      397–400
  hybride processes, 391–394
  membrane types and properties, 384–385
  nanofiltration, 389–390
  pervaporation, 388–389
  pharmaceutical processes, 385–387
      extractive membrane bioreactor, 395–396
      product recovery, 397
      volume minimization, 396–397
  reverse osmosis, 387–388
pulp and paper industry, residuals, 655
recent technologies, 383–431
supercritical emissions, zero discharge
      techniques, 406–414
  carbon dioxide dyeing, 412
  carbon dioxide polymerization, 414
  cleaning operations, 412–413
  corrosion challenge, 410–411
  metal recovery applications, 411
  paint industry VOCs, 413
  supercritical water oxidation, 409–410
  supercritical water properties, 408
  textile dyeing discharge, 412
ultraviolet disinfection, 416–430
  absorbance/transmission, 423
  disinfection standards, 428–430
  efficiency parameters, 422–427
  equipment maintenance, 427
  flow rate, 424–425
  germicidal efficiency, 427–428
  germicidal mechanisms, 417–418
  hardness properties, 426
  iron concentrations, 425–426

  lamp components, 418–420
  North American case studies, 416–147
  open channel modular systems, 420–422
  particle size distribution, 424
  suspended solids, 423–424
  wastewater source characteristics,
      426–427
zero water discharge systems, 227
Water allocation network (WAN), process
      pollution prevention, electroplating in-
      process techniques:
  dynamically switchable design, 484–488
  steady-state design, 483–484
Water balance, zero water discharge system,
      captive power plant, 228–230
Water discharge systems. *See also* Aqueous phase
      waste streams
  zero discharge technology, 225–238
      advantages and disadvantages, 226–227
      applications, 225–226
      cement plant case study, 228–230
      DaimlerChrysler wastewater treatment plant
          case study, 230–234
      database guidelines, 226
      design principles, 227–228
      zero effluent system case study, 234–238
Water gas shift (WGS), fuel cell processing,
      345–346
Water pinch analysis, water optimization and
      integration, 622–623
Water pollution prevention:
  groundwater quality, 522–542
      aquifer contamination, 522–523
      arsenic removal technologies, 524–542
  metal finishing industry, 463–475
      economic benefits, 474
      environmental benefits, 472–473
      MERIT partnership for metal
          finishers, 470
      Picklex case study, 470–472
      strategic goals program, 464–466
      technology transferability, 475
      zero discharge goal planning and
          implementation, 466–469
  wastewater recycling, 432–517
      brine concentrators, 432–449
      electroplating in-process techniques,
          476–490
      foundry oxidation technologies, 491–517
      membrane-electrode process, 450–462
      metal finishing industry, 463–475
  wastewater treatment, 383–431
      membrane-based applications, 384–403

supercritical emissions, zero discharge techniques, 406–414
ultraviolet disinfection, 416–430
Water vapor, dielectric barrier discharges, 306–308
Wave energy, sustainable economics and development, 205–206
Weighting techniques, life cycle impact assessment, 52
Weir construction, ultraviolet disinfection, wastewater treatment, 422
Wet acid deposition, health risk assessment, 108–109
Wilson-Formosa Agreement, zero effluent systems, 234–238
Wind power, sustainable economics and development, 203–204
Wood-plastic composites, pulp and paper industry pollution technologies, 657
Worker safety, ultraviolet disinfection performance evaluation, 73–74
World Commission on Environment and Development (WCED), sustainable economics and development, 191–193
Worst case scenarios, ecological risk assessment, two-dimensional chemical model, 142–144

Xonon technology, best available control technology, pollution prevention economics, 173–174

Zero discharge:
brine concentrators, wastewater recycling, 444–448
payback calculations, 449
eco-industrial parks model, 252–259
basic characteristics, 255
development guidelines, 255–256
India, mini-case studies, 256–259
industrial ecology, 252–255
gas turbine nitrogen oxide emissions case study, 238–246
ammonia slip/ammonium-sulfur salts, 244–245
cost analyses, 244
nitrogen oxide control technologies, 240–244
regulatory background, 239
spent catalyst, 245
sulfur-bearing fuels, 245–246
turbine emissions characteristics, 239–240
industrial applications:
conversion technology development, 22

dematerialization, 26–27
design wastes, 22–23
emissions methodology, 20–23
industrial cluster construction, 21–22
inventory input-output, 20–21
investment recovery, 27
Kalundborg case study, 14–15
limits of, 31–32
material-product recycling and reuse, 23–28
opportunity development, 29
paradigm change, 28–30
process engineers and engineering firms, 16–20
process innovations, 28
productivity of raw materials, beer to mushroom case study, 19–20
regulatory reform, 23
research background, 11–12
return on investment, provision for, 30
separation technology, 30–31
sustainability, ecology and emissions, 12–15
sustainability components, 15–16
throughput analysis, 20
toxicity reduction of materials and chemicals, 27–28
transition strategies, 23–28
metal finishing industry, water pollution prevention, 466–469
paradigm, 4
pollution prevention, 2–4
poultry litter, catalytic steam gasification, 595
pulp and paper industry:
chlorine-free processing case study, 672–673
closed-loop effluent management, 673–676
clarification process, 674
concentration, 675–676
evaporation, 675
incineration, 676
stripping, 676
design constraints, 684–685
emission recovery and control, 679–683
greenhouse gases, 680–682
solid waste management and energy conservation, 682–683
sulfur recovery, 679–680
environment discharges, 678
implementation case study, 676–678
individual entrepreneurs, 684
international implementation, 683–684
regulatory issues, 684
research background, 671–672
safety issues, 678–679
science community involvement, 684

Zero discharge (*Contd.*)
  regulatory policy issues, 246–247
  research overview, 1–2, 5–7
  smokestack emissions, 247–252
    Clear Skies Initiative, 248
    Ebara process, 249–252
  supercritical emissions, wastewater treatment,
      406–414
    carbon dioxide dyeing, 412
    carbon dioxide polymerization, 414
    cleaning operations, 412–413
    corrosion challenge, 410–411
    metal recovery applications, 411
    paint industry VOCs, 413
    supercritical water oxidation, 409–410
    supercritical water properties, 408
    textile dyeing discharge, 412
  waste pollution, 2
  water discharge systems, 225–238

advantages and disadvantages, 226–227
applications, 225–226
cement plant case study, 228–230
DaimlerChrysler wastewater treatment plant
    case study, 230–234
database guidelines, 226
design principles, 227–228
zero effluent system case study, 234–238
Zero effluent system, Formosa Plastic
    Manufacturing case study,
    234–238
Zero emissions coal (ZEC), carbon sequestration
    technology, 379
Zero Emissions Research Initiative (ZERI), 4
  industrial development goals, 11–12
  zero discharge technology by, 20–23
Zero liquid discharge, zero water discharge
    system, DaimlerChrysler case study,
    233–234